AF737355

ice
Institution of Civil Engineers
publishing

# Coastal Management
## Changing coast, changing climate, changing minds

Proceedings of the International Conference organised by the Institution of Civil Engineers and held in Amsterdam, The Netherlands, from 8–9 September 2015.

Edited by Alison Baptiste

Co-sponsored by:

Royal HaskoningDHV
Enhancing Society Together

Van Oord
Marine ingenuity

Boskalis Westminster

TenCate

APS
Advanced Precast Systems

moffatt & nichol

xbloc®

bam

BESIX

Bonar
partners in performance

TOPCON

**Organising Committee**:

Alison Baptiste (Chair), Environment Agency
Heidi Burgess, University of Brighton
Kevin Burgess, CH2M
Tony Camilleri, Mackley
Bryan Curtis, Independent consultant
Greg Guthrie, Royal HaskoningDHV
Pippa Van Kuijk, Royal Haskoning DHV

**International Technical Committee:**
Robert D. Brumbaugh, The Nature Conservancy
Sam Hettiarachch, University of Moratuwa
Nabiil Ismail, Arab Academy for Science, Technology and Maritime Transport, Egypt
Shie-Yui Liong, National University of Singapore
Professor Robin McInnes, Coastal and Geotechnical Services
Denise Reed, The Water Institute of the Gulf, USA
Marcel Stive, TU Delft, The Netherlands
Bruce Thom, University of Sydney, Australia

Related titles from ICE Publishing:
*From Sea to Shore - Meeting the Challenges of the Sea (Coasts, Marine Structures and Breakwaters 2013)* (2 volumes). W. Allsop and K Burgess (eds.). ISBN 978-0-7277-5975-7.

*Innovative Coastal Management.* A. Schofield (ed.) ISBN: 978-0-7277-5749-4.

*Coastal Management.* ICE Publishing. ISBN 978-0-7277-3568-3.

ISBN 978-0-7277-6114-9

© Thomas Telford Limited 2016

ICE Publishing is a division of Thomas Telford Ltd, a wholly-owned subsidiary of the Institution of Civil Engineers (ICE).

All rights, including translation, reserved. Except as permitted by the Copyright, Designs and Patents Act 1988, no part of this publication may be reproduced, stored in a retrieval system or transmitted in any form or by any means, electronic, mechanical, photocopying or otherwise, without the prior written permission of the Publishing Director, ICE Publishing, 1 Great George Street, London SW1P 3AA.

This book is published on the understanding that the authors are solely responsible for the statements made and the opinions expressed in it and that its publication does not necessarily imply that such statements and/or opinions are or reflect the views or opinions of the publishers. While every effort has been made to ensure that the statements made and the opinions expressed in this publication provide a safe and accurate guide, no liability or responsibility can be accepted in this respect by the authors or publishers.

Commissioning Editor: Laura Balchin
Production Editor: Rebecca Norris
Market Development Executive: Elizabeth Hobson

Typeset by Manila Typesetting Company
Printed and bound in Great Britain by TJ International Ltd, Padstow

# Contents

# Foreword

These are the proceedings of the Institution of Civil Engineers' 8th International Coastal Management Conference held in Amsterdam on 8 and 9 September 2015.

It was the first time the conference has been held outside of the UK. The Netherlands was chosen as it is highly active in coastal zone management and offers many leading examples of innovation, technology and general management.

The conference brought together engineers, scientists and coastal management professionals from across the world, who focused on current issues, research and practical applications in relation to the overall theme of Changing Coast, Changing Climate, Changing Minds - Sharing International Experience on Practical Solutions to Coastal Management.

We had a record number of submissions from over 30 countries. Following a rigorous peer review, the best 60 papers and 20 posters were presented over the two days of the conference. The papers covered the full range of coastal management topics including planning, policy, engagement, forecasting, modelling, design, construction, monitoring and management, covering coastal evolution, flood risk, climate change, environmental enhancement, highly complex and 'low-tech' solutions, all tailored to meet the needs of the communities that live, work and play at the coast.

I would like to thank the Organising and Technical Committees for the rigorous peer review that they have conducted, as well as all the hard work that they have put into making this conference a success. A special thank you also goes to Jan Hendrik Dronkers, Director General, Rijkswaterstaat, The Netherlands for officially opening the conference.

Finally I would like to invite you, the reader, to think about submitting an abstract to the next International Coastal Management Conference which will be held in 2019.

***Alison Baptiste***
*Director, Strategy and Investment, Environment Agency*
*Conference Chair*

**Coastal Management: Changing coast, changing climate, changing minds**
ISBN 978-0-7277-6114-9

ICE Publishing: All rights reserved
doi: 10.1680/cm.61149.001

# A New Vision on the Marine Dikes' Structure

**Gabriel Tatu**, Department of Hydraulics and Environmental Protection, Technical University for Civil Engineering in Bucharest, Romania
**Alexandru Dimache**, Department of Hydraulics and Environmental Protection, Technical University for Civil Engineering in Bucharest, Romania
**Iulian Iancu**, Department of Hydraulics and Environmental Protection, Technical University for Civil Engineering in Bucharest, Romania
**Andreea Coveianu**, Department of Hydraulics and Environmental Protection, Technical University for Civil Engineering in Bucharest, Romania
**Oana Ciugulea**, Department of Hydraulics and Environmental Protection, Technical University for Civil Engineering in Bucharest, Romania

## Introduction

The structure of the dikes built on the Romanian Black Sea coast, both to protect seaports and/or beaches, are illustrated in figure 1 and 2. They are made of massive blocs of natural stone with different weights, the smallest ones weighing more than 3 kgf.

For the protection of seaports, the structure of these dikes has been shown to be effective in dissipating the wave energy during storms and provide a calm harbor, but from the point of view of the stability of the dike itself, there were many cases of collapse and dike damage despite their massiveness and coating with tetrapods or concrete slabs weighing tons force.

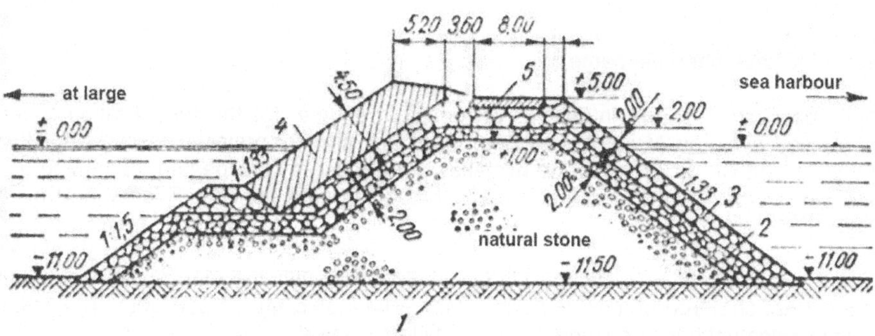


Figure 1: Dikes with slopes protected with tetrapods (Manoliu, I., 1967)
1 - natural stone under 2 tf/piece; 2 - natural blocs 2-10 tf/piece; 3 - natural blocs 10-12 tf/piece; 4 - tetrapods 20 tf/piece, in 2 layers; 5 - concrete slab.

As for the protection of the beaches, this type of structure has proved to be totally inefficient, being completely ineffective from all points of view. Measurements and aerial photographs were made, but it can be seen with the naked eye that whether dikes were transversal (T shape normal on the beach) or longitudinal (parallel to the shore) the beach erosion was not even stopped but rather it has continuously progressed. The shoreline continuously retreated, leading to disastrous situations, when the foundations of important buildings and structures are threatened.

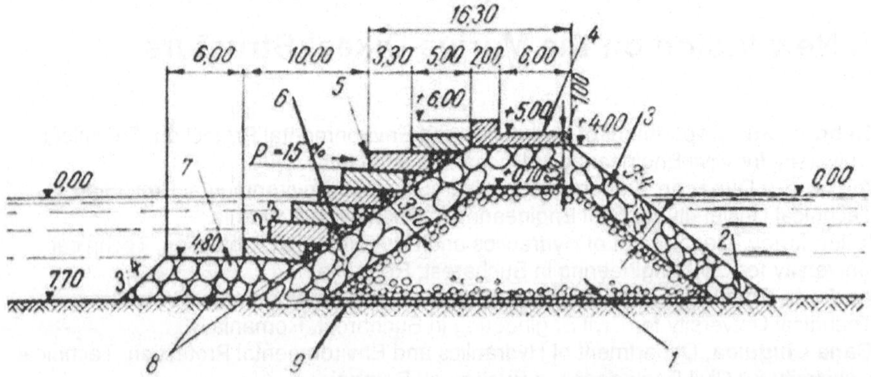

Figure 2: Dikes with embankment with slopes protected with artificial blocs in a regular alignment (Manoliu, I., 1967)
1 - mixed embankment 3-100 kgf/piece; 2 - natural blocs 1-3 tf/piece; 3 - natural blocs 3-5 tf/piece; 4 - concrete platform; 5 - artificial blocs 40 tf (5,00 x 1,75 x 2,00); 6 - natural blocs 5-10 tf/piece; 7 - natural blocs 3-5 tf; 8 - natural blocs 1-3 tf; 9 - allowance for settlement.

In addition, many failures were registered (crashes, namely excessive settlements) that cannot be explained if considering the extremely fast and well protected structure of the dikes (Iulian, C., 1994, 2006, Lazar, P.D., 1996, Luca V.O. et al., 1998).

One can say, with regrets of course but there is an evidence impossible to be contradicted of their effect and their behavior in time, that those dikes represented, just in fact, "money thrown into the sea" (with or without them the result would probably have been the same).

Nowadays, there is the intention, welcomed of course, for the rehabilitation of the protection systems of some beaches; there are already projects designed for this purpose and we do not want the mistakes of the past to be repeated, namely to use the same type of solutions that already proved in time their inefficiency and even futility.

The first steps in this direction are to elucidate and identify the real causes that made the current structure of marine breakwaters to be inefficient, as part of the protection systems against beach erosion and this article aims that.

More specifically, this paper makes an attempt to evaluate the parameters of the flow, namely the water velocity through the large gaps of the structure, because the water velocity may determine the transport of the sand through the dike, including its displacement under its basement or even, on stormy weather, the displacement of the stone pieces of smaller sizes through the dike's body; the compaction of the dike and its diving is then a logical result considering also the possibility of a "cascade" phenomenon development.

## The Mathematical Model

The large gaps in the dike structure, made of rock fill with relatively large size, were modelled as an annular network of pipelines having the dike form, more precise for the core of the dike where the rock fill has the smaller size and the higher hydraulic resistance (figure 3). For simplicity, all network pipelines had the same diameter and the same length.

We considered a strip (plane problem) with the thickness equal to the diameter of the network pipelines and thus we could evaluate the "porosity" *p* of the structure as the ratio of total interior volume of the arteries and the total volume of the strip. By appropriate choice of the network's parameters (length and diameter of the arteries) three porosities were defined, with the size order of the real ones, namely: *p=15% ; p=7.5%; p=5%.*

For simplicity also, calculations were carried out on a steady flow, for the case of a "snapshot"in which, due to the existence of waves, between the two sides of the dike a hydraulic charge *H* is created. More precise, from the assumed relatively constant level on the "back" of the dike, on its seaward side the water level may be above or below with a difference of level equal to the amplitude of the wave. In calculations, we considered a wave amplitude, respectively a hydraulic load *H=1.5 m* but also a much lower value *H=0.5 m* corresponding to that imposed by the presence of a sea current.

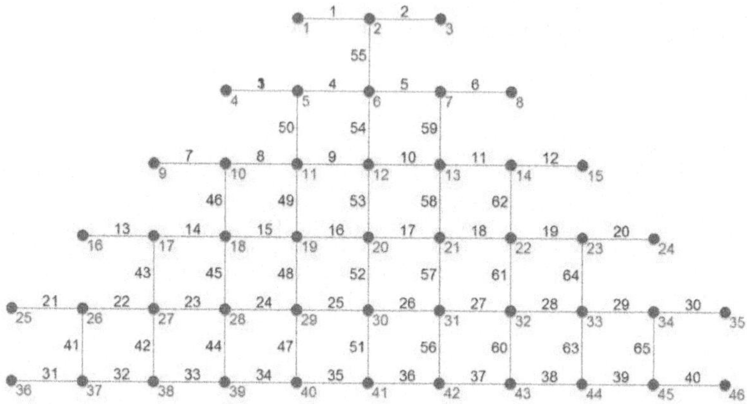


Figure 3: The mathematical model

For modelling the hydraulic head *H*, in the edge nodes of the network it was considered the existence of "tanks with constant water level"; so for tanks of nodes 3, 8, 15, 24, 35 and 46, the constant imposed level was greater with the value *H* (1.5 m or 0.5 m) than the constant imposed level in tanks of nodes 1, 4, 9, 16, 25 and 36.
The calculation results, more precisely the direction and the size of velocity through the arteries, which give approximately the size of the real velocity of flow through the

dike's structure gaps are given in figures 4 ... 9. Regarding the direction of the flow, if the head loss *H* is created by the presence of waves, it successively changes the sign and the flow directions are reversed in the arteries (i.e. in the dike), the speed values remaining the same.

## Results Analysis, Conclusions and Recommendations

All calculations show that the speed of water flow through the holes in core of the dike, in the classical constructive solution, is appreciable even at small hydraulic loads and may become much higher with its growth. The authors consider this as being particularly important, i.e. having important practical implications.

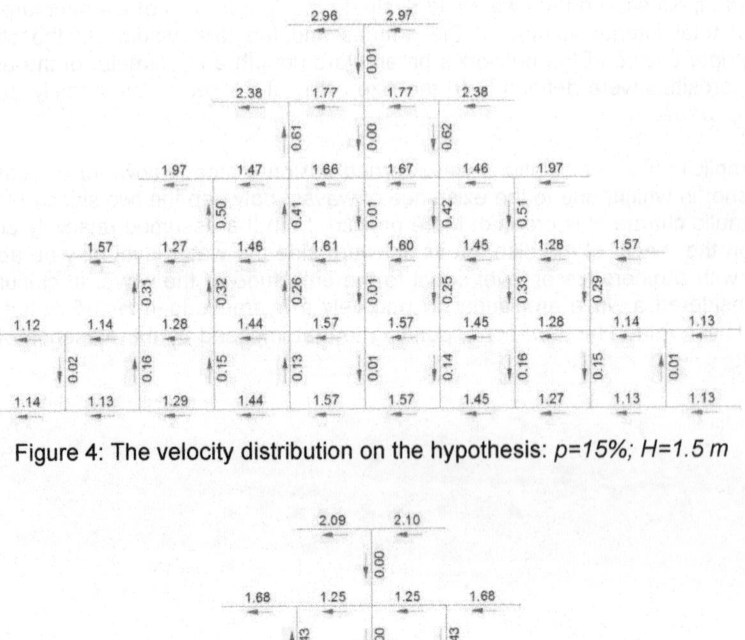

Figure 4: The velocity distribution on the hypothesis: *p=15%; H=1.5 m*

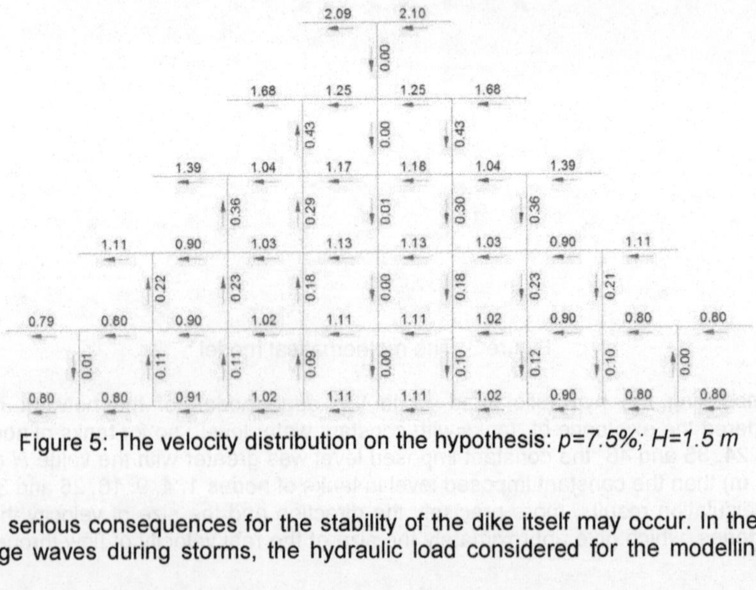

Figure 5: The velocity distribution on the hypothesis: *p=7.5%; H=1.5 m*

Thus, serious consequences for the stability of the dike itself may occur. In the case of large waves during storms, the hydraulic load considered for the modelling can easily reach (+/-) 4 m and that means that through the gaps the velocity may exceed 2 m/s while the flow direction will change cyclically in the rhythm with the waves' period.

Should also be noted that these are average speeds and, in fact, depending on the variable size of the gaps, local real speeds might be higher.

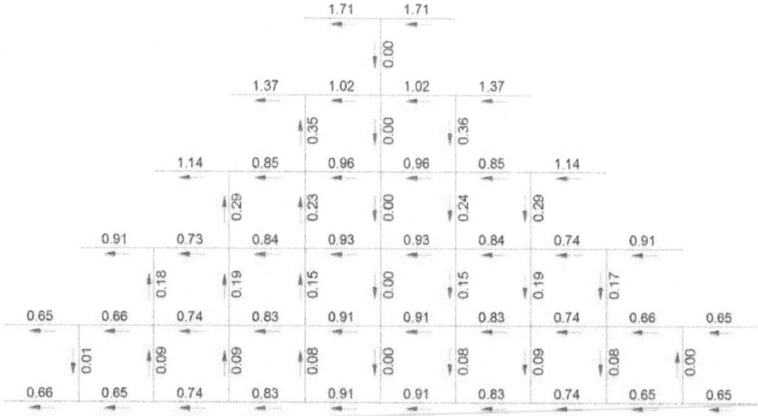

Figure 6: The velocity distribution on the hypothesis: *p=5%; H=1.5 m*

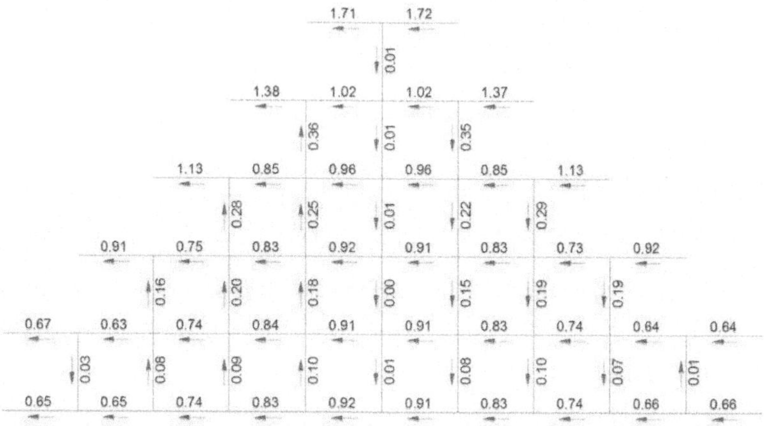

Figure 7: The velocity distribution on the hypothesis: *p=15%; H=0.5 m*

At such speeds, not only sand is moved but also the rock fills with a smaller size can be deployed; the oscillatory character of the movement helps for their dislocation and then their movement through the gaps of the structure; in time, a "cascading" type phenomenon may be generated, i.e. increasing the gaps together with the increasing of speeds.

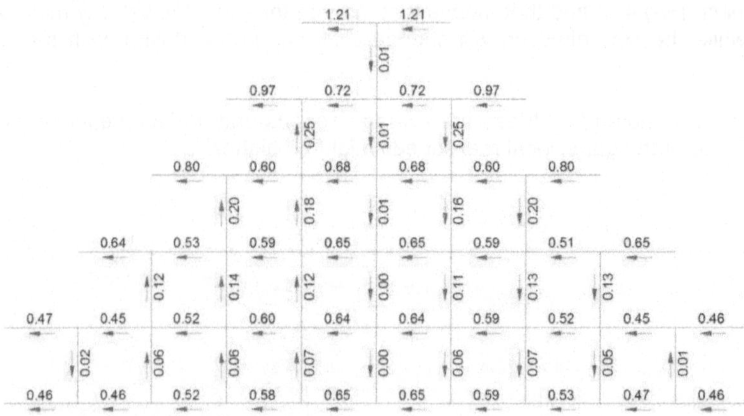


Figure 8: The velocity distribution on the hypothesis: *p=7.5%; H=0.5 m*

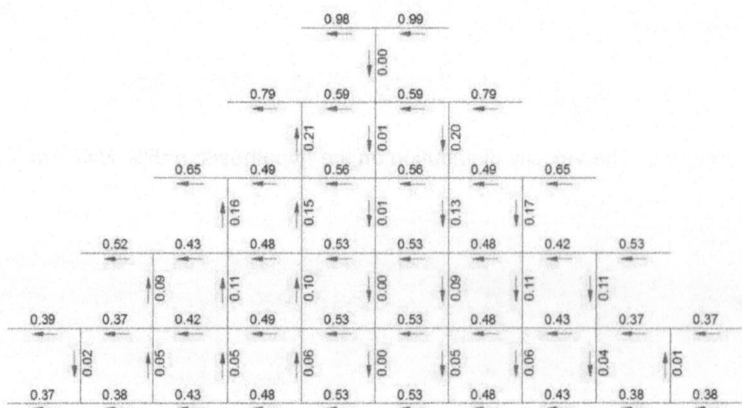


Figure 9: The velocity distribution on the hypothesis: *p=5%; H=0.5 m*

This “mechanism” could explain the spectacular failures of dikes, such as the “breaking” of a part of the dike in the seaport of Constanta, during a storm, when the dike crashed despite their overprotection with tetrapods and blocks that weighted tens of tons force each.

For sure, these phenomena explain the excessive and the fast settlement and even diving of many dikes, such as the longitudinal ones in Mamaia.

More precise, the authors consider that not the dynamic impacting force of the waves has led to such accidents. The “armour” of protection made from tetrapods and large blocks has “done its duty” to dissipate the wave energy but could not prevent aggressive flow phenomena through the dikes and this one caused their collapse from inside.

If this explanation is accepted, the solution comes by itself and consists in building perfect tight dikes both in their body (to prevent the phenomenon of "cascading" presented and incriminated above) and in the adjacent foundation soil (to prevent entrainment of sand from the base of the dike and the phenomenon known as subduction). The constructive solutions depend of course of many other conditions, but overall, it is likely to be able to reduce the whole cost of the works by reducing the volume of protection "armour" from the seaward slope that is currently over-sized.

High permeability of sea dikes has also negative consequences for their effectiveness regarding the beach protection but here the involved phenomena and problems are much more complicated and require further explanation for the particular case of the Black Sea coastal zone.

It is proved and unquestionably that the waves reaching the shore have only an erosive action more or less intense depending on their amplitude and local situation. They just "dig" the shore and never "fill it" in the lack of a source of supply that should come from the sea (Luca, V.O. et al., 1998, Luca, V.O. et al. 2002, Tatu, G. et al., 2014). For a long time, the Romanian coast of the Black Sea, didn't have such sources: no more silt brought by the Danube river; the local "biology" has changed and the sand produced by marine shells' isn't created anymore; there are no other cliffs that could fall down and supply the beach (on the contrary, catastrophic situations have been reached where the water line has got to the foundations of buildings or even several entire streets collapsed).

Sand, once dislocated from the shore, is transported offshore by the many nearby shore currents (their origin, cause, intensity, direction etc. are very various but it is not the place here to discuss about (Kuroji, K. et al., 2006)).

There is the solution of artificial supplying with sand, especially for the beaches, solution that is currently in use nowadays in many places but in these conditions it is necessary for the sand brought with great costs to be detained and prevented from being moved forward into the sea.

The above modelling showed that marine breakwaters do not practically represent a "brake" for marine currents and therefore we believe that the existing dikes did not prevent the currents to "wash off" the sand beach eroded by the waves, this being also the cause of the total inefficiency of these works for protecting the beaches (commented also in „Introduction")

Thus, the logical solution is also that of sealed dikes but, in addition, they shall be arranged to form enclosed areas, for playing the role of "silt traps". Also, to allow continuous access of fresh water, these dikes have to be submerged, giving them, this way, additional advantages such as not affecting the marine landscape and marine sports (water skiing, light boats etc.)

This are the basic ideas of the Romanian Patent highlighted in "References" – Lazar, D.P., et al., which we support wholeheartedly because we believe that this solution is logical, efficient and, by far, the cheapest one for protecting the marine beaches.

## References

Kuroki,K., Goda,Y., Ochi,Y, Morphological Analysis along Southern Romanian Black Sea, JICA Study, June 2006

Iulian,C., Combaterea Eroziunii Marine (Marine Erosion Control), FREDD Ed., Bucharest 2005, ISBN 973-0-03924–0

Iulian,C.,Soluții noi de protecție pentru litoralul marin (New solutions for the protection of the seashore), Ph.D. Thesis, Technical University of Civil Engineering of Bucharest, 1994

Lazăr,P.D.,Contribuții la hidraulica disipării și captării energiei valurilor mici cu aplicație la protecția litoralului românesc al Mării Negre (Contributions to hydraulic dissipation and wave energy capture, with application to protect the Romanian Black Sea coast), Ph.D. Thesis, Technical University of Civil Engineering of Bucharest, 1996

Lazar,D.P., Luca,V.O., Iulian,C., Tatu,G., Patent No. 120279, Construcție Hidrotehnica Uşoara pentru Protejarea si Refacerea Plajelor Marine (Light Hydro Construction for protecting and restoring marine beaches), State Office for Inventions and Trademarks, Romania, 2006

Luca,V.O., Luca,B.A., Construction Hydraulics, “Environment Engineering Hydraulics”, “Orizonturi Universitare” Ed., Timișoara, 2002

Luca,V.O., Tatu,G., Environmental Impact of Free Surface Flows. Evaluation and Protection, “Environment Engineering Hydraulics”, “Orizonturi Universitare” Ed., Timișoara, 2002

Luca,V.O., Tatu,G., Petrescu, V, Hydrodynamics of watercourses, Technical University of Civil Engineering of Bucharest, 1998

Manoliu, I. – Harbours, E.D.P., Bucharest, 1967

Tatu, G., Dimache, Al., Iancu, I., Coveianu, A., Mathematical model for calculating wave runup phenomenon, „Dorin Pavel” Conference, 23-25 May 2014, Politehnica University, Bucharest

**Coastal Management: Changing coast, changing climate, changing minds**
ISBN 978-0-7277-6114-9

ICE Publishing: All rights reserved
doi: 10.1680/cm.61149.009

ice
Institution of Civil Engineers
publishing

# Dealing with Uncertainties in Long-term Predictions of a Coastal Nourishment

**Anna Kroon**, Svašek Hydraulics, Rotterdam, The Netherlands
**Bas van Leeuwen**, Svašek Hydraulics, Rotterdam, The Netherlands
**Dirk-Jan Walstra**, Deltares, Delft, The Netherlands
**Gerard Loman**, Boskalis, Papendrecht, The Netherlands

## Introduction

In coastal engineering, the wave- and tide-driven alongshore sand drift and the resulting shore erosion or accretion can be fairly well predicted with numerical 2DH models. Despite recent advances in process-based morphodynamic modeling, application of such models is inevitably associated with considerable uncertainties from various sources.

This paper describes a comprehensive approach identifying and quantifying individual sources of uncertainty and integrating these into a single probability density function (PDF) of aggregated model outcomes. The approach was developed for the tender of a Design, Build and Maintenance (DBM) contract to upgrade a sea dike at the Holland coast. The reinforcement comprises a sand beach and dune, constructed at the foreshore of an existing 6 km long hard sea defense: the Hondsbossche and Pettemer Zeewering" (referred to as HPZ from here onwards). The main challenge was to design the nourishment so that it meets the functional requirements whilst minimizing the maintenance costs over the 20 year post-construction period.

The above DBM tender was jointly issued in September 2012 by the regional water board "Hoogheemraadschap Hollands Noorderkwartier" and the public works department "Rijkswaterstaat" of the Dutch Ministry of Infrastructure and Environment. In October 2013 this DBM contract has been awarded to a contracting joint venture of Van Oord and Boskalis (the authors formed part of the winning tender team). The designed reinforcement requires 26 million m$^3$ of sand and about 14 million m$^3$ of sand to compensate for expected losses (settlement, hydraulic erosion and aeolian erosion) during the construction and maintenance period. Because of the functional requirements, about 85% of this total volume has been placed during the construction period in 2014. After completion early 2015, the remaining 15% is expected to be necessary for maintenance during the 20 years post-construction period (in 2-3 maintenance campaigns). Given the relatively long maintenance period, an objective quantification of the uncertainties of the nourishment design was imperative for the lump sum DBM contract.

### Project Area and History

Prior to construction of the nourishment, the HPZ was the only sea dike in the predominantly sandy dune sea defense of the Dutch province of North-Holland, see

Figure 1. According to the Dutch Water Law this sea defense is to protect the low-lying area behind against storm surge flooding's having a return period of 1 in 10.000 years. Every 5 years, sea defenses undergo a mandatory evaluation. Due to updated hydraulic design conditions the HPZ did not pass the safety evaluation in 2005. Following a comprehensive Environmental Impact Assessment the responsible public authorities have selected the best possible option of required strengthening of this 6 km long hard dike section to be a sandy beach and dune on the foreshore of the sea dike, with sand type similar to the adjacent dune coast, and having a smooth outcropping waterline.

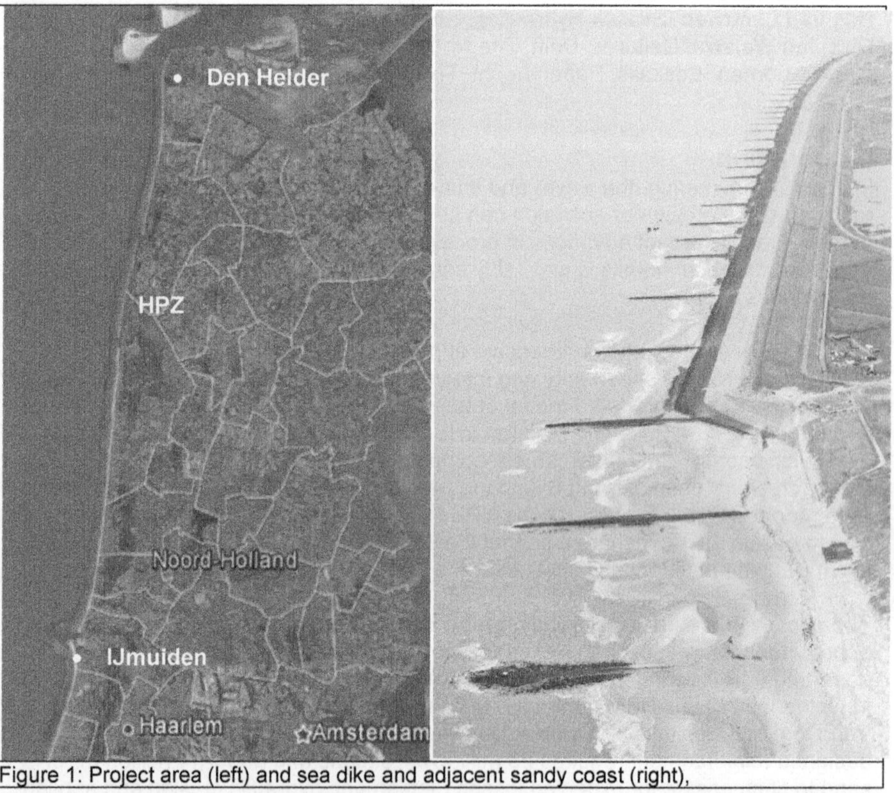


Figure 1: Project area (left) and sea dike and adjacent sandy coast (right),

## Predicting Sediment Loss with a Hydro-morphological 2DH Model

**Approach to Determine Sediment Loss and Modelling Uncertainties**

The main objective is to predict and minimize the hydraulic sand losses for the nourishment, and to underpin the design with a sound and comprehensive estimate of the uncertainties of the expected sand losses from the nourishment during the 20 year post-construction period. The numerical 2DH model package (Finel2D) was applied to simulate the relevant tide-, meteo- and wave-driven alongshore sediment transport and losses. This 2DH model package can also reasonably reproduce the effects of processes, such as rip currents and large scale eddies on local sediment transport and morphology. Finel2D, however, does not account for three dimensional processes such as undertow and asymmetric wave motion. By considering an equilibrium cross-shore profile, cross-shore processes are assumed to be less relevant for the estimation of offshore sand losses.

The reliability of Finel2D was extensively verified for the relevant wave- and tide-driven alongshore transport processes. The calibrated model is subsequently used in numerous sensitivity calculations with the purpose of establishing probability density functions (PDFs) for the expected losses in relation to several uncertainties within the model and its underlying assumptions. By means of a Monte Carlo simulation all the PDFs are combined into a single joint PDF and its corresponding cumulative probability density function. Herewith, the chance of exceeding the predicted sediment losses was determined.

**Description of 2DH Model Package**

For simulating the above 2DH flow use has been made of the numerical flow module of Finel2D (Svašek Hydraulics, 2012), see Figure 2. In this flow module the shallow water equations are solved by finite-element schematization and by explicit time-stepping. The advantage of the finite element method in comparison to for example finite difference methods is the flexibility in triangular grid generation. This allows the user to apply a high resolution in the area of interest and a low resolution in the far-field. This allows for a very efficient grid design that minimizes the required CPU-time and enables us to perform long-term morphodynamic modelling (20 years). Even though the existing breaker zone width under predominant wave conditions requires a spatial resolution of at least 25 m.

As shown in Figure 2, the flow module is linked to the wave module of Finel2D. This wave module contains the spectral wave model SWAN (Delft University of Technology, 2012). The wave module provides the forcing that induces the (breaking) wave driven flows as well as the stirring of sediment. Bed level, water level and current conditions from the flow module are used as input for calculating the spatially varying wave conditions. These calculated wave and flow conditions are used as input of the morphological module, see Figure 2. In this module, the local sediment transport and its spatial gradients are calculated from which the bed changes are derived. Based on previous modelling experience in this region, the empirical sand transport formula of Soulsby-Van Rijn (Soulsby, 1997) is adopted because of its suitability for predicting the wave and current induced sediment transport. To limit computational time, the bed change calculations are conducted with a morphological acceleration factor of about 50. The wave calculation interval (the time between consecutive wave calculations) is 3 hours. To further limit the simulation time, the wave climate is represented by a limited number of wave conditions. The number of wave conditions is reduced in such a way that the remaining conditions accurately represent the annual net and gross alongshore sediment transport (Walstra et al., 2013). It was found that 18 wave conditions were sufficient to represent the measured 20-year time series of wave parameters at the nearby station “Meetpost Noordwijk”.

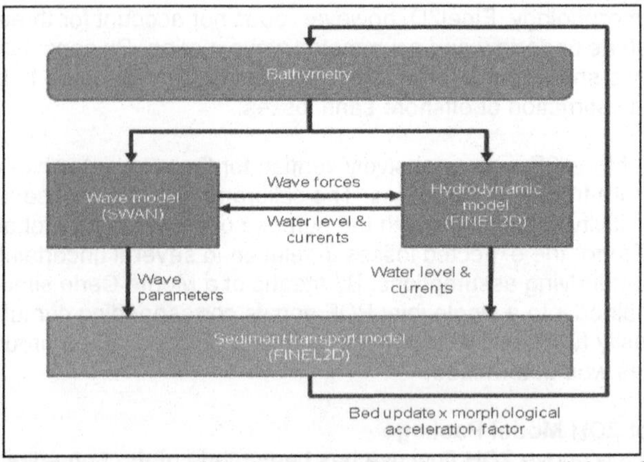


Figure 2: Schematic overview of flow, wave and bed change modules of Finel2D model

**Modelling Assumption: Use of Equilibrium Foreshore Profile**

The foreshore of the new sand coast was designed as a Dean equilibrium profile (Dean, 1977). This profile ranges from the dune foot (3 m+MSL) to the "closure depth" (estimated at 7 m-MSL). The Dean profile and the closure depth were derived from the 20 annual profiles surveyed at a nearby coast section (unaffected by groins and sand nourishments). The cross-shore profile evolution towards a (dynamic) equilibrium was estimated with Unibest-TC (Walstra et al., 2012) to establish the depth of closure and to assess the influence of cross-shore sand-bar migration on the short-term fluctuations of the sediment budget. With the incorporation of the equilibrium profile it is assumed that the long-term offshore sediment losses can be neglected (the natural sand bars that occur in this area do not influence the long-term sediment budget, see also Wijnberg en Terwindt, (1995).This assumption justifies a 2DH approach.

## Calibration and Validation of Hydro-morphological 2DH Model

**Calibration and Validation of Wave and Flow Modules**

The tidal water level of the flow module has been calibrated and validated for the main astronomic constituents of five closest tide gauge stations using the HPZ project-model (North Sea model). The validated M2 amplitude has a maximal deviation of 4% of the measured amplitude. Comparison of predicted and measured current velocities at a.o. "Stroompaal IJmond" shows a slight deficit of tidal asymmetry. This quantified effect of these deviations onto the sediment transport is accounted for in the PDF of model uncertainties, see hereafter.

The prediction of reliable alongshore wave-driven currents have been investigated with the HPZ process model (straight uniform coast section). Here the results of Finel2D are compared with two well-established numerical 1DH models (the analytical model of Thornton and Guza, 1986; and the numerical models WaTrUC & SedTrUC,Hydronamic, 2012). Figure 3a shows the results of the wave decay prediction due to depth-induced wave-breaking of oblique shore-wards travelling waves. Figure 3b shows the predicted profiles of the resulting wave-driven alongshore current. On this barred foreshore profile, the wave height profile of the Finel2D model showed good agreement with the other profiles.

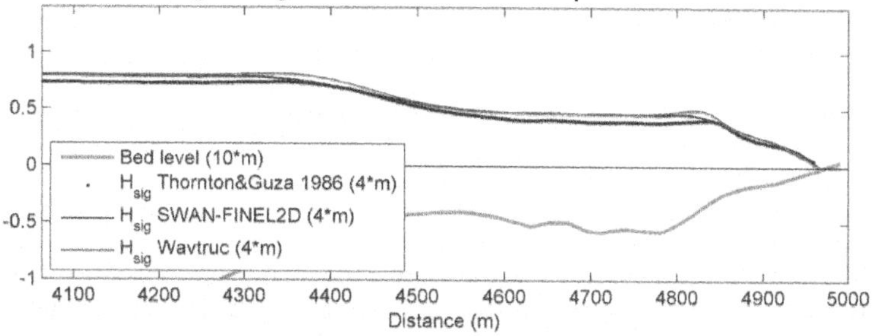


Figure 3a: Predicted wave height decay due to depth-induced wave-breaking.

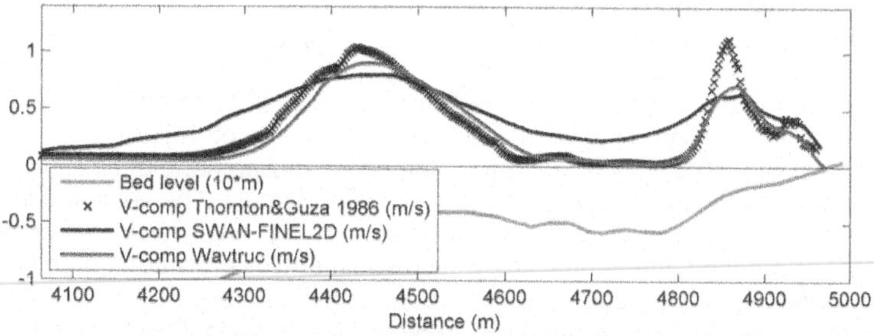


Figure 3b: Predicted wave-driven alongshore current profile for above decay

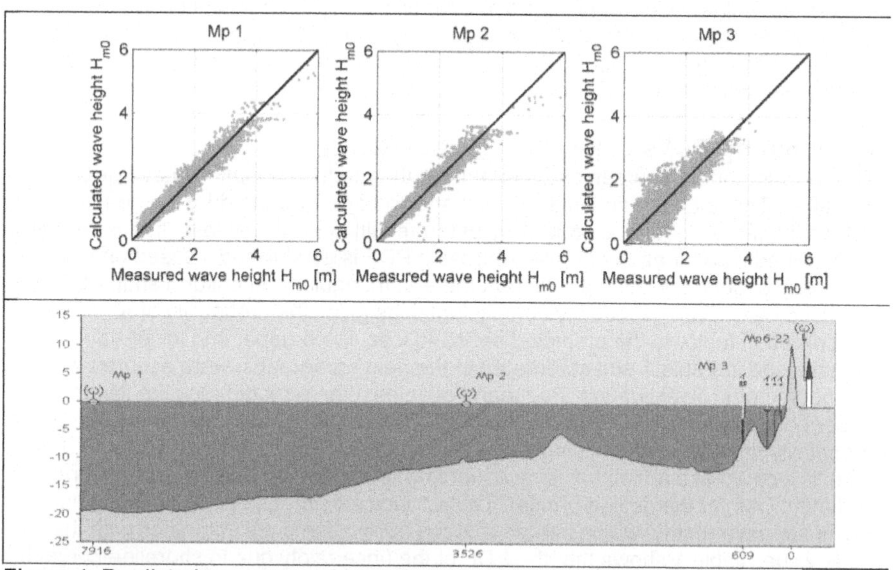


Figure 4: Predicted vs measured wave heights (year-round) at HPZ nearshore locations Mp1 (top left), Mp2 (top middle), Mp3 (top right) and their location in cross shore profile (bottom).

The Finel2D profile predicted somewhat more diffusion in the cross-shore distribution of wave-driven longshore current, compared to the other 2 profiles. This deviation has been included in the PDF of model uncertainty, see hereafter.

The wave module was calibrated and validated with measurements of three wave buoys located along a typical near- and foreshore profile at the HPZ, see Figure 4. The wave conditions have been measured here during several years by "Rijkswaterstaat" at o.a. 20 m water depth (MP1), at 10m water depth seaward (MP2) and landward (MP3) of the outer breaker bank. Scatter plots of the significant wave height show a narrow band of predicted versus measured significant wave heights without a systematic deviation, see Figure 4. The scatter deviation has been included in the PDF of model uncertainty, see hereafter.

**Calibration and Validation of Sediment Drift and Bed Change Module**
The morphological module of Finel2D has been calibrated and validated for two dominant processes. First, the average annual long-term wave- and tide driven sediment loss in the foreshore was predicted with the HPZ project-model and compared with analyzed long-term profile measurements of "Rijkswaterstaat" (Van Oord & Boskalis, 2012). As a calibration, the coast section north of HPZ was used to estimate the scaling factor of the Soulsby-Van Rijn formula. The sediment loss in the coast section south of HPZ was used as validation. This finally yielded a scaling factor of 0.2. Second, the annual tide-driven sediment loss has been predicted with the similar MV2 project-model. Here the predicted large scale depth-increase is due to the contracting tidal flow around the seaward "Maasvlakte 2" extension of the Port of Rotterdam. The predictions have been compared with available measurements. This finally yielded a scaling factor of 0.3 for predicting the dominant tide-driven sediment transports at water depths deeper than about 5 m-MSL. As a final calibration setting the weighted averaged scaling factor of 0.28 is applied in order to represent the losses from both the breaker zone and deeper waters.

## Determining PDFs of Modelling Uncertainty

The most important sources of uncertainty in the model calculations have been identified. These uncertainties are stochastic and are accounted for by means of estimating the exceedance probability of the resulting predicted loss. In this chapter it is explained how for each of these sources a PDF is established. Based on risk-modelling experience with several parameters, triangular and uniform shaped PDFs are used here. Each probability density functions gives the probability of a multiplication factor to the predicted net sand loss. Each upper and lower bound, summarised in Table 1 and elaborated in the next paragraphs, were estimated from a thorough model investigation. By means of at least three model runs the lower and upper bounds of each PDF are established. Of most PDFs the peak of the triangular probability density function coincides with a factor 1. Only for one triangular PDF the peak is located at a factor 1.1 (to account for the under-prediction of alongshore sediment loss for the design profile). Table 1 indicates that most of the triangular PDFs are asymmetric, which implies that the 50% value does not match with the peak value. Table 1 shows that the PDF of the uncertainty due to shoreline curvature has been estimated as uniform. In Figure 5 an example is shown of the resulting PDF for wave-driven sand loss.

Most uncertainty ranges in Table 1 are based on model outcomes of sensitivity calculations. However, several of the identified uncertainties are not directly quantifiable by means of sensitivity calculations and follow from uncertainties in measurement equipment for example or expert judgement. In the next paragraphs the methodology of estimation of each of the uncertainty ranges is explained further.

Table 1: Overview of uncertainty ranges

| Uncertainty | Lower bound | Peak value | Upper bound | PDF-shape |
|---|---|---|---|---|
| Wave-driven sand loss | -50% | 0% | +100% | Triangular |
| Tidal-driven sand loss | -40% | 0% | +40% | Triangular |
| Model uncertainty: curved coastline | -30% | 0% | +50% | Uniform |
| Model uncertainty: 'seeding' | -15% | 0% | 0% | Triangular |
| Model uncertainty: spin-up | +10% | +10% | +25% | Triangular |
| Model uncertainty: groins | -33% | 0% | +16% | Triangular |
| Model uncertainty: tidal asymmetry | -22% | 0% | +22% | Triangular |
| Yearly variation in wave forcing | -25% | -2% | +27% | Triangular |
| Cross-shore profile variations | -10% | +7% | +20% | Triangular |

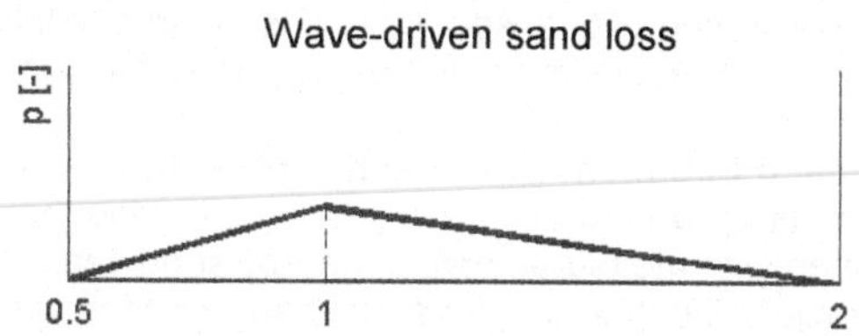


Figure 5: Probability density functions for the considered uncertainties

**Uncertainty in Measured Sand Loss Data Used for Calibration and Validation**

The observed trend of sand volume loss in the adjacent coastal sections of the HPZ contains uncertainties due to measurement errors and seasonal variation in profile. But most important uncertainty is caused by the method by which the trend of sand volume loss is established. Which outliers are disregarded and whether the average or maximum trend is chosen. The resulting trend in sand volume loss varies between -50% and +100%, see Table 1for the uncertainty range related to "wave-driven sand loss". The observed sand losses in the large scale scouring around "Maasvlakte 2" is estimated to vary between +-40% by measurement errors only, see Table 1for the uncertainty range related to "tidal-driven sand loss" (Van Oord & Boskalis, 2013).

**Uncertainties in Modelled Sediment Drift for Model Settings and Schematisation Choices**

The HPZ project-model was extensively calibrated and validated. As a result, the systematic error of prediction is expected to be minimal. However, various sources of uncertainty remain that originate from the schematization choices such as the order in which the selected wave conditions are imposed (chronology effect) and the schematization of the existing groins. The remaining uncertainties from this model calibration and validation are summarised in Table 1. These concern, the presence of groins, the asymmetry in the tidal-induced currents, the coupling frequency between the wave model and the hydrodynamic model, the parameter settings of the Soulsby-Van Rijn transport formula, the "seeding" order of wave conditions and the uncertainty due to a curved shoreline. It is not assumed that by investigation of the sensitivity to the mentioned settings and choices all uncertainties in the model are captured. To account for these unaddressed model uncertainties an additional uncertainty for uncaptured processes and non-linear model behaviour on a curved coastline is assumed which has a range of -30% to +50%.

**Uncertainties in Modelled Sediment Drift due to Duration of Wave Exposure**
The schematized wave climate is based on a selection of wave conditions, which generate the 20-year average alongshore transport. Due to this schematization choice the model is unfit to estimate the variation in alongshore transport due to annual fluctuations in the wave forcing. As such, the effect of the year-to-year variation in wave forcing is investigated using a calibrated 1D coastline model (CERC formula type of U.S. Army Corps of Engineers) and the 2DH model SWAN. Based on 20-years wave measurements it is found that in the "worst" year the expected loss can be up to 60% higher. For longer maintenance periods worse years (high wave energy) and mild years (low wave energy) smoothen the cumulative uncertainty which is found to be -25% and +30% for the planned 10 year maintenance interval. Year-to-year variation in wave forcing does not only influence the surf zone transport but the tidal transport as well due to enhanced wave stirring. Hence a similar analysis is performed on the wave stirring of tidal sand drift. It is found that the year to year variation is comparable to the results found for the surf zone transport, see Table 1.

**Uncertainties in Modelled Sediment Drift due to Cross-shore Profile Variations**
The design profile of the near- and foreshore is shaped following the equilibrium profile of Dean, calibrated with measurements of nearby sand coast section. This long-term averaged profile will evolve into a natural barred beach system. The alongshore sediment drift, predicted for this design profile, may vary for natural deviations from the theoretical profile. As such longshore drift predictions have been made with the HPZ process-model for the 15 years of measured profiles at a comparable nearby location. These calculations revealed that the annual sediment transports can vary with -10% to +20% compared to that for the design profile, and that the average transport on the natural profiles is on average 10% larger than the transport on the design profile. It is assumed that these sediment transport uncertainties also apply to the alongshore sediment losses, see Table 1.

## Determining the joined PDF of sediment loss prediction

It is assumed that all the PDFs derived in the previous chapter are statistically independent. With a Monte Carlo simulation a joint PDF of the predicted net volume loss is determined. By means of 5 million random samples of each of the individual PDFs a joint probability density function has been established. The joint PDF and its cumulative probability of exceedance are presented in Figure 6.

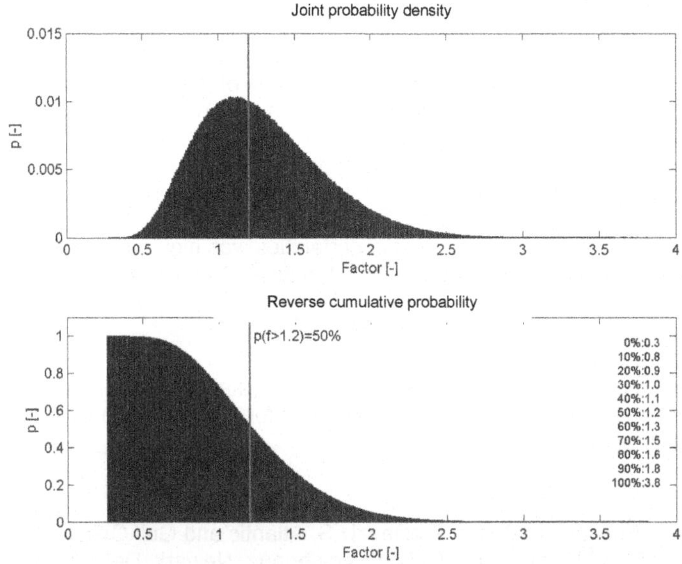


Figure 6: Joint PDF and probability of exceedance for predicted net sand loss

Since most individual PDFs are asymmetric the resulting joint PDF is asymmetric as well. It has its maximum (highest probability density) around a multiplication factor of 1.1, and a long high-end tail that ends at a multiplication factor of about 4 times the model prediction. The corresponding cumulative exceedance probability clearly shows the most important result, that is that the prediction with a well-calibrated and validated model needs to be multiplied with a factor 1.2 for representing the 50% exceedance chance. And not a factor of 1.0, which is often assumed, corresponding here with 70% exceedance chance. The joint exceedance curve of Figure 6 further shows - as an example - that the predicted sand loss over the 20 years has to be multiplied by a factor 1.8 for representing the 10% exceedance chance.

## Conclusion

The presented model approach and results were established within a 12 months lasting tender phase. In absence of such a thorough uncertainty analysis, it is common practice in coastal engineering to apply a multiply factor of 1.0 to the outcome of a calibrated hydro-morphological 2DH model for a 50% chance of exceedance. And to consider uncertainty factors of 2.0 and 0.5 for representing the bandwidth in between 10% and 90% chance of exceedance. The here presented comprehensive method results in deviating uncertainty factors: 1.2 for 50% chance of exceedance and 1.8 and 0.8 for 10% and 90% chance of exceedance. With these outcomes it was possible to perform an in-depth risk analysis of the 20-years nourishment volume as part of the financial DBM bid.

## Discussion

It is shown that triangular PDFs for the various investigated uncertainties give a non-triangular joint PDF. Important is the a-priori definition of the minimum and maximum values of these triangular PDFs. Here, the absolute maximum and minimum values have been estimated. Alternatively, the 10% and 90% exceedance values could also have been estimated and applied to asymmetric triangular PDFs or asymmetric PDFs with asymmetric Gaussian distribution tails. However, other uncertainty analyses found that such results do not differ much from the results of the present practical method as regards to the ranges that were of interest in the presented application.

The outlined method is practically applicable and provides an as objective as possible and reproducible approach to account for model prediction uncertainties in which the uncertainties associated with a 20 years maintenance period were included. This comprehensive estimation of uncertainties was high valued by the client and we believe has made a substantial contribution to the contract award to Van Oord and Boskalis.

## Acknowledgements

The authors want to thank the Dutch Government (Rijkswaterstaat) and the Water Board Hoogheemraadschap Hollands Noorderkwartier for their kind approval on publication of this research.

## References

Dean, R. G., 1977. Equilibrium Beach Profiles: U.S. Atlantic and Gulf Coasts. Department of Civil Engineering, Ocean Engineering. Newark, Delaware: University of Delaware, Report No. 12, 45p.

Delft University of Technology, (2012), "SWAN Scientific and technical documentation", SWAN Cycle III, version 40.91

Hydronamic, (2012), Personal communication with Gerard Loman (Boskalis) about use of tried & tested models WavTrUC (Wave Transformation along Uniform Coast) and SedTrUC (Sediment Transport along Uniform Coast)

Soulsby, R.L., (1997), "Dynamics of marine sands: a manual for practical applications", Thomas Telford, London

Svašek Hydraulics, (2012), "Technical Manual FINEL2D", version 7.13, Rotterdam

Thornton, E.B. and R.T. Guza (1986). Surf Zone Longshore Currents and Random Waves: Field Data and Models. *Journal of Physical Oceanography* 16: p 1165-1178.

U.S. Army Corps of Engineers. (2002). Coastal Engineering Manual. Engineer Manual 1110-2-1100, U.S. Army Corps of Engineers, Washington, D.C. (in 6 volumes).

Van Oord & Boskalis (2013), "Zwakke Schakels Noord-Holland, Versterking en Onderhoud, Kustdeel RSP17.00 – RSP28.32, Dialoogproduct Technische Kwaliteit, Deel A Zandberekening", *DBM bid document*

Van Rijn, L.C., (1995). Sand budget and coastline changes of the central coast of Holland between Den Helder and Hoek van Holland. WL|Delft Hydraulics H2129.

Walstra, D.J.R., Reniers, A.J.HM., Ranasinghe, R., Roelvink, J.A., Ruessink, B.G., 2012. On bar growth and decay during interannual net offshore migration. Coastal Engineering 60, 190–200. doi:10.1016/j.coastaleng.2011.10.002.

Wijnberg, K.M., Terwindt, J.H.J., 1995. Extracting decadal morphological behavior from high-resolution, long-term bathymetric surveys along the Holland coast using eigen function analysis. Marine Geology 126 (1–4), 301–330. doi: 10.1016/0025-3227(95)00084-C.

**Coastal Management: Changing coast, changing climate, changing minds**
ISBN 978-0-7277-6114-9

ICE Publishing: All rights reserved
doi: 10.1680/cm.61149.019

# Network Rail – a New Approach to Coastal Asset Management – North Wales Case Study

**Graham Kenn,** JBA Consulting, Haywards Heath, UK
**Kevin Giles**, Network Rail, Cardiff, UK

## Introduction

Network Rail in partnership with JBA Consulting have recently developed a new methodology to assess the risk posed by coastal processes to the Network Rail portfolio of coastal, estuarine and river flood and erosion defence assets, and the rail track infrastructure that they protect. This paper discusses the context of the system in relation to the latest guidance on effective asset management, the new coastal process risk assessment methodology that was established, the tools developed to deliver the system, and the findings from the pilot study which considered 29 Coastal, Estuarine and River Defence (CERDs) assets located across the north Wales coast (see Figure 1). The CERD assets owned and maintained by Network Rail include a variety of structures including sea walls, rock revetments, masonry pitching, breakwaters and softer elements such as beaches, salt marsh and dune systems.

Network Rail are responsible for managing and maintaining many hundreds of CERD assets across England and Wales. Following a recent review issued by the Office of Rail Regulation (ORR) (2013) a number of suggestions were made regarding future developments Network Rail could consider to improve their system for managing the risks posed to CERD assets from coastal processes. This report recommended that Network Rail review existing asset management policies at a national level and ensure that assets are being managed through a risk based approach that takes into consideration whole life-cycle costing. The new methodology developed addresses the recommendations made within the ORR report and includes improved understanding with regard to:

- The number and type of CERDs managed by Network Rail
- The condition of these assets
- Developing whole life-cycle costing analysis for management options
- Providing tools that allow Network Rail to deliver robust Coastal and Estuarine Asset Management Plans.

These requirements were brought into sharp focus by the coastal storms and subsequent CERD asset damage experienced during the 2013/14 winter.

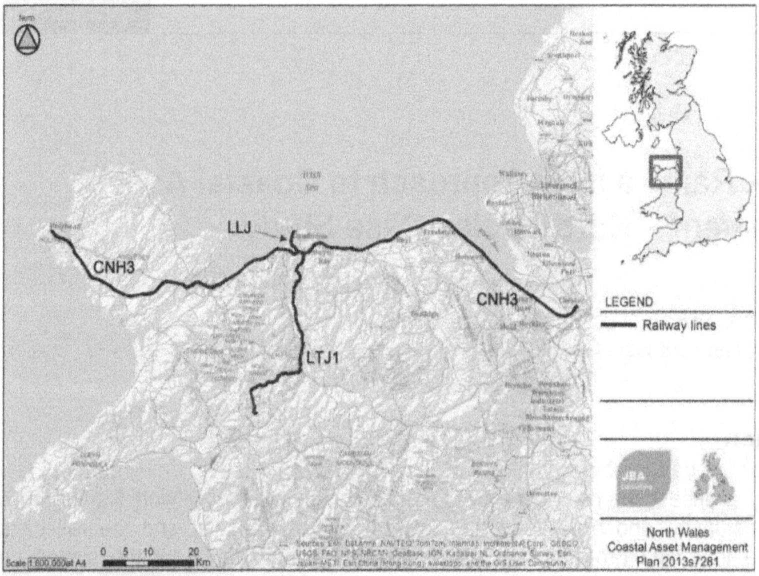


Figure 1: North Wales rail lines

**Wider Context**

Network Rail are by no means alone in the drive to improve their approach to asset management and investment. The recent ICE Guiding Principles of Asset Management report (ICE, 2013) highlighted that 'The UK's approach to infrastructure investment has in general been timid, uncoordinated, incremental and wasteful in its procurement and insufficiently targeted to supporting balanced and sustainable growth in the economy'. Many of Network Rail's CERD assets were constructed over 100 years ago, with only minor modifications and maintenance since this time. This poses a significant challenge to asset management teams, who are charged with developing a management plan that will deliver an effective strategy for the replacement or repairs required to extend the life span of their assets, in order to ensure a safe and efficient railway network is maintained over the coming years.

This is coupled with the increasing physical pressures exerted on these assets by the effects of climate change. A further quote provided by Network Rail in the ICE Guiding Principles report (ICE, 2013) states that effective asset management supports the delivery of a promise to "deliver the timetable, so that trains run safely, punctually and reliably now, and in increasing numbers in the future".

**Requirements for the Proposed Asset Management System**

The principles of the Network Rail CERD asset management system followed latest guidance and best practice documented in ISO 55000:2014, Asset management – Overview, principles and terminology (ISO, 2014). This ensured the most up-to-date methodology and concepts were utilised to help develop the system. ISO 55000 states that the core elements of a best practice asset management plan should, amongst others considerations, attempt to:

- Recognise the associated risks of the assets;
- Inform evidence-based decision making in a system wide and whole asset lifecycle approach;
- Understand the true costs of ownership and the implications of deferred interventions;
- Future proof the asset management plan to consider long term goals; and
- Turn data into useful information.

These five core elements are discussed in more detail in the following pages.

## CERD Asset Management Methodology

Traditionally Network Rail CERD asset management has focused on highlighting individual asset defects and then developing a plan to rectify these defects, often on a site by site basis. The new methodology developed as part of this study considered not only the physical condition of the asset but also the wider risks posed by coastal processes to the asset and the impacts to the rail infrastructure they protect, with the aim of developing a system that used consequence based risk evaluation. The coastal processes deemed appropriate for inclusion within the new methodology were as follows:

- Risk of coastal erosion
- Risk of wave overtopping
- Risk of Still Water Level flooding

In order for the asset management plan to be effective over an extended time frame these risks were analysed in the present day but also for three future time epochs 2025, 2055 and 2115. By analysing the future risk profile, it allowed Network Rail to start developing a long term view of their likely liabilities both in terms of the increasing coastal process risks but also the likely future maintenance and capital costs.

### Erosion Risk

A primary function of many CERDs is to protect the shoreline against wave action, which in-turn could lead to coastal erosion and damage to track infrastructure. The rate of erosion is dictated by a number of factors but primarily the wave height, wave period, resistance of the bedrock/drift deposits against which the waves are acting, and the distance between the CERD and the rail infrastructure.

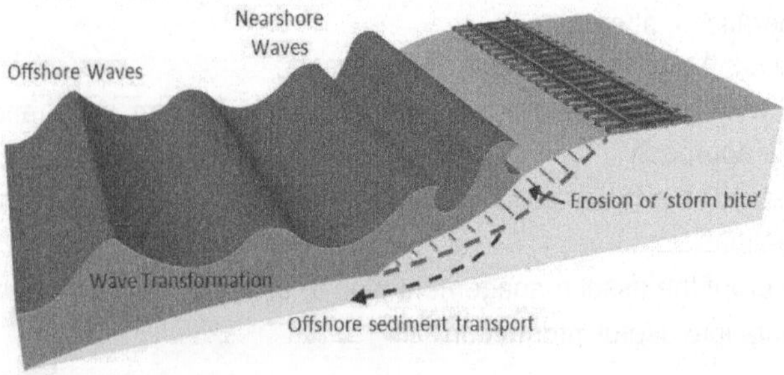


Figure 2: Components of coastal erosion risk to the track

The system developed considered four component factors, each of which were then assigned a score (low to high) dependent on the individual assessment of risk. The factors considered were:

- Condition of the CERD
- Relative influence of waves
- Track distance
- Relative erodibility

These scores were then combined to provide an overall assessment of an individual CERD assets' exposure to coastal erosion risk.

**Wave Overtopping Risk**
Risk of wave overtopping is a critical factor when considering the coastal process risk posed to a CERD and the rail infrastructure located behind the asset.

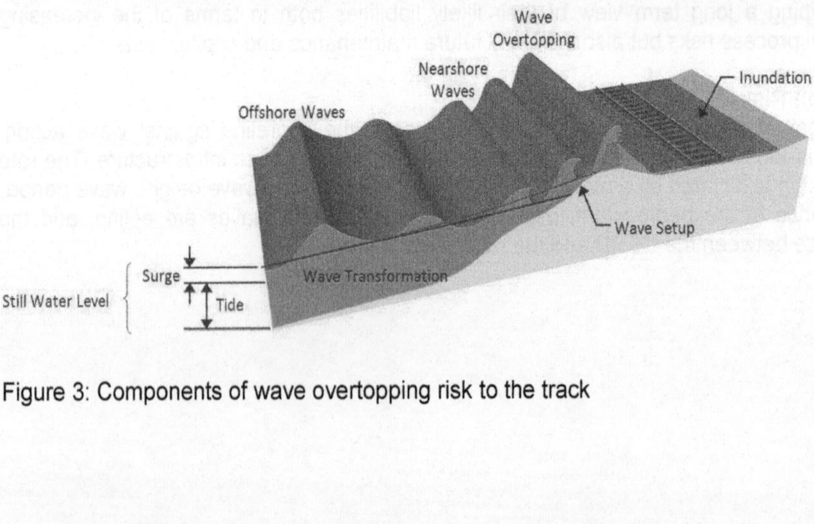


Figure 3: Components of wave overtopping risk to the track

The system developed considered two component risk factors. As with erosion risk, these factors were assigned a score dependent on the assessment of risk. The factors considered were:

- Overtopping rate at the CERD
- Track distance

These scores were then combined to provide an overall assessment of an individual CERD assets' exposure to wave overtopping risk.

**Still Water Level Flood Risk**

The final factor considered critical to the assessment of CERD asset and rail infrastructure risk was the threat posed by Still Water Level (SWL) flooding. A primary function of many CERDs is to protect the shoreline against high still water levels, which if not protected against could lead to inundation of the rail track behind. The location of the track infrastructure either within, or above, the undefended floodplain then determines the probability of Network Rail service interruption due to coastal flooding.

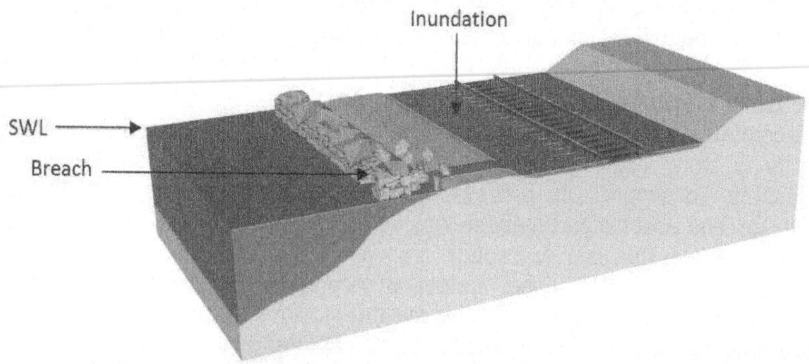


Figure 4: Components of flood risk to the track

The system developed considered two component risk factors. As with the previous categories these factors were assigned a score dependent on the assessment of risk. The factors considered were:

- Condition of the CERD
- Relative location of the line within the floodplain

**Overall Coastal Risk Assessment**

Following the assessment of the risk posed from the individual coastal processes the scores derived were combined via an algorithm to give an overall assessment of the combined risk posed by coastal processes to any one CERD asset and rail infrastructure. This combined score allows asset managers to consider their portfolio of

CERDs and rank them according to their overall coastal process risk, or the risk posed from any one of the individual component factors within the system.

## Climate Change and Future Impacts

The risk assessment was further broken down into two return period events, allowing comparison of risk against a more frequent storm event and an extreme storm event, thus providing a further insight into the risk profile of each CERD asset. This was enhanced with the use of four time epochs, with all data analysed for the risk posed by coastal processes in the present day and the years 2025, 2055 and 2105. This aligns with the time epochs used within the Shoreline Management Plans governing the study areas. As part of the structure condition assessment, projections for asset deterioration were undertaken using industry standard methodologies (Defra/Environment Agency, 2009). This was an important component of the final analysis as it allowed estimates of a structures residual life to be made. This could then be combined within the assessment of climate change impacts on factors such as sea level and wave height. This provided an overall assessment of how a structure will respond in the future when faced with a decreasing structural condition (if limited maintenance is undertaken) and increasing physical pressures and loading, caused by climate change impacts. This component of the analysis is used by the asset manager to help determine the optimum time for intervention, be that maintenance activities or large scale capital expenditure.

## Innovative Data Capture and Visualisation Tools

ISO 55000 recommends the use of sophisticated technology to provide asset managers with the ability to store, capture, analyse and exploit information across the entire portfolio of assets. To simplify the process of capturing the asset condition data a bespoke iPad App and associated database was developed (AssetCoast) as part of the system. The App allows the user to capture a range of site condition data including functionality for geo-referenced asset defect photographs and compatibility with topographic survey equipment. The use of this system provided significant time savings when compared with more traditional methods of data capture.

Following collection of the site inspection data a range of tools were used to assess the risk components discussed above, these included wave transformation modelling, wave overtopping modelling and GIS interrogation. Once all data were collected the outputs were presented via a series of visualisation tools, including:

- An asset condition interactive pdf
- A risk mechanism matrix
- An asset investment prioritisation tool

### Asset Condition Interactive pdf

The asset condition interactive pdf allows Network Rail to see the location of all the assets that were included within the assessment and provides three layers of information that can be interrogated. Hovering the pointer over an individual asset provides a summary box of asset information. This can then be clicked to pull up a detailed plan of the asset showing colour coded and geo-referenced asset defects. A further click takes the users through to a detailed inspection report. This is an extremely powerful tool for the Network Rail asset manager effectively collating all their asset data information in one place; thus allowing them to interrogate the system and find any information they require, all provided within a single pdf file that is portable and can be used in the field as well as the office.

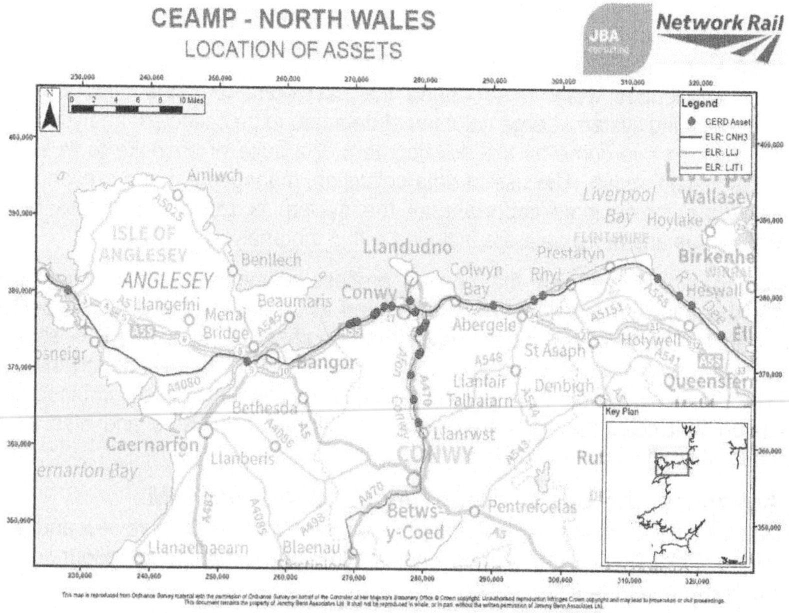


Figure 5: Asset overview window (Note: Some information has been removed for confidentiality)

### Risk Mechanism Matrix

The risk mechanism matrix pulls together all the analytical data that was collated (asset condition, erosion risk, overtopping risk and still water level risk) for all CERDs. The matrix is an extremely powerful tool and allows the user to sort, rank and filter a huge volume of data and return information on whichever component the asset manager is interested in. For example, this means that at a touch of a button the asset manager can find out the relative risk posed to all his assets from wave overtopping for a 1 in 200-year storm in the year 2055.

### Asset Investment Prioritisation Tool

The final tool included within the AssetCoast system is an asset investment prioritisation index. This index provides a summary of all the data gathered during the process and presents recommendations for asset investment over the four time epochs considered. The index is derived from the raw data analysed, the results from the algorithms and also engineering judgement, to provide a suggested schedule for both maintenance and capital expenditure. It makes suggestions on the timeframes when maintenance intervention will no longer be effective and capital investment should be considered to ensure that a continued sufficient standard of protection is provided.

The asset investment prioritisation index meets all the requirements of an asset management plan as outlined in ISO 55000 and so can be considered a vital tool to allow Network Rail to holistically manage their entire asset portfolio.

## Challenges Faced

A number of challenges were faced during the development of the coastal asset management planning system. Large volumes of data had to be collected, analysed and presented. This was overcome by the development of a suite of bespoke tools which have been discussed above. The use of data collection, management and presentation tools will become increasingly important as the system is rolled out across other Network Rail routes and more data are included within the analysis.

Ensuring comparability between the coastal process elements included within the system posed a particular challenge. It was agreed at an early stage that no specific coastal risk should be weighted more heavily than another. To overcome this issue a simple scoring system was used and applied to all elements considered during the risk analysis. The use of algorithms then allowed the separate elements to be combined and scores derived that could be compared between different risk areas and also as a combined total without including any bias.

## Future Advances, the Use of 'Real Time' Updates and BIM

The system was developed in response to the requirements of the ORR review and was trialled in North Wales. However the intention is for it to be appropriate for rollout across the UK wide Network Rail coastal route.

As discussed previously the system utilises a number of innovative data capture, analysis and visualisation tools to provide the asset manager with all the information they require to make informed decisions to address the coastal risk posed to their assets. Currently the system is used to gather historic data, current condition data and future projection data to provide a snap shot of the performance of coastal assets against the pressures of coastal processes both now and in the future.

For future iterations of the AssetCoast system an increasing component will be delivered as 'real time'. This will allow the asset manager to update information about their assets. For example, if repair works have recently been completed the asset manager will have the ability to upgrade the asset condition score. This will then be fed into all the algorithms and outputs to show how this intervention has affected the overall risk posed to the asset and also how this affects the asset ranking when compared with other assets in the portfolio. This will be a useful tool not only for post intervention updates but as a means of assessing which future interventions would be most effective.

The ICE Guiding Principles of Asset Management (ICE, 2013) states that 'A single version of the truth needs to be constructed in place of the plethora of separate and sometimes contradictory data sources that often exist today'. Future iterations of the system will seek to bring together an increasing amount of information that the asset manager requires, adopting the principles of BIM with the aim of ultimately providing a 'one stop shop'. For example there will be increased functionality to include historic asset information and a live record of maintenance works.

Currently Network Rail operate a live coastal flood warning system that provides twice daily updates of the wave height and water level risk posed to coastal assets. This system could be combined with the asset management system providing a single portal for data. This would be particularly powerful when a flood warning is issued. The asset manager could easily access information regarding the condition of the asset and any particular structure weaknesses noted, allowing a much more focused response to be put in place.

## Cross System Resilience

Although the system was developed for use by Network Rail it was also developed to facilitate improved communication and collaboration with other agencies responsible for coastal asset management. In many instances a Network Rail CERD asset will be adjacent to, or even interconnected with a local authority, Environment Agency or Natural Resources Wales coastal asset. There are obvious efficiencies if works to one asset can be considered in the context of what the impact maybe to an adjacent asset, or whether there are efficiencies to be made by delivering a multi asset response. The asset scoring system developed is compatible with the widely used Environment Agency Condition Assessment Methodology (Environment Agency, 2006). In this way Network Rail can communicate much more effectively with their counterparts in other organisations, which helps to facilitate their 'memorandum of understanding'[1] and will hopefully lead to an increased collaborative approach to coastal asset management in the future.

## Conclusions

The Network Rail coastal asset management system was developed in response to recommendations within the ORR review (Office of Rail Regulation, 2013). It was intended to provide Network Rail asset manager's with a consequence based risk evaluation methodology compliant with the latest advice within ISO 55000. Table 1 highlights the key advice within ISO 55000 and demonstrates how the Network Rail system achieves these best practice objectives.

Table 1: Evidence of best practice in asset management

| Element of best practice | Provision in asset investment prioritisation index |
|---|---|
| Recognise the associated risks of the assets | Calculated through an algorithm |
| Inform evidence-based decision making in a system wide approach | Using the calculated risks to prioritise the investment to CERD assets |

[1] Memorandum of understanding between Network Rail and the Environment Agency

| | |
|---|---|
| Understand the true cost of ownership | Costing for recommended improvements |
| Future proofing the asset management plan | Using the changes in risk profile through time to identify timescales for required CERD improvements. |
| To turn data into useful information | Through the use of this tool in combination with the others developed, the asset manager has everything needed to make informed decisions, all presented in a clear, concise format with data easily accessible. |

The system is now in use across the North Wales coastline and is currently being rolled out around all other coastal routes in Wales. It has allowed Network Rail to clearly identify which of its coastal defence assets are most at risk from coastal processes in the present day, under a range of storm conditions. It shows how these risks will change over future time horizons and how their portfolio of assets rank against one another. Armed with this information Network Rail have been able to better estimate the most appropriate timeframes for asset repair or replacement, plan and allocate the appropriate budgets and assess the likely increase in risk that would occur if intervention is delayed.

Network Rail have a sizeable and extremely high value portfolio of coastal defence assets across the UK. The system has been developed with the intention of wider rollout which is likely to be seen over the coming years. The rollout will be accompanied by further technological advances with a move towards 'real time' risk analysis and reporting. With the increasing usage of BIM methodologies the system will be developed to include further streams of data, which will allow an increased number of users to gain benefit from the information. The ultimate objective is a coastal asset management system that not only allows whole organisation alignment with asset management but cross system resilience which facilitates effective dialog between all coastal asset owners.

## References

Defra/Environment Agency (2009) Guidance on determining asset deterioration and the use of condition grade deterioration curves. PAMS Project, MSF9.1

Environment Agency (2006) Condition Assessment Manual. Managing flood risk. Document reference 116_03_SD01

ICE (2013) Realising World Class Infrastructure – Guiding Principles of Asset Management

ISO 55000:2014 (2014) Asset management – Overview, principles and terminology

Office of Rail Regulation (2013) 'Review of Asset Policies' (Draft), Reporter Mandate A0.030

**Coastal Management: Changing coast, changing climate, changing minds**
ISBN 978-0-7277-6114-9

ICE Publishing: All rights reserved
doi: 10.1680/cm.61149.029

ice
Institution of Civil Engineers
publishing

# Contribution of Sand Filled Geotextile Tubes to Decrease Carbon Footprint Emissions in Building Marine Structures

**M. ter Harmsel**, TenCate Geosynthetics Netherlands bv, Nijverdal, The Netherlands

## Introduction

Carbon footprinting as an approach is relatively new and has been developed from Life Cycle Assessment (LCA), which has been around since the late 60s. Both methods take a systematic view of the supply chain from raw material extraction through to the final disposal of the product. This approach crucially prevents decisions being made which may shift the environmental burden up and down the supply chain. The impact can be quantified as a total or can be broken down to present the results as its constituent sub-systems. The latter can be used to identify priority areas for improvements.

A product carbon footprint is an assessment of the global warming potential of a product and is also known as embodied carbon. This is often measured as a cradle to gate assessment, which includes all greenhouse gas (GHG) emissions up until the point where the product leaves the factory gate. For example raw material extraction, transportation at all stages, refining, processing and fabrication up to the product leaving its final factory gate. The boundaries of cradle to site also include the transportation up to the site of use for the product. Finally the boundaries of cradle to grave are the most holistic and include all lifecycle stages. This covers the cradle to site, usage (including operation and maintenance) and finally the end of life stage (recycling, reuse, and disposal). This study considered the cradle to grave carbon footprints of breakwater systems and sludge dewatering systems.

A carbon footprint is a measure of the impact that our activities have on the environment, and in particular climate change. It is the measurement of all the greenhouse gases generated by human activity including construction works, measured in units of tonnes of carbon dioxide equivalent. The lower the carbon footprint, the lesser the impact construction works have on the environment. Engineering solutions are not just compared purely on economic terms, but are beginning to be compared on carbon footprint as well. Therefore engineering solutions that protect and improve the environment are increasingly favoured as opposed to those that have a negative impact to the environment. This paper describes and compares the carbon footprint of sand filled geotextile tubes alternatives versus conventional engineering solutions for marine structures.

## Geotextile Tube Solution

A geotextile tube is a closed ended tubular formed fabric unit tailored with regular filling ports. Its circumference and length may be sized specific for each project and are limited only by constraints of handling practicalities and site conditions. The geotextile tube contains the slurry mixture of solids and water that is pumped in, allows water to dissipate through the permeable fabric skin and retains solids within the geotextile tube. Geotextile tube solutions include use as structural units for marine and hydraulic engineering applications and use as containment and dewatering units for municipal, industrial, mining, agricultural and environmental dewatering applications.

## Marine and Hydraulic Engineering Applications

Geotextile sand filled tubes are used in the following marine and hydraulic applications: revetments, offshore breakwaters, protection dykes, containment dykes and groynes.

Figure 1: Geotextile tube applications

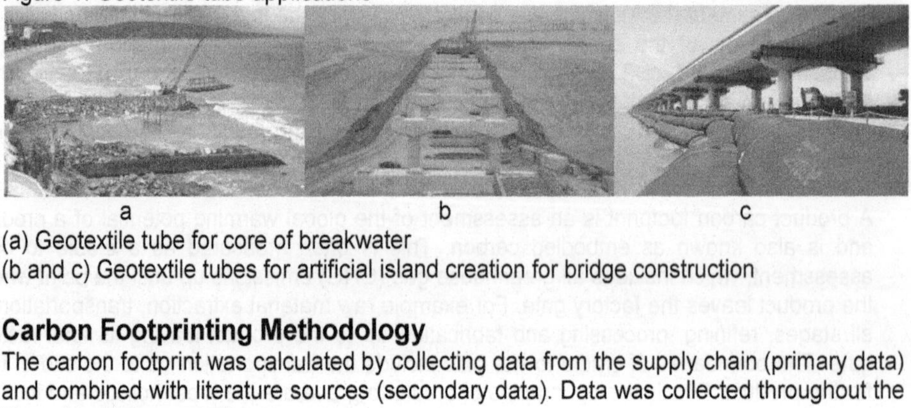

a b c

(a) Geotextile tube for core of breakwater

(b and c) Geotextile tubes for artificial island creation for bridge construction

## Carbon Footprinting Methodology

The carbon footprint was calculated by collecting data from the supply chain (primary data) and combined with literature sources (secondary data). Data was collected throughout the lifecycle which covered:

- Production of raw materials
- Transport of raw materials
- Manufacturing of the geotextiles
- Transportation to final customer
- Use
- Transport to disposal
- End of life

End of life was determined to be negligible in this study. The method used is called QuickSteps and is built upon the PAS 2050:2011 method of carbon footprinting, which is the most robust carbon footprint method to date. The main difference between these two methods is the different requirements to collect primary data from the supply chain and first tier suppliers. However the underlying principles and method requirements are otherwise the same.

The carbon footprint is measured in $CO_2$ equivalents ($CO_2e$) and all IPCC direct GHGs were included in this assessment and converted to $CO_2$ equivalents ($CO_2e$) using the latest IPCC (2007) global warming potentials (GWP). This study excludes capital goods (e.g. manufacturing of vehicles, roads, buildings, machinery etc.); human energy inputs to processes; transport of employees to and from the place of work; animals providing transport services and offsetting of emissions. These exclusions are in line with accepted international standards (ISO 14040:2006 and ISO 14044:2006, and the PAS 2050:2011). The most recent data for primary data collection were used, covering a period of the calendar year in 2010. The period of GHG assessment (i.e. the temporal boundary) is 100 years, which is in line with PAS 2050:2011 and all global warming potential factors are based on a 100 year timeline. These global warming potentials include carbon dioxide, methane, nitrous oxide, hydroflurocarbons (HFC), perflurocarbons (PFC) and sulfur hexafluoride.

The exclusion from the carbon footprint is in line with accepted international standards (ISO 14040:2006 and ISO 14044:2006, and the PAS 2050:2011). This study excludes:

- Capital goods (e.g. manufacturing of vehicles, roads, buildings, machinery etc.)
- Human energy inputs to processes
- Transport of employees to and from the place of work
- Animals providing transport services
- Offsetting of emissions

The most recent data for primary data collection were used, covering a period of the calendar year in 2010. The period of GHG assessment (i.e. the temporal boundary) is 100 years, which is in line with PAS 2050:2011 and all global warming potential factors are based on a 100 year timeline.

## Carbon Footprint Comparisons

Carbon footprint calculations are project specific. In many instances, geotextile tube options result in lower carbon footprints when compared with conventional solutions. A proprietary Carbon Footprint Calculator was developed for the purpose of calculating carbon footprint of and comparison between conventional and geotextile tube solutions. One hypothetical example of a breakwater comparison between geotextile tube option against conventional rock solution in marine and hydraulic engineering application is provided. A project case study is also described whereby the geotextile tube solution resulted in significant carbon footprint savings over the conventional solution.

## Breakwater Example

In this hypothetical example a comparison is made between a breakwater built using only rock material and a breakwater built using a core consisting of stacked geotextile tube construction covered with rip-rap. This example is based on conditions applicable to the Netherlands. Figure 2(a) shows the conventional rock breakwater while Figure 2(b) shows the breakwater option with geotextile tube replaced core. Both have the same overall geometrical cross-sectional dimensions. The breakwater height is assumed as 3.3 m, with a base of 16.2 m, crown of 3 m and side slopes of 1:2. This breakwater geometry is not untypical of an inland breakwater application in the Netherlands. In Figure 2(b) three identical geotextile tube filled to height of 1.5 m are used as the core of the breakwater, replacing rock. A geotextile protection layer is used to cover the geotextile tubes before rip-rap is placed on. Both design options involved the use of a basal geotextile layer for the breakwater.

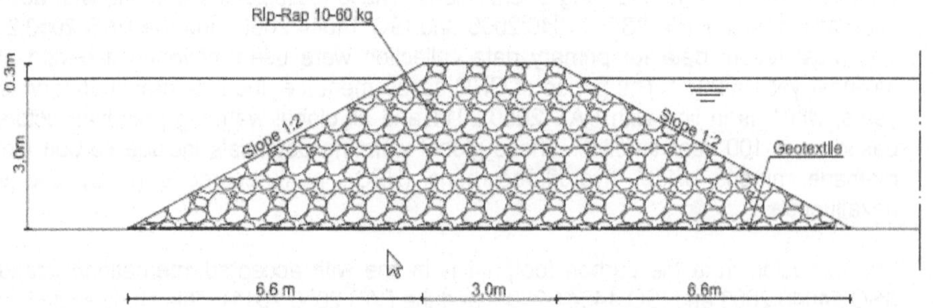


Conventional rock breakwater

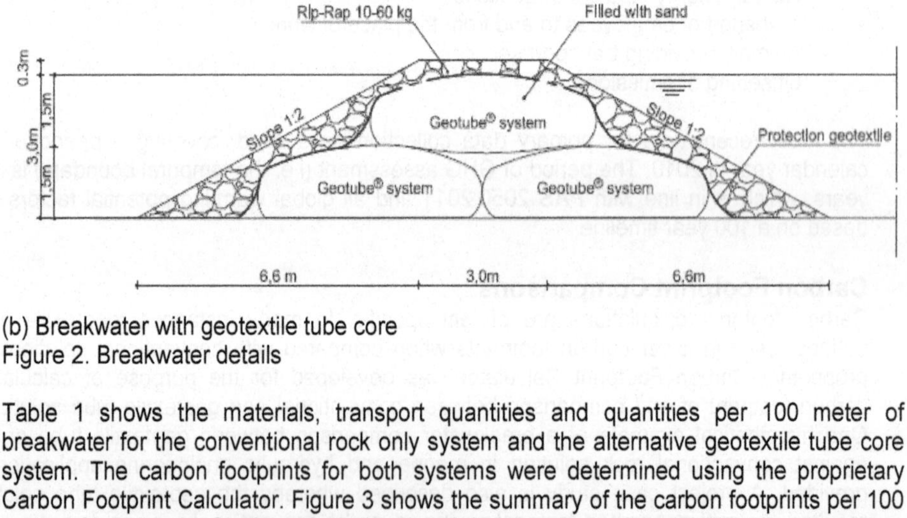


(b) Breakwater with geotextile tube core
Figure 2. Breakwater details

Table 1 shows the materials, transport quantities and quantities per 100 meter of breakwater for the conventional rock only system and the alternative geotextile tube core system. The carbon footprints for both systems were determined using the proprietary Carbon Footprint Calculator. Figure 3 shows the summary of the carbon footprints per 100 meter of breakwater for the conventional rock breakwater and geotextile tube alternative. Figure 3 also shows a savings of 95 tonnes of $CO_2e$ in carbon footprint per 100 meter when the geotextile tube system is used in replacement of the conventional rock breakwater system.

These results show that, in the context of this example, the carbon footprint of the geotextile tube solution was lower because of the lower transport emissions. This is largely a result of the lower quantity of rocks used when compared with the conventional total rock solution, due to replacement of core with geotextile tubes filled with sand. The sand was dredged onsite for this case example and the pumping energy from this operation has been included in the study. However if the sand was imported to the site the transport distance would likely be small. This is because sand is typically sourced from a local resource.

Table 1. Materials, transport distances and quantities for conventional and geotextile tube solutions for a hypothetical inland breakwater in the Netherlands.

| Materials | Transport distance | Quantities per 100 meter of breakwater | |
|---|---|---|---|
| | | Conventional | Geotextile Tube |
| Geotextile tube (4 m diameter) | 150 km by road | N.A. | 300 m |
| Sand to fill geotextile tube | Site available | N.A. | 4600 tonnes |
| Riprap (10 – 60 kg) | 1200km by sea + 50km by road | 7300 tonnes | 1800 tonnes |
| Protection geotextile (200 g/m$^2$ nonwoven) | 150 km by road | N.A. | 2300 m$^2$ |
| Basal geotextile (300 g/m$^2$ nonwoven) | 150 km by road | 1650 m$^2$ | 1650 m$^2$ |

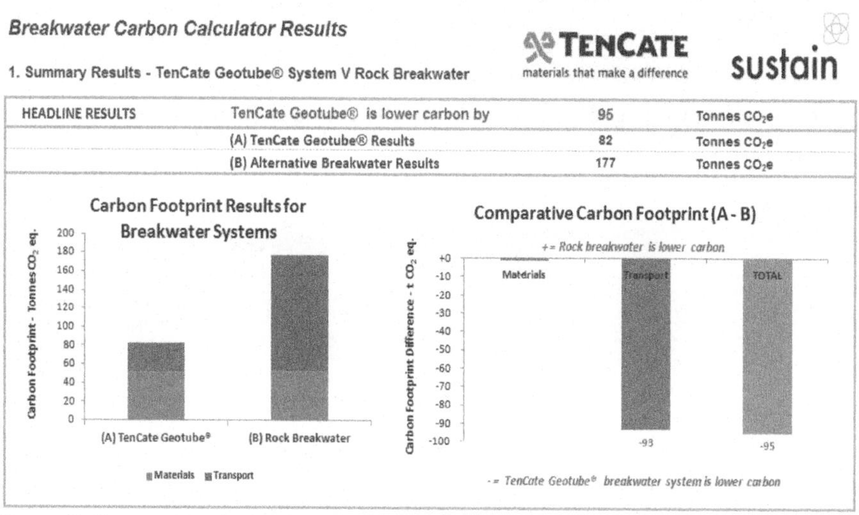


Figure 3. Summary output using proprietary Carbon Footprint Calculator

Case Study Construction of a Polder Dike in Saemangeum, South Korea

The 33.9 km long Saemangeum Sea Dike in Korea links Gunsan in the north to Buan in the south. As of now it is the world's longest sea dike. Before the dike was constructed, Mangyeon River and Dongjin River discharged directly into the Yellow Sea. When the dike was completed, a 400 km2 reservoir was formed and is drained into by both rivers. Future development would involve land reclamation within the formed lake for agricultural, industrial, business, residential, wetland and ecotourism purposes. This paper concerns the land reclamation works for one of the development packages. The Polder Dike that serves as a land reclamation dike during the construction period and as a flood protection dike for the longer term is constructed. The polder dike consists of a sandfill core with rock revetment for erosion protection on both sides of the dike. A road pavement is provided on top of the Polder Dike. For the original design of the Polder Dike rockfill berms are used to contain the sandfill core during construction of the Polder Dike. As an alternative to the original design, geotextile tubes were used to replace the rockfill berms for the construction of the Polder Dike. More than 26 km of geotextile tubes were used for this project. The geotextile tube alternative was more economical than the rockfill berm design. The geotextile tube alternative also help save up to 7 months in construction time. The geotextile tube alternative was also more environmentally friendly, giving a smaller carbon footprint when compared with the rockfill berm design.

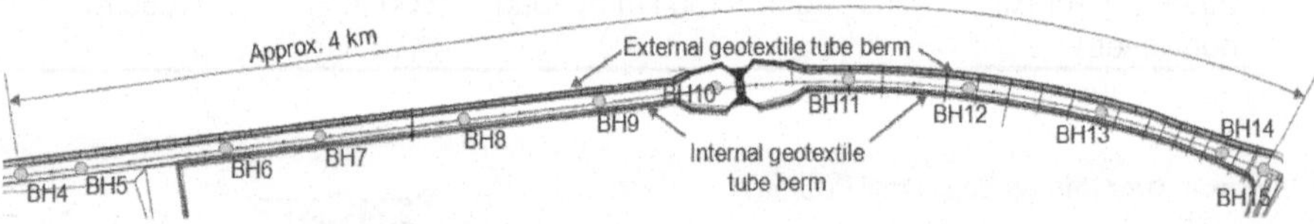


Figure 4. Plan view with geotextile tube berms on both sides of Polder Dike.

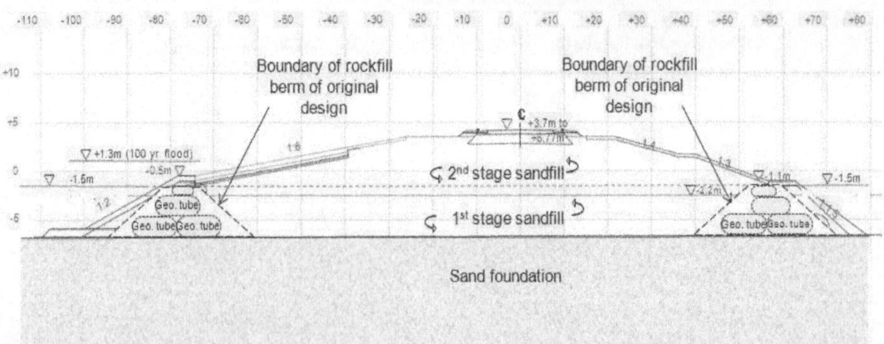


Figure 5. Typical cross-section of the Polder Dike for alternative design with geotextile tube berm (the original design rockfill berm is indicated).

## Cost Saving of Geotextile Tube Berm Alternative Design

Figure 6 shows the berm boundary used to compare quantities of rockfill berm with the equivalent geotextile tube berm. Within the defined boundary, it should be pointed out that the sum of rockfill and sandfill for both designs should add up to the same number. The material quantity differences for the entire Polder Dike are shown in Table 2.

The cost saving of the geotextile berm alternative design over the rockfill berm original design was USD 6.2 million, based on actual tender prices.

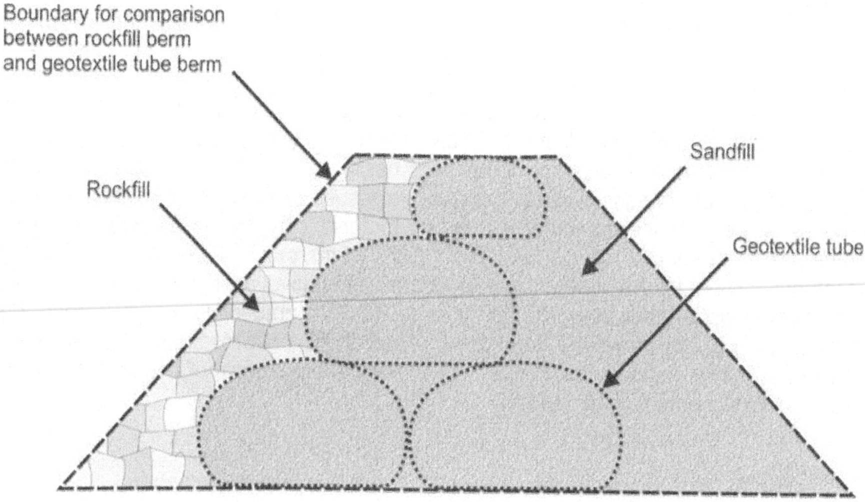


Figure 6. Berm boundary to compare quantities of rockfill berm with the equivalent geotextile tube berm.

| Item | Unit | (X) Rockfill berm | (Y) Geotextile tube berm | (X-Y) Difference |
|---|---|---|---|---|
| Rockfill | $m^3$ | 837.000 | 387.000 | +450.000 |
| Sandfill | $m^3$ | - | 450.000 | -450.000 |
| Geotextile tube | m | - | 26.123 | -26.123 |
| Cost saving | USD | +6.200.000 | | |

Table 2. Material quantity differences between original rockfill berm design and geotextile tube berm alternative design.

Carbon Footprint Comparison

Carbon footprint calculations are project specific. For comparison of carbon footprint savings of the geotextile tube berm alternative design over the rockfill berm original design, likewise to the cost saving comparison, only the difference in quantities between the two berm designs are compared (see Table 2). The transportation distance between the source location of the rockfill and the project site include a road journey of 50 km and a barge journey of 4 km. The transportation distance between the manufacturing location of the geotextile tubes and the project site include road journeys of 500 km and a sea journey of 3,000 km.

For the carbon footprint of the rockfill berm original design the energy consumptions involved in the quarrying of rock, in the transportation of the rockfill, that of mechanical equipment in transferring the rock from dumper trucks onto barges and that involved in the placement of rockfill at site are determined. For the geotextile tube berm alternative design, the carbon footprints of the geotextile tubes used (based on cradle to site life cycle) and that of the sand dredging and filling works involved are determined. In the comparison exercise, the carbon footprint of basal geotextile is not included because it is common for both options. Figure 7 shows the results of the Carbon Footprint Calculator. Figure 7(a) shows the summary for carbon footprint comparison between the geotextile tube berm alternative design and the rockfill berm original design. The total carbon footprint saving for the geotextile tube berm alternative design over the rockfill berm original design is more than 230,000 tons of CO2e, representing a 52% carbon footprint saving. Figure 7(b) shows the breakdown of carbon footprint results for the geotextile tube berm alternative design while Figure 7(c) shows the breakdown of carbon footprint results for the rockfill berm original design.

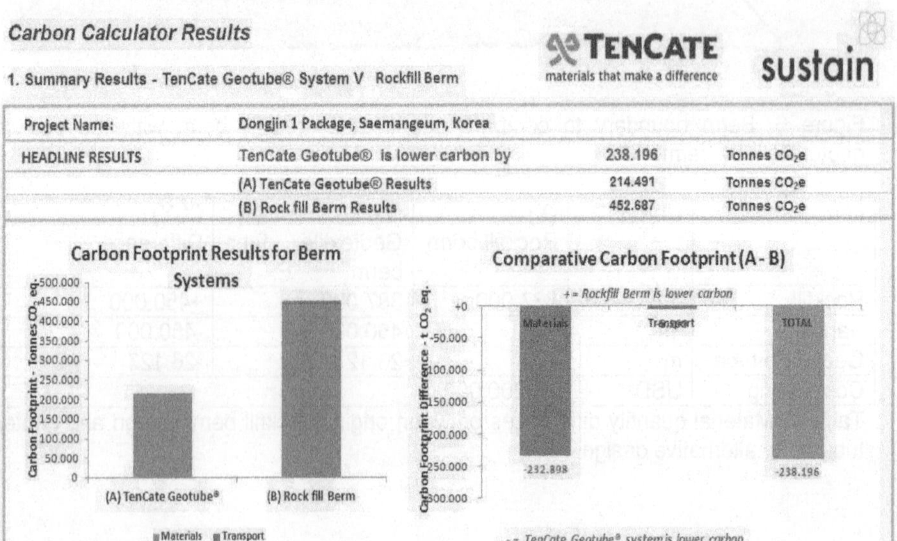

a

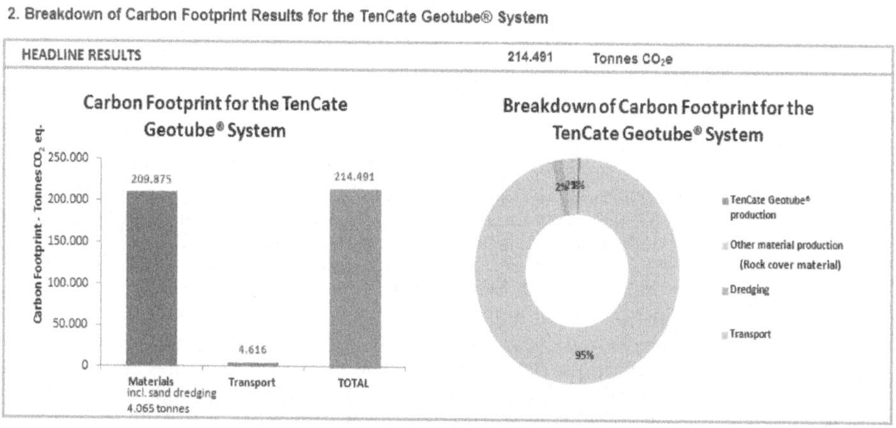


b

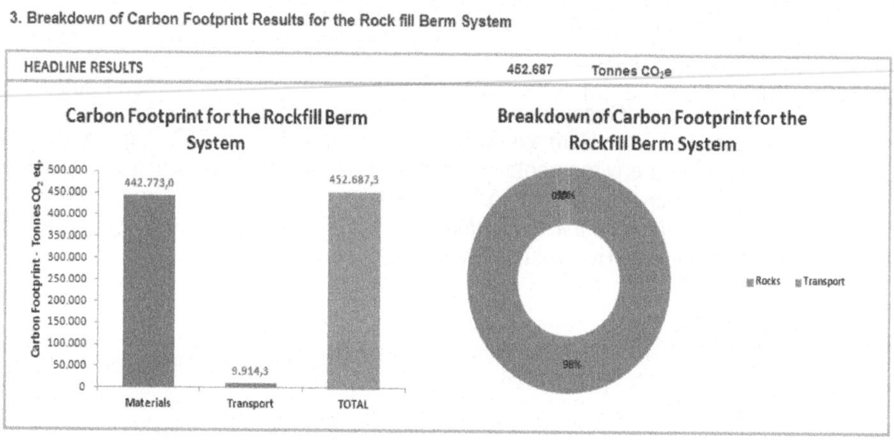


c

Figure 7: Results of Proprietary Carbon Footprint Calculator (a) carbon footprint comparison between geotextile tube berm and rockfill berm (b) carbon footprint of geotextile tube berm (c) carbon footprint of rockfill berm.

## Conclusions

This paper described the methodology for carbon footprinting of geotextile tube solutions and the conventional systems they replace. A case study involving the use of geotextile tubes as economical and environmental replacement of rock for the construction of polder dike for the Dongjin 1 Package in Korea has been presented. The geotextile tube berm alternative design resulted in cost saving of USD 6.2 million and carbon footprint saving of more than 230,000 tons of CO2e or 52% over the rockfill berm original design. The geotextile tube berm alternative also helped shorten the overall project duration by 7 months.

Geotextile tube solutions appear to have more favourable carbon footprints over conventional systems in the cases presented.

## References

Bray, R.N., Bates, A.D. & Land, J.M. (2001). Disposal of dredged material. *Dredging: A Handbook for Engineers*: Chapter 5, 89-111. Butterworth-Heinemann, Oxford, UK.

DEFRA (2010). 2010 Guidelines to Defra/DECC's GHG Conversion Factors for Company reporting, The department for Environment, Food and rural Affairs (DEFRA).

IPCC (2007). Climate Change 2007, Intergovernmental Panel on Climate Change Fourth Assessment Report, Geneva, Switzerland.

ISO 14040:2006. Environmental management – Life cycle assessment – Principles and framework, *International Organization for Standardization*, Geneva, Switzerland.

ISO 1404:2006. Environmental management – Life cycle assessment – Requirements and guidelines, *International Organization for Standardization*, Geneva, Switzerland.

PAS 2050:2011. Assessing the lifecycle greenhouse gas emissions of goods and services, *British Standards Institution*, London, United Kingdom.

Wortelboer, R.J.M., ter Harmsel, M., Zengerink, E., Westerhof, E.J. and Yee, T.W. (2012). Green environmental solutions with geotextile tubes – two case studies: Wack-Wack, Philippines and Zutphen, Netherlands, *5th European Geosynthetics Conference*, Valencia, Spain, (in print).

Ter Harmsel, M., Jones, C.I., Yee, T.W., Environmental Benefit of Carbon Footprint Using Geotextile Tubes, Geosynthetics 2013, Long Beach, California, USA.

Yee, T.W., Lim, L.K., Ter Harmsel, M., Choi, J.C., and Hwang, S.P. "Geotextile tubes as economical and environmental replacement of rockfill for construction of polder dyke in Saemangeum Project, Korea." *Proceedings of the Western Dredging Association (WEDA XXXIII) Technical Conference and Texas A&M University (TAMU 44) Dredging Seminar*, Honolulu, Hawaii, August 25-28, 2013.

**Coastal Management: Changing coast, changing climate, changing minds**
ISBN 978-0-7277-6114-9

ICE Publishing: All rights reserved
doi: 10.1680/cm.61149.039

# Managing Liquefaction Risks in Tidal Systems

**Ben Rijneveld**, Fugro GeoServices B.V., Nieuwegein, The Netherlands
**Martin van der Meer**, Fugro GeoServices B.V., Nieuwegein, The Netherlands
**Hans Waals**, Waterschap Hollandse Delta, Ridderkerk, The Netherlands

## Introduction

To protect the Netherlands against flooding, flood defence systems are present along the coast, estuaries and rivers. In the underwater slopes of the levees in tidal areas often loosely packed sand layers are present, which can be susceptible to liquefaction. Liquefaction can result in a large flow slide and levee failure.

For the occurrence of liquefaction a 'trigger' is necessary. From experience the effect of erosion (steepening) of the underwater slopes is generally regarded as the most important 'trigger'. This effect is especially relevant for the Delta area in the Southwest of the Netherlands. Also other phenomena, such as induced seismic activity, can act as a 'trigger'. This effect is a recent point of attention in the North of the Netherlands, where large gas exploration fields are present that cause seismic activity. In this publication the focus will be on the effect of liquefaction due to erosion, also referred to as static liquefaction.

## Historic Developments

The Southwest of Holland is a tidal system, in which the whole area has changed drastically during the last century. In figure 1 the urbanisation of the delta area in the last centuries is presented. The second map also shows the impact of the disastrous flooding in 1953.

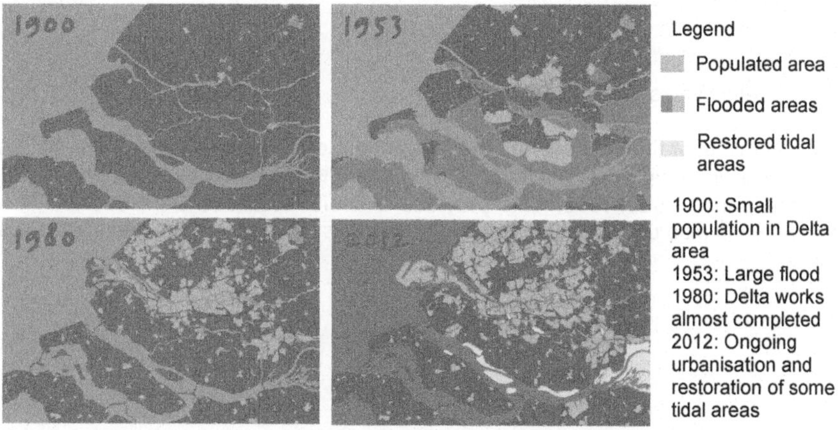


Figure 1: Historic development in the last century, DIND (2014)

Another important change in the tidal system is the construction of the Delta Works. This is a series of storm surge barriers, which were constructed after a devastating flood which occurred in 1953. An overview of these barriers is given in figure 2.

With the completion of the Delta Works in 1986, the protection of Southwest Holland against flooding from the sea increased tremendously. Part of the Delta Works is the closure of several estuaries. However not all of the Delta waters were closed, which changed the tidal flow in some delta branches like Spui and Oude Maas. These changes in the tidal flow causes an on-going erosion process, that after 30 years still has not reached an equilibrium.

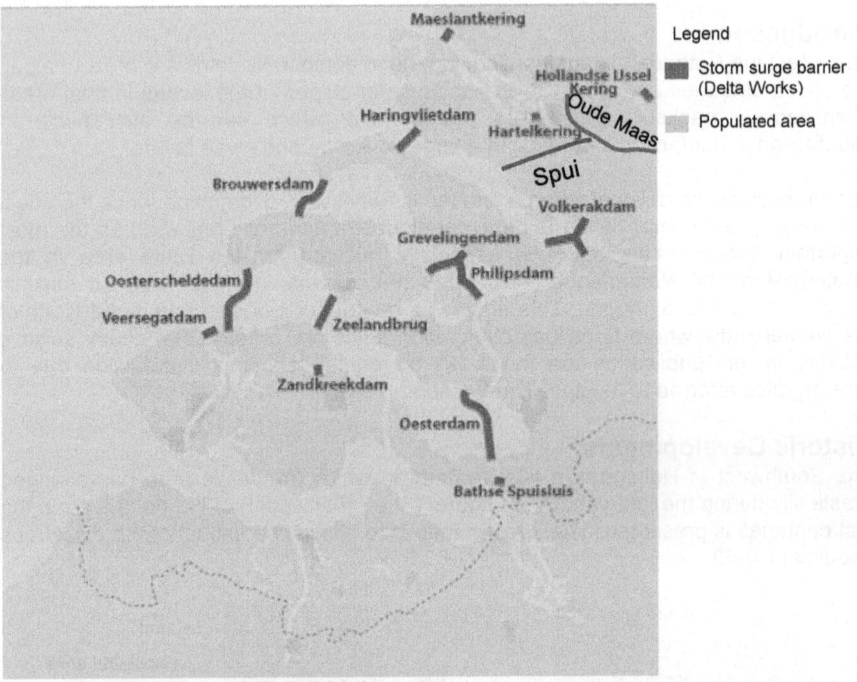


Figure 2: Delta Works, DELT (2014)

In the Southwest of the Netherlands a large amount of flow slides have occurred. An example is given in figure 3. Wilderom (1979) made an inventory of 1129 instabilities during the period 1800–1978. This provides a useful database with reported flow slides, which is also used for the quantitative risk assessment elaborated later in this paper.

Figure 3: Leendert Abrahampolder, Oosterschelde, march 1966.

The traditional solution to build a new dike or levee behind the collapsed embankment is not suitable any more in the heavy populated delta area. Therefore the importance of a proper monitoring and risk assessment has become more and more important.

## Risk Assessment

According to Dutch legislation the safety of all the primary flood defences have to be assessed and reported periodically. One of the failure mechanisms which have to be assessed is static liquefaction. For the assessment of static liquefaction a stepwise approach is prescribed in MVW (2007). This approach is summarised in figure 4.

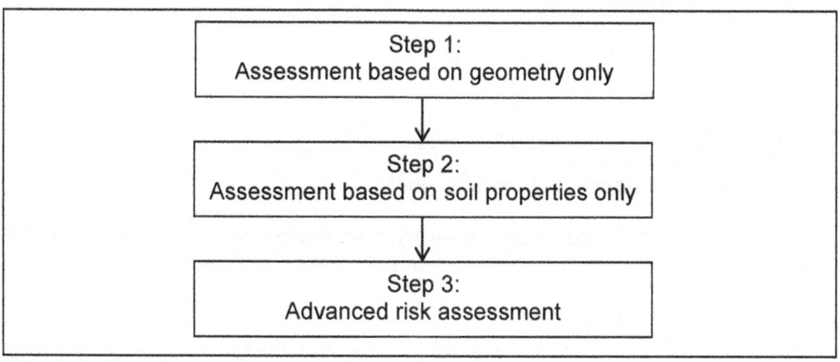

Figure 4: Stepwise assessment of static liquefaction risks

In the first step an assessment based on geometrical properties can be performed. If the slopes of the river banks are gentle, a wide foreland is present and/or the riverbed is shallow enough, static liquefaction will not endanger the stability of the flood defence. If the levee does not meet the requirements of step 1, in step 2 the soil properties have to be analysed. In this step the occurrence of soil layers susceptible to liquefaction are identified. If these soil layers are present, an advanced risk assessment should be performed.

In this chapter this stepwise approach will be illustrated for a case study at the river Spui in the Southwest of the Netherlands.

**Assessment Based on Geometry**

Two aspects are relevant when the geometry is considered. If the river banks are very gentle, static liquefaction will not occur since the static shear stresses in the soil are too low. Secondly, if a flow slide occurs, this slide will not reach the levee if a sufficiently wide foreland is present.

For the river Spui relatively steep river banks are present, so liquefaction cannot be ruled out based on this criterion. For the assessment of the second criterion a post failure slope of 1:15 has been assumed. For the assessment a 1:15 surface was plotted starting from the toe of the levee towards the river bed. At locations where the surface level is below this post failure surface damage to the levee by a flow slide cannot be ruled out. An example of such a geometrical check is presented in figure 5.

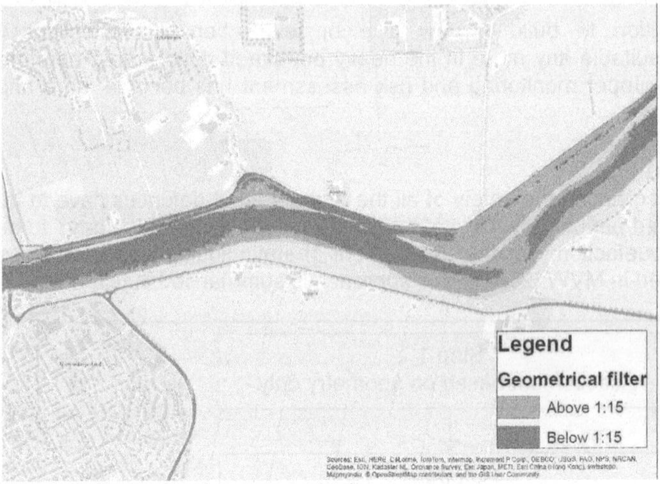


Figure 5: Example of a geometrical assessment

For the areas where damage to the levee by a flow slide cannot be ruled out the liquefaction susceptibility has been investigated.

**Liquefaction Susceptibility Maps**

Along the river soil investigation was performed to establish the regions where liquefiable soils are present. This soil investigation mainly consisted of Cone Penetration Tests with a spacing of approximately 100 m. For every CPT the sand layers were identified and the relative density of the sand layers has been determined. To determine the relative density the correlation with the cone resistance according to Baldi (1986) has been applied. CPT's with loosely packed sand layers with a relative density below 66% and a thickness larger than 3 m are classified as 'very susceptible'. If the relative density is below 66% and the thickness is between 1 and 3 m the CPTs are classified as 'susceptible'. The results are presented graphically in figure 6.

From figure 6 it can be seen that a lot of critical soil layers are encountered along the river. This is a typical result for the river beds in the Southwest of the Netherlands. Based on the first two steps of figure 4 a lot of levee sections did not fulfil the requirements, and as a result a large part of the levee should be strengthened. To optimise the costs, an advance risk assessment was performed so an informed decision on the scope of the levee reinforcement could be made.

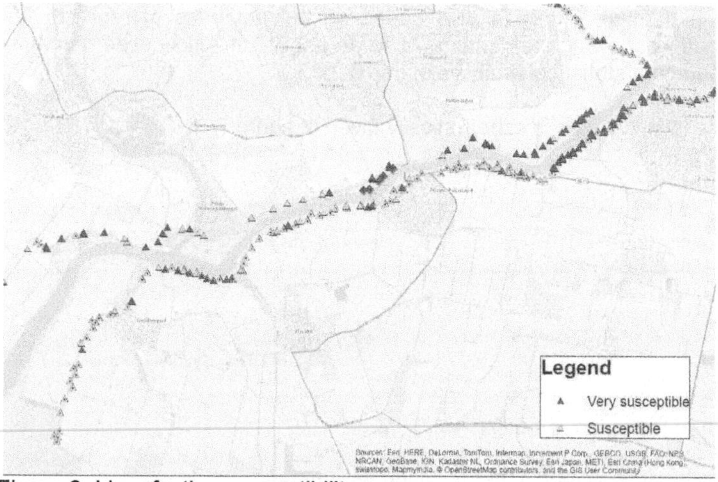

Figure 6: Liquefaction susceptibility map

**Advanced Risk Assessment**

For the advanced assessment a probabilistic analysis has been performed, which is based on MVW (2012). In this analysis the probability of damage to the levee was determined and compared to the standards for acceptable probability defined in the Dutch law. The probability of damage to the levee is the product of three (conditional) events:

- Probability that a flow slide occurs
- Conditional probability that the flow slide reaches the levee, given that liquefaction occurs
- Probability of a high water situation before the damage due to a flow slide is repaired

How these probabilities were determined is elaborated below.

<u>Probability that a flow slide occurs</u>

The calculation of the probability that a flow slide occurs is based on a database of reported flow slides in the Southwest of the Netherlands Wilderom, N.H., (1979). Based on this database and model calibration of the available data the probability that a flow slide occurs can be determined according to MVW (2012):

$$P(zv) = \left(\frac{H_r}{25m}\right)^2 \times \left(\frac{3.5}{\cot \alpha_r}\right)^6 \times \left(\frac{1}{10}\right)^{10\times(R_e-0,4)} \cdot 0.1 / km / year \qquad \text{(-/km/year)}$$

In which:

| | | |
|---|---|---|
| $H_r$ | = fictive height of the slope | (m) |
| Cot $\alpha_R$ | = cotangens of the fictive slope angle | (-) |
| $R_e$ | = lowest value of the average relative density over 3 m | (-) |

<u>Probability of a flow slide reaching the levee, given a flow slide</u>

A flow slide is only relevant for the water retaining function if it reaches the levee. The estimation of the damage profile is also based on the database of reported flow slides in the Southwest of the Netherlands. The probability of a flow slide damaging the levee is based on a statistical analysis of this data.

The length of the failure profile is schematically given in figure 7.

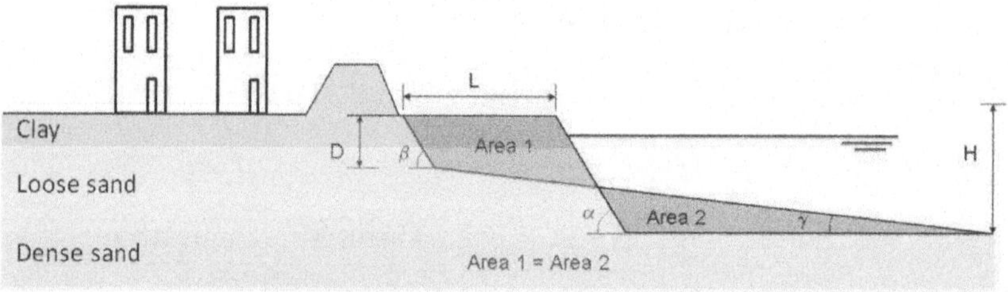


Figure 7: Length of the failure profile

The failure profile is determined by the folowing parameters:

| | | |
|---|---|---|
| cot(α) | = slope of the profile before the slow slide | (-) |
| cot(β) | = upper slope of the profile after the flow slide | (-) |
| cot(γ) | = lower slope of the profile after the flow slide | (-) |
| c | = ratio between Area1 en Area2 (A1=cA2) | (-) |
| H | = depth of the river | (m) |
| D | = height of the steep part of the failure profile | (m) |

Of the four unknown parameters c, β, γ and D, the influence of the uncertainty of the parameter γ on the calculated extend of the flow slide is dominant. Therefore only the uncertainty in this parameter is used for the calculation of the probability of exceedance of the damage profile. For the other parameters deterministic values are applied. Based on statistical analysis of flow slides in the Southwest of Holland, a normal distribution is adopted for cot(γ), with $E_{cot(\gamma)}$ =15,9 and $\sigma_{cot(\gamma)}$ =4,6.

The reliability function can now be defined:

$$Z = L_{allowable} - L$$

In which:

| | | |
|---|---|---|
| L | = length of the damage profile | (m) |
| $L_{allowable}$ | = allowable length of the damage profile | (m) |

The probability of a damage profile which is larger than acceptable, given a flow slide, can be determined according to:

$$P(L > L_{allowable} | ZV) = P(Z < 0) = \Phi(-\beta) \quad (-)$$

Probability of inundation

The occurrence of a flow slide is not related to a high water situation. On the contrary, from the study of historic flow slides it seems that these are more likely when a drop in water level occurs. Therefore, only if a high water situation occurs before the damage due to a flow slide has been repaired this will lead to inundation of the hinterland. Therefore the probability of inundation due to a flow slide can be determined according to:

$$P(Inundation) = P(ZV) \cdot P(L > L_{allowable} | ZV) \cdot P(H > H_{damageprofile}) \quad \text{(-/year)}$$

In which:

| | | |
|---|---|---|
| P(inundation) | = probability of inundation of the hinterland due to a flow slide | (-/year) |
| P(ZV) | = Probability of a flow slide | (-/year) |
| $P(L>L_{allowable}\|ZV)$ | = Probability of a damage profile reaching the levee, given a flow slide | (-) |
| $P(H>H_{damage\ profile})$ | = Probability of a high water, higher than the remaining levee height after a flow slide, within the repair time of the levee | (-) |

For the determination of the probability of inundation the repair time of the levee after a flow slide is important. In the analysis a repair time of about 1 month has been applied. Since the damage due to a flow slide is generally extensive (typically in the order of 100 m) a calamity plan is necessary to make sure that the repair can be performed in this limited time.

In figure 8 the results of the advanced assessment is presented, where the dark/coloured red areas do not meet the safety criteria.

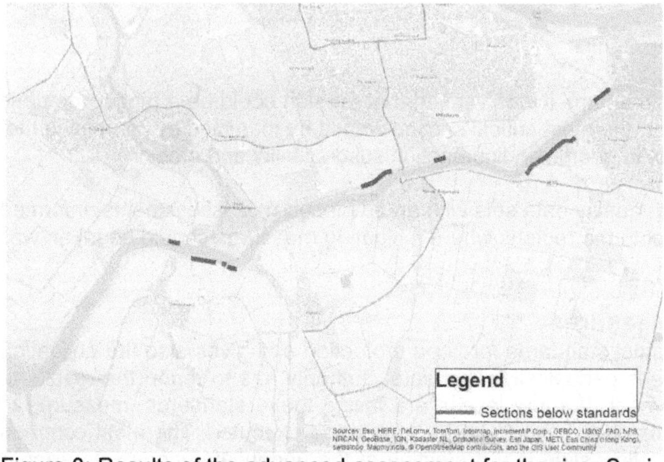


Figure 8: Results of the advanced assessment for the river Spui

For the sections where the safety criteria were not met, an extensive and costly reinforcement program was established. The urgency of these measures is substantiated with erosion maps.

**Erosion Maps**

As stated before the most important 'trigger' for static liquefaction is erosion of the river banks. Therefore the erosion speed is an indication of the urgency of the problem.

Based on periodic soundings of the river bed, locations with significant erosion of the river banks were identified. An example of such an erosion map is presented in figure 9.

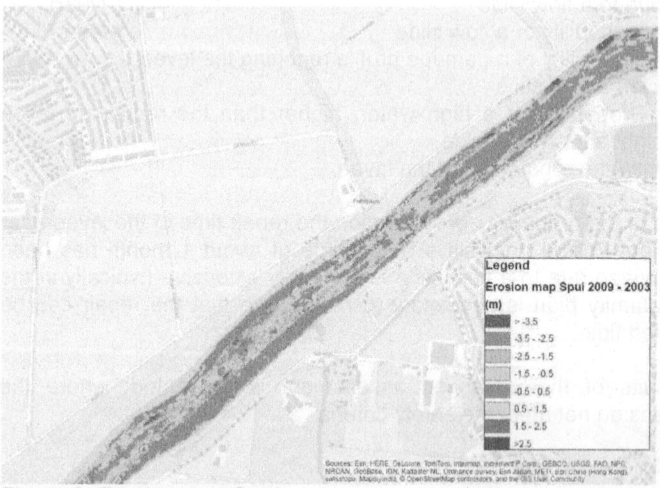


Figure 9: Erosion map Spui

Based on these erosion maps it was verified that erosion could be a trigger for static liquefaction. Moreover, the most critical sections could be identified by combining the data sets with respect to geometry, liquefaction susceptibility and erosion.

Combining these high quality data sets with an advanced risk assessments, informed decisions with respect to the regions where mitigating measures should be taken was made.

## Risk Mitigating Measures

The Dutch law sets strict standards for flood protection and gives also the obligation to assess all the levees periodically. The water authority has to report the results to the National Government. If some levees are failing these standards, measures to improve the flood protection have to be designed and executed. The most common measure to prevent a flow slide is a protection of the underwater slope with armour stone revetments. In this way the erosion process is stopped and the most important 'trigger' for static liquefaction is eliminated.

For the Spui river this measure was developed in a cost effective way. Only the river bed sections, where the calculated inundation probability is larger than the allowed probability, are stabilized with armour stone revetments that are shipped in from Germany. Figure 10 shows the carefully depositing of the stone revetments by a crane on a barge. This project is executed in the years 2014–2015. In the coming years the development of the improved sections will be monitored intensively to establish the effectiveness of this measure.

Figure 10: Carefully depositing the stone revetment for stabilisation of the Spui river bed

## Research and Developments

Flow slide caused by static liquefaction is a difficult mechanism to describe, due to the many uncertainties in river bed properties and erosion processes. Therefore a large uncertainty in the calculation methods is present. In the risk analysis described above these uncertainties are taken into account by using casuistry from flow slides in the past. Beside this very valuable historic data, a more fundamental research program is started in the Netherlands to better understand this failure mechanism.

An example of this is a full scale flow slide test in the Scheldt estuary in the Southwest of the Netherlands. This test took place in September 2014. In this test several flow slides were triggered by dredging at the toe of the underwater slope. During this large scale test a flow slide at real scale was recorded for the first time. Valuable information is obtained with respect to the developments of the geometry and moment of initiation of a flow slide are recorded. At this moment detailed analysis of the tests results are performed and turned into calculation methods for engineering practise.

From analysis of the historic flow slides it appears that these were not only caused by static liquefaction, but that also by the mechanism breach flow. This is a retrogressive mechanism in which an initial disturbance in the underwater slope results in the development of a sand-water mixture flow along the slope. This process continues until the slope reaches equilibrium, generally at a very gentle slope. Currently this mechanism is added to the periodic assessment of the levees in the Netherlands. Knowledge from the large scale tests is directly applied in this assessment procedure.

## Conclusions

A large part of the tidal system in the Southwest of Holland is prone to erosion. In combination with the geology of the area, flow slides are a relevant threat for the flood safety of this densely populated area.

Flow slides caused by static liquefaction are a failure mechanism which is characterised by its sudden occurrence, large extent of damage and large uncertainty in the calculation methods. Moreover, measures to prevent flow slides are costly. Therefore decisions should be based on a thorough risk analysis so the risks and costs can be optimised.

Due to the large uncertainties, a probabilistic approach is appropriate. For the Southwest of Holland a large database of historic flow slides is documented and statistically analysed. This information is crucial for a quantitative risk assessment.

In this paper an advanced risk analysis has been presented, which makes it possible to make informed decisions based on high quality data sets. The geometry as well as the geology of the area should be mapped and combined for the assessment. The geometry can be mapped with land and water surveys. The geology of the area can be explored with Cone Penetration Tests and bore holes.

Application of this method for the river Spui resulted in an optimisation of the scope of reinforcement works.

## References

Baldi, G., et al., (1986), Interpretation of CPT's and CPTU's, *Proceedings of the Fourth International Geotechnical Seminar Field Instrumentation and In-Situ Measurement*, p. 143-156, Singapore

DELT (2014), Deltawerken [Online] available from http://www.deltawerken.com/Deltawerken/16.html [accessed 20 January 2014]

DIND (2014), Dwalen door de tijd [Online] available from http://www.deiningindedelta.nl/dwalen_door_de_tijd/ [accessed 20 January 2014]

MVW (2007), *Voorschrift Toetsen op Veiligheid Primaire Waterkeringen*, Ministerie van Verkeer en Waterstaat

MVW (2012), *Handreiking Toetsen Voorland Zettingsvloeiing t.b.v. het opstellen van het beheerdersoordeel (BO) in de verlengde derde toetsronde*, Ministerie van Verkeer en Waterstaat

Wilderom, N.H., (1979), *Resultaten van het vooroeveronderzoek langs de Zeeuwse stromen*, Nota 75.2, Rijkswaterstaat

**Coastal Management: Changing coast, changing climate, changing minds**
ISBN 978-0-7277-6114-9

ICE Publishing: All rights reserved
doi: 10.1680/cm.61149.049

# Anoxia Monitoring System for Transitional Waters – The Case of the Marano and Grado Lagoon (Italy)

**Isabella Scroccaro**, ARPA FVG, via Cairoli 14, 33057, Palmanova (UD), Italy
**Federico Pittaluga**, ARPA FVG, via Colugna 42, 33100, Udine, Italy
**Giorgio Mattassi**, ARPA FVG, via Colugna 42, 33100, Udine, Italy

## Introduction

In the framework of the SHAPE project (Shaping an Holistic Approach to Protect the Adriatic Environment between coast and sea – IPA Adriatic European Cooperation Programme), ARPA FVG, the Environmental Protection Agency of Friuli Venezia Giulia Region - Italy, realized a pilot action in the Marano and Grado Lagoon. ARPA FVG implemented an anoxia/hypoxia monitoring system, since anoxia/hypoxia monitoring may represent one of the key tools of the decisional process concerning planning and integrated management of the Marano and Grado lagoon. In fact, the application of integrated coastal zone management principles is based on a sound understanding of all the environmental components (geomorphology, hydrodynamics, biology, ecotoxycology, etc.) of coastal ecosystems such as coastal lagoons. This implies an updating of the monitoring methods and an improvement of the monitoring systems. Indeed, a better understanding of the Marano and Grado lagoon will allow decision makers to make better choices regarding the management of the area. In fact, the Marano and Grado Lagoon is a stretch of coast characterized by complex conflicts and environmental problems.

From an environmental point of view the presence of nutrients and pollutant substances is one of the biggest threats. These products come from agriculture and urban/industrial activities, that are spread by rivers, mainly the Cormor and Corno river (Saccon et al. 2011). This phenomenon occurs in an area where sediments have been extensively contaminated from the Idrija mercury mine (Slovenia) through the suspended load carried inside the lagoon by tidal fluxes of the Isonzo River. Furthermore, there is some concern on the risk posed by the potential harmful effects of mercury on the trophic chain, due to the fact that mercury in the sediments tends to remobilize (MeHg) when hypoxia and high temperature values occur (Brambati, 1996; Covelli et al., 2001; Covelli et al., 2008).

During the years, ARPA FVG has been carrying out institutional monitoring of the Marano and Grado Lagoon, collecting water quality data monthly in 19 water bodies, according to the WFD 2000/60/CE and Italian laws. In recent years, ARPA FVG has increased the number of studies on ecology and chemistry, also enhancing the monitoring of fishery products and making hydrodynamic studies. However, despite these efforts, the monitoring activities have always been limited by the conditions of access to the sub-basins. In fact, tidal flats are difficult to reach in descending tide period, furthermore the hourly conditions (no measures at night) and the sampling frequency are limited. Therefore data can be considered only partially representative of the average condition of the lagoon. At most an assessment of the evolution of the seasonal cycles may be obtained (Mattassi et al., 2006). The standing monitoring system composed by the probes is the only way to understand how different parts of the lagoon function during night time, while the collected data are important to know good practices to follow upstream the lagoon.

## Lagoon Ecosystem Description

The Marano and Grado Lagoon is situated along the northernmost stretch of the Adriatic sea, between 45°40'33" and 45°45'39" north latitude and 13°05'08" and 35°08'25" east longitude. It covers more than 16000 ha, extending for more than 32 km from east to west and 5 km from north to south (Mattassi et al. 2006). Its surface falls under the jurisdiction of the municipalities of Grado and Marano.

Through the lagoon inlets of Lignano, S. Andrea, Porto Buso, Morgo, Grado and Primero, seawater flows inside the lagoon. In the last decades, man-made works have been carried out to grant access to boats and merchant ships (in the case of Porto Buso). Various tributary rivers discharge freshwater in the lagoon: Stella, Turgnano, Cormor, Zellina, Corno, Aussa and Natissa. The main one is the Stella river. The Natissa river is the only one flowing into the eastern part of the lagoon (near Grado) and this is the reason why the area has significantly more marine features than the lagoon of Marano. River inputs are the main source of contamination and eutrophication of the lagoon (channelling of rivers has also significantly influenced the self-purifying capacity), as proved by the total nitrogen and total phosphorus concentrations in water. N and P can be found primarily close to the river mouths and show a decreasing gradient towards the sea-inlets (Mattassi et al. 2006). Data collected in 2006 show that the total mean value of nitrogen was 1545 µg L-1 in Marano and 526 µg L-1 in Grado. Total phosphorus mean values ranged between 35 µg L-1 in Marano and 9 µg L-1 in Grado (Mattassi et al. 2006).

Intense inflows usually occur during autumn or spring, when rainfall is more frequent. The Marano and Grado Lagoon is nowadays one of the major centres for the production of shellfish in Italy, especially of *Tapes philippinarum*. At the same time, it is an ecosystem subject to a constant anthropogenic pressure by economic and social actors which have settled in the area and whose activities tend to come into conflict with each other:

- industrial zones (Aussa-Corno area);
- tourist settlements of Lignano and Grado;
- ports for fishing vessel of Marano and Grado;
- commercial port (Porto Nogaro) and marinas;
- agricultural areas at the edge of the lagoon.

Further the Marano and Grado Lagoon represents a primary community site of European interest (SCI – IT3320037) and includes two sites that were declared wetlands of international importance by the Ramsar Convention, since they play a key role in the maintenance of biodiversity and the protection of international migration of wildlife.

## Materials and Methods

### SMATCH Probe

The SMATCH Probe is a multiparametric probe measuring depth, temperature, conductivity/salinity and dissolved oxygen. It is composed by two parts: one is represented by the floating system, while the other is represented by the probes themselves. The floating system is a buoy composed by two half-cylindrical compartment and a central tubular structure built to accommodate the probe (all made by polyethylene). The floating system has been developed specifically for the lagoon environment in order to protect the probe and allow it to lay down on the bottom during low tides.

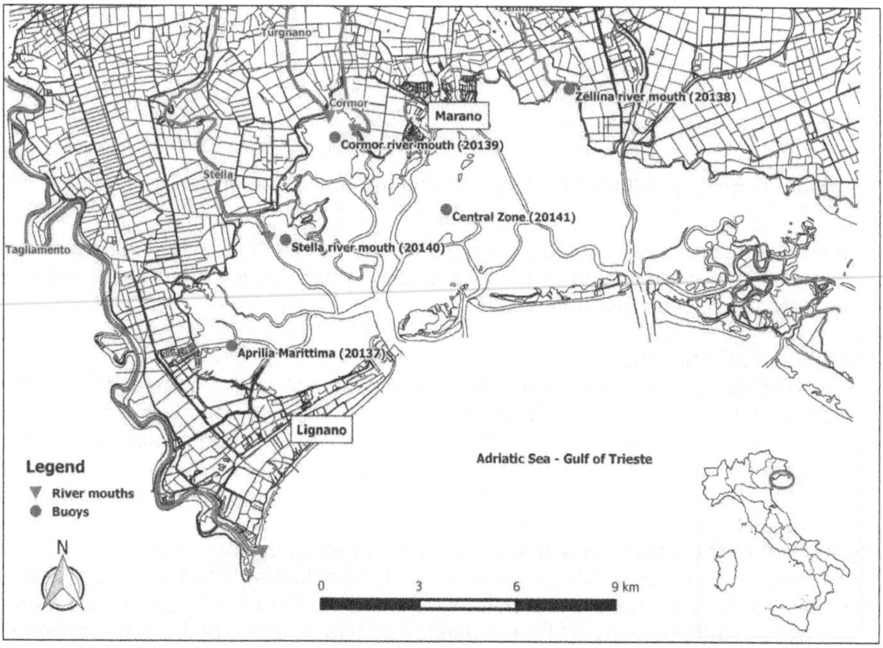


Figure 1 Map of the Marano Lagoon with the monitoring sites.

The probe has a battery pack providing the energy needed for several months. During this period, sensors are protected against biofouling using a localized chlorination system connected to 3 chlorination grids placed around them. The grid activates itself every 20 minutes and extracts Chlorine from water, concentrating it in the sensor zone and thus inhibiting biological activity. Finally, the antenna installed on the top of the probes is a GPS/GPRS, providing a radio connection with a few tens of meters range, used to establish pc-probe communication and set the probes configuration.

Figure 2. Picture of the floating system and scheme of the probe hosted inside.

The probes record measurements every 30 minutes and, once a day, the data stored on the data logger are sent to an e-mail address. An automatic system reads the mail and stores the data on ARPA servers.

**Installation and Set Up**

ARPA FVG decided to install the whole instrumentation in the western side of the lagoon (the Marano lagoon). Therefore the monitoring system was set up in a specific and more limited area with great economic and environmental importance, where safeguard of fishing and shellfish farming activities represent a primary interest. The choice where to locate the anchorage of the equipment was driven by the purpose of monitoring different environments that may be encountered in the lagoon and their critical issues. In order to achieve this goal, three parametric probes were installed in the proximity of the mouths of rivers flowing into the lagoon: Stella, Cormor and Zellina. These three sites correspond also to inner areas, where water exchanges can be particularly limited during low tide. Another probe was positioned near a draining pump exhaust in the proximity of the Aprilia Marittima's channel port. The last probe was located in a central zone of the Marano Lagoon, relatively close to the mouths of the lagoon, more influenced by marine water (Figure 1).

Due to the difficulties of navigation, ARPA FVG used its special vessel "Pontoon Boat", which is able to move even in very shallow water. Operators chose to launch the floating system directly at the Aprilia Marittima harbor where the boat was moored. This choice was made taking into consideration both the considerable weight of buoys (whose weight with the probes amounts to over 120 kg) and because the boat was not provided with a hoist. From the harbour buoys were towed, two at a time, to the installation.

Once the prearranged sites were reached, every buoy was anchored using a 10 meters steel chain (6 mm ϕ) and a 10 kg *Britany* anchor. In the end, a photosensitive signal light was fixed to the protection structure, in order to be seen from boats during the night. After these works, the probes were inserted in the buoy and sensors parameters were set by means of the radio connection with the notebook. After this, the probes were activated.

**Weather Conditions**

Meteorological data have been obtained from 3 different survey stations of the Regional Meteorological Observatory (OSMER FVG). These survey stations better reflect the conditions of the Marano Lagoon. Among all weather parameters, rainfall – temperature - wind – solar radiation are the most important to complete the set of data given by the probes. The rain can increase the amount of nutrients spread from the agricultural areas and also modify the salinity of the lagoon waters. High temperature and wind create opposite effects because wind can mix almost completely the lagoon waters while high temperature can create a stratification. Lastly solar radiation promotes photosynthesis activities in both microplancton (in the water column) and macrophyte/benthonic organisms (on the bottom). July and most of August 2013 have been characterized by the almost absence of precipitation. During September, on the contrary, weather was cloudy and 2 main events bring about 200 mm of rain in the area. Main precipitation events can be summarize in:

- Rainfall July 29$^{th}$ (12-18 mm);
- Rainfall August 14$^{th}$ (10-20 mm);
- Rainfall from August 22$^{th}$ to August 27$^{th}$ (~40 mm);
- Rainfall September 10$^{th}$ (60-80 mm);
- Rainfall September 29$^{th}$ (80-160mm).

Usually strong winds are related to the summer rainstorms but the maximum values of summer 2013 were registered because of a Bora wind event on September 17$^{th}$ (90 km/h), while it normally reaches average values of 10 km/h. Temperature increased during July, reaching maximum values during the first days of August. Between July 4$^{th}$ and August 19$^{th}$ temperature was always over 30°C. After the rain event of the last days of August temperature drops of about 6° and keeps slowly falling during September.

## Results and Discussion

The high frequency of sampling (one measure every 30 minutes) allowed to follow the trend of variables with high temporal accuracy all over the 24 hours for 3 months (July-August-September 2013). In order to search for contingent dystrophic situations of the ecological system, dissolved oxygen is the main indicator that was checked on. In general, maximum values of 150-160% saturation were found, representing microalgal blooms or strong macroalgal activity. Minimum values were observed between deep night and dawn. In particular, dissolved oxygen values for probe 20138, located near Zellina river mouth, reached even 275% saturation during the afternoon fluctuations in July, then values decreased down to 8% in the night (Figure 4). The absolute value of the daily fluctuations of dissolved oxygen are generally less large during August and September, but the minimum is even smaller. Probe 20139 nearby Cormor river mouth registered values of 4% in August and 2% in September, while in the previous days maximum values reached respectively 215% and 150% (Figure 5).

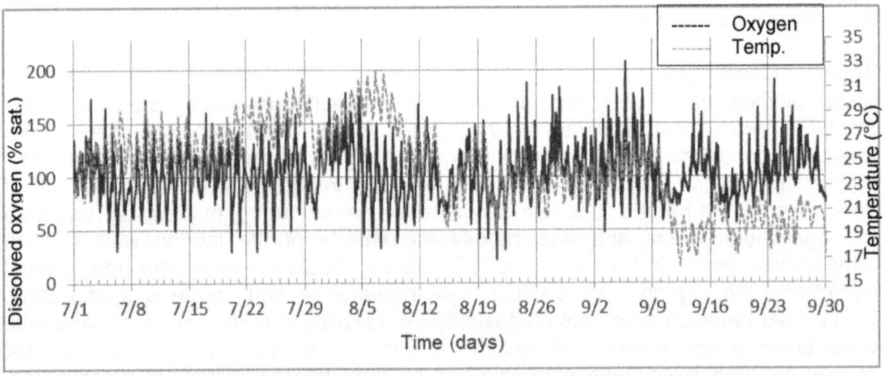


Figure 3. Time series of dissolved oxygen (in black) and temperature (in grey) for probe 20137 (near Aprilia Marittima site) during summer 2013.

Despite of these extreme values, hypoxia phenomena stand at most for some hours, then the values tend rapidly to invert the course.

Unexpected values sporadically appear in the three inner sites located nearby Stella, Cormor and Zellina river mouths (probes 20138-20139-20140). They registered peaks of dissolved oxygen that interrupt the usual decreasing trend of the oxygen saturation in the first hours of the morning (Figures 4, 5 and 6). These outliers are probably related to low tide, when the instrumentation could be rested on the bottom of the lagoon. Such a theory is based on the fact that these phenomena occur usually during the most powerful low tides, as mainly showed by the data of August 2013. Right now it's hard to give a precise explanation of these outliers. However, since the oxygen sensor is located near the bottom of the buoy, it could acquire strange values due to the increased proximity (or even contact) to seagrasses during their photosynthetic activity.

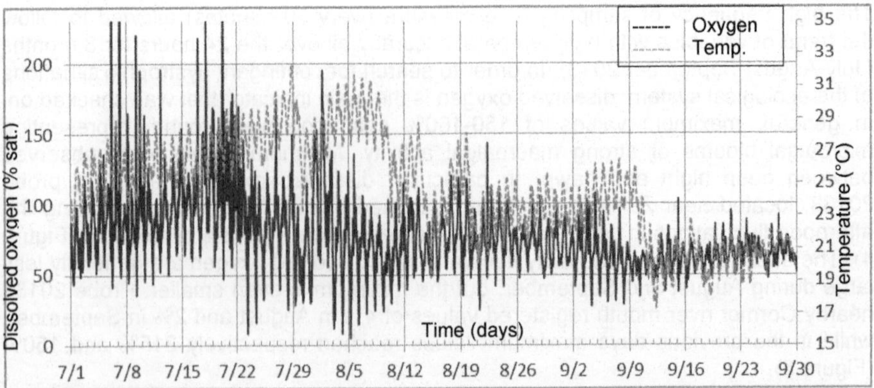


Figure 4. Time series of dissolved oxygen (in black) and temperature (in grey) for probe 20138 (near Zellina river mouth) during summer 2013.

In every site it was observed that dissolved oxygen values lie inside a range between 140% and 60% for the most of the time. In particular the average value during July and August is near the saturation (102% and 93% respectively), while in September there was a slight decline, and average water values did not reach the saturation (78 % approximately). The average values are lowered mostly because of two sites: the Zellina site and mainly the Cormor site where oxygen stood under 50% after the rainfall and the drop of temperature of September 10th.

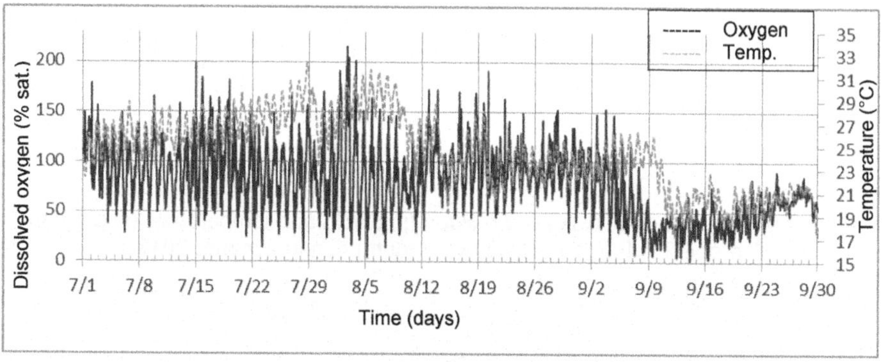


Figure 5. Time series of dissolved oxygen (in black) and temperature (in grey) for probe 20139 (near Cormor river mouth) during summer 2013.

Temperature emerged to be the triggering factor influencing the intervals of extreme dissolved oxygen values both over and under the saturation. Between half July to the first days of August (the hottest days of summer 2013) and in the first days of September, high average temperature and strong solar radiation promoted fito-biological activity in the lagoon. This situation pushed oxygen values to fluctuate in a wide range, between supersaturate and successive hypoxia conditions, also severe (Figure 4 and 5). Inner areas were the most affected and among all the sites on Zellina river mouth and Cormor river mouth (probes 20138 and 20139) displayed often critical values. On the contrary the probe placed in the central area near the lagoon inlet (probe 20141) turned out to have the most stable values and smallest range between supersatured and undersatured conditions (not shown).

On the daytime span, other sensors generally revealed that temperature and pH values are directly connected to the tide too. Indeed their behavior is similar to that showed by dissolved oxygen. This is not completely true for the salinity which seems to have a more complex and less predictable behavior, even if it's tide related for sure.

September's data differ from the general behavior of the summer season. Low levels of dissolved oxygen were registered during the most of the month. They were probably related to a large discharge of water in the system. The large amount of organic matter driven by rivers combined with a very cloudy period (solar radiation decrease) shifted the lagoon equilibrium, making the metabolic processes prevail on photosynthetic ones. This situation created a quite long period of undersaturation, but the resilience of the lagoon system seems quite high, able to withstand stress rather intense even on relatively long periods.

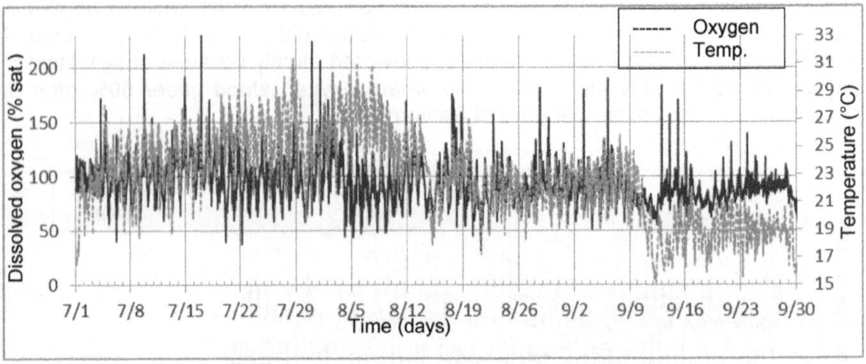


Figure 6. Time series of dissolved oxygen (in black) and temperature (in grey) for probe 20140 (near Stella river mouth) during summer 2013.

Figure 7. Encrusting organisms on the sensor's area of the probe.

Finally, salinity data were maybe altered by a massive grown of encrusting organisms, probably triggered by the large amount of nutrients discharged into the lagoon from the Stella and Cormor rivers. Thick biomasses belonging to *Serpuloidea*'s family (Figure 7) covered the sensors' zone of the probes in just a couple of days, ignoring the chlorination grid that could not work properly due to the general low salinity in the water (there was not enough Chlorine to be extracted). Thus these biomasses modified the electrical resistance of the conductivity sensor, with the result that all values almost fell to zero until probes were cleaned.

## Conclusions

The standing monitoring system implemented in the Marano lagoon by ARPA FVG is a step forward compared to the classic sampling approach usually carried on in the lagoon. This monitoring system overcomes the limits imposed by the conditions of access to the sub-basins, by the hourly conditions and the frequency of sampling.

The installation sites were chosen to meet both the need to analyze specific areas and to allow, practically and technically, the installation and maintenance of the probes' system. This permitted to observe the behavior of the lagoon, focusing on critical areas. Also real ecological problems that occur in this ecosystem, characterized by low turnover of water and high nutrients discharge, were highlighted. Although the analysis covers only a limited period, some important observations on the lagoon behavior have been carried out. Generally speaking, what was observed in the past was confirmed. The lagoon showed an average condition of diurnal dissolved oxygen values in the amount of 91.2% of saturation (over the whole period). Events of supersaturation with values over 200% were registered in coherence with what already seen along the years by ARPA FVG monitoring (Mattassi et al., 2004; Mattassi et al., 2007). The critical areas have been confirmed to be mostly the river mouths of Stella and Cormor, which drive in the lagoon significant nutrient inputs related to manure and urban wastewater (Mattassi et al., 2004; Mattassi et al., 2007; Saccon et al., 2011). Indeed in these areas algal blooms were observed during late spring in the years 2000-2001-2002-2003.

The most problematic situation is related to episodic hypoxia occurring regularly in the early morning within the inner zones nearby the river mouths of Stella, Cormor and Zellina. Very high temperature values can push the system to reach extreme dissolved oxygen values. Values fall up to 5% in the course of the night and early morning (severe hypoxia), while exceeding 200% oxygen saturation during the daylight (afternoon). When hypoxia episodes occur, pH values fall too. PH showed a synchronized trend compared to oxygen. This confirms that most of dissolved oxygen comes from photosynthetic activity that modifies the carbonate equilibrium due to consumption of $CO_2$ present in the water (Martelli et al., 2004). Furthermore, while hypoxia get more and more severe the pH rapidly decrease (min. 6.81), probably due to anaerobic processes that establish and free acid compounds in the water. In any case pH levels never exceeded the tolerance range of the aquatic species that lies between 6-9. The continuous occurrence of extreme oxygen values, however, did not lead to evident damages to local aquatic fauna (e.g. fish mass deaths). Therefore it has to be considered that only limited damage was caused. It should also be remembered that these values were only found in the inner parts of the lagoon and do not seem to affect the entire basin.

In conclusion, the standing monitoring system represented an important action towards the implementation of Integrated Coastal Zone Management policies in Friuli Venezia Giulia. Anyway, a lot of work has still to be done in order to fully comprehend the lagoon system behavior. The implementation of a biogeochemical model coupled with a hydrodynamic model of the Marano and Grado Lagoon was put in place, too. Moreover, in the next years, this instrumentation will support the monitoring activities laid down by the European directives 2000/60/CE WFD regarding transitional water.

## Acknowledgements

The authors wish to thank all the boat pilots and the personnel of the Osservatorio Alto Adriatico (OAA), the Department of Latisana and the SOS Sviluppo e Innovazione Tecnologica e Funzionale of ARPA FVG that have technically supported this work.

## References

Brambati A. (1996), *Metalli pesanti nelle lagune di Marano e Grado, Piano di studi finalizzato all'accertamento della presenza di eventuali sostanze tossiche persitenti nel bacino lagunare di Marano e Grado ed al suo risanamento*; Regione Autonoma FVG, Direzione dell'Ambiente – Servizio Idraulica 1996.

Covelli S., Faganeli J., Horvat M. and Brambati A. (2001), *Mercury contamination of coastal sediments as the results of long-term cinnabar activity (Gulf of Trieste, Northern Adriatic Sea).* Applied Geochemistry 16: 541-558.

Covelli S., Faganeli J., De Vittor C., Predonzani S., Acquavita A., Horvat M. (2008), *Benthic fluxes of mercury species in a lagoon environment (Grado Lagoon, Northern Adriatic Sea, Italy);* Applied Geochemistry 23 (2008) 529–546.

Martelli F., Nocciolini S., Cavalieri S. (2004). *La Laguna di Orbetello e il monitoraggio dell'ARPAT.* Atti Convegno APAT. Ottobre 2004 Venezia.

Mattassi G., Daris F., Decorte E., Suraci C., Zanello A. (2006), *Le Lagune di Marano e Grado: classificazione di qualità mediante utilizzo di macrodescrittori chimico-fisici 1987-2004.* Atti Conv. Lincei 222: 183-194.

Mattassi G., Rossin P., Giacomich P. (2007), *Le lagune di Marano e di Grado: quadro sinottico e problematiche di applicazione della WFD 2000/60. Linee guida e casi di studio per la gestione dei siti della rete Natura 2000 in ambiente di transizione:* atti del workshop tenutosi a Grado, Palazzo dei congressi, 7-8 giugno 2006, Trieste, EUT Edizioni Università di Trieste, 2007, pp. 296-308.

Saccon P., Leis A., Marca A., Kaiser J., Campisi L., Savarino J., Böttcher M.E., Eisenhauer A., Sültenfuss J., Erbland J. (2011), *Assessment of potential nitrate pollution sources in the Marano Lagoon (Italy) and its catchment area using a multi isotope approach*; IAEA-CN-186/134 (2011): 473-481.

**Coastal Management: Changing coast, changing climate, changing minds**
ISBN 978-0-7277-6114-9

ICE Publishing: All rights reserved
doi: 10.1680/cm.61149.059

# Examples of the Software Tool MorphAn Applied in Coastal Management along the Dutch Coast

**Pieter van Geer**, Deltares, Delft, The Netherlands

## Introduction

Effective data management and visualisation is becoming increasingly important for coastal managers as the effect of sea level rise and changing environmental conditions due to climate change pose increasing challenges. In search for answers to these challenges that fulfil criteria of economic cost-benefit and social values, the need for software tools that can couple data analysis with model predictions is increasing. In particular, the ability to assess coastal safety in both the short term and the long term is expected to greatly increase the effectiveness of coastal management schemes.

In this paper we present a coastal management tool called MorphAn that enables easy assessment of the morphodynamics of coasts. MorphAn allows coastal managers and engineers to import, view and analyse coastal data, such as the measured and predicted development of coastal profiles over time, in a user-friendly GIS-environment. MorphAn is currently used by the Dutch Ministry of Infrastructure and the Environment to underpin its nourishment strategy.

A fundamental improvement of MorphAn over other GIS-based products is the ability to link empirical, behavioural and process-based models to data in the MorphAn environment in order to analyse coastal developments and predict future change. In this paper we introduce MorphAn and highlight the use of empirical dune assessment rules, as well as the process-based storm impact model XBeach (Roelvink et al., 2009) in MorphAn to assess nourishment schemes and dune safety near the Dutch coastal town Egmond.

## Description of MorphAn

MorphAn (Morphological Analysis) is a new GIS based desktop application that supports assessment of sandy coasts. First MorphAn provides basic tools for analysis of measured data. Furthermore, MorphAn supports calculations needed for assessing the safety of a sandy dune coast against flooding during an extreme storm event. Typically such an event has a time frame of several hours to days. The application also supports assessment of the development of a coastline on a longer timescale (years to decades). The necessary steps of these assessments included in MorphAn are:

- Data management
- Analysis of the measured data
- Calculations with various models

- Analysis of the calculation results
- Export of the results to various formats

The software intends to facilitate both coastal (zone) managers and consultants in the field of coastal engineering in their daily work. To serve that purpose, MorphAn is freely available at http://morphan.deltares.nl.

MorphAn is built as a plugin in the Delta Shell framework (Donchyts et al. 2010). This framework offers a large variety of tools and components that can visualize data, setup and couple numerical models, visualize measured data and numerical model input and –output and export modeled and measured data. Besides tools to import data in different formats, the framework also includes open-source GIS libraries, used to define, visualize and transform data, including time navigation and the possibility to add other data layers (e.g., shapefiles, ascii files, Web Map Service).

The MorphAn plugin adds a definition for coastal measurements (cross-shore profiles) and additional data (nourishments and other input and output of the calculation models). This data can be visualized either on a map or several representations in cross-shore orientation. Furthermore it includes models that can be used to assess dune safety and development of the coastline. Each model visualizes its input and calculation results on a map and in dedicated cross-shore views. When navigating through the model results, the user can examine calculation results of a specific cross-shore location by clicking on the result in a map. The associated figures will then automatically be shown.

For the last three years, MorphAn has been used to analyze the development of the coastline to underpin a nourishment scheme aimed at maintaining the coastline position in The Netherlands. To that end it includes a model that calculates the trend in the development of the coastline based on measurements. This can be used to identify locations where the coastline retreats.

In the near future, MorphAn will also be incorporated in the software instruments that are provided by the Dutch Ministry of Infrastructure and the Environment (Rijkswaterstaat) to carry out a legal assessment of the safety of the Dutch water defenses against flooding. This functionality has already been used in various projects related to strengthening of the coastline, determination of legal boundaries or maintenance and management of the coastline by Dutch water boards and projects in Belgium, the UK and Denmark. To support this type of assessment MorphAn incorporates models and a workflow as prescribed in Dutch assessment rules, but also an implementation of the XBeach model (Roelvink et al. 2009) that is widely applied around the world to study resilience of a dune coast or barrier island to (extreme) storm conditions.

**Data Management**

The Netherlands has a long history of collecting cross-shore profile measurements along its coast. Consequently a database that describes the morphological changes of the coastal zone exists. This database consists of profiles measured yearly every 250 meter along the coast going back to 1965. Supplemented by a large database of gridded measurements (both bathymetry of submerged areas as well as topography measurements), this provides a good starting point for several types of assessment of the development of the Dutch Coast. MorphAn includes the mentioned database with yearly profile measurements. It also allows import of shapefiles and (2D) gridded measurements (in several well-known GIS formats). This data can be used to correct or supplement the yearly measurements. Together with an editor that allows batch operations on the measured transect (for example to account for the effect of sea level rise or structural erosion of the beach or dune) this makes MorphAn a powerful tool to quickly assess the coastal zone including the effects of possible changes in the future.

Besides access to data, MorphAn also offers its users the possibility to manage the imported dataset. Within the program it is possible to: Add, remove, rename or duplicate sets with measurements; import, change or delete individual measurements and to change single coordinates.

In 2015 plans for further development of MorphAn include an increased possibility to also link to databases that are online available. In that way it is possible to also include public available data.

**Analysis of the Measured Data in MorphAn**

MorphAn provides coastal managers a variety of possibilities to visualise the development of profiles in time. This ranges from a GIS based map to a detailed view of individual cross-shore measurements. The data analysis functions of MorphAn allow coastal managers to quickly examine the effect of nourishments and other developments of the measured beach profiles in time.

Single measurements

MorphAn includes two views that represent individual measurements (Figure 1). The side view shows individual measurements by highlighting thee different sources of the measurement: altimeter data, bathymetrical survey and interpolated values to connect those sources. Next to the side view, MorphAn also offers a view that visualizes the difference between two measurements at the same location, but with a different time stamp.

Visualization of measurements aggregated in time

As shown in Figure 2, the software also offers three views that represent measurements aggregated in time per location:

- The time stack representation (left upper figure) helps highlighting the development of parts of the measured profile in time. This allows for a quick analysis of the development of tidal channels near a beach or the development of dunes or breaker bars in time. The history view indicates the last moment
- The history view (right of the figure) gives an indication of the mobility of sediment at certain layers. In this case the red colours indicate recent profile measurements did not exceed that level. Green colours represent locations where the most recent measurements all exceeded that part of the profile.
- The comparison view allows the user to plot multiple measurements in the same figure in order to compare them. Various colour scales can be applied to help identify the profile development as well.

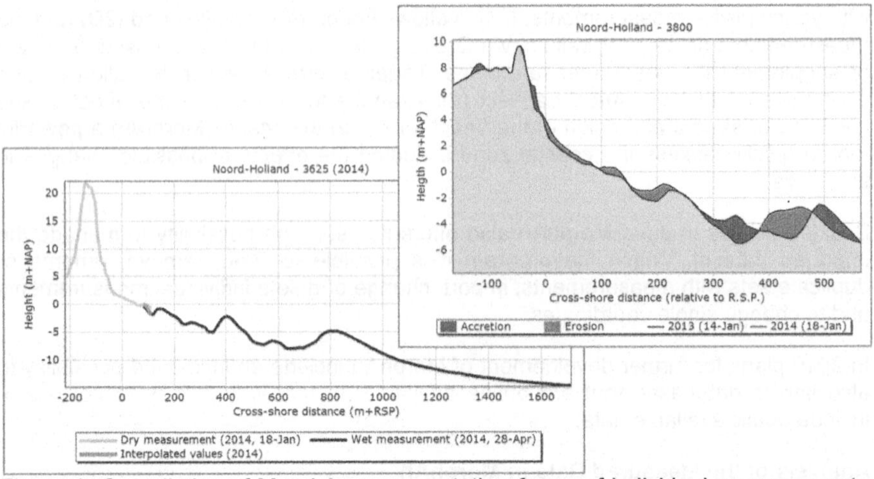


Figure 1: Compilation of MorphAn representation forms of individual measurements. The side view (left lower image) shows details of a single measurement. The difference view (right upper part) shows the difference between two measurements.

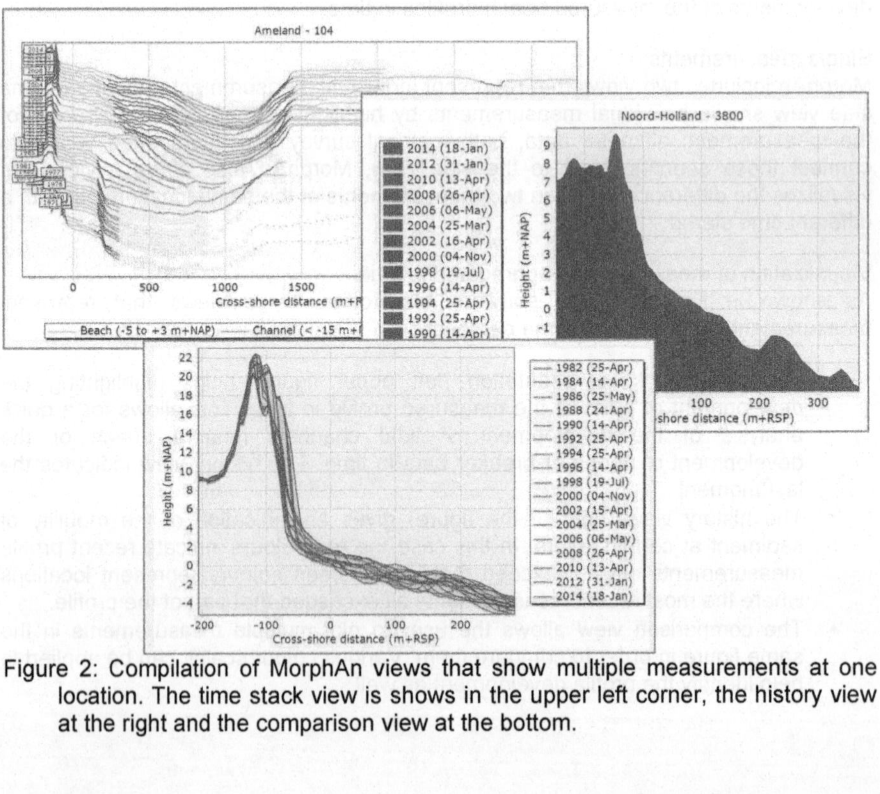


Figure 2: Compilation of MorphAn views that show multiple measurements at one location. The time stack view is shows in the upper left corner, the history view at the right and the comparison view at the bottom.

Map representation of the data

A map is the most efficient way of presenting te data in time and space. The map included in MorphAn (Figure 3) is coupled to a time navigator. Each measurement is indicated by blue and brown lines that represent the source of measurement similar to the side view. By scrolling through time with the time navigator the spatial and temporal extend of the data is shown.

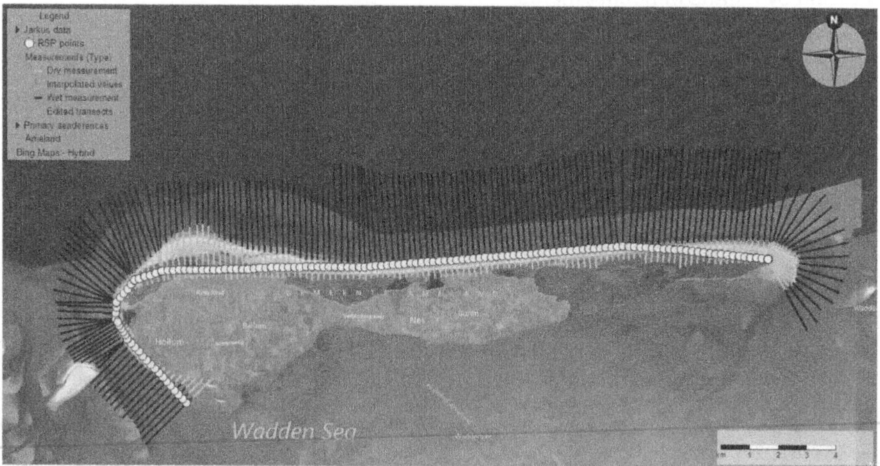


Figure 3: Map representation of the measurements of the beach of Ameland in 2014 as produced by MorphAn. In the background two WMS layers show the location of the primary sea defences[1] and an aerial photograph[2] of the area.

**Coastal Development and Nourishment Strategies in MorphAn**

MorphAn contains several models to calculate coastal development. This includes calculation of a representative coastline for a cross-shore profile measurement (also called momentary coastline, or MKL), but also the extrapolated development in time towards an expected coastline in the future (also called TKL). This tool provides insight into the development of the coastline on a time scale of year to decades. Comparison between the computed future coastline and a reference coastline (BKL) provides input for the scheduling of nourishments.

MorphAn calculates the MKL based on the volume between two horizontal reference levels (Figure 4). The upper level is assumed to best represent the dune foot position. The lower boundary is chosen such that the mean low water line lies in the middle of the two reference levels. Dividing this volume by the height difference between these levels gives a measure that represents the momentary coastline or MKL.

For determination of the trend in the development of the MKL values in time it is important to take into account nourishments that have been applied. Often nourishments are trend breaking events and one would typically not include such an event while determining the trend in the development of the MKL (see also Figure 5).

[1] http://geoservices.rijkswaterstaat.nl/primaire_waterkeringen_rws?
[2] http://www.bing.com/maps/

As a third step MorphAn includes a model to calculate the trend in the development of the MKL values at one location and extrapolates this trend towards the future, as visualized in Figure 5. Besides the expected coastline positions (TKL values) the model also calculates the moment of intersection between this trend and the reference coastline at that location (BKL). This moment is an important parameter when determining a nourishment strategy to hold the coastline at a predefined location by means of nourishments.

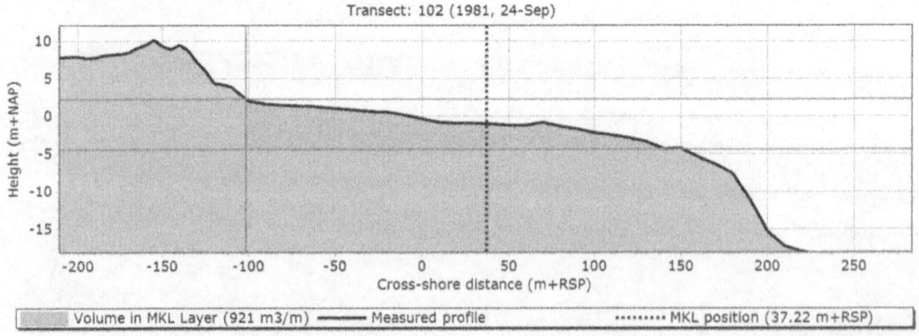


Figure 4: MorphAn representation of the calculated momentary coastline position (MKL) at location 102 at Ameland.

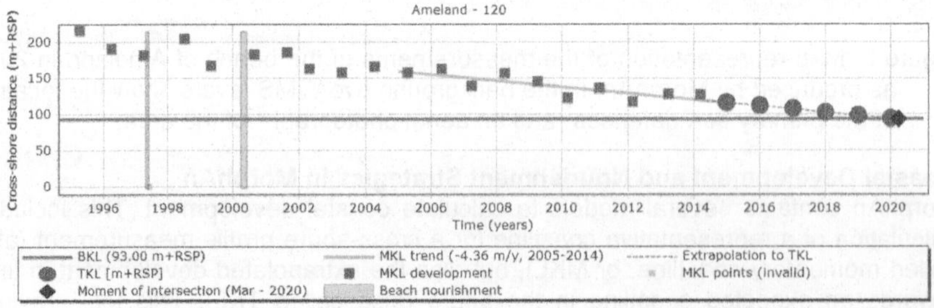


Figure 5: A representation of the development of the MKL in time at Ameland (location 120) as represented by MorphAn. The figure also includes the fitted and extrapolated trend in the development as well as the expected coastline positions (TKL) until 2020.

**Coastal Safety in MorphAn**

Next to coastal development, MorphAn also includes models to assess the response of the coast during storm events. It integrates the Dutch assessment rules as well as a GUI for the process-based storm impact model XBeach, with the possibility to add support for calculations on gravel coasts using XBeach-G (XBeach-Gravel; McCall et al, 2014, Masselink et al.).

Figure 6 shows the result of a legal safety assessment of a single measurement. According to the assessment rules, the effect of dune erosion during an extreme storm should be estimated with the empirical Duros+ model (Van Gent et al. 2008). After that storm a small amount of sediment should still be present to prevent flooding of the hinterland ("Boundary profile"). The assessor checks this by comparing the most landward point of the erosion profile (Retreat-point or R-point) with the “Landward boundary of sea defence” indicated by the green and blue dots in the figure. This assessment needs to be done for each of the 15 most recent profile measurements to incorporate temporal changes of the beach profile and dunes.

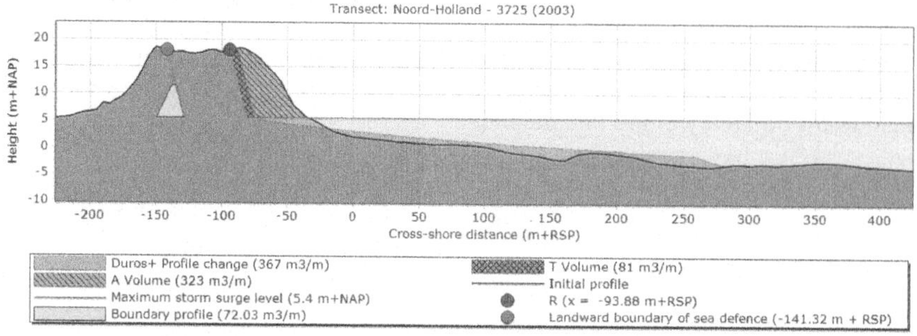


Figure 6: MorphAn representation of a calculation result of the Dutch legal safety assessment model Duros+

MorphAn also allows automated setup of 1D-XBeach models based on a selection of cross-shore profile measurements or results of the legal assessment model. Instead of an empirical description of the final erosion profile, XBeach is a process based model that calculates development of water levels, waves, currents and consequently profile changes in time. The GUI offers a simple view on the models output, but also allows thorough investigations of the model results. Both models allow coastal managers to quickly assess future coastal safety and the impact of nourishment strategies or planned dune coastal strengthening projects thereon.

**Other Tools**

Several tools that improve navigation through results, analysis of the data and results and performing additional operations or calculations need to be mentioned:

- Navigation through the results is made easy by allowing the user to click on particular results causing other output views to automatically select corresponding results.
- All views that contain time dependent data are coupled to a time navigator. This utility can be used to navigate (or animate) through time, showing measurement data or results.
- The possibility of writing python scripts or toolboxes that have access to all data and tool underlying the interface opens a large variety of possibilities. This enables personalizing plots by adding additional data or automatic changing colour schemes or fonts. Furthermore additional calculations based on existing results can be done. Also scripting can help importing from and exporting to file formats that are not supported by default.

## Safety of Egmond

Egmond is one of the coastal towns along the Dutch coastline. Traditionally it was founded some centuries ago near a low part of the dune to facilitate local fisherman to pull their boats over the dunes when done fishing. Typically these towns are therefore most likely to be weak spots when it comes to safety of the hinterland.

Figure 7 shows the result of a safety assessment of the town of Egmond according to the safety assessment rules. The red line indicates the "landward bioundary of the sea defence" that should not be exceeded by the green line, which represents the most landward retreat point at each location in the 15 most recent measurements. The blue area indicates the range of all calculated retreat points. Although several buildings and roads are at risk during an extreme event, flooding of the hinterland is less likely to occur, since the retreat points are still some 20 to 60 meters away from the landward boundary of the sea defence.

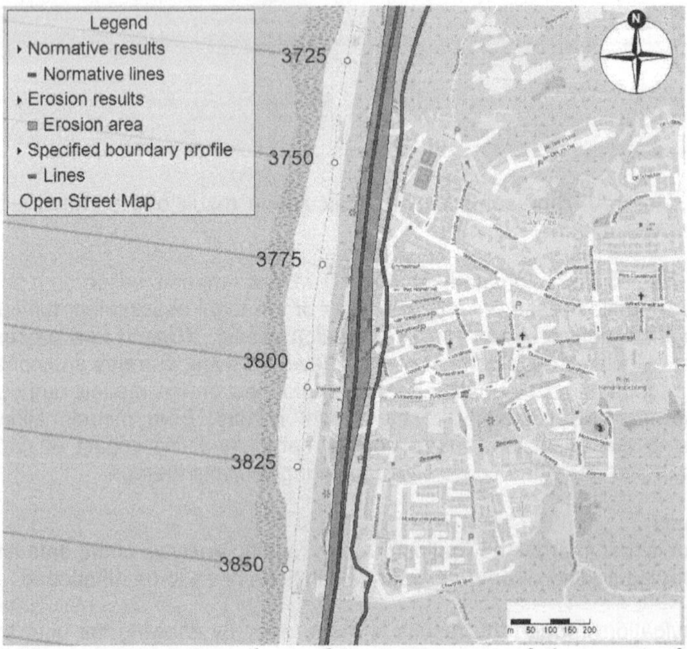


Figure 7: Map representation of a safety assessment of the coast of Egmond generated by MorphAn.

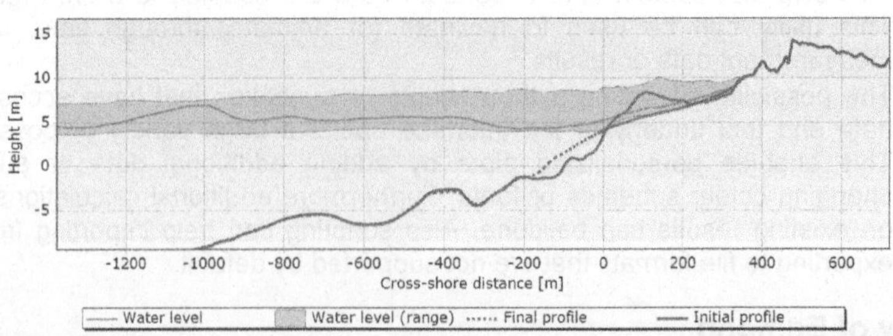


Figure 8: XBeach model result after a 5 hour storm under normative conditions near Egmond

However, near location 3800 a passage to the beach crosses the dunes. The dune is here at its lowest and this was probably the reason for founding the town at this location in the first place. We used MorphAn to extract another cross-shore profile from grid measurements at the exact location of the passage and used that to simulate dune erosion using XBeach 1D. Figure 8 shows that at this location instead of a retreat point around 50 meters (cross-shore location), waves overtop the first dune row flooding the plane behind it. This is in fact the entrance of the main street as can be deduced from Figure 7. Although the dunes will probably not breach (because of the existence of a second, higher dune row) overwash of the first dune row could cause significant flooding during an extreme storm event at this location.

## Nourishment Scheme at Ameland

An interesting location to assess the development of the coastline is Ameland. Ameland is one of the barrier islands enclosing the Wadden Sea in the Northern part of The Netherlands. As such it has tidal inlets both at the Western and Eastern tip of the island. Tidal movement combined with waves that mostly come from the North West, cause ebb-tidal deltas to migrate westward.

Figure 9 shows the result of an analysis of the coastline development at Ameland done with MorphAn. The coloured bars indicate the magnitude and sign of the calculated trend. A seaward pointing bar indicates the coastline is moving seaward, a landward oriented bar indicates retreat of the coastline. Furthermore, red colours show that the calculated TKL position is already landward from the BKL. This can also be derived from a comparison between the calculated TKL (green line) and BKL (blue line).

Migration of a former ebb tidal delta is clearly present at the western part of the island. At the tip of the island a tidal channel already approaches the island, causing a strong landward directed development of the coastline. More to the East a seaward movement of the coastline can be observed. Although this former ebb-tidal delta will keep moving eastward, in the middle of the island presently the TKL approaches and even crosses the BKL. This caused the authorities to schedule a beach nourishment in 2015 for the middle part of this island, since the location where the BKL is not met is quite close to the most important town on the island. Not holding the BKL line at the western part of the island is considered less of a problem, since not many people live in that area and also because the movement of the tidal channel is hard to stop.

## Conclusions

Recently we developed a new software tool called MorphAn. This tool includes several models that can calculate erosion profiles as a consequence of extreme storms or (estimated) coastline positions. This allows for interactive assessment of the coastline position, underpinning a nourishment strategy, and assessment of the safety of dunes. These features are showcased in two field cases.

From assessment of Egmond it follows that during an extreme event it is likely that several buildings and roads will be demolished due to dune erosion. These objects however are not in the official protected area. Erosion under these conditions will be limited to the part of the dune that is considered to be the sea defence. A preliminary XBeach calculation however shows that flooding of the beach passage could cause significant damage.

Figure 9: Map representation of an assessment result of the development of the coastline of Ameland as presented by MorphAn.

Analysis of the development of the coastline at Ameland shows the effect of a migrating (former) ebb tidal delta. Part of the beach is widened, where another part has a retreating coastline. At the western part of the island this is not considered a large problem, whereas in the middle of the island the results of such an assessment lead to nourishments scheduled in 2015. MorphAn is freely available through the website http://morphan.deltares.nl. Although this paper shows two field cases in The Netherlands, it can be applied anywhere around the world.

## Acknowledgments

The development of MorphAn was initiated and funded by the Dutch Ministry of Infrastructure and the Environment (Rijkswaterstaat) with an additional contribution of the collaborating Dutch water boards (STOWA) for development of the transect editor.

## References

Donchyts, G., Jagers, B., (2010): "*DeltaShell - an open modelling environment*", Proceedings of 2010 International Congress on Environmental Modelling and Software, Ottawa, Canada

Masselink, G., McCall, R.T., Poate, T.G, and van Geer, P.F.C.(In Press): "*Modelling extreme storm response on gravel beaches using XBeach-G implemented through a graphical user interface.*" Maritime Engineering (Proceedings of the ICE)

McCall, R.T., Poate, T.G., Masselink, G. Roelvink, J.A., Almeida, L.P., Davidson, M. and Russell, P.E., (2014): "*Modelling storm hydrodynamics on gravel beaches with XBeach-G*". Coastal Engineering Volume 91, 231–250

Roelvink, Reniers, Van Dongeren, Van Thiel de Vries, McCall and Lescinski, (2009): "*Modeling storm impacts on beaches, dunes and barrier islands*". Coastal Engineering Volume 56 (11-12) 1133-1152.

Van Gent, M.R.A., J.S.M. Van Thiel de Vries, E.M. Coeveld, J.H. de Vroeg, J. van de Graaff (2008). "*Large-scale dune erosion tests to study the influence of wave periods*", Coastal Engineering Volume 55(12). 1041-1051.

Coastal Management: Changing coast, changing climate, changing minds
ISBN 978-0-7277-6114-9

ICE Publishing: All rights reserved
doi: 10.1680/cm.61149.069

# Coastal Management and Environmental Guidelines Related to Coastal Defence Works

**Paola La Valle**, Institute for Environmental Protection and Research (ISPRA) - Via V. Brancati 60, 00144, Rome, Italy
**Daniela Paganelli**, Institute for Environmental Protection and Research (ISPRA) - Via V. Brancati 60, 00144, Rome, Italy
**Stefania Ercole**, Institute for Environmental Protection and Research (ISPRA) - Via V. Brancati 60, 00144, Rome, Italy
**Iolanda Lisi**, Institute for Environmental Protection and Research (ISPRA) - Via V. Brancati 60, 00144, Rome, Italy
**Corrado Teofili**, Federparchi-Europarc - Via Nazionale, 230, 00184, Rome, Italy
**Luisa Nicoletti**, Institute for Environmental Protection and Research (ISPRA) - Via V. Brancati 60, 00144, Rome, Italy

## Introduction

The Mediterranean coastal zone is characterized by landscapes of outstanding natural value and by a large number of particularly important habitats, in terms of biodiversity and functional complexity. At the same time, coastal environments are extremely sensitive and vulnerable transitional zones, strongly influenced by the increased urbanisation (ports, industries, tourism and infrastructures) and human activities that, together with poor coastal defence policies, have directly or indirectly turned coastal erosion from a natural phenomenon into a problem of growing intensity (Airoldi et al. 2005; Martins et al. 2009).

Currently about twenty thousand kilometers of European coasts, corresponding to around 20%, face serious impacts. Most of the impact zones (15,100 km) are actively retreating, some of them in spite of coastal protection works (2,900 km) (EUROSION 2004; Sutherland 2010). As a response to the growing need to reduce erosion and to protect building and infrastructures, defence works have become common features of coastal landscapes (EUROSION 2004). Despite their purposes, it is recognised that they produce undesirable effects on coastal and marine environments, especially when protected or sensitive habitats and species are present. For this reason the problems associated with an intense and rapid coastal erosion have raised the attention on shoreline protection not only in terms of preserving economic and social resources, but also in terms of protecting and preserving biodiversity and ecological resources, in accordance with the Integrated Coastal Zone Management (ICZM).

In this framework within the European (MED) projects COASTANCE related to "Regional action strategies for coastal zone adaptation to climate change" and COASTGAP "Coastal Governance and Adaptation Policies in the Mediterranean", ISPRA was charged by the Lazio Region to develop specific guidelines to support environmental studies related to the coastal defence works.

The guidelines propose a matrix-system, named "structure/impact" *vs* "habitat/species", which allows to know in advance, for different coastal defences, the potential impact types expected on marine and coastal environments, and in particular on protected habitats and associated flora and fauna species (Paganelli et al. 2013; 2014). This paper describes the matrix-system, highlighting its effectiveness as a simplified tool for the management and conservation of marine-coastal environments and for the drawing up and the evaluation of environmental studies related to coastal defense works.

## Creation of the Matrix-system

The matrix-system is organized in 9 "structure/impact *vs* habitat/species" matrices, referring to the specific categories of coastal defences as reported below: seawalls and dikes; nearshore breakwaters (emerged and submerged) and artificial reef; groynes (permeable and impermeable); composite groynes; artificial headlands; beach nourishment; beach drainage systems; dune reprofiling; windbreak fences, dune grass planting and dune access management (Paganelli et al. 2014). Each category includes all the defence works that potentially produce the same physical and biological effects on marine and coastal environments, resulting from their interference with coastal processes (such as hydrodynamic regime, littoral transport, morphodynamics changes, etc.).

The matrices are organized in a hierarchical system of classification of the marine and coastal environments, taking into account both emerged and submerged environments. This hierarchical system, with increasing detail, allows the identification of the protected habitat and species that may be potentially affected by impacts. In order to classify the protected habitats we referred to the habitat types reported in the Habitats Directive (92/43/EEC). The matrix-system, assigning a key role to the habitat types of the Directive, allows us to pursue one of the objectives of the Directive itself, i.e. satisfying the conservation state of natural and semi-natural habitats for the wild flora and fauna species.

Analysing the Italian landscapes, a total of 38 marine-coastal habitats of European interest have been identified and are included in the matrices with the same codes of the Directive (Natura 2000 Code), comprising submerged and dry habitats, low sandy shores, and rocky shores (Table 1). Given the great geographical and ecological variety of the Italian environments, the identified habitat types are reasonably representative of the whole Mediterranean basin. Based on morphogenetic, lythomorphologic and pedological homogeneity criteria, the protected habitat types have been grouped into 11 environmental units called "physiographic categories".

The use of physiographic categories allows a wider and more flexible applicability of the methodology both when available information do not allow to classify the habitat types according to the Directive and in presence of non-protected habitats.

The 11 physiographic categories were thereafter classified in 4 main macro-environments: marine habitats, wetlands and halophytic habitats, dune habitats and cliff habitats (Table 1).

Table 1: A hierarchical classification of the protected marine-coastal habitat types *sensu* Habitats Directive (92/43/EEC) in physiographic categories and macro-environments. Priority habitats are reported with an asterisk (*) following the Natura 2000 code.

| PROTECTED MARINE-COASTAL HABITAT TYPES (NATURA 2000 CODE) | PHYSIOGRAPHIC CATEGORIES | MACRO-ENVIRONMENTS |
|---|---|---|
| 1110,1160 | Marine waters, soft bottoms | MARINE HABITATS |
| 1110, 1170, 8330 | Marine waters, hard bottoms | |
| 1120* | Posidonia oceanica beds | |
| 1130, 1140 | Estuarine and tidal systems | WETLANDS AND HALOPHYTIC HABITATS |
| 3120, 3130,3140, 3170* | Standing waters, temporary lakes and ponds | |
| 1150*, 1310, 1320, 1410, 1420, 6420 | Coastal brackish and saline lagoons | |
| 1210, 1310 | Dry beach | DUNE HABITATS |
| 2110,2120,2130 | Embryodune and avandune | |
| 2130*, 2160, 2210, 2230, 2240, 2250*, 2260, 2270*, 6220 | Avandune continental side, fixed dune and stabilised sands | |
| 1410, 1510*, 6420 | Interdune and backdune humid depressions | |
| 1240, 1430, 5320, 5330, 5410, 5420, 5430, 8210 | Rocky shores and cliffs habitats | CLIFF HABITATS |

Once defined the allocation of protected habitat types to macro-environments, the potential physical effects and biological impacts generated by each coastal defence category have been listed for each macro-environment and physiographic category. Furthermore, for each physiographic category, each potential impact was matched with the possibly affected protected habitat types and the associated flora and fauna species. In particular, for the protected wild flora species the eco-physiological adaptation criterion was used. This is possible because the species-habitat relationship is very strong in these environments (Van der Maarel & Van der Maarel-Versluys 1996; Van der Maarel 2003).

For associating protected fauna species with physiographic categories the habitat use criterion was identified; it describes how an individual uses the physical and biological resources of a given environment. Taking into account the most sensitive phases of the species' life cycle, the following 8 habitat use categories were identified: Residents and Sessiles (RS), Larval Recruitment and Settlement (LRS), Feeding (F), Nursery Area (NA), Reproduction and Mating (RM), Nesting and egg Deposition (ND), Temporary Stops (TS), Prolonged Stops and Migration (PSM).

## Use of the Matrix-system

The matrix generally identifies, for each macro-environment, which are the main expected physical effects and potential impacts. It indicates the physiographic categories, the habitat types and the habitat use categories for the fauna species potentially involved by each impact type.

In order to clarify the interpretation and use of the matrix-system, a brief description of the matrix related to "nearshore breakwaters (emerged and submerged) and artificial reefs (offshore)" is reported as an example (Table 2), analysing some expected physical effects.

The main physical effects caused by nearshore breakwaters are: loss of substrate, substrate variation, turbidity, eutrophication, trampling and noise. Loss of substrate, for example, may be linked to both the construction of the structure (affecting both marine habitats and wetlands and halophytic habitats) and erosion phenomena linked to changed hydrodynamic conditions (affecting dune habitats). This effect always results in habitat loss and/or changes with possible impacts on the associated flora and fauna species.

The matrix shows how this impact type can potentially affect different habitat types (and associated flora species): habitat types 1160 and 1110 in marine habitats (M); 1130 and 1140 in wetlands and halophytic habitats (W); 1210, 1310, 2110, 2120, 2130*, 2160, 2210, 2230, 2240, 2250*, 2260, 6220* in dune habitats (D).

Regarding the fauna species, it is important to note that the matrix does not directly provide the species list, but it identifies the habitat use categories. For example, in marine habitats (M), the fauna species potentially involved by substrate loss are residents and sessile species (RS), species that use this habitat for recruitment, reproduction, nesting and deposition (LRS, RM and ND), and for feeding (F) and the nursery areas (N/A). In wetlands habitats (W) in addition to the habitat use categories mentioned above, species that use the area of interest for temporary and prolonged stationing (TS and PSM) are also involved. Finally in dune habitats the following habitat use categories are concerned: RS, LRS, RM and ND, F, PSM.

It is important to underline that a same fauna species can be associated with different physiographic categories and the same category, in different environments, can be associated with different species.

With regard to turbidity, for example, it can be observed that the effect is temporary (during the construction phase, C) and present in marine and wetlands habitats. Turbidity can produce impacts on flora (e.g. decrease of photosynthetic capacity) and on fauna (e.g. decrease of predatory capacity) species. The habitat types and the related flora species that can suffer this impact are marine habitat types 1110, 1160, 1170, 8330, 1120* and wetlands habitats 1130 and 1140. As for the fauna species, the habitat use categories potentially involved are: RS, LRS, F, PSM, RM and NA.

A peculiar situation occurs for noise, that is a temporary effect closely associated to the construction phase. Noise does not have effects on the habitat (intended as "the environment or series of environmental factors in which a species or community develops", from Biondi et al. 2006) and on the associated flora. For this reason, the 'noise' impact has been indicated in the matrices as "non present" (n.p.) - in the column "habitat type and associated flora species".

Noise can induce however significant impacts on wild fauna; for example in the case of dune and cliffs habitats (D and C), the habitat use categories potentially involved are RS, F, RM, ND, TS and PSM. In presence of marine habitats, the RS, F, NA, RM and PSM habitat use categories are involved; in wetlands habitats (W), the use categories involved are the same as marine habitats plus ND and TS.

Table 2: The "structure/impact *vs* habitat/species" matrix related to nearshore breakwaters (emerged and submerged) and artificial reefs.

| MATRIX: NEARSHORE BREAKWATERS (EMERGED AND SUBMERGED) AND ARTIFICIAL REEFS (OFFSHORE) | | | | CATEGORIES, HABITAT AND SPECIES AFFECTED BY IMPACTS | | |
|---|---|---|---|---|---|---|
| MACRO-ENVIRONMENTS | MAIN PHYSICAL EFFECTS | Phases (C/O) | MAIN POTENTIAL IMPACTS | Physiographic categories | Habitat types and associated flora species | Habitat Use Categories (fauna species) |
| MARINE HABITATS (M) | Loss of substrate linked to structure placement operations | C/O | Loss and/or variation of habitat and of flora and fauna species | M1 | 1110, 1160 | RS LRS F ND RM NA |
| | Substrate variations linked to the changed hydrodynamic conditions | O | Habitat loss and/or variations, with effects on the flora and fauna (e.g. changes in species composition) | M1 | 1110, 1160 | |
| | | | | M3 | 1120* | |
| | Turbidity and suspended load, linked to movement of sediments | C | Effects on the flora (e.g. a decreased photosynthetic ability) and on the fauna (e.g. a decreased predatory ability) | M1 | 1110, 1160 | RS NA F LRS RM PSM |
| | | | | M2 | 1160, 1170, 8330 | |
| | | | | M3 | 1120* | |
| | Eutrophication linked to the reduced water exchange | O | Effects on the flora and on the fauna (e.g. algal bloom and anoxia phenomena) | M1 | 1110, 1160 | RS LRS NA F ND RM PSM |
| | Trampling | C | Effects on the flora and fauna | M1 | 1110, 1160 | RS LRS ND |
| | Noise | C | Effects on the fauna (e.g. disturbance in fish and marine reptile species) | M1 | n.p. | RS F NA RM PSM |
| | | | | M2 | | |
| | | | | M3 | | |
| WETLANDS AND HALOPHYTIC HABITATS (W) | Loss of substrate linked to structure placement operations | C/O | Loss and/or variation of habitat and of flora and fauna species | W1 | 1130, 1140 | RS LRS F ND R NA TS PSM |
| | Substrate variations linked to the changed hydrodynamic conditions | O | Habitat loss and/or variations, with effects on the flora and fauna (e.g. changes in species composition) | | | |
| | Turbidity and suspended load linked to movement of sediments | C | Effects on the flora (e.g. a decreased photosynthetic ability) and on the fauna (e.g. a decreased predatory ability) | W1 | 1130, 1140 | RS NA F LRS RM PSM |
| | Eutrophication linked to the reduced water exchange | O | Effects on the flora and on the fauna (e.g. algal bloom and anoxia phenomena) | W1 | 1130, 1140 | RS LRS NA F ND RM PSM |
| | Trampling | C | Effects on the flora and fauna | W1 | 1130, 1140 | RS LRS ND |
| | Noise | C | Effects on the fauna (e.g. disturbance in bird, reptile and mammal species) | W1 | n.p. | RS F NA RM ND TS PSM |
| | | | | W2 | | |
| | | | | W3 | | |
| DUNE HABITATS (D) | Loss of substrate linked to possible down-drift erosion phenomena | O | Habitat loss and/or variations, with effects on the flora and fauna (e.g. changes in species composition) | D1 | 1210, 1310 | RS LRS F ND RM TS PSM |
| | | | | D2 | 2110, 2120, 2230 | |
| | | | | D3 | 2130*, 2160, 2210, 2230, 2240, 2250*, 2260, 6220* | |
| | Trampling | C | Effects on the flora and fauna (e.g. invertebrates, birds and, in D1, marine reptiles) | D1 | 1210, 1310 | RS LRS ND |
| | | | | D2 | 2110, 2120, 2230 | |
| | | | | D3 | 2130*, 2160, 2210, 2230, 2240, 2250*, 2260, 6220* | |
| | | | | D4 | 1410, 1510*, 6420 | |
| | Noise | C | Effects on the fauna (e.g. disturbance in bird, reptile and mammal species) | D1 | n.p. | RS F RM ND TS PSM |
| | | | | D2 | | |
| | | | | D3 | | |
| | | | | D4 | | |
| CLIFF HABITATS (C) | Noise | C | Effects on the fauna (e.g. disturbance in bird species) | C1 | n.p. | RS F RM ND TS PSM |

The above shows how the analysis of a specific matrix allows to know, before the realization of coastal works and before carrying out environmental studies, the list of the expected potential effects/impacts, the list of protected habitat types (and associated flora species) and the habitat use categories that are subject to these impacts.

For obtaining also the list of the protected fauna species, a detailed study should be carried out by experts to identify which are the species really affected by the impacts according to habitat use categories.

## Conclusions

The proposed matrix-system aims to identify the main impact types generated by coastal defence works and to know in advance the critical elements, in terms of protected habitats and species, that may be affected by these impacts.

The matrix provides a qualitative list of the physical effects and potential impacts on habitats and species, but it does not quantify their entity. The quantitative estimation of the impact requires in fact that both environmental aspects of the project area (i.e. littoral transport system, conservation status of habitats and species etc.) and technical-design elements (i.e. form, texture, source and utilization of materials etc.) are adequately considered early in the planning process.

An important aspect of the proposed matrix-system is represented by its hierarchical structure, that assigns a key role to the habitat types of the Habitats Directive and that makes it an effective tool to assess the cause-effect relationships between "structure/impact" and "habitat/species".

This system can be used at different levels of detail and analysis, according to the available knowledge on the area of intervention and the different objectives to be pursued. In fact, critical elements (protected habitats and species) resulting from the application of the matrix to a real-case, can be used as a reference list for the drawing up and the evaluation of environmental studies related to coastal defence works.

Furthermore, the use of physiographic categories allows a wider and more flexible applicability of the methodology both when available information do not allow to classify the habitat types according to the Directive and in presence of non-protected habitats.

It is important to emphasize that to identify protected habitat types and species really occurring in the area of interest, an expert-based evaluation is essential (such as, for example, botanists, zoologists and geomorphologists).
Based on the above, we believe that the matrix-system can be a useful tool for the management and conservation of marine-coastal environments and for the Environmental Impact Studies, also in an extra- Mediterranean context.

### References

Airoldi L., Abbiati M., Beck M.W., Hawkins S.J., Jonsson P.R., Martin D., Moschella P.S., Sundelo A., Thompson R.C. & Aberg P. (2005) *An ecological perspective on the deployment and design of low-crested and other hard coastal defence structures'*, Coastal Engineering, Vol 52, pp. 1073-1087.

Bellan-Santini D., Bellan G., Bitar G., Harmelin J.G., Pergent G. (2002) *Handbook for interpreting types of marine habitat for the selection of sites to be included in the national inventories of natural sites of conservation interest*, RAC/SPA edit., UNEP publ., pp. 217, December 2002.

Biondi E., Blasi C., Burrascano S., Casavecchia S., Copiz R., Del Vico E., Galdenzi D., Gigante D., Lasen C., Spampinato G., Venanzoni R. & Zivkovic L. (2009) *Manuale italiano di interpretazione degli habitat della Direttiva 92/43/CEE*, Ministero dell'Ambiente e della Tutela del Territorio e del Mare'. Available:

http://sweb01.dbv.uniroma1.it/cirbfep/pubblicazioni/pdf/habitat_ita.pdf [23 January 2015].

EUROSION (2004) *Living with coastal erosion in Europe: Sediment and Space for Sustainability - PART I*, Major findings and Policy Recommendations of the EUROSION project. Available: http://www.eurosion.org/reports-online/part1.pdf [23 January 2015].

Martins G.M., Amaral A.F., Wallenstein F.M., Neto A.I. (2009) *Influence of a breakwater on nearby rocky intertidal community structure.* Marine Environmental Research, 67, 237-245.

Paganelli D., La Valle P., Ercole S., Teofili C. & Nicoletti L. (2013) *Assessing the impacts of coastal defense structures on habitat types and species of European interest (92/43/EC): a methodological approach*, Journal of Coastal Research (S.I.), Vol 65, pp. 1009-1014.

Paganelli D., La Valle P., Ercole S., Lisi I., Teofili C. & Nicoletti L. (2014) *Linee guida per gli studi ambientali connessi alla realizzazione di opere di difesa costiera*, ISPRA, Manuali e Linee Guida 105/2014. Available: http://www.isprambiente.gov.it/it/pubblicazioni/manuali-e-linee-guida/linee-guida-per-gli-studi-ambientali-connessi-alla-realizzazione-di-opere-di-difesa-costiera [23 January 2015].

Sutherland J. (2010) *Guidelines on beach monitoring for coastal erosion*, Concepts and Science for Coastal Erosion Management - CONSCIENCE, Deliverable D15. Available: http://www.conscience-eu.net/documents/deliverable15-guidelines-on-beach-monitoring-for-coastal-erosion.pdf [23 January 2015].

Van der Maarel E. (2003) *Some remarks on the functions of European coastal ecosystems*, Phytocoenologia, Vol 33, pp.187-202.

Van der Maarel E. & Van der Maarel-Versluys M. (1996) *Distribution and conservation status of littoral vascular plant species along the european coasts*, Journal of Coastal Conservation, Vol 2, n.1, pp. 73-92.

***Acknowledgments***

The present work was funded by the Lazio Region local authority within the European (MED) project COASTANCE related to “Regional action strategies for coastal zone adaptation to climate change” (www.coastance.eu). Authors would like to thank Antonella Ceccarelli (ISPRA) for the useful support and careful reading of the manuscript.

**Coastal Management: Changing coast, changing climate, changing minds**
ISBN 978-0-7277-6114-9

ICE Publishing: All rights reserved
doi: 10.1680/cm.61149.077

# Determining the Downcoast Control Point for the Parabolic Bay Shape Equation

**Jonathan Kemp,** Engineer-Advisor, International Marine and Dredging Consultants, Antwerp, Belgium
**Thomas Eccleshall,** Graduate Structural Engineer, Hyder Consulting, London, UK
**Richard Simons,** Professor, Department of Civil and Environmental Engineering, University College London, UK

## Introduction

For coastal development projects, artificial beaches are an attractive feature and large scale coastal developments are being planned and constructed throughout the Middle East at a remarkable rate. In the design of any artificial beach it is desirable to avoid, or minimise, any future maintenance commitments by designing the initial beach planshape in such a way that it remains relatively stable under the incident wave climate experienced. Headland bays, or embayments, where a sandy beach is held by one, or between two erosion resistant headlands, tend to evolve to a stable beach planshape with little movement of the beach contours over time. The creation of artificial stable beaches has mainly been achieved through the application of the static equilibrium bay theory which is based on empirical equations of natural headland bay beaches.

Much of the research to date discusses the assessment of the stability of existing beaches and the introduction of headland control structures to address unstable beaches. There is little published work on the design of new artificial beaches for coastal development projects. If there is an existing beach, locating the downcoast drift point is simpler and the parabolic bay shape equation can be used to assess the stability of the beach. However, when there is no existing beach the process of selecting a suitable downcoast point becomes more problematic. It is more usual for a developer or coastal manager to define, or require, a minimum beach width either for amenity or coastal defence purposes.

This paper describes the results of a 3-month joint research project between HR Wallingford and University College London in 2009 as one part of a master thesis. The main aim of the research was to investigate the uncertainty surrounding the placement of the downcoast control point required for the application of the parabolic bay shape equation and to try and define a method for locating it.

## History and Development of Stable Bay Theory and Equations

Complex geomorphological features can be found along many coastal strips around the world. Sandy beaches in various shapes and sizes, from a long straight beach to a short highly curved one in coves, are generally very attractive to public users and as such are usually incorporated into the design of new coastal developments. Most curved beaches have distinct compartments constrained by rocky outcrops or headlands and, according to Short and Masselink (1999), these headland bay beaches or embayed beaches represent around 50 % of the world's coastline.

The equilibrium beach planshape of headland bay beaches was identified as early as the 1900's and is a generally widely accepted concept today. A summary of the chronological development of the theory and key equations is presented below:

- 1906 – Geologists and geographers first noted the beach planshape of headland bay beaches.
- 1944 - Krumbein examined beach process of Half Moon Bay, California and suggested the approximation of a logarithmic spiral.
- 1960 – Silvester initiated research on the stability crenulated shaped bays from a coastal engineers point of view with results reported from a physical model study showing an initially straight beach eroding to a stable curved bay.
- 1961 – Mashima first used the term parabolic to describe the geometric shape of stable coasts in Japan.
- 1965 – Yasso applied a log spiral equation (Figure *1*) to four natural headland bay crenulated beaches in the US.
- 1970's – Further work by Silvester and Ho advocated headland control for coastal engineering purposes, leading to the development of crenulated-shaped equilibrium beach planshape theory.
- 1987 – Hsu *et al* showed that the log spiral does not follow the complete periphery of the bay from the model tests of Silvester and Ho (1972), therefore suggested a different equation is required.
- 1989 – Hsu and Evans revisited the parabola equation and developed the parabolic equation (Figure *1*) for engineering purposes from fitting the planshape of 27 different beaches in static equilibrium. The origin of the equation, unlike the log spiral, is the wave diffraction point around the headland and the results is a beach planshape in static equilibrium. This makes the equation ideal for engineering purposes. It is also important to note that water levels were not believed to be important if the shoreline shape is considered at MHW.
- 1993 – Silvester and Hsu proposed that in terms of beach stability, headland bay beach can be classified as being in: *(1) Dynamic Equilibrium;* with continual sediment supply the shoreline maybe further seaward of the static equilibrium position and the apparent stable shoreline position is maintained by the balance between sediment input and output. *(2) Static Equilibrium;* is a bay with almost negligible sediment input or output and is stable when the tangential section of downcoast is parallel to the wave crests. Waves will break simultaneously around the whole bay, meaning there is no longshore

component of breaking wave energy and hence no littoral drift in the embayment. *(3) Unstable;* occurs when the beach is in an eroding state because of a reduction in sediment supply. In 2008 Hsu *et al.* proposed an additional classification: *(4) Natural beach reshaping;* as a result of the repositioning or extension of an updrift headland or coastal structure from which erosion occurred accompanied by accretion.

- 1997 – Silvester and Hsu advocated the use of headland control as a naturally functioning and preferable means of shore protection through the use of the parabolic equation.
- 1999 – Moreno and Kraus developed the hyperbolic tangent equation (Figure *1*) through the analysis of 46 beaches of mixed beach stability not exclusively beaches in *static equilibrium*. This equation can fit the shape of any bayed beach, regardless of whether it is in *static* or *dynamic equilibrium* or *unstable*.
- 2001 – Gonzalez and Medina provide guidance on the locating the downcoast control point.
- 2002 – Parabolic equation is included in the Coastal Engineering Manual.
- 2004 – Overview of the applicability of the empirical equations delivered by Benedet *et al.* suggests that the preferred method is the parabolic equation.
- 2008 – Hsu *et al.* introduces the additional *natural beach reshaping* classification and compares the three main empirical bay shape equations.
- 2010 – A review of the latest research and summary of static bay beach concept is provided by Hsu *et al.*
- 2010 - The sensitivity of locating the downcoast control point was studied by Lausman *et al.* and indicated that the placement of the control points is a subjective exercise.

**Empirical Equations**

The three major empirical equations (log-spiral, parabolic and hyperbolic) shown in Figure *1* differ in mathematical functions, coordinate systems, origins and controlling parameters related to wave direction and bay geometry. The overview of the applicability of the empirical equations delivered by Benedet *et al.* (2004) suggests that the preferred method is the parabolic equation. This is due to the use of the wave diffraction point as the centre of the coordinate system, thus allowing the effect of altering the wave diffraction point (by installation of engineering structures) to be assessed. Benedet *et al.* (2004) goes on to suggest that the other equations have a purely geometric relationship without physical insight of wave diffraction, which therefore deems them inappropriate for use in assessing the impact of headland control on the bay curvature. Hsu *et al.* (2008) concurs with these findings on the wave diffraction point and goes further by stating that the log-spiral or hyperbolic tangent models provide no means for determining the stability (i.e. in static or dynamic equilibrium) of existing beaches. Hsu *et al* (2010) states that this is because the database used to develop the log-spiral and hyperbolic equations come from beaches of mixed stability and not exclusively from beaches in static equilibrium. Therefore, the beach planshape these two equations prescribe can be of a beach in any state, unlike the parabolic equation which was developed only using beaches in static equilibrium.

Only the parabolic bay shape equation of Hsu and Evans (1989) directly links the change of shoreline to the point of diffraction, which is a fixed point that physically exists (either a natural headland or a coastal structure). Because the physical location of the wave diffraction point is used as the origin of the coordinate system for the parabolic bay shape equation, the effect of its relocation, by various engineering means, can be easily assessed using the new diffraction point. The parabolic bay shape equation is currently widely adopted for the assessment of headland bay beach stability and the effects of coastal structures on headland bay beaches.

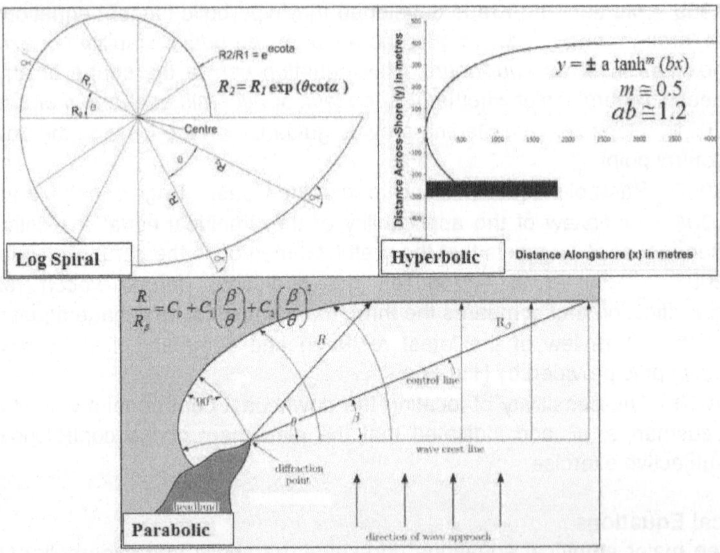


Figure 1: Definition sketch for empirical equations

## Development of Technique for Determining the Downcoast Point

Most of the research to date focuses on the assessment of the stability of existing beaches and how to manage existing beaches; there is comparatively little material on the design of new artificial beaches for coastal development projects. The location of the downcoast control point has always been difficult to assess (Silvester and Hsu, 1993) even if there is an existing beach. However, when there is no existing beach, the process of selecting a suitable downcoast point becomes even more problematic. Work undertaken by Lausman *et al.* (2010) also concluded that locating the downcoast point is a difficult issue and recommended more research in deriving a general relationship between the boundary of the curved and straight sections of the beach. His work, which involved expert elicitation, showed that if the user could not see the resulting beach planshape after selecting the wave diffraction point and the downcoast point, large discrepancies occurred. One could therefore hypothesise that a similar large error could occur if no existing beach existed i.e. in the case of a new coastal development.

It can therefore be considered that the main limitation with the application of the parabolic equation is locating the downcoast control point and the main limitation with the existing hyperbolic equation is the lack of a tangible wave diffraction point (and it can be fitted to any beach regardless of its stability). The work reported here has set out to determine a relationship linking the existing hyperbolic tangent equation with the wave diffraction point and wave obliquity and then use this equation to locate the downcoast control in the parabolic equation. This would then reduce the ambiguity of locating the downcoast control point associated with the parabolic equation.

**Determining the Wave Diffraction Point in the Hyperbolic Equation**

As one part of the short MSc research study, the hyperbolic equation was applied to a variety of different headland bay beaches from around the world. The case studies were selected based on several requirements: they show a crenulated bay shape, they are under the influence of at least one headland and, most importantly, there was information on the predominant wave conditions. One of the restricting parameters in finding suitable case studies at the time of the study was being able to find relevant information, in the short space of time, on the predominant wave direction. It was hoped to obtain the information used in the development of the parabolic equation (around 40 different bays) but this was not forthcoming and equally it was difficult to find locations where full wave climates were available. This led to only the following seven case studies being available (see Figure 2): Acre, Israel; Folkestone, UK; Waterfront 5, Kuwait; Felixstowe, UK; Elmer, UK; Jumeriah Beach, UAE; and Vivonne Bay, Australia. Clearly this is not a robust solution but it provides a start for further research. A summary of the dominant wave directions for each case study site is presented in Table 1

Table 1: Predominant wave direction

| **Name** | **Predominant wave direction (°N)** |
|---|---|
| Folkestone, UK | 186 |
| Elmer, UK | 215 |
| Jumeriah Beach, UAE | 305 |
| Acre, Israel | 294 |
| Kangaroo Island, Australia | 200 |
| Felixstowe, UK | 120 |
| Waterfront 5, Kuwait | 340 |

To determine a relationship linking the geometric origin of hyperbolic tangent equation with a wave diffraction point it was applied to the seven case studies. Through the fitting of the hyperbolic tangent equation to each of the seven case studies, by varying the values of the coefficients a, b & m, a clear pattern emerged: the diffraction point occurred in a similar location.

Based on these results, analysis was carried out to determine the location of the diffraction point relative to the origin of the hyperbolic tangent equation. This analysis involved investigating the coordinates of the diffraction point on an axis where the "origin" of the hyperbolic tangent shape equation is at (0, 0). The definitions of the different parameters are shown in Figure *3*.

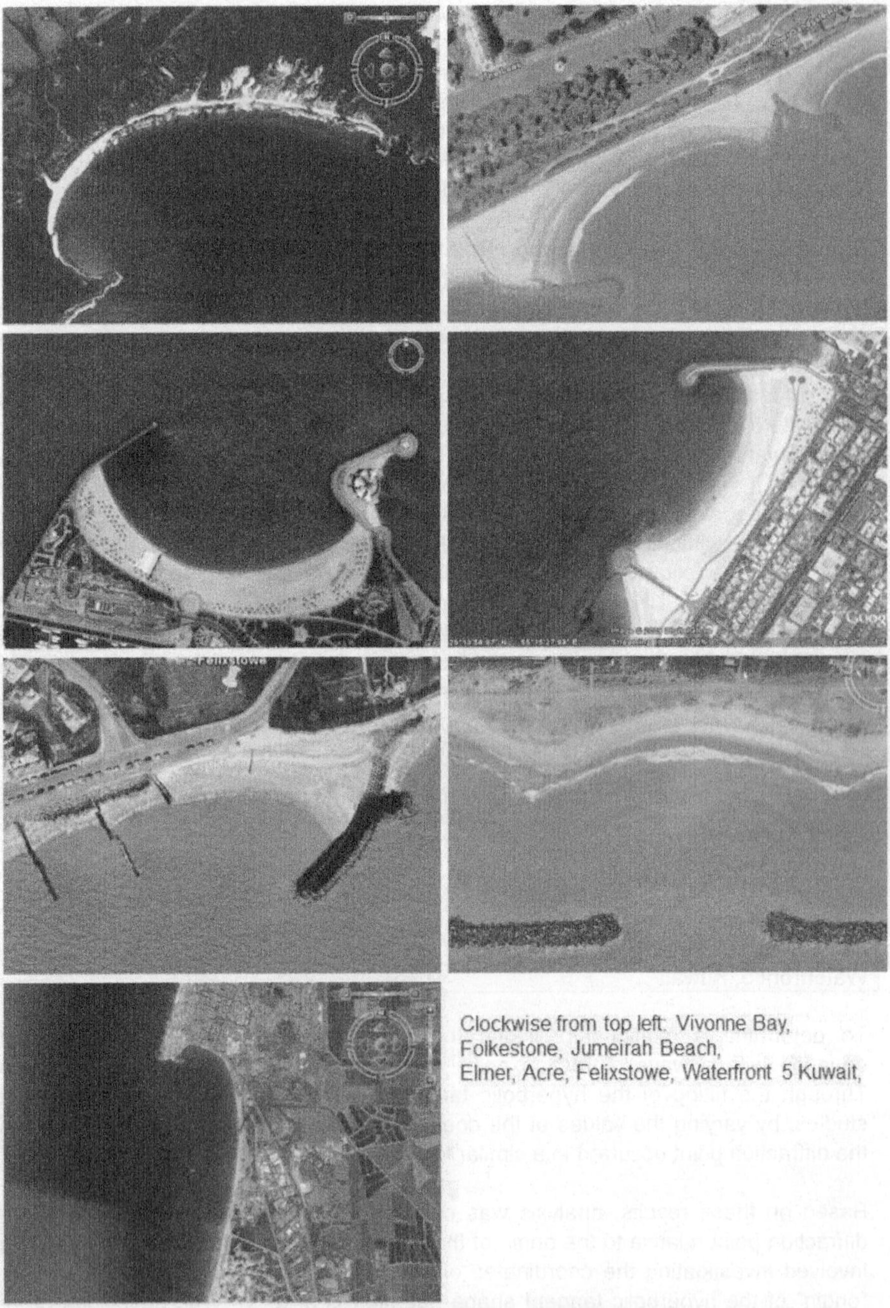


Clockwise from top left: Vivonne Bay,
Folkestone, Jumeirah Beach,
Elmer, Acre, Felixstowe, Waterfront 5 Kuwait,

Figure 2: Location of case studies.

For any modification of the hyperbolic tangent equation to be applicable for designing new beaches or assessing the influence of a headland on a beach, the coordinates of the diffraction point needs to be determined as non-dimensional expressions. For the purpose of analysis, the coordinates of the diffraction point are defined as '*c*' and '*d*' (shown in Figure 3). This means they can be used for any scale of headland bay. To achieve this, a scaling distance was measured for each bay. The coordinates of the diffraction point were then divided by this scaling distance, thus producing a non-dimensional expression that can be applied to different sized bays. The measurement which was determined to be the most easily obtainable was the value '*a*' (shown in Figure 3). This was the distance between the $x = 0$ line and the straight section of the beach. In designing new beaches the width of beach, from the back of the beach to the waterline, is often given as a design criterion (e.g. a 25m wide beach). The values of dimensionless coefficients resulting from the analysis are shown in Table 2. The values of the coefficients for the seven beaches investigated support the findings of Moreno and Kraus (1999) that $ab \approx 1.2$ and $m \approx 0.5$.

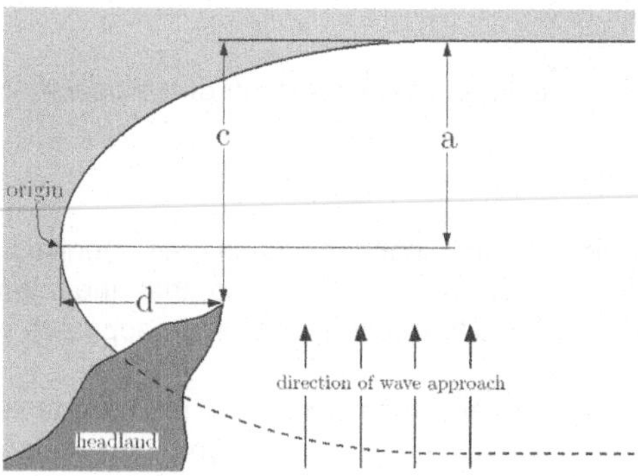


Figure 3: Definition sketch of parameters

Table 2: Values of non-dimensional coefficients

| Name | ab | m | c/a | d/a |
|---|---|---|---|---|
| Folkestone, UK | 1.20 | 0.55 | 1.22 | 0.42 |
| Elmer, UK | 1.32 | 0.65 | 1.29 | 0.86 |
| Jumeriah Beach, UAE | 1.20 | 0.50 | 1.20 | 0.68 |
| Acre, Israel | 1.10 | 0.55 | 1.21 | 0.48 |
| Kangaroo Island, Australia | 1.24 | 0.55 | 1.39 | 0.53 |
| Felixstowe, UK | 1.29 | 0.55 | 1.26 | 0.51 |
| Waterfront 5, Kuwait | 1.46 | 0.41 | 1.22 | 0.46 |

The analysis of the seven case studies led to the following relationships, which was used to identify the location of the coordinates of the wave diffraction point:

$$c/a \cong 1.22 \text{ for the } x \text{ coordinate} \quad (1)$$

$$d/a \cong 0.49 \text{ for the } y \text{ coordinates} \quad (2)$$

**Modified Hyperbolic Tangent Equation**

Using the results for determining the diffraction point, a modified hyperbolic tangent shape equation was developed. The origin of the Cartesian system of the modified hyperbolic tangent shape equation occurs at the diffraction point, in contrast to the hyperbolic tangent shape equation where the origin was a nondescript point. The modified hyperbolic tangent shape equation can also be defined in terms of the 'beach width' (which is a function of *c* in Figure 3), so eliminating the ambiguity in identifying the downcoast control point. The following modified hyperbolic tangent shape equation is proposed for a bay to the:

Right of the headland: $y = (\pm(0.82c)\,tanh^{0.55}\,((1.464/c)x + 0.588)) + 0.18c$ (3)

Left of the headland: $y = (\pm(0.82c)\,tanh^{0.55}\,((-1.464/c)x + 0.588)) + 0.18c$ (4)

where:
*y* = Distance cross shore [m]
*x* = Distance along shore [m]
*c* = Orthogonal distance between straight section of the beach and the point of wave diffraction [m]
(0,0) = origin at diffraction point

As with the hyperbolic tangent shape equation, the modified hyperbolic tangent shape equation empirical formula is rotated so that the straight section of the equation produced is parallel to the wave crests of the approach wave.

The modified hyperbolic tangent shape equation also produces a curve which is symmetric with the x-axis and produces two asymptotes. These are found at y = ±0.82*c*. The line *y* = +0.82*c* indicates the location on the shoreline which is no longer under the influence of the headland. As the hyperbolic tangent shape equation produces a curve which is symmetric with the x-axis and produces two asymptotes (where the line y = +a indicates the location on the shoreline that is no longer under the influence of the headland) it removes the difficulty in locating the position of the downcoast control point. Only the 'c' position is required, which can be directly related to the beach width.

## Application of Modified Hyperbolic Equation

In order to streamline the application of both the parabolic bay shape equation and the modified hyperbolic tangent shape equation a new tool was developed in ArcGIS. The tool calculates the beach planshape of a headland bay beach based on the parabolic and modified hyperbolic equations. To identify the location of the downcoast point the user defines the diffraction point by the clicking on the headland, defines the wave direction in the tool and then defines the beach width by clicking on the appropriate section of the beach. The tool then plots the result of the modified hyperbolic equation graphically within GIS as a shapefile, overlaying an image or map of the existing beach or proposed development, along with any other data layers (Figure 4). The resulting beach planshape of the modified hyperbolic equation shows where the beach becomes straight and therefore allows the user to clearly define the downcoast point for the parabolic equation. The process is repeated in the tool with parabolic equation to produce the stable beach planshape. This allows the stability of a headland bay beach to be assessed visually by comparing the existing shoreline periphery with the static equilibrium planshape, as well as being used in the design of proposed artificial beaches (Figure *4*).

It offers an intuitive environment with simple inputs and instant visualisation of the results and because the output files are in a standard GIS format they can be distributed to other users (e.g. coastal managers, clients or developers) to input into their own GIS. The tool also provides information in a text format for all the output parameters associated with each equation. As the tool is embedded within GIS it can also be used in the estimation of the volume of sand required for beach nourishment and through the using the other GIS functions a 3d view of the beach can be generated. By combining the parabolic bay shape equation and modified hyperbolic tangent shape equation into GIS allows a simple method in the design of proposed artificial beaches, offering an intuitive environment with simple inputs and instant visualisation of the results.

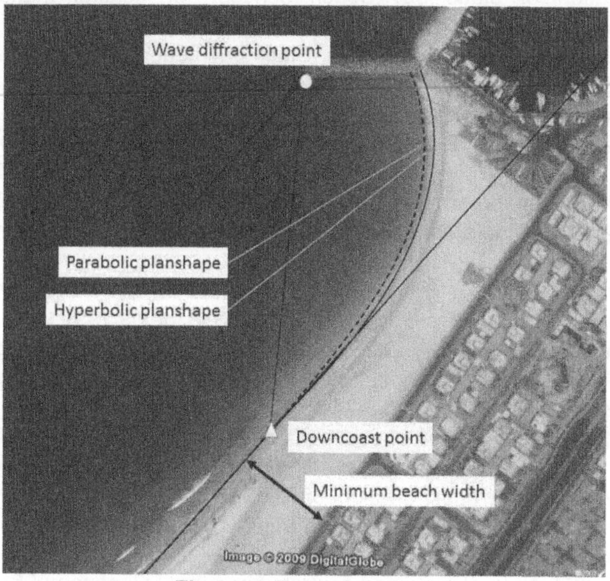


Figure 4: Example output

## Conclusions and Recommendations

Although it is recognised that the short research study had limited data available it has managed to link the hyperbolic equation to the wave diffraction point, which is a crucial step in improving this equation. It allows the user to simply define a beach width and the tool will then plot the beach planshape. As it has been stated that the hyperbolic equation was not developed only for static equilibrium beaches, the parabolic equation has to be applied afterwards. However, the location of the downcoast point (where the shoreline becomes straight) is easily identifiable from the results of the hyperbolic planshape. Further research is required, using more case studies, to verify the location of the wave diffraction point in the hyperbolic equation.

## References

González M and Medina R (2001), On the application of static equilibrium bay formulations to natural and man-made beaches. *Coastal Engineering*, 43, 209-225.

Halligan GH (1906), Sand movement on the New South Wales coast. *Proceedings Limnology society, New South Wales,* 31: 619-640.

Hsu JRC, Silvester R and Xia YM (1987), New characteristics of equilibrium shaped bays. *Proceedings of 8th Australian conf. on coastal and Ocean Eng.*, 3986-3999.

Hsu JRC and Evans C (1989), Parabolic bay shapes and applications. *Proceedings of Institution of Civil Engineers*. Part 2, vol 87. Thomas Telford, London, 557-570.

Hsu JRC, Benedet L, Klein AHF, Raabe ALA, Tsai CP and Hsu TW (2008), Appreciation of static bay beach concept for coastal management and protection. *Journal of Coastal Research*, 24, 198-215.

Hsu JRC, Yu MJ and Benedet L (2010), Static bay beach concept for scientists and engineers: a review. *Coastal Engineering*, 57, 76-91.

Krumbein WC (1944), Shore processes and beach characteristics. Technical Memorandum, vol 3. Beach Erosion Board, U.S. Army Corps of Engineers.

Lausman R, Klein AHF and Stive M (2010), Uncertainty in the application of the parabolic bay shape equation: part 1. *Coastal Engineering*, 57, 142-151.

Mashima Y (1960), Stable configuration of coastline. *Coastal Engineering in Japan,* 4:47-59.

Moreno L and Kraus NC (1999), Equilibrium shape of headland-bay beaches for engineering design. *Proceedings of Coastal Sediments 1999*.

Short AD and Masselink G (1999), Embayed and structurally controlled beaches, in: Short, A.D. (Editor), Handbook of Beach and Shoreface morphodynamics. New York: John Willey; & Sons, 230-249.

Silvester R (1960), Stabilization of sedimentary coastlines. *Nature,* 188:467-469.

Silvester R (1970), Development of crenulate shaped bays to equilibrium. *J. Waterways and Harbours division*, 96, 275-287.

Silvester R and Ho SK (1972), Use of crenulated shaped bays to stabilize coasts. *Proceedings 13th ICCE*, 2, 1347-1365.

Silvester R (1985), Natural headland control of beaches. *Continental Shelf Research*, 4, 581-596.

Silvester R and Hsu JRC (1993), Coastal stabilisation: Innovative concepts. Prentice-Hall, 578p.

Silvester R and Hsu JRC (1997,) Coastal Stabilization. Singapore: World Scientific Publishing, 578p.

Yasso WE (1965), Plan geometry of headland bay beaches. *Journal of geology*, 73, 702-714.

**Coastal Management: Changing coast, changing climate, changing minds**
ISBN 978-0-7277-6114-9

ICE Publishing: All rights reserved
doi: 10.1680/cm.61149.087

# Tottering on the Edge: a Composite Approach to Analysing the Causes and Mitigation of Coastal Slope Instabilities: a Case Study from Totland, Isle of Wight, UK

**Lucy Wiggins**, Mott MacDonald, Croydon, UK
**Elizabeth Ward**, Mott MacDonald, Croydon, UK
**Peter Marsden**, Isle of Wight Council

## Introduction

This paper uses the case study of Totland on the Isle of Wight, UK (Figure 1), to examine the suitability of the current FDGiA approach to funding for coastal instability schemes in the UK.

On the 26th December 2012 and for the following couple of weeks the seawall fronting the cliffs along Totland Bay was pushed 20m seaward due to a 120m wide landslide in the sea cliffs behind the existing defences. This resulted in the closure of the promenade and a resultant economic loss for businesses along the frontage. The future risk to cliff top property and assets that remain along the frontage was also in question.

Figure 1: (a) Location Map of Totland on the Isle of Wight (Streetmap 2013) (b) Local site plan of the frontage (Google Earth Pro (Licenced) Image, Dated January 2004)

Consequently a study (Mott MacDonald, 2013) was undertaken to establish the cause and propensity for failure along the frontage, the future risk of failure and the economic viability of mitigating against this risk and reinstating the coastal footpath. To allow the footpath to re-open, and to stabilise the cliffs against further failure and retreat, coastal defence works, including rebuilding the seawall and promenade around the toe of the landslide are required, in addition to cliff stabilisation works. This combination of works would help protect the coastal slope from wave attack at the toe as well as stabilising the slopes to reduce the residual risk of failure of both the slopes and ultimately the coastal defences.

This paper will examine the interdependent relationship between shoreline and clifftop recession rates and assess the shortfalls in available funding when the retreat of the cliff is affected by composite failure mechanisms as a result of coastal erosion and sub-aerial processes rather than solely shoreline retreat.

## Case Study Description

The Isle of Wight is approximately 40km by 20km off the south coast of England (Figure 1). The island is plagued by landslides as a result of a combination of conditions (geology, hydrology, hydrogeology, topography, sea level rise to name a few) meaning it is particularly susceptible on the steep coastal slopes. Hence, there are many historical landslide complexes as well as those that are currently active, such as the Totland Landslide in Totland Bay (Figure 2). The Totland landslide destroyed the sea wall and promenade in the landslides' immediate vicinity, opening up numerous gaps in the seawall, and allowing material behind and beneath the wall to be eroded, causing further ongoing collapses.

The geology of the cliffs in Totland Bay comprise the Headon and Osbourne Beds which form the lower part of the Headon Hill Formation, part of the Solent Group (Insole et al, 1988) and are known to be prone to failure (Hutchinson and Bromhead, 2001). The Headon Hill Formation consists of thinly bedded marls and clays with subordinate bands of limestone and sand. Marker beds visible in the cliff face identify that that the dip is between 1.2 and 1.5 degrees northwards with the effective dip at the coast being near horizontal (Card Geotechnics Limited, 2003). This gentle northwards dip promotes seepage erosion and landsliding (SCOPAC, 2004).

The frontage is exposed to both tidal currents and modified open sea waves, including swell. Waves can propagate directly from the west on to the frontage, while currents run parallel to the shoreline in an approximately north–south direction. As the bays are relatively closed systems, they receive sediment inputs only from erosion of local cliffs. Much of the material yielded is too fine to remain on beaches and is transported seaward (SCOPAC, 2004).

In front of the sea cliffs is a concrete promenade approximately 5m wide. There are a variety of seawall sections with rock and timber groynes and parts of the concrete seawall have additional rock armour along the toe to provide increased scour protection. In the 1960s, to account for landsliding and undermining, it is recorded that 4m piles were driven into a 'firm clay base' beneath the beach deposits and a reinforced concrete encasement of the existing sea wall face was completed (Report of County Surveyor, 1966). Cliff drainage has been present at selected points since 1925 to prevent or reduce future cliff top instability (Lewis and Duvivier, 1973; Posford Duvivier, 1993; HR Wallingford, 1999).

Figure 2: (a) Failure of the landslide (Mott MacDonald, 2013; HM Coastguard, 2013) (b) Geomorphological map of the landslide indicating the location of Cross-Sections A and B, highlighting the different failure mechanisms within the landslide (Mot MacDonald, 2013).

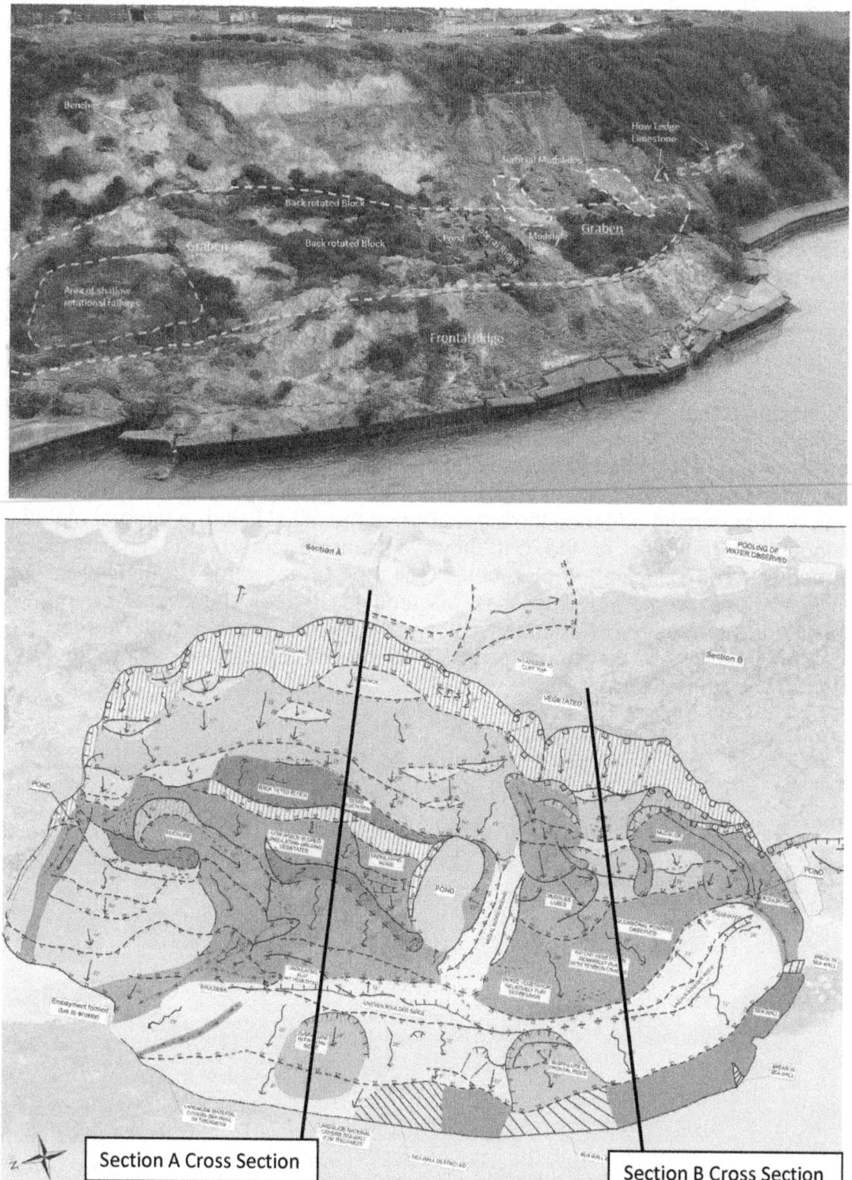

**Modes of Landslide Failure at Totland**

There are two main modes of failure at Totland: (a) through sub-aerial processes and (b) through coastal processes. Transects of the cliff were undertaken in both the landslide and the rest of the frontage to identify geomorphological features such as the backscarp, tension cracks, rotated blocks and mudslides. These were geomorphologically mapped, along with noting the geological features observed in the cliff face, to produce a GIS based geomorphological map (Figure 3) which provided the main basis for developing the failure mechanisms.

The Recent Large Scale Failure

In formulating any remedial measures for potential coastal instability and future landslides it is imperative to develop a model understanding of the failure mechanisms. The landslide has been separated into Sections A and B based on different failure models. Figures 3 and 4 show cross sections through Sections A and B (Figure 3) of the deep seated landslide which have been constructed based on the geomorphological mapping observations, historical ground investigation borehole data and topographical data created from LiDAR images and laser scans of the cliff by Channel Coastal Observatory (2013).

It is considered that Section A failed along a slip plane in the clay layers beneath the piles of the seawall. The resultant translational movement along the failure plane lead to the formation of a graben at the base of the cliff. The removal of support at the toe of the cliff led to the failure of the upper section through multiple sub-rotational blocks – in an almost domino effect up the slope. During the initial failure a ridge of material developed at the base of the cliff; however tension cracks and small failures observed within this indicate the graben is still moving seawards. Continual erosion at the cliff base by marine processes will tend to steepen the overall cliff profile increasing the probability of a further major slope failure. There is expected to be a critical threshold where continued recession of the cliff toe cannot pass without the whole cliff failing in response.

Figure 3: Failure Model for Section A of the Landslide

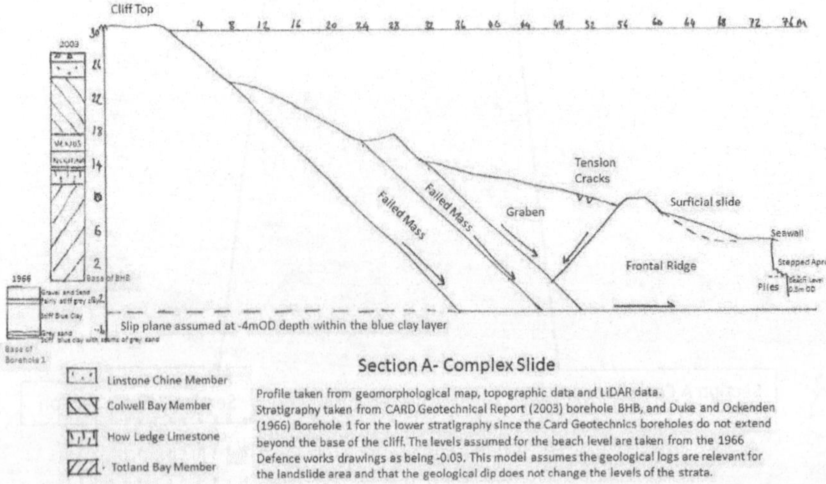

The geomorphological mapping highlighted that Section B has slightly different surface features. There were no multiple rotational blocks observed and the main failure is considered to be a complex failure with both rotational and translational elements. The failure plane is considered to be along the same clay layer as identified in Section A. However, it was observed that the geology of this section contained more clay in the cliffs than Section A and there were seepage lines observed in the backscarp. Owing to the increased clay content and the presence of seepage lines, shallow surficial mudslides developed.

The main cause of failure is considered to be the increased groundwater along key stratigraphic horizons related to the wet autumn in 2012. In addition to ground water, the pooling of surface water and spring lines has led to the shallower surficial mudslides following the main landslide event.

Figure 4: Failure Model for Section B of the Landslide

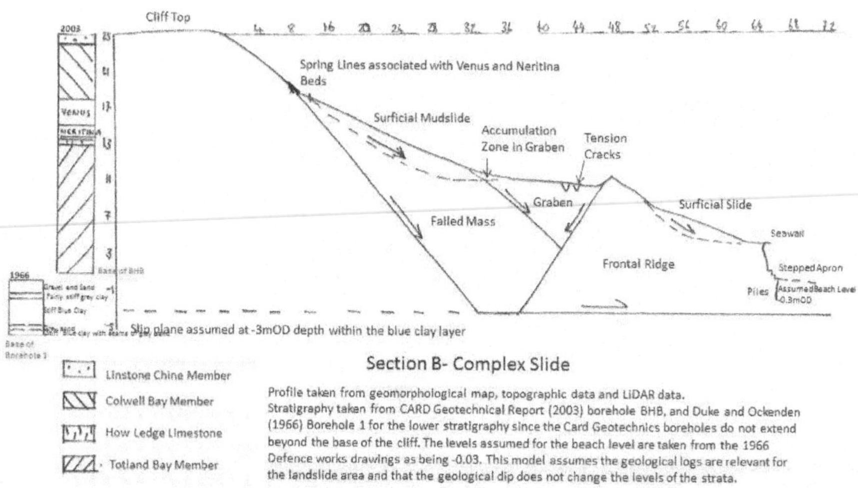


Stability of the Whole Frontage

Along the rest of the frontage there is no evidence of large deep seated landslides, only superficial mudslides were noted. The backscarps of these mudslides originated along seepage/spring lines and there was evidence of pooling water and the development of tension cracks. Mudslides are actively occurring along the frontage and therefore should be considered a high hazard. The mudslide hazard is exacerbated by the scouring at the toe of the cliff as a result of wave overtopping of the seawall. This will eventually lead to a large scale failure, similar in scale to the recent landslide, if no intervention is put in place. The mode of failure is outlined in Figure 5.

Figure 5: The Failure Model for the Rest of the Frontage

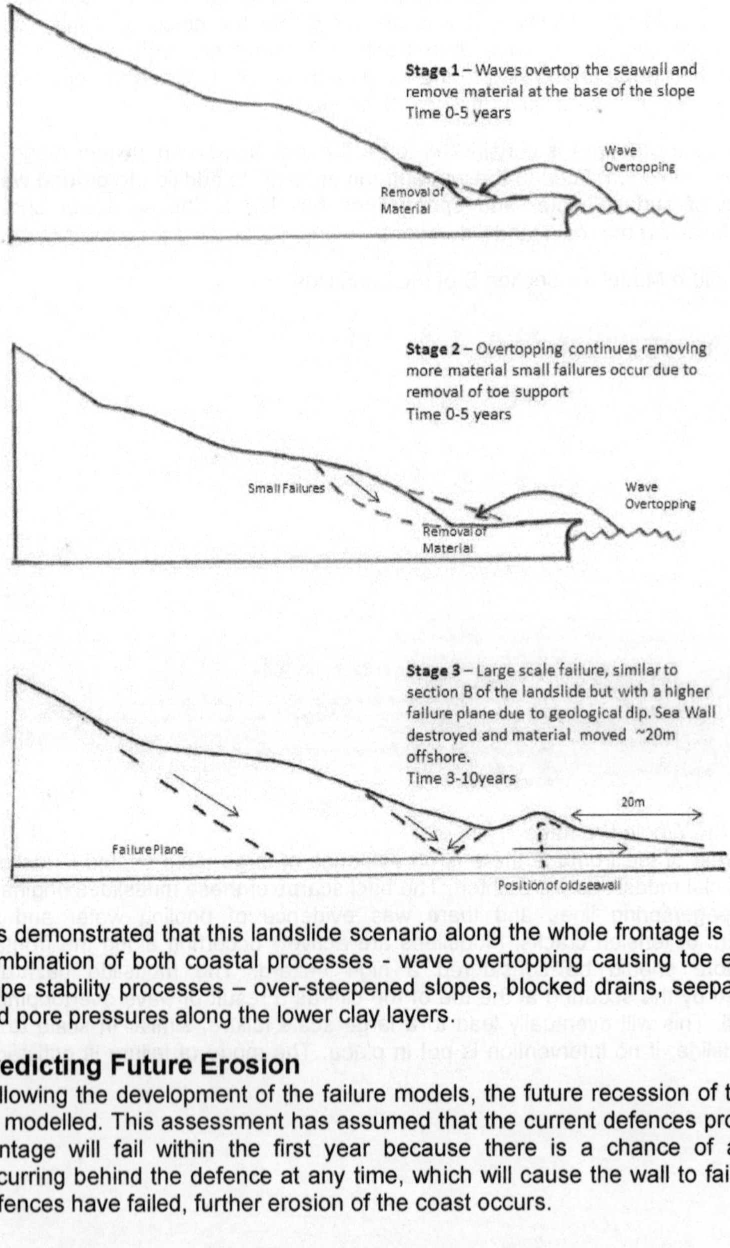


It is demonstrated that this landslide scenario along the whole frontage is driven by a combination of both coastal processes - wave overtopping causing toe erosion and slope stability processes – over-steepened slopes, blocked drains, seepage erosion and pore pressures along the lower clay layers.

## Predicting Future Erosion

Following the development of the failure models, the future recession of the cliff can be modelled. This assessment has assumed that the current defences protecting the frontage will fail within the first year because there is a chance of a landslide occurring behind the defence at any time, which will cause the wall to fail. Once the defences have failed, further erosion of the coast occurs.

Coastal erosion/retreat has been projected into the future using Historical Trend Analysis Rule (HTAR), and is based on retreat of the cliff under two recession scenarios:

- Shoreline erosion resulting from coastal erosion; and
- Cliff top erosion resulting from mass failures and landsliding.

**Historic Trend Analysis Rule (HTAR)**
Specific rates for shoreline erosion have been identified using the HTAR. The HTAR is a model relating the rate of shoreline retreat to the rate of sea level rise (NRC, 1997; Leatherman, 1988, 1989; Bray et al, 1992).

Equation 1: The HTAR equation

$$R_2 = \left(\frac{R_1}{S_1}\right) S_2$$

Where:
$R_1$ = Historic Retreat Rate $R_2$ = Future Retreat Rate $S_1$ = Historic Sea Level Rise $S_2$ = Future Sea Level Rise

For the whole study area the rate of historical sea level rise (S1) has been taken to be 2.26mm/year based on mean sea level data from 1997-2012 at recorded at Bournemouth (Permanent Service for Mean Sea Level, 2013).

Rates of future sea level rise (S2) are based on the UKCP09 medium scenario data outputs (95th percentile) as recommended in the Environment Agency guidance (EA, 2011) (Table 1).

Table 1: Projected sea level rise at Totland over the next 100 years (medium emission scenario)

| Year | Cumulative Sea Level Rise (Medium Emission Scenario 95%ile) (m) |
|---|---|
| 2015 | 0.119 |
| 2040 | 0.267 |
| 2065 | 0.441 |
| 2090 | 0.642 |

Source: UKCP09 User Interface, accessed November 2013.

The historical retreat rate (R1) of the two different recession scenarios is used to calculate the two different future retreat rates (R2).

- **Shoreline erosion resulting from coastal erosion** - A value 0.5m/year has been taken as the historic retreat rate under the coastal erosion scenario. This is an average of all the suggested rates of retreat for the Totland area, using figures obtained in the SCOPAC (2004) report.
- **Cliff top erosion resulting from mass failures and landsliding** – using the cliff failure models developed previously it was postulated that there is likely to be 20m of retreat every 10 years due to significant mass failures as a result of destabilisation of the cliff.

Key assumptions in the erosion assessment:

- It is assumed that the first 5 years of erosion after defence failure are the greatest as the cliff tries to regain equilibrium.

- It is assumed that within the whole section the defences will fail within the next year because there is a possibility of the landslide occurring at any time.

**Projected Retreat Rates**
The projected cumulative retreat rates outputted from the HTAR equation have been mapped in GIS to create a map of potential erosion extent over the next 100 years (Figure 6). It can be seen that using the shoreline recession scenario the amount of land potential lost is significantly less when compared to the cliff recession scenario.

Figure 6: The Retreat Rates (Mott MacDonald, 2013)

## Impacts on Future Mitigation of Failure and Management of the Coast

Based on the projected rate of retreat, the year of impact on assets including property, buildings, roads, utilities and environmental features is modelled.

The erosion benefits or assets at risk from erosion over 100 years has been calculated for each of the retreat scenarios and is shown in Table 2 below. Property values are calculated using the average price of property in the post code area (2015 valuations).

Table 2: The number of properties lost under projected rates of erosion for the next 100 years

| | Coastal Retreat | | | Cliff Retreat | | |
|---|---|---|---|---|---|---|
| Year | 0-20 | 20-50 | 50-100 | 0-20 | 20-50 | 50-100 |
| Number of properties | 18 | 41 | 112 | 97 | 144 | 314 |
| Total | 171 | | | 555 | | |

The models indicate that an additional 384 properties are at risk under the cliff retreat scenario than the coastal retreat scenario. Therefore, the value of the benefits, and ultimately the availability of funding is significantly larger in the cliff retreat than the coastal retreat scenario.

In the UK, funding for coastal erosion protection is through the Flood Defence Grant in Aid (FDGiA) process using Environment Agency Guidance (Flood and Coastal Risk Management Appraisal guidance (FCRM-AG), 2010). The economic assessment considers the value (cost) of the options and whether investment in any option is worthwhile against the benefits. In order to compare the different options it is useful to consider the benefit cost ratios for each option. Using these benefit cost ratios Partnership Funding scores can be developed which will outline the potential percentage of the scheme which can be funded and the further contributions to the scheme that are required. Table 3 highlights the economics for a management scheme based on a rock revetment and cliff stabilisation along the whole of the frontage.

Table 3: The Results of the Economic Assessment for the Different Retreat Scenarios Considered

| Scenario | Discounted Value of Benefits over 100 years | Benefit Cost Ratio | Partnership Funding (PF) Score | Further Contributions Required |
|---|---|---|---|---|
| Coastal Recession | £7.5 million | 1.8 | 28% | £3 million |
| Cliff Recession | £111.8 million | 26.4 | 226% | - |

From these results it can be seen that the different retreat scenarios lead to significant differences in the amount of FDGiA funding available and further contributions required.

Under Environment Agency guidance it is difficult to obtain funding based on the cliff retreat scenario as it is not directly related to managing the risks from coastal erosion or flooding. Consequently there will be significant disparities between funding availability depending on whether if cliff retreat is or is not accounted for in economic assessments; resulting in significant differences for future management of the frontage. If all the cliff recession was to be caused by purely coastal recession at the toe of the cliff, the Totland frontage would be eligible for potentially 8x more funding for a new seawall and promenade installed via capital maintenance funding.

## Conclusions

Using Totland as an example of a coastal cliff system it is clear that the UK's guidance on assessing the funding potential for coastal management schemes is not always suited to coastal areas where the retreat of the cliff is governed by both coastal retreat and cliff retreat associated with cliff instability and landsliding.

Environment Agency funding guidance does acknowledge that coastal retreat is often due to a combination of different processes in addition to direct wave action, and the fact that if cliff stabilisation works are not undertaken there is often a risk of the coastal defence structure failing as a result of landsliding behind the defence itself. Although the guidance understands the risk of mass failures and landslides, it does not allow the assets that are at risk of failure through landsliding to be valued in economic assessments. Therefore, the value of benefits is significantly lower than it could be and the viability of implementing a larger, more effective scheme is significantly reduced.

Consequently, it is proposed that a new method of calculating the value of assets at risk is determined which would incorporate both coastal and cliff recession. This will become an increasingly important discussion point in the future due to climate change and the increasing the value of assets at risk. Not only will predicted sea level rise lead to increased erosion at the toe of the cliff but predicted increased rainfall will lead to increased magnitude and frequency of landslide events, putting more assets at risk.

## References


Bray *et al.,* (1992) Sea level rise and global warming: scenarios, physical impacts and policies.

Card Geotechnics Limited, 2003. Fort Warden, Totland, Isle of Wight- Report on Cliff stability. Geotechnical Investigation.

Environment Agency, 2010. Flood and Coastal Erosion Risk Management appraisal guidance

Environment Agency (2010) Assessment of Coastal Erosion and Landsliding for the Funding of Coastal Risk Management Projects – Guidance Notes.

HR Wallingford, 1999. Coastal Impact Study for Production Licence Application, South West of the Isle of Wight. Report EX3935, Report to ARC Marine Ltd.

Hutchinson, J.N., Bromhead, E.N., 2002. Keynote Paper: Isle of Wight Landslides. In: McInnes, R.G.,

Insole, A., Daley, B., Gale, A., 1998. Geology of the Isle of Wight. Geologists' Association Guide

Mott Macdonald, 2013. Totland to Colwell Bay Landslide Assessment

Lewis and Duvivier, 1973. West Wight - Coastline from Totland Bay to Norton, West Wight

PSMSL (2012) Service for Mean Sea Level [online] *Available from* www.psmsl.org

Posford Duvivier, 1993. Coast Protection: Totland Seawall. Report to South Wight Borough Council, 4pp.

SCOPAC, 2004. http://stream.port.ac.uk/environment/scopac5/nwiow/nwiow.htm URL accessed 07/10/13

UKCIP (2014) UKCP09: UK Climate Projections www.ukcip.org.uk/ukcp09/

Zoopla (2014) Current House Prices *Available from* http://www.zoopla.co.uk/

Coastal Management: Changing coast, changing climate, changing minds
ISBN 978-0-7277-6114-9

ICE Publishing: All rights reserved
doi: 10.1680/cm.61149.097

# A Proposal for Qualitative and Quantitative Analysis Methods for Vertical Evacuation from a Tsunami in Coastal Cities

**Mauricio Reyes,** University of Valparaíso, Chile
**Fusanori Miura,** Yamaguchi University, Japan

## Introduction

A method for qualitative and quantitative assessment of horizontal and vertical evacuation aspects in coastal cities threatened by tsunami is proposed. It is based on the collection of data from Talcahuano city, that was affected by the tsunami in February 27th, 2010, data from Valparaíso and Viña del Mar cities, both with large population and also prone to be affected by tsunami, and Iquique city, in similar situation.

The first stage of the study is the evaluation of horizontal evacuation routes at Valparaíso and Viña del Mar cities using several criteria, both qualitative and quantitative, in order to determine a zone where vertical evacuation can reduce the vulnerabilities associated to Human Suffering in case of tsunami. For the definition and calculus of vulnerability indicators, the methods of Project Cycle Management (PCM) and Analytical Hierarchy Process (AHP) will be applied. Then, the zones where vertical evacuation is needed will be analyzed in terms of geographical information, in order to utilize Google Street View © for a fast evaluation of buildings located at the boundary area where vertical evacuation is needed, and with the aim of finding its availability to be used as a vertical evacuation building. As a basis for this method, the damages from the earthquake and tsunami at Talcahuano city will be considered as a reference, occurred in February 27th 2010.

The outcome of this research includes maps with vulnerability zonation and radar charts indicating the characteristics of horizontal evacuation routes, together with a set of zones and buildings that the local authorities can study to be considered for vertical evacuation planning in each city.

## Contextualization

Chile has a particular position and orientation that makes it prone to be affected by the tectonic plate's behavior. The country stretches from north to south between the volcanoes line of the Andes Mountains and the Pacific Ocean, and thus parallel to the subduction trench in almost its whole extension. So, Chile is located just over the zone where Nazca and South American plates converge and generates earthquakes, some of them tsunamigenic. Since 1562 and based on historical data, more than 31 near-field tsunamis have ravaged the coasts of Chile, producing even the devastation of entire cities like Arica in 1604 and Concepción in 1751 (Winckler, Reyes, & Sepúlveda, 2010). Taking into account the paleogeological data can be seen that some giant tsunamis shows a larger recurrence, thus introducing a shadow of uncertainty over the statistical approach for earthquakes occurrence (Goldfinger, Ikeda, Yeats, & Ren, 2013). Generally speaking the majority of earthquakes in Chile meet the conditions of being tsunamigenic, so tsunami generation scenarios must be always considered in the coastal regions of Chile. The Figure 1 shows the tectonics configuration of the southeast Pacific Ocean.

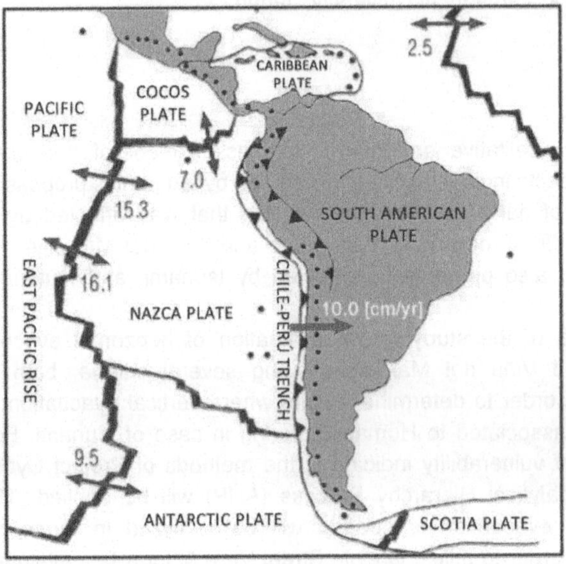


Figure 1. Tectonic situation of Chile. Modified from (Winckler, Reyes, & Sepúlveda, 2010).

## Methodology

The risk assessment methodology proposed by the authors is applied for a specific aspect of Tsunami Disaster Risk, namely Human Suffering. The core idea of the method is to generate a measurable basis for risk control, using qualitative and quantitative variables.

### Risk Conceptualization

Risk is a problematic concept, not only because it is abstract: it represents something that seems unreal, because it anticipates events that only exist in the future, and also it has elements of randomness (Cardona, 2003). If there are certainties then there is no risk, so it is intimately linked with individual or social psychology, even when there are attempts to create a sense of objectivity, as Cardona (2003) proposed.

Risk theory has been applied in several disciplines, with different hazards; with such a diversity that has led to many risk's definitions and/or methods for its calculation. Among a diverse range of risk research application, there is disaster management. However, there is no agreement on terminology, and several definitions of risk can be found, depending on the type of disaster, technical approach or scientific theory, and all these factors often generate confusion in risk management processes, as Pliefke et al. (2007) stated.

As this study is a continuation branch of the previous work of the authors (Reyes, 2013), the mathematical definition they used for risk, has also been used for this study:

$$R = P \times I \tag{1}$$

Where:

R: Risk

P: Probability of occurrence of Hazard scenario

I: Impact or level of consequences, depending on vulnerability and exposure.

Beside that practical mathematical definition must be a wider conceptualization, which includes the basis for qualitative analysis. In this research the core concept of Risk taken into account is *'The combination of the probability of an event and its negative consequences.'* (UNISDR, 2009). That broad concept allows introducing qualitative analysis for consequences, something very useful when information is scarce or hard to be measured.

**Hazard Assessment and Hazard Scenario Hypothesis (HSH)**

The cases of study are the coastal cities of Valparaíso and Viña del Mar, neighbor urbanizations very interconnected. Both have been subject of recurrent earthquakes with magnitude higher than 7.8Mw, as the years 1647, 1730, 1822, 1906 and 1985 (D. Comte, 1986). However, there are uncertainties in the historical information, that allow to think about the occurrence of a big earthquake, similar to the estimations for 1730 earthquake, with estimated magnitude 8.5Mw and maybe more according to recent studies (Carvajal, 2015), and the only one from the series that triggered a major tsunami. For disaster scenarios definition, the 1730's earthquake will be considered, supposing that can occur within the next 100[yr], with the probability of occurrence factor equal to 4, as shown in Table 1. Similar hypothesis has been assumed by the Hydrographic and Oceanographic Service of the Chilean Army (SHOA), to generate the official inundation maps for Chilean coastal cities.

For disaster scenario definition, the hypothetical earthquake can be described as follows:

1. Occurrence of at least one earthquake with magnitude higher than Mw=8.5 and rupture zone extending under the eastern border of the South-American plate, probably reaching beyond the shore line, where Valparaíso and Viña del Mar are located. So can be expected a strong shaking in coastal towns, with

long duration (over 4 minutes), and large horizontal and/or vertical displacement of the shoreline can be expected.

2. Hypothetically, this earthquake would release the slip accumulated between coupling plates, thus generating a large displacement of the sea bottom in the Chile-Peru trench, which would in turn trigger a major tsunami. The arrival time of destructive waves would depend on the characteristics of the rupture, but in the worst case they may arrive in less than ten minutes.

Table 1. Scale of probability of occurrence for hazard evaluation. Modified from (Reyes, 2013).

| **EXPECTED TIME FOR OCCURRENCE** (One event will strike once within the next...) | **PROBABILITY OF OCCURRENCE FACTOR (P)** |
|---|---|
| 10[yr] | 5 |
| **10 to 100[yr]** | **4** |
| 100 to 500[yr] | 3 |
| 500 to 1000[yr] | 2 |
| 1000[yr] or more | 1 |

**Vulnerabilities Assessment and Vulnerabilities Selection**

A revision of the basic Project Cycle Management (PCM) procedure that was applied by the author in previous study (Reyes, 2013) has been performed, in order to discuss the systemic understanding of Tsunami Disaster Risk (TDR). The Figure 2 shows the systematization of impact (vulnerabilities) for a coastal city.

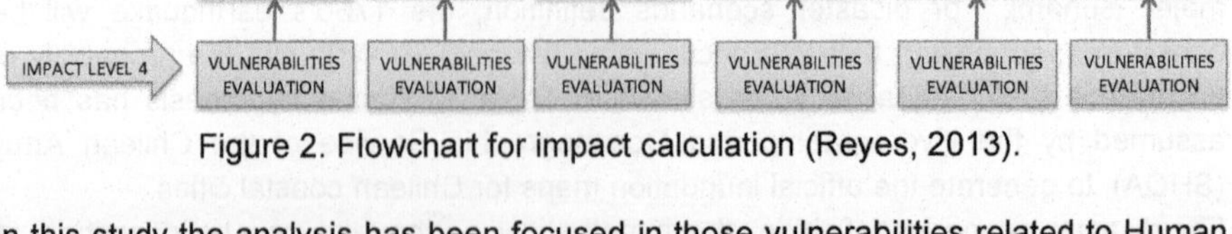


Figure 2. Flowchart for impact calculation (Reyes, 2013).

In this study the analysis has been focused in those vulnerabilities related to Human Suffering (HS). For impact level 4, the following vulnerabilities are considered components of that Human Suffering (Reyes, 2013):

- Human casualties due to building collapse (HS01).
- People that needs assistance during disaster (HS02).
- People trapped under debris (HS03).
- People swept away by tsunami (HS04).

With similar criteria, the vulnerability HS04 can be systematized in multiple components, as follows:

- Characterization of the occupancy or exposed population (OEP) to tsunami hazard (HS04-01).
- Consideration of time of the day and/or season in hazard scenario for OEP characterization (HS04-02).
- Reliability of formal or informal Tsunami Early Warning systems (HS04-03)
- Reliability of evacuation infrastructure, planning and drills (HS04-04)
- Reliability of tsunami-swept-away people estimation and characterization (HS04-05)
- Characterization and enhancement of previous tsunami experience or tsunami education and/or training (HS04-06).
- Professional tsunami rescue teams, equipment, facilities, training, availability, access, and/or coverage (HS04-07).
- Communitarian tsunami rescue teams, equipment, facilities, training, availability, access, and/or coverage (HS04-08).

The core objective is the Tsunami Disaster Risk reduction, which can be achieved by the reduction of damage in a cause-effect chain that configures a hard-to-manage expansion of the vulnerabilities tree from Figure 2. For purposes of this research, the vulnerability HS04-04 is the target for deeper analysis, concerning to the susceptibility and reliability of horizontal evacuation routes, as follows:

- Susceptibility of the evacuation route to be blocked by…
    - … debris.
    - … traffic jams.
    - … fires and/or explosions.
    - … channels or rivers overflow.
    - … crowds due to narrow places.
    - …crowds due to rush hours or massive events.
    - …crowds during summer season.
    - …electric lines collapse.
- Reliability of the evacuation route in terms of…
    - … presence of elements that obstruct people's displacement.
    - … displacement time from inundated area to safe zone.
    - … useful width for displacement.
    - … slope (steep or mild).
    - … tsunami evacuation's signage available along the route.
    - ….handicapped people's evacuation.
    - ….evacuation during night time.

Such systematization was achieved through a brainstorming session between four thesis students and the professor, as an amplification of the PCM already performed in (Reyes, 2013), as mentioned above.

**Impact Estimation**

Both susceptibility and reliability were associated to impact by means of qualitative scales, as shown in:

Table 2. Qualitative scale for impact (susceptibility). Modified from (Reyes, 2013).

| QUALITATIVE SCALE FOR IMPACT | |
|---|---|
| No susceptibility | 1 |
| Low susceptibility | 2 |
| Medium susceptibility | 3 |
| High susceptibility | 4 |
| Very high susceptibility | 5 |

Table 3. Qualitative scale for impact (reliability). Modified from (Reyes, 2013).

| QUALITATIVE SCALE FOR IMPACT | |
|---|---|
| No reliability | 5 |
| Low reliability | 4 |
| Medium reliability | 3 |
| High reliability | 2 |
| Very high reliability | 1 |

The evaluation of impact must be performed by several persons under the HSH assumption, thus introducing diversity of criterion. In this case, each route has been evaluated with the mean value from the respective team members valuation.

**Risk Calculation**

The matrix proposed in the referenced study (Reyes, 2013) has been utilized for risk calculation, as it is shown in Table 4.

With that, risk can be calculated for each vulnerability component, but to obtain an unique value of risk at some level of vulnerabilities, it is required to define a weighted summation. One powerful tool is the Analytical Hierarchy Process (AHP) proposed by Saaty (1980), which has been applied in this research project: a survey to introduce several specialist's opinion has been conducted.

Table 4.Risk matrix (Reyes, 2013).

| RISK MATRIX | | | IMPACT LEVELS | | | | |
|---|---|---|---|---|---|---|---|
| | PROBABILITY OF OCCURRENCE (FACTOR) | | NEGLIGIBLE | MINOR | MODERATE | SERIOUS | SEVERE |
| | | | 1 | 2 | 3 | 4 | 5 |
| EXPECTED TIME FOR OCCURRENCE (T) | 1000[yr] ≤ T | 1 | LOW - ACCEPTABLE | LOW - ACCEPTABLE | LOW - TOLERABLE | LOW - TOLERABLE | MEDIUM - TOLERABLE (UNDER CONDITIONS) |
| | 500[yr] ≤ T < 1000[yr] | 2 | LOW - ACCEPTABLE | LOW - TOLERABLE | MEDIUM - TOLERABLE (UNDER CONDITIONS) | MEDIUM - TOLERABLE (UNDER CONDITIONS) | HIGH - INTOLERABLE (UNSTABLE) |
| | 100[yr] ≤ T < 500[yr] | 3 | LOW - TOLERABLE | MEDIUM - TOLERABLE (UNDER CONDITIONS) | MEDIUM - TOLERABLE (UNDER CONDITIONS) | HIGH - INTOLERABLE (UNSTABLE) | HIGH - INTOLERABLE (CRITICAL) |
| | 10[yr] ≤ T < 100[yr] | 4 | LOW - TOLERABLE | MEDIUM - TOLERABLE (UNDER CONDITIONS) | HIGH - INTOLERABLE (UNSTABLE) | HIGH - INTOLERABLE (CRITICAL) | HIGH - INTOLERABLE (CRITICAL) |
| | T < 10[yr] | 5 | MEDIUM - TOLERABLE (UNDER CONDITIONS) | HIGH - INTOLERABLE (UNSTABLE) | HIGH - INTOLERABLE (CRITICAL) | HIGH - INTOLERABLE (CRITICAL) | HIGH - INTOLERABLE (CRITICAL) |

## Cases of Study

The idea of the applied method is a virtual determining of riskier zones for horizontal evacuation. For that purpose, the first stage of the study was performed in the coastal and neighbor cities of Valparaíso and Viña del Mar, located at the middle part of Chile (33°S), 115[km] away from Santiago, the Capital. With a stable population close to 279000 and 289000 inhabitants respectively (INE, 2012), and a floating population during summer season that can reach more than one million people, these cities are the most important urban zones in the coast of central Chile. However, they are very different in configuration: while Valparaíso is an old port city, with narrow streets and old buildings, Viña del Mar is a modern city, with wide ways and tall buildings. Unfortunately, both cities are too young to have internalized memories about tsunamis: in 1730 Valparaíso was just a little town, barely affected by the tsunami, and Viña del Mar was build massively after the 1906 earthquake which destroyed Valparaíso, without destructive tsunami.

## Evacuation Routes *In Situ* Assessment

On August 29 2014 a group of 18 undergraduate students and the professor (aged between 18 and 39 years old) were divided in teams for each evacuation route analysis, assuming role-playing as during a real disaster scenario. One person by group was designated as handicapped person (blind), with the aim of measurement the associated displacement time. The rules were:

1. Don't use mobile or digital maps.
2. Try to choose your way according to the information on the route.
3. Don't run, walk fast simulating a real evacuation.

The starting points were situated as close as possible to the shoreline, and as far as possible from the official evacuation points through some official evacuation route. That does not mean the shortest way. In fact, the situation of Viña del Mar city is very particular, since it has an open channel for a river, where tsunami waves can be

accelerated and arise faster and sooner to the city. So the shorter ways to get a high point are forbidden and the evacuation routes design considers it (ONEMI, 2012). The image shown in Figure 3 allows observing a general view of the analyzed city.
The case of Valparaíso is very different, due to its urban configuration. In spite of the shorter length of the evacuation routes and larger number of evacuation points compared to its neighbor city, there are other aspects that configure a different composition of vulnerabilities. The Figure 4 indicates the study zone at Valparaíso.

## Comments

The results obtained are showing a clear characterization of the evacuation routes in each city. Some vulnerabilities are very different if both cities are compared, so the risks associated to tsunami disaster scenario can be determinant for the way of management of the emergency and the recovery process. Using the information of time and pedestrian velocities, among other detected factors, a set of criteria for definition of riskier areas for horizontal evacuation will be proposed, and then applied in the cities of Iquique and Talcahuano, validating in this way the remote use of this method for planning purposes. The next stage of this work is the virtual work with Google Street View ©, intended for definition of zones where vertical evacuation is needed, buildings selection and fast analysis of them, looking for those able to be used for vertical evacuation, using Multicriteria analysis. However, as it is known that there are not buildings designed for vertical evacuation, the demonstration of the existence of an unsatisfied demand for vertical evacuation should trigger actions from authorities to find solutions based on more detailed technical assessment, and not only visual characterization.

Figure 3. Inundation zone and evacuation routes at Viña del Mar. Several hotpots are indicated. Modified from (Google Earth, 2015a).

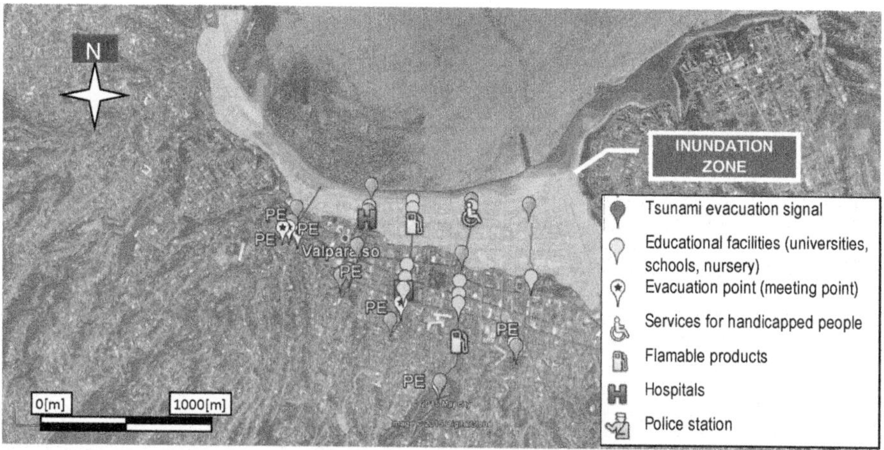


Figure 4. Study zone at Valparaíso. Modified from(Google Earth, 2015b).

## Results

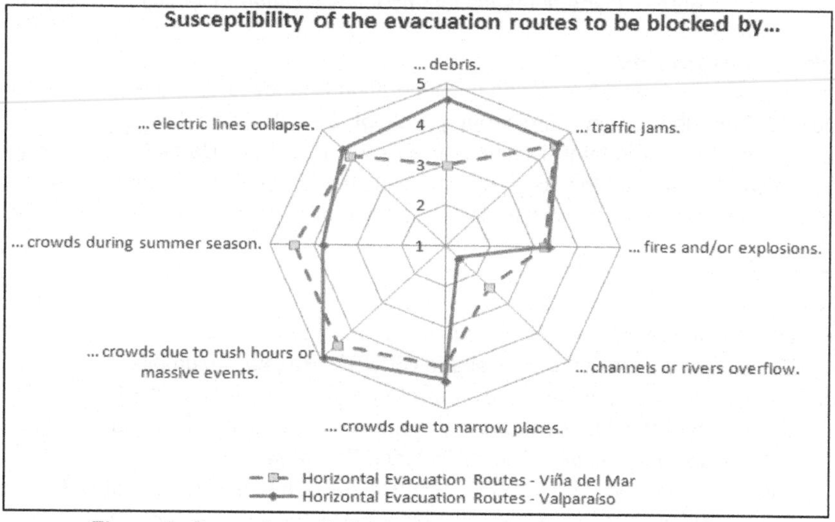


Figure 5. General results for impact associated to susceptibility.

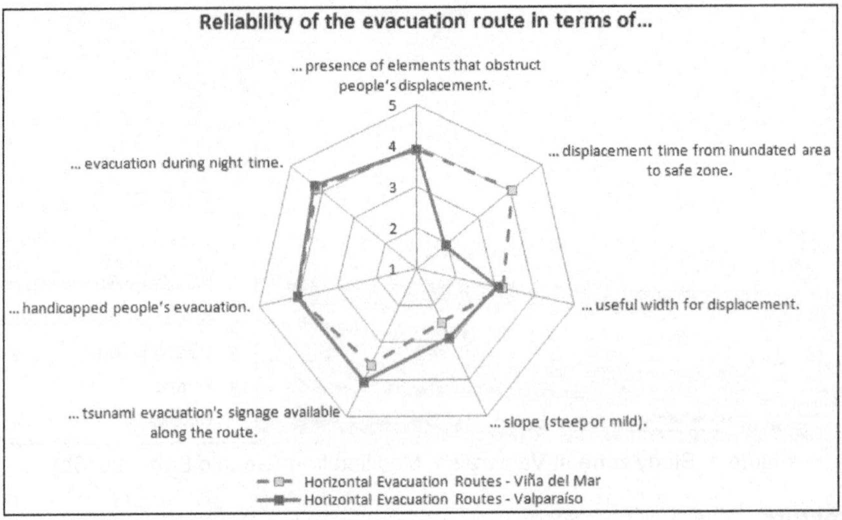


Figure 6. General results for impact associated to reliability.

## Acknowledgments

Thanks to the Japanese International Cooperation Agency (JICA), the undergraduate students from the Ocean Engineering School of Valparaíso University, and my friends from the Tsunami Diploma from the Pontifical Catholic University of Valparaíso. And special thanks to my thesis students: Jazmín, Constanza, Alex and Eduardo, who worked very hard for this, but never enough until they really finish it.

## References

Cardona, O. (2003). *The need of re-thinking in a holistic way the concepts of vulnerability and risk (in Spanish).* Universidad de los Andes, Centro de Estudios sobre Desastres y Riesgos CEDERI, Bogotá, Colombia.

Carvajal, M. (2015). *Tamaño del Terremoto de Chile Central de 1730, inferido de evidencias históricas, geológicas y modelación numérica del tsunami resultante.* Valparaíso: Universidad Técnica Federico Santa María.

D. Comte, A. E. (1986). The 1985 Central Chile Earthquake: A Repeat of Previous Great Earthquakes in the Region. *Science, Vol 233 N°4762*, 449-453.

Goldfinger, C., Ikeda, Y., Yeats, R., & Ren, J. (2013). Superquakes and Supercycles. *Seismological Research Letters*, Vol. 84 Nr.1.

Google Earth. (2015a). Viña del Mar, 33°01'13"S, 71°32'30"W eye elvation 100m.

Google Earth. (2015b). Valparaíso, 30°2'40" S, 74°37'15"W', eye elevation 1000m.

INE. (2012). *Instituto Nacional de Estadíasticas.* Recuperado el 20 de January de 2015, de www.ine.cl

ONEMI. (April de 2012). *Repositorio Digital Oficina Nacional de Emergencias del Ministerio del Interior, Chile.* Recuperado el 5 de January de 2015, de http://repositoriodigitalonemi.cl/web/handle/123456789/1662

Pliefke, T., Sperbeck, S., Urban, M., Peil, U., & Budelmann, H. (2007). A

Standardized Methodology for Managing Disaster Risk - An attempt to Remove Ambiguity. (págs. 283 - 294). Ghent, Belgium: Taerwe and Proske.

Reyes, M. (2013). *A Proposal of Tsunami Risk Assessment Method for Iquique City, Chile.* Tokyo: National Graduate Institute for Policy Studies (GRIPS).

Saaty, T. (1980). *The Analytic Hierarchy Process: Planning, Priority Setting, Resource Allocation.* McGraw Hill.

UNISDR. (2009). *UNISDR.* Recuperado el 15 de January de 2015, de http://www.unisdr.org/we/inform/terminology#letter-r

Winckler, P., Reyes, M., & Sepúlveda, I. (2010). *El Tsunami del 27-02-2010 en isla Robinson Crusoe, Archipiélago de Juan Fernández.* Valparaíso: Universidad de Valparaíso.

**Coastal Management: Changing coast, changing climate, changing minds**
ISBN 978-0-7277-6114-9

ICE Publishing: All rights reserved
doi: 10.1680/cm.61149.109

# Mapping Salt Marsh Margins – a Methodological Comparison

**Beth Riley,** Mott MacDonald Ltd, London, UK
**Iris Möller,** Cambridge Coastal Research Unit, University of Cambridge, Cambridge, UK

## Introduction

Mapping historical coastal change is vital for understanding the processes driving such change and thus for projecting shoreline changes into the future, therefore helping shape policy decisions along the coastline. Studies in dynamic environments such as are found at the coast are often hindered by methodological problems and challenges associated with data availability, quality and thus difficulty of interpretation, which then affects conclusions drawn from the results. This can include (although is not limited to): problems in defining the shoreline, inaccurate detection of the shoreline, thresholds and non-linear changes associated with processes, and low quality (or lack of) historical data.

This paper focuses on the process of shoreline detection and mapping, which forms the basis and often drives the accuracy of historical change studies. It aims to highlight the problems arising from the subjectivity involved in shoreline detection and calls for the development of a more robust method for detecting and mapping the shoreline. Furthermore, it then aims to compare the accuracy of different methods of shoreline detection and mapping, assessing the accuracy and success of a suggested methodology by means of a historical change study which was undertaken for the Stiffkey Marshes, North Norfolk (Riley, 2012).

### Monitoring Open-coast Marsh Edge Dynamics

Environmental studies often aim to identify, detect and map boundaries from remote sensing imagery to study landform or land cover changes over time. Two dominant methods are used within literature: manual tracing (visual interpretation) and automated mapping (e.g. Boak and Turner, 2005). Manual mapping to create a boundary involves the use of a vector shapefile to trace around a visible boundary edge from an aerial or satellite image. The main disadvantage of this method is the subjectivity and the reliance on the knowledge and skills of the researcher (Morgan et al., 2010). An automated approach uses classifications to identify differences, which can be computed automatically using computer algorithms (unsupervised classification) or through the identification of different categories by the user (supervised classifications). For an automated approach, the accuracy of the classification is paramount, and therefore pre- and post-processing techniques are often employed to improve the mapping of the boundary.

In dynamic coastal environments, such as salt marshes, historical change studies often opt for the manual approach (e.g. Cooper et al., 2001; Burd, 1992; Freiss et al.,

2012). The problem of subjectivity thus still perseveres in studies of landform change in coastal environments; it is particularly acute when classification methods are applied to vegetated areas such as salt marshes (e.g. algae cover, fragmented edges, submerged vegetation which can look similar in aerial images). Error assessments are often difficult to include in historical mapping studies due to a lack of ground truth data in addition to the reliance on methodologies which rely heavily on the skills of the researcher. A more objective approach, with associated accuracy assessments, is therefore required to improve historical shoreline mapping in dynamic coastal environments.

## Case Study – Stiffkey Marshes, North Norfolk

Salt marsh ecosystems are very sensitive to environmental changes; both to natural changes, such as alterations to sediment supply and natural fluctuations in climate, as well as human induced changes such as anthropogenic climate change or local interventions such as reclamation and, in recent decades, the managed realignment of sea defences (Allen and Pye, 1992; Allen 2000). However they also form an important part of coastal management policy and provide a number of services including (but not limited to): flood protection, areas of increased biodiversity, important sinks and sources of nutrients and key bird habitats. The juxtaposition of the recognition of their value to society against the rising pressures on these ecosystems, calls for efficient and sustainable management of these ecosystems with reliable monitoring and prediction of future salt marsh change. This requires a comprehensive understanding of the drivers of salt marsh change, alongside accurate methodologies to monitor change.

This paper draws on research undertaken to assess the historical changes in the salt marsh at Stiffkey, Norfolk, UK (Figure 1) which employed the method we have developed and are presenting in this paper. The uncertainty surrounding the current trajectory of marsh evolution at Stiffkey and the potential requirement for management over the medium to long term (20-100 years) (EA, 2010) makes this location a suitable case study site for this comparative study of marsh edge mapping methods. Furthermore, aerial images are available for the site from as early as 1955 and on an annual to bi-annual frequency from 1991 onwards, thus providing a good catalogue of data.

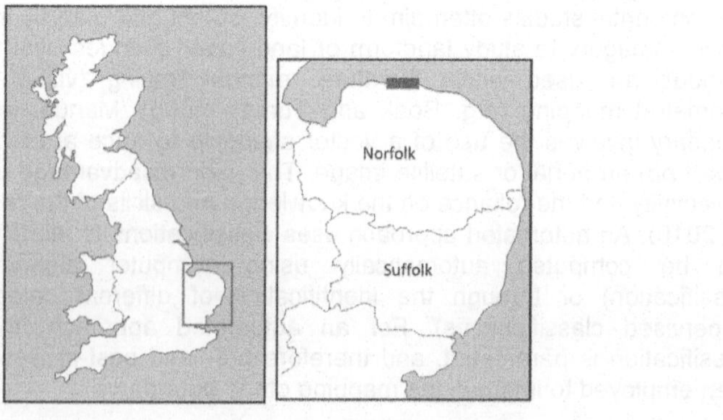


Figure 1: Location of the study area at Stiffkey Marshes, Norfolk, UK.

The term 'marsh edge' as used in this paper refers to the seaward edge of an open coast marsh. Furthermore, 'marsh edge change' refers to the two-dimensional change in marsh edge position and not to specific morphological changes such as the steepening or fragmentation of the marsh edge.

## Method – Digitising the Marsh Edge

A series of aerial photographs were used in this study with imagery post-2002 supplied in colour format (hereafter 'RGB') and all other photographs in black and white format (hereafter 'BW'). Aerial images which were sourced in hard copy were scanned and georeferenced in ArcMap by selecting a series of fixed features on the images as ground control points. Global Navigation Satellite System (GNSS) points of the present day marsh edge were collected during a field visit and were recorded with a Leica smartnet GNSS station to provide locational information at an average accuracy of 0.014m 3DCQ for the accuracy assessment of image-derived location information.

Several steps were used to digitise the marsh edge (summarised in Figure 2), as described in the following sections. The method of digitising the marsh edge presented here is hereafter referred to as the 'classification method' as it relies on a classification of the marsh into a number of different categories identifiable in the image.

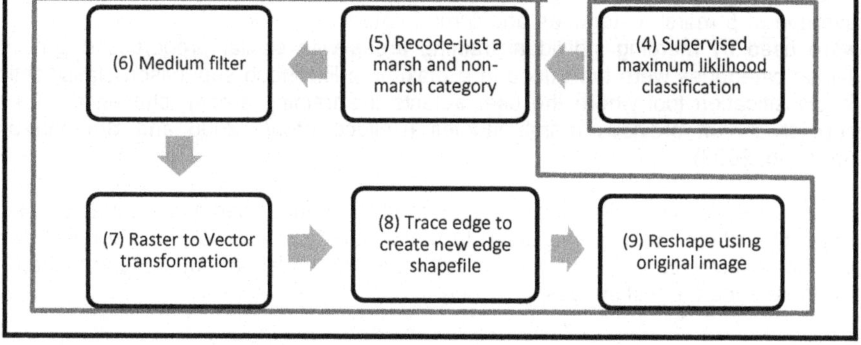


Figure 2: Flow chart to show stages in the classification method.

Grouped into: Pre-processing stages (1–3), classification (4) and post-processing stages (5–9).

**Pre-processing of Images (Stages 1–3)**

Pre-processing of the images involved making small changes to the aerial photographs to improve the accuracy of the classification of the image. This included creating a subset of the image to reduce the variety of environments to classify, and improving the contrast within the image with stretch and filters. Before classification, a gaussian contrast stretch and a 3 x 3 edge enhance spatial domain filter were used to enhance the difference between salt marsh vegetation and mudflat.

Spatial domain filtering uses neighbouring pixels to modify the middle pixel (Rees, 2001). This is generally used to either smooth, or average the values, or highlight differences in the values. An edge enhancement filter (hereafter 'edge' filter) has been used in this study to highlight areas where there is a large difference between adjacent pixels, with the aim of highlighting boundaries in the data (Morgan *et al.*, 2010).

Contrast stretches alter the frequency distribution of the pixel values to enhance particular features, by increasing or altering the tonal variation in the image (Morgan *et al.*, 2010). A gaussian stretch applies a gaussian relationship to the histogram and therefore increases the amount of pixel values present and hence increases the variety of tones in the image.

**Classification of the Marsh Edge (Stage 4)**

The decision was taken to undertake an unsupervised classification for the BW images, and supervised maximum likelihood classifications for the RGB photographs from results of a visual assessment (using randomly selected areas of the marsh edge) of classifications of 3 RGB and 3 BW images. To run a supervised classification, an average of 50 training areas (areas of the image selected to represent the required categories) were used per image, roughly half on vegetated surfaces, and half on non-vegetated surfaces (shortened as 'marsh' and 'mud' respectively). The mean plots for each category were then analysed and split into further categories depending on features. Over the nine RGB photographs, an average of 5 marsh categories and 5 mud categories (hence 10 categories in total) were used by merging individual training areas with similar properties (Figure 3). These categories were then used in a maximum likelihood supervised classification (a classification tool where the user selects the training areas), chosen due to its popular use in previous marsh studies (Belluco *et al.*, 2006 and Shuman and Ambrose, 2003).

The unsupervised classifications used on the BW images used 4 categories as visual inspection (three images were used to trial a variety of category numbers) showed that an accurate separation between salt marsh and tidal flat categories resulted on the three separate test images.

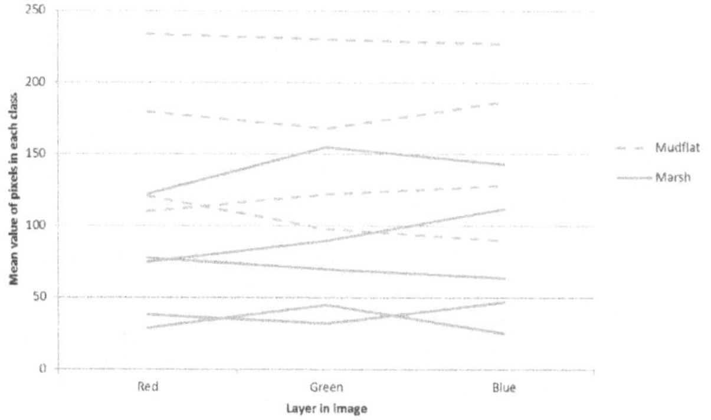


Figure 3: Mean plot of pixel values from merged classes (2010 image).

**Post-processing of Images (Stages 5–9)**

After classifying the images, they were recorded so that only two categories remained: vegetated and non-vegetated areas. A medium filter (5x5 kernel and median value) was then applied to each image to eliminate some of the smaller details within the marshes, hence allowing easier extraction of the dominant edge transition as a feature.

The post-processed images were then transformed into polygon feature classes. A polyline shapefile was subsequently created and the marsh edge substracted. After the trace was completed, the original image was loaded into ArcMap to check whether any areas had been mis-classified. These areas were often large creeks with standing water and therefore very easy to manually correct (Figure 4).

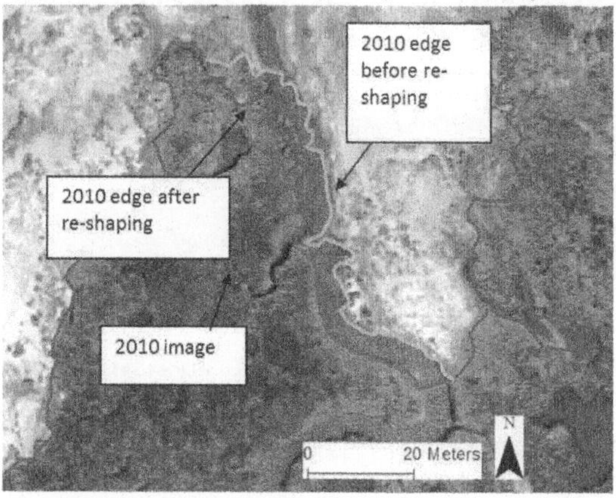


Figure 4: Screen shot to show 2010 digitised edge before and after the image had been manually corrected within ArcMap to correctly define a creek with water present.

**Comparison of Marsh Edge Mapping Methods**
To fully evaluate the accuracy and therefore application of the classification method, two other methods were tested on the 2011 image and an accuracy assessment was undertaken. These are summarized in Table 1.

Table 1: Other methods used to compare results with the classification method

| Method | Description |
|---|---|
| Manual Trace (hereafter the 'trace' method) | The 2011 aerial image was loaded into ArcMap and a polyline shapefile was used to manually trace around what visually appeared to be the marsh edge. |
| Near Infrared Radiation (NIR) assessment (hereafter the 'NIR' method) | A Spectral Enhancement was used to create an image displaying the Normalised Difference Vegetation Index (NDVI), which calculates pixel values using Equation 1. This resulting image was subsequently classified using an unsupervised classification. This classification was post-processed using the same techniques of recoding and filtering as described for the classification method above. |

$$\mathrm{NDVI} = \frac{a-b}{a+b} \qquad \text{Equation 1}$$

Where NDVI = normalized difference vegetation index, a=near-infrared spectral measurements and b= visible (red) spectral measurements.

**Marsh Edge Accuracy Assessment**
GNSS points were collected during fieldwork in 2012, providing an average accuracy of the field data to 0.014 m. The surveyed points were loaded into ArcMap and overlayed on the 2011 digitised edge. Transects were used to calculate the distance between the different digitised 2011 marsh edges and the field results (labelled as the Shore-Normal Distance Error (SNDE)) (see Figure 5).

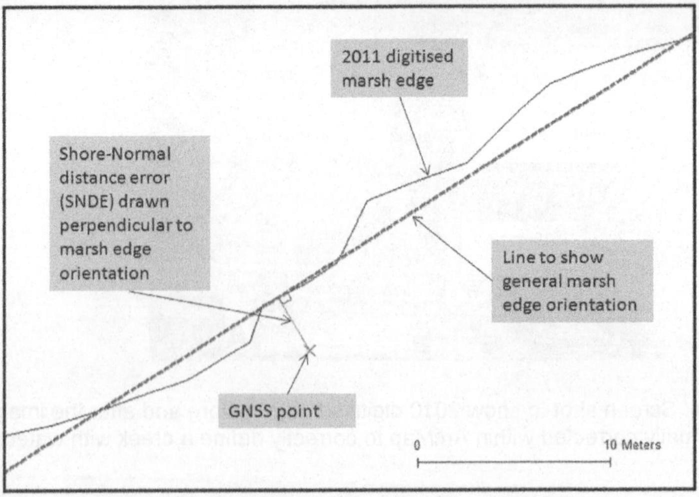


Figure 5: Diagram to show calculation of the SNDE

The SNDE calculations were then averaged and compared to the perimeter of the marsh edge to calculate the Shore-Normal Area Error (SNAE). This process was then repeated for the trace method and the NIR method.

## Results

The accuracy in marsh edge position varies spatially along the marsh edge and occurs in both onshore and offshore directions (note: onshore/offshore refers to the position of the GNSS points relative to the digitized line) (Figure 6). The accuracy assessment shows a maximum SNDE of 9.98 m in the seaward (offshore) and 6.55 m in the landward (onshore) direction. A histogram showing the distribution of errors (Figure 6) shows that the overall SNDE is in the offshore direction (skewness=0.942) and the mean overall estimate of error is calculated as 1.05 m in the offshore direction. The standard deviation of the SNDE is 2.64 m.

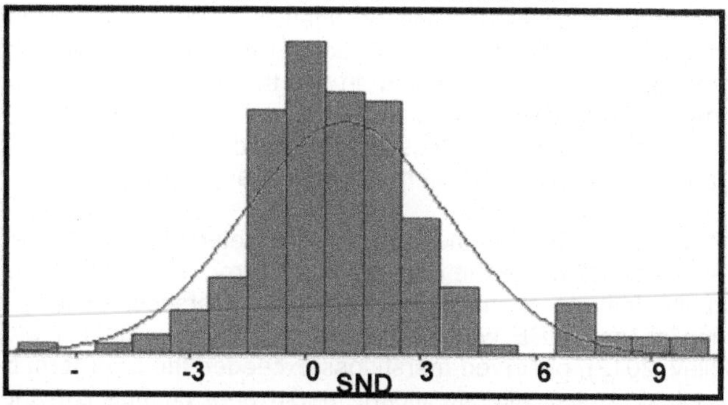


Figure 6: Histogram to show the distribution of SNDE and a normal distribution curve.

Figure 7 displays the difference in distribution between minimum and maximum errors between the different methods for marsh edge detection applied to the 2011 image. The median errors are all relatively similar, with the classification method showing the smallest value and they are all in an offshore direction. However, the NIR method shows a much larger range of values. The interquartile range is much greater for the trace method than for the NIR or classification method.

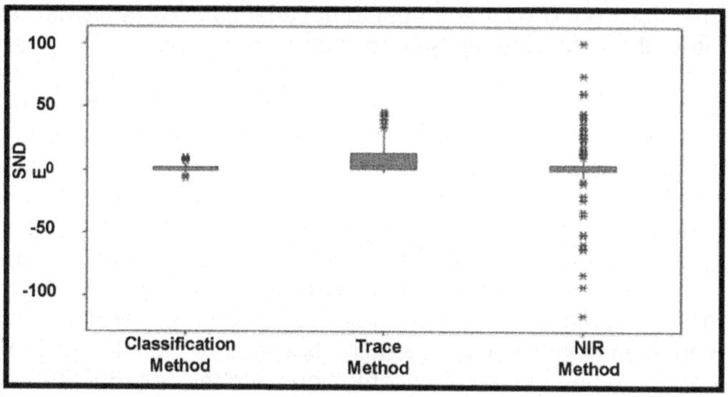


Figure 7: Boxplots to show the SNDE distribution for the three different marsh edge mapping methods averaged over the whole marsh.

## Discussion

This paper has presented a semi-automated and largely objective methodology to detect the edge of a salt marsh which was developed as part of a wider study to assess historical marsh change at Stiffkey, North Norfolk. There are key elements of the methodology and its application that can be considered for wider application at salt marsh sites and other dynamic coastal environments. However, further work is recommended to apply the classification method to other marsh edge types to see whether it performs with a similar consistency of error.

### Application of the Classification Method at Marshes

The dominant advantage of the classification method presented in this paper is that it is possible to achieve a robust classification of areas where the edge of the marsh is very complex and fragmented. A more detailed assessment of the results spatially across the marsh showed similar results between the different methods in areas where there was a clear defined cliffed marsh edge. However in instances where the edge was fragmented, variability resulted in all the method, with the smallest variability occurring within the classification method. The classification method therefore provided a more consistent (less variable) error in defining the marsh edge which is particularly important for detection of marsh change over time.

Furthermore the accuracy of the method can be determined and the changes observed can be discussed within the context of this error. This method was applied to research into the historic changes in the Stiffkey Marshes. In this study, the standard deviation of the SNDE was calculated as 2.64 m. Within a wider marsh change study (Riley, 2012), observed marsh loss exceeded the error term by a factor of 4.5. In this context, statistically significant marsh area change can thus still be observed, in spite of the error term associated with the classification method.

When this error estimate is applied to the smaller annual change in marsh extent obtained from mapping the marsh edge using the classification method, patterns of change are less certain. However, the standard deviation of the SNAEs is likely to be an over-estimation of the mapping error due to the impact of the extremities on the overall standard deviation, and errors are more likely to be systematic and consistent over different images. This suggests that this method is appropriate for mapping changes to marsh extent over the annual to decadal timescale, provided the focus is on relative rather than absolute marsh extent change.

In recent decades, the development of remote sensing technologies and the increasing availability of satellite and LiDAR imagery have led to less reliance on aerial photography for the mapping of coastal change. The relatively recent introduction of this technology, however, means that historic change still often relies on the analysis of aerial photography. This study has shown that RGB and BW aerial images can be processed in ways that provide a reliable way of automating the mapping salt marsh margins, albeit with a consistent offset between the mapped and actual marsh margins. In future, longer historical records of NIR imagery offer great opportunities for clearly identifying vegetation because of the interaction between the quantity of leaf tissue and pigment concentrations and the red (600-700 nm) and NIR (800-1100 nm) parts of the spectrum (Hardisky et al., 1986). This has also been highlighted by this study, however, caution is needed when using NIR imagery in coastal environments as classification schemes will be sensitive to detecting seasonal growth of seaweed and algae thus misclassifying tidal flat as salt marsh. For inter-annual comparisons of imagery acquired in seasonal climates, it is thus critical that successive imagery is acquired at the same time of year. RGB or multi-spectral data is therefore likely to continue to be needed alongside NIR data to accurately map marsh extent through time

**Application of the Classification Method to Shoreline Change Studies**
The results of this study support the use of classification methods more widely for the study of landform change, and particularly salt marsh margin change, in dynamic coastal environments.

Pre-processing and post-processing techniques, such as presented in this paper, can be relatively quickly and easily employed to allow the classification method to be applied in a way that allows reliable inter-annual seaward marsh edge mapping and thus aerial marsh extent change to be assessed. While systematic errors in marsh edge position occur when using this method, these errors are consistent within and between images, such that relative change can reliably calculated. Alternative ways of mapping marsh edge change, such as manual edge tracing, have been shown in this study to lead to a less predictable error term.

The classification method thus reduces the unpredictable nature of errors in edge mapping that are, e.g. associated with manual methods affected by subjectivity. In reality, historical studies will frequently need to make use a range of data, (e.g. RGB and BW aerial photographs). Researchers must remain flexible in the application of pre- and post-processing techniques and adapt the specific methods accordingly. The results of this study also imply that an accurate comparison of individual studies is only possible if the methodology employed has been clearly described and is consistent between studies.

## Conclusions

### Development of the Classification Method

- A classification method for shoreline protection requires pre and post processing techniques to improve the accuracy of the classification.
- The method cannot be fully automated and post-processing requires re-profiling in specific areas.
- However the classification method does provide a semi-automated method which, for long lengths of coastline, could also reduce the time required to digitise the marsh edge.

### Applying the Classification Method to Historical Marsh Edge Change Studies

- The classification method was successfully used in a historical salt marsh change study at Stiffkey, North Norfolk.
- In this study, the standard deviation of the SNDE was calculated as 2.64 m. This was used successfully to observe marsh erosion in a wider study (Riley, 2012).
- Other methods employed at the Stiffkey Marshes resulted in larger errors and would have potentially not enabled specific conclusions to be drawn regarding marsh edge change at Stiffkey Marshes.

- This method has real potential to be applied at other locations and further studies should be undertaken to assess the consistency of error over difference marsh types.

**Application of the Classification Method to Wider Coastal Environments**

- Historical shoreline changes are fundamental in developing policies for future, long term coastal management.
- The classification method, by employing pre and post processing techniques, could be employed over a wider range of environments and studies than it currently is.
- This may improve the accuracy of historical shoreline studies and provide greater consistency.
- The method presented here, including pre- and post-processing techniques would not require a large amount of training or investment in time and could therefore be applied in a wide range of organisations and studies.

## References


Allen J. R. L., (2000) Morphodynamics of Holocene Salt Marshes: A Review Sketch from the Atlantic and Southern North Sea Coasts of Europe. *Quaternary Science Reviews* **19:** 1155-1231

Allen J.R.L. and Pye K., (1992) *Salt marshes: morphodynamics, conservation and engineering significance.* Cambridge University Press: Cambridge

Belluco, E. *et al.,* (2006) Mapping salt-marsh vegetation by multispectral and hyperspectral remote sensing. *Remote Sensing of Environment.* **105 (1):** 54-67

Boak, E. H. and Turner, I. L., (2005) Shoreline Definition and Detection: A Review. *Journal of Coastal Research* **21(4):** 688-703

Cooper, N. J., *et al.,* (2001) 25 years of salt marsh erosion in Essex: Implications for coastal defence and nature conservation. *Journal of Coastal Conservation.* **7:** 31-40

Environmental Agency (EA)., (2010) North Norfolk Shoreline Management Plan. East Anglia Coastal Group.

Freiss, D. A. *et al.,* (2012) Remote sensing of geomorphological and ecological change in response to salt marsh managed realignment, The Wash, UK. *International Journal of Applied Earth Observation and Geoinformation.* **18:** 57-68 Hardisky, M. A. *et al.*, (1986) Remote Sensing of Coastal Wetlands. *BioScience.* **36(7):** 453-460

Morgan, J. L. *et al.,* (2010) Aerial Photography: A Rapidly Evolving Tool for Ecological Management. *BioScience* **60(1):** 47-59

Pethick, J. S. (1980) Salt-marsh initiation during the Holocene transgression: the example of the north Norfolk marshes, England. *Journal of Biogeography.* **7:** 1-9

Rees, W. G., (2001) *Physical principles of remote sensing.* Cambridge: Cambridge University Press

Riley, E.R. (2012) Open-Coast Marsh Edge Dynamics: 60 years of change in Stiffkey Marshes, Norfolk. *M.Phil. Thesis, University of Cambridge.*

Shuman, C. S., and Ambrose, R. F., (2003) A Comparison of Remote Sensing and Ground-Based Methods for Monitoring Wetland Restoration Success. *Restoration Ecology.* **11(3):** 325-333

Van de Koppel, J., *et al.,* (2005) Self-Organisation and Vegetation Collapse in Salt Marsh Ecosystems. *The American Naturalist.* **165(1):** E1-E12

**Coastal Management: Changing coast, changing climate, changing minds**
ISBN 978-0-7277-6114-9

ICE Publishing: All rights reserved
doi: 10.1680/cm.61149.119

# Impacts of Climate Change on Extreme Wave Climate Along the Western Coast of Sri Lanka

**R. M. J. Bamunawala**, PhD Fellow, Department Water Science and Engineering, UNESCO-IHE, Delft, The Netherlands
**S. S. L. Hettiarachchi**, Senior Professor, Department of Civil Engineering, University of Moratuwa, Moratuwa, Sri Lanka
**S. P. Samarawickrama**, Professor, Department of Civil Engineering, University of Moratuwa, Moratuwa, Sri Lanka
**P. N. Wikramanayake**, Senior Lecturer, Department of Civil Engineering, Open University of Sri Lanka, Nawala, Sri Lanka
**Roshanka Ranasinghe**, Professor of Climate Change Impacts and Coastal Risk, Department of Water Education, UNESCO – IHE, Delft, The Netherlands

## Abstract
Climate change and climate change driven impacts are most widely argued topics among contemporary researchers and scientists. Broadly there are two schools of thought that present entirely contrasting perceptions about the overall concept of climate change and its impacts. While one of the perceptions state on frequently varied climate change and occurrence of extreme weather events, others rely on the perception that there is nothing called climate change, and it is only the indifferences in occurrence of climate events. The title of this study itself implies the fact that this research study supports the perception of climate change and its probable impacts, where detailed studies and numerical modelling have been carried out to determine the climate change induced impacts on the wave climate along the western coast of Sri Lanka.

Extensive concentrations of green-house gasses emitted to the earth's atmosphere and vast amounts of aerosols govern the majority of anthropogenic causes for climate change, while many of the natural causes such as changes in solar radiation also contributes immensely to earth's climate change (Houghton, et al., 2001). Absence of detailed studies carried out on investigating probable impacts on wave climate due to possible changes in climate is one of the major drawbacks in managing the unique coastal systems in Sri Lanka. As a country where coastal resources play a major role in its development, it is extremely essential to ensure a clear understanding about the probable impacts that could arise on its coastal areas. This detailed investigation provides vital information on probable impacts on the wave climate along the western coast of Sri Lanka, under the influence of global climate changes.

The presented study is a part of the collaborative research project carried out with UNESCO – IHE, Delft, The Netherlands, which investigates the Impacts of Climate Change on Small Tidal Inlets. This particular research was carried out in support of the objectives of the said research objectives, where WAVEWATCH III and SWAN wave models were used as a nested hybrid model to determine the future wave conditions along the western coast of Sri Lanka.

Wave modelling was carried out for three different time periods, year 1981 – 2000 (past wave climate), year 2041 – 2060 (first future wave climate) and year 2081 – 2100 (second future wave climate). The past wave climate was used to compare the variations of the two future wave climates considered. Out of the 6 hourly wave data obtained at desired locations along the western coast of Sri Lanka, average wave climate and extreme wave climate conditions were analyzed accordingly and the results related to the extreme wave climate are being presented.

The outcomes of this study indicate that there is a prominent threat to the coasts in Sri Lanka, due to a considerable increment in severity and frequency of occurrence for extreme wave climates. This probable extreme wave climate of the western coast of Sri Lanka should be considered very carefully, as many of the socio-economic activities are directly related to the coast, and hence could be subjected to significant disruptions, under the influences of this extreme wave climate events.

The results from this study enhance the knowledge on both episodic and chronic hazards along the western coast of Sri Lanka and provide information for strategic planning and risk management measures. The outcomes have wide applications for the region, as it is a ground-breaking research of this kind for the entire Indian Ocean region.

## Significance of the Research Study

Climate changes of the earth's atmosphere and its possible impacts are of the most widely discussed contemporary scientific problems. These climate changes may occur due to both internal variability within the climate system and due to external factors, which can be either anthropogenic or natural. Increasing concentration of atmospheric greenhouse-gases tends to warm the earth's surface and its lower atmosphere, while an increase in certain forms of aerosols tends to cool it (Houghton, et al., 2001). Natural factors, such as changes in solar output or its explosive volcanic activities, can also cause radioactive forcing and hence influence the earth's climate. Complex climate models are required to provide detailed estimates of feedbacks and regional features in the climate system. Although confidence in the ability of these models to provide useful projections of future climate has improved on a range of space and time-scales, the present-day climate models cannot yet simulate all aspects of climate (Houghton, et al., 2001). For example, there are particular uncertainties associated with clouds and their interaction with radiation and aerosols. Besides, climate model simulations are a combination of a forced climate change component together with internally generated natural variability. The internal variability of the global and regional climate system adds a further level of uncertainty in the evaluation of a climate change simulation (Christensen, et al., 2007).

As an important element of the climate system, ocean wave heights (among many other ocean surface characteristics) could be affected by anthropogenic forcing. However, ocean wave heights are not directly available from the output of global climate models. Useful projections of future wave height climate need to be produced through dynamical or statistical "downscaling" approaches, just like other regional climate change information. Therefore, there are various sources of uncertainty in the generation of ocean wave height climate change projections (Hamer, et al., 2010a).

Within the context of climate change, one of the recurrent questions is how this change could impact waves and thus the wave-dominated coasts. As an example, the western coast of Sri Lanka is bounded by the Indian Ocean and it is characterized by intense human activities, such as sea transport, fishing, coastal shipping, ports, seaside resorts, touristic sandy beaches and surfing areas. Since the said coast is completely open to the ocean, its wave climate is characterized by swells and storms generated by strong winds in the Indian Ocean. In the work done by Wang & Swail, 2006, it was mentioned that the changes in wave conditions could even modify the coastal morphology and hence have significant impacts on human activities. In the context of global warming, significant climate changes at the oceanic basin scale could modify the wave climate. General Circulation Models (GCM) indeed project atmospheric changes such as a poleward shift of storm tracks (Yin, 2005) or a decrease of the total number and intensity of cyclones in the Northern Hemisphere (CATTO, et al., 2011). These changes can then impact the resulting wave climate, in terms of wave height, period and direction.

Concerning the future wave climate, work done by Christensen et al., 2007 have highlighted the fact that there is a vast lack of information on potential changes in regional wave climate. However, a significant work is yet to be carried out with respect to the possible climate change driven wave climate changes in the Indian Ocean, which will indicate the possible impacts to the Western coast of Sri Lanka. The aim of the study carried out was to provide more rigorous projections of the offshore ocean wave climate of the Western coastline of Sri Lanka, and to provide a suitable dataset, which can be used to assess the possible coastal impacts of climate change in the region, under the forcing of future climate scenarios.

## Objectives and Methodology of the Research Study

The major objectives of the study are as the following.

1) Set up and calibrate a wave model that is capable of predicting the off-shore (shallow water) wave climate around Sri Lanka, given the wind field over the relevant portion of the Indian Ocean.
2) Application of the said wave model for different scenarios of climate change and environment.

The research methodology of the presented study is as the following.

1) Selection of an appropriate wave generation model.
2) Possible models are MIKE SW, SWAN, WAM and WW3, and out of these WW3 and SWAN models were used as the model.
3) Acquisition of necessary wind data sets for wave modelling process.
4) Possible sources for actual (historical) wind data are re-analysed wind data from global models (ECMWF, NOAA), Satellite measurements (Quickscat, Adeos) and Wind data from NOAA was used in the study as the historical wind data set.

5) Predicted wind data (CCAM wind data) was provided by the CSIRO, Australia. The same wind model was used to generate the wind data for the Thailand study as well.
6) Acquisition of necessary deep-water wave data for the research domain to calibration and validation of the wave model.
7) The possible sources are Satellite altimetry (TOPEX/POSEIDON, JASON), SAR measurements by satellites (ERS), Wave hindcasts from global models (ECMWF, NOAA) and limited deep-water wave measurements obtained off Sri Lanka.
8) Analyses of predicted wind data sets and determine the bias correction procedure, if necessary.
9) Set up the wave model to cover an appropriate domain around Sri Lanka, relevant to the data availability and the needs of the study. (As shown in Figure 1)
10) For this study $30^0$ x $30^0$ area will be taken as the domain (Latitude $-8^0$ to $22^0$ and Longitude $65^0$ to $95^0$)
11) Calibration and validation of the model
12) Run the model with the predicted future wind field to predict the future wave climate for year 2050 and 2100.
13) Analyzing the differences in the wave climates for the current and future scenarios and assessing their impacts on various uses of the coastal areas.

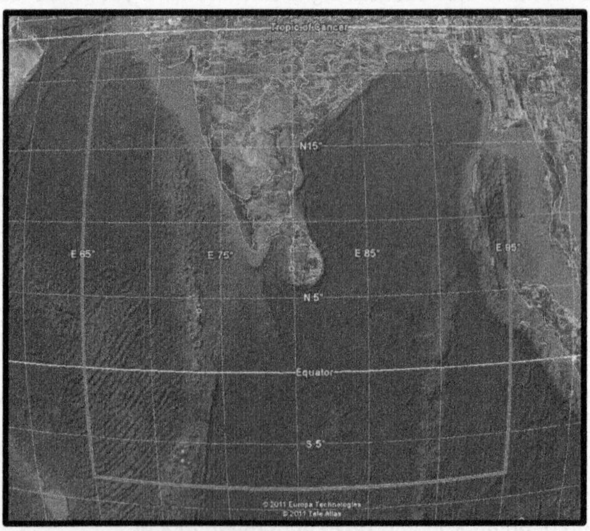


Figure 1: Domain area of the study

## Approach to the Research Problem

One of the most critical issues in wave modeling with future climate change projections is the uncertainty related to the projection of Regional Climate Changes. McInnes et al., 2007 has pointed out the fact that there are three main sources of uncertainty to be considered when producing projections of global warming-induced climate change for a region for a given year in the future.

1) The uncertainty in the future evolution of greenhouse gas concentrations in the atmosphere.
2) The uncertainty in how much the global average surface temperature will respond to increases in atmospheric greenhouse gas concentrations.
3) The uncertainty in how changes to the climate as a result of global warming will vary spatially and hence how the climate of the region under consideration will respond to an increase in global average surface temperature.

The first uncertainty can be addressed by considering different plausible storylines of future global demographic, economic and technological change. The Intergovernmental Panel on Climate Change's Special Report on Emission Scenarios has provided greenhouse gas emission scenarios associated with a suite of such storylines. The second uncertainty can be addressed by considering the future rates of future global warming in simulations of different climate models forced with increases in atmospheric greenhouse gas concentrations arising from emission scenario corresponding to the Special Report on Emission Scenarios (SRES). The third uncertainty, the uncertainty in the response of the regional climate to a given global warming value, can be addressed by considering the response of the climate of the region of interest to global warming in multiple climate models (McInnes, et al., 2007).

Although the work done relate to wave climate with future climate changes are rare to find for the Sri Lankan context, there are number of successful benchmark attempts that have been made in various parts of the world to address the said issue. Almost in all such research work carried out, the outcomes claim a significant relationship between the future wave climate and its forcing conditions that are to be altered by future climate change scenarios.

The work done by Hamer et al., 2010a emphasize that available wave climate projections are highly insufficient to assess the effects of probable climate change impacts on the erosion of the world's coasts. According to the said work, at present, a considerable research effort is placed into regional ocean wave projections, with forcing conditions derived from a few selected emission scenarios, arising from selected GCMs. It has identified several shortcomings of such an approach, as the following.

1) Limitations of statistical confidence in the projections.
2) Additional effort requirements in modelling due to repeated model runs.
3) Laps in the global coverage, including the crucial areas that are most likely to be at risk due to changing wave conditions.

The propose solutions given in the said study for this problematic scenario was to shift to global projections. The aim should be to use different wave models and statistical downscaling approaches to produce ensembles of wave projections that correspond to climate projections from different climate models for different emission scenarios. When combined, a distribution of projections will be available, so it will allow an assessment of all three levels of uncertainties associated with forcing, climate models, and downscaling methods, respectively, presenting projections within acceptable statistical confidence intervals. The proposed design will provide suitable data on a global scale for carrying out surface ocean wave projections, focusing on mid- and late-twenty-first-century time slices, to service the increasing demands of the coastal impacts community. As a result, future studies could include dynamic coupling of wave processes into coupled ocean–atmosphere global climate models (Hamer, et al., 2010a).

According to work done by Swail & Wang, 2005, results of ocean wave climate change scenarios for the northern hemisphere oceans for the twenty-first century show that significant changes can be anticipated in both the North Atlantic and the North Pacific under all the three forcing-scenarios. The rate and sign of the projected future wave height changes are not constant throughout the 21st century and in some regions, these appear to be very much dependent on the forcing conditions as well. The rate of change appears to have a positive relationship with the rate of increase in the greenhouse-gases forcing.

## Use of WAVEWATCH III and SWAN Models to Model Future Wave Climates

The basic scientific philosophy of SWAN is identical to that of WAM (Cycle 3 and 4) and it uses the same formulations for the source terms. On the other hand, SWAN contains some additional formulations, primarily for shallow water. Moreover, the numerical techniques are very different. WAVEWATCH III not only uses different numerical techniques but also different formulations for the wind input and white capping (The Wave Model Development and Implementation Group, 1988).

When SWAN is nested with WAVEWATCH III, it was noted that the boundary conditions for SWAN provided by WAVEWATCH III are not consistent, even the same physics are being used. The potential reasons are manifold such as the differences in numerical techniques employed and implementation for geographic area (i.e. spatial and spectral resolutions, coefficients, etc.) (Chawla & Tolman, 2007, Tolman, 2010, and Cavaleri, et al., 2007).

To overcome this issue, the deep-water boundary of SWAN nest was located in WAVEWATCH III, where shallow water effects do not dominate. This is primarily important to avoid the existence of large discontinuities between the two models. At the same time, the spatial and spectral resolutions were adjusted so that it will not differ largely from one another, in order to avoid additional modifications of the model layout (Cavaleri, et al., 2007, and Chawla & Tolman, 2007).

Since SWAN in coded in such a manner, so that the scientific findings with another wave model could be shared with SWAN. Similarly, SWAN could be readily nested in WAVEWATCH III for collaborative modelling of wave climates from Deep Ocean to shoreline. In this research study, the said similarities have been extensively exploited, in order to couple the WW3 and SWAN together to model the necessary outputs. (Cavaleri, et al., 2007, Chawla & Tolman, 2007, The SWAN Team, 2010, The SWAN Team, 2011 (c), and The Wave Model Development and Implementation Group, 1988).

## Outputs of the Wave Model Developed

The outputs obtained via wave modelling are of three different time spans (year 1981–2000, 2041–2060 and 2081–2100). In order to match with the overall project requirements, outputs were obtained for a point near Colombo. Comparisons for the change in wave climates were made with respect to the past wave climate obtained (year 1981-2000). Mean wave height (Hs) and direction (Dir) are the wave properties considered in the presented study.

Table 1 consists of projected changes of 99th percentile wave climate for two future time periods, compared with the projected wave climate for past time period. The values related to 50th percentile are also provided as a reference to the proportions of magnitude changes during projected extreme events. It indicates a significant increment during the first future time period (8.5%) and reduces to a slightly lesser value (about 7%) during the second future time period considered. Contrastingly, the wave direction corresponding to the said wave heights indicate increments in both the future time periods, where the increment for the second time period (9.5%) is comparatively smaller than that for the first time period (about 8%).

Table 1: Variation of 99th percentile wave climate properties

| | 1981 – 2000 | 2041 – 2060 | 2081 - 2100 |
|---|---|---|---|
| HS (50%) | 0.93 | 0.99 | 1.01 |
| HS (99%) | 2.71 | 2.90 | 2.93 |
| Mean HS (HS > 99%) | 3.06 | 3.32 | 3.28 |
| Mean Dir (HS > 99%) | 259.17 | 279.92 | 283.83 |
| Dir (99%) | 323.00 | 347.04 | 345.60 |

Similarly, extreme wave climate data corresponding to wave heights exceeding a threshold value of 3.0 m are shown in Table 2. The values related to 50th percentile are also provided as a reference to the proportions of magnitude changes of the projected extreme events.

Table 2: Variation of wave climate exceeding threshold wave height of 3.0 m

| | 1981 – 2000 | 2041 – 2060 | 2081 - 2100 |
|---|---|---|---|
| $H_S$ (50%) | 0.93 | 0.99 | 1.01 |
| Mean $H_S$ (for $H_S$ > 3.0) | 3.06 | 3.32 | 3.28 |
| Mean Dir (for $H_S$ > 3.0) | 259.17 | 279.92 | 283.83 |

Although, there is no conclusive evidence to indicate a major deviations in extreme wave climate patterns, the number of extreme wave climate records recorded during the three time periods considered have to be noted. Table 3 indicates the number of predicted extreme wave climate records (i.e. number of records where the wave height exceeded its considered threshold value of 3.0 m) for three different time periods considered.

Table 3: Number of modelled extreme wave records

| | 1981 – 2000 | 2041 – 2060 | 2081 - 2100 |
|---|---|---|---|
| Number of times where $H_S$ > 3.0 m | 392 | 731 | 976 |

These numbers clearly indicate the fact that there is a clear trend of occurrence of extreme wave events at a much higher frequency. In fact, it is an increment of about 87% for the first future wave climate period and that value increases up to about

150% for the second time period considered. These values clearly indicate that there exists a very high possibility of having severe storm conditions in future.

Since the modelled wave climate is of 6 hourly intervals, this could be interpreted as high frequency occurrence of extreme wave climates of longer time durations (i.e. there would be a high number of severe wave conditions, lasting for longer durations)

## Conclusions

The high increments of extreme wave climates need to be investigated carefully, as its frequencies indicate very high increments for both the time periods considered. This means that both the severity and frequency of occurrence for extreme wave climates may be increased very significantly, causing major disruptions to the coastal systems. The directional increments also indicate growths in both the time periods considered, valued about $21^0$ and $25^0$ respectively. Therefore, the combined effects of predicted extreme wave climate could be very severe.

The domain area was restricted by the project requirements. However, if a more detailed study to be carried out for the Indian Ocean region, relevant wind data should be available for a larger domain, covering a larger area of the Indian Ocean. This research study was not carried out at such a high detailed level of the entire Indian Ocean, thus the availability of wind data was sufficient only for this particular research study.

## Bibliography

Church, J. A., Woodworth, P. L. & Aarup, T., 2010. *Underestanding Sea-Level Rise and Variability.* s.l.:Wiley-Blackwell Publishing Ltd.

Hamer, M. A., McInnes, K. L. & Ranasinghe, R., 2009. *Future Projections of the East Australian Wave Climate.* Halifax, Canada, s.n.

CATTO, J. L., SHAFFREY, L. C. & HODGES, K. I., 2011. *Northern Hemisphere Extratropical Cyclones in a Warming Climate in the HiGEM High-Resolution Climate Model,* s.l.: American Meteorological Society.

Cavaleri, L., Alves, J. H. G. M. & Ardhuin, F., 2007. Wave Modelling - The State of the Art. *Progress in Oceanography,* Volume 75, pp. 603 - 674.

Charles, E. et al., March 2012. *Climate change impact on waves in the Bay of Biscay, France,* s.l.: s.n.

Chawla, A. & Tolman, H. L., 2007. *Automated grid generation for WAVEWATCH III,* s.l.: U. S. Department of Commerce.

Christensen, J. H. et al., 2007. *Regional Climate Projections,* s.l.: Cambridge University Press, Cambridge, United Kingdom and New York, NY, USA.

GoogleEarth, 6., 2010. [Online].

Hamer, M. A., Wang, X. L., Church, J. A. & Swail, V. R., 2010a. *Coordinating Global Ocean Wave Climate Projections.* s.l.:American Meteorological Society.

Hemer, M., McInnes, K. & Ranasinghe, R., September, 2010b. *Investigation of Climate Driven Variations in Offshore Wave Climate Along the NSW Coast,* s.l.: s.n.

Houghton, J. T. et al., 2001. *Climate Change 2001: The Scientific Basis,* Cambridge: Cambridge University Press.

Kalnay, Kalnay, Kalnay & Kalnay, 1996. *The NCEP/NCAR 40-year reanalysis project.* s.l.:Bull. Amer. Meteor. Soc..

Madduma Bandara, C. M., 1989. *A survey of the coastal zone of Sri Lank.* Colombo: Coast Conservation Department.

McInnes, K., Macadam, I. & O'Grady, J., 2007. *Climate Change Projections for the Wooli Wooli Estuary and Batemans Bay,* s.l.: New South Wales Department of Environment and Climate Change.
Scheffer, H. J., Fernando, K. R. M. D. & Fittschen, T., May, 1994. *Directional Wave Climate Study; South-West Coast of Sri Lanka,* s.l.: s.n.
Swail, V. R. & Wang, X. L., 2005. Projection and Analysis of Extreme Wave Climate. *JOURNAL OF CLIMATE,* Volume VOLUME 19, pp. 5581-5605.
The SWAN Team, 2010. *SWAN Programming Rules,* s.l.: Delft University of Technology.
The SWAN Team, 2011 (c). *SWAN Implementation Manual,* s.l.: Delft University of Technology.
The Wave Model Development and Implementation Group, 1988. The WAM Model - A Third Generation Ocean Wave Prediction Model. *Journal of Physical Oceanography*, Volume 18, pp. 1775 - 1810.
Tolman, H. L., 2010. *WAVEWATCH III Development Best Practices,* s.l.: U. S. Department of Commerce.
Wang, X. L. & Swail, V. R., 2006. Climate Change Signal and Uncertainty in Projections of Ocean Wave Heights. *Climate Dynamics,* Volume Volume 26, pp. 109 - 126.
Yin, J. H., 2005. *A Consistent Poleward Shift of the Storm Tracks in Simulations of 21st Century Climate,* s.l.: Geophysical Research Letters.

**Coastal Management: Changing coast, changing climate, changing minds**
ISBN 978-0-7277-6114-9

ICE Publishing: All rights reserved
doi: 10.1680/cm.61149.129

# Possibility of Tsunami Risk Map Based on Extreme Value Analysis with Historical Records

**Hisamichi Nobuoka**, Coastal engineering Ibaraki University, Hitachi, Japan

## Introduction

Though Japan has prepared tsunami hazard map about last 10 years, 'The 2011 off the Pacific Coast of Tohoku Earthquake' ( called Tohoku Earthquake in this paper after here), caused bigger tsunami than that in distributed hazard maps in many area. This is the main reason why we had much number of victims. In case we know the much information of a plate at and around epicenter near target area, we can estimate the tsunami risk from the probability of fault slips directly. If there are small numbers of the information of that, we need another risk analysis method. This paper shows a method and results of a tsunami risk map employed by statistical extreme value analysis with historical records of tsunami heights.

## Methodology of Tsunami Risk Map by Historical Record Heights

Ibaraki prefecture is between Tokyo and Tohoku and the south edge of the tsunami source of Tohoku Earthquake as shown in Figure 1.

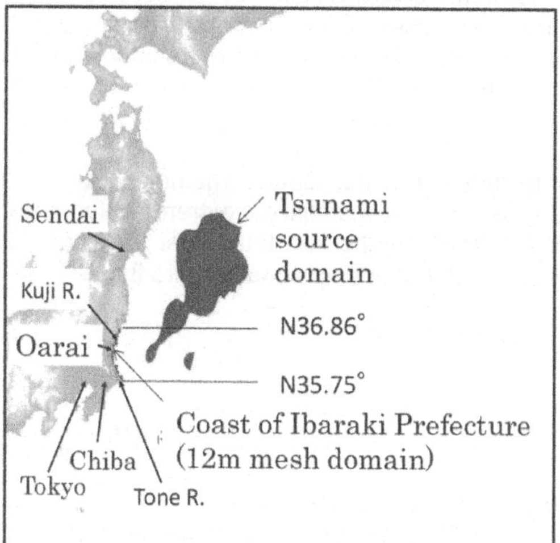


Figure 1: Location and Calculation Domain

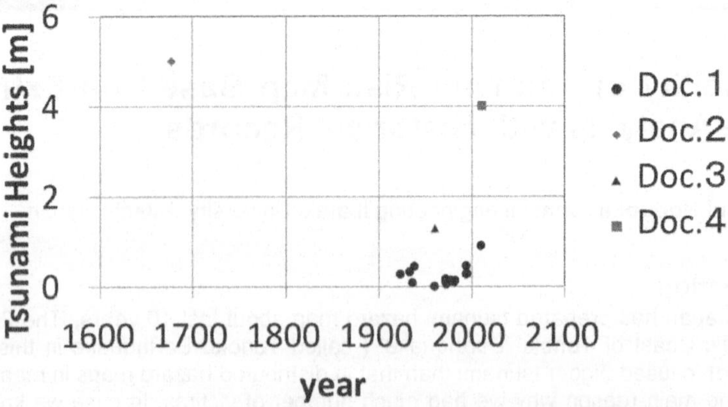


Figure 2: Tsunami Records in Ooarai, Ibaraki Prefecture
Doc.1: Watanabe(1998), Doc.2 Takeuchi et. al.(2007), Doc. 3. Mito local Meteorological office(), Doc.4, Japan Meteorological Agency(2012).

**Extreme Value Analysis on Tsunami Heights**

Ooarai port and town has largest number of tsunami records in Ibaraki prefecture; 15 records except less than 0.1 m heights of small tsunamis, as shown in Figure 2. Maximum heights was 5.0 m at the year of 1677 by Empo-off-Boso Earthquake. This height was estimated from the disaster information in documents by Takeuchi et. al.(2007). After this tsunami there were no records of documents till the year of 1923. The 4.0 m of tsunami height measured by tsunami gage of Japan Meteorological Agency (2011, 2012) at the time of Tohoku Earthquake is seconds highest in these records.

As the limited number of those records, Goda's method (1988, 1990) on extreme wave heights was employed for present statistical extreme value analysis. He choose 9 distribution functions for small number of data; First Tippett-I, First Tippett-II (4 types) and Weibull (4 types) distributions as equation 1 to 3.

$$F(x) = \exp(-\exp[-(x-B)/A]) \quad (1)$$

$$F(x) = \exp[-(1+(x-B)/kA)^{-k}]$$
at $k$ is 2.5, 3.33, 5.5 or 10.0 (2)

$$F(X) = 1-\exp(-[(x-B)/A]^{k})$$
at $k$ is 0.7, 1.0, 1.4 or 2.0 (3)

The k, A and B are the shape parameter, scale parameter and location parameter. Because in case of less 100 data of population, it is hard to estimate best distribution function so that the 9 types functions are selected by Goda (1988, 1990) .

All of the distributions which satisfied DOL and REC citations were used for present risk analysis in this paper as uncertainty. Because it is also hard to choose the best distribution from only less than 30 data. Sampling data with periods of the year also influence the results much by extreme value analysis. Table-1 shows the 11 types set of sampling data examined in this study.

Table 1: Sampling Data Set

| CASE No | Type | Including Huge Tsunami | Taking 2010 Chile | Sampling term |
|---|---|---|---|---|
| 1 | Max in one decade | Enpo* | Yes | 350 years |
| 2 | | Enpo*, Tohoku** | | |
| 3 | | | No | |
| 4 | | Tohoku** | Yes | |
| 5 | | | No | |
| 6 | | | Yes | 100 years |
| 7 | | | No | |
| 8 | POT Method | Enpo* | Yes | 350 years |
| 9 | | Enpo*, Tohoku** | | |
| 10 | | Tohoku** | | |
| 11 | | | | 100 years |

* Empo : Enpo-bousou-oki Earthquake (1677)
** Tohoku : Tohoku Earthquake (2011)

After null hypothesis test of a distribution function by the DOL standard and the REC standard, each adopted extreme value function was improved by use of the Jackknife method proposed by Tanaka and Takara (1999) against ununiform distribution as the large number of small tsunami heights. A range between maximum and minimum tsunami heights of a return period of each improved function is called uncertainty width in this paper. The representative function is also defined in this paper as minimum variance by using SLSC (Standard Least Squares Criterion) proposed also by Tanaka and Takara (1999). The best data set from 11 sets was selected comparing the return period of the Tohoku Earthquake Tsunami near Sendai where huge tsunamis have comes frequently.

**Making Tsunami Risk Map**

Numerical simulation of tsunami propagation and flood in the land were implemented by use of vertical integrated type of shallow water equations. For the solution of the equations, ADI method of finite difference method was employed with upwind differencing for advection terms. Calculation domain was same as largest flame of Figure-1 and nesting 4 domain from 2592 m to 96 m grid sizes. For risk analysis, two more nesting was set up, of which grids sizes were 48 m and 12 m. Integrated time was only 75 minutes before second tsunami arrived a coast.

The fault slip length was adjusted to fit the calculated height and target maximum and minimum extreme heights of those satisfied distribution functions respectively. To find the slip that the difference between simulated tsunami heights and the height of one return period by extreme value analysis becomes less than 0.1m, the slip distance of fault for tsunami source was varied and simulated tsunami repeatedly. Though a area which holds many number of fault slip may be able to calculate probabilistic fault slip directly, the presented study area has only two records of comparatively large slips so it is hard to calculate directly.

Each return period has flooding maps called probabilistic tsunami flood maps. One return period has several flooding maps as like maximum, minimum and representative flood area because of the uncertainty heights generated by the small number of the records on extreme value analysis. Two kind risk maps were examine; one is damage risk map of houses and buildings and the other is human life risk map. Causality and damage of houses were calculated from the calculated flood depth by use of assessment methods presented by Central Disaster Prevention Council (Japan) (2003).

## Result of Extreme Tsunami Heights

### Sampling Data Set and Time Length

The unbiased estimations were changed about 1.5 times at the maximum by the Jackknife method. This result suggested that the method worked much in the bias to low height.

Figure-3 is the estimation results of return periods on 4 m tsunami height at Ooarai in 11 cases shown in Tabel-1. This 4m height is same as the maximum measured that at Ooarai in Tohoku Earthquake by Japan Meteorological Agency. The bar and horizontal line in red show the maximum, minimum and representational height in distribution functions those satisfied DOL standard and the REC standard. These results have variety of heights too much; the ranges of return periods are 90 to 580 years. Three longest cases are data set which length is 350 years and except legend tsunami, Empo-off-Boso Earthquak in Ibaraki in 1677.

We have no other records of type of Earthquake as Empo-off-Boso Earthquak which occurred southern outside of Tohoku Earthquake. Frequency of huge earthquakes as Tohoku Earthquake is once in 400 to 800 years (The Headquarters for Earthquake Research Promotion (Japan), 2011). Case-10 of sampling data, of which return

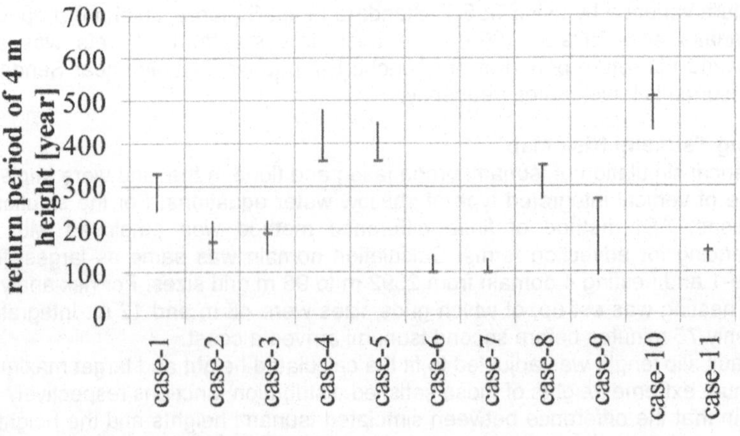


Figure 3: Return Periods of Year for 4m Height at Ooarai

periods are 430 to 580 years, is in the range of Tohoku Earthquake and is chosen in this paper for risk maps. Figue-4 shows the results of return period by the satisfied four distribution functions in Case-10. The representative distribution function became First Tippett-II of k=0.25 which calculate highest value in longer return periods . The range of tsunami heights in 1,000 and 10,000 years return periods are large as 4.8 m to 5.4 m and 9.5 to 15.6 m respectively.
The number of case which calculated the maximum height under than 15.6 m in 10,000 years is only two in all cases. The maximum height in Case-4 and Case-10 were 14.0 m to 14.4 m. The number of minimum height under 9.5 m was four cases.

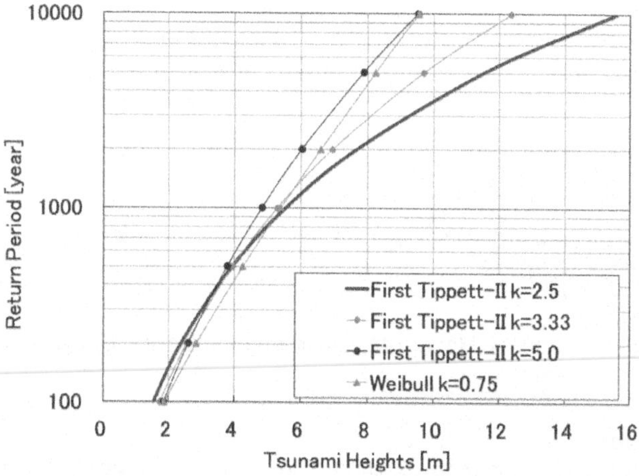


Figure 4: Tsunami Heights and Return Periods in Case-10 at Ooarai

**Probabilistic Tsunami Flood Maps**
The computed maximum and minimum flood ranges in 100, 1,000 and 10,000 years of return period in Case-10, it means six flood maps, are combined in Figure-5, where is small delta of Tone River which is southern border of Ibaraki Prefecture to Chiba Prefecture. A tsunami comes from right of figure, Pacific ocean side. The area painted green in the figure is no flood of land. The others painted in red to light blue are the probabilistic flood areas.
The minimum and maximum ranges in 100 years return period painted in red and orange are very narrow band along a part of coast, where ground elevation is lower than 3 m. It is hard to distinguish maximum and minimum ranges. The area in 1,000 years painted dark and right purple expanding in delta and plane area in Chiba Prefecture. In the north area in the figure, as there are sand dunes which height is 7 to 8 m along the coastal lines, tsunami floods stop at the dune, so the maximum and minimum range is not so different. In this type, we can recognize uncertainty of tsunami height at coast is large but uncertainty of those flood range is small, and can specify hazard range of 1/1,000 risk is maximum that for a government in this area. These are only one example because it depend on the relation of a local tsunami height and ground elevations. The range in 10,000 years painted in dark and right blue expanding to landward. A uncertainty is large as the maximum range is much

wider than minimum area. In this type, we should care the uncertainty for making a disaster making policy.

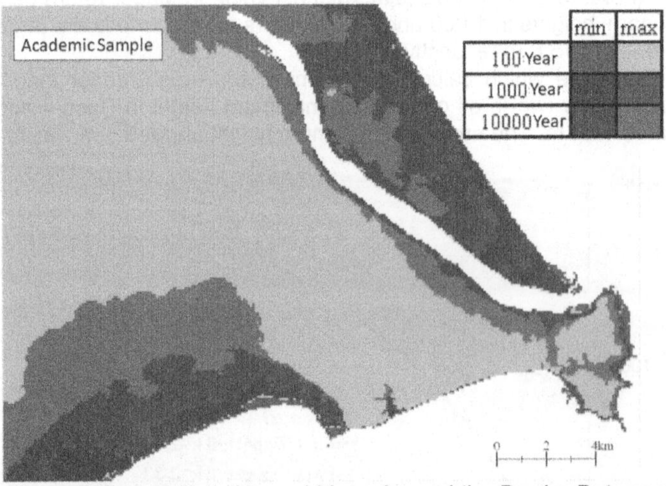


Figure 5: Probabilistic Hazard Maps Around the Border Between Ibaraki and Chiba Prefecture.

**Tsunami Risk Maps**

Figure-6 (a) is the probabilistic map of tsunami inundations located at plane area of Kuji River mouth where is north region of Ibaraki Prefecture. There is one nuclear power station at the just bottom-outside of this map. To show the many probabilities results, the results by representative tsunami heights in each probability only were painted in this figure. The fault slip of tsunami was discounted 75 % for the range in 500 years of return period becomes as same as flood in Tohoku Earthquake (Ibaraki Prefecture 2011). Local government has regulated new hazard range after Tohoku Earthquake(Ibaraki Prefecture, 2012), of which range closed to the range of 1,000 years return period in this figure.

The gross area of inundation in each return period is shown Fiure-6 (b). The flood range expands much from the 2,000 years to 5,000 years of return period. Figure-6 (c) shows function of the number of damage houses. The numbers of half and complicate broken houses increase during 500 to 2,000 years of return period. These distribution can be found from the house damage map in Figure-6 (d) and (e). The function of number of causalities shows in Figure-6(f) in the case if any people would not evacuate. The number of lost life increase much from 1,000 to 10,000. From these examples of risk map and function, we should recognize large potential damage remain outside of hazard range which had shown by local governments, as almost same as the results in 500 years of return period in these results. We should make constraint conditions and priority or specific target.

## Remarkable Summary

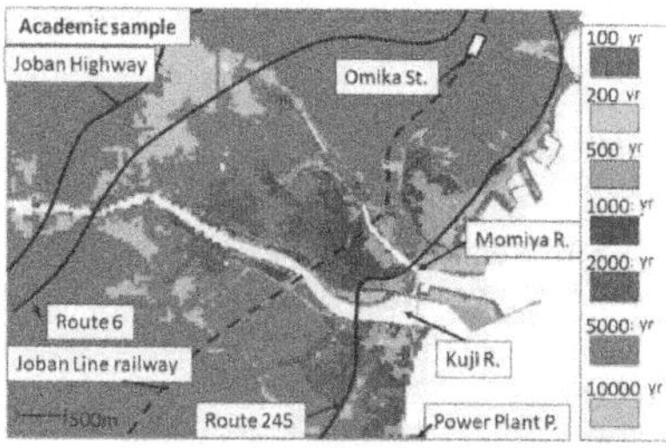


(a) Probabilistic Hazard Map and Return Period

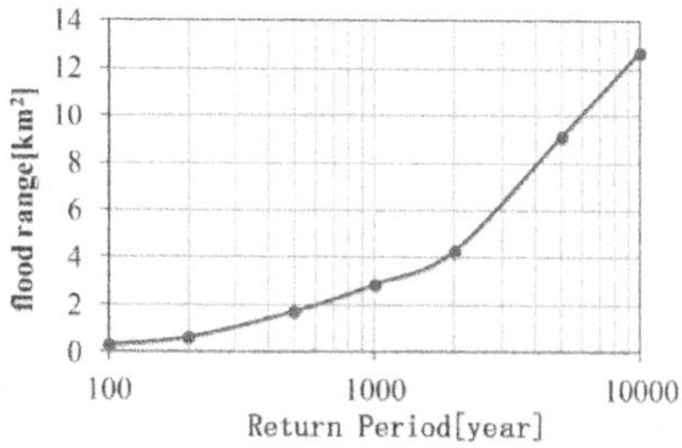


(b) Function of Flood Range

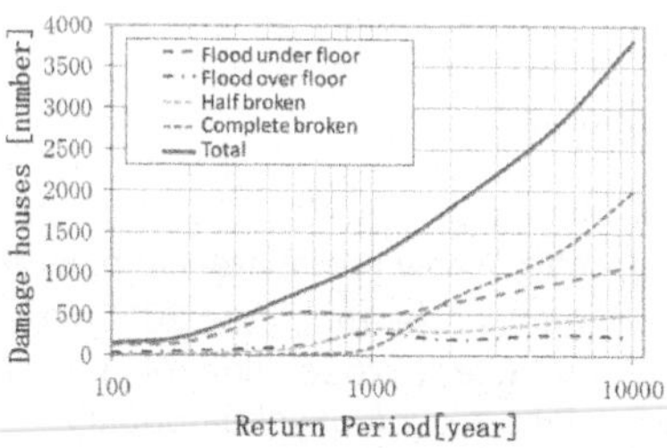


(c) Function of Damage Houses

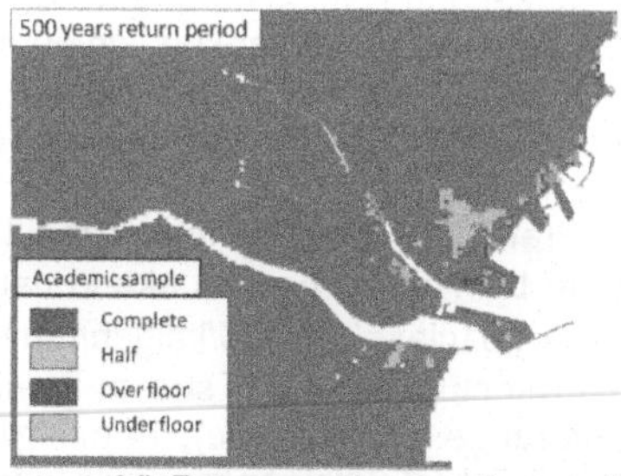


(d) Damage House Map on 500 yr Return Period

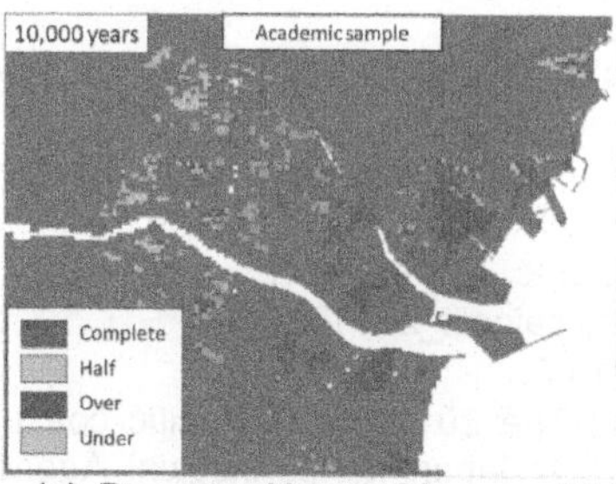


(e) Damage House Map on 10,000 yr r.p. and Return Period

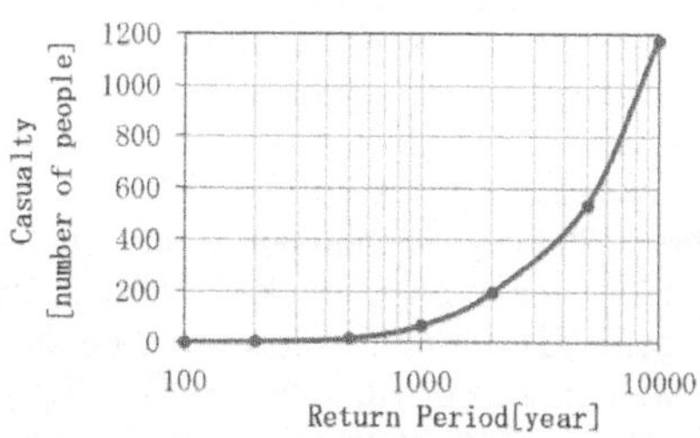


(f) Function of Causality

Figure 6. Tsunami Risk Map Around Kuji River, Ibaraki Prefecture

Though the problem about a setup of the period length of a sampling data for extreme analysis should be solved and an uncertainty and error which cannot be disregarded will remain, tsunami risk maps used extreme statistics analysis based on tsunami quantity records, have a possibility of being utilizable for measuring disaster reduction and a safe target on long-term plans.

Although improvement for long term is effective to reduce the unexpected phenomenon and those risk, it is important to share a risk in citizen. For that purpose, we should be clear the scientific relation between the present method and a tsunami hazard based on the probability of earthquake motions which has lack of information.

## Acknowledge

This work was supported by JSPS KAKENHI Grant Number 25350503 for Scientific Research (C). Local data of topography, bathymetry, houses and population were provided by Ibaraki Prefecture.

## Reference

Central Disaster Management Council (2003) Tou-Nankai Nankai Jishin ni kansuru Senmon-Iinkai (Paper on the assessment-of-damage technique concerning the Tonankai and Nankai earthquakes), http://www.bousai.go.jp/jishin/tonankai_nankai/pdf/sankou_siryou.pdf, reference date 2013-04-30.

Fujii Y., K. Satake, S. Sakai, M. Shinohara, and T. Kanazawa (2011): Tsunami source of the 2011 off the Pacific coast of Tohoku Earthquake, Earth Planets Space, 63, pp.815–820.

Ibaraki Prefecture (2011) Ibarakiken-Tsunami-Shinsui-Jisseki-Zu (Tsunami flood range by Tohoku Earthquake in Japanese), http://www.pref.ibaraki.jp/bukyoku/doboku/01class/class06/kaigan/tunamisinnsui.html, reference date 2015-01-05.

Ibaraki Prefecture (2012) Tsunani-Shinsui-Soutei-ni-tuite (Expectation of Tsunami flood range, in Japanese), 14 p. http://www.pref.ibaraki.jp/bukyoku/doboku/01class/class06/kaigan/tsunamisinnsui/l2shinsui.htm, reference date 2015-01-05.

Goda Y. (1988) On the Methodology of Selecting Design Wave Height, ICCE, No21, chapter 67, pp.899-913.

GodaY.(1990) Kouwan Kouzoubutu no Taiha-Sekkei (Design of Maritime Structures on wind waves in Japanese), Kashima Press, 333p.

Japan Meteorological Agency (2011) Saigai-ji JishinTsunami Sokuhou ;2011 Tohoku-chiho Taiheiyo Oki Jishin (Quick Report on Earthquake and Tsunami; The 2011 earthquake off the Pacific coast of Tohoku) . http://www.jma.go.jp/jma/kishou/books/saigaiji/saigaiji_201101/saigaiji_201101.pdf, reference date 2015-01-05.

Japan Meteorological Agency, (2012) Report on The 2011 of the Pacific coast of Tohoku Earthquake, Technical Report of The Japan Meteorological Agency, No.133, TOKYO (in Japanese), 479p.

Mito local Meteorological Office(1996): Ibraki no KIsho Hyakune (Weather Records during 100 years in Ibarak, in Japanese), 337 p.

Takeuchi. H., R. Fuji, N. Mimura, F. Imamura, K. Satake, Y. Tsuji, K. Hochi and T Matsuura (2007): Survey of run-up height of Empo Boso-oki Earthquake Tsunami on the coast from Chiba Prefecture to Fukushima Prefecture, Rekishi-Jishin (Historical Earthquake Studies), 22, pp.53–59. (in Japanese)

Tanaka S. and Kaoru Takara(1999) Goodness-of-Fit and Stability Assessment in Flood Frequency Analysis, Annual Journal of Hydraulic Engineering, 43, pp.153-156. (in Japanese)

The Headquarters for Earthquake Research Promotion (Japan) (2011) Long term evacuation of Seismic Activity from the off of Sanriku to Bousou (Second report)

, http://www.jishin.go.jp/main/chousa/kaikou_pdf/sanriku_boso_4.pdf (in Japanese), reference date 2015-01-05.

Watanabe, H.(1998) Nihon higai tsunami sōran (Comprehensive list of tsunamis to hit the Japanese islands in Japanese), 238p.

**Coastal Management: Changing coast, changing climate, changing minds**
ISBN 978-0-7277-6114-9

ICE Publishing: All rights reserved
doi: 10.1680/cm.61149.139

# Blue is the New Green – Harnessing Urban Coastal Infrastructure for Ecological Enhancement

**S. Perkol-Finkel**, SeArc – Ecological Marine Consulting LTD, Tel Aviv, Israel, and ECOncrete Tech LTD, Tel Aviv, Israel
**I. Sella**, SeArc – Ecological Marine Consulting LTD, Tel Aviv, Israel, and ECOncrete Tech LTD, Tel Aviv, Israel

## Introduction

With two thirds of the human population concentrated around coastlines (Creel, 2003), accelerated coastal development and changes to natural coastlines are inevitable. Throughout the years natural coastlines undergo severe changes, and in many cases are completely overtaken by man-made infrastructure such as ports, coastal defense measures, power plants, industrial facilities, and residential properties. The proliferation of hardened shorelines is also fuelled by processes related to global climate change, sea level rise and ongoing increase in the frequency and magnitude of extreme weather events.

As most marine flora and fauna reside in coastal areas, anthropogenic changes to coastlines are one of the key reasons for loss of coastal habitats and for changes in species assemblage, richness, and biodiversity (Dugan et al., 2011). While coastal infrastructure such as seawalls or breakwaters add significant amounts of hard substrate open to colonization by marine organisms, these man-made structures do not support similar species assemblages to those of natural coastal and marine habitats, and are often associated with nuisance and invasive species (Firth et al., 2014b). These differences are greatly associated to design and material features related to high inclination, low structural complexity, high homogeneity, and different artificial substrate properties, all of which are rarely found, or do not exist in natural habitats (Perkol-Finkel and Sella, 2014).

Concrete is one of the main construction materials globally, and in the marine environment it commonly accounts for over 50% of coastal and marine infrastructure (Kampa E and Laaser C, 2009). This widely used material is known as a poor substrate for biological recruitment and is often considered toxic to marine organisms, mainly due to unique surface chemistry which impairs the settlement of various marine larvae (Lukens and Selberg., 2004). The ecological enhancement measures presented hereby is based on the use of innovative ecologically active concrete technologies developed by ECOncrete® Tech Ltd that harnesses biological processes for creating environmentally and structurally improved infrastructure. These technologies increase the ability of concrete based coastal infrastructure such as seawalls or pier piles to supply enhanced ecosystem services, while improving their structural integrity and durability. This is achieved by slight modifications to the composition, surface texture, and macro-design of concrete elements (Perkol-Finkel and Sella, 2014).

To date, coastal and marine infrastructure have been designed and built with little or no consideration to marine life developing on them. As a result, the ability of these structures to provide ecosystem services similar to those offered by natural habitats is severely compromised. One of the key reasons that limits the ability of coastal infrastructure such as seawalls, riprap belts or breakwaters, to serve as surrogates to natural coastal habitats is the absence of design and substrate features which create environmental heterogeneity and water-retaining features (Firth et al., 2014a). Recent studies indicate that incorporating water retaining elements that functionally mimic natural rock pools have the potential to positively effect biodiversity in built environments (Brown and Chapman, 2014, Firth et al., 2013). We propose an innovative approach, of designing and retrofitting coastal and marine infrastructure using environmentally sensitive design and construction technologies that enhance their ability to provide valuable ecosystem services, while also elevating their structural integrity and longevity.

The overall goal of the enhancement projects presented below was to integrate ecological considerations into the design and construction works of an active waterfront infrastructure, thus decreasing its ecological footprint. The paper summarizes two case studies implementing ecological enhancement at the Brooklyn Bride Park (BBP) waterfront, (New York, NY, USA). The first is an example for structural repairs of aging infrastructure, i.e., deteriorated pier piles, by applying environmentally sensitive technologies for concrete encasement which provides required structural repair while generating valuable habitat and ecosystem services. The other is an example for increasing the ecological performance of a constructed riprap waterfront, by integrating precast tide-pools as a part of the riprap arrangement, thus adding water retaining habitat feature that are lacking from standard coastal infrastructure. Both strategies are examples for integration of principles of Reconciliation Ecology (Rosenzweig, 2003) in the design of coastal infrastructure that serve a structural/societal goal on one hand, and an ecological function on the other. Results are discussed with respect to ecological and biological benefits, structural performance and associated regulatory perspectives.

## Description of Case Studies

Brooklyn Bride Park (BBP) located at Brooklyn, NY, USA, is an 85-acre post-industrial waterfront site stretching 2 kilometers along Brooklyn's East River edge on a defunct cargo shipping and storage complex. The ambitious park design sought to transform a previously environmentally hostile site into a thriving civic landscape while preserving the experience of the industrial waterfront. Today, the park, composed of 6 renovated/retrofitted piers (Piers 1-6) and surrounded areas, provides green space for active and passive uses including playing fields, sport courts, playgrounds, lawns, a roller rink and a greenway and boat launches for non-motorized vessels. Unlike other waterfront parks, where visitors remain perched above the water, BBP encourages close interaction with the water.

As a part of the holistic environmental approach of the park, two different enhancement projects have been incorporated into the park's renovation plan. The first is an example for structural repair of aging infrastructure, i.e., deteriorated pier piles at Pier 6 (Fig. 1). This was done using an innovative technology of ecological concrete encasement which creates valuable habitat and ecosystem services. The other is an example for increasing the ecological performance of a constructed riprap waterfront, at the Pier 4 (Fig. 1), by integrating precast concrete tide-pools as a part of the riprap arrangement, thus adding water retaining habitat feature that are lacking from standard coastal infrastructure.

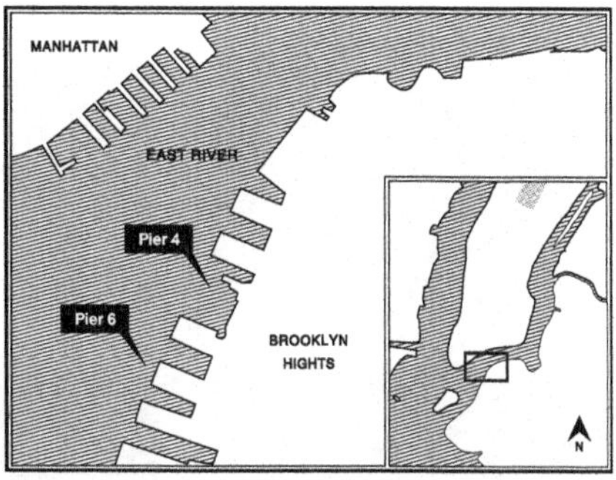


Figure 1: Location of the two enhancement projects at Brooklyn Bridge Park, NY, USA

**Ecological Pile Encasement – Pier 6**
Aging wooden pier piles, deteriorated by marine borers (mainly gribbles and shipworms like *Teredo* and *Limnoria*), require a structural repair in the form of concrete encasement or "jacket", for maintaining the load bearing structural properties of the pile. As opposed to standard pile encasement, utilizing Portland cement and left-in-place fiberglass forms (susceptible to long-term deteriorate), the ecological pile encasement utilizes an innovative concrete mix that enhances growth of marine flora and fauna (Perkol-Finkel and Sella, 2014). In addition, textured forms are applied and stripped after casting, imprinting a rough texture onto the surface of the concrete jacket, which inducing rich marine growth.

During June 2013, a total of 18 piles underwent concrete encasement using ecological pile encasement technology in BBP's Pier 6 nine piles at the south face of the pier and 9 on the face (Fig 1.). On-site, under water casting was executed by D'Onofrio General Contractors Corp. and Walker Diving - Underwater Construction. The ecological jackets illustrated in Fig. 2 provide all the functional and structural support required from a standard concrete encasement, yet with an added biological and ecological value.

All of the ecological jackets as well as three standard control jackets (Portland based concrete with fiberglass form) at each face of the pier, were monitored 3, 10, and 14 months post deployment (MPD). Monitoring included underwater photography and a careful on-site survey for generating a comprehensive species list (performed when water clarity was sufficient: 3 and 14 MPD), as well as sampling for biomass (performed 10 and 14 MPD). The latter was performed using small concrete tiles that were mounted upon the jackets using a strap (1.5 meter below the top of the jacket) allowing the tiles to be detached for laboratory analyses. The tiles were weighed prior to deployment and were reweighed after submersion (dry weight after 48h in 60°C) so that the amount of accumulated organic and inorganic weight could be measured accurately. Similarly, detachable fiberglass tiles were mounted onto control jackets and all recruited cover was analyzed for organic and inorganic weight.

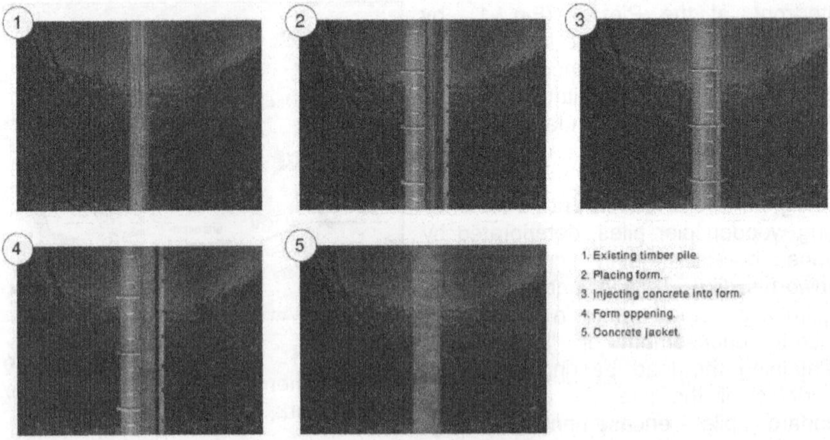


Figure 2: Schematic illustration of the ecological pile encasement system

Community data was analyzed using PERMANOVA tests (permutational analysis of variance - a non-parametric equivalent to univariate statistical tests) conducted on Bray Curtis Similarity matrix applied on raw data, while biomass data were conducted on Euclidean Distance Similarity matrix applied on raw data. Tested factors in both cases included Site (Random Factor: North vs. South), Treatment (Fixed Factor: ECOncrete vs. Control), Row (Fixed Factor: 1, 2, 3), and Season (Random factor: Spring vs. Summer). For community data, a non-parametric Multi Dimensional Scaling (MDS) was conducted in order to illustrate trends within the data. Each dot in the MDS represents a single sample, where samples with similar community assemblages cluster closely on the plot, while ones with different communities are spread further apart. Vectors representing the relative contribution of taxa to the similarity between samples were superimposed on the plot. All analyses were performed using the PRIMER V6.1.13 & PERMANOVA+ V1.0.3 programs (Anderson et al., 2008, Clarke et al., 2008).

**Riprap Enhancement – Pier 4**

Constructed beaches, revetments and breakwaters are often composed of boulders of quarry or dredged rock ("riprap") placed at the waterline to protect the coastlines and/or prevent erosion. While composed of rock, a natural material, such coastal defense measures are not surrogates to natural rocky habitats and often provide limited ecological value to the surrounding (see Dugan et al., 2011 and references therein). In order to enhance the ecological value and increase the biological productivity of riprap zones, a precast tide-pool that integrates into coastal defense structures such as breakwaters or riprap beaches was developed and tested (Fig. 3). The unit takes the place of standard armoring unit/rocks and fills up with water during high tide thus creating a stable habitat that provides shallow water retaining, moist niches that are absent from standard coastal infrastructure. Floating debris capturing is minimized due to an internal passive intake system.

Seven designed tide-pools were integrated during autumn-winter 2013-4 into the newly constructed Pier 4 beach at BBP within the mean higher high water (MHHW)

zone. Each pool retains a volume of 13 gallon and creates a submerged habitat disconnected from the open water at low tide.

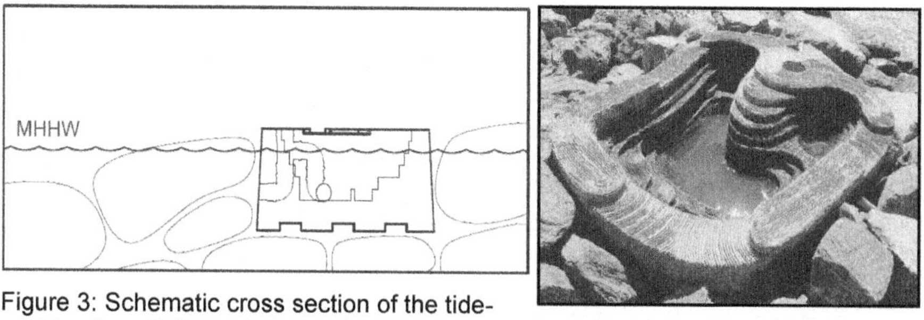


Figure 3: Schematic cross section of the tide-pool in place (left), and photo of the tide-pool on site (right)

A biological survey of flora and fauna recruited onto the tide-pools was conducted during the spring and late summer of 2014. Results are discussed in comparison to marine life recruited onto the riprap adjacent the pools.

## Results

The applied enhancement measures at BBP contributed to the park's ecological value by increasing the availability of bio-enhanced substrate capable of sustaining rich benthic communities and by adding new ecological niches, i.e., water retaining elements that are not to be found in the parks' infrastructure. Specific results of each of the case studies are detailed below.

### Ecological Pile Encasement

Already at the 3 MPD sampling the ecological jackets had live cover of 70-100% and a number of blue crabs (*Callinectes sapidus*) were observed mating on them. Control fiberglass jackets that present a smoother surface, exhibited only scattered colonization (20-50%). Similar results were obtained in the 10 MPD survey, with a dominance of filter feeding species (tunicates, barnacles, sessile polychaetes, sponges, and bivalves), and habitat formers (barnacles and sessile polychaetes) that contribute to biogenic buildup onto the substrate by calcium carbonate deposition. A shift in the community structure was noted on the final, 14 MPD monitoring, where ecological jackets presented a more diverse community than prior samplings (Table 1). Assemblages included coralline algae, sponges, gastropods, barnacles, colonial and solitary tunicates, bryozoans, and sessile polychaetes. Live cover on the jackets was 90-100% with the lower cover at the upper tidal area (Fig. 4). Control jackets presented scattered colonization with live cover of app. 40-85% with the lower cover at the upper tidal area (Fig. 4).

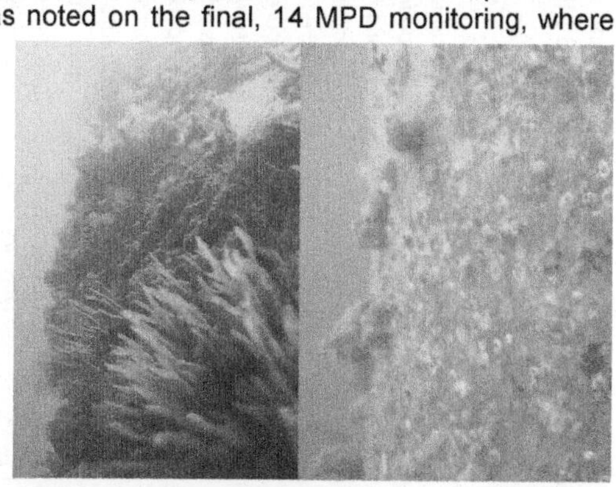

Figure 4: Ecological jackets (left) and control - fiberglass jacket (right), 14 MPD

Table 1: Species list of taxa appearing on the jackets 3, 10 and 14 MPD. Note that different sampling efforts were conducted at the different time points (visual census 3 MPD, on-shore tile analysis 10 MPD and both visual and tile analysis 14 MPD).

| PHYLUM | CLASS | SPECIES | 3 MPD visual census | | 10 MPD tile analysis | | 14 MPD visual & tile | |
|---|---|---|---|---|---|---|---|---|
| | | | *ECOncrete* | *Control* | *ECOncrete* | *Control* | *ECOncrete* | *Control* |
| **Porifera** | Desmospongiae | *Microciona prolifera* | + | | + | | + | |
| | | *Halichondria bowerbanki* | | | | | + | |
| | | *Haliclona loosanoffi* | | | | | + | |
| **Cnidaria** | Anthozoa | *Bunodactis stella* | | + | | | | |
| | | *diadumene leucolena* | | | | | + | + |
| **Bryozoa** | Gymnolaemata | *Schizoporella unicornis* | | | + | + | + | |
| | | *Membranipora membranacea* | | | | | + | |
| **Mollusca** | Bivalvia | *Mytilus edulis* | + | + | | | + | + |
| | Gastropoda | *Urosalpinx Cinerea* | | | | | + | |
| **Annelida** | Polychaeta | *Spirorbis* sp. | + | + | + | | + | |
| | | *Hydroides dianthus* | + | | + | + | + | + |
| | | Family: Sabellidae | + | + | + | + | + | |
| **Arthropoda** | Decapoda | *Callinectes sapidus* | + | | | | | |
| | Maxillopoda | *Balanus improvisus* | + | + | + | + | + | + |
| | Infra: Cirripedia | *Balanus amphitrite ?* | + | + | + | | + | + |
| | Isopoda | | | | + | + | | + |
| **Chordata** | Ascidiacea | *Botryllus schlosseri* | + | + | + | + | + | |
| | | *Molgula* sp. | | | + | | + | + |
| | | *Molgula citrina* | | + | + | + | + | + |
| | | *Molgula manhattensis* | + | + | + | + | + | + |
| | | *Didemnum* sp. | + | + | | | + | |

Communities on the control jackets consisted mainly of sessile polychaetes, barnacles, as well as colonial and solitary tunicates, A total of 19 species were noted in the final survey, of which 18 were found on the enhanced jackets and 9 on the control jackets (Table 1). These findings point to a general trend of maturing community with higher number of species with time, especially on the ecological jackets.

Statistical analyses of results from the tile sampling 14 MPD indicated of a significant difference between ecological concrete tiles and control fiberglass tiles (Site x Treatment interaction, P=0.001). Differences were slightly stronger at the South site (post-hoc comparisons). These differences are illustrated by the MDS for the 14 MPD sampling, where tiles of ecological concrete are clustered on the right side of the plot, while most of the control fiberglass tiles are clustered on the left. North jackets are clustered on the upper side of the plot, and South jackets are clustered on the bottom (Fig. 5). The superimposed vectors that represent the relative contribution of various taxa to the similarity between the samples, illustrate that fiberglass control tiles had a more dominant appearance of soft bodied colonial tunicates and tube worms (Sabellidae Polychaets), while ecological tiles had a more diverse community structure dominated by solitary tunicates, bryozoans, isopods, barnacles, and hard bodied Tube worms (Serpullidae polychaets).

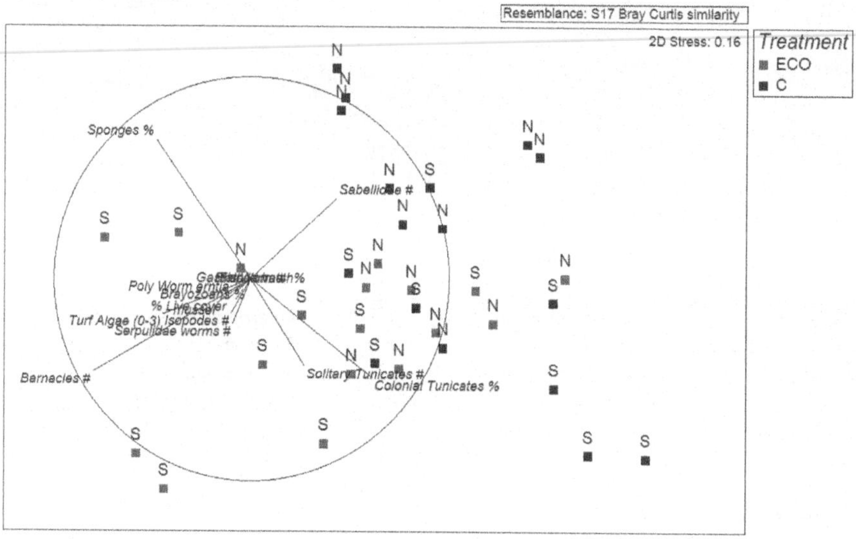


Figure 5: MDS of community data from tiles sampling conducted 14 MPD. Vectors are indicative of the relative contribution of the taxon to the similarity between the samples scattered in the plot. Ecological concrete tiles (ECO) are marked blue, Control (C - fiberglass) tiles are marked black. N = North Jackets, S = South Jackets

Biomass of the sampled tiles significantly differed between ecological and control fiberglass tiles (P=0.001 both 10 and 14 MPD). After 10 months, ecological concrete tiles had an overall average biomass accumulation of 0.083 gr/cm$^2$, more than a tenfold compared to control fiberglass tiles that had an average of 0.007 gr/cm$^2$. This trend persisted at a slightly lesser magnitude in the 14 MPD sampling during which ecological concrete tiles had an overall average biomass accumulation of 0.072 gr/cm$^2$, 3.5 folds greater than control fiberglass tiles that had an average of 0.021 gr/cm$^2$.

**Riprap Enhancement**

During August 2014, ap. 9 months post installation, a monitoring event of the 7 tide-pools was conducted. Tide-pools showed a live cover of 89-100% of the water retaining portions of the pools. Both tide-pool clusters presented a community composed mostly of filamentous green algae, branching brown algae, copepods, amphipods, isopods, as well as *sabellidae* and *Spirorbis* Worms (Table 2, Fig 6.). In addition, two individuals of the Harris Mud Crab (*Rithropanopeus harrisii*) and 17 individuals of an identified juvenile/post larval fish were noted.

Figure 6. Pier 4 tide-pools, BBP, NY August 2014

Table 2: Invertebrate species that colonized the tide-pools.

| PHYLUM | CLASS | TAXON (FAM/GENUS/SPECIES) |
|---|---|---|
| **Chlorophyta** | Chlorophyceae | *Enteromorpha* sp. |
| **Heterokontophyta** | Fucales | *Fucus* sp. |
| **Annelida** | Polychaeta | *Spirorbis* sp. |
| | | Sabellidae |
| **Arthropoda** | Malacostraca | *Rithropanopeus harrisii* |
| | Maxillopoda Subclass: Copepoda | *Acartia* sp. |
| | Maxillopoda Subclass: Amphipoda | *Ampelisca* sp. |
| **Chordata** | | Unidentified Juvenile fish |

The control area surveyed included the rocky area surrounding the pools at a distance of 1-2f from the edge of the pools. This area (at the MHHW) was very poor in biological findings, with only a few patches of algae, mostly at lower part of the rocks, where moist areas were noted (Fig. 7).

## Discussion

Results support the notion that ecological enhancement of hard coastal infrastructure is feasible, even in a heavily urbanized setting. The two case studies demonstrate the ability to harness active infrastructure such as pier piles and coastal protection measures for biological and ecological purposes without compromising their functional purpose, showing an increase in ecosystem services provided by the structure.

Figure 7. Pier 4 tide-pools. Red line marks the upper algae line. Yellow line marks the upper frequently wet tidal area. Arrows mark tide-pools

The enhanced jackets made of ecologically active concrete with a rough surface texture and complex design indeed enhanced the recruitment of marine organisms onto the jackets, creating a richer and more diverse habitat compared to control fiberglass jackets that offer very limited habitat value. Species richness on the enhanced jackets was double then the one found on the control jackets (18:9 respectively).The majority of the species recruited onto the enhanced jackets were filter feeders like tunicates, sponges, and bryozoans, capable of contributing to water quality and clarity in the area. In addition, many of the species dominating the enhanced jackets were habitat forming species such as barnacles, bryozoans and sessile polychaetes that add to the complexity of the habitat with time, provide food and shelter to fish and motile invertebrates such as crabs, which used the ecological jackets as nursing grounds.

Habitat forming species that deposit calcium carbonate ($CaCO_3$) skeletons onto the substrate like barnacles and serpulid worms are also considered ecosystem engineers (Jones et al., 1994), contributing to the modification of the habitat by means of biogenic buildup. The latter, also serves to protect the infrastructure via Bioprotection, sheltering the structure from weathering and erosion (Coombs, 2011, Coombs et al., 2013). It is important to note, the above mentioned biological assemblages that developed on the enhanced jackets did not interfere with the concrete encasement performance, and a hands-on (Level II) inspection of the encasements a year post deployment found hard and sound concrete (Dinos, 2014).

Initial results from the designed tide-pools indicate that the designed precast units successfully mimic conditions offered by natural rock pools, and thus are capable of extending the upper line of biological activity in the riprap area. The pools widen the wet tidal habitat while also extending the distribution of water dependent organisms in the area. We expect this trend to become even more significant with time, as community develops further and progresses into advanced successional stages.

## Recommendations

- As coastal population is expected to grow in the future, and with constant threats of sea level rise and increase in severity and frequency of extreme weather events, coastal development and application of coastal defense measures is bound to increase. In order to cope with global climate change and maintain healthy and productive coastal environments, it is crucial to integrate ecological considerations into the planning, design, and implementation of future hard coastal infrastructure and management schemes.

- Ecological enhancement of coastal and marine infrastructure is a feasible, scalable and effective means for reducing the ecological footprint of hard coastal infrastructure, by increasing their ecological value and biological performance. The case studies presented demonstrate the ability of implementing principles of ecological engineering (Bergen et al., 2001) for enhancing ecosystem services in urban marine environments.
- While "green" building standards such as the LEED system (Leadership in Energy & Environmental Design) are becoming more and more common world-wide, similar guidelines for "blue" coastal and marine infrastructure are only now spurring (see:Envision™, 2012, WEDG, 2014) calling for further research and development of innovative environmentally sensitive technologies.

## References

ANDERSON, M., GORLEY, R. & CLARKE, K. 2008. *PERMANOVA+ for PRIMER. Guid to Software and Statistical Methods. PRIMER-E, Plymouth UK, 214pp*

BERGEN, S. D., BOLTON, S. M. & FRIDLEY, J. L. 2001. Design principles for ecological engineering. *Ecological Engineering,* 18, 201-210.

BROWN, M. A. & CHAPMAN, M. G. 2014. Mitigating against the loss of species by adding artificial intertidal pools to existing seawalls. *Marine Ecology Progress Series,* 497, 119-129.

CLARKE, K. R., SOMERFIELD, P. J. & GORLEY, R. N. 2008. Testing of null hypotheses in exploratory community analyses: similarity profiles and biota-environment linkage. *Journal of Experimental Marine Biology and Ecology,* 366, 56-69.

COOMBS, D. W. 2011. *Biogeomorphology of Coastal Structures: Understanding Interactions Between Hard Substrata and Colonising Organisms as a Tool for Ecological Enhancement.* University of Exeter, UK.

COOMBS, D. W., NAYLOR, L. A., VILES, H. A. & THOMPSON, R. C. 2013. Bioprotection and disturbance: Seaweed, microclimatic stability and conditions for mechanical weathering in the intertidal zone. *Geomorphology,* 202, 4-14.

CREEL, L. 2003. Ripple Effects: Population and Coastal Regions *Population Reference Bureau, MEASURE Communication* [Online]. [Accessed September].

DINOS, G. 2014. BBP - Pier 6 ECOncrete Encasement Inspection. CH2M HILL PA, Report # 201997.

DUGAN, J. E., AIROLDI, A., CHAPMAN, M. G., WALKER, S. & SCHLACHER, T. 2011. Estuarine and coastal structures: environmental effects. A focus on shore and nearshore structures. *In:* ELLIOTT, M. & DUGAN, J. E. (eds.) *Treatise on Estuarine and Coastal Science* New York Elsevier

ENVISION™ 2012. A Rating System for Sustainable Infrastructure. The Institute for Sustainable Infrastructure Washington, DC

FIRTH, L. B., SCHOFIELD, M., WHITE, F. J., SKOV, M. W. & HAWKINS, S. J. 2014a. Biodiversity in intertidal rock pools: Informing engineering criteria for artificial habitat enhancement in the built environment. *Marine Environmental Research,* 102, 122-130.

FIRTH, L. B., THOMPSON, R. C., BOHN, K., ABBIATI, M., AIROLDI, L., BOUMA, T. J., BOZZEDA, F., CECCHERELLI, V. U., COLANGELO, M. A., EVANS, A.,

FERRARIO, F., HANLEY, M. E., HINZ, H., HOGGART, S. P. G., JACKSON, J. E., MOORE, P., MORGAN, E. H., PERKOL-FINKEL, S., SKOV, M. W., STRAIN, E. M., VAN BELZEN, J. & HAWKINS, S. J. 2014b. Between a rock and a hard place: Environmental and engineering considerations when designing coastal defence structures. *Coastal Engineering,* 87, 122-135.

FIRTH, L. B., THOMPSON, R. C., WHITE, F. J., SCHOFIELD, M., SKOV, M. W., HOGGART, S. P. G., JACKSON, J., KNIGHTS, A. M. & HAWKINS, S. J. 2013. The importance of water-retaining features for biodiversity on artificial intertidal coastal defence structures. *Diversity and Distributions,* 19, 1275-1283.

JONES, C. G., LAWTON, J. H. & SHACHAK, M. 1994. Organisms as ecosystem engineers. *Oikos,* 69, 373-386.

KAMPA E & LAASER C 2009. Heavily Modified Water Bodies: "Information Exchange on Designation, Assessment of Ecological Potential, Objective Setting and Measures" - Updated Discussion Paper. *Common Implementation Strategy Workshop Brussels, 39 pp.*

LUKENS, R. R. & SELBERG., C. 2004. Guidelines for Marine Artificial Reef Materials. *A Joint Publication of the Gulf and Atlantic States Marine Fisheries Commissions.*

PERKOL-FINKEL, S. & SELLA, I. Ecologically Active Concrete for Coastal and Marine Infrastructure: Innovative Matrices and Designs. *In:* ALLSOP, W. & BURGESS, K., eds. From Sea to Shore - Meeting the Challenges of the Sea, 2014. ICE Publishing, 1139-1149.

ROSENZWEIG M.L. 2003. *Win-Win Ecology How Earth's Species Can Survive in the Midst of Human Enterprise,* New York, Oxford University Press, Inc.

WEDG 2014. Waterfront Edge Design Guidelines. The Metropolitan Waterfront Alliance.

**Coastal Management: Changing coast, changing climate, changing minds**
ISBN 978-0-7277-6114-9

ICE Publishing: All rights reserved
doi: 10.1680/cm.61149.151

# Engineered Shoreline Protection – Case Study from the Jersey Shore

**A. J. Chew**, Arthur Chew Consulting, Linwood, NJ, USA
**C. Timpson**, TenCate Geosynthetics, Commerce, GA, USA

## Introduction

Shoreline protection is an important but confusing and costly endeavor for small municipalities. Which material should be used? What are the financial costs related to the project? What are the environmental impacts of the project? Can the project be permitted by the federal government and state government? These are some of the questions that every shoreline protection project must tackle.

This paper details the process that Ocean City, New Jersey went through to select a shoreline protection project for a vulnerable stretch of beaches on the north end of the island. The shore protection project was completed in the spring of 2011 almost 18 months prior to Hurricane Sandy. Ocean City is located south of Atlantic City and is 11 miles south of the landfall of the eye of Sandy. The paper will detail the impacts of Hurricane Sandy on the shoreline protection project.

## 1. Problem

The beaches of Ocean City have always eroded during the storms of the fall and winter. The eroded sand moves off shore then during the spring and summer the sand is pushed back onto the beaches with the gentle waves of summer. Periodically, the fall and winter storms will erode some of the dunes and this sand replenishes the beaches. This is a common cycle that occurs in all barrier island communities. When a series of storms occurs in quick succession, the shape of the barrier island changes as Mother Nature moves the sands. This would not be a problem if the barrier islands were left in their natural state but human desire to enjoy the beach and ocean has led to the development and construction of infrastructure on the barrier islands that must be protected.

In Ocean City, if all of the dunes are eroded, either bulkheads or stone revetments protect most of the island's ocean front properties, structures, roads and utilities except one area at the north end of the island. The 750 foot length of beach known as Waverly Beach which is located south of Surf Road and north of the Beach Road – East Atlantic Boulevard intersection does not have any structure between the beach and East Atlantic Boulevard as shown in Figure 1.

Figure 1: Google Earth Image of East Atlantic Blvd

From August to November of 2009, the City of Ocean City had been subjected to a series of five (5) storms (Hurricane Bill August 20, 2009, Hurricane Danny August 29, 2009, September 11, 2009 storm event, October 17-19, 2009 double nor'easter, and the November 12-14, 2009 nor'easter). These storms eroded almost all of the dunes on Waverly Beach as well as approximately a 500 foot long section of beach parking area. The November 2009 storm undermined approximately three to six (3-6") inches of a 100 foot long section of East Atlantic Boulevard.

## 2. Cause

While the beaches on the northern end of Ocean City have always been more prone to erosion and Waverly Beach was subject to significant erosion in the late 1970s and early 1980s, Waverly Beach has been subject to more significant erosion since 2000. One possible explanation for this erosion is that the beach is located in close proximity to the beach fill borrow area off the north east coast of Ocean City. The City has expressed this concern to the Army Corp of Engineers, who designs and constructs the beach fill projects for the area, and the Corp has been studying the problem. In the 2013 beach fill, the borrow area was expanded further away from Ocean City. In addition, requirements have been put on the contractor to remove less sand from the western edge of the borrow area. These two actions hope to reduce the rate of erosion on the beaches by leaving a natural sand bar off the coast to trip up waves that come from the north east.

The City of Ocean City recognized the problem of a lack of protective structure between the beach and East Atlantic Boulevard so in early 2006 geotextile containment units were installed as a core for the dunes on the 677 foot wide section of Waverly Beach. The geotextile containment units were gradually uncovered during the series of storms in 2009. When the tubes were revealed, it was clear that the geotextile containment units had been damaged in the three years since the installation. In many areas the geotextile containment units were cut and torn by a mechanically powered track vehicle or other mechanically powered vehicle. Indications are that the bulldozers from the 2007/2008 beach fill drove over the containment units and the blades cut the structures. Adding to the damage to the structure, the units were undermined and in one area pilings from an old wood groin punctured the geotextile containment units. The series of 2009 storms emptied the geotextile containment units of sand and the units became useless. The geotextile containment units were damaged beyond repair due to the random nature of the cuts in the units as well as the series of punctures.

Figure 2: Damage Due to Scour and Old Wood Groin Underneath

### 3. Initial Actions

The City needed to act to protect the valuable infrastructure (parking area, paved road, gas lines, water lines, sanitary sewer systems, storm water systems, overhead electric, cable, and telephone lines as well as million dollar homes) in this now very vulnerable section of the City. The Ocean City Public Works Department placed harvested sand along the former dune line of Waverly Beach to hold back high tides from entering East Atlantic Boulevard. Sand was collected from the streets as well as harvested from accretion areas on the northern and western beaches of the island. This temporary measure would not adequately protect the road from future storms but would prevent further undermining of the road surface and potential damage to the public utilities within the road from the twice daily advances of the high tide ocean waves. Failure to add more permanent protection in a timely fashion would have resulted in the loss of the road as well as the utilities within the road. This cost could have been up to $500,000 depending on the extent of the erosion as well as utilities being lost. This estimate does not include costs related to clean up. Since this damage would probably only occur in the event of another significant storm event, it is difficult to predict the likelihood of this damage occurring. It should be noted that the rate of erosion slowed dramatically when the erosion hit the compacted gravel of the parking area and roadbed.

Ocean City had to navigate the process for selecting the best shoreline protection method. The City engineering group had a wealth of local knowledge of the shoreline and ocean impacts upon it. In addition, the group had experience with geotextile products, stone revetments, and bulkheads of various materials. The area in question previously was protected by a geotextile product. The area immediately to the north of the area in question is protected by a stone revetment. The area to the south of the area in question is protected by a steel bulkhead. The southern end of the island has several miles of ocean front structures protected by a timber bulkhead.

The City's design professionals weighed the pros and cons of several products including rock groins, gabion baskets, rock mattresses, bulkheads of various materials, geotextile products and combinations of the aforementioned. Initial financial costs, life cycle costs, aesthetic impacts, length of permitting process, impacts to beachgoers, and environmental impacts including the projects' carbon footprint were all considered. During the selection process it was assumed that the beach would be restored by an Army Corp of Engineers beach fill project. The north end of Ocean City is part of a fifty (50) year beach fill project that receives beach fills every three years with a fortuitously previously planned fill scheduled to start in January of 2010 less than 2 month after the storm event. Below is a summary of the analysis by the design team. Estimates were based on similar projects within the state using prevailing wage rates. Pricing from other projects was obtained from New Jersey Department of Environmental Protection (NJDEP) construction office, Army Corps of Engineers (ACE), neighboring municipalities as well as manufacturers and suppliers of materials.

## 4. Solutions

Geotextile Containment Units

| | |
|---|---|
| Description: | Polyester or polypropylene geotextile product. A large fabric bag filled with sand. |
| Cost: | $666.67/LF - $500,000 |
| Useable Life: | 25 years |

Permit: No additional permitting needed. Existing beach and dune maintenance permit was used to install initially and can be used again. Install location above high tide line so no ACE permit needed. NJDEP Coastal Area Facility Review Act (CAFRA) rules permit up to 300 linear feet of geotextile containment units to be installed as shore protection.

Pros: Cost effective. Quickest solution as it can be completed under existing permit. One manufacturer offered the City of Ocean City an experimental armoring for the tubes at no additional cost to the City to defray concerns of damage to the structure. Product proven effective in Ocean City, Sea Isle and Atlantic City. Environmentally friendly since only one truck will be needed to haul material to site then use of onsite sand can be used to fill the units.

Cons: Lead time on materials could delay material supply. Lack of sand in intertidal zone due to beach erosion may delay project and project may not be able to start until the upcoming beach fill or sand can be trucked in from an inland quarry but this would increase cost significantly. Product can be damaged by track vehicles.

<u>Bulkhead</u>

Description: Vinyl or steel wall installed to elevation 9 NAVD88 to retain soils.

Cost: \$2,000/LF - \$1,500,000 plus engineering and permitting costs - \$1,750,000 total cost

Useable Life: 50 years

Permit: Install location above high tide line so no ACE permit needed. Under the NJDEP permitting environment at the time, it was and still is very unlikely that a permit could be obtained.

Pros: 50 year plus solution

Cons: Costs to build a bulkhead heavy enough to hold up to the ocean wave action will be significant. Bulkhead will cause increased scour at bulkhead tow if beach and dune are eroded. Larger carbon footprint due to manufacturing of materials plus trucking to the site.

<u>Gabions</u>

Description: Wire cages filled with softball to watermelon sized stones. Cages are one foot cubic and stacked in a pyramid construction.

Cost: \$333.33/LF - \$250,000 for a stack with a 6 foot wide base, then 5 foot wide second level, 4 foot wide third level, and a 3 foot wide top level. NJDEP bid results (\$500/LF) are for sub-

| | |
|---|---|
| | aqueous installation and thus significantly higher than our project. |
| Useable Life: | 25 years |
| Permit: | Install location above high tide line so no ACE permit needed. An emergency NJDEP permit would need to be obtained prior to bidding or starting of work. An emergency permit takes approximately 1 month to obtain. Then the City must apply for and obtain a full NJDEP CAFRA permit. |
| Pros: | Quick and easy to install |
| Cons: | Wire cages not proven to hold up in heavy surf. Metal cages will deteriorate over time in salt-environment. Cages can be damaged by mechanical equipment. If cages fail, rocks will be spread by wave action and this could lead to a difficult clean up. Potential hazard (trip and fall, foot cuts) if gabions become uncovered. It is unknown if an emergency permit would be considered with the pending beach fill the situation may not be considered an emergency. Public may not accept as this is an unknown product in Ocean City. Larger carbon footprint due to manufacturing of materials plus trucking to the site. |

<u>Stone Revetment</u>

| | |
|---|---|
| Description: | A jetty parallel to the shore line similar to those installed on Surf Road, 5th Street and 9th Street. |
| Cost: | $1,000/LF - $750,000 NJDEP bid results ($3,500/LF) are for sub-aqueous installation and thus significantly higher than our project. |
| Useable Life: | 100+ years |
| Permit: | Install location above high tide line so no ACE permit needed. An emergency NJDEP permit would need to be obtained prior to bidding or starting of work. An emergency permit takes approximately 1 month to obtain. Then the City must apply for and obtain a full NJDEP CAFRA permit. |
| Pros: | Most permanent option, proven effective, already exists in immediate area so public and lifeguards are familiar with product and how to manage interaction. Could be combined with geotextile products, gabions baskets, or rock mattresses to reduce costs. |
| Cons: | Not many experienced contractors to construct. Expensive. Potential hazard (trip and fall, foot cuts) if stones become uncovered. It is unknown if an emergency permit would be considered with the pending beach fill the situation may not be considered an emergency. Larger carbon footprint due to significant trucking of materials from Pennsylvania to the site. |

## 5. Decision

The City gave careful thought to all the pros and cons that each possible shore protection structure offered. Initially the elected officials were leaning towards installing a stone revetment since its life cycle costs were the lowest. Also, the stones were viewed as a permanent solution to the problem. Weighing in against the stone revetment was the desire to remove all hard structures from the beach. For the last decade, Ocean City has been working to remove storm water outfalls and other hard structures from the beach to make the beach safer for beachgoers and bathers. In addition, the city has had a significant green push with installing numerous solar panels, becoming one of the leading communities in New Jersey for biking, and building a new visitor welcome center that utilizes geothermal heating and cooling as well as solar energy. The lower carbon footprint of the geotextile units was viewed very favorably.

Ultimately Ocean City selected to install geotextile containment units with the mayor noting safety and environmental concerns. The project limits were expanded from just Waverly Beach. To the south the project would extend three hundred and fifty feet to connect into a steel bulkhead protecting a residential property. To the north the project would be set behind an existing stone revetment that was built in the 1940's but built to an elevation approximately five feet below the desired protection level. The north extension of the project would include an additional five hundred linear feet of tubes that would connect into a planned steel bulkhead by a residential property owner.

## 6. Design and Construction

During the design process there were serious concerns about damage to the initial geotextile tube installation and how to prevent future damage. The previous installation of tubes was damaged by track equipment during beach fill operations. In addition pilings from a previously installed wood groin were exposed from erosion during the storm and pierced the tubes.

To prevent damage from above during beach maintenance operations or debris hitting the structures in storm events, the geotextile tubes were covered in a protective shroud. The shroud consisted of a second layer of geotextile material. This extra layer of material not only would protect against damage from physical objects but also from the damaging effects of the sun if the tubes were not covered with sand. In addition to the protective shroud on the geotextile structure, the location of the structure was previously surveyed so their location could be added to future beach fill project plans. The location was also physically marked on adjacent bulkheads and stone groins so inspectors could easily delineate the location of the structures under the sand to keep beach maintenance equipment away from the area.

The other damaging factor to the first geotextile project was erosion under the previous tubes when the tubes settled and rolled to lower elevations. As the geotextile structures lowered in elevation, they were pierced by previously unexposed wood piling. After being punctured, the ocean wave action emptied the tubes of their sand filling. The specifications for the new installation required the contractor to excavate five feet below the elevation of the geotextile tubes for twenty feet on either side of the tubes to remove all debris and obstructions.

A scour apron was also included in the project and installed with a downward slope on the seaward edge of the geotextile structure to prevent erosion and undermining during storm events. The scour apron provided an additional five feet of vertical protection to the structure. A three geotextile tube pyramid was designed with four feet of sand dune installed on top. A cross section of the typical geotextile shore protection structure for this project can be found in Figure 3. Each of the tubes was designed to be six feet in height after being filled with sand so in total a seventeen foot high structure with the sand dune would provide twenty one feet of protection for the infrastructure behind it. During normal conditions, the beach would look and function as a natural recreational beach that would appear no different than the other beaches in the community but during extreme erosion and storm events the geotextile structure would become exposed and provide protection for the infrastructure and properties landward of the structure.

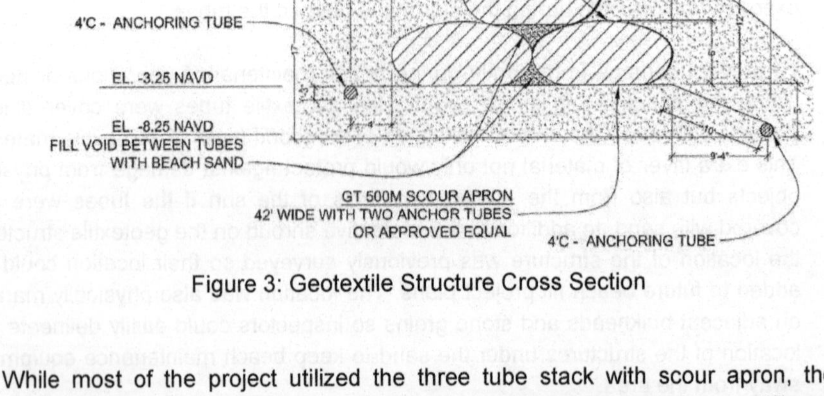


Figure 3: Geotextile Structure Cross Section

While most of the project utilized the three tube stack with scour apron, the northern most portion of the project on Surf Road Beach had to utilize a different approach. Surf Road Beach already had a stone revetment that was installed in the 1940s. The revetment was not providing the desired level of protection for the infrastructure landward of the revetment due to the low elevation of the top of the revetment. The location of the revetment prevented the installation of the three tube stack without removing the stone revetment or significantly changing the beach profile.

There were concerns that if the tubes were to be set directly on the revetment that wave action may cause holes to be worn in the geotextile material from movement during wave action. If this occurred the tubes would empty of sand and be rendered useless in protecting the island. Once again a scour apron was utilized. The sand on the landward side of the revetment was removed and the apron was set six to nine feet below the top of the revetment. Then a single tube was installed ten feet landward of the stone revetment with the bottom of the tube set one to three feet below the top of the stone revetment. The scour apron and anchor tube would encapsulate the sand below the apron and between the stone revetment and geotextile containment unit. Thus creating a combined stone geotextile shore protection structure (Figure 4).

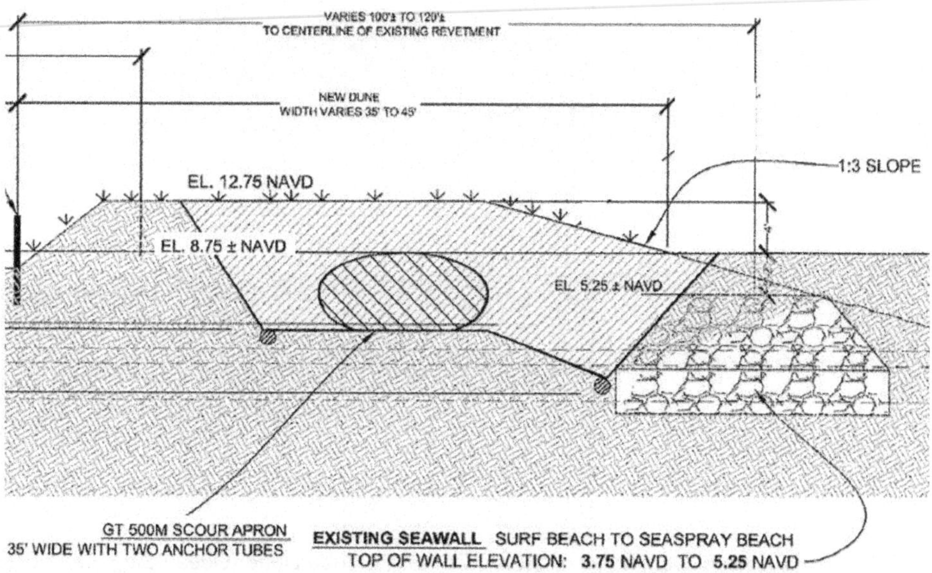


Figure 4: Combined Stone Geotextile Shore Protection Structure Cross Section

Another area of concern during the design was the tie ins to the two stone jetties that ran perpendicular to the shore line as well as the two storm water outfall pipes that were within the project limits. The design utilized smaller geotextile containment unit called pillows. While the tubes measured between 200 and 350 feet in length, the pillows were 10 feet in length. The geotextile tubes were designed to stop short of the jetties and outfall pipes then the pillows would be utilized to fill the remaining gap to the stone and under the pipe. A geotextile scour apron was laid directly on top of the stone revetment to reduce wear on the pillows. The pillows were utilized for the potential scenario where a hole develops in the geotextile then the pillow could quickly, easily, and cost effectively be replaced.

## 7. Results

After the November 12 to 14, 2009 storm, an ACE beach replenishment project began in January of 2010 which was less than sixty days after the storm event. The beach replenishment project was completed in June of 2010. The shore protection project was put out for public bid in September of 2010 with a notice to proceed being issued on November 12, 2010, exactly one year after the storm hit Ocean City.

Figure 5: Completed, Covered Geotextile Tube Dune (Ocean Side, Land Side)

The project was completed in the spring of 2011 almost eighteen months prior to Hurricane Sandy making landfall on October 29, 2012. Ocean City is located south of Atlantic City and is 11 miles south of the landfall of the eye of Sandy. While Ocean City did not see the storm surge that communities north of the eye saw, there was still a high water mark over six feet above mean high tide as well as seven to ten foot storm surge. Both were records for Ocean City. As with the November 2009 storm, Waverly beach suffered severe erosion. All of the dune on top of the geotextile structure was eroded and the top tube of the structure was left exposed after the storm.

Figure 6: Uncovered Dune after Storm

The condition of Waverly beach prior to Hurricane Sandy was very similar to the condition of the beach prior the 2009 storm, but in 2012 the infrastructure fared much better. The November 2009 storm was equivalent to a ten year storm and destroyed a parking area plus undermined the adjacent road surface. Hurricane Sandy was equivalent to a fifty year storm for Ocean City but did not damage the same parking area except for leaving it covered in sand and debris.

Figure 7: Performance of Engineered Dune

The geotextile tubes were not damaged during the storm event despite being bombarded with significant debris including numerous large rocks from an adjacent stone groin. The City was very happy with the performance of the geotextile product and would not hesitate to use the product again in the future.

**Coastal Management: Changing coast, changing climate, changing minds**
ISBN 978-0-7277-6114-9

ICE Publishing: All rights reserved
doi: 10.1680/cm.61149.163

# Adaptation to Coastal Cliff Instability and Erosion and Property Loss: Case study into the Knipe Point Cliff Retreat Pathfinder Project

**Robin P. Siddle**, Scarborough Borough Council, Scarborough, North Yorkshire, UK
**Stewart Rowe**, Scarborough Borough Council, Scarborough, North Yorkshire, UK
**Roger Moore**, CH2M HILL, Birmingham, England & University of Sussex, Brighton, UK

## Introduction

This paper presents a case study into the issues and role of adaptive management being implemented at Knipe Point, Cayton Bay, North Yorkshire, England, to counter the threat of coastal cliff instability on cliff-top property. Contrary to the wishes of the residents affected, construction of coastal defences and cliff stabilisation measures was deemed unaffordable and environmentally unacceptable at the site. As a result, Scarborough Borough Council (SBC) has sought to develop an 'adaptation' approach involving relocation of residents from at risk properties to a new site through a concept of partnership land banking and property development. A summary of the method of adaptation undertaken is explored along with the obstacles encountered and key lessons learnt.

The geomorphological setting and effects of past, present and future climatic influences is presented to provide context on the natural events and processes that have shaped the site in the past and into the future.

## Site Location & Background

Knipe Point is a small community located on the coast c.5km south of Scarborough and c.10km north of Filey in North Yorkshire (Figure 1). The settlement was developed on the northern headland of Cayton Bay in the 1980s. Pre-1980, Knipe Point was the site of a National Association of Local Government Officers (NALGO) holiday camp and consisted of temporary, mainly wooden chalets that were popular in the 1950's. In the early 1980's, 56 bungalows were constructed and given permission to be used as non-permanent holiday accommodation. A change to planning permission seeking permanent occupancy was subsequently submitted by the Knipe Point Management Company to SBC and was refused, but later successfully overturned on appeal. The site is predominately occupied by retired residents and a large majority of the bungalows are second homes. The 170ha woodland on seaward face of the site has been designated a Site of Special Scientific Interest (SSSI) by Natural England due to its geological and ecological features (English Nature, 1999).

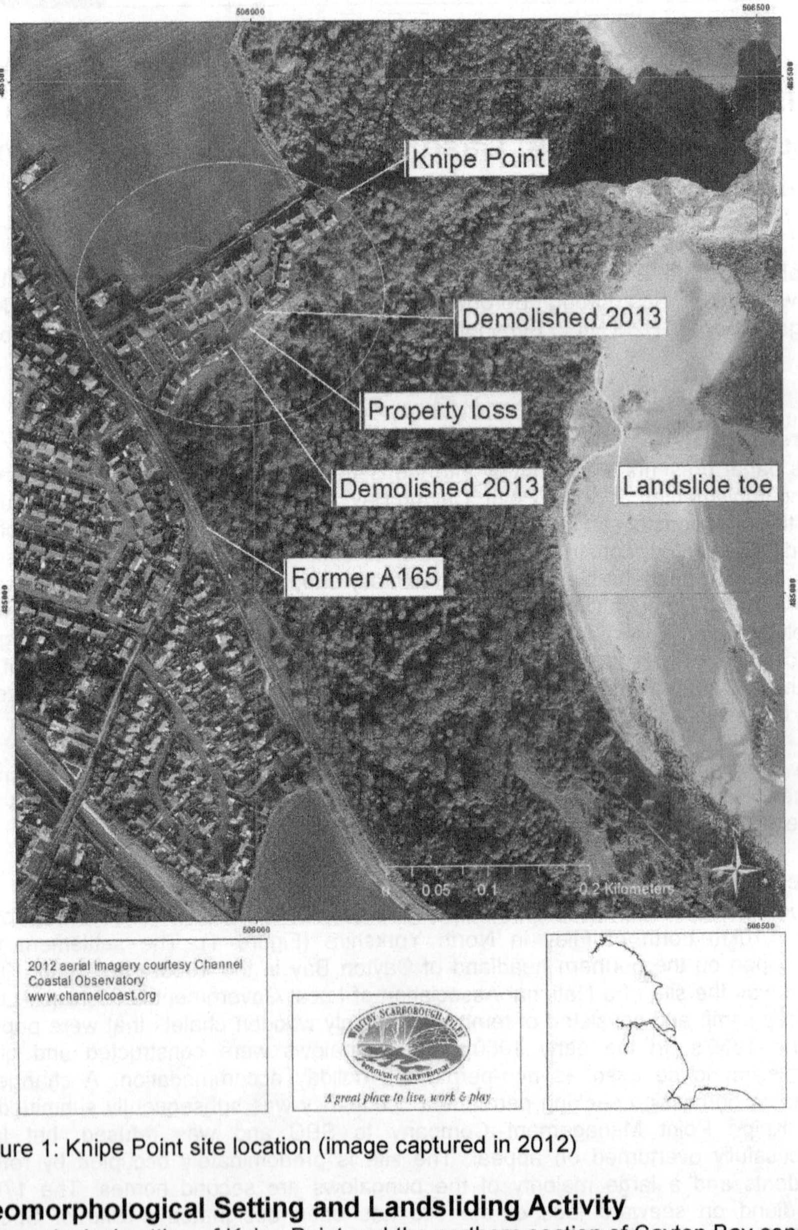


Figure 1: Knipe Point site location (image captured in 2012)

## Geomorphological Setting and Landsliding Activity

The geological setting of Knipe Point and the northern section of Cayton Bay consists of glacial sediments (till) of the Filey Formation overlaying faulted Jurassic rocks of the Lower Calcareous Grit and Oxford Clay Formations (Johnson et al, 2012).

Cayton Bay and subsequently the Knipe Point area has had a long history of both recorded and inferred landslide activity. The earliest historically recorded landslide events as noted by Johnson et al (2012) date from 1850-1890 when headscarp subsidence occurred near Osgodby, which is within the vicinity of the presently named Knipe Point estate. Episodic movements, tension cracks, mudslides and rear scarp recession have been recorded over the historical period and continue to the present day.

Two main types of landsliding have been identified that are presently affecting the community at Knipe Point. The first type of landsliding categorised was an active large mudslide complex in the superficial glacial deposits recognised through geomorphological mapping and interpretation (Fish et al, 2006). Coastal erosion at the toe of the mudslide complex was also noted. The second type of movement identified was a reactivation of a large deep-seated translational landslide system in the winter of 2007/08 covering an area 55m in width and 70m in length (Johnson et al, 2012). The date of the initial deep-seated landslide is unknown but it has been proposed that it possibly dates 'back to the mid-Holocene around 3000 years ago, when sea-level reached its approximate current position' (Fish et al, 2006, pg9). At this time of renewed sea-level incursion, rapid coastal erosion of the Oxford Clay and overlying Calcareous Grits and tills will have occurred, and coupled with seasonal excess ground water pressures, will have resulted in major landslides and development of the cliffs as seen today.

**Deep-seated Landslide Reactivation Winter 2007/08**

In early 2008, concerned residents notified Scarborough Borough Council of cracking and landsliding activity in the communal gardens fronting the seaward properties of the site (Figure 2). SBC inspected the site and subsequently employed Halcrow Group Limited (a CH2M HILL company) to conduct a Cliff Stability Assessment and Ground Investigation Report in order to appraise possible engineering stabilisation options.

Figure 2: Winter 2008 Image of Knipe Point (courtesy of SBC) showing the start of landsliding, which led to 3 properties being demolished

Prior to the completion of the report three residential properties were inspected by North Yorkshire County Council's building control and were deemed to be dangerous structures. These were subsequently demolished in the interests of public safety (Figure 3). It should be noted that the cost associated with the demolition of dangerous structures in the UK is met by the owner of the property. Recently however, and not in time for the demolition of properties at Knipe Point, DEFRA have made a sum of £6,000 available to aid covering the cost of demolition of individual properties affected by coastal erosion under the Coastal Erosion Assistance Grant.

Figure 3: Winter 2008 Image of Knipe Point (courtesy of Oakley Plant Hire Ltd) showing demolition of the first 3 properties

The Halcrow Group Limited report (2009) highlighted the main features and impacts of the deep-seated landslide reactivation and confirmed the depth of failure coincident with the base of the Oxford Clay strata. This included cliff top retreat, large tension cracks, major ground movements and uplift of the beach fronting Cayton Cliff (Figure 1). It was also revealed that human provoked effects on stability were doubtful and that it could 'be fully explained as a natural consequence of the unique ground conditions, groundwater regime and episodic nature of deep-seated coastal landsliding' (Halcrow, 2009, pg48)

Two independent water tables are active below Knipe Point, the first of which is a perched water table within the overlying glacial sediments and the second is deep within the bedrock. Slope stability modelling undertaken during investigations in 2009 identified through sensitivity testing that both the deep-seated and shallow landslides are especially sensitive to increases in groundwater levels. It was found that a 1m increase of groundwater in the upper water table is sufficient to reduce the factor of safety from marginally stable to unstable. The deep-seated failure requires a rise of groundwater in the lower water table of 3m to move the factor of safety to unstable. It was concluded that the deep-seated slide was highly sensitive to fluctuations in short term and seasonal groundwater levels (Halcrow, 2009).

UK Climate Projections 2009 (UKCP09) for changes in precipitation within Yorkshire and Humber under the medium emissions scenario with a 90% probability demonstrate that by the 2050's Knipe Point could expect to see the following:

- +25% increase in winter precipitation
- +1% increase in summer precipitation

UKCP09 also present a wider range scenario to account for the highest and lowest values of change seen in the low, medium and high emissions cases and all three probability levels of 10, 50 and 90%. The predictions given indicate:

- 0% to +27% increase in winter precipitation
- -37% to +9% increase in summer precipitation

(Murphy, et al, 2009)

Rainfall records collected prior to the landslide at Knipe Point, during the summer of 2007 demonstrate a 13% increase in precipitation compared to the previous 12 year average. This led to above average ground water levels during the winter of 2007/2008, which in itself, experienced an above average rainfall of +89% in January and +28% in March 2008 (Johnson et al, 2012). The increased amount of ground water is therefore hypothesised as being the main trigger for deep-seated and shallow movement at Knipe Point. With predictions for increased summer and winter precipitation under the medium emissions scenario due to climate change as highlighted in the UKCP09, it is highly likely that more frequent ground movement and landslides will continue to occur at Knipe Point putting more properties and lives at risk.

In the winter of 2012/13, three more properties at Knipe Point had to be demolished following a sustained wet summer and early winter period resulting in a build-up of antecedent rainfall triggering reactivation of the landslides (Figure 4).

Coastal erosion is also affecting the cliffs below Knipe Point along with the landslide complex as a whole through cut-back and lowering of the landslide toe on the beach. The primary cause of the landslide reactivation in 2008 has been attributed to excess groundwater acting within the cliffs. Although not directly a trigger of the landslide, coastal erosion acting over many tens, hundreds and thousands of years in the past will have prepared the cliffs for failure. With the current rate of coastal erosion and taking account of sea level rise, due to climate change, it has been predicted that within the next 50 to 100 years toe erosion of 40m is possible (Halcrow, 2009).

Figure 4: Left image showing the properties demolished at Knipe Point in 2013, right image showing where the properties once stood after demolition was complete, images courtesy of SBC

## The Department for Environment, Food and Rural Affairs (Defra): Pathfinder Programme

In 2009 Defra put out a call for bids under the Coastal Change Pathfinder Programme to test 'new and innovative approaches to planning for and managing coastal change' (Regeneris Consulting, 2011, pg1). Scarborough Borough Council on behalf of the community of Knipe Point, submitted a concept bid to the programme and were successful in securing £1,022,500 of funding to trial the proposal. Previous attempts by SBC to develop and promote an engineering solution to the problems were unsuccessful due the high cost of a potential solution outweighing the asset benefits. This was compounded with the judgement that any proposed scheme would have had a negative impact on the environment and SSSI.

The Pathfinder bid submitted involved sourcing and purchasing land as a form of land banking, in order to relocate residents and properties to a site within the proximity of Knipe Point, but not at risk of coastal erosion or land instability. The concept of the land bank solution was devised in order to address an imbalance within property insurance. Currently in England, insurance is only offered against physical property loss and not the land on which it occupies. This form of insurance works well in cases such as property loss due to flooding or fire, as it is possible to rebuild upon the same plot, however with cliff retreat due to erosion and landsliding the land is partially or wholly lost so a rebuild is either not possible or desirable given the risk. It is important to note that it is not possible to insure against coastal erosion in England.

The Pathfinder grant received was bound to a concept of proving replacement land and services for affected residents to then use their insurance pay-outs to reconstruct their homes. The partnership concept of land banking adaptation and homeowner rebuilding has not previously been attempted in England.

The land bank adaptation solution has involved the valuing, sourcing and negotiations to purchase a large plot of land to relocate the 15 most immediate at risk properties identified within the Halcrow 2009 report. The land, when sourced, was originally planned to be purchased by SBC with Pathfinder monies, and then conditional terms and conditions imposed on the site and its developments to stop potential profiteering. It is important to note that this method is not seen as compensation of loss to homeowners but instead as a way to adapt to the impacts of coastal change and instability, which is set to worsen with the effects of climate change.

The Knipe Point Pathfinder project was undertaken in a manner that allowed flexibility and development of the original scope in order to bring the adaptation concept into fruition. The initial bid was only intended to encompass properties that were the sole residence of the homeowner and not a holiday home or letting property. Upon consultation with the Knipe Point Residents Association (KPRA) this was overturned as it was deemed unreasonable, as if a coastal defence scheme had been acceptable this would have benefited all homes including second homes.

The first stage of the project involved an independent valuation of the land on which the Knipe Point properties were situated. This gave a land valuation of £45,000 per plot in order to gauge the sum available to purchase a new site. A demolition sum of £5,000 per property was also made available to those not covered for this, within their insurance policies.

The second stage involved the sourcing of land available within the locality of Knipe Point through the Council's Housing Allocation Development Plan, taking account of the sum available for land purchase and the servicing requirements. Unexpectedly, only 3 sites out of 49 were found to suitable as most were either too small or not for sale. Of these 3 sites a preferred site was selected by the Council and offered to the participating residents. This offer was subsequently rejected by the Resident Association on the grounds of an unwillingness to be part of a large housing estate, which included social housing that was too close to a school and too far from Knipe Point and the local hospital. Following this rejection SBC tasked the Residents Association to source a site that would be suitable for them. Three sites were proposed by the Residents Association, all of which had issues of suitability from the Council's perspective. One site was at risk of coastal erosion and another was considered an unacceptable expansion of Scarborough Town by the planning department. The third site, Saxon Park, which was not without planning obstacles, was deemed to warrant further investigation.

Upon supplementary negotiations with the Knipe Point Residents Association they agreed to meet the cost of trying to change the planning constraints on Saxon Park from one which only allowed holiday accommodation not exceeding one month, to a 12 month permanent residential occupancy permission. This was a time consuming process that involved many meetings with the local planning authority. Given the plight and mitigating circumstances the community was facing, an agreement was made with the planners. A permanent occupancy was granted to the identified at risk properties, however a caveat was imposed whereby should the new property be sold the occupancy would revert back to the original condition, unless it was bought by another resident of Knipe Point following the loss of property.

SBC agreed that an independent solicitor appointed by the Knipe Point Residents Association would handle the negotiations of purchasing and serving the site with the land owner. The purchasing discussions took much longer than expected and have caused some delay to the project. Completion of the purchase of the site was undertaken in late January 2015. It is anticipated that identified residents will begin to move to the site once insurance pay-outs have been made. As of April 2015 no residents have commissioned or begun construction work at Saxon Park. Residents have been made aware of lead in times for the construction of new builds and have been advised by SBC to rebuild them as soon as reasonably practical. New homes built prior to the loss of property at Knipe Point will alleviate the need for the Council to provide emergency accommodation upon loss, therefore reducing cost to the Council and freeing up property for others in need.

It has been considered and agreed by SBC that should the residents rebuild now in advance of loss, it will be acceptable to rent their properties out at Saxon Park, in the short-term until loss of their property at Knipe Point. This was considered reasonable by the Council due to the reinvestment of their insurance pay-outs to rebuild their properties.

**Adaptation in Practice: Obstacles and Lessons Learnt**

A variety of issues and lessons have been learnt through this case study. While some of these will reflect local conditions, many will be relevant for other sites around the UK facing similar impact of cliff retreat on property and communities. A number of obstacles have been encountered and overcome through the development and progress of the Pathfinder project. The summary below highlights in no particular order a number of lessons and obstacles faced by the project:

- Land to build new properties on is not readily available or affordable, especially when looking for a plot to build 15 new properties. Recommendation that Planning Departments/Councils allocate and set aside land in advance of property loss for at risk properties as the process is very time consuming.
- Homeowner reluctance to accept the risk of property loss and not wanting to move. Possibly due to the age of the residents and the fact that most properties are second homes.
- Negative media reporting, with relation to causing and adding distress to other areas of risk within the Borough where the Pathfinder project is not in place. Also negative media coverage of project progress, causing a lack of public support.
- Offer of a free plot of land being turned down by residents due to them being unhappy with the selected location. Threat of legal challenge by the Residents Association over site location if not changed.
- An elderly community not wanting the hassle and stress of having to rebuild property came to light.
- Debates over which owner should get what plot on the new site, and some residents noting they should get a larger plot or even two plots due to the size of their current land holding. Signs of residents feeling they are owed and had a right to the land emerged throughout the process.
- Issue around community cohesion and difficulties between the Council and residents with respect to residents being unhappy with the perceived slow progress of the project (Regeneris Consulting, 2011).
- Second homes should be included in any future land bank project in line with DEFRA's concept of better protecting homes against coastal erosion, therefore avoiding discrimination.
- Keeping the community together was a goal of the Council, however upon consultation with the community they were not opposed to being split up. This was possibly due to the large number of second homes.
- Ability for the Council to liaise directly with the Residents Association was important, however having also carried out private meetings with residents it was found that the Association did not always express the views of everyone in the community. Often individual residents felt intimidated by the Association and did not want to speak up in disagreement through concern over social exclusion.
- The appointment of an impartial solicitor to negotiate the purchase and serving of the proposed new site has aided trust and better relationships between the Residents Association and the Council.
- Residents have different insurance policies and pay-out amounts, as a result some companies pay-out before loss and some after loss.

- Too many options were given to the community, which led to project scope creep. In hindsight, keep it simple.
- Over time residents felt that the Council was inflicting something on them and wanted someone to blame. Some even felt offended by the sum of money offered to them to buy a new plot of land.
- Originally the Council placed many restrictions on the land that would be allocated to residents; however these were subsequently dropped, due mainly to the fact that residents were contributing the majority of the cash needed to rebuild their property.
- Increased awareness of coastal erosion and cliff instability within the wider community.

## Conclusion

Current and predicted increases in precipitation and sea level rise leading to increased water tables and coastal erosion are affecting many coastal settlements within the UK and are set to get worse in the future. The experimental Pathfinder project demonstrated at Knipe Point clearly shows it is possible to adapt to the impacts of cliff instability and erosion through other means than solely hard engineering. The partnership land bank solution has taken a long time to deliver, but will ultimately remove 15 properties from an at risk area with no hope of defences to a new safe site.

Many obstacles and lessons have been learnt, developed and overcome throughout this project. It is notable that the residents at Knipe Point are able to contribute to the rebuilding of their properties, through insurance pay-outs on land granted to them, which would not be possible if the problem was solely coastal erosion driven due to insurance limitations. In principle this method of adaption would also be viable elsewhere in the UK, if people in at risk areas could privately fund the development on land provided to them through a government grant. Unfortunately this approach to rollback has yet to be accepted and sponsored by Defra.

The ability to source available and suitable land to relocate at risk properties was seen as one of the most challenging aspects of the project. Within the Borough of Scarborough, building land is a very scarce commodity and attracts a premium price. It is recommended that should land banking become an approved and grantable policy, land is set aside in advance of property loss by the local planning authority. This will speed up the process of relocation and reduce the distress on affected residents and remove complications around site suitability.

## References

English Nature. 1999. *Saltburn to Bridlington Coastal Natural Area Profile.* www.naturalareas.naturalengland.org.uk/.../natural/profiles%5CnaProfile

Fish, P., Moore, R., Carey, J.M. 2006. Landside Geomorphology of Cayton Bay, North Yorkshire, UK. *Proceedings of the Yorkshire Geological Society*, 56(1), 5-14.

Halcrow Group Limited. 2009. *Cayton Bay Cliff Stability Assessment Ground Investigation and Appraisal of Engineering Stabilisation Options.* Technical Report Prepared for Scarborough Borough Council.

Johnson, R.M., Fish, P.R., 2012. Reactivation of the coastal landslide system at Cayton Bay, North Yorkshire, UK. *Proceedings of the Yorkshire Geological Society*, v.59; p77-89.

Murphy, J.M., Sexton, D.M.H., Jekins, G.J., P.M., Booth, B.B.B., Brown, C.C., Clark, R.T., Collins, M., Harris, G.R., Kendon, E.J., Betts, R.A., Brown, S,J., Howard, T.P., Humphrey, K.A., McCarthy, M.P., McDonald, R.E., Stephens, A., Wallace, C., Warren, R., Wilby, R., Wood, R.A. 2009. *UK Climate Projections Science Report: Climate change projections*. Met Office Hadley Centre, Exeter.

Regeneris Consulting, 2011. *Coastal Pathfinder Evaluation: An Assessment of the Five Large Pathfinder Projects*. Report Prepared for DEFRA.

Coastal Management: Changing coast, changing climate, changing minds
ISBN 978-0-7277-6114-9

ICE Publishing: All rights reserved
doi: 10.1680/cm.61149.173

# Sea Level Rise Impact Assessment and Mitigation Alternatives Development for Balboa Islands, City of Newport Beach, California

**Ying Poon,** Everest International Consultants, Inc., Long Beach, USA
**Robert Stein,** City of Newport Beach, Newport Beach, USA

## Introduction

Newport Beach, incorporated in 1906, is an affluent city in Orange County, California. The population was 85,287 at the 2010 census. Newport Beach is also home to Newport Harbor. Newport Harbor is a semi-artificial harbor that was formed by dredging Newport Bay estuary during the early 1900s. Several artificial islands were built, including Balboa Island and Little Balboa Island (islands) which are now covered with private homes.

Figure 1: Site location

The City of Newport Beach (City) has been dealing with localized flooding for several decades due to lower-than-optimum seawall and land mass heights in various locations around Newport Harbor. Flooding of streets and walkways has occurred on Balboa Peninsula, Balboa Island and other localized areas as a consequence of high tides and waves. The challenge of flood control is compounded by City storm drain lines that empty by gravity into the Bay and therefore do not provide flood relief when the bay water level is high. To a lesser extent, distress in the form of concrete cracks and construction joints allows water to breach the protection of these walls and contribute to localized flooding. To prepare for sea level rise and potential flood damages, the City initiated a study to assess the potential flood impact to Balboa Island and Little Balboa Island. The study includes: 1) an evaluation of the islands' seawall condition, 2) prediction of flood water overtopping and resulting inundation due to sea level rise and storm events over the next 100 years, and 3) evaluation of flood hazard mitigation alternatives. Details of the study can be found in Everest (2011).

Using the findings and recommendation of the study, a key City committee composed of City Council members and interested residents appointed by City Council, asked City staff in 2011 to frame the challenges and suggest options for protecting property and facilities around Newport Harbor. Meeting agenda topics systematically explored various sea level rise mitigation options and progressively delved more deeply into the more promising engineering and operational options. Once a general consensus had been reached with committee members, a series targeted outreach programs with affected residents was launched.

## Balboa Island and Little Balboa Island

Balboa Island and Little Balboa Island, adjacent islands located in at the eastern end of Lower Newport Bay (Figure 1), are encircled by concrete seawalls that provide protection from flooding during periods of high tides and waves. Wave overtopping and flooding do occasionally occur on Balboa Island during high tide and high wave events but in general do not cause damage to residences, businesses, vehicles or infrastructure. The higher seawalls around Little Balboa Island protect it from seawater overtopping.

Figure 2: City personnel pumping storm water

During high water events, the City closes the storm drain outlets into Newport Bay to prevent sea water from flowing back through the storm water outlets into the streets. With the storm drain outlets shut, storm water is trapped in the street. Figure 2 shows City personnel pumping out water that was collected at the storm drain inlet located at a bay-front street during a high water, high wave event in December 22, 2010. During high water and high wave events, there are four areas on Balboa Island that are potentially susceptible to flooding: the Balboa Island Ferry Boat landing and its surroundings, and access points to the three bridges on the island at Marine Avenue, Park Avenue and Collins Island. Extension or reconstruction of the seawalls may require design modifications at these four areas to prevent them from acting as openings in an otherwise solid seawall fortification around the island.

## Flood Modeling

An unstructured, variable grid, Godunov-type finite volume model, based on the shallow water equations, BreZo, was applied to Newport Harbor to map present-day (2010) and future flood zones on Balboa Island and Little Balboa Island. The numerical analysis considered projected sea level rise for 2025, 2050 and 2100 along with various protection measures. The model was developed to account for both wave- and tide-driven sea wall overtopping. Moreover, overtopping flows are predicted to spread in accordance with local topography and friction effects. Detailed descriptions of the Brezo model used for this study can be found in Sanders (2008) and Schubert *et al.* (2008).

A variable-resolution computational mesh was developed for Newport Harbor. The mesh was locally refined on Balboa and Little Balboa Islands for flood mapping precision, an intermediate resolution was used within the embayment, and a coarser resolution was used offshore. The finest and coarsest resolution corresponds to approximately 10 feet and 1000 feet, respectively. In total, the mesh consisted of 282,935 cells. Several datasets were relied upon to implement BreZo including topographic and bathymetric data, seawall and house first-flood elevation data, ocean tide height data (including sea level rise projections), and wave-driven overtopping rates.

## Water Level Description

Extreme high tide scenarios are needed to support hydraulic modeling of coastal flooding and plan for sea level rise. The hydraulic model uses Equation 1 to simulate the rise and fall of the ocean height, *h(t),* which in turn causes the tide to rise in the embayment and potentially flood low lying areas.

$$h(t) = H_0 + a\cos\left(\frac{2\pi(t-t_1)}{t_2}\right) + b\,\mathrm{sech}^2\left(\frac{t-t_3}{t_4}\right) \qquad (1)$$

In this equation, $t$ represents time, $H_0$ represents mean sea level elevation, $a$ represents the amplitude of a periodic wave with phase $t_1$ and period $t_2$, and $b$ represents the amplitude of a solitary wave with phase $t_3$ and duration parameter $t_4$. Equation 1 allows the hydraulic model to account for increases in sea level and the probability of an extreme high tide. The value of $H_0$ is varied by year to account for higher sea levels projected for the future. The remaining tide-height parameters (a, b, and time constants) are specified to simulate either a 10 percent or 1 percent annual exceedance probability tide. Two sources of information were merged to parameterize Equation 1: hourly tide measurements and sea level rise projection.

Over 80 years of hourly tide measurements at Los Angeles (NOAA Station ID: 9410660) were analyzed for this study. Using the hourly record, the mean and maximum value of the tide height from each year was computed and a linear model was least-squares fitted to each of the time series. These tide heights and trends are shown in Figure 3.

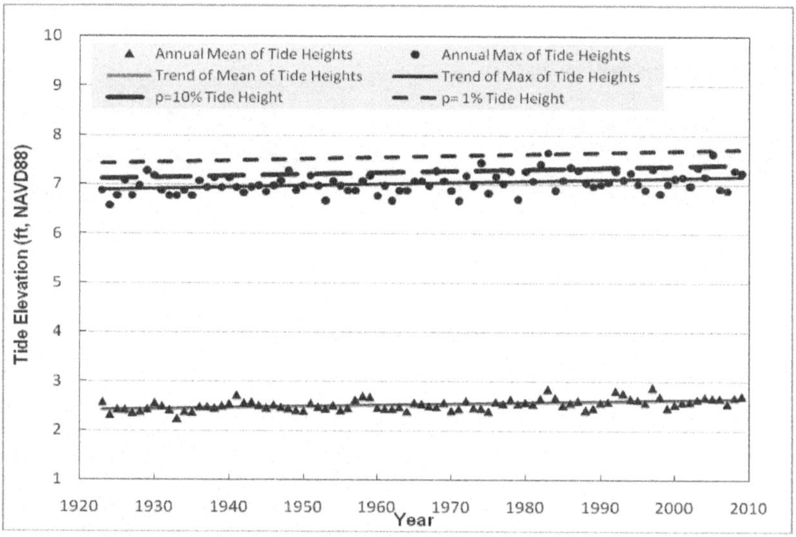

Figure 3: Mean and annual maximum tide elevations, and trends of mean, maximum, 10% and 1% annual exceedance-probability tide heights, Los Angeles, 1923 – 2009

As shown Figure 3, the maximum tide heights exhibit considerable inter-annual variability on the order of one foot. For this study, this variability is treated by considering the extreme tide height to be a random variable and probability analysis was performed to obtain the 1% and 10% extreme high tide trends shown in the figure. Based on the analyses of the historical data, mean sea level for 2010 is 2.65 feet NAVD88, and the 10% and 1% probability of occurrence for extreme tide heights for 2010 are 7.41 and 7.71 feet NAVD88, respectively.

Future mean sea levels are estimated by applying the methodology recommended by the U.S. Army Corps of Engineers (USACE 2009). For conservative planning purposes, the *high* rate of future sea level rise, SLR (in feet), is appropriate. SLR between years $y_1$ and $y_2$ is calculated as follows:

$$SLR(y_1, y_2) = 0.0056(y_2 - y_1) + 3.2964 \times 10^{-4}\left((y_2 - 1986)^2 - (y_1 - 1986)^2\right) \qquad (2)$$

Figure 4 show projections of future tide heights based on the sea level rise trends given by Eq. 2 assuming the trend for extreme high tide follows the trend for mean sea level. Projections point to a rapid increase of sea level over the next decade compared to the historical rate of rise. The projections also show that the differences between 10% and 1% extreme tide events is very important in the near term but less important in the long term relative to sea level rise.

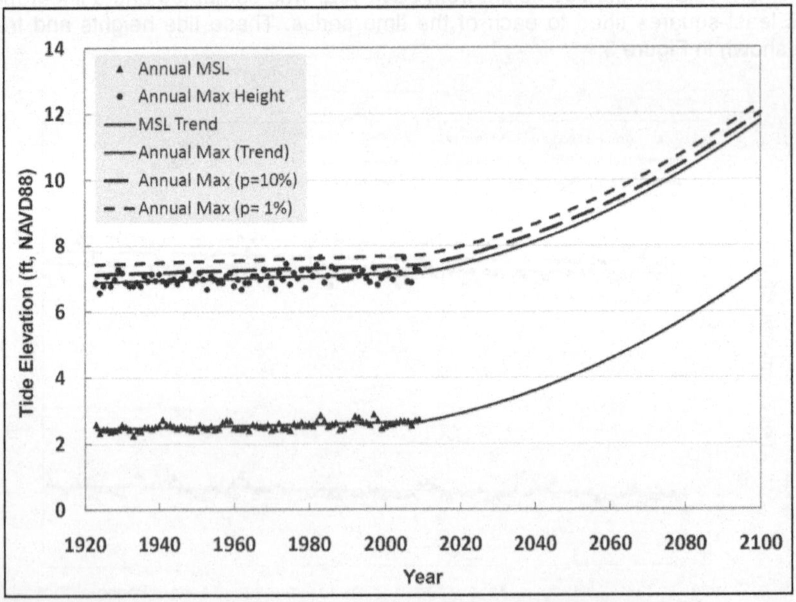


Figure 4: Projections of mean sea level and extreme tide heights through Year 2100

## Wave-driven Overtopping Rates

Waves are an important driver of embayment flooding when the tide level approaches the crest of a flood barrier. Wave action provides the necessary energy for water to rise up above the tide level and spill over the barrier. The rate at which this occurs is dependent on many factors such as the wave height, period, water depth at the toe of the wall, the height of the tide, and the height of the wall. Figure 5 shows an example of wave overtopping a seawall at Balboa Island during the December 22, 2010 flood event.

Figure 5: Waves Splashing over the Balboa Island Seawall (December 22, 2010)

It can be expected that as the tide height, $h$, rises, the rate of wave overtopping will commence once the tide reaches a height that allows the first drops of water to spill over the wall. As the tide continues to rise, the rate of overtopping can be expected to increase until the tide height surpasses the height of the wall. Thereafter, weir-like flow can be expected to dominate the overtopping rate and wave-driven flow can be ignored. BreZo was modified to account for the mode of wave overtopping described here. The following formula was used to specify the volumetric flow rate along sea walls where wave overtopping is expected:

$$q_{wo} = \begin{cases} q_{wo}^{\max}\left(\dfrac{h - h_{q0}}{h_{wall} - h_{q0}}\right) & if \quad h_{q0} \le h \le h_{wall} \\ 0 & otherwise \end{cases} \tag{3}$$

where $q_{wo}^{\max}$ represents the maximum rate of wave overtopping based on all possible water levels, $h_{wall}$ represents the height of the wall, and $h_{q0}$ represents the tide height corresponding to the onset of overtopping.

Wave overtopping rates were calculated using the ACES program developed by the U.S. Army Corps of Engineers. The parameters controlling the onset of wave overtopping and the overtopping rate include structure type and slope, beach slope, water depth, and wave characteristics (height and period). Based on the wall types, mudline elevations, and wave conditions for Balboa Island, wave overtoppings were calculated for three different reaches. Reaches 1 and 2 correspond to the southwest and south sides of Balboa Island, respectively, and Reach 3 corresponds to the south side of Little Balboa Island. All other sides of these islands were ignored, relative to wave overtopping due to a negligible wind fetch.

## Model Scenarios and Results

The model was used to compare flood zones based on present-day seawall heights, a future sea wall raised uniformly to an elevation of 9.8 feet NAVD88, and an interim extension of the sea wall height by 6 to 8 inches (replacing existing seawall cap with a taller cap, adding an extension to the existing cap, or deploying sandbags). A total of 17 different modeling scenarios were considered to account for the varied effects of high tides, waves, sea levels, and sea wall heights. These model scenarios and model results are shown in Table 1. In the table, the number of flood impacted buildings for each model scenario is estimated by comparing the modeled flood water depth at each parcel with a randomly generated first floor elevation for that parcel based on the first floor probability distribution.

Table 1: Model Results for 17 Scenarios

| Scenario | Year | Tide Height (Annual Exceedance Probability) | Wave Scenario | Average Flood Depth (ft) | Impacted Parcels (Number) | Impacted Parcels (%) | Impacted Buildings (Number) | Impacted Buildings (%) |
|---|---|---|---|---|---|---|---|---|
| Existing Condition Scenarios | | | | | | | | |
| 1 | 2010 | 10% | No Waves | 0.26 | 61 | 4.0 | 3 ± 2 | 0.2 |
| 2 | 2010 | 10% | Wind Waves | 0.26 | 61 | 4.3 | 3 ± 2 | 0.2 |
| 3 | 2010 | 10% | Ocean Swell | 0.29 | 514 | 36.5 | 24 ± 5 | 1.7 |
| 4 | 2010 | 1% | Wind Waves | 0.36 | 324 | 23.0 | 22 ± 4 | 1.5 |
| 5 | 2025 | 10% | Wind Waves | 0.48 | 681 | 48.3 | 66 ± 7 | 4.7 |
| 6 | 2025 | 10% | Ocean Swell | 0.79 | 1,176 | 83.4 | 235 ± 13 | 16.6 |
| 7 | 2025 | 1% | Wind Waves | 1.16 | 1,179 | 83.6 | 420 ± 14 | 29.8 |
| 8 | 2050 | 10% | No Waves | 1.84 | 1,410 | 100.0 | 894 ± 17 | 63.4 |
| 9 | 2050 | 1% | No Waves | 2.15 | 1,410 | 100.0 | 1047 ± 15 | 74.3 |
| 10 | 2100 | 10% | No Waves | 5.02 | 1,410 | 100.0 | 1410 ± 1 | 100.0 |
| 6-inch Extension Scenarios | | | | | | | | |
| 11 | 2010 | 1% | Wind Waves | 0.03 | 0 | 0.0 | 0 | 0.0 |
| 12 | 2025 | 1% | Wind Waves | 0.12 | 12 | 0.9 | 0 - 1 | <0.1 |
| 10-foot Seawall Scenarios | | | | | | | | |
| 13 | 2010 | 1% | Wind Waves | 0 | 0 | 0.0 | 0 | 0.0 |
| 14 | 2025 | 1% | Wind Waves | 0 | 0 | 0.0 | 0 | 0.0 |
| 15 | 2050 | 1% | Wind Waves | 0 | 0 | 0.0 | 0 | 0.0 |
| 16 | 2050 | 10% | Wind Waves | 0 | 0 | 0.0 | 0 | 0.0 |
| 17 | 2100 | 1% | Wind Waves | 5.30 | 1,410 | 100.0 | 1410 ± 1 | 100.0 |

The model results show that Balboa Island is presently vulnerable to extensive flooding because of low points in the seawall at the Balboa Ferry Ramp, the Collins Island Bridge crossing, and the relatively low height of its seawall. The sea wall around Little Balboa Island is higher than the Balboa Island sea wall and therefore offers better protection against present-day flood risks. Modeling shows that present-day flooding may occur from a combination of high tides and high waves, and the southwest boundary of the island is particularly vulnerable to overtopping due to its exposure to southerly and westerly winds. Modeling also shows that a combination of high tides and ocean swell may cause overtopping of the southern seawall due to its orientation towards the harbor mouth. Figure 6 shows an example of flood model results for the existing seawalls with and without the effects of wave overtopping under a 10% exceedance tide (Scenario 1 versus Scenario 3).

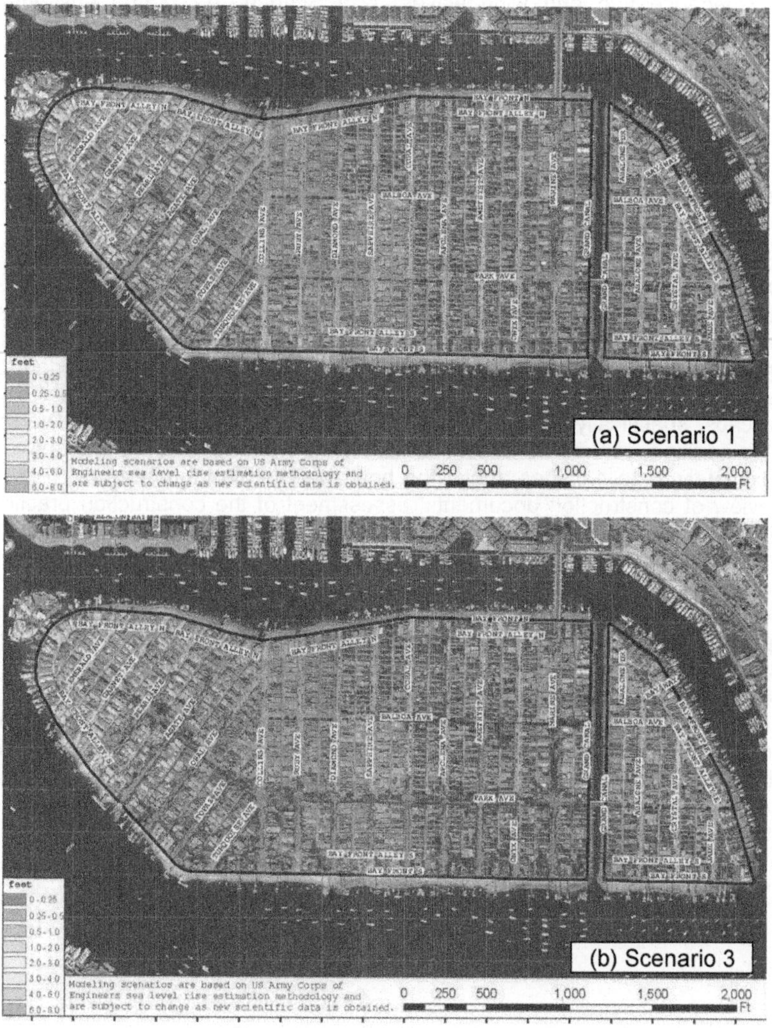


Figure 6: Modeled flood depths for existing condition (a) Scenario 1 – without waves, and (b) Scenario 3 – with waves

Modeling also shows that by extending the existing seawall cap by 6 inches and raising the height of the Balboa Ferry Ramp and Collins Island Bridge crossing to the elevation of adjoining sea walls, both islands would be protected through 2025 from a flood event caused by the combined effects a 1% exceedance probability storm tide and 10% exceedance probability wind waves. However, the islands would not be protected far past 2025 based on current sea level rise projections. The effectiveness of extending the seawall cap by 6 inches is illustrated in Figure 7, which compares the flood model results for Scenario 7 (existing seawall, year 2025, 1% exceedance tide and wind waves) with Scenario 12 (same environmental condition as Scenario 7 but with 6-inch extension of seawall).

Modeling also shows that a seawall constructed to an elevation of 9.8 ft NAVD88, uniformly around the islands with appropriate barriers at the ferry and bridge access points, would provide protection from flooding for the 1% storm tide and 10% wave scenario through 2050. Based on the USACE's high rate of projected sea level rise, the islands' seawalls would need to be raised 3 to 4 feet to protect the islands after 2050.

## Flood Inundation Mitigation Solutions

Review of available documents, reports and drawings revealed that most of the existing concrete seawalls were constructed in the 1930s. Site observations determined that an undocumented extension of 6 to 12 inches was added to the seawall cap around Little Balboa Island. Field investigations found universal distress in the caps, specifically multiple cracks, coinciding with the locations of the soldier piles. In addition, common and continuous cracks were observed along the seawall approximately 2- to 4-inches above the boardwalk behind the seawalls. Based on the review of construction documents, assessment of the construction practices in the 1930s, and site observations of the existing seawall conditions, it is estimated that the remaining useful life of the seawalls is between 10 to 25 years.

In the short term, using sandbags or raising the existing cap can be effective in preventing flooding from high water levels and wave overtopping. Given the limited useful lifetime of the seawalls, a replacement seawall will be required in the near future. The new seawall will be designed to allow wall height to be increased 4- to 5-feet of height if rising sea level dictates. For the Balboa Ferry Terminal, solutions include either raising the dock and launch ramp, or shifting the launch ramp further into the main channel so that existing properties can remain unchanged. Finally, to limit future damages, the report recommends establishing an appropriate minimum lowest first floor elevation in accordance with the federal Base Flood Elevation.

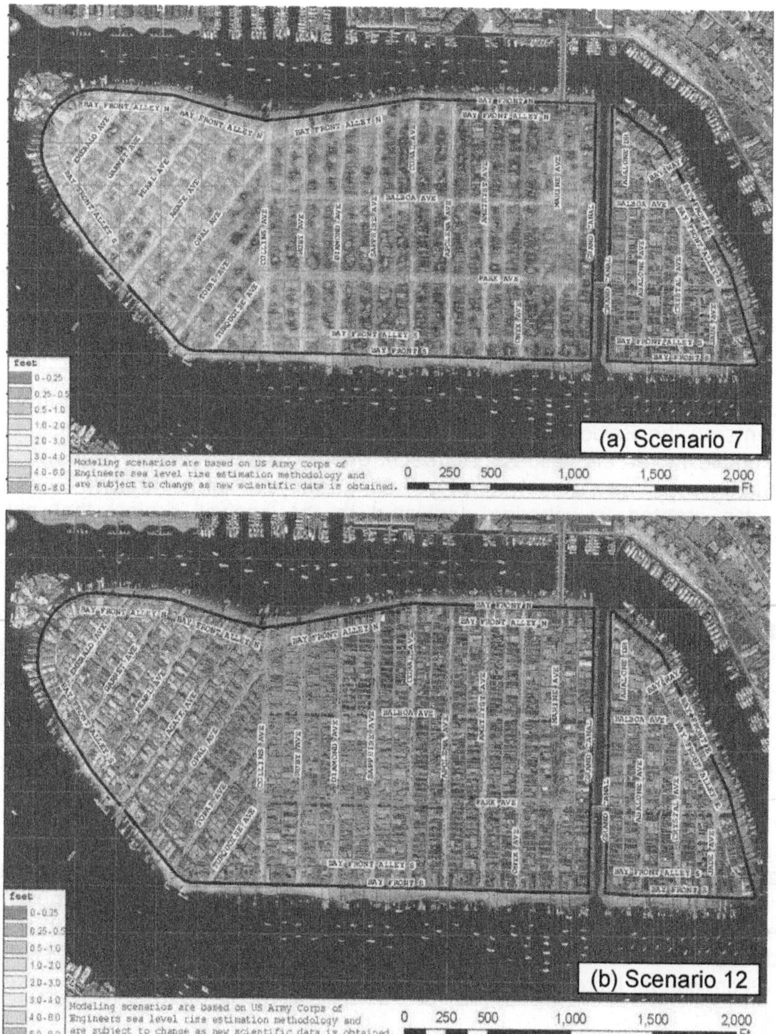


Figure 7: Modeled Year 2025 flood depths for 1% exceedance tide and wind waves (a) Scenario 7 – existing seawall and (b) Scenario 12 – 6-inch extension

## Public Outreach

The Islands' residents are split on their preferences for increased flood protection versus maintaining view planes. Given the uncertainty on the magnitude of future sea level rise, the City is pursuing the following implementation strategy. New walls will be designed assuming a sea level rise of 65 inches by Year 2100 with an ultimate top of wall elevation of about 14 feet NAVD88. However, to essentially preserve view planes, the top of wall elevation will be limited to a maximum of 9.0 feet NAVD 88 when the new seawalls are installed, an approximate increase of wall height of less than 9 inches. Then, as needed, the seawall cap will be periodically raised to provide flood protection from rising sea water. The adjacent boardwalk will also be periodically raised to preserve the strolling experience on the boardwalks around the islands.

The City understands that as sea level rises, so will groundwater. Long term protection of properties on the Islands will likely require that house pad elevations, as well as streets and sidewalks, rise in tandem with increasing top of seawall elevations. The City is considering a reconfiguration of the storm drain collection system to reduce the number of discharge points to the bay, as well as an upgrade to the tide gate system to allow for remote operation. Pumps and forebays can be created to help minimize street flooding when tide gates are closed at high tides. Finnally, the need for a groundwater extraction system will be examined.

A community outreach program, structured to reach a broad spectrum of the residents on the Balboa Islands, was launched to communicate with the residents the complexity of the issues. Outreach efforts included visual displays, formal community meetings, casual social gathering, and special information events. The City also created a webpage dedicated to the seawall project that not only addressed frequently asked questions but provided a wealth of scientific information with links to climate change reports and related journal articles.

## Summary

Given the complexity of the emerging scientific understanding of sea level rise, it was incumbent that City staffs distill the essential scientific information and define a few, distinct engineering solutions that reasonably covered feasible flood protection options. These options were presented in an objective manner to the community along with supporting information with a special effort made to reach as many of the islands' residents as possible and solicit their comments and advice. While unanimous consensus cannot be reasonably expected, City staff's goal is to give all ideas fair consideration so that residents will feel that the final recommendations have been reached based on a balanced and objective process. Final direction from City Council may be decided later this year. If indeed the Islands will need to be raised over the next century, this can be an opportunity to consider rezoning residential and commercial areas in a manner to provide for future flood resiliency as well as provide for increased visitor services and amenities.

## References

Everest International Consultants, Inc. 2011, Assessment of Seawall Structural Integrity and Potential for Seawall Over-Topping for Balboa Island and Little Balboa Island.

Sanders, B.F. 2008, Integration of a shallow-water model with a local time step, Journal of Hydraulic Research, 46(4) 466-475.

Schubert, J.E., Sanders, B.F. Smith, M.J. and Wright N.G. 2008, Unstructured mesh generation and landcover-based resistance for hydrodynamic modeling of urban flooding, Advances in Water Resources, doi:10.1016/j.advwatres.2008.07.012.

U.S. Army Corps of Engineers. 2009, Water Resources Policies and Authorities Incorporating Sea-Level Change Considerations in Civil Works Programs.

**Coastal Management: Changing coast, changing climate, changing minds**
ISBN 978-0-7277-6114-9

ICE Publishing: All rights reserved
doi: 10.1680/cm.61149.183

# Changing a Coastline to Enhance Society: Lessons from Littlehaven Seawall Realignment, UK

**Nick Cooper**, Royal HaskoningDHV, Newcastle-upon-Tyne, UK
**Gregor Guthrie**, Royal HaskoningDHV, Peterborough, UK
**Simon Wilson**, South Tyneside Council, South Shields, UK
**Tony Hanson**, South Tyneside Council, South Shields, UK
**Steve Burdett**, Royal HaskoningDHV, Newcastle-upon-Tyne, UK
**Colin Toovey**, Royal HaskoningDHV, Haywards Heath, UK
**Simon Howard**, Royal HaskoningDHV, Haywards Heath, UK

## Introduction

The Littlehaven seawall realignment scheme, completed in 2014, has transformed the beach, promenade and sea defences along a stretch of the seaside town of South Shields' seafront in northeast England, UK (Figure 1), turning a previously neglected length of coastline into a desirable visitor destination.

The project involved the demolition of a pre-existing, but failing, seawall, the onsite re-use of excavated material from backing land in landscaping, and construction of a 500m length of seawall and promenade along a new, landward and more sustainable, embayed alignment. High-quality concrete finishes, bespoke public art and breath-taking streetscape were used to ensure this striking scheme captured the public imagination and revitalised the seafront.

The project has been the subject of previous published papers focusing on key technical, environmental and economic aspects of the appraisal, design and construction phases (Cooper, Wilson and Hanson, 2013) and on the scheme's post-construction performance and award-winning features of distinction (Cooper *et al*, 2015).

It is not, therefore, the intention of this paper to repeat those aspects. Rather, by focusing on the themes of the conference, this paper considers how the realignment has influenced, or been influenced by, *"changing coast, changing climate and changing minds"*.

## Changing Coast

The coast at Littlehaven has long been subject to both natural and anthropogenic change, as captured by a series of historic maps covering the frontage since the mid-19$^{th}$ Century. These depict the one-time significant importance of the role of the coal industry, and its subsequent demise, through the railways on the backing land that were used to transport coal that was mined from local collieries to timber loading staithes on the nearby River Tyne estuary for export around the world.

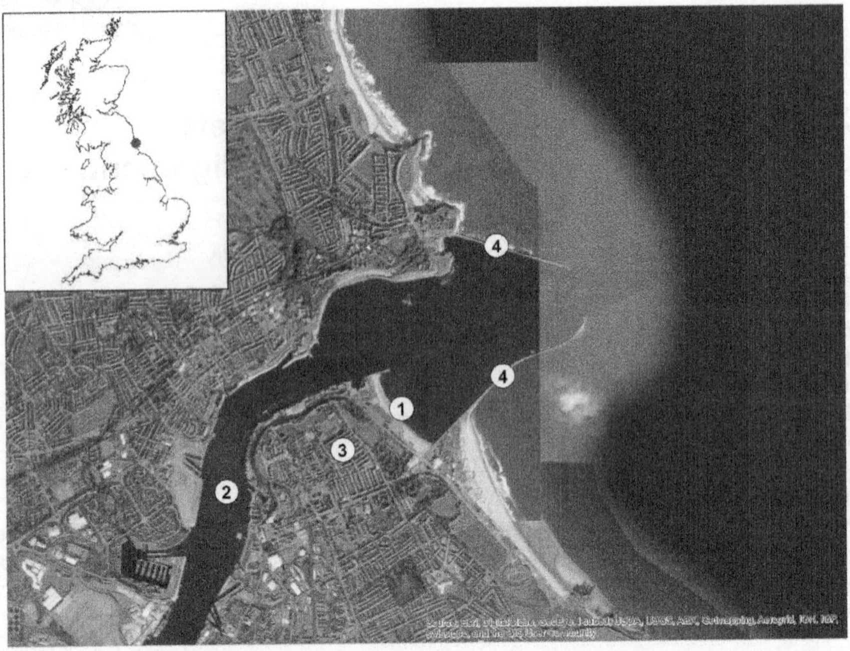

Figure 1: Scheme location
(1) Littlehaven; (2) River Tyne; (3) South Shields; (4) Harbour Piers

Changes during wartime are also evident, with Littlehaven having been used during the First World War as a Royal Naval Air Service station for seaplanes, which were launched from the beach by means of a timber slipway, and during the Second World War as a barrage balloon station, intended to help protect the shipyards on the River Tyne against aerial bombardment from the Luftwaffe.

Meeting the needs of the frontage at this time, a sea wall was constructed sometime in the early 1940s, extending seawards across an area of inter-tidal foreshore and enabling its reclamation from the sea. However, this resulted in the seawall protruding seawards to directly intercept the mark of high water in the centre of the Littlehaven frontage.

Over subsequent decades, and especially since the turn of the 21st Century, storm wave action caused notable deterioration in the structural condition of the original seawall at this location, as well as both beach lowering at its toe and frequent wave overtopping of its crest along its length. The area was developed as a car park to meet the changing needs of the frontage. Works were undertaken to enhance the attraction of the area to the west, including the construction of a new hotel in the early 2000s. However, all this worked within the constraints imposed by the existing defences and those defences were failing. The use of the car park was severely limited due to the risk of flooding. This led to a lack of investment, creating an area whose very appearance deterred the vision for more fundamental change.

It was recognised whilst preparing the Shoreline Management Plan (SMP) for the frontage that a change in coastal management approach was needed (Royal Haskoning, 2007). The SMP is a document which assesses the risks to people, property and both the natural and historic environments from sea flooding and coastal erosion over the forthcoming century, and presents a suite of shoreline management policies to manage those risks in a sustainable manner. It was at this time that the concept of managed realignment of the seawall was first introduced as a means of providing a more sustainable management policy than the previous approach of maintaining a deteriorating asset along its existing alignment.

Then, recognising the under-utilised amenity potential of the Littlehaven frontage, it was considered that the realignment scheme could not only provide much-improved coastal defences for the town of South Shields but could also provide an opportunity for enhancement in environment and society by stimulating the frontage through inclusion of a variety of public art and streetscape features that would add amenity value to the core coastal defence functionality and help to revitalise the frontage.

To this end, the scheme was developed within the context of a strategic 'vision for change' along the frontage. This incorporated the philosophy of managed realignment as a sustainable policy to provide the required functional coastal defence (Figure 2), together with a wider amenity beach and an aspirational promenade design that serves to set a high quality design standard for the area and provides the local community and visitors to the frontage with a new way to celebrate and enjoy this part of the coast. Complementing this, a new stepped apron leading from the promenade to the beach was designed to provide increased access and connectivity between the land and marine environments to the benefit of visitors.

Figure 2: Coastal defence

The high quality design standard has been achieved through the inclusion of bespoke public art, promenade furniture and contemporary lighting to complement the hard engineering of the coastal defence. These enhancements were designed within the context of the scheme's strong sense of place and maritime history, as dictated by its physical setting where the River Tyne meets the North Sea. Recognising that the changing coast would both provide opportunities for public realm enhancement and create new important maritime vistas, two major art features were designed and constructed to capitalise on this.

The first art feature, the Littlehaven Sail (Figure 3), provides an inviting gateway at one end of the promenade and provides the area with a unique sense of identity and a focal point for public congregations. The shape of the sail is designed to represent that of the fishing cobles, native to the north east of England, that were used by early pilots of the River Tyne to board large vessels and navigate their safe passage. The second art feature, the Littlehaven Eye (Figure 4), encourages visitors to ascend to the crest of the re-landscaped area behind the new seawall to gain new, elevated perspective, views. In particular, the Littlehaven Eye has been located so that through its central opening an unrivalled view of ships can be obtained as they pass through the piers at the mouth of the River Tyne. Complementing this theme of maritime voyages, the opening couplet from *'Blow the Wind Southerly'*, a traditional Northumbrian folk song which tells of a woman desperately hoping for a southerly wind to blow her lover back home to her from over the ocean, is emblazoned around the opening of the Littlehaven Eye.

Figure 3: Littlehaven Sail

Figure 4: Littlehaven Eye

The maritime theme is continued through the concrete finishing detail and street furniture design along the promenade. The concrete pigment was closely matched to the colour of the indigenous beach sand to give the appearance of a seamless transition between the natural environment and the man-made promenade, whilst various sea shanties and motifs have been sand-blasted into the concrete seawall. Seating along the length of the promenade has been designed to reflect the shape, and evolution, of sand dunes, whilst the seating in the areas of public realm open space are designed to mimic a selection of shapes of marine diatoms. Feature lighting has also been used to wash the promenade with a range of colours and change the atmosphere of the frontage at night-time to an area where people want to walk and meet for public events.

With an appropriate strategic 'vision for change' and particular attention to design and finishing detail, a 'conventional' coastal defence scheme can be used as a vehicle to deliver an enhancement to society by stimulating increased amenity and economic use of a previously under-utilised resource. The Littlehaven realignment scheme exemplifies this and has resulted, in a positive and proactive manner, in a changing coast that is fit for both coastal defence and amenity purposes over the next century.

## Changing Climate

### Environmental climate

In order that the Littlehaven realignment scheme provided a sustainable shoreline management solution, it was necessary to consider not only contemporary problems facing the original seawall, such as its poor and deteriorating condition, its undermining and the frequent wave overtopping, but also the likely effects of future climate change, especially in terms of sea level rise. It was considered that this environmental change may affect the scheme's future performance in two potential ways.

First, a change in sea level needed to be taken into account when designing the deck level of the promenade and the crest level of the backing seawall. This was achieved by making an allowance for the rates of sea level rise projected for this frontage within wave overtopping modelling assessments which were undertaken using the AMAZON model (Hu *et al.*, 2000).

Second, sea level rise was considered likely to affect the beach behaviour over time, in keeping with the classic 'Bruun Rule' of sandy beach response to sea level rise whereby the rising tidal frame would, in effect, cause the landward transgression of the shore profile (Bruun, 1962). Any fixed structure on a beach is likely to limit this landward transgression, reducing the width and scope for sustaining the beach. With this in mind, the design of the set-back (landward) position and alignment of the new seawall was therefore made with sufficiently increased beach width to accommodate such natural response to sea level rise over the next century in an uninhibited manner (Figure 5).

The sea level rise values in these assessments were derived from the UK's presently most authoritative source of climate change projections for the next century, namely the United Kingdom Climate Projections 2009 (UKCP09) 'Marine and Coastal projections' (Lowe *et al.* 2009). Values under the 'low', 'medium' and 'high' future greenhouse gas emissions scenarios were used to assess beach sensitivity to sea level rise.

The Littlehaven realignment scheme is believed to be the first coastal defence project to pro-actively remove a concrete seawall and reconstruct a defence along a new landward position in this manner for the specific purposes of sustainable coastal management, creating different perspective for realignment, rather merely one focusing on retreat.

Figure 5: Managed realignment
(black line = original seawall alignment; white line = new seawall alignment)

**Economic and Funding Climate**

Shortly after publication of the SMP in 2007, advocating that the policy of managed realignment at Littlehaven should be implemented within a few years due to the poor and deteriorating condition of the existing seawall, the UK economy entered a period of recession in 2008-09, with a subsequent period of austerity in the public sector economy initiated by the UK coalition Government which came into power in 2010. Coupled with this, the Government announced in 2010 that the Regional Development Agencies would be abolished in England by 2012 with a view to reducing the government deficit. These were non-departmental public bodies established for the purpose of, primarily, economic development. Other, relatively recently-completed, coastal defence schemes with added seafront amenity enhancement value had benefited from funding contributions from their respective Regional Development Agencies. These included Redcar Seafront Regeneration in northeast England and Weston-Super-Mare Seafront Enhancements in southwest England amongst others. Abolishment of the Regional Development Agencies before design and delivery of the Littlehaven realignment scheme meant that this funding source was no longer available.

Due to the benefits derived from the coastal defence aspects of the scheme, including the reduced risk of sea flooding and coastal erosion, a proportion of the scheme costs (£936k) were funded by central Government's Flood and Coastal

Erosion Risk Management Grant-in-Aid. Prior to 2011, such Grant-in-Aid would ordinarily have covered the entirety of the eligible scheme costs, but with a change in the funding arrangements the concept of 'partnership funding contributions' had become introduced meaning that the Grant-in-Aid was a reduced proportion and third party funding was additionally required to meet the full scheme costs. Due to this change in the funding climate, further contributions were successfully sourced from the Environment Agency's Local Levy (£200k) and the Local Transport Plan (£150k), but a significant proportion (£4M) of the remaining total overall scheme cost of £5.3M was funded by the maritime Local Authority.

Whilst the total scheme costs represent a relatively large sum of public sector finance during a climate of economic austerity, the maritime Local Authority recognised the value of its seafront and riverside to the local economy and had a strategic 'vision for change' which delivered not only the Littlehaven realignment scheme but other, complementary, capital projects within a long-running programme of regeneration activities which have made the area a popular destination for local residents and tourists alike.

## Changing Minds

### Pre-scheme Activities

During the 1980s and 1990s the emphasis in the Littlehaven area had been to keep the sea wall in place, accepting, or at least attempting to reduce, the overtopping, but effectively still shoring up the existing coastline against erosion. In part this was driven by the fact that, from coast protection risk perspective, this all could be afforded at that time. This reflected the underlying sectorial mind-set of the time; the focus on defence and, particularly in terms of funding, the focus on protecting what was there. Against this attitude, recognising the perceived significance of the frontage, the first SMP, developed for the area in 1998, defined a policy of "Hold the Line". A subsequent coastal management study reinforced this approach, recommending rebuilding of the sea wall on the same alignment at the end of its effective life.

The initial SMPs, undertaken throughout the whole coast of England and Wales by 2000, took an important step forward by promoting the concept of taking a broader, longer term perspective in shoreline management. However, in reviewing this process it was recognised that the initial SMPs had not gone far enough. The fundamental principle of the updated SMP in 2007 built on this concern, looking further forward rather than backwards: looking forward to sustainable solutions over the next 100 years and beyond (Guthrie *et al.* 2009) and critically considering a more integrated approach to management of the frontage.

In the updated SMP, this was reflected in the following core principles:

- To contribute to sustainable development and support and integrated approach to land use planning.
- To avoid damage to and enhance the natural heritage.
- To support the cultural heritage
- To minimise reliance on defence.

Whilst being a non-statutory document, the updated SMP was intended to inform statutory land use plans about the risks from sea flooding and coastal erosion over future timescales, and on the most sustainable policies to manage those risks over the short-term (25 years), medium-term (50 years) and long-term (100 years). The aim, however, was not to impose but to open up opportunities and to encourage thought about possibilities.

This was a very different concept that had to overcome significant barriers, not least in moving beyond the headline policy definition of "Managed Realignment" to an understanding of the intent behind this policy.

Despite working closely with planning authorities during the development of the updated SMP, there was a perception that managed realignment meant abandoning the frontage. In discussion with land use development planners this was explained within the updated SMP document, including the important concepts:

- Specific defence policy for this frontage cannot be determined solely on the basis of risk to existing assets.
- The essential elements for future development have been identified as the need to maintain a high quality promenade with scope for recreational facilities.
- In terms (solely) of sustainable defence, to allow the frontage to retreat to a natural profile would be the preferred policy. This, however, ignores the potential for enhancement. The intent of the preferred policy is to relax the line consistent with an integrated development plan aimed at deriving a balanced enhancement of both the natural and human environment.

This revolutionised the attitudes in the collaboration built between coastal managers and land use development planners. When the updated SMP was completed in 2007, it then became adopted by the maritime Local Authority. Based on this, the key findings and policy recommendations of the SMP were incorporated into the Local Development Framework. This comprises a suite of spatial land use planning documents, based around the Core Strategy – the key compulsory local development document specified in UK planning law which sets out principles regarding the development and use of land in a Local Authority's area of jurisdiction. The Core Strategy also became adopted in 2007.

The SMP became a catalyst for change and a catalyst for collaboration. Even so, much hard work was still required in taking such an approach forward and in developing the mind-set for positive change necessary to deliver the project.

**Scheme Development Activities**

Once the policy of managed realignment had been endorsed by the technical officers of the maritime Local Authority and then debated and adopted by its democratically-elected members, a positive public relations campaign was initiated and run throughout the scheme's development, especially during the construction stage when most disturbance and disruption to local residents, businesses and visitors would inevitably occur. This pro-active and positive approach helped to generate a sense of local ownership to the scheme and a source of local pride in the legacy that was being created.

This positive approach is captured in the strapline *"Hello Tomorrow, Change is Happening"* which, along with an attractive logo, was used to brand the scheme and provide an instant sense of recognition and a kite-mark identity of high associated quality on all publicity material relating to the scheme (Figure 6). This included letter drops to local residents and fortnightly update bulletins to local businesses.

Figure 6: Scheme strapline and logo to assist with 'changing minds'

The developers of the Littlehaven realignment scheme also engaged with interested or potentially affected stakeholders throughout its planning and delivery. This included the establishment of a dedicated Community Liaison Officer as a single point of contact throughout the construction phase and the provision of an on-site Visitors Centre. This provided an exhibition of the scheme's need and reports on progress with its construction, and a comments book for feedback. A site employee was on hand at all times to answer questions from the public.

Presentations were also made to local clubs, charitable societies and specialist interest groups, whilst site visits were hosted on numerous occasions to welcome local primary school pupils, college students, coastal engineers from neighbouring authorities, civil engineers from local firms, top-level officers, elected members and dignitaries from the maritime Local Authority and members of local professional societies, including the Institution of Civil Engineers and the Chartered Institution of Water and Environmental Managers.

In addition to these activities, a project website - www.southtyneside.info/Littlehaven – was developed and maintained throughout the construction phase to keep interested individuals updated, with regular photographs included to document progress with the development.

These engagement approaches generated much positive local media and local community interest and created a positive mind-set in the local community. To further showcase the scheme and commemorate the end of construction, an official opening ceremony and celebration day was held in 2014. The scheme was officially opened by the Mayor of South Tyneside and Her Majesty's Lord Lieutenant of Tyne and Wear. The celebration event aimed to bring the site to life with a fun-filled programme of family friendly activities such as a wing walking display, local dance groups, kite flying, beach activities, and local musicians. The day of events culminated in a fireworks display and light show that were set to music, bring into full effect the amenity enhancement aspects of the project.

## Conclusions

The development and success of the Littlehaven seawall realignment scheme demonstrates the benefits that can be derived from an ongoing commitment to an integrated approach to sustainable coastal management throughout all of the stages linking policy through to scheme delivery. It shows how, with an appropriate 'vision for change' and with a team working pro-actively and collaboratively, a coastal community, local economy and seafront environment can become revitalised and enhanced.

Flying kites has been an essential element throughout this process, from the initial concepts flown and explored within the updated SMP, through the exciting ideas and options considered during planning and design and in finally celebrating the successful outcome. The project highlights the benefits of taking forward new ideas and the importance of pushing boundaries and involving people as we adapt positively to the increasing pressures of a changing coastline – *"Hello Tomorrow, Change is Happening"*.

## Acknowledgements

The authors would like to thank the landscape architects Oobe and the principal contractors Galliford Try for their assistance in the design and delivery of the Littlehaven realignment scheme.

## References

Bruun, P., 1962. Sea level rise as a cause of shore erosion. *Journal Waterways and Harbours Division*, 88(1-3), 117-130.

Cooper, N.J., Wilson, S., Hanson, T., West, D. and Goodall, M., in press. Littlehaven Promenade and Sea Wall, South Shields, UK. *Proceedings Institution of Civil Engineers Maritime Engineering Journal*.

Cooper, N.J., Wilson, S. and Hanson, T., 2013. Realignment of Littlehaven Sea Wall, South Tyneside, UK. *Proceedings Institution of Civil Engineers Coastlines, Structures and Breakwaters*. November 2013.

Guthrie, J.G.L., Cooper, N.J., Rowe, S., Patterson, P., Oakes, T., Thomas, K., Briggs, P., Lombardo, S., 2009. One Hundred and Beyond: Using Shoreline Management Plans. ICE Breakwaters Conference 2009.

Hu, K., Mingham, C.G. and Causon, D.M., 2000. Numerical simulation of wave overtopping of coastal structures by solving non-linear spectral wave equations. *Coastal Engineering*, Vol.41, pp433-465

Lowe, J. A., Howard, T. P., Pardaens, A., Tinker, J., Holt, J., Wakelin, S., Milne, G., Leake, J., Wolf, J., Horsburgh, K., Reeder, T., Jenkins, G., Ridley, J., Dye, S., Bradley, S. (2009), *UK Climate Projections science report: Marine and coastal projections*. Met Office Hadley Centre, Exeter, UK.

Royal Haskoning, 2007. *North East (River Tyne to Flamborough Head) Shoreline Management Plan 2*. Report to Scarborough Borough Council. February 2007.

**Coastal Management: Changing coast, changing climate, changing minds**
ISBN 978-0-7277-6114-9

ICE Publishing: All rights reserved
doi: 10.1680/cm.61149.193

# A New Coastal Change Adaptation Planning Guide for England

**Alan Frampton**, CH2M HILL, Exeter, UK
**Andy Parsons**, CH2M HILL, York, UK
**Jeremy Pickles**, East Riding of Yorkshire Council, Beverley, UK
**Jennifer Kippax**, East Riding of Yorkshire Council, Beverley, UK

## Introduction

The National Planning Policy Framework (NPPF) sets out the Government's land use planning policies for England and describes how these are expected to be applied (Communities and Local Government, 2012). Land use plans developed by local planning authorities are required to consider flood risk and coastal change and to reduce risk to the developed and natural environment by avoiding inappropriate development in vulnerable areas or adding to the impacts of physical changes to the coast.

Within the NPPF, coastal change is defined as physical change to the shoreline through erosion, coastal landslip, permanent inundation or coastal accretion. Coastal change affects many undefended coasts and the rate of change and coastal risks in many of these areas are predicted to increase due to sea level rise and climate change (Rogers *et. al*, 2011). Coastal change can also occur in areas where there is a change in coastal management policy, for example a decision to no longer maintain or replace defences.

In areas near the coast that are likely to be affected by coastal change, the NPPF expects local planning authorities to define Coastal Change Management Areas (CCMAs) and within these areas define what development may be appropriate under what circumstances.

In order to help promote a consistent and systematic approach to the development and implementation of CCMAs, a new Coastal Change Adaptation Planning Guide has been developed. The work to develop the guide was undertaken by CH2M HILL under contract to East Riding of Yorkshire Council which received funding and guidance from England's Department for Environment, Food and Rural Affairs (Defra).

## Background

Coastal change poses a threat to many coastal communities, their infrastructure and their heritage, including complete loss of homes and land. Coastal change also poses a threat to coastal habitats and wildlife, particularly where they are constrained as a result of development, but can also provide opportunities where managed realignment or withdrawal of defence maintenance is proposed.

Past development on the coast took place with less knowledge of natural processes and flood and erosion risks than is now available. As a result of this there is a legacy of development in areas that are now vulnerable to ongoing coastal change (Pontee and Parsons, 2012). As recognised in the Shoreline Management Plans (SMPs) for England and Wales completed between 2006 and 2011 (Environment Agency, 2009), it will not be technically, economically or environmentally sustainable, nor indeed desirable, to defend all locations against future coastal change. With limited public funding for coastal defence, it will also not be affordable to continue defending all areas that currently benefit from defences. These factors, combined with impacts of climate change, will result in increasing need for land use planning to deal with coastal change in future.

Recognising the need to consider adaptation as a means of helping deliver more sustainable shoreline management solutions, Defra led a Coastal Change Pathfinder (CCP) programme as part of a consultation on the development of a national coastal change policy. The CCP programme, which was undertaken between 2009 and 2011, explored new ways of adapting to coastal change. Under this programme, 15 projects involving English local authorities working in partnership with their communities trialled innovative approaches to planning for and managing change (Defra, 2012).

The UK government has adopted the concept of adaptation as a means of delivering sustainable coastal change solutions within the land use planning system for England. As set out in the NPPF, this is to be delivered through CCMAs. Specifically, Paragraphs 106, 107 and 108 of the NPPF state that "*Local Planning Authorities (LPAs) should reduce risk from coastal change by avoiding inappropriate development in vulnerable areas or adding to the impacts of physical change to the coast*". To do so, LPAs "*should identify as a CCMA any area likely to be affected by physical changes to the coast, and be clear as to what development will be appropriate in such areas and in what circumstances; and make provision for development and infrastructure that needs to be relocated away from CCMAs.*" (Communities and Local Government, 2012).

The planning practice guidance that accompanies the NPPF (Communities and Local Government, 2014) provides limited guidance on the considerations for defining CCMAs and, although it cross-references SMPs, it does not give technical details of the process or steps to take.

## Development of New Coastal Change Adaptation Planning Guidance

Building upon the lessons learnt from the 15 CCP projects (Defra, 2012) and other similar projects undertaken by others both around the UK and rest of the world, a new guide on developing approaches to enable coastal change adaptation through planning has been developed. The guide, which has been designed to aid the identification and delivery of CCMAs and suitable adaptation approaches for areas affected by physical coastal change, was developed in two phases.

**Phase 1: Literature Review**

Phase 1 involved an extensive literature review of practical coastal change adaptation projects and approaches, legislation and regulations from the UK and rest of the world. In total, over 100 documents and sources of information were reviewed. These included legislation, regulations, policies, plans, papers and study reports relating to coastal change adaptation. To support this phase, telephone interviews were also conducted with 14 organisations in England which had been involved in coastal change adaptation projects in the recent past, including those which had been involved in the CCP programme as well as others such as The National Trust. These interviews captured first-hand experience of those projects and provided insights and lessons learnt that were not always documented in project reports. Output from this phase is provided in a supporting technical report to the guidance so that it may be used by others in the future.

Phase 1 also reviewed the planning framework for England to identify how the new Coastal Change Adaptation Planning Guide (CCAPG) would sit within that framework. This is illustrated in Figure 1.

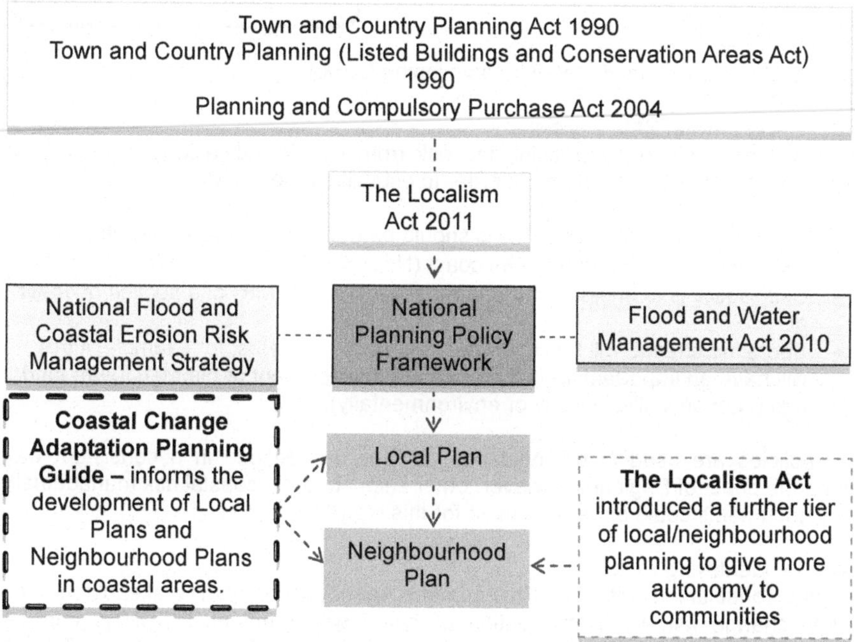


Figure 1: Land use and spatial planning context within which the new CCAPG sits

**Phase 2: Developing the Guidance**

Phase 2 used the information from Phase 1 to develop a new, practical, user-friendly guide that sets out firstly how to identify and define areas at risk of coastal change, and then to establish the types of policies that are most likely to be appropriate in different coastal settings, in a consistent and robust way. The guide identifies a 4-stage process to the development of CCMAs as illustrated in Figure 2 and described in summary below.

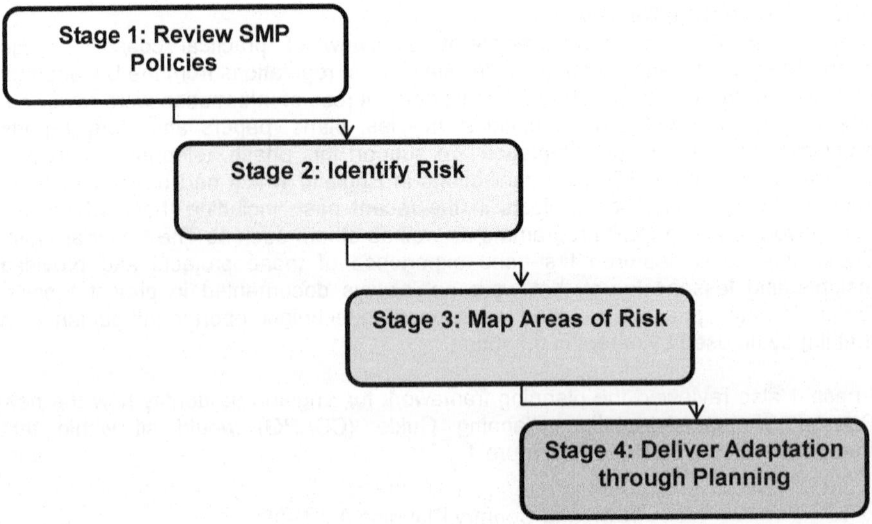


Figure 2: Overview of the 4 stages to developing CCMAs

Stage 1 – Reviewing SMP Policies
This first stage involves appraising the risk from the Shoreline Management Plan (SMP) and interpreting its policies in order to establish potential CCMAs.

The NPPF states that potential CCMAs should be defined in Local Plans where:

- SMP policy is not to defend the coast (No Active Intervention, NAI);
- SMP policy is to implement Managed Realignment (MR) of a section of coast; or
- Shoreline change will be significant over next 100 years; for example, if this will have an important impact on existing development or planned future land use (economically, socially or environmentally).

SMP policies are used in the guidance as they are defined in a consistent way around the coast of England. However, other such strategic coastal risk management policy documents could equally be used for this stage.

Stage 2 – Identifying Risk
Through consultation with the Environment Agency and other stakeholders, the coastal change risk for each section of coast within potential CCMAs can be identified in Stage 2. This will depend on the setting of the coast and factors influencing risk such as the physical setting of the coast (e.g. beach, sand dunes, simple cliffs, coastal landslides, coastal floodplain, estuary, salt marsh etc.) and features and land uses in the coastal zone (e.g. urban/rural, developed/undeveloped etc.).

Stage 3 – Mapping Areas of Risk
The third stage involves mapping the CCMAs and the zones of risk within them. This may be based on erosion risk information contained in the SMP, coastal monitoring

data from the national network of regional coastal monitoring programmes, National Coastal Erosion Risk Mapping (where appropriate and available), expert assessment or any other appropriate relevant studies or predictions from local monitoring data. In addition to the erosion risk area, CCMA boundaries may need to be extended to avoid, for example, cutting through settlements or transport infrastructure, or to allow a zone of land for roll-back, relocation or diversion of infrastructure or to allow for a specific type of habitat creation.

The resulting areas defined as CCMAs may cover (i) lengths of coast from tens of metres to many kilometres in length, (ii) areas of coast from tens of square metres to many square kilometres, and (iii) individual SMP policy units, or multiple SMP policy units, based upon the nature of the coast.

By way of example, Figure 3 provides an illustration of the CCMA zoning concept.

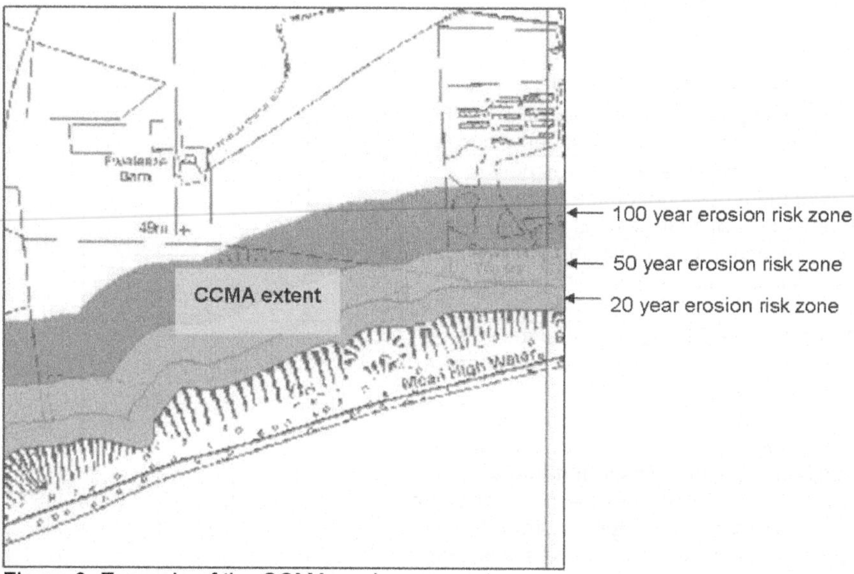


Figure 3: Example of the CCMA zoning concept

Stage 4 – Delivering adaptation through planning

Once the CCMAs and different risk zones within them are mapped through Stages 1 to 3, it is then possible in Stage 4 to identify adaptation requirements for areas at risk from coastal change depending upon the coastal setting, and to determine what type of development should or should not be allowed to occur in each area, feeding into planning policy and development management adaptation approaches. Table 1 provides some examples of the types of approaches that could be employed in different coastal settings. Note that the examples in Table 1 and their suggested applicability are not intended to be all encompassing and exclusive, but are simply examples intended to promote ideas for those developing Local and Neighbourhood Plans.

Case study examples of planning policy and development management measures for coastal change adaptation employed by various planning authorities are also included in the guide.

Table 1: Examples of coastal change adaptation approaches applicable in different coastal settings

| | Coastal physical settings considered for adaptation | | | | |
|---|---|---|---|---|---|
| | Cliffs | Beaches | Sand Dunes | Estuaries | Defended Coastlines |
| **Planning Policy and Development Management Adaptation Approaches** | | | | | |
| Prepare a Rollback policy for the relocation of property, community facilities and infrastructure. | ✓ | ✓ | ✓ | ✓ | |
| Avoid development in vulnerable areas. | ✓ | ✓ | ✓ | ✓ | |
| Identify suitable type of time-limited (temporary) development that could occur. | ✓ | ✓ | | | |
| Implement a policy that ensures new development does not cause adverse effects / transfer coastal change risks to other areas. | | ✓ | ✓ | ✓ | ✓ |
| Use area action plans / neighbourhood plans to manage future development in coastal communities. | | ✓ | ✓ | ✓ | ✓ |
| Implement a policy that encourages new developments to incorporate green space to help manage future risks. | | | | ✓ | ✓ |
| Encourage developers to include resilience measures to property / infrastructure where development is necessary and future relocation will not be possible. To support this, provide advice to developers / property owners / businesses on the measures they can take. | | ✓ | | ✓ | ✓ |
| Implement managed realignment to manage coastal change, working with natural processes and restoring habitat. | | ✓ | ✓ | ✓ | |
| **Other Adaptation Approaches** | | | | | |
| Develop community / individual emergency response plans in case of future coastal change incidents. | ✓ | ✓ | | ✓ | ✓ |
| Develop community / individual recovery plans to respond to a coastal change incident occurring. | ✓ | ✓ | | ✓ | ✓ |
| Improve the response to incidents by the emergency response organisations. | ✓ | ✓ | | ✓ | ✓ |

| | Coastal physical settings considered for adaptation | | | | |
|---|---|---|---|---|---|
| | Cliffs | Beaches | Sand Dunes | Estuaries | Defended Coastlines |
| Introduce flood defences that are flexible to risk changes over time, and resilient to extreme weather events and the longer-term projected impacts. | | ✓ | | ✓ | ✓ |
| Remove defences to restore natural processes, making use of the natural environment. | | ✓ | ✓ | ✓ | ✓ |
| In relation to historical assets, explore other ways to conserve the asset in situ but also record assets to secure the evidence. | ✓ | ✓ | ✓ | ✓ | ✓ |

Other aspects of the CCAPG
This 4-stage process is supported by recommendations on best-practice for systems and structures to aid successful development and then implementation of CCMAs, such as funding, community engagement and administrative arrangements within organisations with responsibility for leading coastal change adaptation.

The guide is supported by a series of 'prompt questions' for users to consider at each of the 4 stages in developing CCMAs, and 9 case studies from around England covering a range of topics from the following areas:

- Alde and Ore Estuaries, Suffolk – example of community engagement and participation in decision making with regards future flood risk management and adaptation.
- Cowes to Gurnard, Isle of Wight – example of local planning policy and development control in practice in an area at risk of coastal landslides.
- East Riding of Yorkshire coast – example of local planning policy and development control, including roll-back of development, in an area at risk of rapid coastal erosion.
- Manhood Peninsula, West Sussex – example of a community-led initiative to develop an integrated coastal zone management plan for the area that includes coastal change adaptation planning and implementation.
- North Norfolk coast – example of local planning policy and development control in an area at risk of rapid coastal erosion.
- Reculver, Kent – example of incorporating coastal change adaptation in a wider Masterplan for regeneration of a local area.
- South Milton Sands, Devon – example of planning and implementing coastal change adaptation in an area of sand dunes including measures relating to access, parking, archaeology, nature conservation and continued protection to coastal community.
- Steart Peninsula, Somerset – example of planning and implementing a managed realignment scheme with extensive community engagement that creates coastal habitat whilst providing continued flood defence and local

access to communities.

- West Dorset, Weymouth & Portland coast – example of coastal risk planning guidance developed for a length of coast with recommendations on planning policy and development control for all sections of the coast, including those where the SMP policy is to continue to defend as well as those where the SMP policy is for no active intervention or managed realignment.

Prior to publication, the final guide was consulted on widely with organisations including LPAs, Environment Agency, Natural England, RSPB, Defra and England's Department for Communities and Local Government. This has ensured that a wide range of views and experience has been captured and that the guide is robust and usable.

It is anticipated that the guide, which is expected to be published online at www.gov.uk in 2015 by Defra, will be used by coastal managers, engineers, planners and professionals involved with implementation of CCMAs. The guide should also help coastal communities to increase their understanding of coastal risk adaptation and help them understand how to get involved.

## Conclusions

The new guide is anticipated to be of benefit to coastal practitioners and local authorities needing to develop and implement CCMAs as it provides:

- A staged, consistent approach to identifying and implementing CCMAs in England
- Sound processes for managing coastal change, all of which can be built into future policy
- A source of coastal change adaptation case studies from across England on a range of topics including:
  - Developing planning guidance linked to coastal change adaptation
  - Working with the community to plan for adaptation
  - Encouraging/incentivising rollback through planning policy
- Access to over 100 sources of further information and identification of organisations around England that have experience of developing and applying adaptation measures at the coast.

Whilst the Coastal Change Adaptation Planning Guide has been developed with a focus on the planning system in England, many of the principles contained within it are universal and could be applied in most situations around the globe where there is a need to proactively plan for and manage adaptation in the face of ongoing climate change impacts.

## References

COMMUNITIES AND LOCAL GOVERNMENT. (2014). *National Planning Policy Framework, On-Line Planning Practice Guidance*. [Online]. Department for Communities and Local Government, March 2014. Available from: http://planningguidance.planningportal.gov.uk/. [Accessed: 4th November 2014].

COMMUNITIES AND LOCAL GOVERNMENT. (2012). *National Planning Policy Framework*. [Online]. Department for Communities and Local Government, March 2012. Available from:

https://www.gov.uk/government/publications/national-planning-policy-framework--2. [Accessed: 4th November 2014].

DEPARTMENT FOR ENVIRONMENT, FOOD AND RURAL AFFAIRS. (2012). *Coastal Change Pathfinder Review, Final Report*. [Online]. Department for Environment, Food and Rural Affairs, January 2012. Available from: https://www.gov.uk/government/publications/coastal-change-pathfinder-review. [Accessed: 4th November 2014].

DEPARTMENT FOR ENVIRONMENT, FOOD AND RURAL AFFAIRS. (2010). *Coastal Change Pathfinders website*. [Online]. Department for Environment, Food and Rural Affairs, April 2010. Available from: http://archive.defra.gov.uk/environment/flooding/manage/pathfinder/. [Accessed: 4th November 2014].

ENVIRONMENT AGENCY. (2009). *Shoreline Management Plans (SMPs)*. [Online]. Environment Agency, March 2009. Available from: https://www.gov.uk/government/publications/shoreline-management-plans-smps. [Accessed: 4th November 2014].

PONTEE N. I. and PARSONS, A. (2012). *Adaptation as part of sustainable shoreline management in England and Wales*. [Online]. Proceedings of the Institution of Civil Engineers' Maritime Engineering 165 September 2012 Issue MA3, p113–130. Available from: http://dx.doi.org/10.1680/maen.2011.35. [Accessed: 4th November 2014].

ROGERS, J., ALLAN, E., HARDIMAN, N., and JEANS, K. (2011). *The National Coastal Erosion Risk Mapping Project – From Start to Finish…and Beyond.* [Online]. ICE Conference on Innovative Coastal Zone Management: Sustainable Engineering for a Dynamic Coast. 15-16 November 2011 in Belfast, ICE Publishing, 2012. Available from: http://www.ice.org.uk/ICE_Web_Portal/media/Events/Breakwaters%202013/The-National-Coastal-Erosion-Risk-Mapping-Project---From-Start-to-Finish----and-Beyond.pdf. [Accessed: 4th November 2014].

Coastal Management: Changing coast, changing climate, changing minds
ISBN 978-0-7277-6114-9

ICE Publishing: All rights reserved
doi: 10.1680/cm.61149.203

# Results from the Dutch Ecobeach Pilot Project

**Bas Reedijk**, BAM Infraconsult bv, Gouda, The Netherlands
**Ad van 't Zelfde**, BAM Infraconsult bv, Gouda, The Netherlands
**Jelle-Jan Pieterse**, Hillebrand, Middelburg, The Netherlands
**Marco van Leeuwen**, VU University Amsterdam, The Netherlands
**John Rae Manaois**, BAM Infraconsult bv, Gouda, The Netherlands

## Introduction

Ecobeach is an innovative vertical passive beach drainage system discovered by accident by the Danish contractor Poul Jakobsen in the early 1990s. The invention was initially called "Pressure Equalizing Module System" (PEM). The inventor indicated a positive effect on natural accretion and reduction of erosion of the beach. The demonstrated positive results at different pilot projects around the world triggered the Dutch contractor BAM to set up a test of the system at the beach of Egmond aan Zee (Figure 2), along the Dutch North Sea coast, in cooperation with Rijkswaterstaat (Dutch Ministry of Public Works). This pilot, which was called Ecobeach, was carried out between 2007 and 2011. The system was installed in 2 sections of 3 km length at the test area North and test area South, separated by an area of 1 km without Ecobeach. These selected areas were located within a coastal zone which has been monitored intensively.

## Dutch Pilot

A field experiment was carried out for the pilot, in which a statistical analysis was executed for coastal state indicators (CSI's). In addition a scientific approach was used to investigate possible working principles of the system. The drainage system consists of vertical drainage tubes with a length of 1.75 m which are installed between the dune foot and average low waterline at 0.25 m below the beach surface (Figure 1). The tubes have filter slots. The tubes are placed with an interspacing of 10 m in cross-shore rows, while the long-shore distance between the rows is 100 m. Over 600 drains were installed in the test areas.

Figure 1: Installation of a drainage tube below the beach surface

Figure 2: Test area Ecobeach system

## Coastal Changes

According to (Pieterse, 2009) the direction of the net long-shore sediment transport along the Dutch coast near Egmond is from South to the North. The Egmond area (test area North) is frequently nourished over time (Figure 3). Test area South is less affected by direct nourishments. The statistical analyses of the sand volumes consider test area South in comparison with Castricum as test area South is adjacent to Castricum. Test area North is separated from test area South by a 1 km area without Ecobeach. Castricum is bounded by Heemskerk to the south. Test area North and Heemskerk are compared as in both areas a beach nourishment was applied in 2005, before the Ecobeach installation.

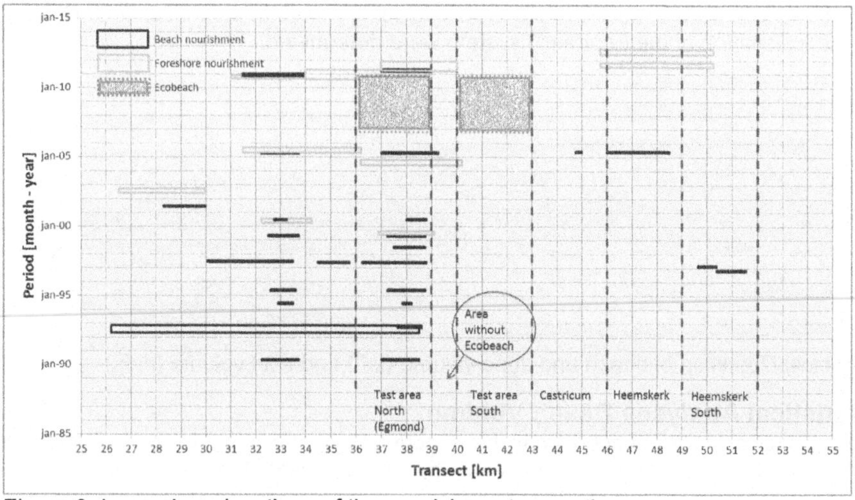


Figure 3: Long-shore locations of the nourishments over time

## Approach

The main objective of the experiment was to investigate potential working principles as well as to measure possible sand volume changes caused by the Ecobeach system. The sand volumes were measured at the project location and the adjacent locations in order to compare effects of the Ecobeach system. In geographical terms from North to South, the considered project locations are test area North (JARKUS transects from RSP 36 to 39), area without Ecobeach (RSP 39 to 40), test area South (RSP 40 to 43), Castricum (RSP 43 to 46), Heemskerk (RSP 46 to 49) and Heemskerk South (RSP 49 to 52). The sand volumes are also compared with the historical data (annual coastal measurements carried out since 1965) in order to be able to draw conclusions on the effect of the Ecobeach system along the coast. The morphological behavior of the coast can be distinguished in three parts: natural variability, impact of nourishments and effects of the Ecobeach system.

The definition of the beach volume describes the layer between the dune foot at NAP + 3 m and the average low water level at NAP – 0.78 m. The boundary extends from the beach surface until a defined reference line shoreward (Figure 4). The volume above the dune foot is defined as the dune volume which is also bounded by the reference line shoreward. The total volume covers the sand volume between the level NAP – 8 m and the dune crest, extending to the defined reference line. The statistical analyses of test area South and Castricum comprise both total volumes and beach volumes. Test area North and Heemskerk consider statistical analysis of beach volumes alone and beach + dune volumes combined. The volumes are derived from the JARKUS data which is the annual survey of the dune, beach and foreshore profiles extending to approximately RSP + 2500 m seawards. The survey data consists of data from 1965 to 2014. Along the Dutch coast near Egmond, the JARKUS transects are located every 250 m (perpendicular to the coastline). The aggregated values of 12 or 13 transects along the length of the areas were used to average the variability of the volumes at which 1965 = 0 $m^3$/m. Both test areas have 13 JARKUS transects. Also in both outer transects of the test areas, drains have been installed. The behavior of the CSI's including trends and confidence limits prior to the installation of the Ecobeach system were determined.

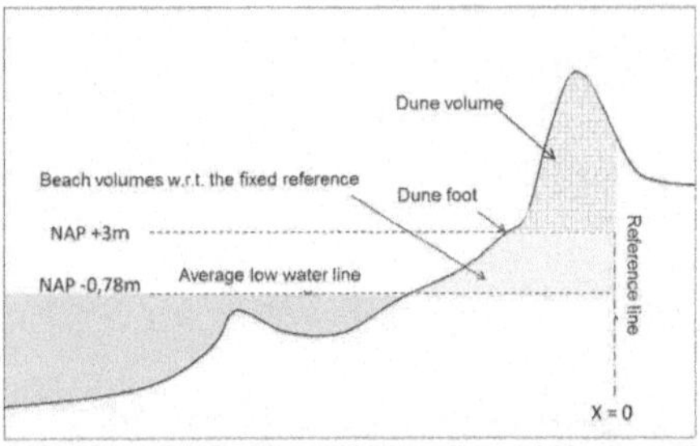


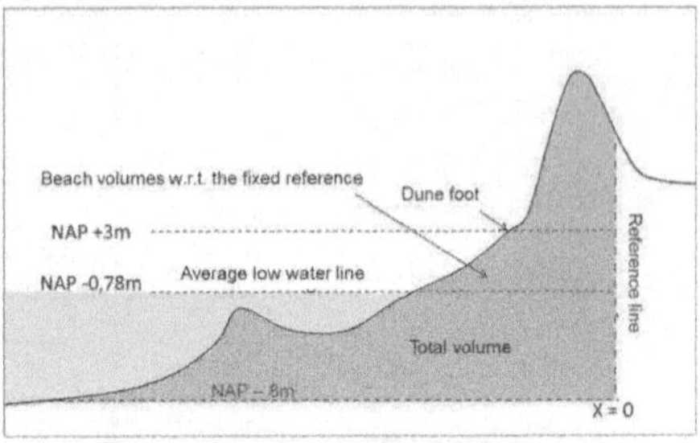


Figure 4: Definition of beach and dune volume (left) and total volume (right)

## Statistical Analyses Beach Volume

The statistical data obtained from the experimental results led to an analysis that includes both the total sand volumes and theories of potential working principles.

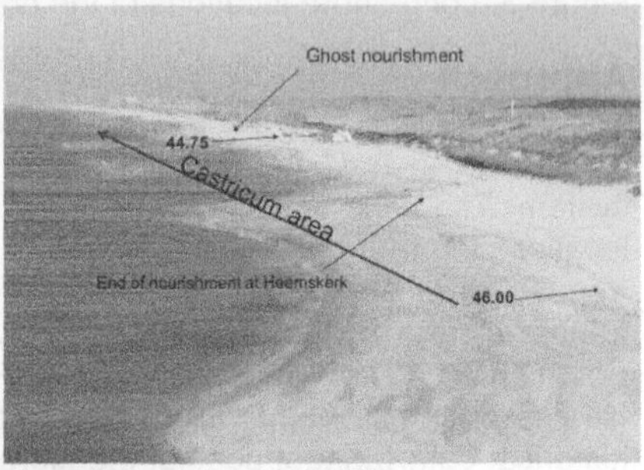


Figure 5: Partly nourished beach of Heemskerk and Castricum (2005)

### Test Area South and Castricum

The beaches at test area South and Castricum are stable and show limited variation of beach volume based on historical data from 1965 to 2007 (prior to the installation of the Ecobeach system) (Figure 6). No significant influence of the approximately 15 years sand bar cycle can be observed on the beach volume. Directly after installation of the Ecobeach system, the beach volume in the southern test area started to grow. It reached the highest measured value since 1965 in January 2011, gaining almost 50 $m^3$/m of beach. Wave conditions during the test period were similar to previous periods (Walstra, 2014). Although the severe storm Tilo occurred in November 2007 with the highest waves ever measured at IJmuiden, and the first storm closure of the Rotterdam storm surge barrier, no decrease in beach volume was observed in 2008. After completing the removal of the Ecobeach system in February 2011, the beach volume decreased, and in 2014 the beach volume was almost back to the value of 1965. After the Ecobeach system was installed in the southern test area, the Castricum area showed a similar growth rate of the beach volume. The beach volume for both the test area South and the Castricum area is outside the confidence limits for the years 2009 to 2011 based on the historical data.

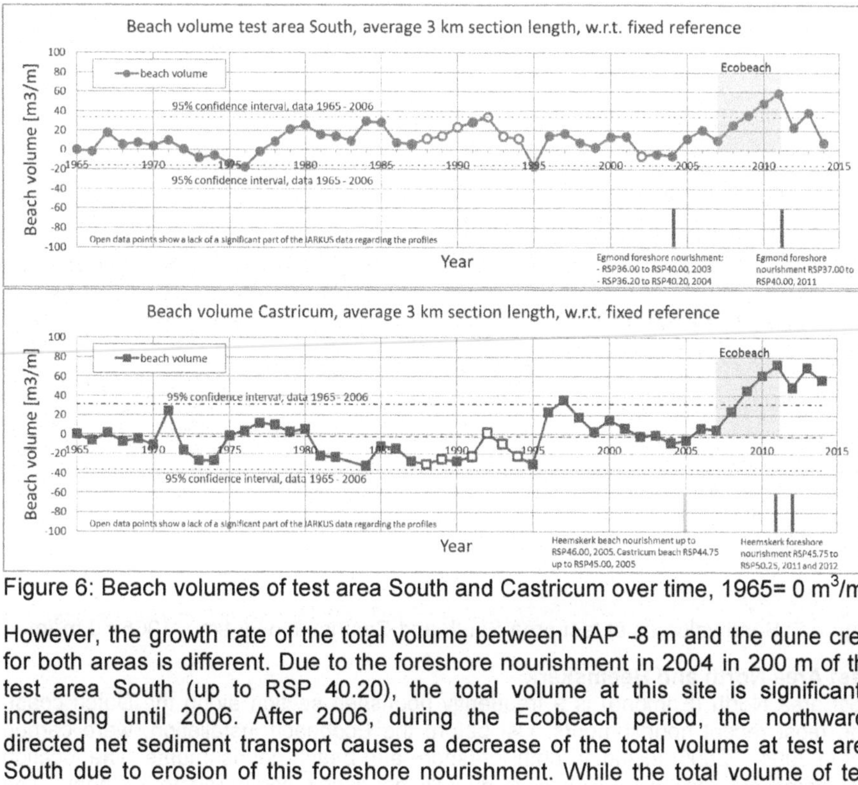


Figure 6: Beach volumes of test area South and Castricum over time, 1965= 0 $m^3/m$

However, the growth rate of the total volume between NAP -8 m and the dune crest for both areas is different. Due to the foreshore nourishment in 2004 in 200 m of the test area South (up to RSP 40.20), the total volume at this site is significantly increasing until 2006. After 2006, during the Ecobeach period, the northwards directed net sediment transport causes a decrease of the total volume at test area South due to erosion of this foreshore nourishment. While the total volume of test area South decreased during the Ecobeach period, the total volume in the Castricum area increased due to feeding from outside this area. Feeding of the Castricum area has occurred by the 250 $m^3/m$ beach nourishment which was placed in the adjacent Heemskerk area in May/June 2005, see Figure 5.

It is rather remarkable that after installation of the Ecobeach system in test area South, both areas show a similar growth in beach volume, while the total volume of test area South decreased and the total volume of the Castricum area increased.
Coarsening of sand particles of the upper 2 m of the beach has been observed in the southern test area (Ekkelenkamp, 2011). This coarsening has led to increased beach stability and reduced longshore transport in the southern test area. The Castricum area is wedged in between the 2005 nourishment at Heemskerk and the southern

Ecobeach test area and will have been affected by both the sand feeding from the 2005 nourishment at Heemskerk and the beach growth and coarsening in the southern test area. It is remarkable that although from 2005 onwards the total volume from NAP -8 m to dune crest of the Castricum area started to grow significantly, only after installation of the Ecobeach system the beach volume started to show a significant growth (Figure 7). It has been observed that after installation of the Ecobeach system an increase of Aeolian transport occurs (Pieterse, 2009). This Aeolian transport is not bounded by artificial human demarcations. Also during severe wave conditions and during storm surges, when dune erosion takes place, sand will move from one area to the other.

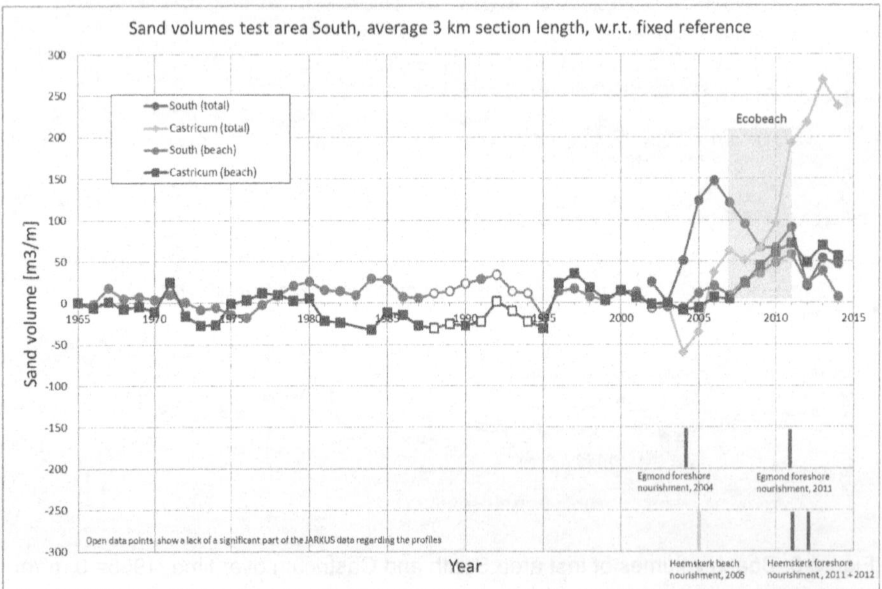


Figure 7: Sand volumes of test area South and Castricum over time, 1965= 0 $m^3/m$

**Test Area North and Heemskerk**

Test area North (Egmond) is a frequently nourished section along the Dutch coast. The most recent nourishments, just before the Ecobeach installation, were carried out in 2004 – 2005, including both foreshore and beach nourishments. The location at Heemskerk, however, is less frequently nourished compared to Egmond. The 2005 beach nourishment was the first nourishment in the Heemskerk area. Both coast sections show receding beach volumes on average based on historical data from 1965 until present. The Egmond area is receding more than the Heemskerk area. Nourishments in the Egmond area started in 1990. From Figure 8, it can be observed that the nourishments at Egmond were able to counteract the beach erosion. However, nourishments erode in the years after placement. The beach volume in 2004 is again below the beach volume of 1999. In November 2006, 1.5 year after the beach nourishment at Egmond, the drains were installed in the northern test area. The drains have also been installed in the beach nourishment.

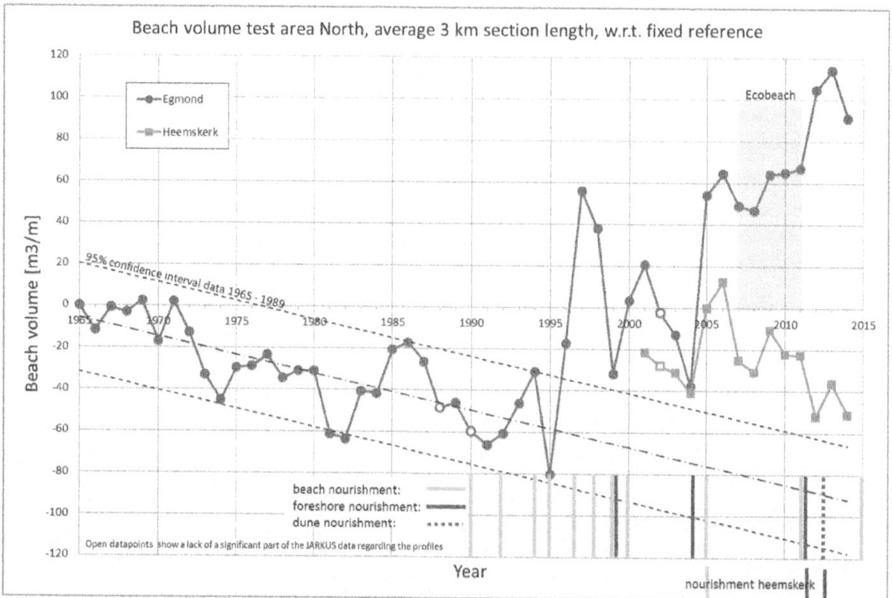


Figure 8: Sand volumes of test area North and Heemskerk over time, 1965= 0 $m^3/m$

Without Ecobeach, a rapid erosion of the beach nourishment would have been expected. However, after installation of the Ecobeach system in combination with the foreshore and beach nourishment, an increase of the beach volume occurred at the northern test area. Also this area shows the highest beach volume ever measured at the end of the test period in 2011. The measured beach volume is higher at the end of the Ecobeach test period in January 2011 than measured in 2006 after the beach nourishment. The growth is even more when the combination of beach and dune volume is considered as shown in Figure 9.

The beach at Heemskerk (3 km south of test area South) is assumed not to be significantly affected by the Ecobeach system. Beach volume variations after the Heemskerk beach nourishment of 2005 are within the historical natural variability from 2007 onwards. Furthermore, the behavior of combined beach and dune volumes during the Ecobeach test period is similar to the behavior before the nourishment of 2005. A significant difference is observed between the nourished beach at test area North with Ecobeach and the coast at Heemskerk without Ecobeach during the test period

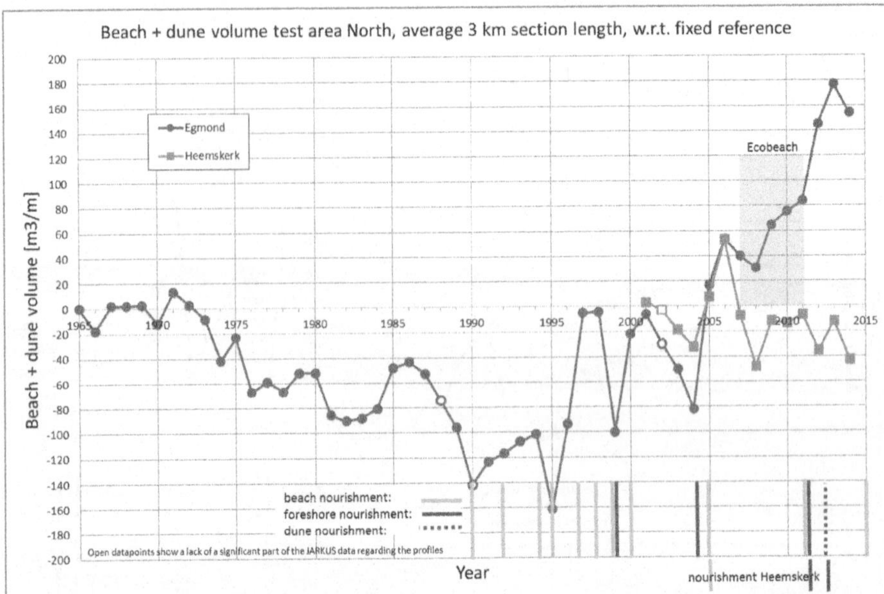


Figure 9: Beach volumes of test area North and Heemskerk over time, 1965= 0 $m^3/m$

## Scientific Research

Scientific research for possible working principles of Ecobeach was carried out by BAM together with universities and scientists from different fields of specialism. The research took place from 2008 to 2014. Hypotheses for possible working principles were formulated in relation with the field of (geo) hydrology, geology, sedimentology and biology.

Coarsening of beach sediments was found in the upper 2 m of the intertidal zone of the southern test area (Ekkelenkamp, 2011), while measurements of ground water pressure and salinity showed the outflow of fresh water from a drainage tube near the low water line (Pieterse, 2009). This triggered the research to focus more on beach sediment.

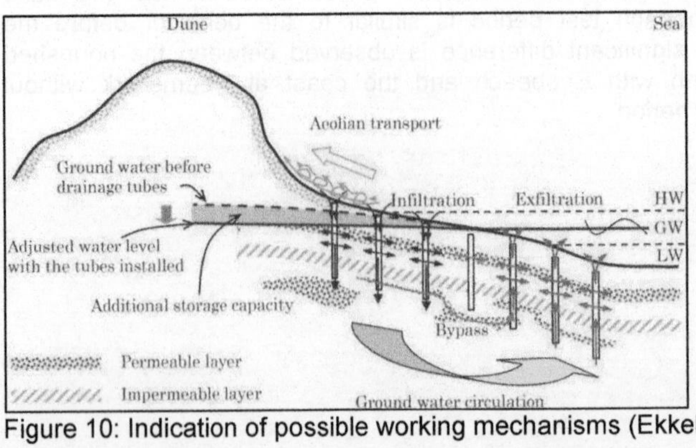


Figure 10: Indication of possible working mechanisms (Ekkelenkamp, 2011)

**Sediment Coarseness Investigation**

Many sand samples were analyzed by using a laser diffraction spectrophotometer. Most important property output for the purpose of this investigation is the grading and the grain shape of the sample. Beach sampling was carried out along 20 km coastline including the Ecobeach test areas. The granulometric analysis comprised the comparison of the sediment data from before, during and after the test. To determine the sediment properties, cone penetration tests and boreholes were made. The results showed a significant coarsening of the sediment in the active zone of the beach at the test area South. The upper 2 m of the beach is defined as the active beach zone. Moreover, soil stratification was observed. The beach consists of horizontal layers with each layer containing its own permeability. The drainage tubes connects the highly permeable layers resulting into drying of the beaches after high water conditions. Dry beaches promote Aeolian transport of fine sediments to the dunes. Therefore, sediments at the beach become coarser leading to a higher permeability. The last measurements in 2014, three years after the end of the test period, resulted in a decrease of the presence of relatively coarse material within the southern test area. Test area North did not show significant coarsening of the sediments during the Ecobeach test period. This may be due to the extensive nourishments of this area with offshore sand and thus absence of coarser fractions. Investigations in Denmark also showed the presence of coarser sand in two Ecobeach test areas along the coast of Hvide Sande which is also located on the North Sea coast.

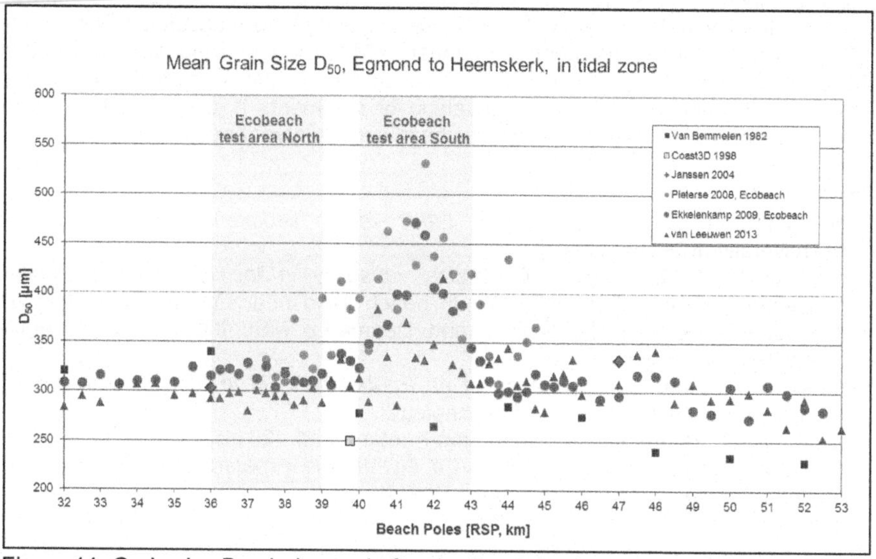


Figure 11: Grain size $D_{50}$ during and after the Dutch Ecobeach pilot

**Working Principle**

Several hypotheses concerning the working principles of the Ecobeach system were investigated. Beach stabilization based on groundwater transport through the drainage tubes is a possible working mechanism. Outflow from fresh water through a drain near the low water line has been observed. The drainage tubes will penetrate through less permeable silt layers and connect more permeable shell layers, leading to a higher vertical permeability of the beach near the drains. As the beach is able to dry quicker, the wind driven sediment transport increases. More fine sand is transported by wind towards the dunes resulting in a coarsening of the active beach zone in the vicinity of the drainage tubes. A coarser beach dries relatively quick, because the infiltration capacity of water is improved and the thickness of the capillary zone is decreased. Therefore a local effect near the drainage tubes can be extended to the rest of the beach. A coarser beach has higher permeability and a higher permeability causes faster infiltration of water. Therefore less backwash of incoming waves is expected leading to less suspension of sediments caused by backwash.

Coarsening of the beach also affects the stability. During stormy conditions, the eroded sand volume at a relatively coarse-grained beach is significantly lower than at a relatively fine-grained beach. Therefore the coarsening of the beach will have a stabilizing effect on the coastline. Moreover, due to the increased wind driven sand transport, the volume of the dunes will increase, creating an additional sand buffer to protect the coastline during a storm.

## Conclusion

After installation of the Ecobeach system near Egmond, significant changes in beach volume trends were observed during the test period between 2007 and 2011. A significant increase in beach volumes was observed during the test period compared to the natural variability before the test, reaching highest measured values since 1965 at the end of the test period in January 2011. Coarser sand was observed in the upper 2 m of the beach at test area South and at test areas in Hvide Sande, Denmark. A potential cause for coarsening of sediments is the increase in wind driven transport of fine sand to the dunes because of a dryer beach.

## Reference

Ekkelenkamp, H.M. (2011), *Drainage tubes versus sediment*, MSc thesis, Delft University of Technology

van Leeuwen, M.G.J. (2014), *Grain-size variability in the Ecobeach test sites* (Temporal land spatial variability of beach sand near Egmond aan Zee and Hvide Sande, the Netherlands and Denmark), MSc thesis, VU University Amsterdam

Pieterse, J.J. (2009), *The influence of an Ecobeach PEM in beach development*, MSc thesis, Deft University of Technology

Walstra, D.J.R., Brière, C.D.E., Vonhögen-Peeters, L.M. (2014), *Evaluating the PEM passive beach drainage system in a 4-year field experiment at Egmond (The Netherlands)*, Coastal Engineering, Elsevier

Coastal Management: Changing coast, changing climate, changing minds
ISBN 978-0-7277-6114-9

ICE Publishing: All rights reserved
doi: 10.1680/cm.61149.213

# Changing Perspectives – Exploring and Implementing the Realities of Sea Level Rise in the Coastal Community of Surrey, British Columbia

**Carrie Baron,** City of Surrey, Surrey, B.C., Canada
**Jeannie Lee,** City of Surrey, Surrey, B.C., Canada
**David Hislop,** City of Surrey, Surrey, B.C. Canada

## Introduction

The City of Surrey is a Canadian coastal community located within the Greater Vancouver area. It has a long history of flood protection and mitigation works starting from pioneering settlers in the 1800s reclaiming estuarine areas for farming purposes to the present where the floodplain is now a significant area for industrial, commercial and agricultural land uses in the region.

The City's floodplain areas cover over 85 sq. km. These floodplains are vulnerable to both coastal and riverine flood processes. Citywide, there are three distinct floodplain areas: the coastal area, the Serpentine & Nicomekl River systems, and the Fraser River floodplain. Flood protection works servicing these areas consist of 115km of dykes, 30 pump stations, 2 sea dams, and over 170 flood boxes.

In 2008, some lowland areas of Surrey experienced flooding during high tide, storm surge events. Further investigation found that rising tidal events, land subsidence and rising groundwater tables combined to cause the recurring problem. The sandy soils are affected by tidal water levels resulting in groundwater emerging at the surface during significant tidal events.

As the City was exploring and studying this new flooding situation, the Province of British Columbia (BC) began a detailed investigation of sea level rise and coastal dyke protection. Since 2008, the Province has published various guidance documents for local governments to use in the planning, maintenance, and operation of sea dykes. The provincial government continues to be involved with climate change science and its integration into designs. In BC, municipalities are responsible for planning, designing, constructing, operating and maintaining all flood protection works.

With the release of these new provincial guidance documents, the increasing flooding issues in the City's coastal areas, and the extensive floodplain area potentially affected by projected sea level rise, the City took the initiative to extensively study the vulnerable coastal and riverine systems throughout Surrey. This report summarizes the City's investigative journey and findings to date and also illustrates how the City is incorporating these results into Surrey's processes for operations and land use planning.

## Rivers Meet the Sea

The Serpentine and Nicomekl Rivers are the main river systems within Surrey. These combined watersheds are over 300 sq.km with a combined floodplain area of 50 sq.km as shown on Figure 1. Both river systems have sea dams located near their mouths. The sea dams prevent salt water from intruding upriver enabling farmers to use river water for irrigation. The sea dams also provide flood protection to upstream properties.

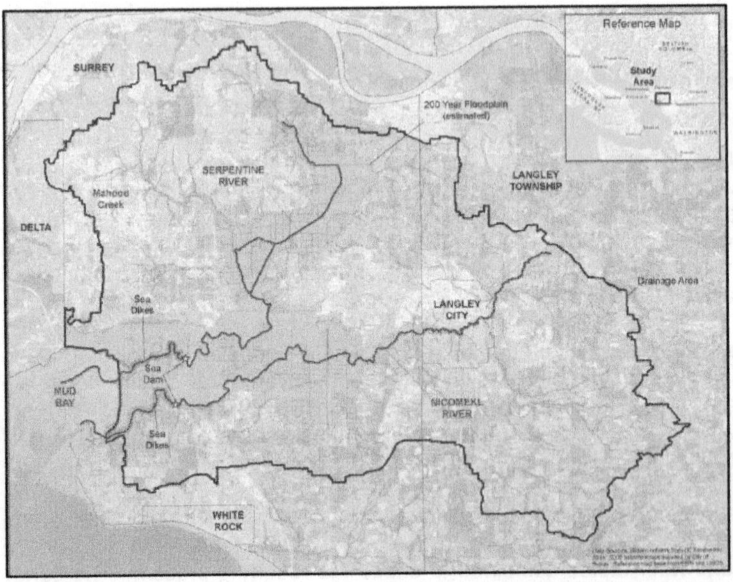


Figure 1: Serpentine and Nicomekl River Watersheds (Northwest Hydraulic Consultants, 2012)

Provincial guidance documents regarding climate change and sea level rise have focused solely on coastal areas and did not account for riverine systems influenced by coastal process. Since there was little available information on these interface areas, the City undertook its own research and studies with the use of engineering consultants and specialists. The goals of these studies were to explore potential impacts to the community, identify vulnerabilities, determine critical timing and develop a new overall floodplain strategy for both the coastal and river systems.

## Climate Change Floodplain Review Phase 1 (2012)

One of the first steps was to conduct a joint probability flood assessment of the coastal shoreline combined with river modelling. The work involved the development of a 50 year continuous simulation model to determine oceanic, hydrologic and hydraulic responses. Historic tide, flow and rainfall records were used to develop the base models.

Modelling consisted of linking the following programs for the continuous simulation:

- River Coastal Ocean Model (RiCOM)
- SWAN Model
- PC Overtopping software
- USEPA HSPF continuous hydrologic model
- US Army Corps. HEC-RAS Model

This was the first study of its kind in the province and has since been replicated in other communities and is to become the published standard for the province as an alternative to current provincial guidance.

Key findings from this study identified:

- The sea dams are vital in reducing flooding in the riverine systems. Without the sea dams, flooding is projected to be greater in terms of both duration and frequency.
- Upgrades are required for all coastal dyke systems. Recommended increases to coastal dyke heights are estimated to be between 1.7 to 3.6 m. An example of the potential impact is illustrated in Figure 2;
- Of the 13 existing bridge river crossings, three are projected to be completely submerged and 7 partially submerged by 2100. None of these bridge structures are currently designed for these conditions.
- Sea level rise will influence over 20 km upstream on both the Nicomekl and Serpentine Rivers.
- Near the sea dams, the current 200 year level of protection is predicted to have less than a 2 year level of protection by 2100.
- River dyke designs need to be reassessed for all flood cell and spillway features as they will be too low in the projected future condition.

Figure 2: North Crescent Beach Dyke Required Upgrades 2010 and 2100

This Climate Change Floodplain Review began to answer many questions regarding potential impacts to the City from sea level rise; however, it also raised many new issues that required further investigation.

Before the City undertook the next phase of our floodplain review, we conducted detailed assessments on outstanding issues such as subsidence and precipitation.

## Subsidence Review (2014)

Floodplain areas are primarily comprised of highly compressible soils. Some initial satellite work by others indicates subsidence rates in the Fraser Valley could be greater than the projected change in sea level rise. For the City, this could result in a 1.5 to 2.0 m net increase in coastal water levels for the floodplain areas. Before starting additional modelling work on the floodplain regimes, the City undertook a detailed subsidence assessment.

The City retained a consultant to analyze the information, and determine the variability of subsidence across the coastal floodplain and the Serpentine and Nicomekl River floodplains, using historic and current satellite data. The influence of pump stations on subsidence was also investigated. Primary goals in this review were to determine average subsidence rates and identify point loads and temporary influences. The study effort was challenging due to the wide distribution of data points and the limited years of data available for analysis.

The report outlines key subsidence areas, estimates rates of subsidence, and develops simplified subsidence contours (Figure 3) that can be easily understood and integrated into future studies of the floodplain.

The City is continuing to review subsidence in our compressible soil areas on an annual basis. This has become more affordable and manageable as satellite data and data processing have improved.

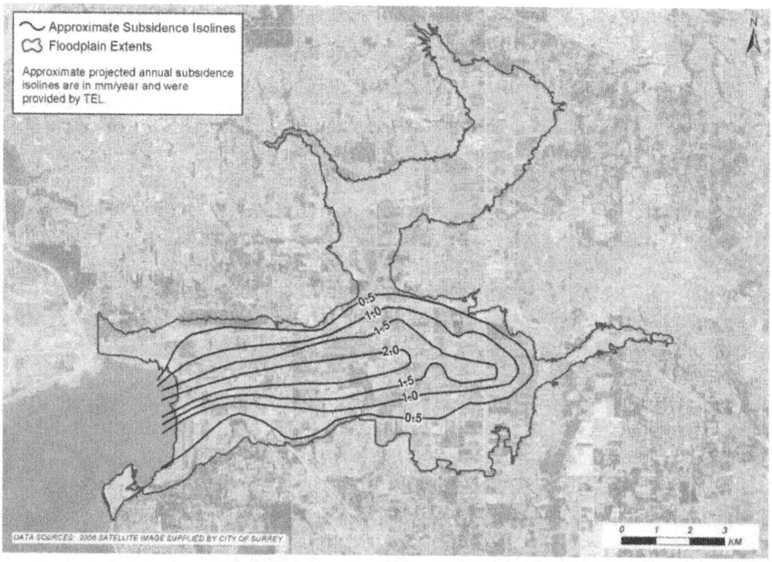


Figure 3: Approximate Projected Annual Subsidence Isolines. Northwest Hydraulic Consultants, 2014

## Precipitation Trending and Downscaling (2013 to present)

The Climate Change Floodplain Review Phase 1 was based on historic precipitation data. The upper reaches of both the Serpentine and Nicomekl Rivers are influenced more by precipitation events than by tidal/sea level rise. Climate change predictions and global climate modelling indicate this region may experience more intense storms and higher annual rainfall totals in the future. However, the large coarse scale of current Global Climate Models (GCMs) may not accurately reflect the changes at the local level. The City wanted to examine the potential impacts of future precipitation trends on the overall City and specifically floodplain response. To better understand how projected precipitation changes may influence Surrey, the City initiated a series of studies. The first was a rainfall trending assessment looking at historic data from rainfall gauges located throughout Surrey, identifying statistically relevant trends, and projecting these into the future. This study found:

- The annual rainfall at all recording station is increasing over the past 47 years.
- Overall monthly rainfalls are increasing with the exception of the month of February where a decreasing trend was identified.
- 12 hr and 18 hr rainfall events in the south part of Surrey have been statistically increasing at significant levels.

With these results, the City updated the Intensity-Duration–Frequency (IDF) rainfall curves used for drainage infrastructure design in upland areas.

The City then commissioned an analysis of a variety of emerging downscaling techniques of the global climate models to project potential future rainfall patterns. This work is underway and it is anticipated it will initiate discussions regarding revisions to design requirements of all infrastructure affected by precipitation in the City.

In conjunction with these studies, the City retained a consultant to develop two future precipitation scenarios for input into future floodplain modelling. This work involved assessing downscaled precipitation projection data prepared by the Pacific Climate Impacts Consortium (PCIC) based on 12 Global Climate Models (GCM). From these alternatives, two GCMs were selected to represent medium and extreme scenarios. The existing Surrey Municipal Hall rain gauge data was then modified to reflect the two projected precipitation regimes. For each future scenario, the precipitation time series was modified to reduce the number of precipitation days while increasing daily precipitation totals to meet future conditions. The return period of each event was also estimated for future conditions based on the two climate models.

The end results are two sets of future precipitation hyetographs representing moderate and extreme cases that can be integrated into flood modeling to examine the hydrologic sensitivities of the floodplain. The moderate scenario resulted in a 33% increase for the 200 year rainfall event compared to present values, while the extreme scenario has a 50% increase in precipitation.

## Climate Change Floodplain Review Phase 2 (2014)

With the information gathered by these previous studies, the City began the next phase of investigations. Our consultant refined the Phase 1 simulation model to include: detailed subsidence conditions; potential future precipitation scenarios; dam and dyke breach scenarios using HEC-RAS2D 5.0 (beta); seismic failure of the dam structures in the year 2100; and a variety of interim sea level rise scenarios for the years 2020, 2040, 2070 and 2200. These various scenarios are to assist with prioritizing vulnerabilities and estimating the overall floodplain extents over time.

Key outcomes from these assessments are:

- The coastal dykes have a rapidly decreasing level of service in the long term. All coastal dykes may be overtopped annually by 2070. Return frequencies overtime are summarized on Figure 4.
- Floodplain extents are not significantly altered over the projected period. However, flood depth, duration of flooding and frequency will significantly increase.
- In the Serpentine and Nicomekl River systems, the current 200 year floodplain will have a return period of less than 2 years by 2100. In the upper watershed, the current 200 year water levels will be influenced by precipitation changes increasing the flood frequency to a 5 to 10 year recurrence by 2100.
- The dyke breach modelling shows that a present day breach would result in extensive damage to the floodplain and infrastructure with the potential for loss of life. In 2100, this same breach would result in catastrophic impacts and probable loss of life due to inundation depths and high water velocities. In 2100, a dyke breach during a significant storm event is predicted to significantly affect properties and infrastructure at least 10 km upstream of the river mouth.

- The extent of infrastructure vulnerabilities is projected to expand in the future. Figure 5 outlines impacts to significant transportation corridors, railways, bridges and dyke infrastructure over time.

| Site | Location | DCE(m) | 2010 | 2020 | 2040 | 2070 | 2100 |
|---|---|---|---|---|---|---|---|
| 1a | Colebrook – Serpentine | 2.84 | 22 | 8 | 2 | 0 | 0 |
| 1b | Crescent Beach East | 2.88 | 220 | 70 | 7 | 0 | 0 |
| 1c | Mud Bay – Serpentine | 3.00 | 28 | 11 | 2 | 0 | 0 |
| 1d | Mud Bay – Nicomekl | 2.98 | 350 | 110 | 10 | 0 | 0 |
| 2 | Colebrook (Highway 99) | 3.15 | 22 | 10 | 2 | 0 | 0 |
| 3 | Crescent Beach North | 2.90 | 2 | 1 | 0 | 0 | 0 |
| 4 | Crescent Beach South | 3.30 | 70 | 22 | 3 | 0 | 0 |
| 5 | BNSF Railway | 3.20 | 7 | 3 | 0 | 0 | 0 |
| 6 | 8th Ave. at Campbell | 2.30 | 4 | 1 | 0 | 0 | 0 |

Figure 4: Total Water Level Return Periods (Years) Associated with the Current Dyke Construction Elevation for Selected Years from 2010 to 2100. Northwest Hydraulic Consultants, 2014

| Asset | 2010 | 2020 | 2040 | 2070 | 2100 |
|---|---|---|---|---|---|
| Bridges | 12 | 13 | 13 | 15 | 15 |
| Sea Dams | 0 | 0 | 0 | 1 | 2 |
| Major roads (km) | 3.2 | 3.3 | 3.8 | 4.4 | 5.1 |
| Dykes (km) | 7.4 | 13.7 | 21.1 | 46.0 | 58.2 |
| Rail lines (km) | 0.3 | 0.3 | 0.4 | 0.6 | 0.7 |

Figure 5: Flooded Assets Over Time During a 200 Year Design Storm Event (Dykes Intact)

## Other Projects Underway and Proposed

The City continues to integrate the findings of climate change studies into various City projects. Some key projects currently under design include:

- Preliminary designs for the replacement of the two sea dams including transition designs for future coastal conditions of 2100;
- Preliminary designs for a portion of Nicomekl River dyke to incorporate river levels predicted for 2100;
- Replacement pump station and dyke reconstruction at the Colebrook Pump Station for current and 2100 needs; and
- Development of an overall flood strategy for coastal protection.

## Integration of Climate Change Adaptation Works into City Bylaws and Policies

Conducting studies to determine the potential impacts of sea level rise and climate change on the City and its residents is only one component of ensuring a safe and informed community. Surrey City Council recognizes the need to assess all the issues that may impact the City. As a result, the City adopted a citywide Climate Change Adaptation Strategy. This plan assimilates the science of climate change with doing business and providing services to our community.

The development of the Climate Change Adaptation Strategy involved all City Departments, staff from various levels of the organization, and key community stakeholders. The Strategy looks at all manners of potential impacts from agriculture and irrigation to changes in vegetative and wildlife communities. Energy issues were included in a separate document: Community Energy and Emissions Strategy.

Sea level rise and future precipitation changes are prominently featured in the Adaptation Strategy and have led to increased support to continuing progress in the development of strategies and infrastructure works to protect the public. The Strategy has also enabled a new funding category to be included in the City's Drainage Utility to help advance climate adaptation works in the City.

Surrey has also recently adopted a new Official Community Plan. Within this Plan, policies such as the newly established Hazard Land Development Permit Areas (DPAs) were created. These DPAs will assist to ensure new developments and re-developments in and adjacent to floodplains are designed to protect public safety and effectively manage the lands within the floodplains. Similar revisions are currently underway to the City's Zoning By-Law in relation to floodplain requirements. Recent examples of climate resilient construction as a result of the new DPAs can be seen as re-development occurs (Figure 6).

Figure 6: Raised Home for Flood Resiliency

Informing the community by educating those who live or work in flood-prone areas is essential to the success of effective floodplain management. Research on educating two demographically different communities was recently completed for the City. From that study and the annual work with one community, development of a broader communications strategy is underway.

## Regional Integration

The City of Surrey, along with other local governments in the region, understands the importance of coordinated floodplain management. Through the Joint Program Committee of the Fraser Basin Council, local governments, provincial ministries, regional governments, First Nations, federal ports and airports, and private and public sector organizations have come together to advance the science and enable a coordinated approach to address the risks related to climate change and sea level rise in the region. These regional works and projects are intended to expose the economic importance of vulnerable infrastructure in floodplain areas.

Other groups are also conducting joint partnership studies to highlight various factors which may be overlooked when floodplain decisions are being made. The BC Agricultural Institute has recently completed works looking at the economic impact to the agricultural sector in the City of Surrey, City of Richmond and Corporation of Delta. This project is critical in highlighting the value of agriculture to the region and the potential impact of flooding to the industry and local food supply.

## Conclusion

Climate science has experienced considerable advancements in a short period of time. The City of Surrey recognizes the importance of integrating current science into both capital and community planning to ensure our residents and industries continue to live in a safe and healthy environment with effective management for current and future needs. Through our works to date, we have advanced climate science and embraced its integration into our community.

## References

Northwest Hydraulic Consultants (2012), Serpentine, Nicomekl & Campbell Rivers Climate Change Floodplain Review

Northwest Hydraulic Consultants (2014), Serpentine & Nicomekl Rivers Climate Change Floodplain Review – Phase 2

Northwest Hydraulic Consultants (2014), Two 21st Century Scenarios of Hourly Precipitation for the City of Surrey, B.C.

Urban Systems Ltd. (2009), Crescent Beach Climate Change Floodplain Review

Climate Action Initiative BC Agriculture & Food (2014), Potential Economic & Agricultural Production Impacts of Climate Change Related to Flooding in the Fraser Delta

City of Surrey (2014), City of Surrey Official Community Plan

City of Surrey (2013), Climate Change Adaptation Strategy

Fraser Basin Council (2014), Introducing the Lower Mainland Flood Management Strategy

Dillon Consulting Ltd. (2013), Rainfall Trend Analysis for the City of Surrey

Dillon Consulting Ltd. (2014), Updated IDF Statistics for the City of Surrey

Thurber Engineering (2014), Surrey Flood Plain Subsidence Statistical Characterization and Geotechnical Review

**Coastal Management: Changing coast, changing climate, changing minds**
ISBN 978-0-7277-6114-9

ICE Publishing: All rights reserved
doi: 10.1680/cm.61149.223

# Coastal Communities 2150: Increasing Resilience to Climate Change Through Community Engagement

**John Gower,** Environment Agency, Worthing, UK
**Andrew Gilham,** Environment Agency, Worthing, UK

## Introduction

Coastal communities across Europe face the prospect of significant risk from future climate change through rising sea levels, increased storm surges, accelerating erosion rates and increased air and sea temperatures. These changes are likely to impact all areas of coastal life and providing increased traditional levels of protection (such as higher sea walls) is not always a realistic option either environmentally, economically or socially.

Climate change is predicted to cause significant increases in sea levels, wave heights and storms well into the next century. By 2100 the highest tide of a year could be over 1 metre higher than currently seen (From Ice to High Seas 2013). Increased global temperatures will continue to impact and it is predicted that sea levels will rise for many hundreds of years even if global carbon dioxide emission reduction targets are achieved. Upper estimates predict a possible extra 9–15 metres of sea level rise in 1000+ year's time.

The Environment Agency's Flood and Coastal Erosion Risk Management Long-term Investment Scenarios (LTIS) 2014 reports that the cost of flood and coastal risk management in England and Wales is likely to exceed £102 billion (plus inflation) over the next 100 years, just to maintain the current levels of protection against high tides and more intense storms. Uncertainties in the impacts from climate change could significantly increase these costs to a point where we may need a fundamental rethink in our approach to planning how we defend coastal communities in the future.

This paper describes work to prepare coastal communities at 10 sites in the south east of England for climate change. It explains the process of engaging local people, and their representatives, to become part of the discussions for future planning and adaptation to long-term change. It gives positive examples of best practice from other regions, reports lessons learnt and shows that by becoming better informed people are more likely to make the right decisions for their future.

## Coastal Communities 2150 INTERREG project

The learning for this paper came from work carrier out by the Environment Agency and Kent and Hampshire County Councils in the Coastal Communities 2150 project (CC2150). This was a multi partner European project co-funded by the 2 Seas INTERREG IVA Programme. Although focusing on the UK work, the findings came from collaborative working through the 'cross boarder partnership' and are equally applicable throughout northern Europe. The objectives of CC2150 were to increase awareness of climate change and to get coastal communities involved and actively engaged in the challenges of adapting to the problems that climate change and rising seas are predicted to bring in the next 150 years. The project was led by the Environment Agency, with partners from Kent and Hampshire County Councils, the University of Wageningen from The Netherlands and two partners from Belgium: the Province of West Vlaanderen and the Maritime and Coastal Services Agency - Coastal Division. The region Nord Pas de Calais from France was also an observer.

## CC2150 Partner Study Areas

### At Risk Communities

CC2150 engaged with vulnerable coastal communities in the partners' priority areas by forming community engagement groups who helped devise, pilot and test climate change communication and awareness raising tools. Partners then helped their communities produce long-term visions and action plans for how the pilot study areas might adapt to and take advantage of future changes. These visions and action plans would then be used by the community to inform future plans, thus ensuring that vulnerable communities had tools and information to reduce their risk. CC2150 ran from January 2011 until June 2014.

Each of the Partners had different priorities for local engagement. By choosing a selection of communities the project achieved a broader spread and understanding of different coastal communities' engagement needs.

## Stakeholder Analysis

All partners conducted a stakeholder analysis that identified local people, organisations and groups and the value of engaging with them during the project. CC2150 developed a new methodology for recording and presenting results. These stakeholder planning and engagement diagrams (see example in figure 2) were then used to target activities and to choose representatives to form community engagement steering groups.

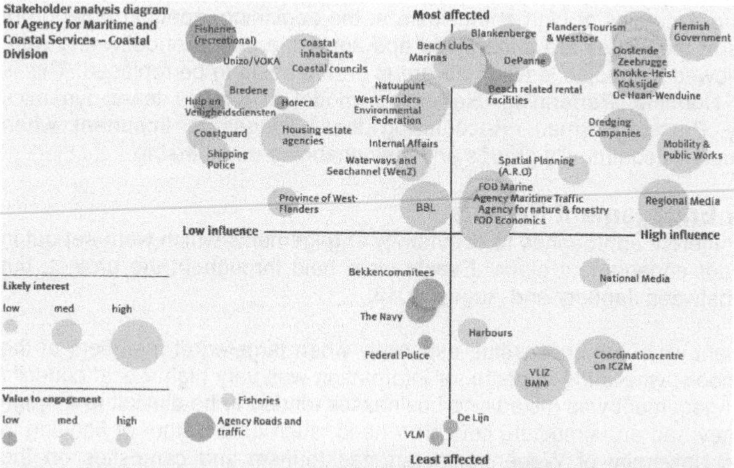


Figure 2: Example Stakeholder Analysis from the Agency for Maritime and Coastal Services – Coastal Division

A copy of the full stakeholder analysis report is available at www.CC2150.eu

## Stakeholder Engagement Planning

Stakeholder engagement plans were developed to decide what communication tools and methods were needed and how they would be used. A lot of interest was found in the English partners' communities who were very enthusiastic to become involved in local engagement steering groups to help with the planning and preparation stages. In Belgium the target audience was Municipalities, Town Mayors and coastal businesses. In The Netherlands the focus was on improving tourism and leisure on the island of Schouwen Duiveland, where many areas, including Natura 2000 sites will be impacted by climate change. There was much less appetite for the English style of local community engagement in Belgium and The Netherlands, but they had more success with engaging with business, especially in The Netherlands. Kent CC investigated the psychological principles and barriers to communicating on climate change through a study by the University of Kent (Sutton et. al. 2012). This work was used to inform the engagement activities.

## Involving Local Representatives – Use of Steering Groups

Working with engagement groups helped steer the project work and was invaluable for getting local support. English partners found high interest from town and parish councillors and members of public who were happy to give their time for planning and local engagement events. In Sussex, the Environment Agency's group took responsibility for signing off the lower River Ouse valley 'Engagement Plan'. This group greatly benefitted from having an independent chair person.

The River Ouse valley steering group was made up of 12 members. This was based on previous experience involving local communities where it is useful to have enough members to spread workload and be representative, but also to keep numbers manageable enough for effective meetings. This community group followed the usual group dynamics of high enthusiasm at the beginning, needing a period of bedding in whilst the issues are understood and agreed, a period of delivering work effectively, followed by people leaving the group and needing to be replaced. This is the 'Forming, Norming, Performing, Reforming' model of group / team dynamics developed by Bruce Tuckman. Recognising these stages is important when providing support for community groups and their changing membership.

## Community Engagement Results

Partners had different approaches to community engagements which were set out in their stakeholder engagement plans. Events were held throughout the project, but concentrated between January and August 2013.

The engagement was very successful, especially when targeted at members of the public and schools, where the appetite for information was very high. Local council / municipality engagement was mixed and businesses tended to be difficult to engage with, unless they had an immediate concern / need, such as a history of flooding in the area. The University of Wageningen targeted tourism and campsites on the Island of Schouwen Duiveland, which suffered badly in the 1953 floods. They had success in helping a privately owned campsite relocate away from coastal defences by working with them over time and allowing planning permission for a new, higher quality site. The business was happy with their better facilities and their old campsite site was used to make an 'environmental corridor' to connect two separate parcels of land to make a bigger more resilient nature reserve.

Figure 2: Photos of using maps and asking the Oostende public how their coast could be improved when new defences are installed

## Engagement Tools

Partners developed and used a series of tools to demonstrate the future impacts of climate change and the benefits of adaptation. There was a lot in common between the partners in the use of the most succesful tools. Images, aerial photos and maps were extremely important. Illustrations and computer visualisations were very successful in communicating the impacts of sea level rise and how land use in coastal zones might need to change in the future.

Table 2: Engagement tools

| *Tool* | *Benefit* | *Cost*<br>*Low, Med, High* |
|---|---|---|
| *Maps & aerial photos* | *High general interest, factual and freely available* | *Low / free* |
| *Images of the past & historical events* | *Factual & reminds how much coastline can change* | *Low / free* |
| *Postcards to the future* | *Fires imagination, links location with the future* | *Low* |
| *Displays at local coastal events* | *Interested audience but requires staff attendance for best results* | *Medium* |
| *Posters* | *Allows messages and concepts to be explained* | *Medium* |
| *Flood markers* | *Helps remind people of past events* | *Low / medium* |
| *Flythrough/ videos* | *Presents complex ideas in simple ways. Can be web based* | *High* |
| *Visualisations / adaptation options* | *Good for presenting future scenarios and adaptations* | *Med / high* |
| *Web sites* | *Useful but not well used* | *Med / low* |
| *Social Media* | *Good for raising awareness but face to face discussion more successful* | *Low* |

Figure 3: Front and reverse of 'postcard to the future'

Kent County Council produced 'postcards to the future', which were cheap, extremely effective and sucessfully used by 3 other partners. Hampshire CC developed a computer aided 'fly-through' based on 'LIDAR' ground levels and predicted sea level rise. The Environment Agency developed this concept and made it interactive by adding a choice of sea level rise and controls to 'draw users in'. This was very effective in explaining climate change impacts for the lower Ouse valley. The fly-throughs were liked by all sections of the communities, from school children through to local authorities and councillors.

We found that many of the communication tools were transferrable between partner communities, with maps, fly-throughs and postcards to the future being especially successful. It is interesting that these are at opposite ends of the cost spectrum, with maps being freely available, postcards being very cheap (€0.5) whereas a fly-through can cost upto €10,000.

Computer enhanced fly-throughs were especially successful in getting the message across about how vulnerable certain coastal communities are to rising sea levels. They were useful in schools, public exhibitions and even for briefing newly elected members of local councils on climate change, sea level rise and flood risk.

Figure 4: Lower Ouse Valley Fly-through allows user to select sea level rise application available at http://www.fusiongfx.com/EA/

## Websites and Social Media

Kent and Hampshire County Councils set up community web sites to share information. Whilst successful, they were not used as much as expected and greater benefits were gained from face to face engagement events. Facebook and Twitter were used, but they were more useful as signposts to actual events, where face to face interaction was the preferred way for communities' to receive information on climate change. Static displays in libraries and similar events without a person to explain the issues and take verbal feed back were less successful than anticipated.

## Community Visions and Action Plans

The 3 UK partners developed new long-term visions and action plans for how the at risk communities could adapt to climate change, whereas the Dutch and Belgium partners used existing plans. Involving communities in adaptation was extremely successful and greatly increased awareness of the risks and opportunities that climate change will bring. Levels of understanding increased throughout the engagement process. This has left an important legacy of improved community knowledge which will help future discussions on short to medium-term proposals to improve local defences.

The River Ouse valley community engagement group has adopted the local CC2150 vision 'waking up to tomorrow' and are continuing to work beyond the life of the project on its main recommendations;

- to continue increasing local awareness of climate change,
- to make the local communities more resilient to flood risk by signing up to the Environment Agency's Flood warning direct telephone warning service,
- to work more with businesses to raise awareness and look for solutions, and
- begin local adaptations to a warmer climate that will have more extremes of rainfall and storms.

Figure 5: Vision and Adaptation Plan for the lower River Ouse valley

## Barriers to Engagement

The acceptance of climate change by communities was mixed through partners' areas but generally easier than expected. Some people did not accept that human activity is causing / contributing to sea level rise and climate change but most people who attended organised CC2150 events did accept this, especially in coastal areas. Lewes, Newhaven and Seaford in the River Ouse valley were wholly accepting without a single climate change denier making a challenge. The project team for this area had prepared carefully including producing background posters on climate change, measured sea level rises and set up a series of explanatory meetings to deal with the science behind climate change. Recent flooding in Lewes (October 2000) and a series of near misses with very high tides at Newhaven meant locals were very aware of coastal flood risk and keen to become involved. The heavy storms of winter 2013/14 also reinforced these messages.

## Recommendations and Benefits of Engaging with Communities

Engaging with communities takes time and it took longer than expected to design and complete the engagement programmes. Climate change is complex and CC2150 found it was best to break the subject down into smaller 'bite size' portions. Community groups reacted well to being given technical presentations over a series of weeks, each dealing with a defined topic. This allowed full understanding and for the group to move on. It is important to be flexible and work at the right pace otherwise people will lose interest. A great benefit from CC2150 was where community representatives became strong supporters and willingly gave their time, both at the steering groups and to help at local engagement events. CC2150 ended in June 2014 and has left a legacy of extensive communication materials, over 40 case studies and 9 local vision and adaptation plans. These will be useful at other locations throughout northern Europe and beyond.

The lower River Ouse group has continued to meet beyond the end of the project and work on greater community involvement in local climate change adaptation measures. This has been of great benefit to the Environment Agency, with additional town meetings being organised by the local community partnership and beach users. The local communities and town council now understand much better how the local sea defence, a popular shingle beach, is maintained to manage flood risk. Relationships with Newhaven Town, Seaford Town and the local parish councils have been greatly improved with a much clearer understanding of the issues. In Newhaven a major project is planned to improve the existing defences. The public consultation will be much easier because of increased knowledge and awareness and the support being shown by community representatives involved in the project. These local residents have now become strong public advocates for adapting to climate change and are playing a key role in communicating important messages that were previously delivered solely by public bodies. This public support is a new and much welcomed benefit.

## Pathway Approach to Adaptation

An important learning point in CC2150 was the use of 'pathways' and breaking complex subjects into smaller, more easily manageable tasks. This is applicable to both the problem of explaining climate change and the need to adapt to it. Figure 6 shows how a better future can be achieved by 'Envisioning' where you want to be in 150 years, and then considering where current actions such as public policy etc. are forecasting you will arrive – the expected future. There inevitably is a gap, often much too great to make in one leap. But by 'Backcasting' and envisioning the smaller steps needed to deal with the big problem, it is possible to set out a pathway that is both manageable and achievable which leads to a more sustainable and prosperous future. It is important that the first, and subsequent steps are on the 'new pathway'. This is sometimes called the path of no or low regrets and illustrates how small amounts of extra investment in the right measures now, can have great benefits in the future. An example would be installing a wider foundation that allowed sea walls to be extended with future sea level rise.

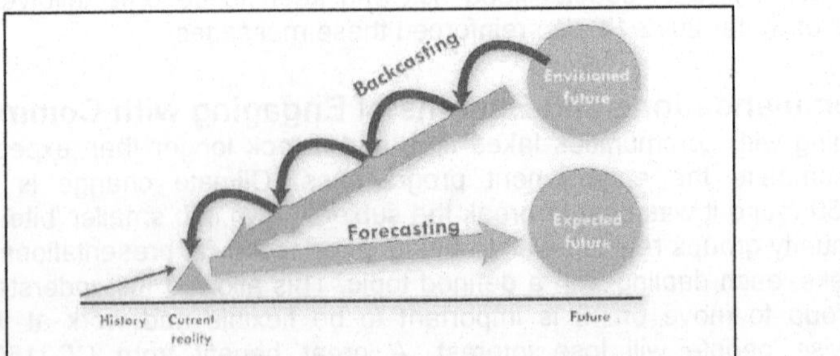


Figure 6: Illustrating the 'backcasting' pathway approach

## CC2150 Key Learning Points

- Communities were more accepting of climate change than expected and wanted to become part of the decision making for deciding their future.
- School children and young people were very interested and not overwhelmed even by extreme climate change and sea level rise scenarios. This was in contrast to warnings in the early stages of not to frighten people.
- There was less public confusion over 'uncertainty' (i.e. how much sea levels and temperatures might rise) than expected.
- Face to face communications worked best. Web sites and social media were more successful at promoting events than starting discussion or raising awareness.
- People respond well to photographs and maps, especially when they show historical coastal changes.
- Illustrated visualisations and concepts for the future help to start discussion about what can be done now, and how they can become a pathway to future adaptation.
- Shared tools and concepts transferred very well across the 3 different countries.
- Informed communities are much more likely to engage with decision-making over adaptation options.
- The CC2150 'Pathway' approach of breaking down problems into smaller easy to deal with chunks is a successful way of dealing with climate change adaptation.
- Local advocates provide huge benefits in explaining current coastal work and future options for adaptation.
- The project has left a valuable positive legacy.

## Concluding Remarks

CC2150 was a very succesful project which has greatly increased the resilience of communities in the pilot areas by making them aware of climate change and the need to adapt. Partners engaged extensively in their 6 areas and developed successful ways for involving community members and local councils, who produced and used long-term visions and action plans. Local steering groups were extremely useful in getting community involvement and preparing adaptation actions for the future. CC2150 identified considerable best practice and communication tools which are transferrable to other areas. It has already left a positive legacy with better informed communities continuing to work, independently, beyond the life of the project. In the lower River Ouse valley community members have continuing to use the vision and principles of long-term adaptive planning and have become advocates and powerful spokespeople for the issues, which will greatly help local councils and the Environment Agency. This and further information can found at www.CCC2150.eu

## References

From Ice to High Seas: Sea-level rise and European coastlines, The ice2sea consortium, Cambridge, United Kingdom, (2013) www.ice2sea.eu

Flood and coastal erosion risk management Long-term investment scenarios (LTIS) 2014 - Environment Agency 2014

Reeder.T and Ranger.N, How do you adapt in an uncertain world? Lessons from the Thames Estuary 2100 project.
R.M. Sutton, et, al, Engaging Coastal Communities in Climate Mitigation and Adaptation Measures – basis for the application of psychological principles Phase 1 and 2 Reports
Tuckman, Bruce W. (1965) 'Developmental sequence in small groups', *Psychological Bulletin*, 63, 384-399.
UKCP09Marine Projections Report
CC2150 Final Project and Communications Report

**Coastal Management: Changing coast, changing climate, changing minds**
ISBN 978-0-7277-6114-9

ICE Publishing: All rights reserved
doi: 10.1680/cm.61149.233

# A Novel Technique for Rapid Flood Forecasting in Coastal Areas

**M. van Ledden**, Royal HaskoningDHV, George Hintzenweg 85, Rotterdam, The Netherlands. Also: Delft University of Technology, Stevinweg 1, Delft, The Netherlands. Email: mathijs.van.ledden@rhdhv.com
**H.W. van den Brink**, Royal Meteorological Institute, De Bilt, The Netherlands
**J. Klein**, University of Twente, Enschede, The Netherlands
**S. Caires**, Deltares, Boussinesqweg 1, Delft, The Netherlands
**J. Groeneweg**, Deltares, Boussinesqweg 1, Delft, The Netherlands

## Introduction

Large-scale coastal flood disasters occur frequently around the world. For example, the North Sea region (see Figure 1) has been battered with many severe storms in the past resulting in significant damage and loss of life. Well-known examples are the storms in 1953 (Netherlands, UK, Belgium) and 1962 (Germany) and very recently the storms of December 2013 and January 2014 battering the French, English and Scottish coastlines. These storms can cause significant damage in the coastal areas around the North Sea. For example, the January 1990 storm caused an economic loss of around US $ 6.8 billion (1992 price level), see Dorland et al. (1999).

Figure 1: North Sea region with numbered locations considered in the validation of this North Sea Storm Atlas (source: Klein, 2015)

Early storm forecasting and issuing warnings are key aspects for mitigating flood risk during storm events. These forecasts have to be timely and accurate in order to give good insight into which locations are at risk and what effective emergency measures can be taken. Furthermore, a key aspect for mitigating flood risk (both in planning/design studies but also in emergency situations) is the understanding of hydraulic loads during extreme storm events. This understanding can be obtained through storm scenario analyses. Therefore, both early storm forecasting and storm scenario analyses tools are paramount for flood warning and mitigation.

Currently, the countries along the North Sea have or are implementing operational storm surge forecasting system based on real-time detailed meteorological and hydrodynamic modeling. These forecasting systems are accurate, but computationally expensive, heavily dependent on the reliability of the ICT infrastructure during extreme events and involve high implementation and maintenance costs. Furthermore, the needed computational time of the detailed models lead to challenges in terms of having sufficient time/flexibility to do long forecasts and uncertainty or scenario analyses. In parallel with these real-time systems, a flexible and fast forecasting tool allowing for longer term forecasting and scenario analysis is, therefore, desirable.

In tropical cyclone prone regions, the need for a tool, allowing for quick longer term forecasting and scenario analysis, in parallel with real-time forecasting based on detailed meteorological and hydrodynamic modeling, led to the development of a *Storm Atlas*. The atlas was developed after Hurricane Katrina for hurricanes around New Orleans. After this hurricane, a suite of synthetic hurricanes, with varying tropical cyclone parametric characteristics (such as storm track, radius and intensity) has been used to force detailed meteorological and hydrodynamic models, resulting in a database of parametric storm characteristics and associated near shore surge and wave conditions. For emergency purposes, these have been put into the so-called "Hurricane Surge Atlas" and the application of this atlas during various new hurricanes (Gustav, 2008; Ike, 2008; Isaac 2012) was very successful (Van Ledden et al., 2013).

Such a storm atlas is, therefore, desirable in coastal regions prone to flooding. However, given the fundamental difference between extreme storms in hurricane prone regions (which are tropical) and at higher latitudes (which are extra-tropical) the parametric approach used in the "Hurricane Surge Atlas" cannot be directly applied in extra-tropical regions. Motivated by these challenges, this paper proposes a technique for quick storm forecasting valid in extra-tropical regions with a pilot application to storm surge forecasting in the North Sea. The objective of this pilot North Sea Storm Atlas is a quick but still reasonably accurate storm surge prediction for 10-day ensemble weather forecasts.

## Methodology

The main idea behind this atlas is the use the seasonal forecasting dataset of the European Centre for Medium-Range Weather forecast (ECMWF), which cumulates to 3500 years, to drive local hydrodynamic models and to locally classify or, more precisely, identify storms. The storm characterization is carried out in a non-parametric way and surges are predicted though the identification of analogues in the created long term data base of storm characteristic and associated surges.

The creation of the atlas involved four main steps, which are briefly described in the following.

**Step 1: Meteorological Data Collection and Computation of Associated Surges**
ECMWF seasonal forecast surface wind and mean seal level pressure (MSLP) data covering the region with longitudes between 16°W and 14°E and latitudes between 42°N and 65°N have been retrieved from January 1981 until August 2012. These seven months long 6-hourly forecasts are available at the start of every month. Furthermore, the forecast data in the period between May 2011 and April 2012 comes from 51 ensemble members per month and the forecast data in the remaining period from 15 ensemble members per month. Each 7 month forecast was concatenated into one fictive time series, leading to a dataset that as a whole resembles a time period of about 3,500 years. These wind and MSLP data were used to force a detailed North Sea 2DH hydrodynamic model (WAQUA), leading to associated storm surges at a number of chosen locations along the North Sea coast.

**Step 2: Storm Characterization**
After an extensive analysis of the data and of ways to characterize storms, the MSLP fields of the retrieved ECMWF dataset were used to define a set of 50 Empirical Orthogonal Functions (EOFs), the first 50, to be used in the (non-parametric) characterization of individual MSLP fields. An EOF analysis describes the variation in the spatial pattern of a given parameter (MSLP in case of the Storm Surge Atlas). As a first step, the mean state (i.e., the climatology) is subtracted.

The EOF analysis uses the dataset to construct orthogonal patterns that optimally describe the temporal variations. The first EOF pattern is the pattern that explains most of the variance in the original time series, etcetera. Depending on the percentage of the variance that has to be explained, less or more EOF patterns are needed to describe the original field. For every time step, all EOF patterns have certain amplitudes ('eigenvalues'), so that summation of the weighted EOF patterns results in (approximately) the original MSLP field.

The resulting first 4 EOF patterns of the MSLP obtained are shown in Figure 2. Especially the 2$^{nd}$ pattern is expected to be important for the surge at the Dutch coast because a large amplitude of this pattern indicates a strong wind from North-North-West directions. Although not shown herein, this is also confirmed by comparing the amplitude of the 2$^{nd}$ EOF pattern with the associated skew surge at Hoek van Holland in the Netherlands. These two parameters are clearly correlated.

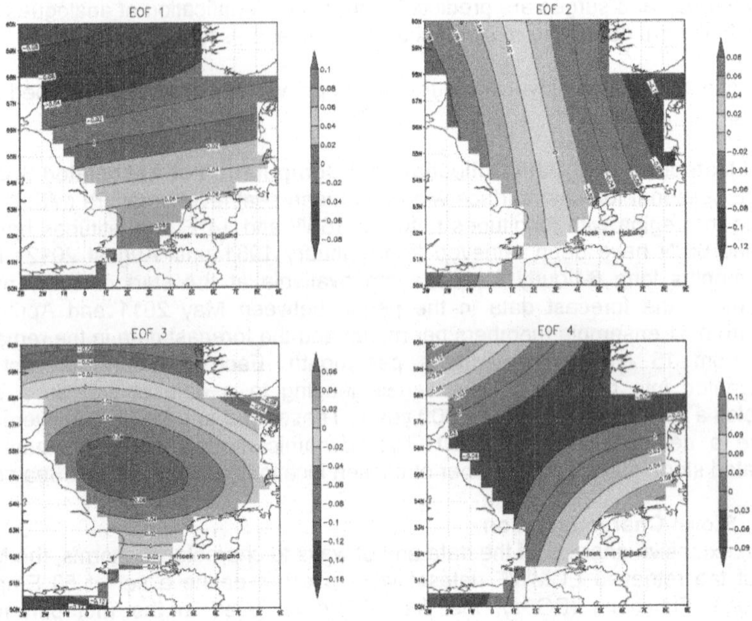


Figure 2: Spatial patterns ('eigenvectors') of the 4 first EOFs

**Step 3: Storm Surge Forecast**

The storm surge forecast is carried out by means of identification of analogues in the complied database. The database consists of the 6-hourly coefficients of the 50 EOF identified in Step 2 and the associated surge levels for the coming 6 hours computed with a 2DH hydrodynamic model of the North Sea (WAQUA).

The analogues are identified as follows:

a) For each forecast MSLP field the coefficients of the 50 EOF identified in Step 2 are computed. The distance $D_{t,T}$ between the MSLP field of time step $t$ of the forecast $f$ and time step $T$ of the database $d$ is defined as

$$D_{t,T} = \sum_{\Delta=0}^{2} W_{\Delta} \sum_{i=1}^{50} \{c_{i,f}(t-\Delta) - c_{i,d}(T-\Delta)\}^2$$

where the $c_i$'s are the 50 EOF coefficients. The distance is computed by summing over the t 50 EOFs, as well as over 3 time steps: the current one ($\Delta$=0), and 6 and 12 hours back ($\Delta$=1 and 2, respectively). These extra time steps guarantee that the analogue has a similar 'history' over the previous 12 hours as the current meteorological situation, which is important for the surge development. The weighting factor $W_\Delta$ can be used to give the previous MSLP fields less effect in the determination of the analogue. We have set $W_{\Delta=0}=W_{\Delta=1}=1$, and $W_{\Delta=2}=0.5$. The associated surge evolution over a six hour period is selected from the MSLP match in the database with the smallest distance $D_{t,T}$ ('best match').

b) The surge prediction may have discontinuities caused by the matching principle. The resampled storm surge basically consists of consecutive 6-hourly pieces of surge. When 2 sequential pressure fields in the weather forecast do not match to 2 sequential pressure fields in the database, the surge predictions may not be continuous. To tackle this problem, a smoothening technique is applied based on an exponential function such that a smooth development of the surge is guaranteed.

Figure 3 shows the scatterplot of the simulated versus the modelled skew surges at Hoek van Holland for all 1194 runs. It shows a good correlation (r=0.85), but a slight underestimation of the most extreme events. For the coastal stations along the North Sea, most of the correlations exceed 0.80. Remarkably, the skill increases while moving from the eastern UK coast to the Belgium/Dutch coast, being highest for Cuxhaven. No explanation is found yet of such spatial pattern. More detailed testing of the predictive performance is presented in the next section.

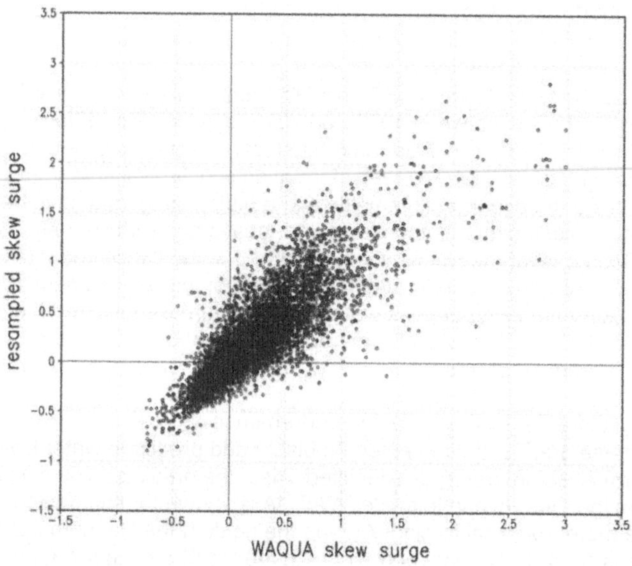


Figure 3: Scatter plot of all simulated skew surges at Hoek van Holland by resampling versus the modelled skew surge by a 2DH model for the North Sea (WAQUA)

**Step 4: Visualization Tool**
The output of the Storm Atlas is visualized through a purpose built web-based interface. Figure 4 shows a screen dump of the interface. The interface presents for each selected station the forecasted surge level based on the weather forecast on a high resolution (in red) and low resolution (in blue) grid. The blue area depicts the bandwidth based on the ensemble runs. The green line is the astronomical tide for the station. These are produced online at synoptic times. The forecasts can also be accessed in the Delft-FEWS environment.

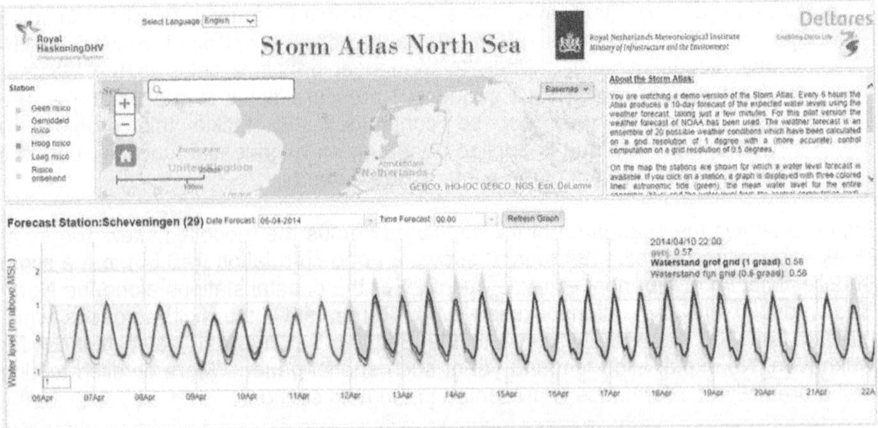


Figure 4: Web-based interface for the North Sea Storm Atlas

## Validation

The predictive performance of the North Sea Storm Atlas has been tested using 33 historical storms in the North Sea between 1979 and 2013 (see for more details Klein, 2015 and Klein et al., 2015). This timeframe for storm selection has been selected since detailed meteorological hindcast information is only available from 1979 onwards. From this database, pressure fields of the historical storms have been used. The performance assessment has focused on 11 locations, of which 6 are located at the North Sea coast of the United Kingdom and 5 at the Dutch coast (see Figure 1 for locations). Three physical parameters were part of this assessment: the peak water level, duration and timing of peak water level. Herein, we focus on the peak surge level only since this is considered the most important parameter for determining the risk of a specific storm.

Figure 5 presents two (extreme) examples of the performance at Hoek van Holland, The Netherlands. In these graphs, the observed (in black) and predicted water level by the North Sea Storm Atlas (in green) are depicted. Also, the predicted water level variation (in blue) from the hydrodynamic model WAQUA is shown for this particular storm using the same meteorological forcing. Around the peak of the December 2003 storm, the water level is predicted almost perfectly with the North Sea Storm Atlas. This is, however, not the case for the November 2007 storm. The peak surge level for the 2006 storm is under predicted with about 1 meter. It is also noted that the result from the hydrodynamic model WAQUA also shows a substantial mismatch (> 0.5 meter below the observed peak water level) with the observations.

To quantify the overall performance of the atlas, the peak water level results of all storms and all Dutch and UK stations have been analyzed for different types of errors: Root-Mean-Square-Error (RMSE) for the total error, bias for systematic errors and standard deviation for random errors. Considering all stations, the RMSE of the peak water level is 0.47 meter with a bias of -0.31 meter and a standard deviation of 0.36 meter. This analysis has also been carried out for Hoek van Holland only. For this station, the bias is slightly less (-0.11 meter), but the standard deviation is not different.

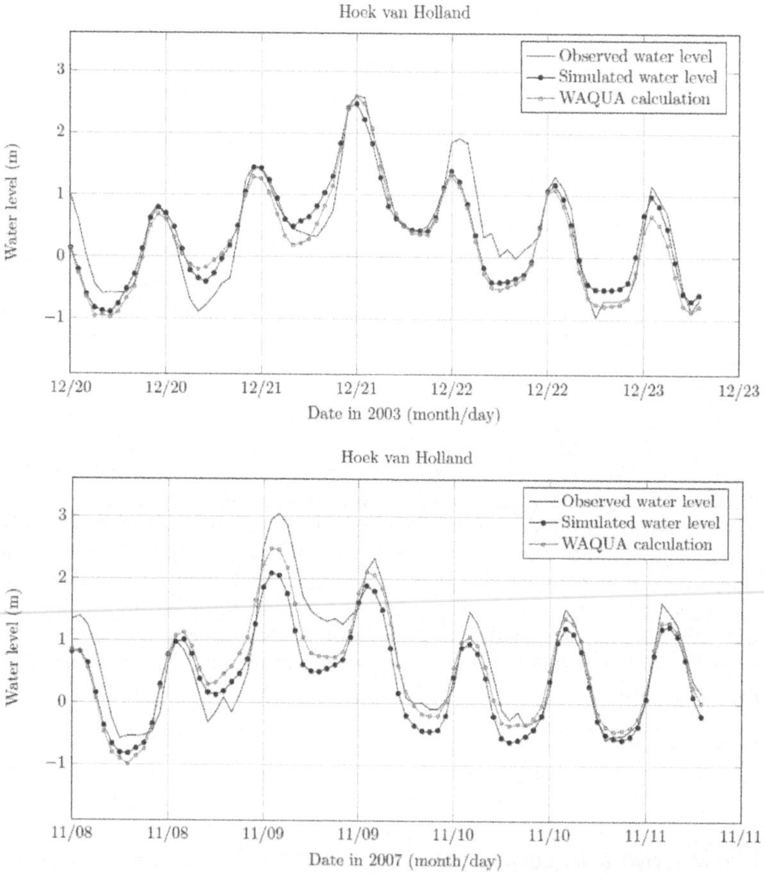


Figure 5: Validation of the North Sea Storm Atlas at Hoek van Holland, The Netherlands for storms in December 2003 (upper panel) and November 2007 (lower panel). The black line is the observed water level, the blue line indicates the simulated water level with the North Sea Atlas, and the green line is the result from the hydrodynamic model WAQUA

The overall result of this performance assessment indicates that the predictive performance is rather good. A random error of a few decimeters is acceptable if mid-term predictions (say 3 – 5 days in advance) are considered. Also, this error is not substantially larger than the typical error for peak water level predictions with real-time surge models. Further extending the database with more storms will probably improve the performance of the atlas for severe storms even further. Another source of uncertainty which needs further investigation is the tidal signal to be used in this atlas.

## Discussion and Conclusions

This paper has presented an algorithm that very quickly (order minutes) determines the surge due to pressure patterns at the North Sea for the coming 10 days at many coastal stations along the North Sea. The storm surge results for the coastal stations along the North Sea have been visualized through a web-based interface. The results from the North Sea Storm Surge Atlas can also be viewed with FEWS Noordzee, the operational forecasting system of the Dutch authorities.

A detailed validation has been carried out using 33 historical storms at the North Sea from 1979 onwards. In total, 11 stations along the Dutch and UK coastlines have been included in this validation. The result shows that the random error in the peak water level predictions with the atlas is 0.36 meter. This error seems acceptable if the mid-term storm surge forecasting of 3-5 days in advance is considered as the primary objective with this tool.

If further testing and validation is successful, it is envisioned that this tool can be valuable next to real-time models. The advantage is that it can produce a quick insight in the potential storm surge hazard for coastal regions without detailed real-time computations. Given the global availability of the ECMWF seasonal forecast data and the generality of the non-parametric approach used, this storm atlas concept is globally applicable and can also be a cost-effective solution for areas with limited resources. Furthermore, in this pilot only surge levels have been considered and it is only applied in forecast mode producing both the forecast and associated uncertainties. The atlas can be extended to be used in scenario analysis and also by adding other relevant flood variables, such as wave conditions, dune erosion, flood extents in the hinterland, etc.

## Acknowledgement

The development of the Storm Atlas for the North Sea has been partly funded by The Collaboratorium Climate and Weather project “North Sea Storm Surge Atlas (TM01)”. The authors gratefully acknowledge the support from this foundation. Also, the project WTI2017 - Hydraulic Loads project commissioned by the Dutch Ministry of Infrastructure and the Environment is acknowledged for matching funds. The performance assessment of the North Sea Storm Atlas has been carried out in cooperation with the University of Twente. The cooperation with prof.dr. Suzanne Hulscher and Dr. Pieter Roos during the M.Sc. project has been highly appreciated.

## References

Dorland, C., Tol, R.S.J., Palutikof, J.P., 1999. Vulnerability of the Netherlands and northwest Europe to storm damage under climate change. A Model Approach Based on Storm Damage in the Netherlands. Climate Change (43), 513-535.

Klein, J., 2015. The North Sea Storm Surge Atlas. Performance Assessment and Uncertainty Analysis. M.Sc. thesis, University of Twente, Enschede.

Klein, J., Van Ledden, M., Van den Brink, H.W., Van den Berg, N.J.F., Roos, P.C., Hulscher, S.J.H.M., 2015. Validation of the North Sea Storm Surge Atlas by hindcasting historical storms, accepted for publication at IAHR conference 2015, Amsterdam.

Van Ledden, M., Kluskens, R., Hillen, M., Agnew, M.E., Van der Sande, B., Spencer, S.G., Turner, R.A., 2013. A digital Hurricane Surge Atlas for New Orleans, Proceedings of HYDRO2013, International, 4 – 6 Dec 2013, IIT Madras, India.

**Coastal Management: Changing coast, changing climate, changing minds**
ISBN 978-0-7277-6114-9

ICE Publishing: All rights reserved
doi: 10.1680/cm.61149.241

# Adaptive Estuarine Management in Micro to Hyper Tidal Environments

**Paul J. Canning**, Atkins Limited, Bristol, UK

## Introduction

Estuaries often represent areas where a range of societal features intersect and are influenced by flood and coastal erosion risks. These can include flood and coastal erosion risk management (FCERM) assets, environmental designations, infrastructure, marine operations and tourism. In the south-west UK, a strategic approach has been taken to planning how FCERM assets are managed, and how they influence and are influenced by other societal needs and physical processes. Three FCERM strategies that cover the different natural and built environments of the Severn Estuary, Exe Estuary and Poole Harbour were recently completed in 2013. These strategies have set out a sustainable plan for managing 400km of FCRM assets, protecting over 200,000 people and £12 Billion of infrastructure over the next 100 years in response to climate change projections. This paper identifies the challenges that were overcome to enable viable strategic plans, emergent issues, and lessons learnt through this process.

## Overview of Existing and Future Challenges

### Existing Situation and FCERM Legacy

The natural and built environments of the Severn Estuary, Exe estuary and Poole Harbour (and how this relates to FCERM) can be considered as predominantly influenced by physical (notably tidal) processes. In overview:

- The Severn Estuary is hypertidal (having a tidal range of 11-12m). The wind wave climate can reach 4m, with swell wave influence towards the Bristol Channel.
- The Exe Estuary is macrotidal (tidal range of 4m). The wind wave climate is constrained to less than 1m (no swell) due to storm sheltering from the estuary mouth sand spit.
- Poole Harbour is micro-tidal (tidal range of 1.4m). The wind and swell wave climate is similar to that of the Exe Estuary for the same reasons.

The Severn and Exe Estuary, and Poole Harbour have also experienced significant anthropogenic modification over centuries, to enable development of infrastructure, communities and agriculture. The physical processes, and to some extent form of development in the reclaimed floodplains, have driven the overall design history of FCERM assets. In overview:

- The Severn Estuary has had the majority of its floodplains reclaimed since at least the Roman period, including the large coastal monolithic floodplains.

- The Exe Estuary has similarly had the majority of its floodplains reclaimed since the 12-13$^{th}$ centuries onwards. The present day form of reclamation is different in detail to the Severn Estuary, in that it is now related to a few, quite different developments. These consist of the now London-Penzance mainline railway on the west bank (and branchline railway to a lesser extent on the east bank), the city of Exeter on the lower River Exe, agricultural practice on the lower River Clyst, and the development of the town of Exmouth at the estuary mouth (partially on what was originally a sand spit).
- Poole Harbour is different to the Severn and Exe estuaries in that a large part of the harbour floodplain is not reclaimed (on the southern shoreline). However reclamation has occurred at the lower rivers Piddle and Frome, northern floodplains (now developed as the town of Poole) and the sand spits (now developed as Sandbanks and managed as Studland Peninsula).

Subsequent to the majority of built environment development and FCERM asset construction, increasing concern over natural environment impacts resulted in the legal designation of the study sites in the latter 20$^{th}$ century for their natural environmental value, described as (paraphrased from the JNCC website):

- Severn Estuary SAC, SPA and Ramsar. The designated features most relevant to this paper are intertidal mudflats, sandflats, saltmarsh and birds (numerous species and assemblages of approximately 93,000 waterfowl).
- Exe Estuary SPA and Ramsar, and Dawlish Warren SAC. Again the relevant designated features are intertidal mudflats, sandflats, saltmarsh and birds (Avocet, Slavonian Grebe and assemblages of approximately 23,000 waterfowl); as well as sand dunes and supported flora.
- Poole Harbour SPA and Ramsar. Relevant designated features are intertidal and terrestrial habitats, and birds (numerous species and assemblages of approximately 28,000 waterfowl).

The distribution of physical, anthropogenic and environmental features of the sites are shown in Figure 1. The combination of development, FCERM asset construction and environmental designation sets out the starting point from where the strategies proceed from. Whilst this is self-evident, there are significant ramifications from this legacy which is often centuries old and, until the late 20$^{th}$ century, was not managed in a clearly coherent manner. This is discussed later in this paper, but in summary the legacy means that future management of the estuaries starts from an arguably less than balanced, sustainable position.

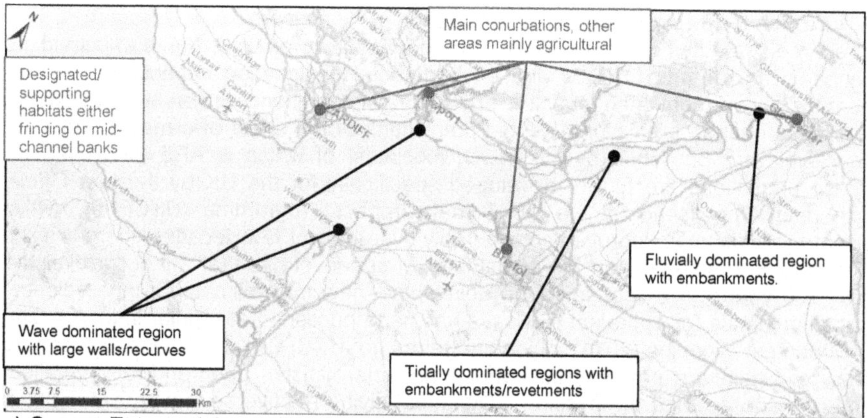


a) Severn Estuary (please note different scale)

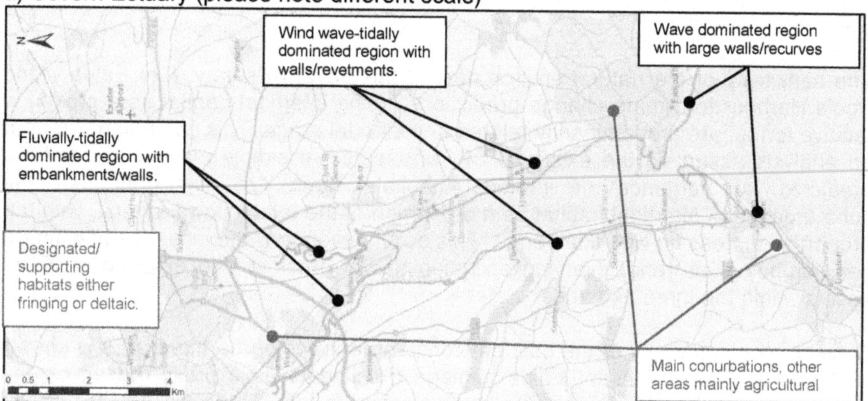


b) Exe Estuary

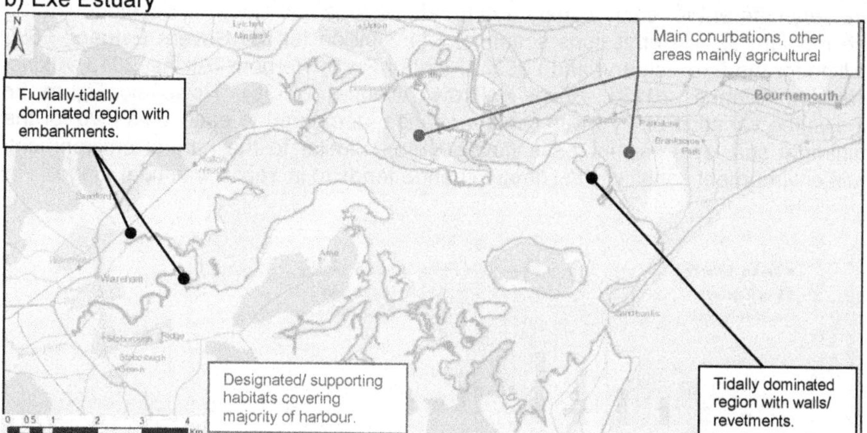


c) Poole Harbour

Figure 1: Overview of physical, anthropogenic and environmental features

**Future Scenarios and Drivers for Change**

The three strategies of interest were all predominantly promoted due to a) immediate concerns as regards FCERM and environmental designation impacts, which flow from the above mentioned legacies, and b) longer term concerns on how to respond to climate change predictions. Global predictions of the effect of climate change on sea levels are published by the IPCC, the latest of which is AR5 (IPCC, 2013). These predictions are further developed specifically for the UK by the Met Office, with results published by the UK Climate Impact Programme (UKCP09). Whilst climate models have improved considerably over the last few decades, there is still a high level of uncertainty on the actual level of sea level rise that will occur over the next 100 years. In response to the developing view from climate change science, formal guidance for appraisal purposes has been published at regular intervals. The Environment Agency (2011) guidance refers to UKCP09 predictions for climate change, which are probabilistic rather than deterministic, geographically detailed, and relate to a range of emissions scenarios, giving relative sea level rise over the next 100 years between 0.4m-1m (low 50%ile to upper end) for the three estuary sites.

The sensitivity of the natural environment of the Severn Estuary, Exe Estuary and Poole Harbour to climate change predictions (using intertidal habitat as a proxy), in relative terms, are predominantly related to their tidal range. This can be shown from an analysis assuming the existing FCERM assets are simply maintained, and the predicted submergence of intertidal habitat. Figure 2a highlights that the submergence of intertidal habitat as a proportion of the total follows a trend, with the proportion increasing with tidal range. This occurs even with the influence of different geomorphological trends (varying accretion and/or erosion) and extent of FCERM assets within the three estuaries.

Similarly, the sensitivity of the built environment features of the three sites is shown in Figure 2b, using Present Value damage (PVd) data as a proxy (based on the scenario of existing FCERM assets being allowed to fail). The absolute PVd values are very different for each estuary due to relative size and extent of development: for the medium 95%ile emissions scenario; £11.3 billion for the Severn Estuary; £655 million for the Exe Estuary; and £352 million for Poole Harbour (Atkins, 2013; Atkins-Halcrow Alliance, 2013a; Atkins-Halcrow Alliance, 2013b). However, Figure 2b shows the variation of PVd with climate change, compared to each sites PVd for the minimum sea level rise increase (low 50%ile). Similar to the natural environment, built environment sensitivity to climate change tends to increase with tidal range.

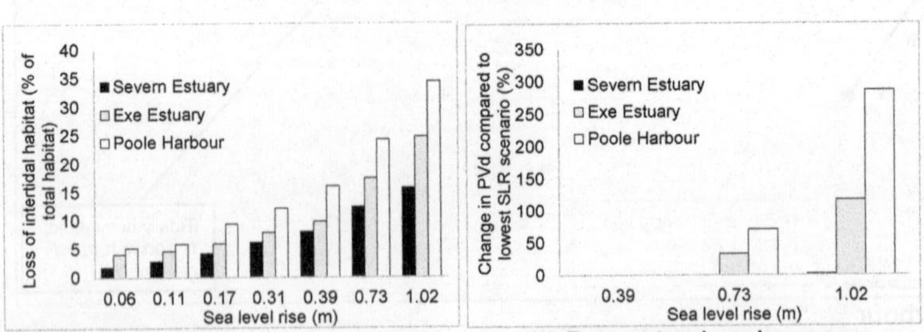


a) Intertidal habitat b) Present value damages

Figure 2: Relative sensitivity of natural and built environment to climate change

## Developing Adaptive Management Approaches for the Natural and Built Environment

### Framework for Responding to Climate Change

Within the UK system, various guidance and legislation effectively sets out the framework that quantitatively, or qualitatively, forms drivers for change in FCERM and, as a consequence, environmental designation management. In response to this guidance and legislation, a consistent approach was taken across all three strategies to enable transparent, auditable decision making. Development and assessment of options was carried out via a range of analytical and numerical tools (Canning et al, 2010 and 2012) at a range of scales:

- Assessment of the potential for compensatory habitat was carried out at an estuary-wide scale. A filtering process was carried out using a range of criteria. This defined potential viable sites, their maximum spatial area, and a broad hierarchy.
- Assessment of the approach to FCERM (alignment, type and SoP). This critically considered and merged SMP2 and CFMP policies. These were qualitatively assessed based on the baseline scenario and high level policy drivers at an estuary-wide level (such as the need for habitat creation to address climate change impacts), and quantitatively assessed for socio-economic viability and preference. This was generally undertaken at a flood cell level, although complex issues could change this (see below).
- Assessment of the most efficient and appropriate design of FCERM assets. This was undertaken at a sub-reach scale, for sections with consistent exposure, form, geometry and condition.

To inform at what scale socio-economic decisions could be made, analysis of the key issues that could have far-field hydrodynamic and/or geomorphological impacts was undertaken, including a) large-scale breach of existing FCERM assets and consequent flooding, and b) breach of large-scale geomorphological features such as sand spits. Testing of large-scale breach of existing FCERM assets found that the Exe Estuary and Poole Harbour would not experience significant far-field impacts (Atkins-Halcrow Alliance, 2013; HR Wallingford, 2010); the Severn Estuary was similarly not affected over the majority of its extent, except in the fluvially dominated region at the upper extremity of the estuary. Analysis of the potential for sand spits to breach found that the Severn Estuary and Poole Harbour were not affected, due to the lack of any relatively extensive pits (Severn Estuary) or the low risk of breach over the next 100 years (Poole Harbour); however the Exe Estuary was found to be influenced by likely breach of the estuary mouth sand spit (Dawlish Warren) within 100 years (Canning et al, 2012). Consequently, FCERM and socio-economic decision-making was carried out estuary-wide in the Exe Estuary, but at a flood cell level for the Severn Estuary and Poole Harbour.

### Developing Adaptive Management Plans

The above analytical and numerical tools, option assessment framework, and extension engagement with stakeholders progressively developed plans for FCERM that were adaptive to future climate change predictions. These are shown conceptually in Figure 3 and 4, and discussed below.

Within the Severn Estuary, application of the above approach found that the vast majority of FCERM assets provided appropriate (determined as the optimal economic approach) Standards of Protection (SoP) for the majority of flood cells, with relatively limited need for immediate improvements FCERM asset systems. The incremental programme of FCERM asset improvements to respond to variable climate change was spatially dispersed, economically viable and technically achievable. Similarly, the immediate and future programme of compensatory habitat as a consequence of the above FCERM programme could be achievable. These potential changes in FCERM asset form, SoP and alignment are relatively small in comparison to the wider FCERM asset system.

The above points suggest that the existing FCERM asset system could be considered already well adapted (or balanced) as there are limited immediate changes required due to their physical and economical (i.e. form and function) appropriateness. The FCERM asset Severn Estuary FRM Strategy could be considered a large-scale, long term asset management plan, predominantly focussed on recommending physically and economically efficient design of FCERM structures; an example of 'incremental adaptation' (IPCC, 2014). This focus on design efficiency results in the potential for larger-scale transitions in FCERM asset form in the future, driven by the dominant form of hydrodynamic influence (particularly either wave and/or tidal) on flood risk. In practice, for some flood cells this would result in the FCERM asset system transitioning from 'soft' to 'hard' forms (such as earth embankments, being inefficient in withstanding wave action, being converted to rock armouring, concrete revetments and/or wave recurve walls).

Similarly, within the Exe Estuary the majority of FCERM assets were found to provide appropriate SoPs, although the previous management of the storm sheltering sand spit was less than optimal. Recommended changes in FCERM at this location (see Canning et al, 2012 for details) influenced FCERM within the wider estuary, spatially and temporally. This consequently required immediate and future FCERM to be considered both at a fully linked estuary-wide scale as well as a local, flood cell scale. Together with the immediate and future programme of compensatory habitat, this forms relatively large changes in asset form, SoP and alignment. Salient points from the Poole Harbour findings include that again the majority of FCERM assets provided appropriate SoP for the majority of flood cells, with relatively limited need for immediate improvements FCERM asset systems. The incremental programme of FCERM asset improvements to respond to variable climate change was spatially localised, economically viable and technically achievable; similarly, the immediate programme of compensatory habitat as a consequence of the above FCERM programme could be achievable, although at the upper end of climate change predictions this may become difficult. In comparison to the Severn Estuary, the above points suggest that the existing FCERM asset systems as a whole in the Exe Estuary and Poole Harbour could be considered less well adapted to existing drivers. In response to this, particularly within the Exe Estuary, the adaptive response to legacy, existing drivers and future climate change could be considered an example of 'transformational adaptation' (IPCC, 2014).

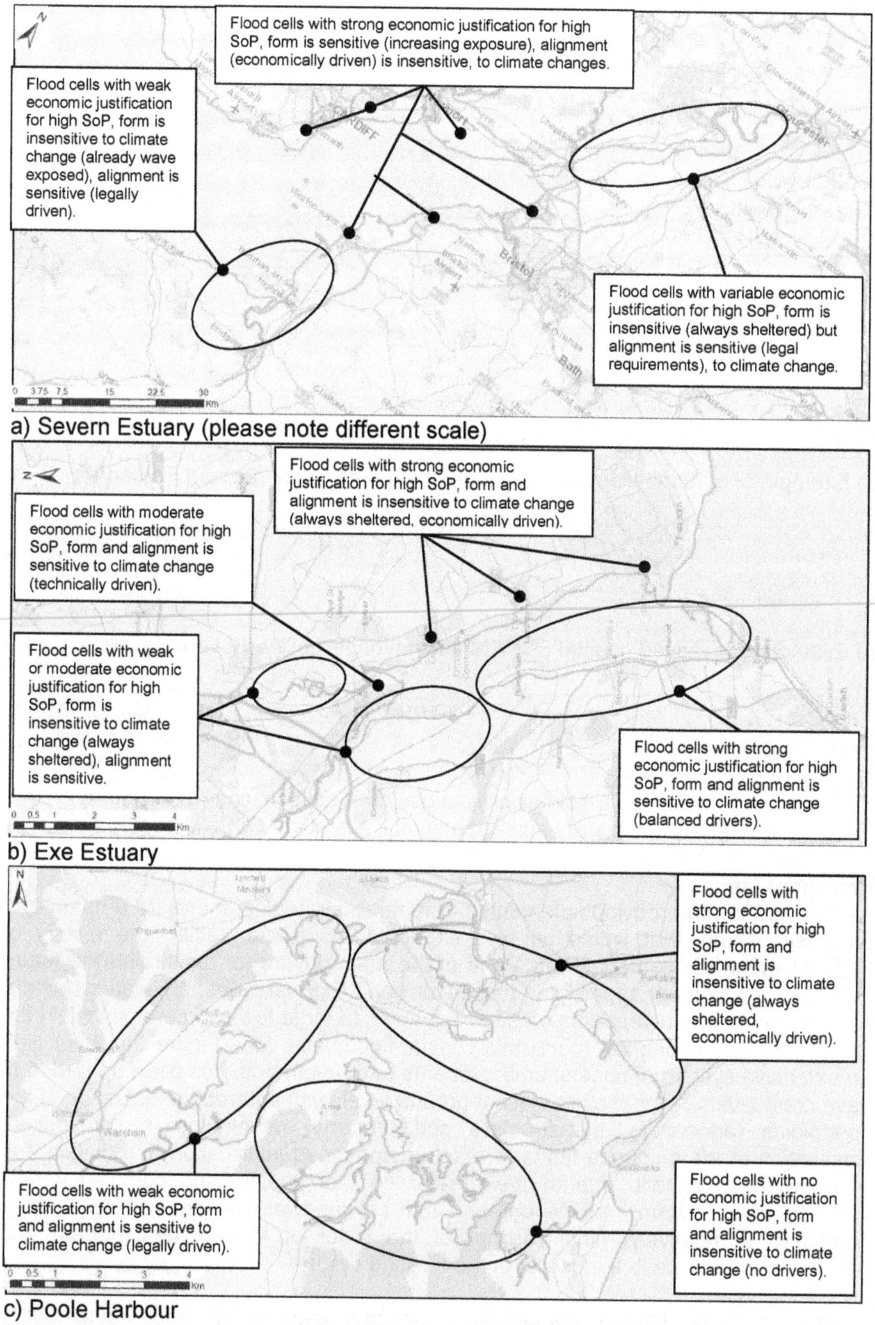


Figure 3: Overview of salient issues within the FCERM strategies

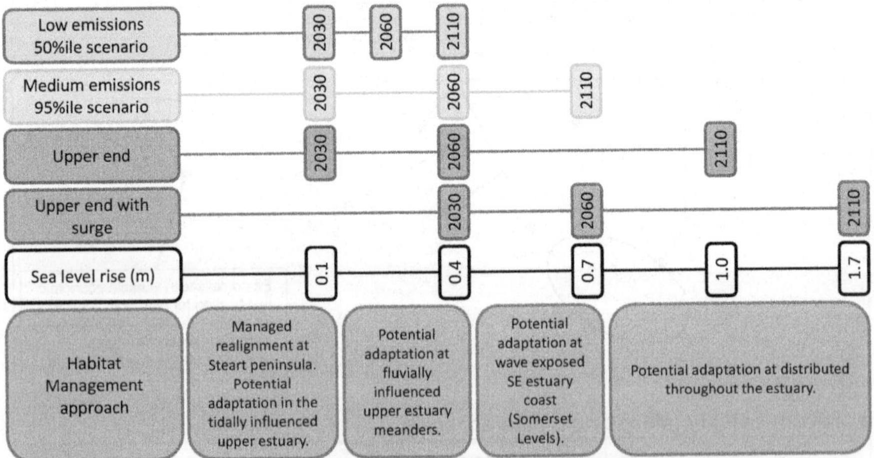


a) Example of estuary-wide compensatory habitat flexibility (Severn Estuary).

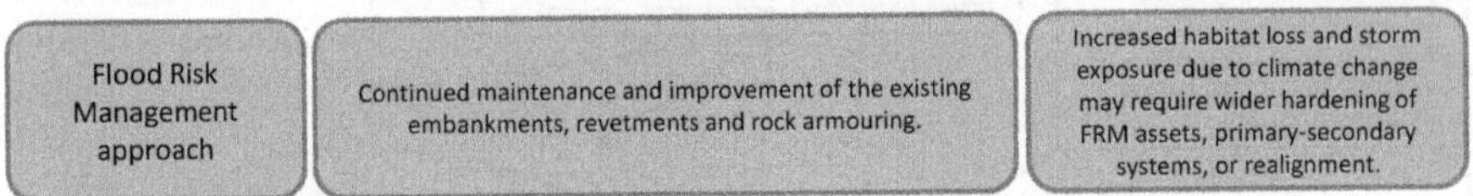


b) Example of exposed coastal FCERM asset type flexibility (Severn Estuary).

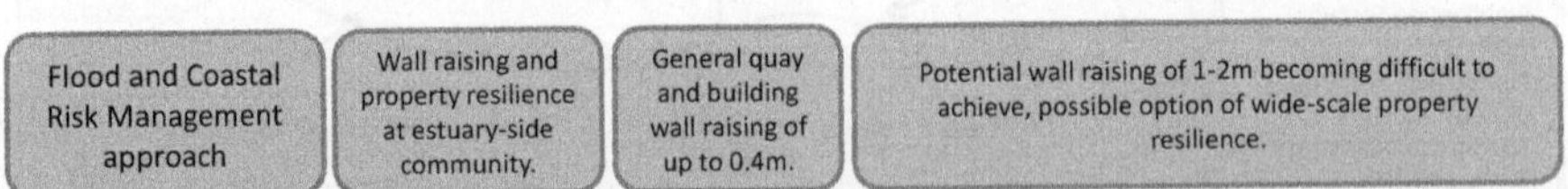


c) Example of limits of FCERM flexibility in constrained locations (Exe Estuary).

Figure 4: Examples of flexibility and adaptation limits in the FCERM strategies

**Emergent Issues**

As well as the more obvious elements of the three strategies, there are a number of emergent issues that would be less expected. As noted earlier, the history of FCERM within the estuaries forms one of the main drivers (or constraints) to future FCERM. This is most apparent in the Severn and Exe estuaries, at locations where embankments and revetments manage flood risk to large low-lying areas. Within the Severn Estuary, along the monolithic coastal floodplains (see Figure 3a), over time an extensive system of coastal embankments and revetments has been upgraded to have crest levels 3-4m above adjacent ground level, with anthropogenic usage of the floodplains (agriculture, infrastructure and property) developing in conjunction. Application of the guidance tends to result in continued justification to maintain and upgrade these assets, due to the greater cost of constructing completely new FCERM asset systems (of the order of metres) landward of the existing system, compared to relatively minor raising (of the order of tens of centimetres) and hardening (due to wave exposure) of the existing system.

Engagement with stakeholders identified concerns over the real world sustainability of this approach. A similar example exists in the Exe Estuary on the west bank, where the London-Penzance mainline railway is located on (and defines) the shoreline. Similar economic decision-making would tend to occur under FCERM-AG, with recent storm damage raising similar concerns from society and the civil engineering profession as regards future sustainability. These examples arguably represent situations where previous investment in infrastructure results in continuation of historic coastal management, with difficulty in fully capturing sustainability issues within existing guidance.

Analysis of the ability of the FCERM asset system (and as a consequence the designated sites) to adapt flexibly to variable climate change was undertaken for the three strategies. The Severn Estuary system was found to be relatively insensitive to climate change, in the sense that changes in form and alignment of FCERM assets were proportionally localised and/or minor relative to its scale. However, the form of the more wave-exposed FCERM assets was indicated to be sensitive, with socio-economic analysis supporting transitioning from 'soft' earth embankments to 'hard' concrete or rock revetments. Consequent future impacts on the designated site ('coastal squeeze') were found to be deliverable within the estuary system.

The Exe Estuary system (aside from the planned change in management to the estuary mouth sand spit) was found to be more sensitive to climate change, in a localised manner. At the upper end of climate change predictions, some FCERM asset systems were indicated to require significant changes (for example transition from wall raising to community-wide property resilience works) that may be difficult to deliver in practice. Consequent future impacts on the designated site were found to be deliverable within the estuary system, although the extensive infrastructure development along the estuary focusses this to a few deliverable sites. The Poole Harbour system was also sensitive to climate change, but in a larger scale manner. This was particularly related to the extent of potential future changes in FCERM system alignment required to address future impacts on the designated site. At the medium and upper end of climate change predictions (as indicated in Figure 4) this would require large scale realignment to compensate for intertidal habitat losses, and also consequentially freshwater habitat to support the integrity of the designated site.

## Conclusions

The comparative analysis in this paper seeks to highlight a number of emergent challenges for FCERM, that predominantly flow from the fundamental physical regime of the estuaries (whether micro, macro or hyper tidal), and the legacy of anthropogenic development that forms the starting point from which future FCERM proceeds. Particular themes that emerge from this analysis consist of:

- FCERM systems that are relatively well adapted now and relatively insensitive to future change. Strategic long term management in this situation then tends to become a large scale asset management plan, focussed on efficient use of resources to achieve government objectives and comply with legislation.
- FCERM systems that are relatively less well adapted now, and are relatively sensitive to future change. Strategic management, over both the short and long term then tends to become challenging. Notably, for the upper end climate change scenarios, the ability of the natural and built environment to be adapted to continue to achieve present day expectations (whether economic, legislative or societal) becomes problematic.

## Acknowledgements

The author would like to acknowledge funding of the FCERM strategies by the Environment Agency. The views expressed in this paper are solely the authors.

## References

Atkins, 2013. Baseline Assessment Report, Severn Estuary FRM Strategy.

Atkins-Halcrow Alliance, 2014a. Baseline Assessment Report. Exe Estuary FCERM Strategy.

Atkins-Halcrow Alliance, 2014a. Baseline Assessment Report. Poole Harbour, Poole Bay and Wareham FCERM Strategy.

Canning, P., Hills, K., Samphier, R., Quarrier, G. and Dunsford, D., 2010. Climate change adaptation at an estuary-wide scale. Environment Agency Flood and Coastal Erosion Risk Management Conference, Telford, UK. 30th June – 1st July 2010.

Canning, P., Fox, D and Pontee, N., 2012. The benefits of managing spit evolution: a case study in the Exe Estuary, UK. Innovative Coastal Zone Management: Sustainable Engineering for a Dynamic Coast, Institution of Civil Engineering Publishing.

Defra, 2010. Flood and Coastal Erosion Risk Management – Appraisal Guidance.

Environment Agency, 2011. Adapting to climate change: advice for flood and coastal erosion risk management authorities.

IPCC, 2013. Climate Change 2013: The Physical Science Basis. Contribution of Working Group I to the Fifth Assessment Report of the Intergovernmental Panel on Climate Change [Stocker, T.F., D. Qin, G.-K. Plattner, M. Tignor, S.K. Allen, J. Boschung, A. Nauels, Y. Xia, V. Bex and P.M. Midgley (eds.)]. Cambridge University Press, Cambridge, United Kingdom and New York, NY, USA, 1535 pp.

IPCC, 2014. Climate Change 2014: Impacts, Adaptation, and Vulnerability. Part A: Global and Sectoral Aspects. Contribution of Working Group II to the Fifth Assessment Report of the Intergovernmental Panel on Climate Change [Field, C.B., V.R. Barros, D.J. Dokken, K.J. Mach, M.D. Mastrandrea, T.E. Bilir, M. Chatterjee, K.L. Ebi, Y.O. Estrada, R.C. Genova, B. Girma, E.S. Kissel, A.N. Levy, S. MacCracken, P.R. Mastrandrea, and L.L. White (eds.)]. Cambridge University Press, Cambridge, United Kingdom and New York, NY, USA, 1132 pp.

**Coastal Management: Changing coast, changing climate, changing minds**
ISBN 978-0-7277-6114-9

ICE Publishing: All rights reserved
doi: 10.1680/cm.61149.251

# Lessons from Strategic Coastal Monitoring

**Andy Parsons,** CH2M, York, UK
**Paul Fish,** CH2M, Birmingham, UK
**Robin Siddle,** Scarborough Borough Council, Scarborough, UK

## Introduction

The strategic management of flood and erosion risks at the coast typically relies on understanding of ongoing coastal processes including the use of records of past shoreline changes and previous extreme events to inform predictions of future coastal risks to assets.

Monitoring of shoreline change data using surveys of beach profiles, aerial photography, topographic and LiDAR surveys is now undertaken regularly as part of the national network of strategic regional coastal monitoring programmes covering the coast of England. Understanding of the drivers for shoreline change is facilitated by the collection of information on forcing conditions including waves and sea level and rainfall data collected both at national and regional scale.

The prediction of trends in coastal change and extreme events requires consistent long term data sets due to the impacts of natural variability in the weather and climate, predictable changes in tides and the possible impacts of longer term climate phenomena such as the North Atlantic Oscillation and forecast changes in climate and sea level.

This paper describes emerging findings from the North East Regional Coastal Monitoring Programme and discusses the implications of the findings of data analyses undertaken so far. Recommendations are also made for future improvements in the programmes to maximise the benefits of future coastal monitoring to long term strategic coastal defence planning.

## Regional Coastal Monitoring Programmes in the UK

Strategic regional coastal monitoring programmes have generally been in place for between 7 and 12 years on the English coast. Exceptions are the strategic monitoring of erosion on the Holderness coast, Cell 2a, which has been ongoing since the 1950s and shoreline monitoring on the coast of the former Environment Agency Anglian region, which has been underway since 1991, see Figure 1. While there are also many examples of sites with longer periods of local coastal monitoring these typically cover scheme specific frontages. A strategic monitoring programme for the Welsh coast has been in development since 2010 (WCMC, 2013). Although the Scottish and Northern Ireland governments contribute to the UK wide monitoring of waves and tides, strategic monitoring of shoreline change is not yet undertaken.

The requirements for monitoring data to inform future coastal management largely depends on the risks related to coastal change and coastal management practices undertaken to manage those risks. This has resulted in the strategic monitoring tasks undertaken differing, particularly in terms of scale and frequency of data collection between the six English regions identified in Figure 1.

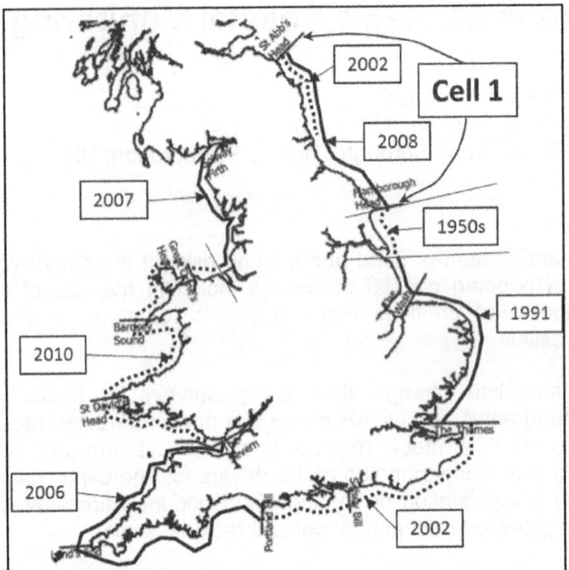


Figure 1: Cell 1 study area and regional monitoring programme inception dates

## National Coastal Monitoring Data

The national network of strategic regional coastal monitoring programmes is complimented by national coastal monitoring funded by Defra and the Environment Agency, including the strategic national tide gauge network and WaveNet programme that form components of the UK Coastal Monitoring and Forecast Service (UKCMF).

The strategic national tide gauge network, see Figure 2, includes 44 tide gauges around the UK coast and was set up to inform flood warnings following the 1953 storm surge which caused widespread coastal flooding on the east coast of the UK. The network now provides consistent and quality controlled medium to long term sea level data that can aid the interpretation of coastal change data and the prediction of extreme events. Two of the strategic national tide gauges, North Shields and Whitby, are located within the Cell 1 frontage.

The North Shields tide gauge site has one of the longest records of sea level in the UK, enabling investigation of extreme events and also trends in relative sea level, with monthly mean data available since 1895 and hourly records available through BODC from 1946. The Whitby tide gauge record has data available from 1980.

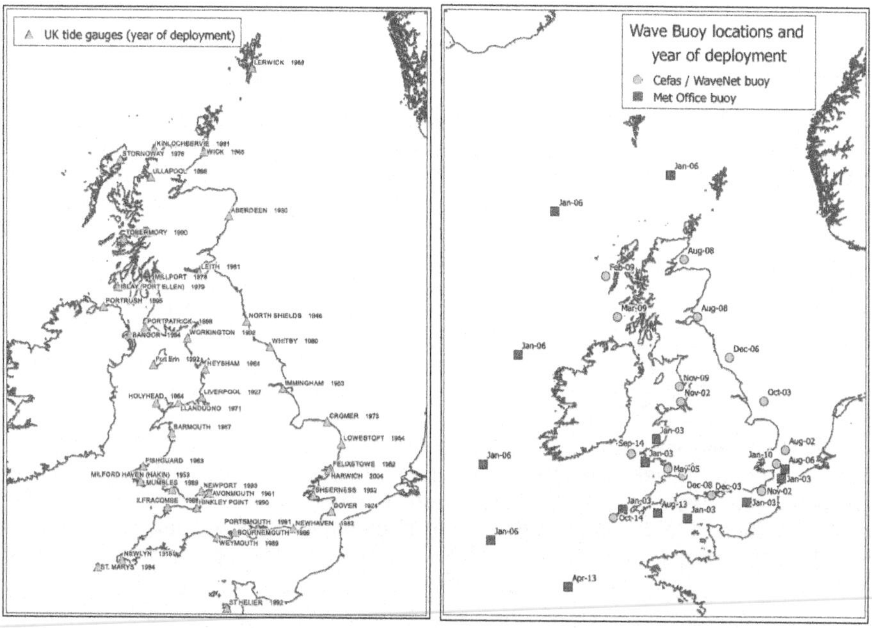


Figure 2: Location of the UKCMF tide gauges and WaveNet wave buoys showing deployment year for each site

The strategic national wave monitoring programme known as WaveNet was set up with funding from Defra in 2002 following a scoping report, Hawkes et. al. (2001) that considered options for siting the buoys taking into account coastal risks. The programme now provides strategic wave monitoring data with durations of 6 to 12 years, see Figure 2. The Cefas WaveNet website, http://cefasmapping.defra.gov.uk/Map also hosts data from UK Met Office wave buoys, which are also shown in Figure 3 together with a large number of other active and historical wave buoys owned by energy companies and wave buoys deployed under the regional coastal monitoring programmes. The WaveNet Tyne-Tees buoy which has data from 2006 is located offshore from the Cell 1 study frontage. In addition to measured data, the Cefas wavenet website also distributes hindcast wave data from the Met Office Wave Watch III model of UK waters, and data covering the period from 1st January 1980 to 31st December 2012 was obtained for the Cell 1 programme.

## The North East Coastal Monitoring Programme

The North East Coastal Monitoring Programme covers the coastline from St Abb's Head at the border with Scotland to Flamborough Head in the East Riding of Yorkshire (Figure 3). This encompasses coastal sediment cell 1 and covers almost 300km of diverse coastline including dunes, hard and soft rock cliffs, periodically active landslides and numerous settlements, harbours and other coastal assets.

The strategic monitoring of shoreline change on the northern part of the Cell 1 frontage from St Abb's Head to the River Tyne, commenced in 2002. The southern part of Cell 1 from the Tyne to Flamborough Head was integrated with the northern part in a combined programme which commenced in 2008 (Cooper et al. 2009) in order to meet the strategic monitoring requirements of the two shoreline management plans that cover Cell 1.

The core component of the Cell 1 programme is the annual survey and analysis of 248 beach profiles, 31 beach topographic surveys and cliff recession monitoring at 14 locations to document change since the previous survey and since the baseline. Repeat surveys at selected locations are also undertaken every 6 months. In addition to these regular topographic surveys, wave buoys have been deployed at three locations along the coastline to provide real time data.

Rapid walk over condition inspections of the cliffs and coastal landslides, engineered assets and are also undertaken biennially to document defence and shoreline change and to update information on coastal cliff instability risk. Aerial surveys, using LiDAR, vertical photography and multispectral techniques are also undertaken every two to three years to document cliff recession rates and patterns of dune evolution. These data sets have also been used to map coastal habitats and investigate the regions archaeological resource. A baseline bathymetric survey, including sediment sampling, has been undertaken at main settlements and a more extensive repeat survey is planned to document offshore sediment movements. All data and analytical reports are freely available to all users on the programme website: http://www.northeastcoastalobservatory.org.uk/

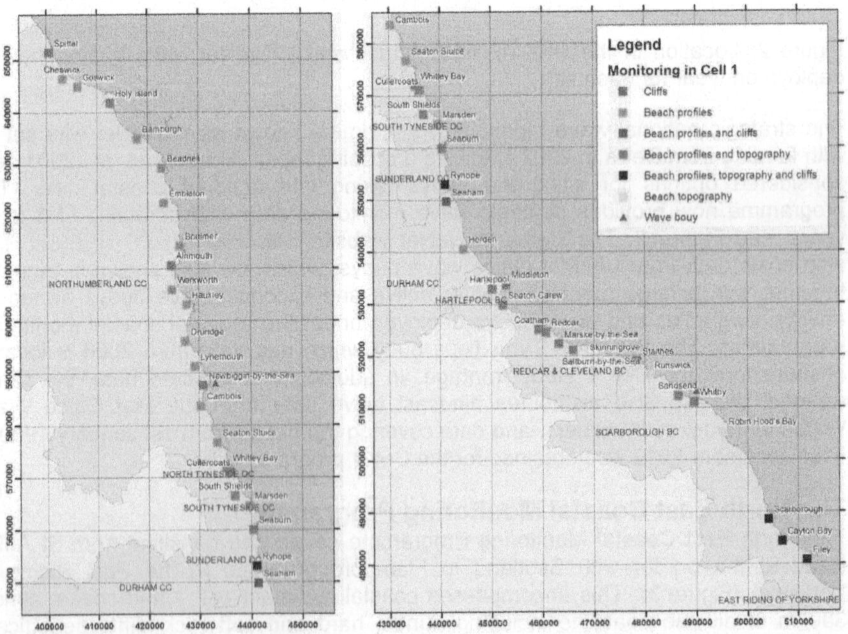


Figure 3: Main data collected under Cell 1 monitoring programme

## Emerging Lessons

### Long Term Data on Mean Sea Level Change

Mean sea level data from North Shields (data provided by PSMSL, 2014) show a long term trend of relative sea level rise of 1.91mm/yr over the last century, see Figure 4. There does not appear to be any clear change in rate of rise over the last 50 years. However, the annual mean data show variability of order of +/- 100mm about the trend line over 10 to 20 years, which may partly relate to tidal and climatic cycles. Analysis of trends in the monitoring data require long term change to be identifiable from short term variability, suggesting that >50 years of data are required for detection of mean sea level trends. The Whitby data period is 32 years, but there are significant gaps in the record with ten missing annual mean data points. Therefore, although the data for Whitby in Figure 5 show a significantly higher rate of rise, calculated as 6.7mm/yr, the trend is likely to be unreliable. However, restricting the analysis at North Shields, to the same data years available at Whitby gave a rate of rise at North Shields of 3mm/yr, less than half that found at Whitby, suggesting that there may be differential land level changes between the two sites. This indicates that further investigation may be warranted and that reliance on regional land level changes such as those used in UKCP09 for predictions of future relative sea level rise could be inadequate.

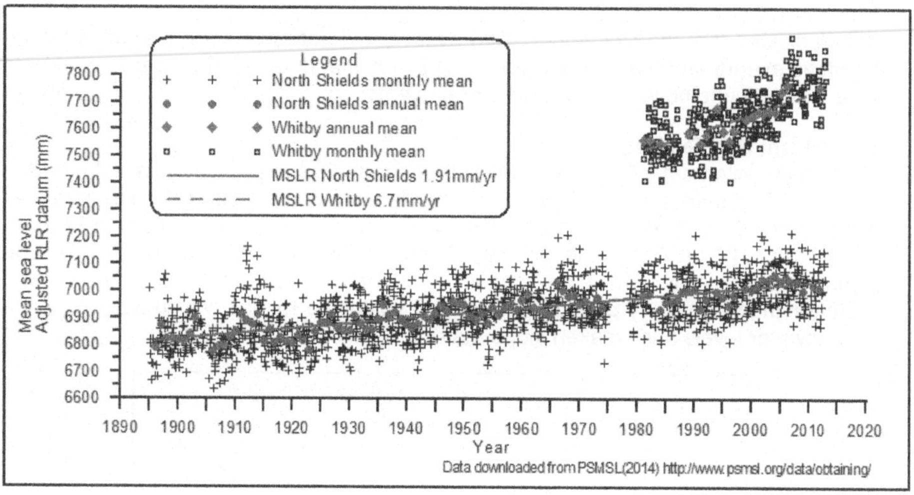


Figure 4: Monthly and annual mean sea level data in Cell 1 from 1895 to 2012

### Extreme Sea Levels

Results from analyses of extreme sea level using gauge data from the national tide gauge data sets from Whitby and North Shields and the Cell 1 programme tide gauge at Scarborough, which has data from 2004, compared very closely with the EA (2011) coastal flood boundary conditions results when data up to the end of November 2013 were analysed. The highest recorded sea levels in Cell 1 occurred during the 5$^{th}$/6$^{th}$ December 2013 storm surge event. At North Shields the peak water level of 3.98mOD, on the 5th December 2013 significantly exceeding the previous maximum recorded sea level of 3.56mOD on 31st January 1953. Based on the preceding extreme sea level statistics the 2013 surge event had an Annual Exceedance Probability (AEP) of about 1 in 500 at North Shields, although accounting for uncertainty the range of could be between 1 in 100 and over 1 in 1000 AEP.

Table 1: Impact of the December 2013 surge on 1 in 100 AEP tide levels in Cell 1

| Location | Extreme Level (mOD) from EA CFB Study (2011) | Confidence intervals (m) from EA CFB Study (2011) | Analysis of North Shields NTSLF data to November 2013 (mOD) | Analysis of North Shields NTSLF data to February 2014 (mOD) |
|---|---|---|---|---|
| North Shields | 3.8 | 0.2 | 3.8 | 3.9 |
| Whitby | 4.0 | 0.3 | 4.0 | 4.3 |
| Scarborough | 4.0 | 0.3 | 4.1 | 4.3 |

The extreme statistics were re-evaluated at the three sites taking into account the recorded data from December 2013 (see Table 1) and it was found that the predicted extreme water levels increase, with the levels for the 1 in 100 AEP event raised by between 0.1 and 0.3m. At North Shields, which has the longest data set with data from 1948 onwards the change is only 0.1m, but at Whitby and Scarborough, where the data records do not go back to the 1953 surge, inclusion of the December 2013 surge has a larger impact on the predicted extremes, which are increased by 0.2 to 0.3m.

This indicates that the December 2013 event could not have been predicted by the previous events and that the EA (2011) CFB national data set should be updated to take into account the winter 2013/14 storm events.

**Long Term Rainfall Data**
Rainfall records from Whitby that cover the period 1962 to present have a long-term average of 575mm per year (Figure 5). The data show a marked increase in rainfall in the last 10 to 15 years with the five wettest years on record occurring in 2012 (883mm), 2010 (839mm) and 2004 (811mm), 2008 (782mm) and 2002 (762mm). These data are likely to have a strong influence on the coastal cliff instability risk that is determined in part by periods of high and sustained rainfall, which act to elevate groundwater levels and weaken slope materials.

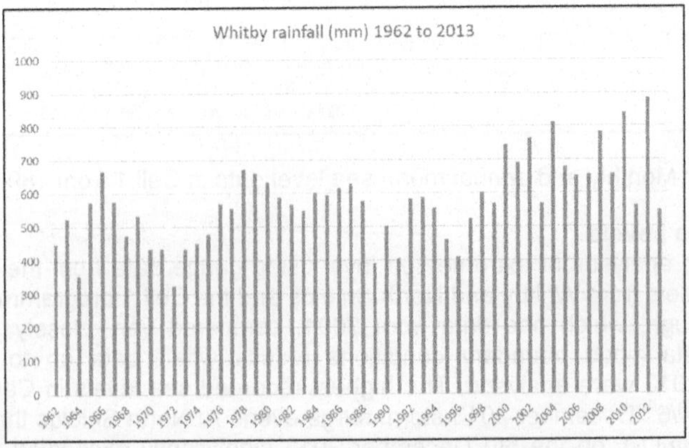


Figure 5: Annual rainfall data for Whitby

**Wave Data Analysis**

Offshore wave monitoring data is available from 2006 from the Tyne Tees WaveNet buoy. The near shore buoys deployed at Scarborough, Whitby and Newbiggin have data periods of less than 3 years, whilst the Tyne Tees WaveNet buoy has data for 8 years. Storm analysis of the wave data has been undertaken to aid interpretation of the beach monitoring data and this is now being incorporated in annual monitoring reports. A wave energy plot in Figure 6 identifies storms with significant offshore wave heights above 4m. The storm analysis shows that there are between 3 and 13 storms each year, with 2010 having most and 2014 the fewest. The three biggest storms on the record occurred over the winter 2007 to 2008.

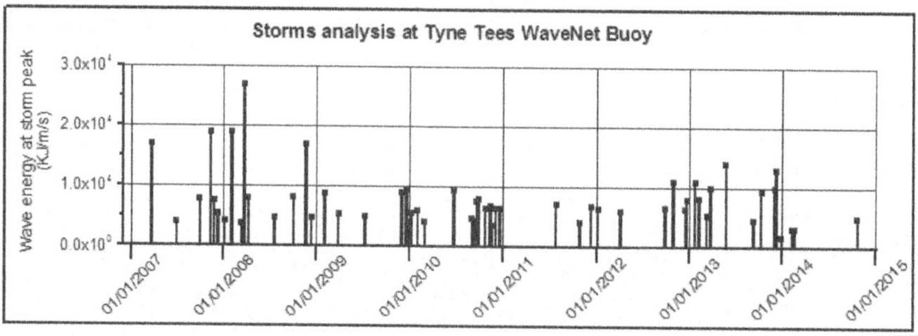


Figure 6: Annual variability of offshore storms using data from 07/12/2006 to 07/12/2014

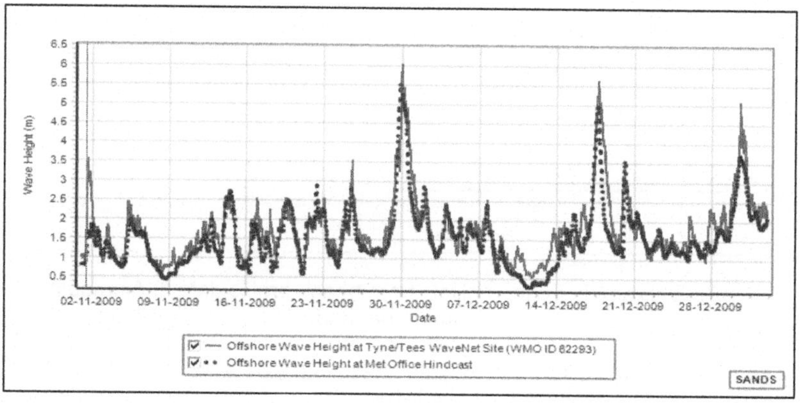


Figure 7: Example plot of hindcast and measured wave data at Tyne Tees, winter 2009

A comparison of the Met Office modelled data at the nearest location to the Tyne Tees buoy found that the hindcast wave data, which is available from 1980 to 2012, show a very similar temporal record to the measured data at Tyne Tees, but the peak wave height on most storms is significantly under-estimated, with peak wave heights often 0.5m or more less than measured, see example plot in Figure 5. This indicates that the model calibration is not good for peak storms in this location and that caution should be used and consideration given to adjusting or calibrating the Met Office hindcast offshore data if it is to be used for boundary conditions in coastal modelling studies.

**Beach Topographic and Profile Monitoring Data**

Topographic and profile surveys of beaches have been undertaken for at least 6 years, with more than 10 years available for some profile locations north of the Tyne, allowing short to medium-term trends to be investigated. However, at all locations where topographic data are available the magnitude and pattern of change between the baseline and most recent surveys is almost identical to that shown when comparing annual surveys. This indicates that seasonal/annual sediment movement is considerable and it masks any underlying trends for erosion or accretion. Collection of longer-term datasets will be required to highlight subtle underlying trends.

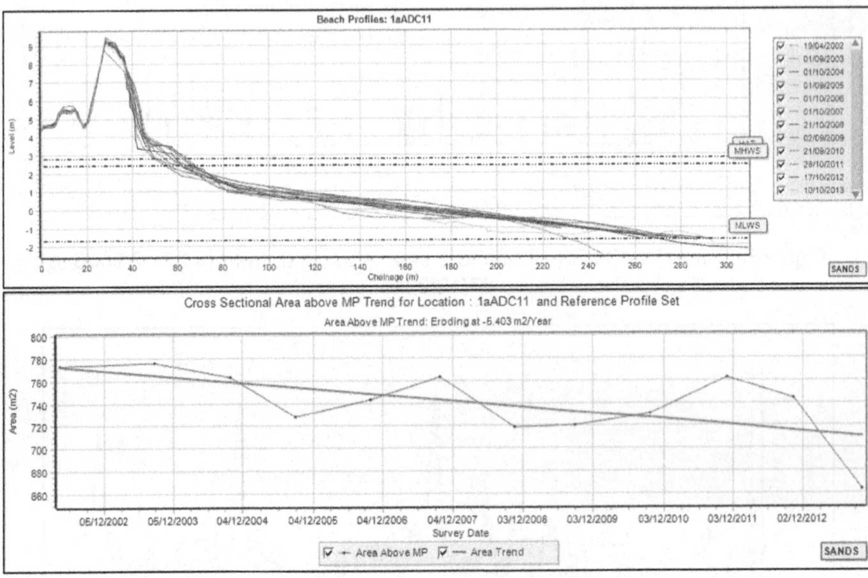


Figure 8: Example trend analysis from beach profile data at Alnmouth, Northumberland

The beach profile data also shows large year to year variability in most locations. However, trends can begin to be established from the data in some locations (see Figure 8). The wave data shows that there were relatively few storms in 2011 and more than normal storms in 2013 (see Figure 6). The data collected to date in Cell 1 are limited to beach levels above low tide, which limits the understanding of sediment movements. In other regional monitoring programmes bathymetric extensions to beach profiles are surveyed, generally less frequently than the land based surveys. Bathymetric surveys have been completed along a limited number of narrow sections already and more comprehensive surveys are planned for Cell 1 in the future. Along with other regional programmes, consideration will be given to the use of innovative approaches to capturing the beach profile below low water and rapid post storm surveys of upper beach topography using remote survey techniques such as

Unmanned Surface Vehicles (USVs), Unmanned Aerial Vehicles (UAVs) and Autonomous Underwater Vehicles (AUVs).

**Cliff Recession Analysis**

Cliff recession monitoring has been undertaken using biannual field survey at a small number of sites and analysis of aerial survey data of the entire Cell 1 frontage captured in 2003, 2008, 2010, 2012/13. Field surveys are prone to errors principally caused by problems in precise identification of the cliff top in the field, particularly where thick vegetation has developed. This has generally led to short-term cliff recession data being unreliable. However, now that a record of at least 6 years' data is available, erosion patterns are becoming clearer and more reliable cliff recession rates can start to be provided. In contrast, aerial surveys can provide recession data for an infinite number of cliff locations and are only limited by the accuracy of the ortho-photos, which is better than ±0.1m with contemporary technologies. Representative cliff profiles can be identified by first mapping the cliff behaviour units (Lee and Clark, 2002). The interpretation of CBUs and recognition of the cliff edge is greatly aided by the use of LiDAR data typically collected simultaneously with digital imagery.

Field survey data collected from Filey Bay between November 2008 and September 2014 are summarised in Figure 9, which shows a record of cliff recession from four profiles out of a total of 28 currently monitored. They highlight that at some locations there has been little or no cliff recession (Profile 16 and 20), but that at other locations, episodic erosion has occurred (Profiles 5 and 7). The episodic erosion is characterised by recession of several metres in a six month period (in fact probably in a single event in a day) separated by a period of years where no detectible cliff recession occurs. The data from Profile 5 shows 3.5m recession in the 6 months between November 2008 and April 2009 and a further 2.1m recession between October 2009 and March 2010. Recession has been much less significant between March 2010 and September 2014, with only 0.8m erosion in this time. From a total of 28 measured profiles in Filey Bay, the average recession rate between 2008 and 2013 has been 0.1m/yr, but rates range from 1.3 to 0.0 m/yr. The data highlight that the most significant cliff recession events occurred between November 2008 to March 2010 (recorded in Profile 5), with other notable failures occurring in the spring/summer of 2011 (Profile 7), spring/summer 2012 (Profiles 7 and 16). The rainfall records from nearby Whitby (Figure 5) show that 2008, 2010 and 2012 were all exceptionally wet years (in the top 5 wettest years since 1962) and these have been plotted on the chart. The data suggest there is a relationship between high rainfall and accelerated cliff recession. Periods of below average beach levels are often associated with accelerated cliff recession (Lee, 2008; Pye and Blott, 2015) and data from nearby profiles are also marked on the chart. The beach data collected from Filey Bay are inconclusive, with periods of low beach level typically unrelated to phases of cliff recession. Collection of longer-term data on cliff recession, rainfall and beach levels will allow these relationships to be further investigated and better understood. In addition, analysis of slope monitoring equipment installed in areas of high cliff instability risk along the Scarborough Borough Council frontage will allow relationships between rainfall, groundwater level and slope failure to be explored.

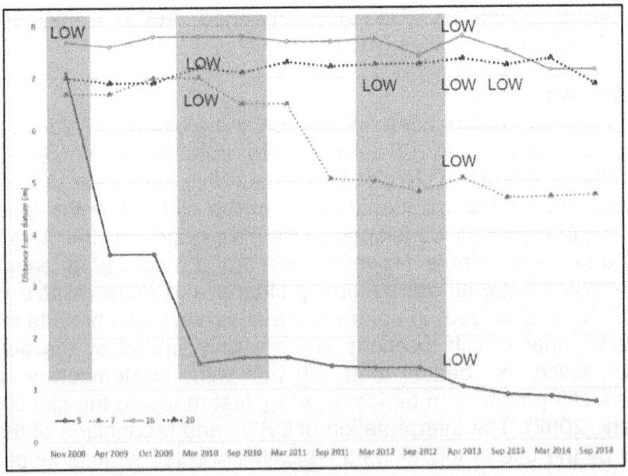


Figure 9: Cliff recession data (distance to cliff edge) measured in Filey Bay. Wet years are highlighted by shaded zones. Periods of lower than average beaches are marked "LOW".

The analysis of aerial survey data covering Filey Bay indicates average recession rates between 2008 and 2012/13 ranged from 0.0m/yr in the composite cliffs at Bempton (till over chalk) to 0.27m/yr in the simple landslides (mudslides in till) around Hunmanby Gap. Data from 2010 to 2012/13 highlights that recession rates can be higher over shorter time periods, reflecting the impact of episodic landslide events and rates of over 2m/yr are recorded in some simple landslide CBUs. The aerial survey dataset has also been used to monitor dune systems that are prevalent along the Northumberland coastline. The data collected to date indicates significant variability in the position of the dune frontage, with periods of both accretion and erosion indicated at rates that are commonly up to 5m/yr, but reaching 10m/yr in certain locations.

In summary, the recession rates derived from aerial survey and field-based techniques are in close agreement and highlight the significant variability in cliff recession rates over time and in different CBUs, which may be missed in smaller datasets covering shorter periods of time or with more limited spatial coverage. However, the relatively short time period covered by the data and the variability over time and in different CBUs means that understanding the relationships between recession, rainfall and low beach levels is uncertain and longer-term changes in the recession rate, potentially in response to climate change and relative sea level rise, are unclear.

## Conclusions

The North East Monitoring programme was designed to improve strategic understanding of coastal risks and ongoing coastal change. Beach and cliff monitoring data is now available for 5 to 10 years and analysis of the data is beginning to show trends and improve understanding of coastal change. Continued monitoring will allow better informed decisions on coastal management to be taken in future. The key lessons and recommendations from analysis of Cell 1 data available so far include:

- Consistent long term monitoring is required to underpin strategic shoreline management decisions. The Cell 1 programme is in its infancy and quality controlled consistent data collection needs to be continued in a manner for at least a further decade to identify trends.
- Analysis of datasets to determine long term coastal change forcing factors, such as sea level rise and rainfall, demonstrate that short term variability can mask longer term trends.
- A relationship between phases of accelerated cliff recession, low beach levels and periods of wet weather is suggested, but longer-term datasets are needed to better understand these complex inter-relationships. The high variability and short-term nature of the currently available data means that relationship between cliff recession and relative sea level rise is unclear.
- Modelled offshore waves have been shown to underestimate peak storm waves offshore Tees Bay, highlighting the need to continue wave buoy deployments and use the data to improve national and regional scale wave modelling in future.
- Relative sea level rise may differ locally from regionally recommended values, but longer term data is required to confirm.
- The extreme sea levels experienced during the December 2013 event could not have been predicted by the data from previous events. The EA (2011) CFB national data set should be updated to take into account the winter 2013/14 storm events.
- The magnitude and pattern of change between the baseline and most recent surveys is almost identical to that shown when comparing annual surveys. This indicates that seasonal/annual sediment movement is considerable and it masks any underlying trends for erosion or accretion. Collection of longer-term datasets will be required to highlight subtle underlying trends

## References

Cooper N, Rowe S, Parsons A, Cooper T. (2009) Cell 1 regional coastal monitoring programme. In proceedings of 44th Defra and Environment Agency Flood and Coastal Management Conference, Telford.

Environment Agency (2011) Coastal flood boundary conditions for UK mainland and islands, R&D project no. SC060064 'Coastal Flood Boundary Conditions'. Report – R&D SC060064/TR3 Design sea levels http://evidence.environment-agency.gov.uk/FCERM/en/Default/FCRM/Project.aspx?ProjectID=F162D56F-87C4-4F14-B77B-A8A3EFDB363F&PageId=a0fe6dfc-506a-452c-9bff-a7ec06b4e6b0

Hawkes, P., Atkins, R., Brampton, A., Fortune, D., Garbett, R., Gouldby, B. (2001) WAVENET: Nearshore Wave Recording Network for England and Wales, Feasibility Study. HR Wallingford, Report TR 122.

http://www.cefas.defra.gov.uk/media/42006/tr122_public.pdf

Lee EM (2008) Coastal cliff behaviour: Observations on the relationship between beach levels and recession rates. Geomorphology, 101, 558-571.

Lee EM, Clark A (2002) Investigation and Management of Soft Rock Cliffs. Thomas Telford, London

North East Coastal Observatory website: http://www.northeastcoastalobservatory.org.uk/

Permanent Service for Mean Sea Level (PSMSL), 2014, "Tide Gauge Data", Retrieved 30 Jun 2014 from http://www.psmsl.org/data/obtaining/

Pye K, Blott SJ (2015) Spatial and temporal variations in soft-cliff erosion along the Holderness coast, East Riding of Yorkshire, UK. Journal of Coastal Research. In press.

Wales Coastal Monitoring Centre (WCMC), 2013, Second Report June 2013. http://www.walescoast.org/wp-content/uploads/2013/10/WCMC-Second-Report-2013.pdf

**Coastal Management: Changing coast, changing climate, changing minds**
ISBN 978-0-7277-6114-9

ICE Publishing: All rights reserved
doi: 10.1680/cm.61149.263

# Understanding the Physical Processes Occurring Within a New Coastal Managed Realignment Site, Medmerry, Sussex, UK

**Heidi Burgess**, University of Brighton, Brighton, UK
**Paul Kilkie**, University of Brighton, Brighton, UK
**Tim Callaway**, RSPB, Hampshire and Isle of Wight Office, Pulborough, UK

## Abstract

In September 2013, the largest Open Coast Managed Realignment scheme in Europe was breached in the southern UK and Medmerry Nature Reserve created. Focusing specifically on the newly created intertidal areas, this paper presents observations and a sample of initial data of the intertidal sediment processes and geomorphological changes occurring within the new managed realignment site. Data shows that all of the sites being monitored are in a state of accretion with the source of sediment coming from both local e.g. through bank failure and distal e.g. off site sources. From the observations suggestions are made as to possible future improvements to both channel and borrow pit design.

## 1.0 Introduction

Part of the remit of the design of a managed realignment (MR) site is that it provides sustainable flood defence as well as becoming a fully functioning ecosystem. To achieve this newly created MR systems also have to be resilient to sea level rise and the increase in storminess. Understanding of the relationship between geomorphological elevation and vegetation is often used as one of the principle tools in the design and construction of these MR sites, but understanding of the geomorphological evolution, response to extreme events and development of abiotic factors in MR sites is incomplete. Although vegetation colonisation may occur rapidly after breach in MR sites, levels of floral bio-diversity are not equivalent to those found in natural intertidal sites (e.g. Garbutt et al., 2006). One of the causes for this may be differences in abiotic factors, where poor drainage resulting from soil compaction from previous land use and / or construction activity (e.g. Spencer and Harvey, 2012; Crooks et al., 2002) leads to anoxia (e.g. Davy et al., 2011) and limited biogeochemical cycling of the sediment. (e.g. Spencer et al., 2008; Mossman et al., 2012)

The pre-breach landscaping of MR sites is designed to encourage the development of varied habitat, both mudflat and saltmarsh, but guidance for both designers and contractors is very limited. Siltation of MR sites, especially those in sediment laden estuaries, has made the creation of sustainable mudflats in these sites difficult as they tend to evolve into saltmarsh (Morris, 2013, Pontee, 2014). Designers have attempted different methods to reduce siltation (e.g. Pontee, 2014) but there is a lack of high definition, long-term data on the geomorphological system response to disturbance / breaching (Viles et al., 2008), the modelling of which would greatly aid design. Therefore to improve design and management of MR intertidal systems we need to improve our knowledge base of how the Erosion, Transportation, Deposition and Consolidation (ETDC) of sediment varies within these newly created MR systems in relation to variations in MR conditions, e.g. size, open coast v estuarine, tidal range, geology, pre-breach land use, construction methods. After the initial breach of a site there is considerable re-distribution of the sediment, and the supply of sediment changes as the geomorphology of the site develops (e.g. Pontee, 2014). This is dependent upon the new hydrodynamics of the system, the created geomorphology, construction methods, and weather events as well as the die back of terrestrial flora. As the system starts to move towards equilibrium and intertidal flora and fauna begin to colonise the movement of sediment starts to stabilise. Therefore the first step in building up an understanding of the system is how the ETDC Cycle functions during the initial period after breach as it changes from terrestrial to intertidal.

Detailed understanding of the ETDC processes for a wide variety of MR sites and sub-environments within MR sites can then be used to improve the design, construction and management and hence the long-term sustainability of these systems. This paper presents data from the first year of monitoring at the largest open coastal Managed Realignment site in Europe, to assess initial post-breach sediment processes and early site geomorphological development. From these data, combined with onsite observations we suggest improvements which may help in future planning, design and management of some MR sites.

## 2.0 The Site and Preliminary Monitoring

Medmerry Nature Reserve (NR) is situated to the West of the Manhood Peninsular on the Southern Coast of England (Figure 1) and, at the time of writing, is the largest open coastal Managed Realignment site in Europe. The site is subject to a semi-diurnal, mesotidal regime. Pre-breach, material taken from a number of borrow pits (BP) was used in the construction of the 7km of flood bunds. The initial design and location of the BPs was related to the objectives of 'winning' suitable material for the construction of the new seawalls, minimising the haul distance (reducing costs) and the extent of site compaction, and contributing to the creation of intertidal habitat by lowering the elevation of areas within the site. However, further onsite investigations and ground conditions during early construction resulted in the need to change the location and design of most of the BPs. Further landscaping within the seaward confines of the bund (e.g. construction of islands and spits and construction of mini-breaches through old banks) was designed in order to encourage the establishment of different types of habitat (Figure 1) compensating for Natura 2000 (N2K) loss elsewhere around the Solent. From the 183 hectare site it was planned that initially 25 hectares would be mudflat and 158 hectares saltmarsh, with the upper marsh initially being available for grazing/arable land. These initial values would adjust in line with sea level rise, providing flood protection for the next 100 years, with the level of protection decreasing from 1:1000 to 1:100, over the century. The site was breached on the 11$^{th}$ September 2013.

The initial monitoring sites were established to encompass a variety of factors which included proximity to breach and to bund, variation in elevation, geomorphologic variation, pre-construction land-use, construction methods and exposure to meteorological conditions. This paper specifically looks at 5 sites in two different constructed landscapes (Figure 1). Sites 1 and 2 are at the extremity of the site close to the newly installed tidal gates, where freshwater enters the intertidal area. Site 1 is up on the banks of the channel and site 2 is situated on a 'beach' area within a channel. Sites 3, 4 and 5 are in the newly created BPs. Site 3 is within a newly created cutting which connects the BP to the main channel. Site 4 is to the back of this same BP, which is near the centre of the Medmerry NR. Site 5, is at the back of a BP towards the breach point.

In order to assess the sediment erosion and deposition and the initial geomorphological development within the Medmerry NR, Sediment Elevation Monitoring (SEM) Pins were installed at the monitoring sites. Installation was staggered as heavy plant were still accessing the interior of the site after breach. Transects of bamboo cane were installed and the length of the cane protruding was recorded periodically during the first year after breach.

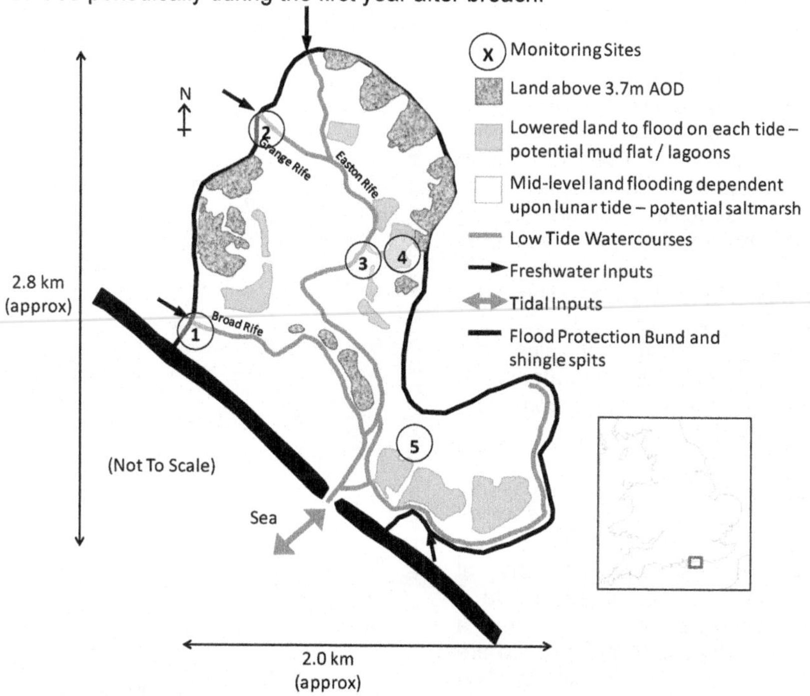


Figure 1: Schematic view of Medmerry Nature Reserve

## 4.0 Site Observations and Preliminary Results

At all sites over the winter period the terrestrial fauna rotted and much was finally washed away. Some of this dead fibrous material was then deposited upon the surface of the newly emerging mudflat areas and was then covered by a layer of sediment. Pioneer halophytic plants tended to establish in the areas where crop stumps and the remnants of other terrestrial plants were still present. This may be attributed to the fact that the propagulaes may have been trapped by the agricultural remnants (c.f. Tollesbury, UK - Garbutt et al., 2006.

SEM pin data for each site was plotted with variation in bed elevation (ΔE) being monitored from the initial bed elevation at the time of installation (t=0) for each pin,

(Figure 2). Generally there was a rise in bed elevation at all sites, with some localised reduction in bed elevation at two sites (3 and 4).

**Site 1**

This site is at the eastern extremity of the MR site, adjacent to a main freshwater channel which enters Medmerry NR through a tidal gate. Protected from predominant South Westerly winds by the shingle spit, locally induced wind waves are generally small. The pins were installed in an upper flat area. This area had previously been semi-improved grassland, used for storing of excavated material unsuitable for bund construction, before being cleared and levelled. The first post breach data was taken in May 2014, eight month after breach. This area floods on most high tides and the sediment in which the pins are situated is continually saturated, with frequent pooling of water, from both tidal and pluvial water.

The pins closest to the channel (Figure 2; 1.1, 1.2 and 1.3) saw an increase of between 60 to 80mm in ΔE during the first eight months of installation. This then reduced to a gradual accretion of around 1mm a month over the summer period. There was less of an increase in bed elevation around pin 1.4, which is situated slightly further from the main channel and its initial elevation was slightly higher than the other three pins, and therefore is subjected to lower frequency and depth of inundation.

It is likely that the ΔE is as a result of the settling of suspended sediment. The source of this sediment is yet to be determined but there are a number of potential sources. Including to the north east of Site 1 a large area to

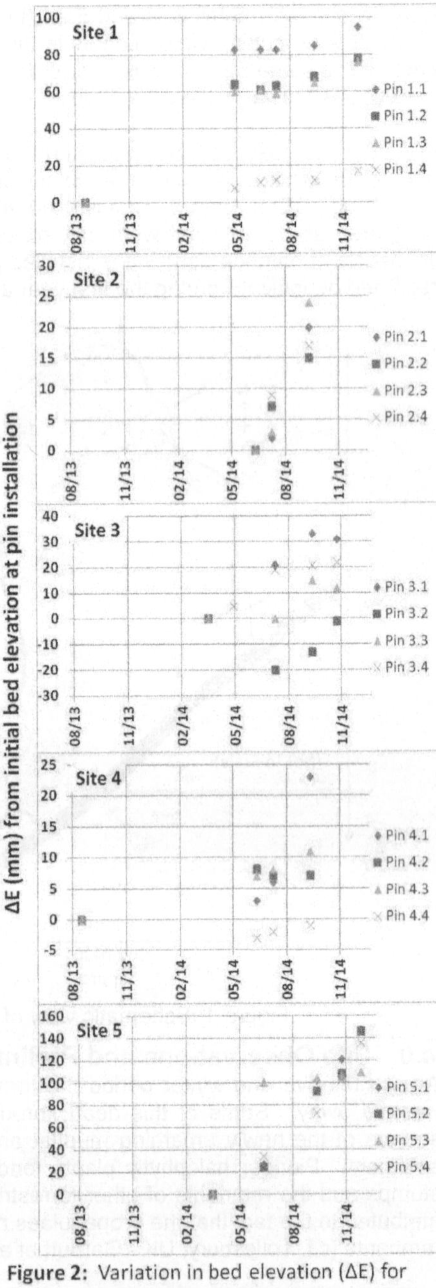


**Figure 2:** Variation in bed elevation (ΔE) for SEM pins. Plots relative to initial bed elevation at the date of pin installation i.e. at t=0, ΔE=0.

where excavated material was placed to create raised intertidal areas, which may also be subject to erosion.

**Site 2**

To the north of site 1, situated in another principle freshwater fed channel, this monitored section of the channel is between the new large culvert containing the tidal gates and an older smaller culvert which was built to allow farm vehicle access. This section was stripped of vegetation, the channel cut and re-profiled to allow transition from the old channel to the new culvert. The pins were positioned on a flattened 'scoured beach' area at the base of the southern bank, but at a slightly higher elevation than the low freshwater flow level (figure 3). During spring tides this whole area, including the upper banks, is inundated.

**Figure 3**: Site 2, (a) July 2013, (b) July 2014. SEM pins installed on the beach area on the right hand side of the picture, position indicated by arrows.

Within this channel the beach area has been raised between 5 and 8mm every month over the monitoring period. This is likely to result from the slumping of sediment from the bank as the channel at this point has widened and the gradient of the banks decreased as the channel converts from its engineered to a more natural profile. This area was subject to considerable machinery activity and modification during construction works. In addition, the bankside vegetation on the steep existing channels, died off rapidly after tidal inundation leading to rapid destabilisation of the channel banks. The site is open to cattle grazing during the summer months and at times the upper banks were heavily trampled. This disturbance potentially provides a further source of sediment as it is transported from the upper banks in the tidal runoff water.

**Site 3**

This site is towards the middle of Medmerry NR at the confluence of the main channel through the site and a newly cut channel which slopes gently up to connect to a large BP (3.035 hectares). Previously overlain by agricultural land, the bed at this site was exposed during construction as up to 1.5 metres of upper soil was excavated (figure 4a). Therefore the initial bed sediment at this site was compacted by both years of soil overburden and heavy plant activity during landscaping. Over high neap tides only the lower section of the cut channel is flooded. During high spring tides tidal water floods the whole channel up into the BP (site 4). In this tidal state the site is exposed to a long fetch, with ≈0.5m wind waves being observed.

It was observed after high rain events of December 2013 / January 2014 that a creek had been scoured into this connecting channel (figure 4b). Because of the limited tidal exchange through this newly cut connection channel it is likely that the creek has formed as a result of a combination of both pluvial runoff from the BP area and of the infrequent tidal discharge from the BP (see site 4).

**Figure 4**: Site 3, (a) October 2013, (b) September 2014. SEM pins on either side of the pluvial scoured (indicated by arrow).

The SEM pins were installed at this site after the formation of the pluvial creek within the newly cut connection channel. Between March and October 2014 there was an average accretion of sediment of between 1.5 and 4.5mm per month. Pin 3.2 initially suffered scouring at the base as a result of weed entanglement. It was observed that the consistency of the bed sediment varied greatly between visits, further indicating the variation in mobility of the sediment at this site. The gradient of channel banks in the new cut has reduced since breach, which will be a source of sediment.

**Site 4**
This BP site is situated to the east of site 3 and was designed to be an intertidal mudflat area. SEM pins were installed 2 months before breach (July 2013) and recorded a comparatively low increase in bed elevation of between 5 and 10mm, over the whole year. Because of the excavation elevation of this site, relative to mean sea level (MSL), it is only inundated on the highest of tides (see site 3). The topographic levelling of this compacted terrain means that often there are puddles of standing water after pluvial and high tide events. The short hydroperiod of this site results in a low delivery of suspended sediment which means that delivery of seeds borne by the intertidal waters will also be low. The limited inundation means that the slopes of the banks of this site remain steep, as constructed, as they have not be subjected to significant hydrological forces.

The combination of compaction and low delivery of sediment and seeds has resulted in the indentation marks of the heavy plant used in construction of the banks still being evident after more than one year, the sediment in this area remaining coarse and highly variable in size, and minimal vegetation cover (figure 5).

**Site 5**

This site is situated directly in the face of the prevailing wind, towards the back of a low elevation BP near the breach in the shingle spit. The low tide water levels in this BP are controlled by the shifting sediment at the entrance channel. The banks of this BP are overtopped on every tide and the remnant barley field (last harvested in the weeks before breach) behind also inundated.

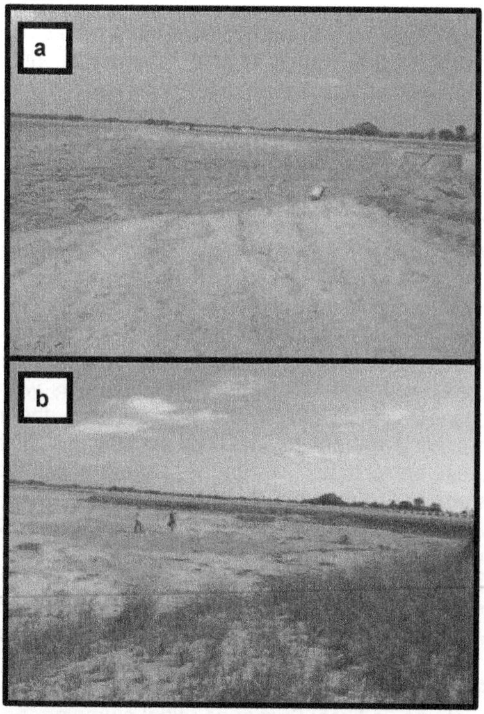


**Figure 5**: Site 4, (a) July 2013, (b) July 2014. SEM pins installed to the right of the picture. Note the lack of vegetation after one more than one year of excavation.

The BP here was subject to significant archaeological excavation which resulted in varied bed topography. Stretches of the northern bank were re-profiled after completion of the excavation of the BPs to reduce the steepness of the banks. As predicted much of the remaining constructed angle of the steep banks of this BP have been reduced considerably by wind wave action resulting in block slumping of the bank. Sediment from the BP banks as well as sediment washed from the remnant barley field has moved forward into the BP area. This combined with input of sediment from other sources within the MR site or from the ocean has resulted in an average of up to 18mm increase in bed elevation per month over the monitoring period. During early tidal inundation events after the breaching of the old sea wall, significant sediment loads were noted on incoming tides (verbal communication with RSPB and EA). Circumstantial evidence of delivery of sediment from external sources was provided by the colonisation by a variety of intertidal plants six months after breach on the higher remnant barley field. The longer hydroperiod together with the remnant crop stumps provide a good environment for the delivery and entrapment of seeds and propagules (see Garbutt et al., 2006).

## 5.0 Summary of Observations

Initial data from the SEM pins indicate that the ETDC processes vary considerably around the site. This initial period of monitoring is not yet long enough to show any seasonal trends, but all of the pin sites, which are all located on relatively flat terrain, show a general trend of accretion. The source of the sediment has been observed to originate from: the re-sculpting of the pre-breach (existing and engineered) geomorphology by tidal action, wind waves and pluvial events. Sediment is being redistributed from newly constructed banks, collapse of banks of the old freshwater channels and from the natural formation of new intertidal drainage creeks. The colonisation by intertidal flora provides evidence that seeds are being imported from an external site, from this circumstantial evidence is drawn that sediment is also likely to be being imported from external sources. These observations raise a few points in relation to construction and early stage management:

**Gradient of the Channel Banks**

There are two types of channel within Medmerry NR, either *'constructed'* as part of the MR site landscaping or *'freshwater'* which previously formed part of the drainage system of the terrestrial system. The constructed channels tended to be devoid of established flora pre-breach and the angle of the banks constructed using the design specification for that of terrestrial drainage channels, rather than natural intertidal creek systems. The freshwater channels were generally over deepened drainage ditches with relatively steep sides with the root system of the freshwater flora providing enhanced stabilisation. With regular tidal inundation the land forming these banks is now continually saturated with tidal water. As the tide drops there is a sharp variation in hydraulic head within the soil, as the compacted soil often remains waterlogged. This then results in slumping of the banks and breakthrough of water lower down in the banks as drainage channels form, again resulting in bank failure. This increased sediment load within the channels can become an issue towards the heads of the channels where sediment tends to accumulate in the calmer water and where the tidal control structures through the bund are situated.

Reducing the constructed angle of the bank slope, as found in natural intertidal systems, is likely to reduce initial slumping, as well as aiding in the drainage of the surrounding land and possibly allow quicker formation of natural intertidal channels (cf. Cornu and Sadro, 2002). In areas higher in the tidal elevation reduced bank slopes would provide a greater area of transition for flora and fauna. Drainage could be further improved by, increasing the length of the channel by increasing the sinuosity of constructed channels, in line with natural marsh creeks.

**Borrow Pits**

The bed elevation data indicates that the BP areas being monitored are all starting to accrete. This is in line with estuarine MR sites, which have shown initial, rapid sediment infill, especially in lower lying, wind protected areas, turning from mudflat to saltmarsh ( Morris, 2013 and Pontee, 2014). However this is a large open coast site as opposed to an estuarine site, with relatively large BPs, often exposed to locally induced wind waves, hence their development may differ. The elevation of the bed of the BP is important relative to MSL but also relative to surrounding land. In Medmerry there are three types of borrow pit, '*a low bed - low bank elevation'*, this type of BP sits low in the tidal range and the banks overtopped on every tide. There is sediment runoff from the surrounding land into the BP with each tide. The banks of the BPs will be also subject to the same hydraulic forces as flooded channel banks (see above). The bed of a *'low bed – high bank elevation'* BP again sits low in comparison to MSL but the banks are higher so may not be over topped on every tide. The *'high bed – high bank elevation'* BP is situated in the upper range of the tidal cycle and the banks very rarely overtopped. The steep sides fail to provide flora and fauna with a gradual transition. The shorter hydroperiod, results in little sediment and seed delivery, again retarding the colonisation of flora and fauna. In addition the bed of these sites will have been subject to compaction resulting both from the years of soil over burden coupled with the movement of heavy plant during the excavation of the upper soil layer. Compaction can lead to problems with water logging and further limit plant colonisation (Spencer et al., 2008; Spencer and Harvey 2012).

The amount of water retained within a BP during low tide is dependent upon the previous tidal level of inundation and the geomorphology of the entrance channel. The gradient of the bed within a BP is generally quite low, slowing surface drainage of both tidal and low tide pluvial water, therefore prolonging water logging. Steeper slopes and, in higher MSL elevation BPs, rudimentary channels would also further aid drainage and possibly creek formation (cf. Cornu and Sadro, 2002), potentially increasing sediment runoff and providing elevation transition for flora and fauna.

More understanding of the roll that depth and fetch limited waves have on the ETDC of sediment in MR environments is needed. This along with consideration at design/construction stage of the size, depth and orientation and angle of BP banks to predominant wind direction could then be potentially be used as a self-sustaining tool in terms of remobilisation or prevention of settling of sediment.

## 6.0 Conclusions

In order to improve and refine design and construction methods so that we are working more in line with nature there needs to be a greater understanding of the complex interconnected processes (physical, biological and chemical) occurring within MR sites. The water and sediment are the fundamental basis of the intertidal systems. Therefore more high resolution data on the sediment ETDC processes are needed, in order to start to understand how the biological and chemical processes vary with sediment, and vice versa. Once the relationship of the sediment ETDC processes are more fully understood in relation to the hydrodynamics, meteorological events, geomorphology and pre-breach use within MR sites, then we can start to understand, and model more accurately, the longer-term, larger scale trends.

With landscaping trying to replicate more closely natural environment there is a potential increase in initial costs. This may be offset in the longer term if a fuller understanding of ETDC cycle and the complex interconnected biological and chemical processes aid us to develop a more holistic design criteria leading to a more diverse and self-sustaining system.

Site conditions and archaeology resulted in the planned intertidal landscaping of Medmerry NR having to be re-designed considerably during construction, within a budget. This work has provided an excellent opportunity to study from a variety of landscaping techniques. Medmerry NR has already proved that it is an effective flood defence site during the winter of 2013/14. The long term bio-diversity of the flora and fauna is another measure of the success of the site and is dependent upon numerous and complex factors, which is beyond the scope of this paper.

Our understanding of the sediment regime and how site morphology can be engineered pre-breach to influence and initiate ETDC processes is a fundamental factor in the future design and construction of all future MR sites. It is essential that funding is built into these projects so that co-ordinated multi-disciplinary detailed and focused monitoring is conducted, starting pre-construction and continuing for the longer-term. MR sites need to be monitored in more depth and findings disseminated so that the lessons learnt and further questions posed by the creation of MR schemes such as Medmerry NR be taken forward for planning, design and management of other similar systems.

## Acknowledgements

Funding for the project monitoring was provided by the Environment Agency. Special thanks to Andrew Cundy, Colin Maplesden and Martin Smith for their critical input into this paper.

## References

Cornu, C.E., Sadro, S., 2002. Physical and functional responses to experimental marsh surface elevation manipulation in Coos Bay's South Slough. Restoration Ecology 10 (3): 474-486.

Crooks, S., Schutten, J., Sheern, G.D., Pye, K. And Davy, A.J., 2002, Drainage and elevation as Factors in the Restoration of Salt Marsh in Britain, Restoration Ecology, 10: 591-602

Davy, A.J., Brown, M.J.H., Mossman, H.L. and Grant, A., 2011, Colonization of a newly developing salt marsh: disentangling independent effects of elevation and redox potential on halophytes. Journal of Ecology, 99: 1350–1357.

Garbutt R.A., Reading, C.J., Wolters, M., Gray, A.J. and Rothery, P., 2006, Monitoring the development of intertidal habitats on former agricultural land after the managed realignment of coastal defences at Tollesbury, Essex, UK, Marine Pollution Bulletin, 53: 155–164

Morris, R. K. A., 2012, Managed realignment: A sediment management perspective, Ocean and Coastal Management 65: 59-66.

Morris, R. K. A., 2013, Managed realignment as a tool for compensatory habitat creation - A re-appraisal, Ocean and Coastal Management, 73: 82-91

Mossman, H.L, Davy, A.J. and Grant, A., 2012, Does managed coastal realignment create saltmarshes with 'equivalent biological characteristics' to natural reference sites?, Journal of Applied Ecology, 49: 1446–1456

Pontee, N., 2014, Accounting for siltation in the design of intertidal creation schemes, Ocean and Coastal Management, 88: 8-12

Spencer, K.L. and Harvey, G.L., 2012, Understanding system disturbance and ecosystem services in restored saltmarshes: Integrating physical and biogeochemical processes, Estuarine, Coastal and Shelf Scinece, 106: 23-32.

Spencer, K.L, Cundy, A.B., Davies-Hearn, S., Hughes, R., Turner, S. And MacLeod, C.L., 2008, Physicochemical changes in sediments at Orplands Farm, Essex, UK following 8 years of managed realignment, Estuarine, Coastal and Shelf Science, 76:608-619

Viles, H.A., Naylor, L.A., Carter, N.E.A and Chaput, D., 2008, Biogeomorphological disturbance regimes: progress in linking ecological and geomorphological systems, Earth Surface Processes and Landforms, 33: 1419-1435

**Coastal Management: Changing coast, changing climate, changing minds**
ISBN 978-0-7277-6114-9

ICE Publishing: All rights reserved
doi: 10.1680/cm.61149.273

ice
Institution of Civil Engineers
publishing

# Integrated Coastal Management Implementation in South Africa – in Practice Not Just Concept

**Tandi Breetzke**, Royal HaskoningDHV, Durban, South Africa
**Gerard van Weele**, Royal HaskoningDHV, Cape Town, South Africa
**Luke Moore**, Royal HaskoningDHV, Durban, South Africa
**Andrew Mather**, eThekwini Municipality, Durban, South Africa

## Overview

Integrated Coastal Management (ICM) in South Africa, as in many other areas of the world, has been a process of exploration and discovery as much as one of definition; from the initial process of establishing policy through to a process of implementation, a process often fraught with inconsistencies. It is in moving from the academic concept of ICM to a cycle of implementation where one encounters the greatest challenges. This is the arena where practical application and sound thinking is required to be applied to the reality of delivery and small victories celebrated.

South Africa can rightly claim to have come a long way in terms of the development of an ICM system. Starting in the 1990s through to the late 2000s, an extensively consultative and acclaimed process led to the promulgation of the National Environmental Management: Integrated Coastal Management Act, (Act no 24 of 2008) and recent National Environmental Management: Integrated Coastal Management Amendment Act, (Act no 36 of 2014) (ICM Act). In conjunction with other statutes related to the conservation of biodiversity and general environmental management, this Act gave hope to coastal champions countrywide. Currently, the South African suite of coastal policy and legislation can be easily compared with its international counterparts and is rated among the best (DEA, 2012). Implementation has, however, been somewhat erratic, differing from agency to agency, place to place and issue to issue. Institutional constraints, regulatory gaps, socio-cultural history and simply the diverse and extensive nature of the South African coastline has led to inconsistent results.

This paper is intended as a thought provoking perspective and proposes to reflect on select implementation processes over the last 5 years. It will detail some of the successes achieved and celebrated as well as highlight challenges faced and lessons learnt. Its main purpose is to contribute to practical implementation and inform the next 5 year cycle of ICM practice in South Africa but also, in considering the range of issues faced by implementation of consistent ICM principles across a varied landscape, offers broader ideas within the international context.

The paper will highlight:

- The various implementation tools included in the ICM Act, namely Coastal Management Programmes, Estuary Management Plans and Coastal Management Lines;
- The process of development and implementation of these tools using select examples; and
- The implementation process good and 'not so good's.

In summary, the paper will highlight how the South African policy journey, which while comparable internationally, suffers from inconsistencies in respect to implementation. This is predominantly a result of differences between implementation agencies (political affiliation and will, institutional constraints, capacity, financial resources, socio-cultural gaps etc.); and differences between places along South Africa's extremely diverse and extensive coastline.

## A Background to Integrated Coastal Management in South Africa

The initial shift towards ICM and the more equitable management of South Africa's coastal zone with its numerous resources came in 1998, with the formulation of South Africa's preeminent environmental legislation, the National Environmental Management Act, Act 107 of 1998 (NEMA). The NEMA's primary objectives included the promotion of a co-ordinated approach to matters affecting the environment by ensuring that co-operative governance, as well as co-ordinating mechanisms and institutions, were implemented as a matter of priority (DEA, 2014). Shortly thereafter in 1998 and 2000, South Africa's first ICM policies emerged in the form of the Coastal Policy Green Paper and the White Paper for Sustainable Coastal Development respectively.

The White Paper was founded on a shared national vision for the coast that included the socially justified and sustainable sharing of benefits as well as a shift towards ICM and away from top-down management (Glavovic, 2006). It recognized the coastal zone as an interconnected functioning system which required coordinated and integrated management (DEAT, 2000). While the White Paper set the foundation for the implementation of ICM in South Africa, the ICM Act, with its emphasis on cooperation and shared management, kicked off its actual implementation. The objectives of the ICM Act include determining the coastal zone, providing a framework for implementation, protecting and securing equitable access to coastal public property, and giving effect to South Africa's international; obligations in terms of international law (DEA, 2014)

## Tools provided for Integrated Coastal Management in South Africa

The ICM Act provides all three spheres of South African government (National, Provincial and Municipal) with various tools to support effective implementation. Tools include Coastal Management Programmes, Estuarine Management Plans and Coastal Management Lines, amongst others.

**Coastal Management Programmes**

In order to facilitate cooperative governance, the ICM Act requires a framework of coastal management programmes (CMPs) to be developed. The National CMP, with its largely strategic focus, provides for coastal management on a national level. Provincial and municipal CMPs are intended to provide for coastal management on a regional and local level respectively. They aim to translate the national goals and objectives into provincial and local outcomes, provide guidance on the alignment of outcomes, needs and issues and formulate priorities, action plans and strategies. These latter CMPs are both operational and strategically focused (DEA, 2012).

CMPs provide a coherent, integrated and co-coordinated framework and directive for ICM and decision-making and provide an applicable vision for the specific portion of coast, they identify priority areas for implementation with accompanying objectives and implementable and measurable required actions. They also include norms and standards or priority management guidelines to manage and utilise the coast and identify specific research priorities (DEA, 2012).

In order to ensure consistency with the development of CMP, the National lead agent for ICM, the National Department of Environmental Affairs, have prepared a Guide for the development of CMPs (DEA, 2012). While this guide is still in draft form it has significantly informed the preparation of CMPs at provincial and municipal levels.

**National Estuarine Management Protocol and Estuary Management Plans**

A further tool detailed in the ICM Act is the development and implementation of a National Estuarine Management Protocol (NEMP) as well as the development of individual estuary management plans (EMPs) for each estuary throughout the country. It is noted that South Africa contains approximately 270 functional estuaries[1] along it's approximately 3200 km of coast line (Van Niekerk and Turpie, 2012). Through these tools, the ICM Act aims to facilitate the efficient and coordinated management of all estuaries by a designated management authorities identified.

The NEMP has since been developed and was gazetted / adopted on the10th of May 2013. It provides the required national policy for estuary management and guides the development of individual estuarine management plans, thus ensuring national consistency. Any interventions and management decisions relating to estuarine ecosystems must be guided by both the NEMP and any EMP developed in line with the NEMP. Estuarine management essentially involves the regulation of human activities, as opposed to the estuarine environment *per se*, and thus public participation and invoking a sense of ownership amongst the local communities is key to the planning phase. The establishment of estuary advisory forums to assist upper government levels is critical to ensuring successful and effective management, as well as the sustainable use of estuarine resources and translating the concerns identified into implementation activities (DEA, 2014).

---

[1] Approximately 79 per cent of the estuarine area in the country is in a threatened state. Fifty four per cent of South African estuaries are in a poor condition. Currently as much as 59 per cent of South Africa's estuarine ecosystem types (83 per cent of total estuarine area in South Africa) are not formally protected (Van Niekerk and Turpie, 2012).

**Coastal Management Lines**

The use of the final tool under discussion, namely coastal management lines (CMLs), is particularly important in the face of the unfolding effects of climate change in the coastal environment, as it involves both a quantification of risks (erosion, flooding, tidal surges, inappropriate development etc.) and pro-active planning for future development. Although the use of CMLs cannot address historical decisions that have locked in development investment along potentially at-risk coastal areas, they can influence how existing development is maintained over time and how new development will be allowed to proceed.

Coastal management lines, as detailed in the ICM Act (as amended), are prescribed boundaries that indicate the limit of development along ecologically sensitive or vulnerable areas, or an area where dynamic natural processes pose a hazard or risk to humans or infrastructure. The ICM Act allows coastal management lines to demarcate areas where authorities can prohibit or restrict the building, alteration or extension of structures that are either wholly or partly seaward of the coastal management line. It is noted that the location of immovable property and the ownership and zonation of vacant land must be taken into consideration when delineating coastal set-back/management lines. The ultimate intention of the coastal management line is to protect coastal public property, private property and public safety; protect the coastal protection zone; and preserve the aesthetic values of the coastal zone (WCG, 2014).

## In Practice

The implementation of the ICM Act and the use of the tools detailed above, in practice, has tended to unfold in a somewhat erratic and inconsistent manner. This irregularity is highlighted using specific case studies.

**In Practice - Coastal Management Programmes**

As detailed, the ICM Act requires all three spheres of government to prepare CMPs within four years of the Act taking effect. The 2014 National CMP, while somewhat overdue, provides a valuable national foundation by unpacking the following key issues:

- The value of the coast;
- The context and purpose of CMPs and the national CMP development process;
- A situational assessment (inventory analysis);
- Roles and responsibilities for coastal management;
- A national vision and priorities; and
- National objectives and actions (DEA, 2014).

The proposed process towards implementation is, however, lacking in substance with the next step identified as 'developing detailed implementation plans' for each of the listed actions. These detailed plans are proposed to include specific target dates, performance indicators, a human resources plan as well as a financial resources plan, amongst others (DEA, 2014). The content of the National CMP also does not follow the requirements as detailed in the draft National Guide to the Development of CMPs in South Africa.

Provincial and local government were also required to prepare CMPs within four years of commencement of the Act. All four coastal provinces had, however, already prepared inaugural CMPs as part of the National White Paper implementation process, the international donor funded Sustainable Coastal Livelihoods Programme, prior to the enactment of the ICM Act. These CMPs, were consequently not in full compliance with the requirements of the ICM Act, and therefore needed to be updated. The Eastern Cape Province already undertook this updating process in 2013 with the Programme having being formally adopted and gazetted. The Western Cape and Northern Cape provinces are currently in the process of updating their CMPs with the KwaZulu-Natal province having completed its status quo assessment in the form of a stand-alone publication entitled “Ugu Lwethu – Our Coast: A profile of coastal KwaZulu-Natal”.

Municipalities, not having undertaken any coastal planning in the past and also required to prepare CMPs within four years of commencement of the Act, were a lot slower to respond. This was primarily as a result of so-called ‘unfunded mandates’ (i.e. the belief that environmental management requirements are forced upon municipalities without the necessary budgetary support from higher tiers of government) and a lack of capacity/financial resources. The slow response was also a result of uncertainty as to who should actually prepare the programmes. In South Africa, Municipal Government is broken into Metropolitan Municipalities; District Municipalities (consisting of multiple local municipalities) and Local Municipalities. While the National White Paper was unclear as to applicable level, the ICM Act identifies Metro and District as the key role-players in the CMP process, even though many of the operational aspects and on the ground requirements are in fact Local municipal responsibilities.

Lessons Learnt

Having either completed, or being in the process of completing, three of the four provincial CMPs as well as three municipal CMPs, the authors are well placed to identify lessons learnt. These are listed as follows:

- Key to the success of CMP development and implementation is buy-in from all parties / stakeholders involved;
- Successful implementation hinges on political support and buy-in;
- Successful development and implementation requires either a dedicated champion or group of champions;
- Adequate and sustainable resources and capacity are required to implement any strategies identified within the plan of action;
- Effective co-operative governance, which includes civil society in cooperative government, will drive successful implementation across sectors and spheres;
- The CMP drafting team, including the client and stakeholders, need to be realistic and propose / adopt an attainable programme of action;
- Success is achieved in small incremental steps as part of an iterative process; and
- Finally, the best available and scientifically defendable information must be used and data gaps identified and processes employed to fill them.

**In Practice - Estuary Management Plans**

Further to the above, the NEMP proposes management standards, details procedures and minimum requirements for EMPs, responsible authorities and approval and review procedures (DEA, 2014). Key objectives of the NEMP include, amongst others, ensuring the sustainable economic and social use of systems without compromising their ecological integrity and functioning; restoring ecological integrity, co-operatively managing systems, promoting awareness and minimizing the potential impacts of predicted climate change (DEA, 2014). No new institutional structures are proposed in the NEMP, however, existing Provincial and Municipal coastal committees established during the development of CMPs, are required to serve as forums for monitoring and reporting on the implementation of EMPs. Any existing and/or planned individual estuary forums are regarded by the NEMP as "*informal additional advisory bodies for facilitating the implementation of the project plans identified in the Estuary Management Plans and to foster continuous stakeholder engagement (DEA, 2014)*".

As a result of estuaries being amongst the most threatened of South Africa's inland aquatic ecosystems (Van Niekerk and Turpie, 2012), the Cape Action for People and the Environment (C.A.P.E.) Estuaries Management Programme, in partnership with the National Department of Environmental Affairs, developed the C.A.P.E. Generic Estuarine Management Plan (GEMP) framework which detailed an integrated management process that predominantly met the EMP requirements of the now adopted NEMP (CSIR, 2009). This framework was implemented to establish EMPs for selected South African estuaries inside the Cape Floristic Region, thereby including estuaries from three of the four coastal provinces. Factors affecting the application of the GEMP primarily included the level of detail that the framework established EMPs on and the scarcity of bio-physical state and socio-economic data. The NEMP recognises this data shortfall and now proposes minimum requirements for the development of EMPs which include a requirement to highlight any major information gaps and make recommendations to address such gaps (DEA, 2012).

Lessons Learnt
While the Royal HaskoningDHV team has completed only four EMPs, these plans have been assessed by the national Department of Environmental Affairs against the NEMP and found to have little or no inconsistencies. Lessons leant from these processes are therefore detailed as:

- The use of scientifically sound local ecological knowledge in conjunction with anecdotal knowledge / evidence ensures the applicability and ultimate successful implementation of the plan;
- Local / neighboring ownership and buy-in (shared responsibility) must be encouraged;
- As with CMPs, knowledge gaps should be acknowledged, identified and efforts made to undertake/contract out the necessary research required. EMP development and implementation is therefore incremental;
- Efforts should be made to partner with research organizations and higher education institutions;
- Sustainable resources (financial & capacity) are required to successfully implement a realistic programme of action;
- Successful implementation includes a programme of monitoring and enforcement; and
- Any strategies, management objectives and/or zonation proposed must promote a balanced and sustainable use of the system – both protection and sustainable utilization.

**In Practice - Coastal Management Lines**

The designation of Coastal Management Lines, as a final ICM Act tool, is provided predominantly for implementation by provincial government. As a result of this, in addition to the concurrent environmental legal competency, the Western Cape Provincial Government (WCG) undertook a process of developing an approach or methodology to determine such lines ahead of any direction or norms and standards developed by National Government. Work undertaken by the authors in respect to the Overberg District represents the third full-scale set of coastal management lines determined by the WCG. It is preceded by a study conducted during 2010 at Milnerton and Langebaan to devise a standard methodology for the delineation of coastal set-backs in the province (DEADP, 2010); the initial pilot project and determination of coastal set-backs along the coastline of the Overberg District during 2011/12 (WCG, 2012); and more recently, the development of coastal management lines for the West Coast District (WCG, 2014) and the refinement of lines for the Overberg (WCG, 2015).

Issues identified and recommendations made though all of these processes include:

- The reality of existing legal coastal development meant that coastal management (set-back) lines have to make provision for existing development and development planning that already extends into the identified hazard zone;
- A management response is required that differentiates between the modeled erosion hazard and pragmatic development control;
- Risk needs to be modeled taking realistic planning horizons into consideration, for example three proposed sea level rise scenarios, namely, low (20 years or 200mm), medium (50 years or 500mm) and high (100 years or 1000mm);
- Decisions regarding land use and development should be based on the nature and value of the proposed development or activity and considered in relation to the risk identified; and
- The use of coastal (set-back) management line regulations, as provided for in the 2008 ICM Act and since amended in the 2014 amendments, are impractical and do not provide space for negotiation, mitigation or discretion. Formal regulations were abandoned in favour of more practical and locally customisable management schemes, specifically the use of overlay zones with accompanying development controls which are implemented as part of a Land Use Management Scheme at a municipal level.

The most recent 2015 approach taken in determining coastal management line(s), accompanied by coastal management risk zones for the Overberg District has been informed by an updated methodology developed for the WCG, implementation thereof in the West Coast District, and work undertaken independently by the City of Cape Town. The approach adopted reflects management principles or criteria suggested by affected authorities as well as the public stakeholders participating in the stakeholder consultation process facilitated by the WCG (WCG, 2015). These principles are:

- Legal development with existing rights (including properties zoned for development) must be allowed to proceed as long as public (i.e. government) liability and the quality of the coastline are not compromised;

- Management controls must allow for private acceptance of liability in risk-prone areas;
- Controls must allow for discretion in decision-making by authorities based on appropriate motivations and information;
- Provision must be made for areas where the local municipalities propose growth, as long as the development proposals are responsive towards coastal risk and ICM Act principles;
- Coastal defences, if constructed, must be constructed and managed in an integrated manner and in accordance with legislative requirements;
- Delineation must avoid the uncertainties surrounding the position of the high water mark; and
- Management control must recognise that 'land use' and physical activities are distinct.

The updated methodology developed for the designation of coastal management lines, is informed by the projections of risk as well as the location and extent of existing development and existing executable development rights. It is intended as a clear guideline for the management of development within risk areas, and the protection of coastal public property and differentiates between areas along the coastline with existing development rights and/or forming part of future municipal development, and those areas that should be left undeveloped due to a high risk from dynamic coastal processes or as coastal public property. This methodology is a refinement of the previous methodology applied and takes into consideration the need for national (National Coastal Committee discussions, i.e. Working Group 8 forum) and local alignment (similar work by the City of Cape Town).

Lessons Learnt

Having undertaken work for both the Overberg and West Coast, the authors are well placed to detail lessons leant in the determination of coastal management lines:

- Local government, both political as well as administrative, buy-in is imperative for effective implementation;
- The drafting of risk zones and the application and somewhat subjective drafting of coastal management lines, should therefore strive to be founded on sound scientific principles;
- Uniformity and consistent application of a negotiated and agreed decision tree significantly reduces potential conflict;
- Decisions made must always strive to be practical, yet responsible and flexible enough to both protect the environment and limit negative impacts on private property rights;
- Government should motivate for funding to expropriate privately owned development rights in high risk areas or consider alternative innovative options, for example, the transfer of development rights;
- Land owners and government agencies are required to implement appropriate defense by 'thinking out the box', living with coastal erosion and wave inundation and adapting to living in a high risk environment; and
- Government agencies must allow for finer scale planning and modeling to be undertaken and thereafter be flexible with solutions and with amending the lines drawn.

## Conclusion

Acknowledgement of local circumstances and conditions is a critical starting point for any implementation scheme and, necessarily, this must be based on the involvement of local actors and authorities in order to ensure their buy-in and to avoid critical mistakes that will undermine the practicality of the management measures. In South Africa, ensuring buy-in on a political decision-making level is equally important, with highly politicised strategic planning determining the scope and direction of executive environmental management.

First round implementation also need not be seen as a once-off event. As evidenced by the development of local scale coastal management tools in South Africa, any number of iterations is necessary to achieve an appropriate level of practical implementation. This allows multiple viewpoints, innovative solutions and progressive contextual changes to align over time, whilst knowledge gaps are identified and filled.

The alignment of different management schemes is a critical factor, as it reduces duplication of effort and conflicting decisions arising from differential mandates. Indications are that a greater reliance on local level implementation and regulation is required, and specifically integrated coastal management that aligns with existing and commonly accepted regulatory schemes such as land use management (Town Planning) schemes. However, when the diversity of land tenure in South Africa is considered, communal ownership also needs to be planned for and addressed on its merits and limitations when coastal management tools are developed for these areas.

The range of scales that these ICM tools must be implemented at also means that different combinations of strategic and/or operational approaches are needed to successfully address the issues facing different areas in space and time. Successful implementation is no mean feat and prospective coastal managers and implementation authorities should pace themselves for the long run, breaking the various tasks down into more implementable phases.

The challenges that face implementation authorities, coastal stakeholders and landowners in South Africa, while extensive, are not insurmountable. Lessons learnt, successful solutions found to challenges faced as well as those situations unable to be resolved, should be shared and collectively applauded. Continued success will depend on a collective response and the ability to address the pragmatic and practical issues of implementation while winning the confidence of all involved in the sensible application of national policy. It is however important to not compromise the need for adaptability in a climate of uncertainty and an environment prone to rapid change. Only then can ideological concepts feed into meaningful and sustainable implementation measures.

## References

Celliers, L., Breetzke, T., Moore, L. & Malan, D. (2009). *A User-Friendly Guide to South Africa's Integrated Coastal Management Act.* The Department of Environmental Affairs and SSI Engineers and Environmental Consultants. Cape Town, South Africa.

Coastal Management Policy Programme (2000). *White Paper for Sustainable Coastal Development in South Africa*, Department of Environmental Affairs and Tourism. Pretoria: Government Printer pp. 137.

DEAT (Department of Environmental Affairs and Tourism) (1998). *Coastal Policy Green Paper: Towards sustainable coastal development in South Africa*. Cape Town, 1998. 155pp.

DEA (Department of Environmental Affairs). (2010). *A Summary Guide to South Africa's Integrated Coastal Management Act.* Department of Environmental Affairs. Cape Town: Cape Town Directorate: Marine and Coastal Management.

DEA (Department of Environmental Affairs). (2012). A Guide to the Development of Coastal Management Programmes in South Africa. Department of Environmental Affairs. Cape Town : Oceans and Coasts Branch, Department of Environmental Affairs, 2012. p. 88.

DEA (Department of Environmental Affairs). (2013).National Estuarine Management Protocol. Department of Environmental Affairs. Pretoria : Department of Environmental Affairs, 2013.

DEA (Department of Environmental Affairs). (2014). South Africa's National Coastal Management Programme. Department of Environmental Affairs. Cape Town: Oceans and Coasts Branch, Department of Environmental Affairs.

DEADP (2010). *Development of a Methodology for Defining and Adopting Coastal Development Setback Lines*, Volume 1: Main Report, 31 May 2010. Western Cape Government Environmental Affairs and Development Planning.

DEAT (Department of Environmental Affairs and Tourism). (2000). *White Paper for Sustainable Coastal Development.* Department of Environmental Affairs and Tourism. Cape Town: Coastal Management Policy Programme. Department of Environmental Affairs and Tourism.

Glavovic, B. (2006). The evolution of coastal management in South Africa: why blood is thicker than water. *Ocean and Coastal Management, 49, 889-904p.*

Van Niekerk, L and Turpie, J.K. *National Biodiversity Assessment 2011: Technical Report. Volume 3: Estuaries Component.* CSIR Report Number CSIR/NRE/ECOS/ER/2011/0045/B. Stellenbosch : Council for Sicentific Research, 2012.

WCG (2012). *The Establishment of Coastal Set-back Lines for the Overberg District,* March 2012. Western Cape Government Environmental Affairs and Development Planning.

WCG (2014). *The Establishment of Coastal Set-back/Management Lines for the West Coast District*, June 2014. Western Cape Government Environmental Affairs and Development Planning.

WCG (2015). *The Establishment of Coastal Management Lines for the Overberg District*, April 2015. Western Cape Government Environmental Affairs and Development Planning.

**Coastal Management: Changing coast, changing climate, changing minds**
ISBN 978-0-7277-6114-9

ICE Publishing: All rights reserved
doi: 10.1680/cm.61149.283

ice
Institution of Civil Engineers
publishing

# Medmerry Managed Realignment; Changing the Minds and Coast of Sussex – a Case Study

**Colin Maplesden**, Environment Agency, Worthing, UK.
**Andrew Gilham**, Environment Agency, Worthing, UK.
**Tim Callaway**, Royal Society for Protection of Birds (RSPB), Brighton, UK.

**With contributions and thanks to:** Pippa Lewis, Environment Agency; Joe Pearce, Environment Agency, Jacqui Bandy, Team Van Oord; Terry Fuller, Jacobs UK Ltd, Deborah Nutt, EC Harris.

## Introduction

The Environment Agency, RSPB and suppliers have worked in partnership at Medmerry to create the largest managed realignment on the open coast in Europe and the first in the UK. Before the new inland banks were completed late in 2013, bulldozers worked throughout winter storms every year to push up the shingle flood defences in efforts to stop breaching and seawater inundation. Even with this work, coastal flooding was expected annually, damaging properties, land and businesses. As well as reducing the risk of flooding, the realignment scheme has created a new 400 hectare (ha) nature reserve. This includes 183ha of intertidal land offsetting tidal-squeeze losses caused by existing and proposed flood and erosion risk management schemes across the Solent.

Early involvement of local people and groups has been essential. This has helped gain support for what had initially been seen as untested and among many local people, unwelcome coastal change proposals. Local representatives worked with the project team helping to change communities' perceptions and promote the scheme's positive contribution in securing the area for the future. The new flood banks protected the area from numerous severe storms in winter 2013/14 and the new nature reserve has already hosted breeding birds among the rarest in the UK.

## Location

Medmerry is on the south coast of England, located on the Manhood peninsula approximately 10km south of Chichester, West Sussex. The land use was historically mainly agricultural, with large holiday park sites interspersed among widely spaced small villages and hamlets. The project area also included the RSPB Bracklesham Bay Nature Reserve and SSSI.

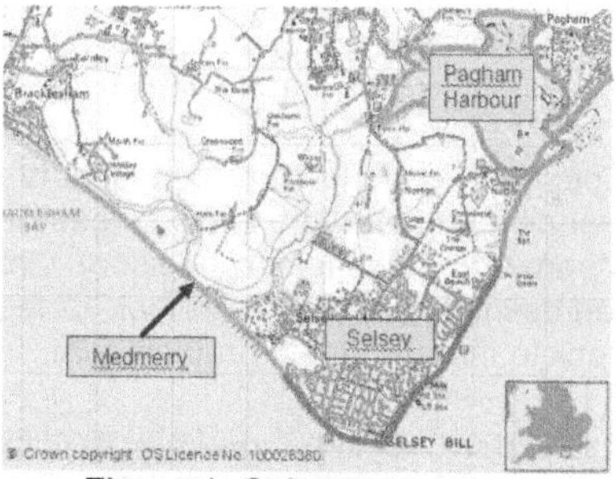


Figure 1: Scheme location

**Background**

Around 350 homes and businesses are at risk from coastal floods in addition to thousands of fixed caravans and holiday chalets. The sole access road and utilities serving the community of 5,000 households in Selsey are also at risk from a breach in the defences at Medmerry. Before the realignment, shingle management throughout each winter was unable to stop regular flooding.

**Strategic Issues**

The 2009 Pagham to East Head Coastal Defence Strategy (PEHCDS) further developed the Shoreline Management Plan (SMP) policy and recommended managed realignment for Medmerry as the most favourable coastal flood risk management option. The recommendation was for new inland embankments to be built from locally sourced earth and clay, with a tidal breach created in the existing coastline.

The SMP considered coastal processes and environmental issues on 70km of coastline, from Medmerry westwards. This included natural harbours within the Solent, which host 4,750ha of intertidal habitat forming part of the Natura 2000 network.

Coastal squeeze caused by rising sea levels is estimated to cause 600ha of habitat to be lost over the next 100 years. Without mitigation through provision of replacement intertidal habitat, coastal flood and erosion risk management schemes could not be implemented across the Solent. This would jeopardise providing improved protection to urban areas including Southampton and Portsmouth affecting 5,000 properties over 10 years and some 25,000 properties in the long term.

## Community Engagement

The delivery partnership between Environment Agency, RSPB and contracted suppliers Team Van Oord, Jacobs and EC Harris were central to engagement with local people, public bodies and businesses. The project team built upon extensive public engagement undertaken for the PEHCDS.

The approach to planning and implementing the stakeholder engagement for the scheme was described by Miller et al. (2010). The project team followed the Environment Agency's 'Building trust with communities' guide, and applied a stakeholder engagement plan that focused on three core objectives:

- Provide information about the multiple benefits offered
- Gain widespread support for the implementation of the scheme
- Community input into the design.

Specialist engagement consultants were contracted to advise on engagement planning and lead the initial stages of implementation. This saw the formation of the Medmerry Stakeholder Advisory Group (MStAG), consisting of self elected local representatives who undertook to feed in the wider communities' views and report back to them from the project team.

Experience during the development of the PEHCDS and from other managed realignment schemes elsewhere showed that ongoing support and involvement of local stakeholders would be essential in promoting the scheme to gain permits and approvals needed for its construction. Many people still view managed realignment as a lesser option for managing flood risk as it brings the sea closer to their properties.

Local communities concerns about the scheme included:

- The ability of the proposed inland embankments to withstand storms
- The loss of agricultural land and recreation opportunities
- The perception of wildlife interests being prioritised over properties and land
- Changes to inland drainage
- Potential loss of access to the coast and ability to walk between local villages
- The effect of changes on holiday parks important to the local economy.

One aspect of concern locally, was the impact on the Bunn Leisure holiday parks situated on low ground stretching inland from the coast at Medmerry. The strategy could not justify continued protection for the whole of the holiday park. This prompted Bunn Leisure to pursue construction of their own defence scheme costing £16.8 million. The works on the 1km Bunn Leisure frontage scheme were designed to complement the Medmerry scheme and were completed at the start of 2013, in parallel with ongoing realignment scheme work.

From the later stages of design through its construction, the project team included a project liaison officer to facilitate communications both with MStAG and with the wider community. Information about the scheme was published regularly on the internet, through email newsletters, social media and press releases as well as through exhibitions, guided walks and presentations.

The trust that the engagement approach helped to forge with local communities had a direct impact on the success of the realignment scheme. It enabled issues to be aired early and quickly resolved.

## Habitat Creation

The new habitat area at Medmerry contributes towards compensatory habitat requirements to offset tidal squeeze losses in the Solent, and will help towards meeting targets for England's biodiversity strategy. However, the construction of the scheme impacted existing habitats and species, meaning that extensive mitigation for water voles, reptiles, ground nesting birds and the RSPB Bracklesham Bay SSSI was also required.

In 2009 a Habitat Creation working group was established to advise on the scheme's habitat creation requirements. This group consisted of Environment Agency, Natural England, RSPB, local wildlife groups and academics. The work of this group ensured that both the mitigation and compensatory habitat needs of the scheme were met, and fully integrated into the flood risk management design. This group also helped balance the recreation aspirations of the scheme, with the need for wildlife refuge areas around the site.

As the Environment Agency's conservation partners for the site, the RSPB brought experience from other realignments such as Wallasea to the design of the habitat.

Their input was integral to the design of bird islands in particular, as well as to proposed land use around the site, including recreational access.

Intertidal habitat comprising mud flat, salt marsh, saline lagoon and transitional grassland requires specific levels and conditions in order to develop. Modelling of the ground levels and expected tidal prism influenced the design of the borrow pits, from which the clay for the new embankments was sourced. Further adjustments were made during construction by the RSPB to maximise suitable conditions for habitat formation. More details of this are given in Burgess et al (2014).

Ecological mitigation, primarily for water voles, reptiles and great created newts, was built into the scheme designs. Drainage channels needed around the landward side of the new embankments were designed to provide 10km of water vole ditch habitat. This was planted with aquatic vegetation 12-18 months before water voles were moved into it. Prior to construction starting areas of reptile habitat were also created, and during the construction period more than 2000 reptiles were moved into these areas.

## Constructing in Partnership

With a scheme the scale of Medmerry and a fast paced programme, it was clear from the outset that new ways of working would be required during the design and construction. The timescales for scheme delivery are shown in Figure 2.

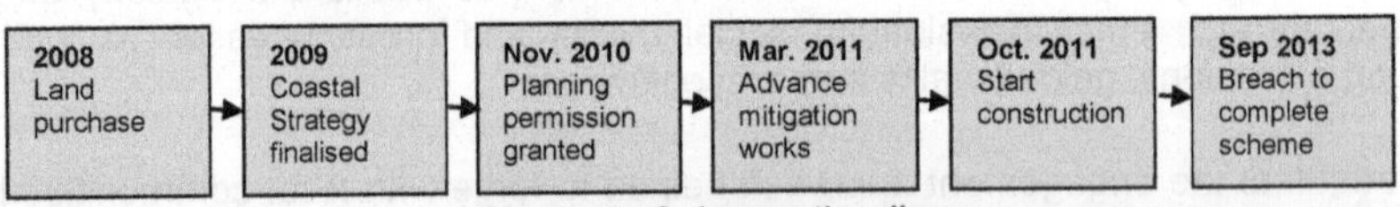


Figure 2: Scheme timeline

During the early stages of the design a co-located team was setup in the designer's office for a minimum of one day a week, enabling quick decisions to be made. The design team adopted the policy of sharing all the important design decisions with MStAG, this ensured that there were 'no suprises' later when designs were tabled elsewhere. In some cases specific workshops were held including value engineering, landscape and recreation, and hazard reduction.

The planning application for the scheme was submitted in July 2010, with permission granted in November 2010. Over 100 letters of support were submitted and unanimous committee vote achieved. Once planning permission was secured considerable work was required prior to the main construction starting on site. This included preparing detailed method statements for traffic, archaeology and ecological management.

Advance environmental mitigation works were mobilised in March 2011. This work cleared the first areas for the construction of the flood banks and planting the areas that would later be used for translocation of reptiles.

**Core Scheme Elements**
The design and construction of the core elements of the scheme are described in this section and shown in Figure 3.

Figure 3: Scheme layout

Flood-bank and borrow pits
With the decision to use all material for the new flood banks from on site, access and quality of material was critical to this element of the work. At times the clay material was too wet and alternative areas had to be opened up. For example, material taken from the flood storage areas could not always be used in the flood bank. Instead this material was used in the creation of viewpoints and landscaping. The contractor led on production of a 3d spatial model of the site. This helped communication between contractor, designer and the project team to ensure that material use was optimised within the site constraints.

As a result of necessary changes the final construction looks quite different in places from the original plans, however this has added to the overall landscape appeal of the site.

Rock ends
To protect the ends of the flood-bank, two large rock structures were constructed at each end of the scheme, requiring 60,000t of rock imported by barge from Larvik, Norway. On arrival at Medmerry it was placed in layers over the flood bank.

The rock structures are designed to allow the existing shingle beach between them to roll back within the realigned area, whilst preventing the sea outflanking the new defences.

The interface at the eastern end of the scheme with the Bunn Leisure frontage was particularly complex due to the aggressive nature of the sea at this location and boundaries of responsibility. The rock end was adjusted on site to extend partway back along the front of the beach. Extensive land raising to the rear using spare material has filled what was previously a low and vulnerable area. This work has greatly reduced the risk of breach of the defence and in keeping with the rest of the scheme has created a large raised viewpoint, giving 360 degree views over the realigned area and out to sea.

Outfalls and drainage

The scheme area contained four streams known locally as 'rifes'. Ordinarily these streams have low flow but in the winter significant flow can be experienced. The primary aim of the Medmerry scheme was to manage the risk of coastal flooding. However, it was also important to ensure that the fluvial drainage of the area was as a minimum not worsened. The capacity of the drainage was maintained by construction of four outfalls and new fluvial flood storage areas. Pre-cast concrete units were used for the outfalls, reducing site safety risks and allowing construction during the winter, saving time on the construction programme.

The outfalls were completed with two gates on each, closing during high tide. Side opening gates were chosen to maximise the ability for fish and eels to pass. The gates required a period of adjustment on installation as tidal water levels within the site developed. The flood storage areas were designed to double up as prime freshwater habitat on the landward side of the flood bank.

Breach

The breach to let the sea in required very careful planning, not least to ensure it was carried out in a safe manner. This area was the lowest part of the site - the existing freshwater drainage outfall was demolished first. Then over 50,000 cubic metres of material was excavated out over a four week period in September 2013. The sea was finally let in to the site!

**Environmental Partnership**

With so many differing needs, objectives, constraints and opportunities across a large and diverse site, the need for a collaborative partnership approach throughout construction was essential. Partnership with the RSPB from the early design and planning phases, maximised opportunities to reduce flood risk and create habitat. Into the future the site will be managed by the RSPB with all the recreation, education and health benefits this will bring.

With a two year construction programme it was crucial for the existing farmland to be maintained. This ensured that land remained productive for as long as possible. This helped protect the populations of birds within the site and in particular on the SSSI, which would form the founder population for the new intertidal areas.

The RSPB were an important part of the construction team and they offered invaluable advice, particularly when building the new freshwater and intertidal habitat areas.

Birds and the push-pull approach

It was identified early in the design stage that there was a significant risk that construction works could be delayed or their efficiency affected by the presence of breeding birds. In addition there was also an unacceptable risk that the works themselves would have a detrimental impact on important and scarce breeding bird populations. Due to the complexity of the works, the sheer amount of materials to be excavated and moved around site and the construction footprint being in excess of 180ha, a detailed programme of works and ongoing monitoring needed to be built in to the overall works programme.

This was achieved primarily by large scale habitat management involving the tenant farmers, backed up by specific activities by the contractor, such as scrub clearance, mowing and brush-cutting and in certain specific incidents the use of bird scaring techniques to discourage nesting birds.

Birds were *'pushed'* away from construction areas using various techniques, including planting of crops known to be unattractive to nesting farmland birds, mowing of ditch and field margins, clearance of scrub and management of freshwater ditches.

Birds were *'pulled'* towards non-construction areas by planting of crops and seed mixes attractive to birds, creating fallow areas and topsoil stripping, cordoning off scrub areas from construction and pumping of water to areas. The *push-pull* technique was very successful in avoiding construction delays whilst maintaining the bird populations on the site.

Water vole mitigation

Of particular importance was the new habitat for the water vole populations at the site. Due to the size of the site, the project team had to ensure that the water vole habitat was dug, holding water and planted in time to allow a growing season prior to animals being moved in to it. The planting of the ditches involved a combination of new grown aquatic plants and translocation of reed clumps from within the site. Prior to the breach being made in the shingle bank to inundate the site, water voles were trapped and moved into the new habitat. Corridors of high ground were left within the site to enable animals not trapped to escape into safer areas of the site. To facilitate this, the breach was undertaken at the lowest tide, so that inundation was gradual over a tidal cycle.

Archaeological finds

Another significant environmental receptor at the site was archaeology. During the early phases of construction it quickly became clear that the Medmerry site contained extensive archaeological finds, which were unknown and unexpected. High levels of preservation from historic siltation, meant that intact elements of the historic landscape was revealed by the excavations. This included several Bronze Age settlements, a Bronze Age cemetery and a medieval wooden structure, several hundred metres in length. It is thought that the wooden structure is a fish weir, the scale of which indicates it would have been used catch fish to sell in Chichester.

The discovery of these finds meant that close working between the archaeological consultants, the contractors and the RSPB was required. This ensured that the important archaeology was recorded, without adversely affecting the construction and avoiding disturbance to ground nesting and feeding birds.

## Construction to Operation

The new flood defence embankments and breach were completed in September 2013. The scheme had become fully operational as a flood defence, with the coastline and habitat ready for natural development over the years to come.

An extensive monitoring programme for the site has been put in place, covering the evolution of the beach, intertidal and habitat areas. This will provide important data informing future coastal evolution and establishment of habitats.

Figure 4: Photo of completed scheme. Copyright John Ackerman 2013

### Scheme Flood Risk Performance

The winter of 2013/14 was one of the stormiest and wettest for 50 years, with a succession of coastal storms putting the new flood banks and drainage outfalls to the test straight away. Ten years of storms were effectively received in the space of one winter. The largest of these storms had waves 1m higher than the 2008 storm when £6m of damage was caused. The success of the new scheme is clear as no coastal or fluvial flooding occurred and risk to life was avoided.

Given the successive nature of the storms it would have been very difficult to repair the previously defended coastline during the winter of 2013/14. Extensive and prolonged flooding would almost certainly have occurred.

### Early Evolution of the Beach

During the storms of winter 2013/14 the existing shingle beach has evolved and 'rolled back' quicker than expected. The beach has flattened and widened to a more natural profile. The breach area especially, has undergone rapid development. Several of the storms moved large quantities of shingle, to form a small spit in the entrance of the breach. However, even within this highly mobile system the breach has remained stable and open.

### Habitat and People Early Wins

The establishment of managed habitats including grasslands, planted reed beds and ditches has been rapid since construction work started in 2011. Since the breach in September 2013 there is clear evidence of intertidal plant colonists moving into the developing mudflats and saltmarsh, however these need time to establish.

During this period not only did key populations of breeding birds remain stable, but breeding numbers of certain species actually increased. In addition, the site has already begun to attract new species. Little ringed plover began to breed for the first time in 2012 and in 2014 a pair of black winged stilts reared an exceptional 3 chicks to fledging. They have only reared young twice before in the UK. The larger intertidal channels and creeks are already attracting both feeding and roosting terns during the summer.

Saltmarsh and shoreline breeding specialist including redshank, avocet, ringed plover and oystercatcher all encouragingly increased their populations in 2014. Wintering water bird numbers have also responded to the establishment of new habitats with a threefold increase in peak counts of total waterbirds using the site.

The monitoring of water voles at Medmerry is being undertaken by the University of Brighton, on behalf of the Environment Agency. Working with a local wildlife group and volunteers, they are undertaking the presence/absence surveys of the new water vole habitat. The most recent of these surveys was undertaken in September 2014, and showed that 75% of the transects on the new perimeter ditch habitat had signs of water voles This means that water voles are successfully using the new habitat.

Feedback from MStAG and the community during and after completion of the scheme has been very positive about the benefits the scheme provides. Comments include *"Normally by now we would've had flooding, but we've had none at all"* and *"Medmerry will enhance the area around Earnley and I am sure it will become a major attraction for wildlife and tourists alike"*.

After the storms of winter 2013/14, many residents have commented how much safer they feel now the scheme is in place. Visitor numbers to the site are very healthy and are expected to increase as the site becomes established.

## Summary of Learning Points

A strong focus on engagement with the community and stakeholders has been essential to enable delivery and success of the completed scheme.

Learning from the *engagement* includes:

- Make engagement and communication a top priority for large and complex flood risk management schemes
- Consider early formation of a self elected stakeholder group and ensure this group is made aware of all important issues first
- Exploring multiple benefits with all parties enables delivery and ultimately completion of an enhanced scheme
- Consider independent facilitation to explore the benefits and unlock difficult conversations
- Ensure the community are involved in the important design decisions and involve them during construction
- Create an environment of trust where issues can be aired and resolved.

A collaborative partnership approach to the design, construction and environmental management of the scheme has been essential.

Learning from the *design and construction* includes:

- Ensure that your conservation partner for managing the site in the long term and other specialists are brought in early. Integrate them in the design, construction and all the important decisions that affect operation
- Form specialist project groups to manage complex areas of work. For Medmerry this included a 'Habitat Creation Working Group' and a 'Communications Working Group'
- Understand how the variation in soil conditions and potential for archaeology may affect the design and construction, integrate this into a 3d spatial model
- Do not underestimate the impact that investigation and discovery of archaeology can have on the construction programme
- Carefully consider the extent of pre-construction investigations and develop contingencies for use of unsuitable material
- Consider the *push-pull* approach for managing birds during construction
- Ensure that there is sufficient time in the programme to manage advance ecology works and creation of habitat mitigation areas

## Conclusion

The completed Medmerry scheme has performed exceptionally well during it's first year of operation. No flooding issues have been experienced despite the extreme weather in the winter of 2013/14. The community feel safe and reassured.

Early indications are that the habitat is developing well, with good populations of birds and other wildlife using the site. Measures of the pre-existing wildlife show that their population has been maintained and enhanced.

The scheme is also delivering wider benefits to the local economy. The nature reserve is attracting more visitors to the area, and the local holiday parks are extending their periods of operation. It is envisaged that the area will become a year round destination for green tourism in the future. Continued monitoring will be essential during the coming years to confirm the operation of the intertidal area and to fully understand the development of the habitats.

The completion of the Medmerry scheme is the next step in history for this area of the Manhood Peninsula coastline. The scheme has been seen as a great success, sustainably integrating flood risk, creation of habitat and the needs of the community.

## References

Burgess H., Kilkie P. and Callaway T. (2015), Understanding the Physical Processes Occurring within a New Coastal Managed Realignment Site, Medmerry, Sussex

Environment Agency (2009), Pagham to East Head Coastal Defence Strategy

Environment Agency (2010), Medmerry Project Appraisal Report

Miller S., Giacomelli, J., Hyam P., Pearce J., Lewis P., Pizer L., and Wilson, G., (2010), Designing the 'right' scheme through community involvement: lessons learned from community engagement on a coastal protection scheme

North Solent Shoreline Management Plan (2010)

Our New Beautiful Beach http://beautifulbeach.bunn-leisure.co.uk

Solent Coastal Habitat Management Plan (2003)

**Coastal Management: Changing coast, changing climate, changing minds**
ISBN 978-0-7277-6114-9

ICE Publishing: All rights reserved
doi: 10.1680/cm.61149.293

# Engineering of an Island Style Breakwater System for the Ft. Pierce Marina

**Jack C. Cox**, Principal Coastal Engineer, SmithGroupJJR, Madison, WI, USA
**Richard E. Czlapinski,** Coastal Engineer Discipline Lead, Tetra Tech, Boynton Beach, FL, USA

## Abstract

The push for creating environmentally more compatible shore and harbor protection often puts the reality of the needs for resilience and strength at odds with the needs of nature. Softer protection schemes such as floating wave attenuators and non-bottom founded protection methods are typically inadequate to resist the ravages of the storm, while the construction of solid ring breakwaters change the texture of the shoreline and affect the biota.

Out-of-the-box thinking was required to find an approach that could both emulate the natural environment and yet still be resilient enough to protect against hurricane events, promote ecology, and offer safe and manageable navigation. The solution was found in building a protection system of artificial islands configured to create wave sheltering, oriented to reduce long term sedimentation effects by controlling tidal flow patterns, and executed to be a living island archipelago with both benthic and terrestrial habitat.

## Introduction

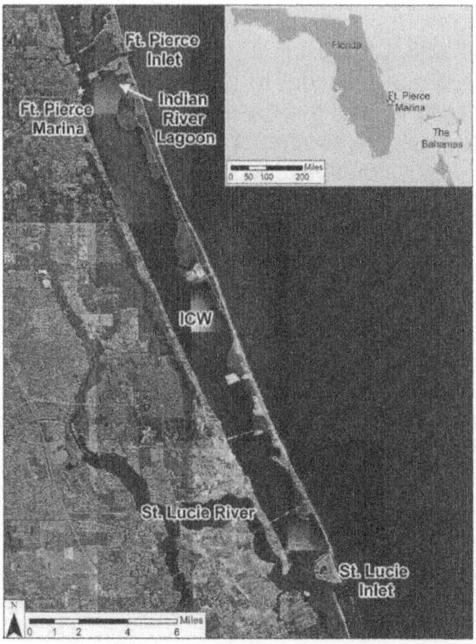


The 280 slip Fort Pierce City Marina is on the Indian River Lagoon, a long and wide estuarine waterbody separated from the Atlantic Ocean by barrier islands on Florida's southeast coast. In 2004, Hurricane Frances made a slow 34 hour approach toward the marina. The hurricane totally destroyed the 138 slip floating dock portion of outer marina as well as 96 of the 97 boats berthed there. Additional damage occurred within the 142 slip inner marina basin. The hurricane caused about $14.9 million in damages to the boats and about $13 million in damages to the marina facilities.

The City applied to the Federal Emergency Management Agency (FEMA) for public assistance funding to cover rebuilding costs that were in excess of the City's insurance coverage.

Since the storm waves that destroyed the marina exceeded the capacity of the original wave attenuator-protected, floating dock system, the City and FEMA quickly realized that a simple replacement of the marina was not a long term solution to the storm losses that the City endured. It became clear that additional protection against storm waves and currents would be needed to keep the marina from being destroyed again in some future hurricane.

## Design Objectives

The design provides for the reconstruction of the outer basin of the marina with approximately 200 slips for vessels ranging from 10 to 40 meters in length. The design objectives are:

- Minimize adverse sedimentation and currents in adjacent areas.
- Promote water quality and flushing in the marina basin similar to the original open water design case.
- Discourage sediment accumulation in the basin and along the approaches.
- Limit wave heights in the basin to less than 0.35 meter during the 50-year event.
- Limit wave heights in the marina entrance to less than 1.0 meter during the 50-year event.
- Provide 3.0 meters of navigable water depth in the marina approach, entrance, and fairways at Mean Low Water (MLW).
- Minimize the length of the breakwater required while encouraging water circulation and promoting basin tranquility
- Minimize scour and deposition damage to adjacent seagrass beds
- Minimize the occurrence of deflected or concentrated flow conditions that may impact adjacent facilities or navigation channels

## Analysis of Alternatives

The Fort Pierce City Marina is located within portions of the flood tidal delta of Fort Pierce Inlet. The site is characterized by strong tidal currents with magnitudes of up to 2 m/s. Any structural alternative therefore, had to have good functional performance in wave and current protection without causing current deflections and concentrations that would pose problems in areas adjacent to the marina site. The basic requirement for an island breakwater system was to provide protection against the largest waves that propagate to the site from the southeast along a 22 km long fetch in the Indian River Lagoon. Shallow water conditions in the Lagoon limit the 100-year $H_{mo}$ wave height to 1.9 m with a corresponding wave period of 5.1 s. Fetch lengths and the corresponding wave heights are smaller in the arc from the southeast to the north. Therefore, the most substantial wave protection needs to be to the southeast.

## Proposed Solution

The proposed solution was to create a permanent wave barrier system configured as a carefully sculpted, but naturally appearing, artificial island breakwater archipelago to protect the marina from the large open water areas. The marina site is within the flood tidal delta of the Fort Pierce Inlet. Persistent depositional features include a linear shoal to the east of the site and a large shoal that becomes exposed during spring tide low water conditions in the middle of the lagoon. The general layout of the island archipelago conforms to these persistent features and therefore is presumed to be sympathetic to and compatible with the local processes rather than in opposition to natural inlet behavior.

The intent was to use much of the dredged sands from the shoaled-in marina basin area as well as supplemental fill from other sources, and place them to form the islands. The islands' construction beneficially utilized sand dredged from the existing approach channel and the marina basin as well as from maintenance dredging operations at Fort Pierce inlet and St Lucie Inlet. A large dog-legged island provides the primary protection from south and southeast wave conditions. The island was to be built high enough so that dunes would provide overtopping wave protection meeting FEMA requirements for the 100-year event. A perimeter dike system forms a containment area that allows the island to be constructed by hydraulic filling techniques in a flowing current environment. The island is then capped with sand to provide a surface compatible in composition, grain size and color with the native sand of the site.

The windward side of the island was to be stabilized using a system of Tee-groins and pocket beaches. The design of the size and locations of the individual Tee-groins followed the procedures outlined by Bodge (1998). The evaluation of the MLLW shoreline for each pocket beach followed the procedures outlined by Hanson and Kraus (2001). The leeward edge of the island was shaped with a bermed ledge to provide a base for oyster reefs and mangrove plantings. Additional vegetative plantings consist of beach grass.

At the lesser exposed northern approach to the marina, a series of 11 small rock islands were added as an archipelago, primarily positioned and shaped to redirect tidal currents away from the mooring basin, and secondarily to provide a maze of detached breakwaters protecting the marina from a secondary fetch direction as well as providing a recreational amenity for canoeists and kayakers. These rock islands also include mangroves and patches of oyster reef substrate materials to environmentally and esthetically enhance the islands while contributing an additional measure of stability during storm events.

## Modeling Study

An initial numerical model effort considered a total of 17 possible island configuration schemes. The modelling explored wave dispersion patterns in and around the proposed islands, sediment transport patterns, and currents and circulation. The objective was to minimize the potential for impacts to sensitive areas such as seagrass beds, avoid shoaling or scour and achieve required berthing tranquility. The evolved version then served as the base configuration in a physical modeling program consisting of two-dimensional (2-D) and three-dimensional (3-D) testing under combined wave and current conditions.

The 3-D hydraulic model testing was conducted to investigate and characterize the following parameters:

- Current speeds in the marina and adjacent offshore areas
- Wave agitation levels in the inner and outer marinas
- Stability of the beach cells and T-groins
- Influence of the floating attenuator constructed at the north end of the marina

- Stability of rock revetments
- Performance of design elements during normal operating conditions as well as during a predicted hurricane strike (scaled condition)

The 3-D testing was conducted in a 25- by 30-meter test basin at the Coastal Engineering Research Laboratory at Queen's University, Kingston, Canada (see Figure 1). The entire model area was designed as a moveable bed that was re-graded to original conditions at the start of each test.

Tests were undertaken at a geometric scale of 1:40 using irregular wave trains synthesized to create a realistic model of conditions at the site. Because tidal currents at the project site also dominate littoral (shoreline) inlet behavior, the testing also included simulated current effects on the marina site.

In the 3-D hydraulic model, the bathymetry of the nearshore area was reproduced so that shallow water wave transformation processes were accurately in the wave propagation from the model boundary to the shoreline. Tests were completed for a range of water levels varying from MLW conditions (-0.15 m) up to the projected 100-year flood level (+1.6 m).

2-D hydraulic model testing was undertaken to evaluate the following parameters in greater detail:

- Scale effects compared with the smaller-scale 3-D tests
- Performance of the beach cells, compared with the performance observed in the 3-D tests
- Design of the geotextile tube/beach cell interface

The 2-D tests were conducted using a fixed bottom profile. The artificial island was modeled in profile with wave attack occurring at normal incidence (perpendicular to the beach).

The 2-D tests were undertaken at a scale of 1:20, thereby providing a better resolution of the wave structure interaction processes and a better opportunity to visually observe the run-up and beach profile development processes.

In addition to the increased scale of the model, the 2-D tests incorporated geotubes (sand-filled tubes) in the cross-section. The structural nature of the island cross-section was under development during the 3-D testing, and as a result, geotubes were not incorporated into the structure for those tests. The 2-D testing afforded the opportunity to observe the profile response to the stabilizing effects of the geotubes.

The preliminary 3-D model tests showed that excessive wave energy was able to penetrate into the marina basin from the northern entrance area. To develop appropriate wave sheltering for the north entrance area, a variety of wave protection options were considered in a supplemental numerical wave modeling program. These options included conventional straight and curved floating attenuators and additional freeform islands. A total of five island configuration options were evaluated in the supplemental study. The analysis of the entrance area options applied the CGWAVE model. Figure 2 illustrates the wave shadowing result of one island arrangement.

Based on the results of 2-D and 3-D testing and on physical and numerical analyses, the recommendations called for the following structures to be used for marina protection:

1. A single dog-legged artificial island constructed of retained sand fill; this will provide the required protection from the largest hurricane waves (100-year/SSE).
2. Beach cells with artificial headland T-groins; this will provide dynamic erosion protection for southerly hurricane waves at all water levels up to the 100-year surge level.
3. A maze of rock reef islands north of the large sand/Tee-groin island breakwater to provide satisfactory wave protection for the marina from NE wave action. These low reef island structures also help redirect high tidal currents out of the marina approach channel and berthing area and divert sediment away from the marina. The islands also offer added marine habitat opportunities.

## Final Design and Construction

The finalized plan is shown in Figures 3. Note the various aspects of the plan including the primary dog legged protection island including Tee-groin headlands and pocket beaches, primary surge protection sand dune, and the small island archipelago directing tidal currents and sediment away from the north approach channel, while buffering the basin from storm waves.

The details of the island construction are shown in Figures 4 through 7. Initially a perimeter dike was constructed of stacked geotextile tubes to define the limits of the main protection island. This created a containment barrier to capture hydraulically placed sand fill. The tubes were then armored with rock-filled marine mattresses to protect against erosion and abrasion. (Figure 6) The tee head groins were built by laying rock-filled marine mattresses on the lagoon bottom, creating a stable footing for the subsequent stone armor placement (Figure 7). Once the island was emergent, the island was overfilled to build surge protecting dunes. On the windward side of the island, the cells were also filled with sand to form pocket beaches. The sand to fill the geotextile tubes as well as to create the large dune on Tern Island (Figure 8) all came from dredging in the marina basin and approach channel as well as from maintenance operations at the adjacent Fort Pierce Inlet as well as the St Lucie Inlet well to the south of the project.

The small island archipelago used a different form of construction. Geotextile tubes were included as the core of the rock islands in the original project design. The contractor developed a value engineering proposal that used a uniform stone size for the cores and armor layers of the islands eliminating the geotextile tubes. A different approach was employed for these, which was to use a foundation mattress and then build the retaining dike from placed rock (Figure 9). The mangroves can adapt well to rooting in a rock environment so soil materials were unnecessary in the island core to support their growth.

## Conclusions

The overall configuration of the breakwater island system is consistent with persistent depositional features in the flood tidal delta area of Fort Pierce Inlet. The multi-island design provides sufficient channel capacity to maintain excellent flushing within the marina while still providing wave protection through the full 180° exposed perimeter of the project. To build the system required 93,000 tons of rock and 90,000 $m^3$. The 6.1 ha breakwater island complex is both structurally and environmentally enhanced with 0.5 ha of oyster reefs, 2.5 ha of artificial reef habitat, 0.9 ha of coastal dune plantings, 0.9 ha of shorebird habitat, 3.3 ha of recruited seagrass beds and 0.6 ha of mangrove plantings. Total cost of the island breakwater system was $19.8 million US.

The demonstration that the prototype achieved the same effectiveness in controlling tidal currents and sedimentation of the marina approaches and basin as predicted in the modeling can be seen in Figure 8. Before the archipelago matrix was even fully completed, the sedimentation diversion was readily apparent. The bypassing of the sediment around the marina, and the drift occurring through the islands is evident.

The habitat enhancements were also obvious even before the completion of construction. The rock of the islands and tee groins quickly became covered with lush algae and fish populations grew quickly from small bait fish size up to very large fish such as a 300 lb. Jewfish that shadowed the marine divers in their inspections of the islands. Naturally floating mangrove propagules became entrapped in the rocks and began to root even before the planned mangrove plantings took place. Shore birds also quickly adopted the dunes of Tern Island while pelicans took advantage of the plentiful fish stocks around the islands.

The community has fully embraced this new form of environmentally sympathetic shore protection, truly blending ecology with technical performance. This evolved design represents a truly "green" approach to new marine development where the components have been purpose sculpted to promote environmental benefit as a first priority. The goal is achieved through using an understanding of the behavior of waves and currents to shape the project into a non-impacting project, emulating the existing lagoon environment and its uses.

## References

Bodge, K. R. Beach Fill Stabilization with Tuned Structures: Experience in the Southeastern U.S.A and the Caribbean, Preprint from Proc. Coastlines, Structures and Breakwaters '98.

Cox, J and M. Wesson (2007). "Manipulating Harbor Geometries to Defeat Long Wave Agitation Problems", ASCE, Ports '07

Hanson, H., and Kraus, N. C. (2001). "Chronic beach erosion adjacent to inlets and remediation by composite (T-head) groins," ERDC/CHL CHETN IV-36, U.S. Army Engineer Research and Development Center, Vicksburg, MS URL: http://chl.erdc.usace.army.mil/library/publications/chetn/pdf/chetn-iv-36.pdf

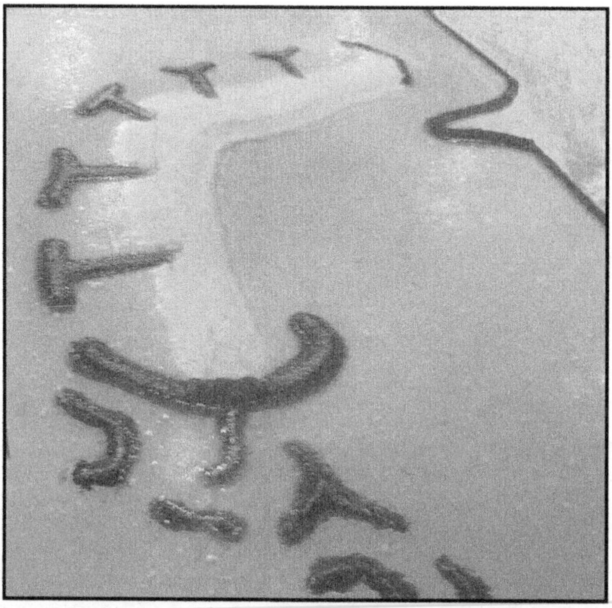

Figure 1 – Large and small islands in physical model tank

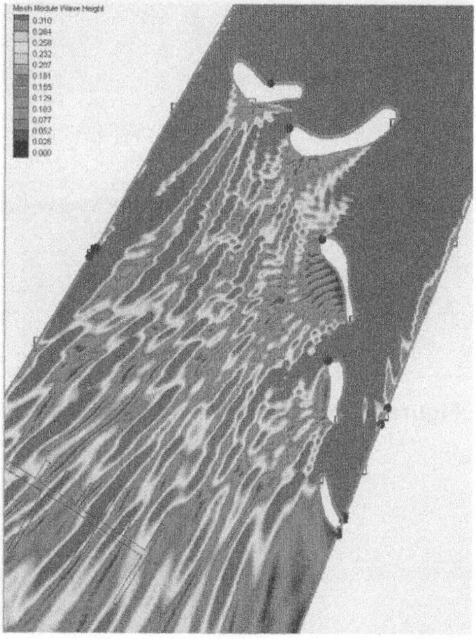


Figure 2 – CGWAVE Model Prediction of Island Wave Shadowing

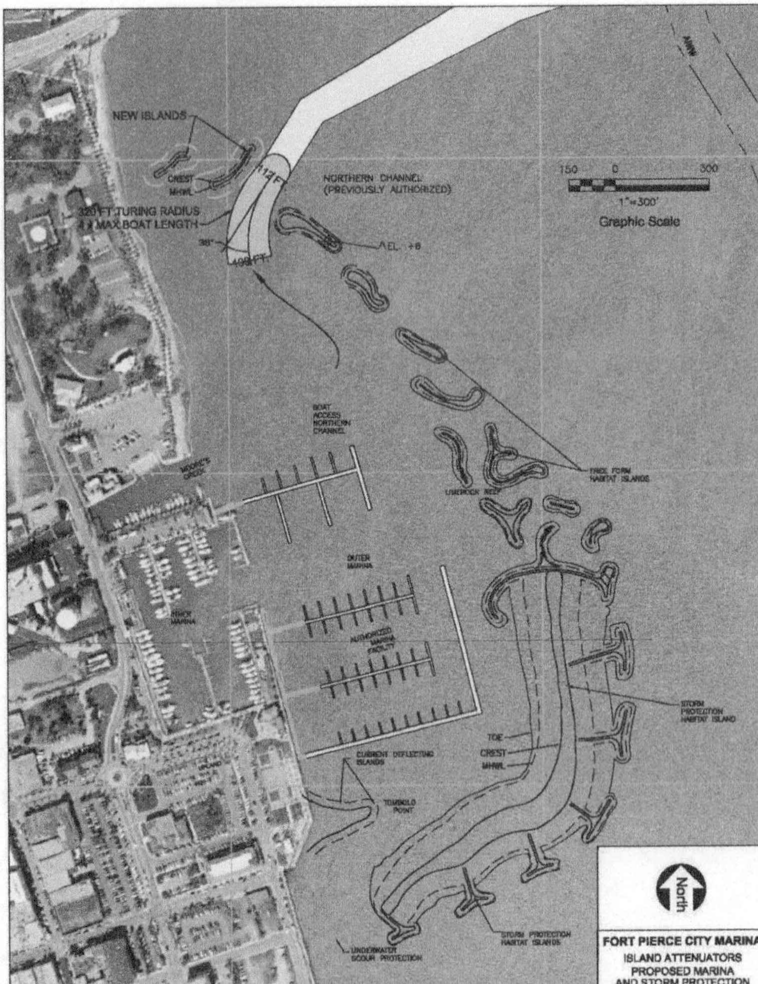


**Figure 3 – Recommended Configuration**

Figure 4 – Construction of the Large (Tern) Island Dike

Figure 5 – Construction of Tee Head Groins

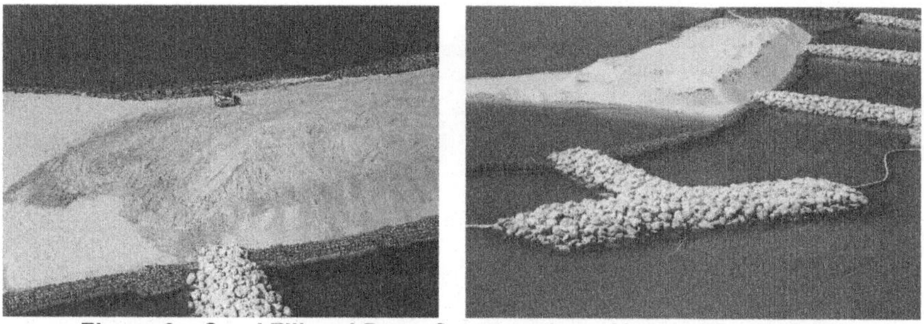

Figure 6 – Sand Fill and Dune Construction of Large (Tern) Island

Figure 7 – Construction of Small Rock Islands

Figure 8 – Effectiveness of sediment and current deflection

**Coastal Management: Changing coast, changing climate, changing minds**
ISBN 978-0-7277-6114-9

ICE Publishing: All rights reserved
doi: 10.1680/cm.61149.303

# Representing and Modelling Coastal Systems Over a Regional Scale for Coastal Management

**James Sutherland**, HR Wallingford, Wallingford, UK
**Robert J. Nicholls**, University of Southampton, Southampton, UK
**Helene Burningham**, University College London, London, UK
**Kate Rossington**, HR Wallingford, Wallingford, UK
**Belen Lopez de San Roman Blanco**, HR Wallingford, Wallingford, UK
**Jennifer M. Brown**, National Oceanography Centre, Liverpool, UK

## Introduction

Coastal and estuarine landforms, such as beaches, dunes, saltmarshes and mudflats, mediate flood and coastal erosion risk, but are part of an evolving coastal system with dynamic properties. Present and future risk simulations generally use static bathymetry/topography and this may give a misleading view of future risk as the coastal system evolves in time and erosion is widespread. Hence, there is a fundamental need to improve and formalize predictions of coastal change at the temporal and spatial scales that are most relevant to coastal management. Such predictions should account for the key geomorphic feedbacks between climate forcing, sediment supply and morphology, including the geomorphic influence of present and future adaptation options.

*Integrating Coastal Sediment Systems* (also known as the iCOASST project -- http://www.icoasst.net/) is addressing this challenge with the overall aim of improving our capability to predict decadal and regional scale morphodynamic change on the open coasts and in the estuaries of the UK to support erosion and flood risk management (Nicholls *et al.*, 2012). This remains a fundamental research challenge where expert judgement is often the main informant to management decisions.

One approach to regional scale morphology develops a conceptual model influenced by: geomorphological understanding, regional and local context, equilibrium concepts and observed behaviour (Whitehouse *et al.*, 2009). Models are run independently, so may not conserve sediment and assumptions may be implicit and inconsistent. Another approach uses process-based numerical models with coupled hydrodynamic and sediment transport modules and parameterised inputs to simulate the evolution of well-constrained problems with limited spatial extents over decades (van der Wegen and Roelvink, 2012).

The alternative approach used in the iCOASST project is to develop and link geomorphic models that do not start at the smallest scale and integrate upwards to the scales of interest for coastal management (van Maanen *et al.* 2015). A key innovation in the iCOASST project has been the development of model compositions that exchange data during run-time and take advantage of recent advances in model-linking technology (Sutherland *et al.*, 2014,

Harpham *et al.*, 2014). These aim to capture the interactions between the open coast, estuaries, and the shallow seabed, which has often been ignored in earlier assessments.

This paper describes the hierarchical approach adopted by the iCOASST project to assess and model coastal morphological systems, and its application to the coastline between Formby Point and Blackpool, including the Ribble estuary. This is within Liverpool Bay, UK (Figure 1) and is part of SMP22 (Halcrow, 2011). Its coastal region comprises protected soft cliffs, sand dunes, beaches, mud flats, salt marshes and hard engineering structures. Key issues raised in the Shoreline Management Plan (Halcrow, 2011) include the risks and uncertainties associated with long-term trends (including climate change) in understanding the sedimentary processes and how to improve the design of management schemes. There is believed to be a net onshore transport of sand towards Formby Point and Blackpool (Halcrow, 2010) so it is important to link offshore sediment sources and their pathways to geomorphic evolution models of the coast.

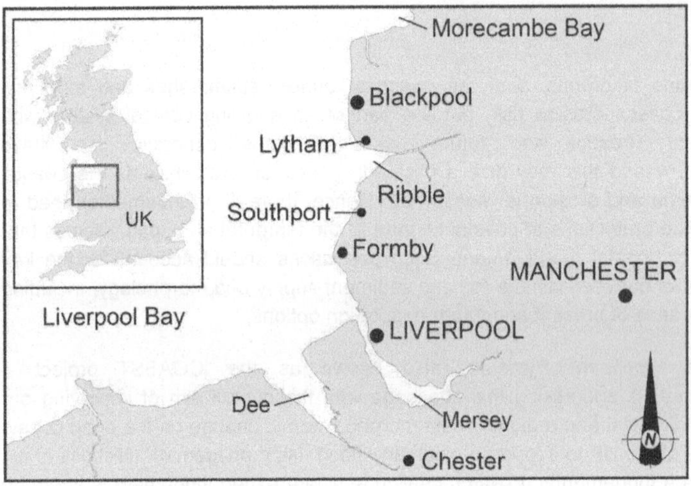


Figure 1: Formby to Blackpool case study area and the wider region

## The iCOASST Approach

A three-tiered approach is being taken:

- The development and implementation of Coastal and Estuarine System Mapping (CESM) which characterises the relationships between coastal landforms and structures;
- Coastal area modelling of inner shelf hydrodynamics and sediment transport to identify sediment pathways and onshore/offshore sediment fluxes; and
- Developing and applying coupled systems (or compositions) of specialised landform-scale models that interact by exchanging data at run-time and represent the coastal zone as a dynamic, evolving system that is, at least partially, self-regulating using feedback mechanisms.

## Coastal and Estuarine Systems Maps

A Coastal and Estuarine Systems Map (CESM) is a conceptualisation of coupled coast and estuary systems that is based upon a set of component landforms and human interventions (Whitehouse *et al.*, 2009, French *et al.*, 2015). It is a means of formalising disparate sources of knowledge, which is implemented in a formal mapping protocol using a set of landform complexes, landforms and human interventions that are common to open coast, estuary and inner-shelf locations. These landforms are linked by sediment transport pathways (solid lines) or influences (dashed lines). A section of the coastal and estuarine system map of the case study area, centred on the Ribble, is shown in Figure 2.

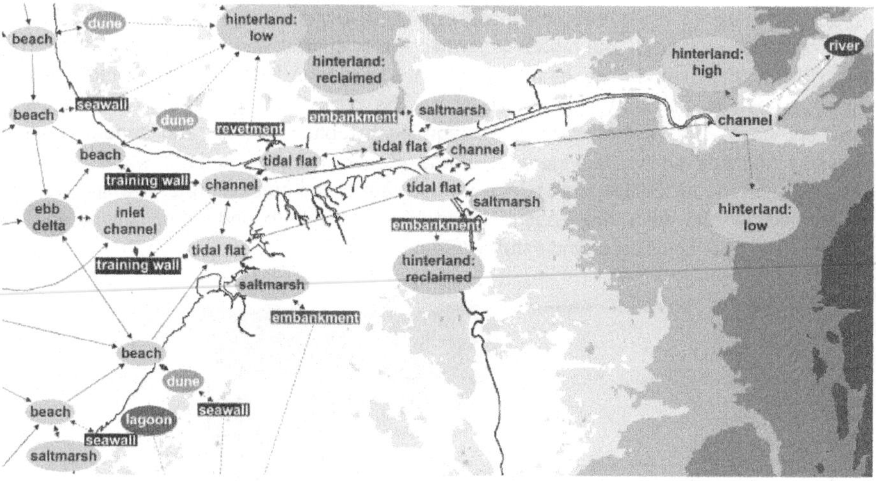


Figure 2: Coastal and Estuarine System Map of the Ribble Estuary

A CESM is a visual representation of sediment pathways and the influence of landforms, structures and other human influences on coastal evolution. The resulting system maps provide a framework for the development and application of coupled systems of reduced-complexity models (which should represent the key landforms and sediment pathways from the CESM) and could act as a repository for the outputs they generate, including flux estimates for the major sediment system pathways. CESMs may find many other applications in both coastal geomorphic analysis and coastal management, which we are investigating (French *et al.*, 2015).

## Coastal Area Modelling

Coastal area models represent flows, waves and sediment transport over a wide spatial scale by discretising processes on a spatial grid. Models range from two-dimensional depth-averaged to fully three-dimensional representations (Stansby 2013) but all provide understanding of hydrodynamic flows and their interactions that drive sediment transport and morphological change. It is not feasible to simulate long time periods with these models at present, so spatially-detailed, physics-based models typically simulate years. They are used in the iCOASST project in this way, to identify sediment transport pathways over large scales and for different sediment sizes. This information informs the CESMs about inner-shelf-coast-estuary interaction and can provide boundary conditions for coupled systems of reduced-complexity models.

The Proudman Oceanographic Laboratory Coastal Ocean Modelling System (POLCOMS) was used in this study. The hydrodynamic component solves the three-dimensional, hydrostatic, Boussinesq equations of motion (Holt and James, 2001) and includes algorithms for wetting and drying. POLCOMS has been implemented within the Liverpool Bay region, including the study area, using a numerical domain with a resolution of ~180 m in the horizontal and 10 vertical sigma levels within the water column.

An annual simulation using realistic atmospheric and riverine forcing (Brown *et al.*, 2015) has been conducted for 2008. The model generated hourly current vectors at each vertical level within the water column. The near-bed current stress was then calculated and used to estimate bedload transport vectors based on the critical Shield's stress for a chosen sediment class. The vectors were time-integrated to obtain monthly and annual residual transport rates as shown in Figure 3. These results indicate the large-scale sediment pathways, inform the CESM and provide an estimate of the potential rate at which sand is transported to and from the coast. Hence, they provide boundary conditions to the composition of the open coast and estuary (Figure 4).

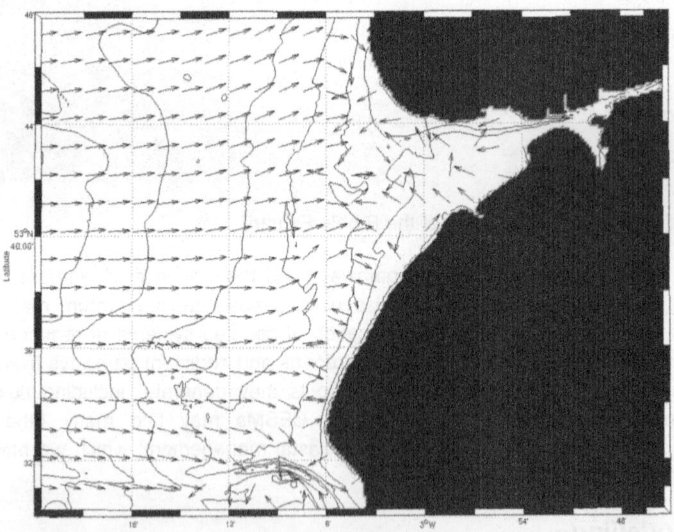


Figure 3: Annual residual bedload transport for 0.25 mm sand in the study area, estimated using POLCOMS Contours are at 5m intervals

## Coupled Systems of Coastal Geomorphic Models

The environment is an interconnected system, which couples many natural processes. The iCOASST project's approach to integrated environmental modelling is to develop a flexible, modular system of model instances, coupled to allow two-way exchanges of data (essentially sediment fluxes in this case) during run-time (Sutherland *et al.*, 2014). The Fluid-Earth implementation (Harpham *et al.*, 2014) of the OpenMI standard for data exchange during run-time is used.

Having defined the CESM, the next step is to define the model components (which are explained below, for this case). The coupling of these models allows us to develop a more complex composition, appropriate to the problem, that links different modelling approaches.

**Forrmby to Blackpool Composition**

For the Formby to Blackpool case study a model composition has been developed (Figure 4) that links the following models instances:

- UnaLinea (one line) model of Formby Point to Southport (U1).
- ASMITA (aggregate estuary) model of the Ribble estuary (A1);
- UnaLinea model of the Blackpool frontage (U2),
- POLCOMS look-up table of Liverpool Bay fluxes (P1).

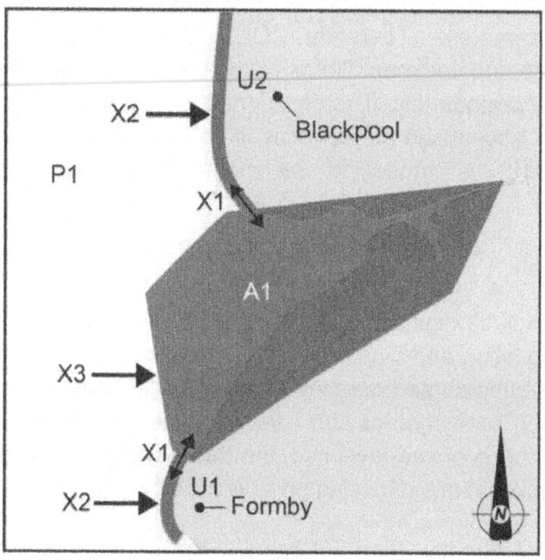


Figure 4: Formby to Blackpool model composition

UnaLinea is a simplified, regional-scale one-line model of wave-driven beach plan-shape evolution. Offshore wave conditions are transformed into the position of breaking at each point along the beach. These breaking wave conditions are used to calculate the longshore drift at each of these points. The change in position of the specified contour is calculated from differences in the wave induced longshore transport. The effects of seawalls and groynes, beach nourishment and sediment feeding to/from offshore may be included (Stripling *et al.*, 2011).

ASMITA is an aggregated model that schematises an estuary into channel, flats, saltmarshes and a delta (Stive *et al.*, 1998, Rossington *et al.*, 2011). Each element has an equilibrium condition when unperturbed, but when it is perturbed (for example by sea level rise, the addition or removal of sediment, reclamation or managed realignment) ASMITA seeks a new equilibrium, which can be defined as a function of hydrodynamic forcing and basin properties.

The following exchanges of data during run-time have been defined:

- UnaLinea – ASMITA exchange (X1);
- POLCOMS – UnaLinea exchange (X2); and
- POLCOMS – ASMITA exchange (X3).

We have assumed that the sediment exchange between UnaLinea and ASMITA (X1) is dominated by littoral drift. We then developed an explicit exchange mechanism that sees UnaLinea's potential littoral drift rates away from the ebb tidal delta as a demand on ASMITA that is modified by sediment availability in the delta. Four options have been identified based on direction of littoral drift on each coast and appropriate modifications to the drift rate have been coded and assigned to the relevant UnaLinea models.

In principle, POLCOMS could become part of the composition, exchanging data with the other models as they run, but in practice this this is not feasible as POLCOMS is much more computationally expensive than the other models. Therefore POLCOMS was used to simulate a representative year and pre-computed sediment fluxes were passed to UnaLinea (X2) and ASMITA (X3) at their offshore boundaries. Therefore, neither UnaLinea nor ASMITA can influence POLCOMS, which is acceptable as the focus is on modelling coastal and estuarine evolution, while POLCOMS is modelling sediment fluxes and not morphological evolution.

### Preliminary application of the composition

Data used to set up the UnaLinea models include digitised shorelines, a 30 year time series of WaveWatch III offshore wave data, the location and types of coastal defences and cross-shore beach profiles (for defining sediment budgets and depth of closure). Data used to set up the ASMITA model includes estuary bathymetries for deriving the volumes of schematised elements (channels, flats, etc) rates of sea-level rise, the dates and volumes of sand extraction and the date and area of reclamations (Rossington *et al.*, 2011).

The composition was set up using the 1904 Admiralty Chart 1904 due to the length of time it takes the estuary to respond to significant reclamations. A 200-year time series of wave conditions was generated by repeating the WaveWatch time series, thereby not allowing for any long-term changes to the wave climate. A sea level rise of 2 mm per year, which is between the measured trends in mean sea level at Liverpool and Heysham (Woodworth *et al.*, 2009), was applied to all years. No reclamations or sand extraction were included beyond 1990 and there was no onshore feed of sediment from POLCOMS to UnaLinea or ASMITA (X2 = X3 = 0).

Results from the ASMITA model run are shown in Figure 5, where the average depth of each of the model elements (relative to mean sea level for that year) is plotted against time. The dates of significant reclamations are shown as vertical lines. Variations during 20$^{th}$ Century are mainly due to reclamations within the estuary although small disturbances in channel volume (from 1950 to 1980) are due to dredging. ASMITA predicts that the Ribble will return to a dynamic equilibrium (rising with sea level) by about 2020, at depths slightly shallower than pre-1950 depths. The ASMITA model run outside the composition shows a much slower (post 1980) return to equilibrium, which it was reaching only by about 2100. The quicker response in the composition is due to the link to the open coast models, which are supplying sand to the Ribble estuary.

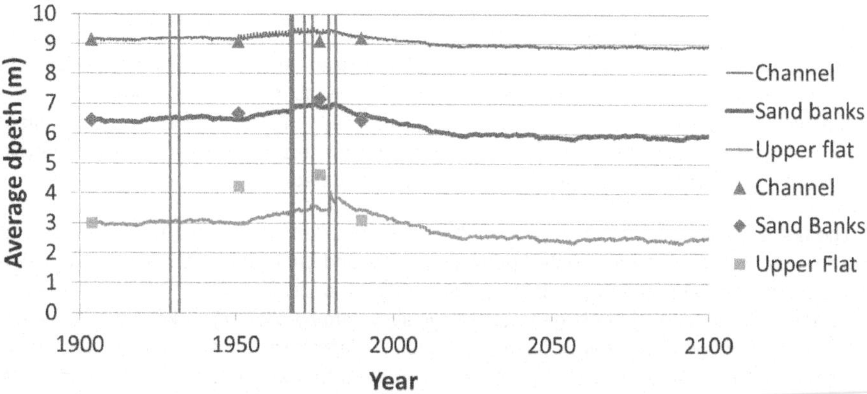


Figure 5: Preliminary changes in average depths of the ASMITA elements in the Ribble estuary over 200 years

**Extension to probabilistic modelling**

The behavioural models in the composition have a relatively short run-time: the 200-year simulation took about 1 hour. The availability of high throughput computing (HTC) facilities means that it is now possible to consider running this composition probabilistically. As there are many uncertain parameters and links, probabilistic simulations will provide us with an ensemble of plausible results, as is commonly obtained in weather forecasting or climate modelling. This will make the communication of the results more difficult, but methods used in other domains will be explored. Sensitivity analysis could also be applied in order to identify the most salient uncertainties, to identify regions of robustness, search for thresholds and test the effect of different assumptions (Epstein, 2008). Potential examples include:

- the sensitivity of the composition to the rate of onshore feed of sediment (X2 and X3) which would enable a more reasoned judgement concerning the appropriate effort to put into regional coastal area modelling;
- the identification of key uncertainties to guide the collection of data or the commissioning of research in order to most improve understanding of that uncertainty;
- an assessment of the variability and trends in beach position (and hence level) at critical structures, which will control the integrity and failure of the structure; and

- the identification of potential thresholds of key coastal state indicators, agreed with local stakeholders.

None of the above are possible using an implicit model of geomorphic change, showing how the iCOASST approach is providing an important improvement in our simulation potential.

## Stakeholder Engagement

It is increasingly recognised that there are benefits in a participatory approach, in terms of capturing stakeholder knowledge and developing shared understanding and solutions. The iCOASST approach lends itself to stakeholder engagement in a number of ways. Stakeholder groups have included representatives of local councils, consultants, Environment Agency, Natural England and local Universities, all of whom have an interest in coastal management at the level of the local SMP, a strategy study or a scheme.

The first engagement was centred around the CESM developed by iCOASST. Stakeholders were able to provide local knowledge on processes and behaviour, comment on the relative importance of links and suggest alterations. CESM could have been used to develop a common understanding of the geomorphological system by allowing stakeholders to develop their own versions, but it was quicker for them to comment on a prepared version. The iCOASST CESM was used to highlight the landforms that should be included in the model composition.

The second stakeholder engagement was centred around the initial results from the model composition. This included discussion on the suitability of the composition to represent the system, the explicit links between the models, the plausibility of the results and suggestions for future scenarios and sensitivity analysis, including different management interventions. By engaging at this stage, we sought to give stakeholders the opportunity to comment on the initial model results and influence the future modelling to be undertaken.

We also plan to engage with stakeholders when we have results from the probabilistic modelling, as learning how to interpret and present these results in a way that is meaningful to stakeholders is challenging. There are other challenges which are likely to emerge as these tools and methods become available. We cannot fully anticipate them and need to recognise that stakeholder engagement should be seen as an iterative process.

## Summary

The traditional approach to analysing morphology at a regional scale develops a conceptual model influenced by geomorphological understanding, regional models, longshore drift models, estuary models, equilibrium concepts and observed behaviour (Whitehouse *et al.*, 2009). Models are run independently, so may not conserve sediment and assumptions may be implicit and inconsistent. Integration is an expert process and is not transparent.

The iCOASST approach has developed a three-pronged approach to the characterisation and prediction of coastal and estuarine sediment systems. CESMs have been trialled with stakeholders, who have engaged positively with the process and wish to remain involved. This has helped to build a consensus over how landforms and interventions interact and has influenced the Formby to Blackpool composition.

The approach also uses detailed, process-based coastal area model runs to identify large-scale sediment pathways and exchanges between the inner shelf and the coastal models (with a boundary at the offshore limit of wave breaking). These exchanges could be replaced by data from a sediment budget analysis, so that the sensitivity to the data source can be tested. The system of coupled models can therefore incorporate information from different domains in a rigorous and transparent way.

Within the Formby to Blackpool open coast-estuary composition, the exchanges between models are explicitly stated and the effect of different assumptions can be tested. The two-way exchange of sediment between the models makes the composition internally consistent (in that sediment is conserved) and ensures that interventions on the open coast will affect the estuary, and *vice versa*.

The Formby to Blackpool composition takes about an hour (rather than days) to run, so this approach lends itself to sensitivity analysis and this will be explored in the near future. This can be used to identify the most salient uncertainties, identify thresholds and test the effect of different assumptions. A set of simulations to explore these issues has been planned. These results will assist more considered coastal management judgements. Even if they do not change the broad decisions in the SMP, they will give greater confidence in these decisions, including subsequent decisions in strategies and schemes.

A model composition can be extended with the addition of additional landforms and processes, as defined by the CESM, such as overwash, breach, dune, inlet/ebb delta, estuary or other models to increase complexity and address specific issues. Experience is being gained on the Suffolk coast within the iCOASST project (Nicholls *et al.,* 2015). This approach could potentially be linked to other models, such as flood risk assessment (Stripling *et al.,* 2011) or habitat / species change to look at economic and environmental effects, as well as morphology. The experience of the iCOASST project will be translated into guidance for coastal managers and coastal stakeholders on the analysis of coastal geomorphic systems and the appropriate use of different methods and tools.

## Acknowledgements

This work was funded by the National Environmental Research Council (http://www.nerc.ac.uk) under grant NE/J00541X/1, with the Environment Agency (http://environment-agency.gov.uk) as an embedded stakeholder.

## References

Brown JM, Amoudry L, Souza A and Rees J (2015). Fate and pathways of dredged estuarine sediment spoil in response to variable sediment size and baroclinic coastal circulation. *Journal of Environmental Management*, 149: 209-221.

Epstein JM (2008). Why Model? *Journal of artificial societies and social simulation*, 11(4) 12. http://jasss.soc.surrey.ac.uk/11/4/12.html (accessed 20/04/2015).

French J, Burningham H *et al.* (2015). Conceptualizing and mapping coupled estuary, coast and inner shelf sediment systems. *Geomorphology,* in review.

Halcrow (2010). *Cell Eleven Tidal and Sediment Transport Study (CETaSS) Phase 2 (ii),* issue 2. Prepared for the North West and North Wales Coastal Group.

Halcrow (2011). *North West England and North Wales Shoreline Management Plan SMP2*, issue 3. Prepared for the North West and North Wales Coastal Group.

Harpham Q, Cleverley P and Kelly D (2014). The FluidEarth 2 implementation of OpenMI 2.0. *Journal of Hydroinformatics* 16(4): 890–906.

Holt JT and James ID (2001). An S coordinate density evolving model for the northwest European continental shelf: 1. Model description and density structure. *Journal of Geophysical Research (Oceans)*,106(C7), 14015-14034.

Nicholls RJ, Bradbury A, Burningham H, *et al.* (2012). iCOASST – integrating coastal sediment systems. *Proceedings of the 33rd International Conference on Coastal Engineering*, Santander, Spain. ASCE.

Nichols RJ, French J, Burningham H *et al.* (2015). Improving decadal coastal geomorphic predictions: an overview of the iCOASST project. *Proceedings of Coastal Sediment 2015* San Diego, USA, ASCE.

Rossington SK, Nicholls RJ, Stive MJF and Wang ZB (2011). Estuary schematisation in behaviour-oriented modelling, *Marine Geology*, 281(1-4): 27-34.

Stansby PK (2013). Coastal hydrodynamics – present and future, *Journal of Hydraulic Research*, 51(4): 341-350, DOI: 10.1080/00221686.2013.821678.

Stive MJF, Capaobianco M, Wang B *et al.* (1998). Morphodynamics of a tidal lagoon and adjacent coasts. *Proceedings of the 8th International Conference on Physics of Estuaries and Coastal Seas*, The Hague. Balkema, 397-407.

Stripling S, Panzeri M, Blanco B *et al.* (2011). Broad scale integration of coastal flood and erosion risk models. FRMRC2 Work Package 4.3 Science Report. See http://web.sbe.hw.ac.uk/frmrc/downloads (accessed 22/01/2015).

Sutherland J, Townend IH, Harpham QK and Pearce GR (2014). From integration to fusion: the challenges ahead. In *Integrated Environmental Modelling to Solve Real World Problems: Methods, Visions and Challenges*. Geological Society, London, Special Publications 408, http://dx.doi.org/10.1144/SP408.6.

Van der Wegen M and Roelvink JA (2012). Reproduction of estuarine bathymetry by means of a process-based model: Western Scheldt case study, the Netherlands. *Geomorphology* 179, 152-167.

Van Maanen, B., Nicholls, R.J., Barkwith, A *et al.* (2015). Simulating mesoscal coastal evolution for decadal costal management: a hybrid integration of multiple, complementary modelling approaches. *Geomorphology*, in review.

Whitehouse R, Balson P, Beech N *et al.* (2009). *Characterisation and prediction of large scale long-term change of coastal geomorphological behaviours: final science report.* Report SC060074/SR1, Environment Agency, ISBN 978-1-84911-090-7.

Woodworth PL, Teferle FN, Bingley RM *et al.* (2009). Trends in UK mean sea level revisited. *Geophys. J. Int.* 176, 19-30.

**Coastal Management: Changing coast, changing climate, changing minds**
ISBN 978-0-7277-6114-9

ICE Publishing: All rights reserved
doi: 10.1680/cm.61149.313

# A Search for Alternative Concepts of Coastal Planning and Future Development of the Port of Zeebrugge

**G. van Holland**, International Marine and Dredging Consultants, Antwerp, Belgium
**B. Verheyen**, International Marine and Dredging Consultants, Antwerp, Belgium
**E. Maes**, Flemish Government, Department of Mobility and Public Works, Maritime Access Division, Antwerp, Belgium
**M. Sas**, International Marine and Dredging Consultants, Antwerp, Belgium

## Introduction

The Maritime Access Division (Flemish Government - Department of Mobility and Public Works) is responsible for the accessibility of the navigation channels to the Flemish Seaports. Extensive maintenance dredging is carried out in the North Sea and Scheldt Estuary. In addition to this, the department is also responsible for capital dredging, and harbour infrastructure and extension.

The Maritime Access Division is specifically looking to facilitate the future growth of the port of Zeebrugge on the North Sea. The port is situated in a sensitive sand dominated natural system, close to several nature protection areas. Important recreational beach areas are adjacent to both sides of the port, which have significant economic value. With respect to accessibility the present layout of the port of Zeebrugge has two disadvantages. Firstly, due to high offshore sediment concentrations and strong tide-induced water exchange, the central part of the harbour requires frequent and intensive maintenance dredging. Secondly, the protruding port causes a contraction of the already strong ebb and flood currents, which has led to the installation of a navigation access window for the larger vessels.

Anticipating future developments, a project was set up in 2011 by the Maritime Access Department. The aim was to develop modelling tools and concepts for the port of Zeebrugge focusing on the hydrodynamic challenges of a possible harbour extension, reduction measures for flow contraction and sediment influx. Meanwhile, on a federal level, two concession areas have been identified for energy storage. The potential energy islands can be integrated in the design of the harbour extension or be part of the design to optimize the accessibility of the port of Zeebrugge.

While in this paper the focus mainly lies on the effect of a range of alternatives on flow behaviour and anticipated morphological response, it is evident that these alternatives are mere sketches that will have to be integrated into a full master plan on the scale of the entire Scheldt River and the Flemish Coast. The first steps to arrive at such a master plan have been taken by the Flemish Government in their

Master Plan Vlaamse Baaien (*literally Flemish Bays*), as presented during the Super Storms Congress on November 26th, 2013 and shown in Figure 1.

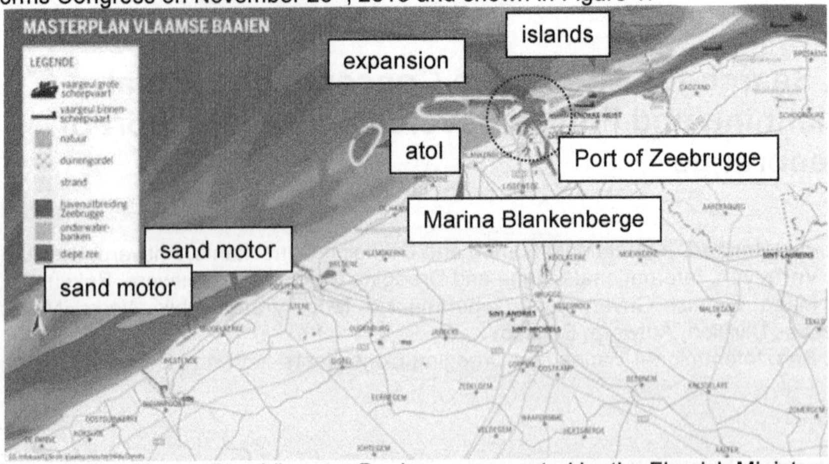


Figure 1: The Master Plan Vlaamse Baaien as presented by the Flemish Minister of Public Works during the Super Storms Congress on 11.26.2013

IMDC nv was commissioned by the Maritime Access Division to develop a numerical modelling tool and to assess various alternative coastal schemes and harbour extensions. The result of this project is a collection of conceptual islands and dams seeking to reduce the flow contraction and sedimentation problem at the port of Zeebrugge, to provide coastal protection, and to create accommodation space for coastal development and future extension of the port area (IMDC, 2015a). The paper gives an overview of the developed model and explores in further detail the various alternatives of dams, islands and extensions of the port of Zeebrugge.

**Elements of the Vlaamse Baaien Master Plan**

In the Master Plan Vlaamse Baaien various elements can be recognized that first were studied separately in the present study:

- Extension of and modifications to the port of Zeebrugge;
- A flow reduction pit or mud trap in front of the harbour mouth;
- An energy atoll (separate island);
- (Island) developments on the east and/or west side of the harbour;
- Preparation for a protected route for shallow draft vessels along the coast;
- Adjustments to the Scheldt estuary or mouth, o.a. by a depth reduction or closing of the maritime route called Scheur-Wielingen and the construction of a navigation channel through the Walvisch Staart (*literally: Whale Tail*).

The alternatives were evaluated with respect to a number of issues:

- For certain types of vessels strong transverse flow conditions in the area of the harbour mouth are problematic and a **navigation window** or tidal window is in place. In the present situation navigation is interrupted during approximately 2.5 hrs per (average) tide for deep drafted vessels (e.g. container ships) and 4 hrs per tide for LNG carriers, which have to follow a stricter criterion.

- **Eddy formations** inside and outside the (existing) port can lead to significant gradients in the cross flow velocity and are therefore potentially problematic for shipping. Currently, only eddy patterns can be observed within the port. Especially during high tide, strong eddies occur with flow rates up to 0.5 m/s. These eddies are important for sedimentation of fine sediments (mud).
- In order to evaluate and compare the sediment trapping efficiency of each scenario a performance indicator for the **sedimentation potential** was developed. This dimensionless indicator is based on the water exchange with the port and the dynamics inside the port and has no physical significance.
- In the present situation, local maximum flow rates of approximately 0.75 to 1.5 m/s and more can be found, leading to seabed changes in front of the port. Changes in the overall flow pattern therefore give a good indication of where **morphological changes** of the seabed can be expected.
- As a result of (historical) changes in the estuary the tidal range has increased in recent decades, and interventions in the mouth may possibly contribute to reducing it. Currently, the tide is amplified from a range just below 5 m in Vlissingen to almost 6.5 m in Antwerp. Scenarios have been evaluated to lower the expected **tidal range** and **tidal amplification** in Antwerp.

### The Numerical Model

For the purpose of the study a 3D model has been developed in TELEMAC. Along the coast, the computational domain extends from the port of Dunkirk (France) to Goeree-Overflakkee (the Netherlands). Cross-shore a semi-circular model boundary was chosen such that the possible impact of the interventions on the water movement lie within the limits of the model. Both the Eastern and the Western Scheldt are included in the model, albeit with a limited resolution. Upstream of Schelle different branches of the Scheldt are schematised through straight channels, so that tide independent upstream boundary conditions could be imposed. The resolution of the computational grid varies gradually from about 15 m in the port of Zeebrugge to about 3 to 4 km along the boundaries. For each of the calculated scenarios, a special computational grid was created, in such a way that the structures or islands were neatly represented by the computational grid. Ten vertical non-uniform sigma layers have been used. Boundary conditions were obtained from a continental shelf model. A non-uniform roughness coefficient has been applied, where it turned out to be important to include the roughness of the breakwaters.

Calibration and validation of the model was carried out using water levels, velocities and discharges measured at different locations in the North Sea and the Scheldt River. Generally the model gave good results with bias and Root Mean Square Error (RMSE) values below 0.05 m and 0.20 m for water levels and 0.10 m/s for velocities (IMDC, 2015b). As the eddy formation in the port turned out to be very sensitive to the precise model settings (Breugem at al, 2013), it was agreed to do a qualitative assessment of the intensity rather than to reproduce the exact position and direction.

## Results

In total, at the request of the Maritime Access department, about fifty **scenario calculations** studying the various elements of the Master Plan Vlaamse Baaien were simulated with the model. In the following sections the conclusions from these calculations are discussed. It should be underlined that no optimisation of the concepts of the modelled scenarios took place. Also, the technical feasibility, as well as the necessity or the political and economic feasibility of the scenarios has not been the subject of this study.

**Modifications to the Existing Port**

Several small adjustments to the existing breakwaters were examined. On the basis of these simulations, it was concluded that minor changes could result in an improvement of the flow conditions or even eliminate the navigation window completely. An (eastward) shift of position and reorientation of the port entrance however, does not improve the navigation conditions and leads to stronger eddy formation in the inner port area. Furthermore, it would require additional manoeuvring, as a straight approach is no longer possible.

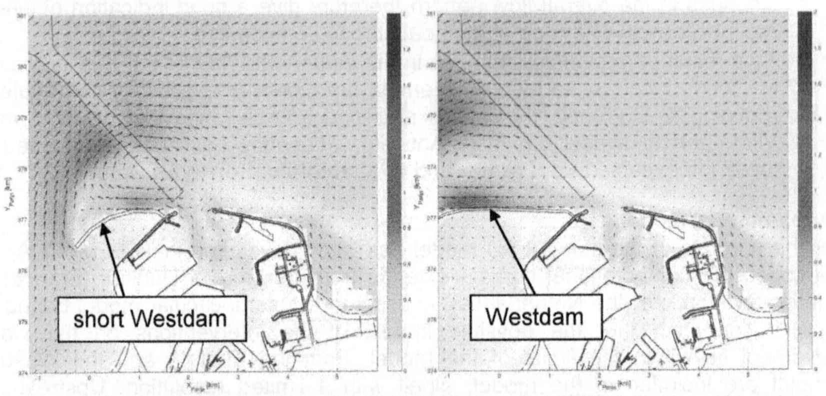


Figure 2: Eddy formation during low tide (2.5 hrs after HW), left: short Westdam, right: Westdam, both without pit

In the current situation the strongest (transverse) flow velocities occur at the harbour entrance. A dam extending towards the west can shift this flow contraction point. This will have a favourable effect on the navigation window, which with an optimal length of the dam (approximately 3 km) could be eliminated completely. At the contraction point the development of a new scour hole is expected. An important point of attention is the eddy that develops 2 to 3 hrs after high tide. Depending on the length of the dam this eddy may extend to the navigation channel (Figure 2).

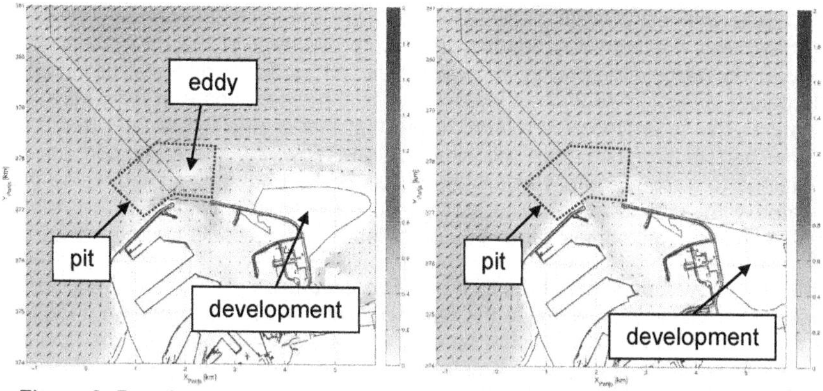


Figure 3: Developments under the lee of the eastern mole, both with pit, flow velocities 1.5 hrs after LW, left: eddy formation present in the access channel

In a number of concepts, land reclamations have been envisaged under the lee of the eastern and western harbour moles, aiming to divert the flow gradually away from the harbour mouth using either soft or hard measures. Doing so might reduce the amount of flow contraction and a more favourable navigation window would be achieved. However, it has been demonstrated that these measures have no significant effect on the flow distribution. Measures furnished with a mild slope (beach) may help to redistribute to some extent the flow field. Moreover, if the reclamations do not connect well to the existing dams, eddy formation may occur. In particular for developments on the east side an eddy may develop during ebb flow (towards the west, Figure 3) that may turn into a stronger eddy inside the port during the flood phase.

**Flow Reduction Pit**

A local deepening of the seabed seaward of the harbour entrance by about 8.5 m has been included in several of the scenarios. The purpose of this pit is to reduce the local flow velocities with minimal intervention and therefore improve the navigation window for the port. Secondly the pit could work as a sediment trap. However, this aspect cannot be examined with the hydrodynamic model only.

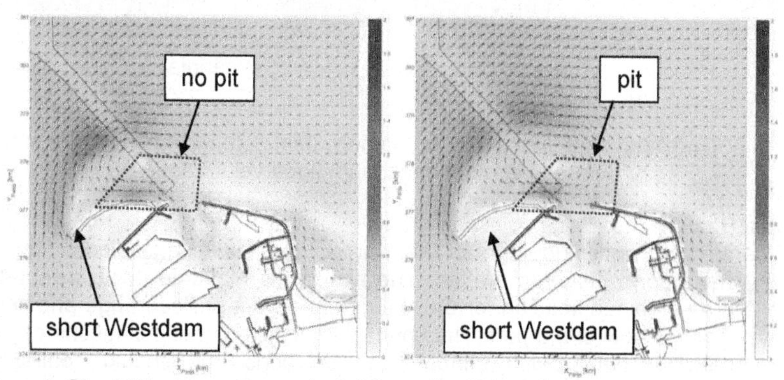


Figure 4: Short Westdam, flow rate 2.5 hrs after HW, left: without pit, right: with pit

The results indicate that in case of a small pit, the local deepening reduces the velocities near the entrance, leading to an improvement of the navigation window. However, for large dimensions, it appears the pit attracts additional flow, reducing this positive effect. In addition, it was observed that scenarios with a pit attract eddy formation. This is presumably related to the fact that due to the greater depth in the pit, the bed friction has less influence on the flow so the eddy can be easily maintained (e.g. see Figure 4 where the centre of the eddy has shifted inside the pit, indicated with a dotted line). Therefore, in case a pit is included in the design, it should be checked if eddies caused by other interventions could be enhanced.

**Additional Flow Openings in the Existing Breakwaters**

Increases in flow velocities inside the harbour might prevent sediment to deposit. For this reason the impact of additional openings in the harbour moles on the flow inside the harbour was studied. Three simulations with an increasing number of openings were made: in the eastern dam just north of the Britannia dock and north of the LNG terminal and in the western dam at the end of the Albert II dock. An additional advantage could be that shipping with limited depth could use these openings to sail eastward.

As expected the additional openings lead to a significant increase of flow inside the harbour, which leads to a decrease of the calculated sedimentation potential. However, the indicator does not include the increased risk of sediment import from the surrounding sand banks and mud flats. Moreover, stronger gradients and eddies inside the harbour will affect the manoeuvring of vessels. The extra openings have no (positive) effect on the navigation window.

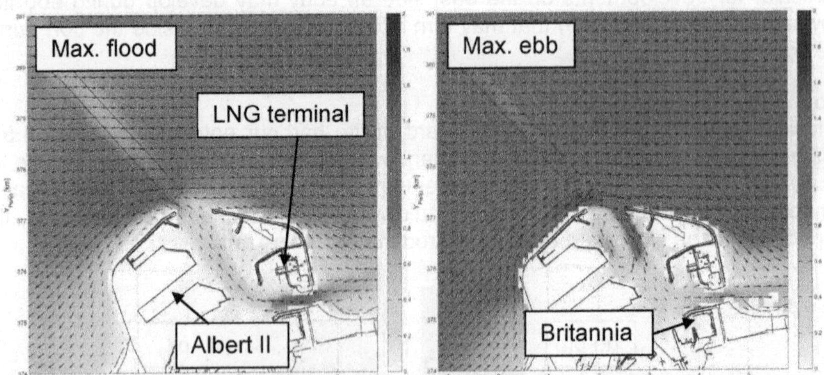


Figure 5: One additional opening in the eastern dam, left: maximum ebb velocities (1 hr before LW), right: maximum flood velocities (HW)

**Vlaamse Baaien**

In the context of the development of the Master Plan Vlaamse Baaien a number of different island concepts were drafted. The aim of the Maritime Access department was to find a combined solution for coastal defence west of Zeebrugge and improve the navigation window of the port. The concepts consisted of combinations of islands and (submerged) dams, with shapes and positions that should contribute to achieving this objective (e.g. see Figure 6).

The simulations indicated that it is possible to reduce the maximum flow velocities along the present coast and harbour mouth through a chain of islands and a dam connecting the last island to the port of Zeebrugge and thus to improve the navigation window. The shape of the islands can help to minimize seaward flow contraction. Structures which extend beyond the continuous line along the islands, will lead to local flow contraction and a potential response of the sea bed. In case of a partially submerged dam a certain flow remains behind the islands reducing the risk of stagnant water and increasing the velocities in front of the harbour mouth.

By means of a series of simulations the distance and location of a single island connected by a dam to the western dam of Zeebrugge was optimized. On the basis of these simulations it was concluded that such an island and dam could have a beneficial effect on the accessibility of the port. The dam could also be viewed as a possible first step towards the expansion of the port.

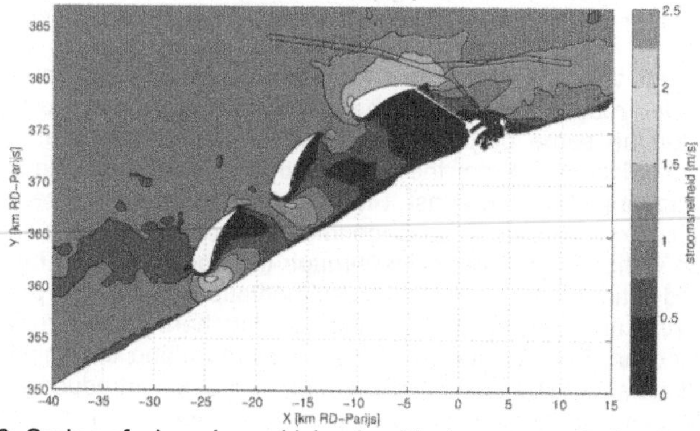


Figure 6: Series of wing-shaped islands with dam connected to Zeebrugge, maximum flow velocities, average tide

It was concluded that a dam with a north-westerly orientation was less beneficial to the navigation window, because the flow behind the dam could recover. On the other hand, an orientation closer to that of the access channel would increase the risk of eddy formation (at the end) of the flood phase. It also appeared that a shorter dam (an island closer to the port) has a less dramatic impact on the overall flow pattern (and thus the morphological adaptation). An island closer to the port does slightly increase the restriction of the navigation window. When the island is integrated into a reclamation siding the western dam the risk of eddy formation is smallest, possibly because of the streamlined design and also because of the deep pit that was included in the scenario (see Figure 7).

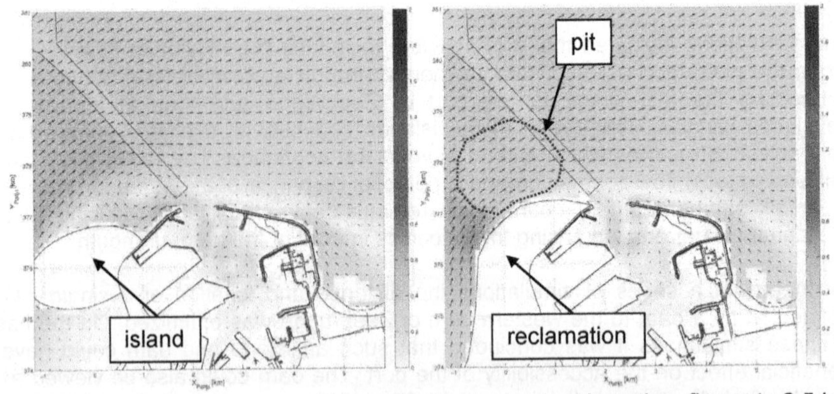


Figure 7: Island of whether or not integrated into a land reclamation, flow rate 2.5 hrs after HW, left: eddy formation will develop, right: with pit

East of Zeebrugge various interventions were devised, including a series of islands to create a protected route for shipping with shallow draft, a submerged dam connected to an island for the same purpose, and various interventions that affect the flow through Scheur-Wielingen. From these simulations it was concluded that by using islands or structures along the coast it is technically possible to create a protected route for shallow draft vessels. A passage in the eastern dam with a lock is required, partly because without a lock velocities through the opening would be too high (and probably would lead to an increase in sedimentation inside the port), and partly because entering and leaving through the existing harbour mouth does not give adequate protection. For the design of the measures along the shipping route, the irregular coastline should be taken into account, as it could introduce a local increase of velocity. For waves, it is necessary to provide measures up to the mouth of the Western Scheldt (Breskens) to allow for the coastal waters to be classified as zone 2 for navigation.

**Expansion of the Port**

The most intuitive way of extending the port of Zeebrugge is the seaward extension of the present harbour, doubling the present length to about 6 km (Figure 8, left). A possible end situation with docks and an eastern passage just above the existing Eastern dam, as well as a construction phase where all dams are submerged, have been simulated with the model. The simulations show that such a seaward extension of the port has a very dramatic impact on the overall flow pattern. In particular it would lead to a deterioration of the navigation window of the (new) harbour mouth.

An alternative concept including a westward expansion of the port (Figure 8) has been developed based on the simulations discussed below. The simulations indicated that a westward expansion would have a less drastic effect on the overall flow pattern and could lead to a complete elimination of the navigation window.

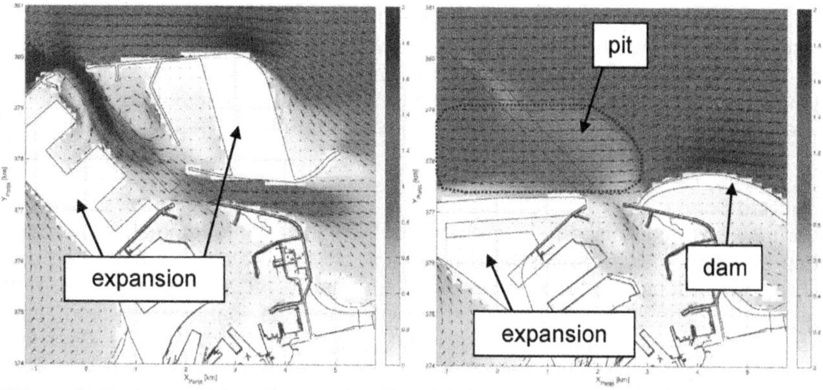


Figure 8: Port expansion, flow during flood, left: seaward, right westward with pit

**Shifting from Scheur-Wielingen to the Walvisch Staart**

An alternative navigation route through the Walvisch Staart would shorten the length of the waterway to the ports in the Scheldt and reduce the length of the estuary. A combination with a depth reduction or blockage of Scheur-Wielingen aims to reduce the tidal range in the estuary. As Scheur-Wielingen is being maintained artificially by dredging it could be brought back to its original depth. However, a simulation where the local seabed was returned to the situation of the 1970's showed no significant changes in velocities in the channel or in the tidal range at Vlissingen.

Simulations with a shoal or blockage by means of one or more islands on top of Scheur-Wielingen (e.g. the large horseshoe island in Figure 9) show that a significant reduction of the tidal range of 25 to 100 cm can be obtained. However, in-between the islands constructed for this purpose an increase of flow velocities should be expected. Since the cross-section of the mouth has been reduced, water level differences induced by the tidal variation lead to strong currents, especially in the new channel through the Walvisch Staart and over the shoal called Vlakte van de Raan. Morphological adaptations will reverse part of the tidal range reduction unless the measures in the mouth are combined with additional measures inside the estuary aiming to reduce the tidal volume passing through the estuary mouth.

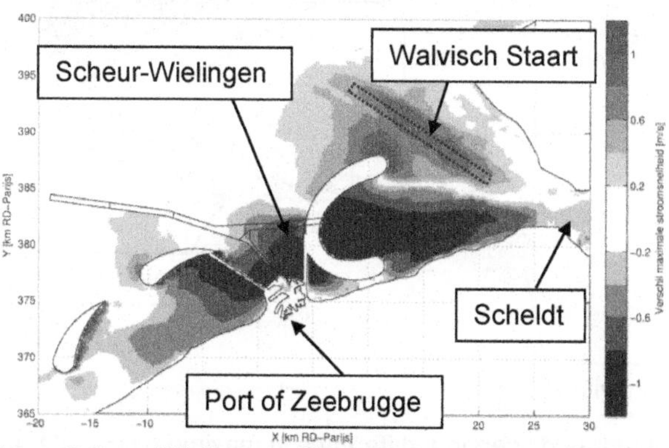


Figure 9: Blockage Scheur-Wielingen, difference in maximum flow rate, average tide

The scenarios as they are drafted and simulated lead to a significant increase in flow velocities in the new channel through the Walvisch Staart (e.g. Figure 9), which is likely to be problematic for shipping and the morphology of the seabed. Further optimization of the scenarios will therefore definitely be needed. However, these scenarios do indicate the order of magnitude of the effects that can be expected of such measures.

**Integrated Scenario**

Several elements discussed above have been combined into an integrated scenario. These elements include a westward extension of the port with an integrated energy atoll, a flow reduction pit and a series of islands on the east side of the harbour (Figure 10). The flow through Scheur-Wielingen is partially blocked by the construction of a shoal (at MSL) which extends to the Vlakte van de Raan to the northwest. A new navigation channel is provided through the Walvisch Staart. Behind the islands a protected route for shipping with limited depth is created. The scenario that was presented during the Super Storms Congress on 26th November 2013 as a potential long-term vision for the Flemish coast was based on this scenario.

As a result of the westward expansion, the contraction point of the flood flow velocities has been shifted to the west and the navigation window is removed. Due to the blockage of Scheur-Wielingen the tidal flow is shifted towards the Walvisch Staart. No eddy formation outside the harbour is expected and also the eddy formation inside the port is removed almost completely. On the Vlakte van de Raan, in the new channel through the Walvisch Staart and in-between the new islands a strong increase of tidal currents is expected. The altered morphology of the estuary mouth has a strong effect on the tide, causing a decrease of the tidal range in Vlissingen and Antwerp of about 1 m. Due to the shelter by the islands against waves a protected navigation route for shallow draft vessels between Zeebrugge and the Scheldt has been created, provided a lock is included in the eastern harbour dam.

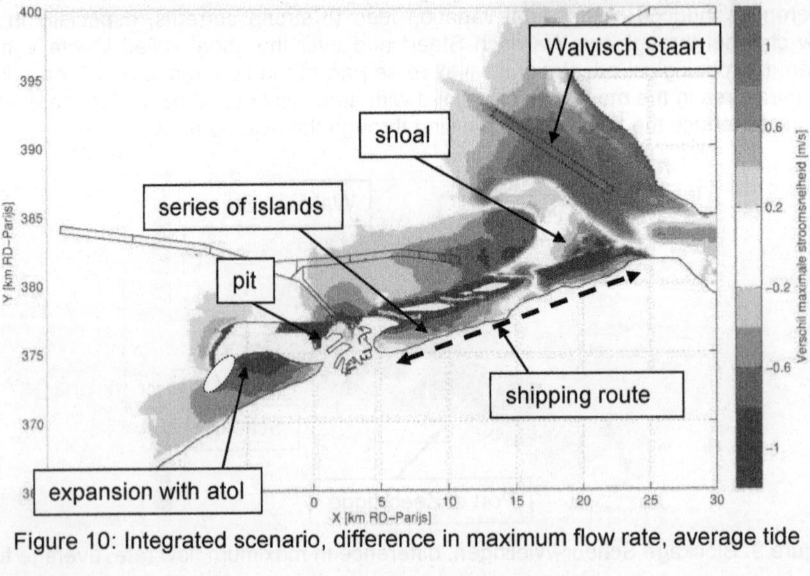


Figure 10: Integrated scenario, difference in maximum flow rate, average tide

## Conclusion

The importance of such an out-of-the-box investigation is paramount: it adds to the social and technical discussion on the feasibility and necessity of large-scale developments, leading in-time to more refined and improved, (possibly) feasible designs to accommodate the future needs of economy, safety and nature. The study has pointed out that measures are perhaps technically feasible but will lead to large modifications in the coastal system so the economic and ecological feasibility is still to be investigated. It also out that, to achieve a successful implementation, one has to work with the forces of nature to shape the concept to a sustainable design.

## References

IMDC (2015a), Analyse impact havenuitbreiding Zeebrugge op onderhouds-baggerwerken vaargeulen Noordzee en voorhaven Zeebrugge - Leveren van ondersteunend numeriek scenario-onderzoek, Synthese rapport, IMDC Rapport I/RA/11401/14.270/VBA, IMDC (in Dutch)

IMDC (2015b), Analyse impact havenuitbreiding Zeebrugge op onderhouds-baggerwerken vaargeulen Noordzee en voorhaven Zeebrugge - Leveren van ondersteunend numeriek scenario-onderzoek, Telemac model - Calibration and validation, IMDC Rapport I/RA/11401/12.242/ABR, IMDC

Breugem A., Decrop B., Da Silva J., van Holland G. & Martens C. (2013). Modelling the Eddy Pattern in the harbour of Zeebrugge. Bundesanstalt fur Wasserbau, Karlsruhe, Germany

**Coastal Management: Changing coast, changing climate, changing minds**
ISBN 978-0-7277-6114-9

ICE Publishing: All rights reserved
doi: 10.1680/cm.61149.325

ice
Institution of Civil Engineers
publishing

# Changing Societies, Changing Coasts and Vice Versa: A Case Study of Bangladesh

**Margriet Hartman**, Royal HaskoningDHV, Amersfoort, The Netherlands

## Aim of this Paper

This paper intends to support the transparent, informed and inclusive decision-making process in planning flood protection infrastructure. Bangladesh is taken as a case study, where considerable research is done on the impacts of flood protection infrastructure on different groups in society. The paper describes first the context of the natural and human environment in Bangladesh. After that, an analysis of the impacts of flood protection infrastructure of embankments, erosion protection and flood shelters follows. Finally a tool is introduced and applied for embankments in Bangladesh, which provides an integrated analysis of costs and benefits of embankments for the different social groups.

## Natural and Social Trends in Bangladesh

The coastal zone of Bangladesh is renowned for its highly complex and extreme interrelated dynamics in its natural and human environment (Cook and Wisner, 2010). The area is a large low-lying deltaic area, where people are highly exposed to natural disasters such as floods, storm surges, cyclones, riverbank erosion and saline water intrusion (Penning-Rowsell, 2013). However, these people show traditionally a high capacity to adapt and respond to these disasters (Elliott et al, 2014).

The region sees a large seasonal and peak variety in meteorological conditions with periods of droughts, heavy rain, storms and cyclones. The hydrodynamic processes (tidal and riverine) and related morphology (erosion and accretion) are characterised by large numbers, like a billion tons of sediment entering the coastal zone annually from the Ganges-Brahmaputra-Meghna Rivers (Islam et al., 1999), a tidal range reaching 8 meters and coastline erosion of one km per year at some locations.

In Western countries ‘flood’ has a unilateral negative meaning. Bangladeshis assign different meanings to ‘floods’, depending on their timing, duration, and magnitude. There are normal monsoon floods (*barsha*), which inundate 20–25% of the country every year and extreme floods (*bonna*) due to extreme rainfall, storm surges or cyclones, which may result in 40–60% of the country under water and occur once every 5 to 10 years (Haque, 1997; Paul, 1997).

*Barsha* are regarded beneficial by the local population as they bring moisture and fresh silt to the soil and enhance fish production. However, *bonna* damage standing crops and may cost human and animal lives.

People learned how to live in the natural dynamics of the delta (Islam, 2001), but they usually could not adjust to the *bonna* floods. People settled on higher parts of land or constructed elevated parts of land (just above the monsoon floods) and put their houses thereon ('dig–elevate–dwell' principle). Furthermore, the agricultural practices were adjusted to the seasonal variations.

However, these days Bangladesh is one of the most densely populated countries in the world. It has nearly 1,200 inhabitants per $km^2$ of land, while the Netherlands (another deltaic country, but in the developed world) has only about 500 (World Bank Data). 75% of the people live in the rural area. The traditional adjustment systems cannot be sustained anymore due to scarcity of land.

Recent socio-economic developments in Bangladesh show an average GDP growth of around 6% annually (World Data Bank) and the Bangladesh Government envisions to reach the status of a middle income country by 2022 (Planning Commission, 2012). Poverty levels dropped by nearly a third since 1992, increasing life expectancy, literacy, and per capita food intake (World Bank, undated). However, in absolute numbers 25 million people lived in extreme poverty in 2010 (Bangladesh Bureau of Statistics, 2011) and this number is still increasing (Rahman, 2011).

Loss of land due to erosion makes many households landless and without livelihood every year. E.g. according to official figures of Bhola Island about 10% (33,000) of the households are landless (Bhola District, 2014, oral communication). People find shelter on government land (like embankments, see Figure 1), slums in urban centres, accretion areas or rent land from other farmers. Most prefer to stay in the area (Penning-Rowsell et al., 2013). Many change their livelihood into fishery, resulting in over-fishing. People in erosion-prone areas consider erosion a much bigger problem than flooding.

Further data show that women, children and elderly suffer most in disasters. Cyclone Sidr in 2007 showed a ratio of female to male deaths of 5 to 1 (World Bank, 2013).

These complex dynamics will be intensified by climate change, population growth and socio-economic development, creating unprecedented challenges to water and environmental management in Bangladesh (IPCC, 2007).

## Flood Protection Infrastructure

### Embankments

Since the 1960s many coastal embankments were constructed by the Government of Bangladesh to enhance agricultural production by irrigation and by reducing saline water intrusion and to control floods (Pal et al, 2011).

Since then, the embankments are regularly upgraded to reach climate change proof levels or to repair breaches and damages. In the last decades, about 20% of Bangladesh' development budget has been spent on water development and flood control projects (Islam, 2001).These efforts continue into present times, for example with the recent World Bank funded 'Coastal Embankment Improvement Project' (World Bank, 2014).

Embankments are not always constructed following design criteria as set by the Government of Bangladesh. Often they are constructed with clay (sometimes more sandy parts) from nearby borrow-pits in the floodplain (see Figure 1). Sometimes the toe of the embankment reaches the borrow-pits. Slopes are often steep to limit the footprint of the embankment.

Figure 1: Embankment with adjacent borrow-pits and in the back landless living on embankment on Bhola Island

**Erosion Protection**
Erosion protection works usually consist of revetment on the upper part and slopes of concrete blocks and/or sand bags under water (see Figure 2). Construction and maintenance need large funds. Often only small stretches of coast line can be protected, increasing the threat to unprotected stretches.

Coastal embankments in erosion-prone areas are highly susceptible to be eroded, when no erosion protection measures are in place at the shoreline. Without proper erosion protection the investment in coastal embankments will have limited sustainability.

Figure 2: Erosion bank protection with sandbags and concrete blocks and house under threat on Bhola Island

**Flood Shelters**
Next to shoreline measures to reduce flood events, there is life-saving infrastructure like the provision of warning and evacuation shelters (Mallick et al., 2011) (see Figure 3).

Figure 3: Multi-purpose flood shelter/school in Noakhali

## Effectiveness of Flood Protection Infrastructure

### Embankments and Erosion Protection

Pal et al. (2011) evaluated the effect of the embankments in Bangladesh on agricultural development and on flood protection. The outcomes show that there is a considerable expansion in irrigation and food production, making Bangladesh today self-sufficient in rice.

However, the study also reveals that the impact on flood control was mixed, with success in regard to normal *barsha* events, while in extreme *bonna* events the impact is much less. For example the massive floods of 1997 and 1998 showed that the embankments did not reduce losses in life and assets.

The result is that people have lost the benefits of the *barsha*, while still experiencing the costs of the *bonna*. Especially, the rural poor and landless are dependent on these natural benefits, which come 'for free'. The fertile sedimentation, irrigation of the fields, recharge of water bodies and groundwater, natural rise of land elevation and entrance of fish are for many poor an indispensable part of their livelihood (Islam, 2001).

On top of that, embankments may even create negative impacts for the rural poor (who usually live on the lowest land). Drainage congestion, filth, longer recession times of flood waters and blockage of streams for navigation and fish migration have most impact on the poor (Islam, 2001 and Kahn, 2008).

Besides these, there is a direct impact for households living and/or working in the footprint area of a proposed embankment. They need to be relocated and will face a hard time to find shelter in the extremely densely populated country and build up a new livelihood.

On the other hand, flood control projects may bring disproportionate benefits to more wealthy farmers and landowners. The wealthy farmers have the opportunity to change to more high-yielding varieties of crops, mechanical irrigation and fertilisers, increasing their production and income. The value of land increases considerably (some say a 10-fold), when the land gets protected by embankments, benefiting the landowners (often from outside the area).

Further, people will develop a sense of security, when they live in an area, protected by embankments. However, the embankments cannot withstand severe cyclones or storm surges. People may start to ignore cyclone warnings. In case of a breach of a coastal embankment, the force and depth of the flood may give people very little time to respond. Often the poor live closest to the embankment (or on it).

Sultana (2010) concludes that poor and landless farmers did not benefit from embankments as much as the non-poor, while they lost more when embankments breached, as their houses are more vulnerable and they live usually closer to the coast. However, it is good to realise that this does not mean poor are less resilient to flood damage. Often they can recover quicker than their non-poor neighbours (Akter, Sonia and Mallick, 2013).

Contrary to the evaluation of embankments, there is a lack of independent evaluation research of the long-term impacts of erosion protection works to make a comprehensive assessment at this moment.

**Shelters**
The capacity of the shelters is still inadequate, but they have helped to reduce the number of deaths considerably (133,882 and 2,380 with the 1991 and 2007 comparable cyclones respectively, Mallick et al., 2011). Also the ratio of female to male death has changed from 14:1 in 1970 to 5:1 in 2007 (World Bank, 2013).
However, the location of flood shelters appears highly influenced by local elites, so that many poor have longer distances to reach the shelters (Mallick and Vogt, 2011).

Still, cultural barriers deter women from entering the shelters. They stay home, as many women are supposed not to leave the house alone, to look after the children, homestead, valuables, animals and garden (Ikeda, 1995).

## Integration of Flood Protection Infrastructure in Societal Needs

The evaluation of the flood protection infrastructure as mentioned above shows that there is a need to assess the positive and negative impacts of the infrastructure on the different groups in society and how the intended benefits of the infrastructure can be more evenly spread over the different groups.

(Foreign) engineers have their background in hydrodynamics, geo-technics and cubic metres and tend to forget about the needs, the knowledge and resilience of the human environment in which their structures will arise. Politicians, consultants and contractors and also international donors have an interest in large-scale infrastructure projects that provide visible results and large funds (Islam, 2001).

In an effort to balance the influence of the technical and financial aspect with social and environmental aspects, often the planning of flood protection infrastructure contains the preparation of an Environmental and Social Impact Assessment (ESIA).

This is required by national legislation and international financiers' guidelines. This document contains an analysis of the social impacts on different groups, although the depth and consistency of this analysis varies (Hartman, in prep). In most cases a genuine integration of the outcomes of the ESIA with the design and financial analysis is limited.

There is a need for a sincere, balanced and integrated analysis of the technical, environmental, social and financial aspects. This is not original and is recognised by many actors in Bangladesh. However, a true 'dialogue' among the different disciplines combining international and local knowledge is still rare (Cook and Wisner, 2010). Early and genuine participation of the affected population, including the most vulnerable groups, in planning and execution will raise the chances of success and awareness of the problems (Elliot et al., 2014).

There are various tools to stimulate integration of disciplines and knowledge. Here I chose the simple tool of Ziller (2012), which consists of an integrated matrix to list financial and non-financial costs and benefits for different social groups.

First a stakeholder analysis will identify the different groups in society. In the case of embankments, we can distinguish: (ultra-)poor, landless, farmers, fishermen, women, children, elderly, businessmen, landowners, politicians, government staff, consultants and contractors. Together with these groups the costs and benefits can be listed. This can be done in different meetings per group, but also in mixed groups to create understanding of the perceptions among the groups.

Table 1 provides the matrix filled for a coastal embankment project using experiences from a number of consultancy projects performed last year, completed with literature research (as described above) in Bangladesh. This table has the purpose to show how this method may work. It has not been validated with the stakeholders in the field.

## Conclusions

The evaluation of the impact of flood protection infrastructure on prevention of flood damage and loss of life in Bangladesh shows a mixed picture. The normal floods (*barsha),* which are generally considered beneficial, can be prevented, while the effects on *banna* floods are often limited.

The application of a tool to analyse the costs and benefits of embankments for different social groups shows the value to stimulate a 'dialogue' among different disciplines involved in the planning of embankments and the experts and the effected people. The table demands to make a complete and balanced overview of the costs and benefits of the proposed infrastructure and how these costs and benefits are distributed over different groups in society. The financial costs and benefits can be monetised and used in the financial analysis. Disproportionate imbalances in how the costs and benefits fall on certain groups are made visible and easy to understand for decision-makers, experts and financiers. Adjustments can be made to the project design to change the balance of costs and benefits.

This method can easily and with limited costs be implemented at the start of the planning of an embankment. It enhances the chance of success of the project, stimulates the participation of all stakeholders and creates transparency in decision-making and understanding and awareness of the complexity of flood protection in Bangladesh and similar countries.

Table 1: Integrated matrix of sustainability and potential costs and benefits of a coastal embankment in a rural area

| Individuals/ groups | Non-financial benefits | Financial benefits | Non-financial costs | Financial costs |
|---|---|---|---|---|
| **All** | Feeling of security behind embankments | Reduced damage and loss of life by moderate floods; Market prices of agricultural products may reduce | Loss of cleansing effect of floods on environment; change in landscape; Degradation of spiritual attachment to river system; Tension among floodplain users | Increased force and depth of a flood in case of a breach: increased risk of loss of life and damage; pumping costs; increased pressure on other areas, where no embankment is present. |
| **(Ultra)-poor** | Opportunity to build homestead on embankment | Temporary labour | Loss of floodplain culture and connection to the river | Loss of floodplain fishery and crustaceans collection, cattle herding, recharge of inland (ground)water bodies |
| **Landless** | Opportunity to build homestead on embankment | Temporary labour | More pressure on available government land with additional resettlers | Loss of floodplain fishery and crustaceans collection, cattle herding, recharge of inland (ground)water bodies |
| **Resettlers from footprint of embankment and borrow-pits** | | Compensation package (not really a benefit) | Loss of homesteads and livelihood; loss of community cohesion; stress for security; risk of HH* break-up; | Resettlement costs; Building-up new livelihood |
| **Fishermen** | | | More competition due to loss of land and people changing to fishery | Decrease fish catch due to blocked fish migration and floodplain breeding grounds |
| **Businessmen** | Improved road connections (when road is built on embankment) | More trade in agricultural produce; supplies to contractor; | | |
| **Farmers** | | Increase in crop production; less crop damage; less salt water intrusion; | | Increase in drainage logging; loss of fertile sediments; loss of natural land elevation; more costs for fertiliser; more costs for pumping |
| **Boatmen** | | | Loss of livelihood and way of living on the water. | Loss of connection between inland and sea navigation (or sluice), old channels may get out of use, berths for boats may disappear |
| **Elderly** | Better access to medical services possible | | Loss of feeling of belonging | |
| **Children** | Better access to school possible | | | |
| **Women** | Better access to medical services possible | Construction labour, supply of food and drinks to work force | Loss of feeling of security; Influx of outside workers | |
| **Local civil servants** | Power to manage tasks and funds | Work, opportunity to obtain additional benefits | Tensions with local population | |
| **Politicians** | Power and respect; visibility | | | Maintenance costs; Debt to international donors. |
| **Landowners** | | Increase in land value behind embankment; income from land sale in footprint; Increase in crop production | | Loss of natural elevation of land by sedimentation |
| **Government Bangladesh** | Power and respect; visibility; | | Loss of natural elevation of land by sedimentation | |
| **International donors** | Visibility | | | Provide large part of the funds for the construction |
| **Consultants** | | Work and profit | | |
| **Contractors** | | Work and profit | Tensions with local population | Contract risks under Bangladeshi law |

HH Household

## Acknowledgements

This paper was the result of experiences and discussions with various experts in a number of coastal development projects in Bangladesh implemented by Royal HaskoningDHV. I like to thank all people involved for sharing their thoughts and opinions. Especially those who live in the coastal zone of Bangladesh and face the natural threats to their livelihood, while having an impressive capacity to adjust and respond to these threats.

## References

Akter, Sonia and Bishawjit Mallick, 2013. The poverty–vulnerability–resilience nexus: Evidence from Bangladesh. *Ecological Economics 96* (2013) 114–124.

Cook, Brain R. and Ben Wisner, 2010. Water, risk and vulnerability in Bangladesh: Twenty years since the FAP, *Environmental Hazards*, 9:1, 3-7

Elliott, Michael, Nicholas D. Cutts, Anna Trono, 2014. A typology of marine and estuarine hazards and risks as vectors of change: A review for vulnerable coasts and their management. *Ocean & Coastal Management.* Volume 93, June 2014, Pages 88–99.

Ikeda, Keiko, 1995. Gender Differences in Human Loss and Vulnerability in Natural Disasters: A Case Study from Bangladesh. *Indian Journal of Gender Studies* 1995 2: 171.

Hartman, Margriet (in preparation), Addressing inequality through Social Impact Assessment: a case study of Bangladesh.

IPCC, 2007. *Climate Change 2007: Impacts, Adaptation and Vulnerability: Summary for Policymakers.* Working Group II Contribution to the Intergovernmental Panel on Climate Change Fourth Assessment Report. IPCC, Geneva.

Islam, Mohammad Rezwanul Syeda Fahliza Begum, Yasushi Yamaguchi and Katsuro Ogawa, 1999. The Ganges and Brahmaputra rivers in Bangladesh: basin denudation and sedimentation. *Hydrological Processes.* 13, 2907-2923.

Islam, Nazrul, 2001. The open approach to flood control: the way to the future in Bangladesh. *Futures* 33, pp. 783–802.

Khan, M. Shah Alam, 2008. Disaster preparedness for sustainable development in Bangladesh. *Disaster Prevention and Management.* Vol. 17 No. 5, 2008, pp. 662-671.

Bangladesh Bureau of Statistics, 2011. Household Income and Expenditure Survey, 2010. Key findings and results. [Accessed on 15 November 2014] http://www-wds.worldbank.org/external/default/WDSContentServer/WDSP/IB/2012/07/12/000386194_20120712025531/Rendered/PDF/709500WP00PUBL0AndExpenditureSurvey.pdf

Mallick, Bishawjit, Khan Rubayet Rahaman, Joachim Vogt, 2011. Social vulnerability analysis for sustainable disaster mitigation planning in coastal Bangladesh, *Disaster Prevention and Management*, Vol. 20 Iss: 3, pp.220 – 237

Mallick, Bishawjit and Joachim Vogt, 2011. Social supremacy and its role in local level disaster mitigation planning in Bangladesh, *Disaster Prevention and Management*, Vol. 20 Iss: 5, pp.543 - 556

Pal, Sudip K., Adebayo J. Adeloye, Mukand S. Babel and Ashim Das Gupta, 2011. Evaluation of the Effectiveness of Water Management Policies in Bangladesh, *International Journal of Water Resources Development*, 27:02, 401-417.

Paul, Bimal Kanti, 1997. Flood Research in Bangladesh in Retrospect and Prospect: a Review. *Geoforum*, Vol. 28, No. 2, pp. 121-131,1997.

Penning-Rowsell, Edmund C., Parvin Sultana and Paul M. Thompson, 2013. The 'last resort'? Population movement in response to climate-related hazards in Bangladesh. *Environmental Science & Policy.* Volume 27, Supplement 1, March 2013, Pages S44–S59.

Planning Commission 2012. *Perspective Plan of Bangladesh 2010-2021, Making vision 2021 a reality*. [pdf] Government of the People's Republic of Bangladesh, General Economics Division, Final Draft, April 2012. [Accessed on 18 January 2015] http://www.plancomm.gov.bd/wp-content/uploads/2013/09/Perspective-Plan-of-Bangladesh.pdf

Rahman, A.K.M., 2011. *Poverty and inequality in Bangladesh*. [pdf] Policy brief, Unnayan Onneshan. [Accessed on 18 January 2015] http://www.unnayan.org/documents/RightsParticipation/policy_brief_poverty_inequality.pdf.

Sultana, Farhana, 2010. Living in hazardous waterscapes: Gendered vulnerabilities and experiences of floods and disasters, *Environmental Hazards*, 9:1, 43-53.

World Bank, *World Databank*, [Accessed in the period between 5 May and 28 October 2014] http://databank.worldbank.org/data/home.aspx Kingdom and New York, NY, USA, pp. 361-409. [pdf: accessed on 21 January 2015] https://ipcc-wg2.gov/AR5/images/uploads/WGIIAR5-Chap5_FINAL.pdf

World Bank, undated. *Bangladesh overview* [0nline, accessed on 18 January 2015] http://www.worldbank.org/en/country/bangladesh/overview

World Bank, 2013. *Improving Women's Odds in Disasters* [0nline, accessed on 23 January 2015] http://www.worldbank.org/en/news/feature/2013/12/12/improving-women-disasters

World Bank, 2014. Coastal Embankment Improvement Project - Phase I (CEIP-I). [accessed: 19-07-2014] http://www-wds.worldbank.org/external/default/WDSContentServer/WDSP/SAR/2014/12/22/090224b082944a2a/1_0/Rendered/PDF/Bangladesh000C0Report000Sequence003.pdf

World Bank, 2012. *Disaster Risk Management in South Asia: A Regional Overview* [pdf] –Washington, DC : The World Bank : GFDRR, 2012. Available at: http://www-wds.worldbank.org/external/default/WDSContentServer/WDSP/IB/2013/07/26/000356161_20130726124954/Rendered/PDF/763020WP0P11400Box0379791B00PUBLIC0.pdf [accessed on 18 January 2015]

Ziller, A. (2012) *The new social impact assessment handbook, A practice guide*. Australian Street Company.

**Coastal Management: Changing coast, changing climate, changing minds**
ISBN 978-0-7277-6114-9

ICE Publishing: All rights reserved
doi: 10.1680/cm.61149.335

# Making Space for Shingle Barriers: the Benefits of Rolling Back

**Uwe Dornbusch**, Environment Agency, Worthing, UK
**Peter Ferguson**, New Forest District Council, Lyndhurst, UK

## Introduction

The winter 2013/14 was characterised by a large number of storms (Met Office, 2014) with the wave height Hs exceeding a return period of between 1 in 1 and 1 in a few tens of years along the southern coastline of England. As a consequence, several shingle beaches have been overwashed leading to a general redistribution of material onto or behind the crest, that is, roll-back of the feature. With anticipated changes to sea levels and potentially to wave environments, "knowledge of gravel-barrier formation, general short-term barrier behaviour, and long-term barrier evolution under changing forcing scenarios are of growing concern."(Orford and Anthony, 2011, p. 42).

It has been demonstrated that swash aligned single barriers respond to sea level rise by rolling back (Orford et al., 1995). Furthermore, it is generally agreed that changes in the environmental conditions (waves, water level and sediment supply) together with the subsurface and back barrier topography will impact on the position of a gravel barrier (Orford, 2011). Sea level rise will lead to barrier roll-back and overstepping as shown for the Holocene transgression in the English Channel off Hastings by Mellett et al. (2012), in more recent historic evidence of Bronze Age oak forests seaward of the present beach position (e.g. Allen et al., 2004), through historical mapping (Nicholls and Webber, 1987) or the internal structure of Chesil Beach (Bennet et al., 2009). Sufficient sediment supply may counterbalance roll-back induced by sea level rise (Bennet et al., 2009), or even lead to significant progradation as exemplified in the long-term development of Dungeness Foreland (Long et al., 2004) or over the last decade at Church Norton Spit as discussed below. Topographic maps show roll-back over the last few centuries under small rates of sea level rise and only small scale changes in sediment supply at many locations in Southeast England. Event based roll-back is documented for example for Hurst Spit (Bradbury and Powell, 1992). Numerical models are increasingly able to represent overwashing and sediment transport onto the crest (Powell, 1990) and eventually to the back of the barrier (Masselink et al., 2014).

Using the assumption that in the future the main pressure along the majority of the mostly swash aligned coastline will be to roll-back, this paper provides evidence of roll-back pressure and its impact during the Winter 2013/14 using 4 case studies. For two of these, roll-back has been managed proactively over different time scales allowing for a first assessment of this management approach.

## Study Area

Shingle beaches in Southern England are composed of a mixture of pebble, gravel and sand (with cobbles locally), where sand makes up ~30% by weight. These beaches have a seaward slope generally steeper than 1 in 10 that intersects with a much shallower gradient sedimentary or rock platform, usually in the intertidal area but at some beaches this junction is subtidal. They are predominantly single ridge barriers and spits but also include fringing beaches often fronting seawalls. Tidal range increases from ~2m in the West to >6m in the East while the 0.05% Hs exceedance decreases from ~5m in the West to <4m in the East (Figure 1).

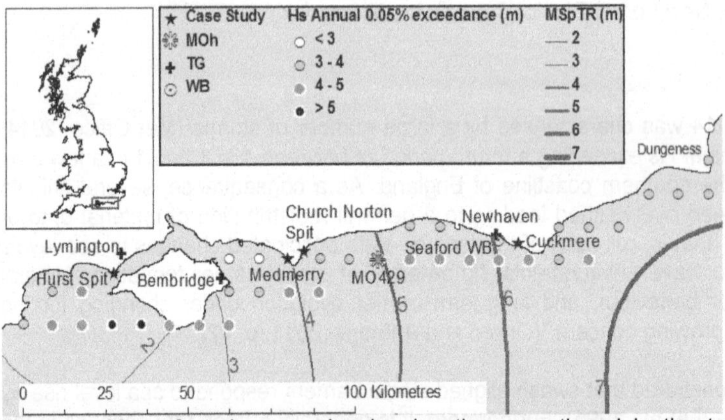


Figure 1: Map of study area showing locations mentioned in the text together with general tide (MSpTR = Mean Spring Tide Range) wave information

## Data

This paper uses survey data collected under the Regional Coastal Monitoring Network (http://coastalmonitoring.org) consisting of LIDAR, ground based Laser scan and GPS surveys, the latter also including surveys carried out for the project Beaches at Risk (BAR, 2006) following the methodology in Dornbusch (2010), together with Met Office wave hindcast data (MOh) (http://cefasmapping.defra.gov.uk/hindcast), buoy and tide gauge data from different locations (Table 1).

Table 1: Wave height (Hs) and water level (h) data in metres for events mentioned in the text. WB = wave buoy, TG = tide gauge, MOh = Met Office wave hindcast

| N° | Location | Date / time | Hs | Hs source | H | H source |
|---|---|---|---|---|---|---|
| 1 | Cuckmere | 24-25/12/1999 | 3.8 | MOh (429) | 3.5 | TideCalc (Newhaven) |
| 2 | Cuckmere | 02/11/13 21:30 | 3.4 | Seaford WB | 2.9 | Newhaven TG |
| 3 | Cuckmere | 03/11/13 22:45 | 2.7 | MOh (429)/Seaford WB | 3.92 | Newhaven TG |
| 4 | Cuckmere | 24/12/14 03:00 | 5.7 | MOh (429) | 2.6 | Newhaven TG |
| 5 | Cuckmere | 14/02/14 22:30 | 4.8 | MOh (429) | 3.48 | Newhaven TG |
| 6 | Hurst Spit | 21/10/89 03:00<br>28/10/89 12:00 | 4.2<br>4.0 | MOh (429) | 1.8<br>1.5 | Bembridge TG - 0.3m |
| 7 | Hurst Spit | 14/12/89 9:00 | 3.5 | MOh (429) | 2.5 | Bembridge TG - 0.3m |
| 8 | Hurst Spit | 14/02/14 21:00 | 5.5 | MOh (429) | 2.3 | Lymington TG |

## Case Studies

### Cuckmere Haven

Cuckmere Haven is an estuary mouth between Chalk cliffs. Groynes were built at various times since the beginning of the 20th century in the area of the present West Beach but only since fixing the river channel through the beach by training the western side in the 1940s and the eastern side in 1971 have the West and East Beach been established as separate beaches. As the tidal prism is too small to keep the channel between the training walls free, removal of shingle from the channel and ebb delta has been ongoing for decades to reduce flood risk further upstream. Any shingle taken was placed on the West Beach building the crest seaward. Longshore transport quickly (often within a month) returned this material into the channel and onto the ebb delta, aided by the deteriorating groynes, resulting in frequent narrowing of the West Beach. Over the last two decades, the beach oscillated between the example position in 2011 and 2013 shown in Figure 2A for profile 66. As the beach sits on a Chalk platform, the equilibrium storm profile has a slope of ~ 1 in 10 so that with a fixed toe position, cut back and overwashing was only a question of when, not if.

The West Beach was overwashed in the night 24-25/12/1999 (see Table 1 for conditions) at a period when the recycling that should have brought the beach crest forward was still underway (Profile 65 in Figure 2A). All overwashed material was bulldozed back into the original position and the battle to keep the shingle in this position continued. The Cuckmere Estuary Strategy (Environment Agency, 2008) proposed to deposit the shingle from the channel to the back of the barrier, but due to a tidal lagoon behind the beach, the extent of such placement was limited and therefore not implemented. On 02 or 03/11/ 2013, during a storm of similar severity to the one in 1999, the beach was again overwashed with 4 overwash fans extending up to 20m landward (Figure 2B). The naturally extended footprint was then used to place 2,700m$^3$ of shingle removed from the Cuckmere channel in that new position, thus rolling back the beach by >20m by the end of December. The following storm events, notably the one on 14/02/2014, which was very similar to that in 1999, did not lead to any overwashing.

Following discussion with Natural England about extending the beach further to accommodate shingle from the Cuckmere, beach material from the channel clearance in December 2014 was used extend the beach further back. At the same time, the crest was lowered to increase the opportunity for overwashing and the beach attaining a more natural appearance and function (profile 13/01/2015 in Figure 2B).

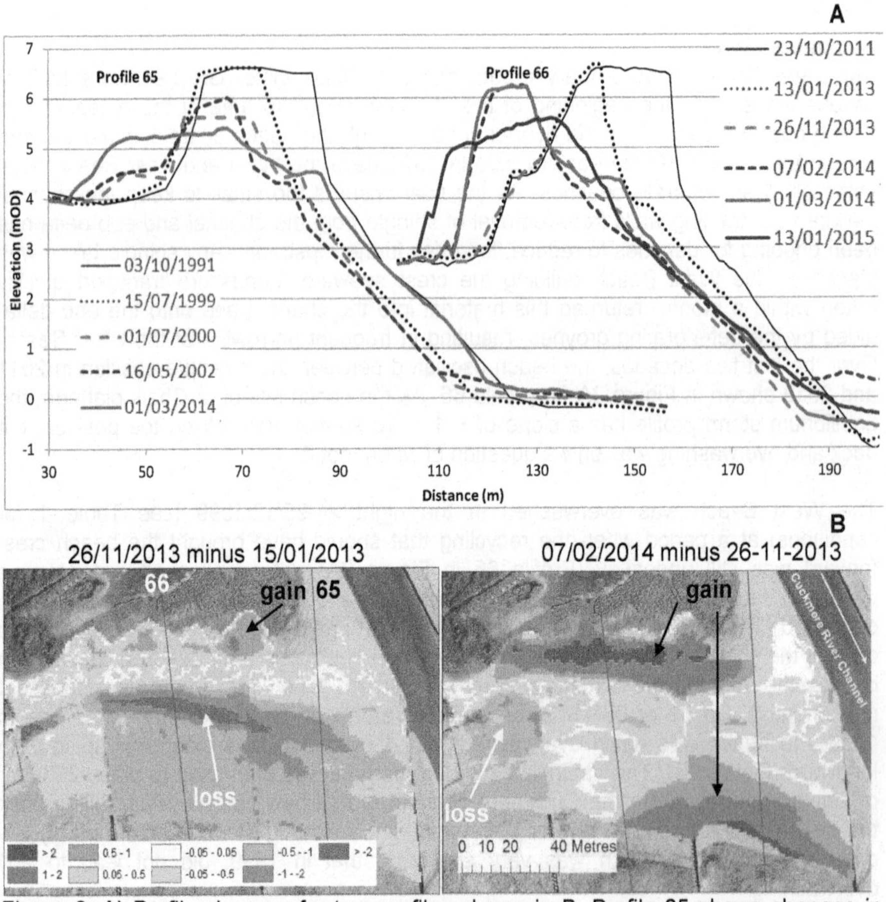


Figure 2: A) Profile changes for two profiles shown in B. Profile 65 shows changes in 1999/2000 based on annual summer profiles that miss the actual roll-back and the final situation in March 2014; profile 66 shows changes during Winter 2013/14 including roll-back and crest lowering (26/11/2013) and recycling to the back (07/02/2014). B) Area changes showing overwash lobes and large-scale beach face erosion (left panel) and deposition of channel dredge material to the back of the barrier (right panel)

## Hurst Spit

Figure 3 shows the same profile used in Bradbury and Powell (1992). Profiles started to be collected in 1987 and very little changed up to the beginning of September 1989. It is unclear what conditions led to the overwashing and roll-back during September 1989, but they left the spit vulnerable to the two storms in October which led to the severe overwashing and flattening of the spit seen in the survey 1/12/1989 and described in the literature as having occurred over a length of 800m with the seaward toe position retreating by as much as 60m and the landward toe by up to 80m with a crest level reduction of as much as 3m (Bradbury and Powell, 1992). As an intermediate measure, the beach was reshaped into a similar form with the crest ~15m landwards and 1m higher by April 1990. Following physical modelling, the beach was rebuilt in 1997 with 300,000m$^3$ of sea dredged material. Apart from creating a higher and wider barrier, the material was added to the back of the 1990 repairs. Continued pressure on the front of the barrier has been dealt with since 2000 by recycling beach material to the backslope, allowing the barrier to migrate under management by ~12m, though loosing width. The storms during 2013/14 up to February had reduced the barrier to a pointy narrow ridge (survey 17/02/2014) when the beach was subject to the worst conditions over the last 25 years during the event on 14/02/2014. The spit was overwashed and lowered locally but the damage was much less than in 1989. Repair over the summer of 2014 again widened the beach to the back. In addition, the natural recovery has created a wide berm between the 25 and 40m location (approximately the 1990 barrier position) forming and altogether more natural barrier.

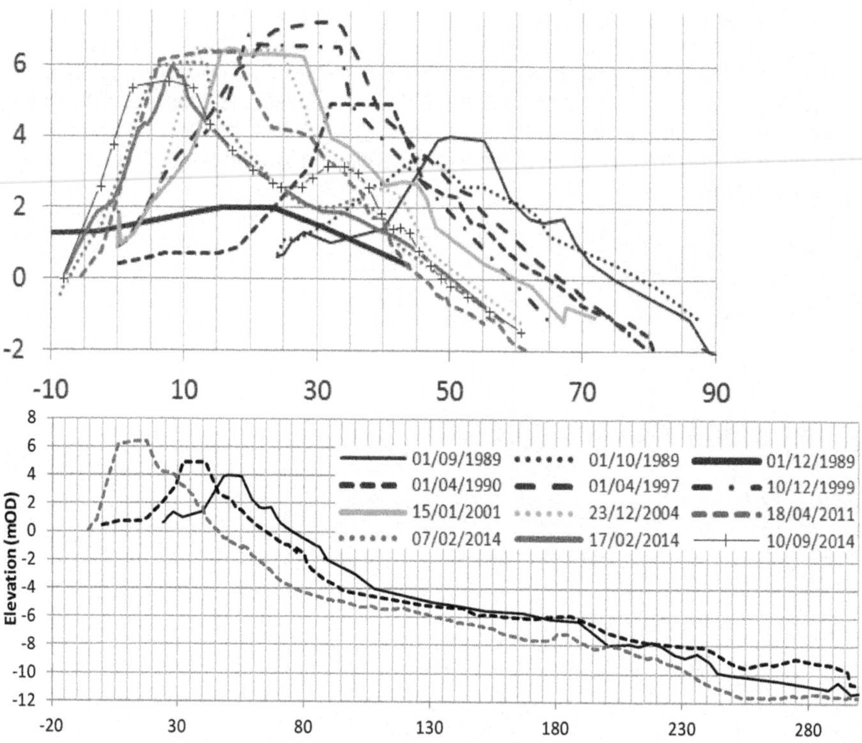


Figure 3: Profile 5f00045 in the centre of Hurst Spit (~450m east of the rock revetment, National Grid Reference: 430347E 90446N)

Figure 3 also shows the changes to the subtidal profile using only 3 of 20 available profiles (those shown are moved in time to the closest topographic survey data for display purposes). Lowering, and thus recession, can be seen along the entire profile down to -12mOD, however, lowering shows a minimum at ~120-130m at elevations of ~5mOD suggesting a more resistant palaeo-surface. Retreat of the shingle barrier means that the profile over which wave energy can be dissipated before impacting the supra-tidal barrier has increased.

**Church Norton Spit**

Church Norton Spit has grown in length by >900m between 2004 and 2014[1] (Figure 4) through continued nearshore and longshore sediment supply. The growth was supported by an extensive ebb-delta complex over which the spit could extend. The winter 2009/10 captures the last period of progradation (Figure 5). Until spring 2012 the barrier gets narrower and the intertidal foreshore lowers. Then, in September 2012, overtopping and roll-back begins, followed in October by a washover and crest lowering (1m) event that moves the back of the crest some 25m landwards. Further significant roll-back happens in the Winter 2012/13 accompanied by crest-raising so that by March 2013 the barrier reaches the same height as the year before but with the front of the crest 25m landward. Over the following Winter 2013/14, it is initially pushed up another metre to its highest position and moved landwards by ~7m. The last storms of that winter lowered the crest again down to ~5mOD with some minor depositions behind the crest. Finally, the first storm of the Winter 2014/15 reduced the crest level further to ~4.5mOD with deposition just behind the crest. This latest period of lowering and overwashing is likely to have been aided by the continuing lowering of the beach toe and foreshore platform.

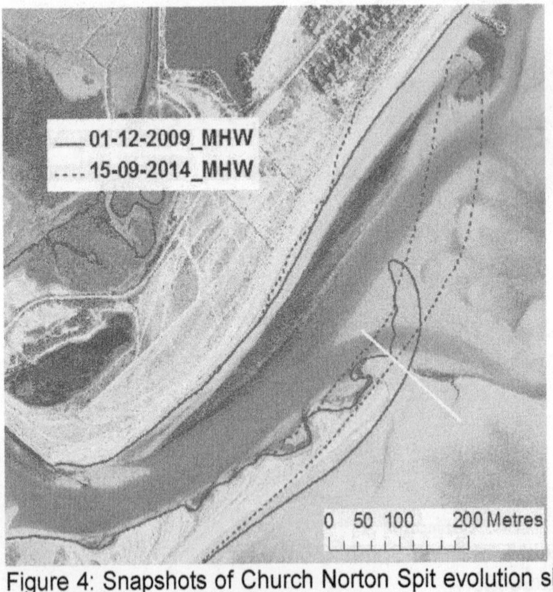


Figure 4: Snapshots of Church Norton Spit evolution showing Mean High Water (MHW) lines for the bracketing dates in Figure 5 over orthophoto from summer 2008. Profile line used in Figure 5 is shown in white.

1 For an animation see: https://www.youtube.com/watch?v=dABo2Chkvco

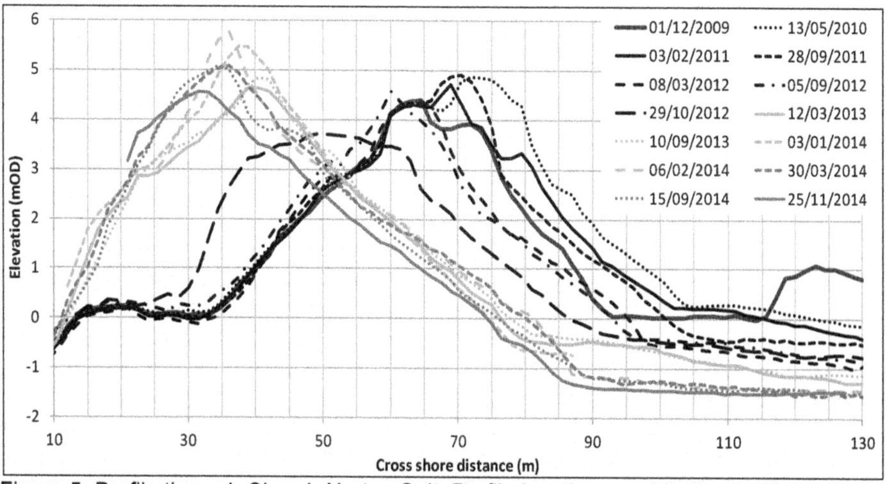

Figure 5: Profile through Church Norton Spit. Profile location as white line in Figure 4

**Medmerry**

The narrow shingle ridge at Medmerry had been a battleground with the sea for decades. While the MHW line retreated ~50m between the 1890s and 1930s along the beach to the north west of the section shown in Figure 6 and ~30-40m along the beach to the south east, the area of present retreat was stabilised with groynes to maintain the outfall structure of the Broad Rife.

Figure 6: Location of the profile shown in Figure 7 at Medmerry (light blue dotted line). Orthophoto shows the situation in the Summer 2008 overlaid with the MHW contours for 3 points in time

Cut off from historic sediment sources through cliff protection at Selsey Bill in the 1940s, pressure on the barrier in this location, but also up-and downdrift, led to the installation of groynes trying to halt the thinning of the barrier through longshore transport. Over the last 60 to 70 years the barrier was kept in the location it inherited in the 1930s and over the last few decades it was overtopped most winters and saw a number of breaches. In these last decades, any material that had been deposited behind the barrier was bulldozed back to its pre-storm position and the ridge kept in the shape shown in Figure 7.

In the autumn of 2013, the barrier was artificially breached following the construction of a flood defence much further inland and the simultaneous creation of intertidal habitat in the Medmerry Managed Realignment project. As a consequence, management of the shingle bank ceased. The first storm in November 2013 cut back the front crest by ~5m, lowered it by ~0.5m and created overwash fans extending for 10 to 20m. The storms in December 2013 overwashed the barrier completely, lowering it by >1m and moving the front crest landward by ~20m and the back of the barrier >50m, thereby exposing the underlying back barrier sediments on the beach face (stepped profile from ~35m in Figure 7). Further retreat of the landward edge was stopped even during the continued roll-back by an existing earth embankment. Despite longshore transport along the Medmerry frontage being in the order of 20-30,000m$^3$ during this winter, it is of note that the cross-sectional area (not counting the eroded marsh strata) of this profile remained almost constant.

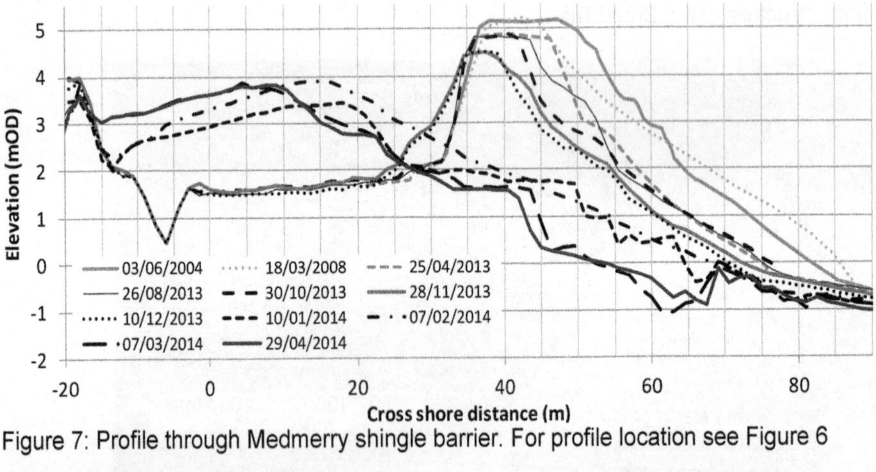


Figure 7: Profile through Medmerry shingle barrier. For profile location see Figure 6

## Discussion

The long-term historic trend for shingle beaches has been to roll back under sea level rise during storm events as shown in the literature and case studies presented in this paper. Although this response can be aggravated by a lack of, or be mitigated through an increased sediment supply, the pressure for beaches to roll back will only increase in the future.

In the past, the principle of maintaining shingle beaches has been to add material to the seaward side, often together with raising the crest. As waves attack and erode the front of the barrier during a storm, maintaining it in a historic position creates narrow, steep and high ridges that are more vulnerable to breaching than lower and wider barriers (Obhrai et al., 2008). Sea level rise will lead to a reduction in freeboard and an increased water depth, allowing higher waves to reach closer to the barrier before being affected by the sea bed. Rolling back onto higher ground will maintain the relationship between barrier geometry and changed marine forcing and is likely to require no additional beach material. Rolling back across terrain of the same level is likely to provide only a temporary solution before lower parts of the profile adjust to this retreat. This appears to be the situation at Hurst Spit. Once the lower part of the profile adjusts, the cross section of the barrier needs to be increased to account for increased forcing through sea level rise.

Where management has pro-actively supported the roll-back process by recycling beach material to the back of the barrier rather than the front to maintain cross sectional area and thus stability of the structure, recent events appear to have had less of an impact than could be expected if the barrier had been kept in place. This conclusion is open to challenge mainly because storm events are not identical in terms of water level, wave height and period, and because the data sources are of different type and quality. However, given the historic and present evidence together, pro-active roll-back can be seen as a precautionary adaptive management intervention and forms the basis for the management at Hurst Spit. Morphological response models like Xbeach-g (Masselink et al., 2014) may be able to support this decision in the future.

This pro-active approach could be seen to require a change of management policy from Hold The Line (HTL) to Managed Realignment (MR), however, it can be argued that the function of HTL is still maintained, albeit in a more landward position: this is the case for Hurst Spit and suggested for implementation in the Barton's Point (Kent) Beach Management Plan. Issues of intentionally covering habitats behind the barrier could potentially be overcome by adopting an opportunistic approach as in the case of Cuckmere Haven or by making an assessment as to the potential extent under natural processes.

We suggest that adapting shingle barrier management by pro-actively rolling them back, will make them more resilient to storm events and potentially more sustainable through reducing longshore transport rate. This may delay the inevitable, more catastrophic roll-back, but unless barrier cross sections are significantly increased, their resilience under rising sea levels has to be questioned.

## References

Allen, M.J., Bone, D., Matthews, C., Scaife, R.J., 2004. Tree trunks, Bronze Age remains and an ancient channel exposed on the foreshore at Bognor Regis, West Sussex. Sussex Archaeological Collections 42, 7–23.

BAR, 2006. Beaches at Risk project [WWW Document]. URL

http://www.sussex.ac.uk/geography/researchprojects/BAR/ (accessed 1.6.15).

Bennet, M.R., Cassidy, N.J., Pile, J., 2009. Internal structure of a barrier beach as revealed by ground penetrating radar (GPR): Chesil beach, UK. Geomorphology 104, 218–229.

Bradbury, A.P., Powell, K.A., 1992. The short term profile response of shingle spits to storm wave action, in: Proceedings of the 23rd International Conference on Coastal Engineering. Presented at the Coastal Engineering 1992, ASCE, Venice, Italy, pp. 2694–2707.

Dornbusch, U., 2010. Ground survey methods for mixed sand and gravel beaches in intertidal environments: a comparison. Journal of Coastal Research 26, 451–464. doi:DOI: 10.2112/08-1134.1

Environment Agency, 2008. Cuckmere Estuary Strategy.

Long, A.J., Plater, A.J., Waller, M.P., Roberts, H., Laidler, P.D., Stupples, P., Schofield, E., 2004. The Depositional and Landscape Histories of Dungeness Foreland and the Port of Rye: Understanding Past Environments and Coastal Change. University of Durham.

Masselink, G., McCall, R., Poate, T., van Geer, P., 2014. Modelling storm response on gravel beaches using XBeach-G. Proceedings of the ICE - Maritime Engineering 167, 173–191. doi:10.1680/maen.14.00020

Mellett, C.L., Hodgson, D.M., Lang, A., Mauz, B., Selby, I., Plater, A.J., 2012. Preservation of a drowned gravel barrier complex: A landscape evolution study from the north-eastern English Channel. Marine Geology 315–318, 115–131. doi:10.1016/j.margeo.2012.04.008

Met Office, 2014. A Global Perspective on the Recent Storms and Floods in the UK [WWW Document]. URL http://www.metoffice.gov.uk/research/news/2014/uk-storms-and-floods (accessed 3.5.14).

Nicholls, R.J., Webber, N.B., 1987. The past, present and future evolution of Hurst Castle spit, Hampshire. Progress in Oceanography 18, 119–137. doi:10.1016/0079-6611(87)90029-2

Obhrai, C., Powell, K.A., Bradbury, A., 2008. A Laboratory Study of Overtopping and Breaching of Shingle Barrier Beaches. Presented at the International Conference on Coastal Engineering, 31/09 to 5/09/2008, Hamburg.

Orford, J.D., 2011. Gravel-dominated coastal-barrier reorganisation variability as a function of coastal susceptibility and barrier resilience, in: Proceedings of the International Conference on Coastal Sediments 2011. Presented at the Coastal Sediments 2011, World Scientific Publishing, Miami, pp. 1257–1270.

Orford, J.D., Anthony, E.J., 2011. Extreme events and the morphodynamics of gravel-dominated coastal barriers: Strengthening uncertain ground. Marine Geology 290, 41 – 45. doi:10.1016/j.margeo.2011.10.005

Orford, J.D., Carter, R.W.G., McKenna, J., Jennings, S.C., 1995. The relationship between the rate of mesoscale sea-level rise and the rate of retreat of swash-aligned gravel-dominated barriers. Marine Geology, Coastal Evolution in the Quarternary: IGCP Project 274 124, 177–186. doi:10.1016/0025-3227(95)00039-2

Powell, K.A., 1990. Predicting short-term profile response for shingle beaches (No. Report SR 219). HR Wallingford.

Coastal Management: Changing coast, changing climate, changing minds
ISBN 978-0-7277-6114-9

ICE Publishing: All rights reserved
doi: 10.1680/cm.61149.345

# Global Approaches to Coastal Erosion and Landslide Management – Understanding the Risks, Empowering Communities and Building Resilience

**Robin McInnes**, Coastal & Geotechnical Services, Isle of Wight, UK
**Roger Moore**, CH2M HILL, Birmingham, and University of Sussex, UK

## Introduction

Over half of the population of the world lives on or close to the coast and in many locations this results in severe development pressures, which impact upon both the physical and natural environments. Within these coastal zones the processes of erosion and landsliding, both natural and man-made hazards, affect communities and infrastructure, and the impacts of climate change are widely expected to further increase risks to society. The expansion of the global population has increased the vulnerability of individuals and assets as demand for housing and other developments has gradually spread onto cliff tops and coastal slopes, which are often of marginal stability.

Even in many developed countries there is a legacy of planned, unplanned or inappropriate development in areas prone to increasing risk from coastal erosion and landslides. There is also reliance, in some locations, on the performance of slope stabilization and coast protection measures, which may not be fit for purpose or are of an age beyond their design life. The impacts of the global recession have also led, in some countries, to a reduction in available funding to support programmes for the replacement or maintenance of necessary defences.

The cost of managing coastal risks can represent a significant burden for national governments, regional or local authorities often with limited financial resources available to ensure an effective, sustainable response. There is now increasing recognition globally that efforts to reduce such coastal risks must be systematically integrated into policies, plans and programmes for sustainable development, and must be supported by states in partnership with the various levels of regional and local administration, and the private sector where applicable (McInnes & Moore, 2014).

This paper promotes an approach that identifies risk areas, appraises the risks, and seeks to avoid or manage them through adaptive planning or appropriate civil or geotechnical engineering measures. The authors recognize that such approaches can only be successful if they are accompanied by appropriate stakeholder engagement alongside education, awareness-raising and encouraging building institutional capacity for change by the affected populations. This is proposed because it will help to ensure that planning policies will shape more sustainable communities that can adapt successfully to the increasing risks likely to be posed by coastal and climate change.

## Background

The physical, social and economic and impacts of coastal erosion and landsliding are of global concern. Increasing vulnerabilities resulting from changing demographic and socio-economic conditions, as well as unplanned urbanisation and development extending into areas of increasing risk, alongside climate variability and geological hazards all present significant challenges. These hazards can have a serious impact across a range of locations and environments - within coastal zones, cities, coastal towns and villages, more isolated communities as well as for coastal infrastructure such as transport links. The costs of emergency action, remediation and prevention can represent a significant burden for the governments or communities affected, often with heavy reliance placed on local or regional authorities with limited financial resources available to permit an effective and sustainable response.

There is a widely held view that the impacts of climate change pose increasing levels of risk to communities in coastal zones susceptible to erosion and landslide activity. In the United Kingdom, for example, it is predicted that up to 700 properties may be lost as a result of coastal erosion over the next twenty years and 2,000 over the next 50 years (Environment Agency, 2013, Moore and Siddle, 2014). On the Isle of Wight, UK erosion and instability within north-western Europe's largest urban coastal landslide complex has resulted in the loss of over 100 properties, businesses and infrastructure in the last century (Moore et al., 2010a) whilst over 25% of the French coast is affected by coastal erosion, parts of which are heavily developed (European Commission, 2004; Deboudt, 2010).

In Australia 80% of the population lives in coastal zones, and on the eastern and southern coasts nearly 300,000 buildings may be vulnerable as a result of rising sea levels and increased rates of erosion (Australian Government, 2009). On the eastern coast of the USA the impacts of Hurricanes Ida in 2009 and Irene in 2011 as well as major storms in 2005 and 2007 devastated coastal communities eroding beaches and dunes, destroying sea walls and hundreds of properties (McInnes & Barter, 2010).

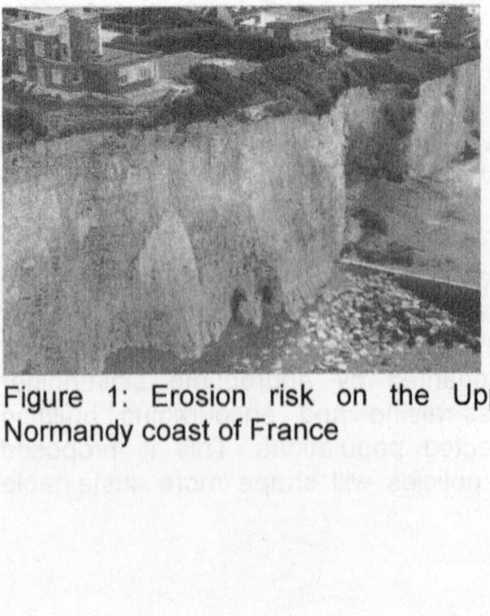

Figure 1: Erosion risk on the Upper Normandy coast of France

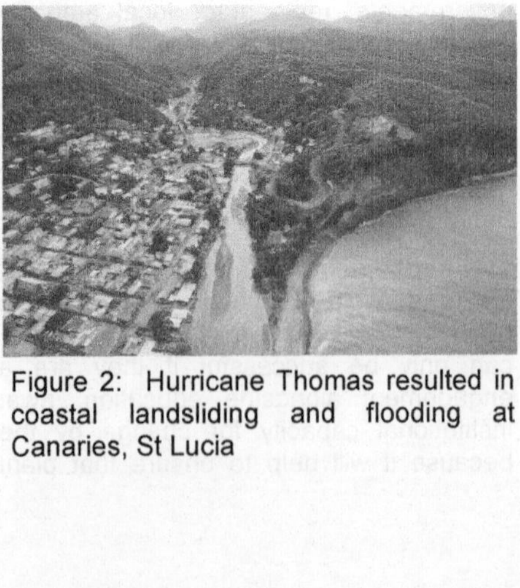

Figure 2: Hurricane Thomas resulted in coastal landsliding and flooding at Canaries, St Lucia

Despite an increasing understanding and the acceptance of the importance of the implementation of management strategies and other risk reduction measures, the impacts of coastal erosion and landsliding continue to pose major challenges in many countries. There is now increasing recognition that efforts to reduce risks must be systematically integrated into policies, plans and programmes for sustainable development, and must be supported by international co-operation and partnerships at the various administrative levels.

## Global Approaches to Risk Reduction

At the international level, the United Nations International Decade for Natural Disaster Reduction (1990-99) recognised the increasing global threat of large-scale natural disasters both at the coast and inland; this initiative was succeeded by the '*International Strategy for Disaster Reduction*' (ISDR). The UN Office for Disaster Reduction (UNISDR) implemented the '*Hyogo Framework for Action 2005-2015*' (HFA) - '*Building the Resilience of Nations and Communities to Disasters*' in 2005 (United Nations, 2007). Since then it has undertaken periodic risk reviews and assessments through its '*Global Assessment Reports for Disaster Risk Reduction* (United Nations, 2011). The UNISDR is now working towards the development of a '*Post-2015 Framework for Disaster Risk Reduction (HFA2)* (United Nations, 2011), which will succeed the current HFA when it completes its ten year term.

Alongside UNESCO's Intergovernmental Oceanographic Commission (IOC), which was established in 1960, another global network, The Global Forum for Coasts, Oceans and Islands, plays an influential role assisting world governments by ensuring that ocean and coastal related issues remain high on political agendas. Such initiatives aim at encouraging, globally, the development of more resilient coastal communities by promoting increased awareness of the importance of risk reduction as an integrated component of sustainable development. Achieving this goal will help to reduce the human, social and economic losses resulting from natural hazards such as coastal erosion and landslides.

Risk reduction requires interdisciplinary action and the United Nations has sought to build partnerships and take a global approach encouraging better co-operation between government authorities and the international and scientific communities, with the aim of reducing risk and improving the quality of life for those living in such areas at risk. Governments must balance coastal and landslide risk management needs against other priorities particularly as a result of the far-reaching consequences of the global economic recession and the need for deficit reduction. Cuts in both budgets for capital and maintenance works as well staff reductions in local government and the key NGOs have been taking place at a time when many parts of the world's coastlines have been experiencing the impacts of severe storm conditions. This emphasises all the more the need for the thorough investigation, monitoring, assessment, long-term planning and management of coastal risks.

## Assessing Coastal Erosion and Landslide Risks

A reliable hazard and risk zoning system for urban planning and development has been recognised as an essential requirement by coastal and landslide specialists, and many nations are working to implement this approach. The actual delivery of solutions to erosion and instability problems often rests with the mayors, local government officials and decision-makers who must deal with, on a day-to-day basis, the impacts of these events. The evaluation of coastal erosion and landslide risk is a judgmental process designed to determine just how significant the estimated risks are, and to establish the best course of action, including the nature of the risk management measures required (Lee & Jones, 2014); this evaluation process then allows risk management strategies to be developed.

Overall responsibility for risk reduction from natural disasters generally rests with the state and the need to reduce risks from natural hazards is usually addressed as part of a national response. Such national strategies, which have the overall objective of protecting people, property, assets and the environment from loss or damage, can only be achieved successfully through involvement of the various administrative tiers including central government departments, regional and local government, non-governmental organisations, the scientific community as well as the key stakeholders.

Although central government often has the overall co-ordinating responsibility, and may be responsible for disaster relief, it is often at the regional and local levels that considerable knowledge and expertise has developed based on practical experience. This knowledge allows regional and local government, working with the key agencies, to develop strategies and respond based on knowledge of the local situation. In the United Kingdom, for example, the predicted impacts of climate change on cliff instability and erosion rates has been explored in considerable detail through both 'National Coastal Erosion Risk Mapping' (Moore et al., 2010b), the development of shoreline management plans for the whole coastlines of England and Wales based on a sediment cell approach (Defra, 2006), a national 'Strategic Coastal Monitoring Programme' (Bradbury, 2007) and a 'National Flood and Coastal Erosion Risk Management Strategy for England' (Defra, 2011).

The investigation and improved understanding of erosion and landslide hazard does, therefore, provide the opportunity to significantly improve coastal land-use planning, with the aim of limiting development to those areas where risks are assessed as likely to be acceptable, and using the higher hazard areas for other low occupancy uses or no development at all. This logical sequential approach to policy, tasks and responsibilities is illustrated in Table 1 (below).

## Planning for Change in Coastal Zones Affected by Erosion and Landsliding

Across the world governments are increasingly committed to ensuring that planning policies help to encourage sustainable communities allowing them to adapt to coastal risks, particularly in the face of climate change. Land-use planning, therefore, has a vital role to play in helping communities to manage risk and adapt to conditions, which may worsen in the future.

Table 1: A Sequential Approach to Coastal Erosion and Landslide Risk Assessment and Planning – Tasks and Responsibilities (adapted from CLG 'Development and Coastal Change Practice Guide' 2010)

| **Coastal Erosion & Landslide Risk and Planning<br>TASKS AND RESPONSIBILITIES** | | | |
|---|---|---|---|
| **POLICY TASK** | **WHAT IS INVOLVED?** | **HOW IS THE TASK ADDRESSED?** | **WHO IS RESPONSIBLE?** |
| 1. Risk appraisal and analysis<br><br>2. Identify risk areas | Undertaking geomorphological mapping, desk studies and reviews of historical data at appropriate scales and levels of detail.<br>Preparing hazard maps illustrating ground behaviour and impacts on existing/future developments.<br>Understanding of the potential impacts of climate change. | By reference to geomorphological, ground behaviour and summary planning guidance maps. | Studies and investigations often commissioned from specialists by regional or local authorities or NGOs.<br>Local Authority technical staff in liaison with their planning department officials. |
| Erosion and Landslide Management<br><br>3A. Risk Avoidance | Ensure through the planning system that new development is prevented in areas at risk of erosion or instability. | Carry out a vulnerability assessment using mapping data.<br>Only consider approval of new developments if is of overriding economic or social importance. | Local Planning Authorities and the technical advisors. |
| 3B. Risk Mitigation | Measures to reduce current and future risks to communities affected by erosion or landsliding. | Improve resilience of communities through civil engineering measures alongside advice and guidance for businesses and residents. | Local Authority technical departments and their advisors or NGOs. |
| 3C. Risk Adaptation | Reducing or eliminating the risk for affected communities. | Identify land likely to become unsuitable for development as a result of increasing levels of risk in the face of climate change.<br>Implement strategies for transition e.g. time-limited planning consents and programmes for relocation of property and assets. | Local Authority technical and planning departments in liaison with businesses and stakeholder groups. |

In order to reduce erosion or landslide risk in the context of land-use planning, assessments of hazard and vulnerability are required as essential tools to support improved standards and practice. Such measures need to be accompanied by appropriate building codes and land-use regulations in order to improve standards and practice. Alongside this, other measures, for example the need to reduce informal urban settlements, to ensure proper siting of critical facilities and infrastructure, zoning hazardous areas and imposing strict development controls in these areas are further essential requirements. Introducing natural hazard considerations into development, land-use and urban planning will, in the medium to long term, reduce the accumulation of risk in both rural areas and in areas of rapid urbanization. Land-use planning supports coastal erosion and landslide risk management because it focuses on modifying the vulnerabilities of developed areas instead of controlling the hazards, thereby supporting mitigation and prevention.

There are numerous examples in both more and less prosperous states where planning and development control systems are in place for coastal zones but the necessary guidance has not always been followed, sometimes with serious consequences. Planning policies in erosion or landslide-prone areas should therefore:

- Ensure that decision-making is based on a thorough understanding of coastal evolution and the potential for change over time, whilst still recognising that uncertainties do exist;
- Prevent new development from being put at risk, first, by avoiding inappropriate development in vulnerable areas, and, second, by encouraging development away from areas that are vulnerable to erosion or landslide risk;
- Ensuring, where possible, that plans are in place to secure the long-term sustainability of settlements in erosion and landslide-prone areas; if this is not sustainable planning should support suitable adaptation measures;
- Implementing policies that promote a strategic risk-based approach, which aims to strike the right balance between economic prosperity and the need for potentially expensive remediation measures. The approach it adopts to achieve this is to carefully assess risks so that they can be avoided or managed.

## Adaptation to Climate Change in Areas Affected by Instability

In many countries a key government objective through the planning process, is to support the transition to a low-carbon economy in a changing climate, taking full account of risks arising from climate change. In order to achieve this objective, the planning system should seek to:

- Minimise vulnerability and provide resilience to impacts arising from climate change.
- Avoid inappropriate development in areas at risk from erosion or instability, by directing development away from areas at highest risk or, where development is necessary, making it safe without increasing the hazard elsewhere.

To achieve these objectives national governments and regional and local government planning authorities should adopt pro-active strategies to mitigate and adapt to climate change (Defra, 2010). For example, in certain situations, coastal or landslide risk may be so problematic that it is necessary to consider the relocation of all or part of a community and or other assets, and to identify and allocate more suitable land within the development plan where a community may be relocated. This may be a case of the development 'rolling back' from the area of active coastal erosion or instability.

Depending on the level of risk, or the timeframe in which implementation is required, the planning authority may be in a position to approve more modest developments which can exist and operate within time-limited constraints. Planning conditions may be applied to such new developments where it is possible to manage the risk to the proposed development during the planned lifetime of the building or structure. Such approaches have been tested in the United Kingdom through a series of innovative 'Pathfinder Project' case studies supported by the government including sites in East Anglia and on the North Yorkshire coast (Defra, 2012; Siddle et al, 2015 this conference).

Time-limited planning permissions that contain conditions relating to the review of that permission in light of the rate and scale of erosion or instability problems, ensuring their eventual removal prior to the impacts of more serious instability aggravated by climate change offer an alternative approach. In France the provisions of the Barnier Law allow the expropriation and compensation by the government of property threatened by natural risks such as instability when remedial works would be more expensive than compulsory purchase. Compensation may be obtained from a state fund accrued through a surcharge of 12% added to all property and motor insurance premiums. Such a mechanism can support the planning system by facilitating the relocation of people and assets away from vulnerable locations.

Adaptive approaches aimed at trying to address potentially increasing levels of risk are part of a process that will, without doubt, be required to take account of worsening conditions aggravated by climate change over the next century. To achieve successful adaptation communities need to be supported and to be involved in the outcomes that will shape the future of their community. There needs to be open discussion on what changes in risk may mean for their community, and agreement on the problems to be addressed in terms of infrastructure, properties and local businesses.

## Building Resilient Local Communities

While consideration of risk should be taken into account at all levels of governance, it is often at the local community level where a coherent and acceptable response to managing the impacts of changing risks can be achieved most effectively. To be successful, this should bring together all sectors of a community including politicians, local government officers, the statutory bodies, local businesses and homeowners. Clearly effective dialogue with community leaders and residents in such affected locations requires a thorough understanding of the nature and the scale of the problem based upon up to date data on hazards and vulnerabilities, and completed risk assessments that are informing the planning process. The findings must be translated into non-technical language to ensure successful engagement with community groups and other stakeholders (eg; Figure 3 below).

Such approaches have proved particularly successful in several key locations affected by landslide risk around the English coast including North Yorkshire, North Norfolk, the Isle of Wight and in West Dorset (McInnes & Moore, 2011, 2014). Residents can play a proactive role in terms of contributing in practical ways to landslide risk reduction strategies through self-help leading to improved management of their own land and property.

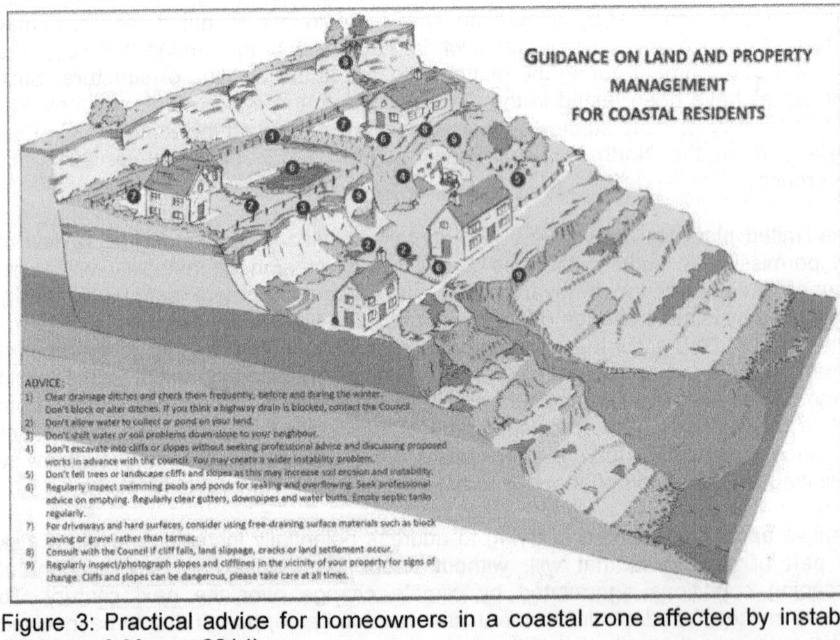


Figure 3: Practical advice for homeowners in a coastal zone affected by instability (McInnes & Moore, 2014)

## Conclusions

1. It is now widely accepted that globally there is likely to be an increase in coastal erosion and landsliding as a result of the impacts of climate change and sea level rise.
2. Whilst research, monitoring, data gathering and increased investment in coastal defences are vital, it is essential also to advance the testing and implementation of further new approaches to adaptive coastal planning in support of sustainable coastal management.
3. At a time of expenditure reductions across public sector functions, including coastal risk management, the statutory planning role is all the more important in terms of supporting the management of changes resulting from increasing coastal erosion and landsliding. The planning process can fulfill a vital role in terms of preventing development in areas at risk, or potential risk, as well as facilitating new approaches to avoidance and adaptation.
4. A thorough review of adaptation techniques being practiced globally is now required and practical guidance should be provided, building on approaches such as Defra's Coastal Change Pathfinder Programme in England and the coastal management initiatives practiced in France and elsewhere.

## References

Australian Government, 2009, 'Climate Change Risks to Australia's Coastline'. First National Assessment. Dept. for Climate Change.

Bradbury, A. P., 2007, Benefits of a Large-Scale, Long-Term Regional Coastal Monitoring programme'. In McInnes (Ed.), Proc. Int. Coastal Mgt. Conf. Cardiff. Thomas Telford.

Communities and Local Government (CLG), 2010, 'Planning Policy Statement 25 Supplement – Development and Coastal Change Practice Guide'. Crown Copyright.

Deboudt, P, 2010, 'Towards Coastal Risk Management in France', Oceans & Coastal Management xxx(2010) 1-13.

Defra, 2006, 'Shoreline Management Plans Guidance'. Crown Copyright.

Defra, 2010, 'Adapting to Climate Change: Developing a Policy and Framework'. Crown Copyright.

Defra and Environment Agency, 2011, 'National Flood and Coastal Erosion Risk Management Strategy for England', Crown Copyright.

Defra, 2012, 'Coastal Change Pathfinder Review'. Crown Copyright.

Environment Agency, 2013. 'Managing Flood and Coastal Erosion Risks in England'

European Commission, 2004 'EUrosion – Living with Coastal Erosion in Europe'. Luxembourg.

Lee, E. M. and Jones, D.K.C, 2014, 'Landslide Risk assessment'. ICE Publishing. 509pps.

McInnes, R.G and Barter, P. 2010. 'Adaptation to Coastal Climate Change'. Report to the Global Oceans Conference, Paris. UNESCO.

McInnes, R.G and Moore, R. 2011. 'Cliff Instability and Erosion Management in Great Britain – A Good Practice Guide'. Halcrow. 88pps.

McInnes, R.G and Moore, R. 2014. 'Living with Ground Instability and Landslides – An International Good Practice Guide'. CH2M HILL. Birmingham. 80pps.

Moore, R, Carey, J.M and McInnes, R.G, 2010a. 'Landslide Behaviour and Climate Change – Predictable Consequences for the Ventnor Underclif, Isle of Wight'. Quarterly Journal of Engineering Geology & Hydrology. 447-460.

Moore, R and Davis, G. 2014. 'Cliff Instability and Erosion Management in Great Britain'. J. Coast Conserv. DOI 10.1007/s11852-014-0359-3.

Moore, R, Rogers, J and Woodget, A. 2010b. 'Climate Change Impact on Cliff Instability and Erosion'. Flood & Coastal Risk Management Conference, Telford.

Siddle, R.P, Rowe, S and Moore, R. 2015. 'Adaptation to Coastal Cliff Instability, Erosion and Property Loss: Case Study into the Knipe Point Cliff Retreat Pathfinder Project'. Int. Coastal Mgt. Conf. Amsterdam. In Press.

UNISDR 2007. 'Words for Action – A Guide for Implementing the Hyogo Framework for Action'. UNISDR. Geneva.

United Nations ISDR, 2011. 'Global Assessment Report on Disaster Risk reduction – Revealing Risk, Redefining Development' (GAR). UNISDR Unit. Geneva.

**Coastal Management: Changing coast, changing climate, changing minds**
ISBN 978-0-7277-6114-9

ICE Publishing: All rights reserved
doi: 10.1680/cm.61149.355

# Ecosystem Based Carbon Footprinting of Marine Engineering Projects

**Jasper Fiselier**, RoyalHaskoningDHV,Amersfoort, The Netherlands
**Bart-Jan Vreman,** RoyalHaskoningDHV, Amersfoort, The Netherlands
**Sander Dekker**, van Oord Dredging and Marine Contractors, Rotterdam, The Netherlands
**Hedwig Thorborg,** Boskalis, Rotterdam, The Netherlands.

## Summary
Dredging projects use large amounts of fuel, which lead to substantial $CO_2$ emissions. So far reducing the carbon footprint of dredging projects mainly looked into the possibilities of dredging schemes and vessels that are more fuel efficient. A reduction in the order of 10 to 20% may be within reach, most of it is a win-win since fuel reduction will also reduce costs. A further reduction in $CO_2$ emission may be possible on a project by project basis but will involve a trade-off between dredging costs and $CO_2$ emissions.

Marine engineering projects also have an impact on primary production and formation of organic carbon and on sedimentation processes and burial of organic carbon in sediments. Both impacts influence carbon sequestration and can be a substantial or even overriding factor in the carbon footprint of a project. For first estimates of the potential role of these ecosystems related impacts, an attempt was made to quantify them using existing information. Information is scarce and the related processes are complex. Nevertheless, first estimations show the relevancy of an ecosystem based approach.

The carbon footprint of a marine engineering project is often larger and more complex than so far anticipated. However, a more comprehensive carbon footprint also shows that a significant reduction in carbon emissions is possible when the designs and dredging and maintenance schemes stimulate sequestration of organic carbon in the wider environment. Examples are steering the release of fine sediments by dredging and locating and creating a sand pit in a way that release and trapping of fine sediments and organic carbon is optimal. Also the use of blue carbon wetlands as part of the design, whether it is for protection purposes or for the beneficial use of dredging sludge, has the potential to reduce the carbon footprint.

## Introduction

Dredging requires enormous amounts of energy and environmental policies require a reduction in related $CO_2$ emissions. Part of the dredging sector aims to reduce $CO_2$ emissions with 20% before 2020. The dredging industry is already working with more energy-efficient vessels that emit less $CO_2$ for every cubic metre excavated. However, maritime engineering projects can be more $CO_2$-efficient by allowing more efficient construction schemes or by creating the right conditions for sequestering $CO_2$ in the ecosystem where the project is constructed. The essence of an ecosystem-based approach is that the design phase takes into consideration $CO_2$ emission and carbon sequestration processes in the coastal ecosystem.

The 'Building with Nature' (BwN) initiative includes the development of a programme that focuses on ecosystem-based $CO_2$ foot printing and design. The Building with Nature initiative is committed to the integration of infrastructure, nature and society in new or alternative forms of engineering (see also www.ecoshape.nl). Steps are currently being taken within BwN to develop a design tool that enables the calculation of the CO-footprint and to steer the design towards reducing it. To that end, a partnership has been formed between two dredging companies (Boskalis and Van Oord), NGOs (The North Sea Foundation and the Wetlands International Foundation), engineering and consulting firms (Arcadis, Royal HaskoningDHV and Witteveen+Bos) and Deltares, a Dutch institute for applied research in the field of water, subsurface and infrastructure. The goal is to be able to develop and construct maritime engineering projects that have a smaller $CO_2$ footprint over their entire lifecycle than they would have with conventional design and construction methods.

## Relevancy of an Ecosystem Based Approach

### The Carbon Content of Dredging Material Is Substantial

In discussions on the $CO_2$ emission by the dredging industry the emphasis so far is on the emission by the equipment used. In most dredging schemes the $CO_2$ emission per $m^3$ dredged may account for 2 to 5 kg $CO_2/m^3$. This is for a dredging cycle that involves transport over 10 to 20 km. During the process of dredging part of the fine sediments and attached organic carbon is released from the sediment. The total organic C content of sand used for nourishment amounts to 30 kg to 120 kg $CO_2$ equivalent. Recent research by Deltares shows that the potential $CO_2$ emissions of dredging sludge is very large (Veld, 2014). So, the potential $CO_2$ emission of one $m^3$ of sand is much larger than the $CO_2$ emission as a result of dredging it. It is therefore important to consider the fate of organic matter in dredged materials. The release of fines including attached organic matter can be steered by the dredging method, which may have a large impact on the $CO_2$ balance of a project. The carbon footprint can also be reduced by carefully selecting the location of the sand pit.

**The Carbon Sequestration Potential of Marine Ecosystems Is Large**

Intertidal wetlands have the ability to sequester carbon. There is a wide range in estimates of how much. It depends on location, stage in succession and other factors such as climate and sedimentation rates. In the long run the total built up of the sediment is the most relevant, since most intertidal areas such as a salt marsh or a mangrove forest will eventually reach a climax state and a biomass maximum. Over a period of 100 years a salt marsh may sequester as much as the equivalent of 400 to 1200 ton $CO_2$/ha, which is much higher than emissions generated by building for example a salt marsh or a mangrove forest as part of a soft defense strategy. So if costs would not be a constraint, a design that uses salt marshes or mangroves usually has a smaller $CO_2$ footprint.

**A Sand Pit Can Also Sequester Large Amounts of Organic Carbon**

Most marine engineering projects involve dredging for which sand is taken out of a pit. The sand pit will trap fine sediments and attached organic carbon if it is deep enough. Depending upon sedimentation rates, the depth and form of the pit a large proportion of the attached organic carbon is permanently trapped in sediments. The carbon sequestration by a sand pit can be significantly larger than that of a intertidal area. So it is an important component of the carbon footprint of a marine engineering project that involves dredging.

**Fine Sediments Enable Carbon Sequestration but Are in Limited Supply**

A sand pit or newly created salt marsh may trap fine sediments that hence are not available to other parts of the intertidal area. If the availability of fine sediments is limited building a salt marsh may lead to the erosion of another one, so the net carbon sequestration of building a salt marsh may be nil. So it is important to consider if this form of substitution may take place. Even if the new salt marsh is not competing with another salt marsh nearby, the fine sediments that are trapped by it, would otherwise bury organic carbon in another deposition area, such as the continental shelve or even further away. Since the availability of fine sediments is often limited, the difference in the way different parts of the coastal system are able to sequester organic carbon in a matrix of fine sediments, is relevant. The efficiency with which different depositional areas use available fine sediments for trapping organic carbon can be expressed in a C/F ratio.

**Phosphorus Is Needed for Formation and Sequestration of Organic Carbon**

Phosphorus is essential for the formation of organic carbon and often the most limiting factor in primary production. Without phosphorus there is no (organic) carbon sequestration. The fate of carbon and phosphorus varies between different deposition areas, a difference which is expressed as the C/P ratio. The C/P varies between 50 and 600 and is often higher in off shore depositional areas and higher in vegetated coastal wetlands. A higher C/P implies a more efficient use of phosphorus in carbon sequestration. So trapping carbon in wetlands can be more but also less efficient than in other parts of the coastal system. Examples exist of newly created salt marshes that trap large amounts of organic carbon due to rapid sedimentation but at low C/P ratios. A design that allows for less rapid sedimentation would lead to higher C/P ratios and potentially much more organic carbon sequestration on the scale of the coastal system.

The aforementioned example shows several ecosystem related impacts that are relevant to the carbon footprint of a marine engineering project. In the example a salt marsh was used, but similar kinds of processes take place in case other types of intertidal areas (e.g. mangrove forests, mud flats) would be created or impacted by a marine engineering project. There are more processes that influence primary production (other nutrients such as nitrogen, temperature, light) and carbon sequestration (e.g. soil forming processes, presence of sulphur and iron). Most of these processes result in specific C/P en C/F ratios that are not influenced by the project (e.g. temperature, light, or presence of iron and sulphur in depositional areas) or are less limiting to primary production, such as nitrogen.

## The Position of the Project in the Coastal Ecosystem

Primary production (e.g. the formation of organic carbon), and sedimentation (e.g. the burial of organic carbon), is key to carbon sequestration. For primary production, the amount of biologically available key nutrients such as phosphorus is essential. Phosphorus is in many environments a limiting factor to primary production and it is critical to assess the impact of a project on the availability of phosphorus.

Similarly, without fine sediments it would not be possible to sequester organic carbon for longer periods in sediments. Sand by itself can contain and sequester organic carbon, but only in limited quantities. A precondition for burial of organic carbon is anoxic conditions that prevent its further decay, a condition that is met in case of a matrix of fine sediments. Fine sediments are often available in limited quantities. The exception is river deltas where large amounts of fine sediments reach the sea. It is therefore important to assess the impact of a project on the availability of fine sediments.

Fine sediments may find their final resting place on the continental shelf, if they are not withheld by marine wetlands. They may even end up in deeper parts of the ocean. Its final destination depends on the characteristics of the coastal system. A project that involves the construction of a salt marsh will sequester organic carbon in this marsh as a consequence of the interaction of primary production (e.g. algae, benthic algae, plant tissue) and sedimentation of fine sediments. It may also sequester fine sediments in organic carbon in the sand pit. However, the fine sediments would without this project contribute to the sequestration of organic carbon elsewhere, perhaps in another salt marsh nearby or on the continental shelf father away. It therefore important to scout for potential substitution processes that involve fine sediments.

### Important Components in the $CO_2$ Footprint of a Marine Engineering Project

A comprehensive carbon footprint of a marine engineering project needs to look at:

- Dredging: The $CO_2$ emission per $m^3$ of sediment is determined by the size of vessel, the way sediment is dredged and placed into its new position and many other factors. The $CO_2$ emission related to the construction of dredging equipment is negligible compared to the fuel consumption. There are several options for optimizing the carbon footprint on this level, with roles and responsibilities for both the contracting authority and the contractor.
- Construction: the $CO_2$ emission by constructing and construction materials. Most marine projects also include the use of harbor dams, groynes etc., for which quarried stone and concrete is used. These building materials represent vast amounts of $CO_2$ emission. The carbon footprint of a project, such as the strengthening of a dike can be greatly reduced if stone and concrete revetments can be reused, which often can be made possible by the design.
- Abstraction area/sand pit: the organic carbon sequestration related to the release and trapping of fine sediments and related organic carbon and phosphorus. The fines released and trapped will lead to changes further down drift in the coastal system, furthering or reducing the formation of intertidal area, the sedimentation rates in depositional areas etc.
- Construction area/project area: often the original sea bed is replaced by a (wider) beach, dunes or land reclamation area. The project may lead to changes in coastal processes, such as long shore sediment transport that may affect and alter down drift areas and change their carbon sequestration capacity.

## Handling the Complexity

Carbon is an important building block of life and is present in all organic life forms. It is present not only in tissue of plants, algae and marine organisms but also in the shells of marine organisms. Carbon is present in different forms, and also the sequestration of carbon has different forms. The sea is an open system. This implies that sand, fine sediments, organic matter and nutrients move freely from one area to another. This movement is restricted for sand, but fine sediments and organic matter, especially in the form of algae, and also (dissolved) nutrients travel considerable distances. This implies that there are regional cause and effect chains that may need to be considered. Carbon cycles are strongly related to primary production and are therefore complex. There are however options to simplify this complexity:

### Short and Long Cycles: Focus on Long Cyclic Sequestration

A distinction can be made between short cyclic forms of carbon sequestration in plants, animals and the humus of soils, and long cyclic forms of sequestration that, in coastal systems, are mainly related to sedimentation. Short cyclic forms show large variations in time and place and are difficult to monitor, to measure and to extrapolate to larger areas and time scales. For instance, the carbon uptake of a salt marsh depends on tides, seasons, salinity, nutrient input, vegetation type and more variables. Most of the variation takes place within a season or a tide, so in essence a short period of time. With time vegetation will reach a climax in which it is fully established and can no longer sequester more organic carbon in organic tissue. Further sequestration by fully grown marine wetlands depends on sedimentation and soil formation. The long term sequestration capacity of, for example, a salt marsh, consists mainly of the organic carbon sequestered in its biomass in its climax state and the carbon sequestered in its sediments over a specific period. Both can be measured and calculated largely independent of the short-cyclic variation.

**For Substitution Focus on Sand, Fines and Phosphorus**

There are many processes that govern primary production and carbon sequestration by sediments. It is proposed to focus on the following processes for regional assessments:

- Sand: sand contains only little organic carbon, but sand is in many coastal systems an essential building block. Most marine engineering projects that take place at sandy shores influence sand budgets and longshore processes and the subsequent erosion and sedimentation may affect coastal wetlands, dunes and beaches as well as the nourishment needs of a project.
- Fines: as argued, fine sediments are needed for the long-cyclic sequestration of organic carbon in soils. Fines are abundant in river deltas but overall fines are only available in limited supply. Fines are also an essential building block for coastal wetlands, such as salt marshes, mudflats and mangroves. Projects change the availability of fines, their distribution and sedimentation.
- Phosphorus: of the nutrients that govern primary production phosphorus is the most limiting. Projects that directly or indirectly alter the availability of phosphorus also alter primary production and therefore carbon sequestration.

The C/P ratio in sediments and soils is influenced by several processes. Over time the C/P ratio increases because phosphorus that is bound to organic matter is released from the soil more effectively than C. The release of phosphorus strongly depends also on the presence of sulphur and iron and is controlled by oxygen and microbiological activity. The organic C/P ratio differs depending on the sedimentation rate, being lower (near 200) in the case of fast sedimentation (>1 cm/year) and at very low sedimentation rates (<0,002 cm/year). Intermediate sedimentation rates can have C/P ratios that are higher up to 600 (see also Ingall E.D. and P. van Cappellen, 1990). Very low sedimentation rates occur in deep sea environments. Very high sedimentation rates can be expected on the coastal shelf, near delta's and in estuaries, but also in sand pits. Frequent resuspension of bottom sediments lowers C/P ratios as well.

**Simplify by Reasoning**

The carbon footprint of a project is complex, but it is possible that the most complex parts are not relevant for a particular case. For example, when comparing two nourishment strategies, annual and periodical, one may argue that mainly the $m^3$

efficiency of the nourishment scheme and the $CO_2/m^3$ efficiency of the equipment used are relevant if the same shallow sand pits are used in both schemes.

Regarding substitution there are basically three different situations:

1. In coastal systems with potentially full replacement and substitution, the emission caused by dredging and constructing will be the paramount factor in the carbon footprint and in comparisons between alternative designs and dredging schemes. This is the simplest level of assessment.
2. In coastal systems with little risk of replacement and substitution, the dredging and construction schemes can be supplemented by the sequestration characteristics of sand pits and marine wetlands, if being part of the project.
3. More complicated are those coastal systems where only partial substitution is expected. Here all processes need to be taken into account and the carbon footprint, will have a considerable bandwidth since especially offsite substitution is very difficult to assess and can often only be done qualitatively.

Furthermore, one may need to focus on those changes in the coastal system that endures over longer periods. So the emphasis should be on expected long-term equilibrium states of the coastal system.

## Initial assessments of an ecosystem based carbon footprint

### Proposed initial calculation scheme

The BwN programme aims to develop a tool that allows the calculation of a more comprehensive carbon footprint that helps to identify the potential to reduce the carbon footprint by altering design, dredging schemes, pit location etc.

Figure 1 shows a simplified scheme for a carbon footprint for marine engineering projects. It distinguishes the direct $CO_2$ emission due to dredging and construction, including the emission related to the production of construction materials. Furthermore, a distinction is made between the abstraction site/pit site and the construction/project site. An important element is the substitution that may take place by following the fate of and impact on sand, fines and phosphorus.

Focus is on long cyclic sequestration and on the role of fines and sedimentation processes. The far field substitution that takes place related to fines and phosphorus is the most complex. So far C/P, C/F and other variables were collected from literature in order to have a first impression of the sequestration capacity and efficiency of different types of depositional areas. Data are scarce and almost none existent for sand pits. The plan is to collect more field data to fill in the gaps. Another information gap is the proportion of the fines that is trapped by depositional areas.

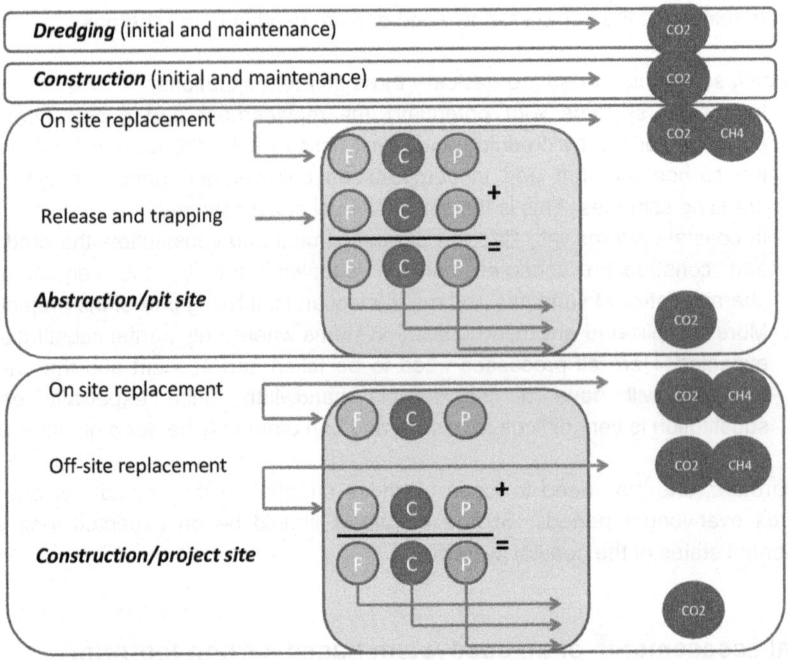


Figure 1: Calculation scheme

**First Assessments**

This scheme was used for a first set of calculation sheets. Using these calculation sheets different cases have been considered, such as the Sand Engine on the coast of Delfland and a new BwN pilot, the mud engine that is meant to stimulate salt marsh development locally in the Wadden Sea using dredged material.

The conceptualization of a project over a 100 year period is an important and daunting step that requires extrapolating trends and also coastal policies. The issue of substitution is strongly governed by long-term and regional assumptions regarding the sand and fine sediment balances and of the availability of phosphorus. The complexity and uncertainty on the long term is very large, so it is proposed to consider intermediate stages for assessment.

The example of the sand engine shows that a design that enables the use of larger dredging equipment (due to sheer volume) and comparatively simple dredging schemes (e.g. with substantial bottom release) may lead to a reduction in the carbon footprint of coastal maintenance projects. It also shows the potentially important role of the sand pit in the overall carbon footprint. Since its location and form influence the release and trapping of fines and related organic carbon and phosphorus, the sand pit merits more attention.

More understanding is needed at the sediment-water interface that determines C/F and C/P ratios in different kinds of depositional areas and marine habitats. For this sediment sampling and analysis is necessary with an emphasis on long-cyclic sequestration processes, so also of deeper sediment layers. Of great interest are the sedimentation characteristics of different forms of sand pits. Also different forms of organic matter need to be distinguished that have different decay rates.

Also the long-term hydrodynamics of various potential carbon sequestration sites is critical in order to define the permanency of long-cyclic carbon sequestration. The carbon sequestration of many years can be lost during an occasional fierce storm, not only in the form of salt marsh erosion but also by resuspension of settled fine sediments out of sand pits and from depositional areas on the continental shelf.

## Next Steps

Based on these first results additional steps are taken by the programme in order to fill in relevant information gaps, validate the calculation scheme and to refine it. Furthermore, specific formats will be developed for different types of projects in different types of coastal environments. More information is gathered for cases that appear to offer opportunities to positively alter the carbon footprint.

## References

Arzayus K.M. and E.A.Canuel,2004. Organic matter degradation in sediments of the York River estuary: Effects of biological versus physical mixing. Geochim.Cosomochim.Acta 69: 445-463.

Dauwe B, P. M. J. Herman, C. H. R. Heip (1998). Community structure and bioturbation potential of macrofauna at four North Sea stations with contrasting food supply. Marine Ecology Volume 173.

Dekker, S, J.van der Klooster, J.Fiselier and H.Thorborg (2014). Pressing the CO2 buttons: towards ecosystem based CO2 footprinting for maritime engineering projects. In: Terra and Aqua.

Emeis, K.C, U. Struck, T. Leipe, F. Pollehne, H. Kunzendorf, C. Christiansen (2000). Changes in the C, N, P burial rates in some Baltic Sea sediments over the last 150 years—relevance to P regeneration rates and the phosphorus cycle. Marine Geology 167 (2000) 43–59

Ingall E.D. and P. van Cappellen (1990): Relation between sedimentation rate and burial of organic phosphorus and organic carbon in marine sediments

Geochimica et Cosmochimica Acta, vol. 54, Issue 2, pp.373-386.

Jansen D.L., D. P. Lundqvist, C. Christiansen, L.C. Lund-Hansen, T. Balstrøm, T. Leipe (2003). Deposition of organic matter and particulate nitrogen and phosphorus at the North Sea – Baltic Sea transition– a GIS study. OCEANOLOGIA, 45 (1), 2003. pp. 283–303.

Laffoley D. and G.Grimsacter (2009): The management of Natural Coastal Carbon Sinks. IUCN report.

Mills M, T. Elsey-Quirk, D.J. Velinsky, D. Kreeger, Angela Padeletti, Martha Maxwell-Doyle (2002): Above- and Belowground Biomass Densities at Salt Marsh Sites in Delaware and Barnegat Bays, New Jersey. (https://s3.amazonaws.com/delawareestuary/sites/default/files/P10 Mills.pdf)

Pidgeon E. (2012): Carbon Sequestration by Coastal Marine Ecosystems: Important Missing Sinks and Sources, Ph.D. Conservation International. Presentation at Blue Carbon Conference in Brussels, 2012.

Volkman J.K., D. Rohjans, J. Rullkötter, B.M. Scholz-Böttcher, G. Liebezeit (2000): Sources and diagenesis of organic matter in tidal flat sediments from the German Wadden Sea. *Continental Shelf Research, Volume 20, Issues 10–11, 1 July 2000, Pages 1139-1158.*

Veld H. (2014): De CO2-belasting van baggerwerken Prognose van emissie op basis van de karakterisering van organisch materiaal. Deltares report.1209965-000.

Zee C. van der, C.P.Slomp, W. van Raaphorst (2002):Authigenic P formation and reactive P burial in sediments of the Nazare Canyon on the Iberian margin (NE Atlantic). Marine Geology 185 (2002) pp. 379-392.

**Coastal Management: Changing coast, changing climate, changing minds**
ISBN 978-0-7277-6114-9

ICE Publishing: All rights reserved
doi: 10.1680/cm.61149.365

# The Wallasea Island Wild Coast Project: Working with Nature and People at a Landscape-scale

**Colin Scott**, ABP Marine Environmental Research (ABPmer), Southampton, UK

## Introduction

The Wallasea Island Wild Coast Project is the RSPB's 'flagship' marine wetland restoration initiative and the most ambitious coastal environmental project in Europe. It is taking place on Wallasea Island at the confluence of the Crouch and Roach Estuaries near to Rochford in Essex (see Figure 1) and will involve a 'Managed Realignment' of the island's coastal defences.

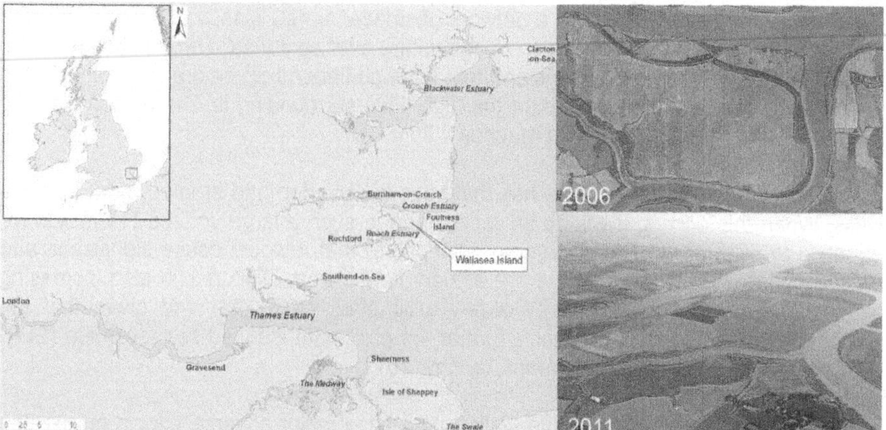


Figure 1: Location of Wallasea Island and pre-construction aerial photographs

Managed Realignment is pursued, where appropriate, along flood-sensitive coastlines where there are often deteriorating and vulnerable seawalls. It typically involves the construction of new, higher quality, seawalls setback from the old defences. These old defences are then breached to allow controlled tidal flooding of the land between the two walls. This creates new shoreline habitat and substantially improves the levels of coastal flood protection to hinterland properties.

The practice of managed realignment is now very familiar and 117 such projects have now been completed across Northern Europe (63 in the UK alone) (source www.omreg.net). These previous projects are often much smaller than Wallasea.

However, in recent years, relatively large projects have been undertaken including the 300ha and 400ha realignments at Medmerry and Steart (as completed in 2013 and 2014 respectively). One comparatively large realignment ('Allfleet's Marsh') has also already been undertaken on the north bank of Wallasea Island by the UK's Department for Environment Food & Rural Affairs (Defra). This project was completed in 2006 (see Figures 1 and 2) and created 118ha of marine habitat (85ha mud and 33ha marsh).

This past experience of realignment (at international, national and local levels) was used to scientifically underpin the Wallasea Wild Coast Project and inform the scheme's design and the methods proposed. When completed this site will cover 668ha and will be managed alongside the adjacent Allfleet's Marsh. Therefore, the full extent of restored marshes and muds will be 786ha. This equates to over 1300 football pitches.

This is coastal habitat restoration at a scale that has not been seen before in Europe. It involves the first use of terrestrial sediments to raise the ground levels of a large island and reshape its full landscape. This will return the island back to the condition it was in prior to being claimed from the sea several hundred years ago (see following section).

## Reasons for the Project

Historically, Wallasea Island was a network of smaller marsh islands and was subject to human encroachment for conversion to agriculture over centuries. Through this process, it has become a single enwalled island that is much flatter than before and has sunk to an elevation that is much lower than the high tides surrounding it. The land is typically 1m ODN while a mean spring tide reaches 2.85m ODN.

In this condition there is a severe risk that tides will overtop and break through the sea walls. When this occurs, it will do so rapidly and in such a large volume (11 million $m^3$ equating to 67% of the Roach Estuary tidal prism) that it would cause significant, and potentially irreversible, changes to the estuary flow regime. This risk is also increasing as sea levels continue to rise by a few millimetres each year and given that the Government's advice is to plan for a further 1m of climate-induced sea level rise (SLR) over the next 100 years (Environment Agency, 2011).

Such an unmanaged breach of the walls would have major adverse effects on existing erosion/siltation patterns and flood protection levels in other areas. It would also affect the water quality and oyster fisheries in the estuary and have effects on recreational and commercial activities. For these reasons, managed realignment of the island is the strategic long-term flood management policy (Environment Agency 2010).

In addition to addressing the flood risk threat, this project will provide major biodiversity gains and enhance fisheries. It will also offset losses of marshes and mudflats that will occur in front of maintained sea defences in other areas of the region due to sea level rise. For this reason, the Environment Agency has made a substantial funding contribution to the project to deliver 155ha of compensatory habitat for such intertidal habitat losses. This project is therefore fundamental to the long-term sustainable management of coastal defences in the region and has major benefits for the protection of coastal properties elsewhere along the coast as well as at Wallasea.

The site will also provide many public amenity benefits. Providing opportunities for public participation is a major element of the RSPB's management approach to this project. Therefore, involving the local community and understanding their concerns was essential for the planning application and in developing project design (which includes extensive public access areas and regular public engagement events on site).

## Designing the Project

Although managed realignment was the defined management strategy for the island, simply altering the position of the seawall and breaching the existing defences would not solve the problems facing this site. This is because a tidal island in its current form would still result in large volumes of tidal water being exchanged to the Roach Estuary.

Therefore an innovative solution was needed. Initially, the idea of introducing structures to create a large-scale Regulated Tidal Exchange (RTE) area (also called Controlled Reduced Tide (CRT)) was considered (based on smaller projects such as Lippenbroek in the Scheldt Estuary (Temmerman et al 2013)). However, a more sustainable long-term solution of restoring the island's landform was made possible by the availability of large volumes of excavated materials from separate National Infrastructure Projects.

The new landscape design uses 7 million $m^3$ of such materials to raise the island's ground level by up to 4m. The landscape was also sensitively re-shaped to reintegrate it hydrodynamically with the Roach Estuary. The new shape (see Figure 2) includes four separate intertidal areas (Called Cells 1 to 4) to allow the project to be implemented in a sequence of controlled, iterative and monitored stages. These areas are divided by seawalls (5m ODN in height) that have large expanses of high-level ground on either side (called an 'SLR Adaptation Zone') that are above the highest astronomical tide levels and into which the tide can move as sea levels rise in the future.

Three of these areas will be managed realignments (Cells 1, 2 and 4) with six separate breaches providing open tidal exchange to the estuary. The breaches have been placed to dissipate energies sensitively through the Roach. The island has also been shaped to ensure that the highest flows are directed through the breaches in the outer Roach to minimise the spatial extent of the minor hydrodynamic changes arising.

Cell 3 at the centre of the site is a Regulated Tidal Exchange area (connected to Cells 1 and 2) that will become a lagoon habitat and flood accommodation zone. A fifth zone (Cell 5) on the landward side of the final defences includes freshwater and terrestrial habitats (see Figure 2).

The new landscape change will reduce the future volumes of tidal exchange with the estuary to a maximum of just over 2 million $m^3$. Once breached, however, the site will accrete with sediments and the tidal volumes will decrease over time. This accretion is expected to occur at a rate which outpaces sea level rise and is expected to be around 3cm/year during the first few years (then reducing over time as bed levels rise).

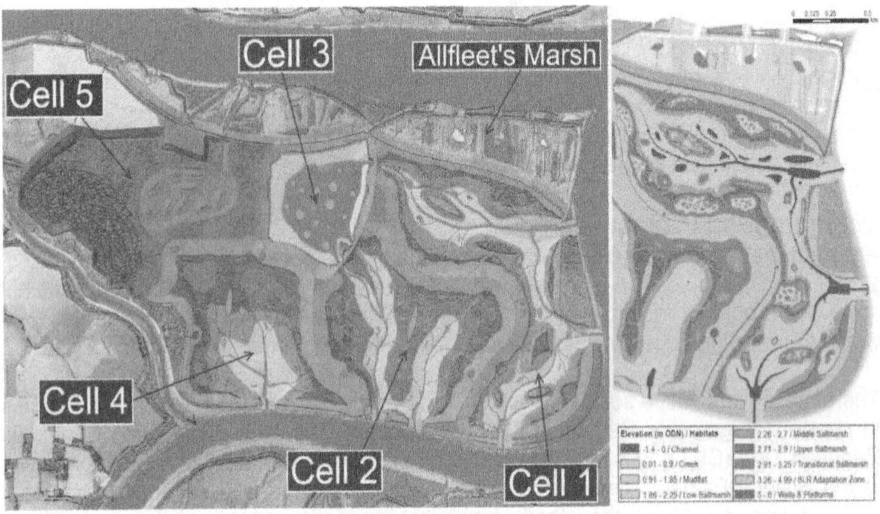


Figure 2: Design of the Wallasea Wild Coast Project with detail of habitats in Cell 1

The design includes shallow profiles and dendritic creeks to ensure that tidal water spreads across the site as fully as possible and in a manner that dissipates the tidal energies as evenly as possible. The functionality of this design and the changes that will occur in the estuary were verified by a series numerical modelling analysis (ABPmer 2008) as well as by drawing upon past project examples.

Among many aspects, this modelling indicates that flow speed increases of 0.2m/s (or 0.2 knots) will occur across outer sections of the Roach from the extra volume of water being drawn into the realignment site (maximum flows are currently around 1.5m/s or 2.9 knots). These increases occur as transient events around the time of peak flood and ebb flows and provide assurances that the scheme will not result in major changes to flow conditions in the wider estuary.

As part of the design, particular consideration has been given to avoiding significant erosion at the breaches where the main flows will be experienced. Empirical models were used to define their appropriate 'regime-shape' based on the pattern of flows that will occur throughout a tidal cycle (as dictated by the morphology of new intertidal areas that are being created) (Townend et al., 2010). As part of the design work there was also a careful refinement of the scheme to avoid impacts to key oyster cultivation areas.

## Collaboration with Crossrail

This unique approach of raising and reshaping a large island has only been possible through partnership working. In particular, it is being realised by beneficially recovering waste soils from separate infrastructure projects. To deliver the first phases of the project, the RSPB has worked jointly with the separately-consented Crossrail project.

Crossrail is a new railway in London which links Reading and Heathrow Airport in the west with Shenfield and Abbey Wood in the east. It is Europe's biggest infrastructure project and includes 26 miles of new tunnels and 10 new stations. The Crossrail tunnelling work will provide some 1.65 million $m^3$ of the landscaping materials that are needed for the Wallasea project to be realised.

This collaboration was encouraged by the Centre for Environment, Fisheries and Aquaculture Science (CEFAS) the UK department who advise on licences for the use of sediments in the marine environment. CEFAS were able to see very early on in the planning and consenting process that this beneficial link could occur.

By using the tunnelling material, as well as other sediments sourced on Wallasea Island itself, the Crossrail team will be completing the whole of the Cell 1 realignment (176ha). They will also be undertaking the bulk of the land excavation and sluice construction work for Cells 3 (67ha) and 5 (153ha) (see Figure 2). To complete Cells 2 and 4 sources of materials from other key infrastructure projects are being explored by RSPB.

## Delivering the Materials

The Crossrail tunnelling materials were transported from London by sea to the Crouch Estuary using vessels and landing infrastructure provided by Crossrail. This work was undertaken by Bam Nuttall Ltd. and Van Oord UK (a joint venture working for Crossrail and with the RSPB and their advisors). This project team were responsible for managing the full sediment handling process from the vessel loading work at two London locations (the Docklands Transfer Station at Barking Riverside and the berth at Northfleet) to the unloading and landscaping work at Wallasea.

From the excavation locations within central London, the materials were delivered by rail to Northfleet and by tipper lorry to Barking. At each of these two quaysides, the materials were then transferred into 1300$m^3$ capacity vessels for transfer on to Wallasea. A maximum of six vessels (more typically five) were then in transit between the two London berths and Wallasea at any given time and a maximum of four vessels (approx. 5,000$m^3$) were unloaded at Wallasea each day.

The process of delivering these materials, and then also removing them for re-use at Wallasea, was strongly influenced by the composition and cohesiveness of the sediments that were available. For a large project like Crossrail this sediment composition varied depending upon the location and nature of the excavation work.

Materials delivered to Barking were from the bulk excavation of stations, shafts and portals or from the sprayed concrete lining (SCL) mined tunnels and caverns (Figure 3). These materials were generally dry and friable though they would contain some large pieces of concrete that needed processing through a screener and then crushing. The materials from Northfleet were predominantly from the western tunnelled section and were excavated using a tunnel boring machine (TBM). Here quality-approved polymers and water were used to condition the material and assist with the tunnelling.

The result of these different locations and methods was that, in general, comparatively dry sediment was delivered to Barking while the material reaching Northfleet tended to have a higher clay component and was more cohesive. The former was relatively easy to handle while the latter required more time and effort to transfer. This variability in the sediment composition was also influenced by the amount of rainfall experienced during the transfer and landscaping processes.

At the Wallasea unloading site, the preferred option for transferring from the vessels involved use of a conveyor which allowed material extracted from the vessels by excavators to then be moved 800m into the site for redistribution. The use of conveyors was preferred here to minimise the noise generated at the site and minimise the disturbance to birds using the adjacent environment (mainly the Allfleet's Marsh managed realignment that was created by Defra on the north bank of the island).

Conveyors are typically used to transfer drier granulated sediments and so it did not work as efficiently for the adhesive clays from Northfleet. To address this, the project team modified the conveyor and also mixed the materials prior to shipping. They also set up a process whereby the loads were taken in 50:50 ratios from both Barking (1,000 tonnes dry) and Northfleet (1,000 tonnes wet clay).

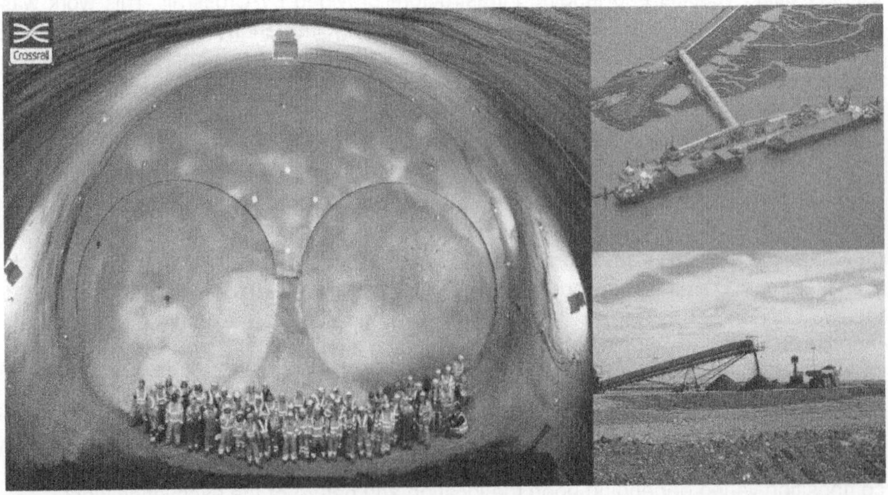


Figure 3: Crossrail SCL-mined tunnel and material unloading (via conveyor) at Wallasea

To optimise efficiency in the transfer process, the vessels operated on fixed schedules and were not allowed at berth for more than an allotted 12 hours working window (even if on occasions that meant a residual proportion of the load was not removed). This method of running ships allowed for five ships to operate around the clock seven days per week and deliver between 25-35,000 $m^3$ of landscaping materials each week (with inherent variation due to weather conditions).

Another element of the optimisation process was to ensure vessels reached the Barking site first where there was a 4-6hr high-water tidal access window. Northfleet was visited second (not tidally constrained) before the ships returned (9-10 hour transit) to Wallasea. Following this optimisation process the rate of delivery increased from 1 million tones (approx. 625,000 $m^3$) within the first 62 weeks to the same volume in almost half the time (32 weeks) thereafter. In total the delivery work will have taken just under 3 years from August 2012 to July 2015 (when breaching of Cell 1 is scheduled).

This work was undertaken in compliance with appropriate legal and planning conditions which included quality testing of the sediment to ensure they met approved criteria and measures to mitigate and monitor other possible effects on site. These measures included surveys of noise levels which indicated that they remained below pre-determined thresholds (and with no complaints arising from the local community) (ABPmer 2014). Also a shipping management plan was implemented to mitigate for any effects on recreational navigation. This identified critical sailing race times and established procedures for limiting ship movements at these periods as well as setting out a framework for good communication between all parties using the estuary.

## Creating the New Landform

Once the material had been delivered to the site by sea it was transferred through the conveyor to the material handling area on a '24 hours/day 7 days/week' basis. Then, during standard construction working hours, the material was loaded into dump truck and delivered across the site to create the new landform. This work involved 20 staff and drivers using 32 pieces of plant including 20 dump trucks (capacity around 16$m^3$), six excavators, 5 dozers and a grader.

To achieve the landscaping work the dozers and grader were GPS-guided and linked to a digital ground model that was prepared by AECOM from the original design concepts created by RSPB and ABPmer. This meant that once the bucket had been placed on the ground, a screen in the cabin provided guidance in the required landscape elevations (i.e. whether bed raising or lowering was required).

This approach allowed the drivers to have a full understanding of the requirements on site and meant that the whole project could be run very efficiently with two supervising engineers and without the need for on-site banksmen. This was important for safety reasons given that there were 20 dump trucks operating at any one time across the site. Using GPS-guided plant was also valuable for shaping those parts of the new habitats that were particularly complex morphologically. This included particularly the network of ditches that were need within Cell 5 to create new mitigation habitat for water voles.

## Protecting Species on Site

As part of the project, work has been undertaken to ensure that there is no adverse effect on any legally protected or vulnerable species using the island. The relevant species that are present include adder, common lizard, water voles, corn bunting, badger and eel. These were distributed across the site and especially alongside the 5 mile field drain network within the site.

Mitigation habitats for these species were created within Cell 5 of the site. This included a 4ha water vole area with 10 miles of new ditch, a 20ha reptile site with 20 hibernacula, a new badger sett and an area planted with seed-bearing crops for corn bunting. A range of measures were undertaken to trap and translocate the reptiles and water voles to the new mitigation habitat 'receptor sites' or to encourage them to move of their own accord through habitat displacement (e.g. by drawing down water from the ditches).

In total over 8,200 common lizards, 50 adders, 200 water voles and a number of badger setts were relocated. Tagging and radio tracking exercises were also undertaken to learn about the effectiveness of these methods. Initial post-mitigation monitoring surveys have indicated that these measures were successful and have shown, for instance, that the water vole population on the island has increased by 75% (CH2MHill 2014) and that badgers are using the artificial sett (Bam Nuttall/Van Oord 2014).

## Costs and Benefits

The overall cost of this project to date has been quoted at £100 million (Reuters: Doyle and McNeill 2014) of which the protected species protection cost £4 million. These are large fees which reflect the technical complexity of the project but it is also important to recognise the multiple long-term benefits that this work brings (see Figure 4). In particular it delivers major 'win win' solutions for nature and people because it:

- avoids flooding through a future unmanaged breach of the island's sea defences;
- creates coastal habitats with benefits for ecology and social amenity;
- provides increased public access to the coast;
- promotes sedimentation to address sea level rise and trap carbon and contaminants;
- provides a beneficial reuse alternative to landfill disposal of the source materials;
- facilitates the sustainable transport of material by sea rather than by road; and
- maintains sustainable populations of important terrestrial species; and
- reduces the costs to society and Government for coastal protection.

## Conclusions

Prior to the Wild Coast Project, Wallasea Island, its adjacent estuaries and properties faced an impending risk of significant damage from a flood surge event. However, this was a site where just allowing the coast to adapt naturally was not possible and where improving the existing sea walls would be excessively costly and unsustainable in the long-term. It would also represent a missed opportunity to alleviate a flood risk, restore lost habitats, beneficially use sediment and create an area of value to the community.

This project therefore illustrates how it is possible, and necessary, to develop innovative and bespoke solutions to address local/regional flood management challenges and deliver multiple environmental, social and economic benefits. As it is the first time ever that such a landscape-scale reshaping of an island has taken place, a number of new technical and logistical challenges have been encountered. The lessons learned have enhanced scientific understanding about marine wetland creation and especially about the suitability of different terrestrial materials for marine habitat restoration.

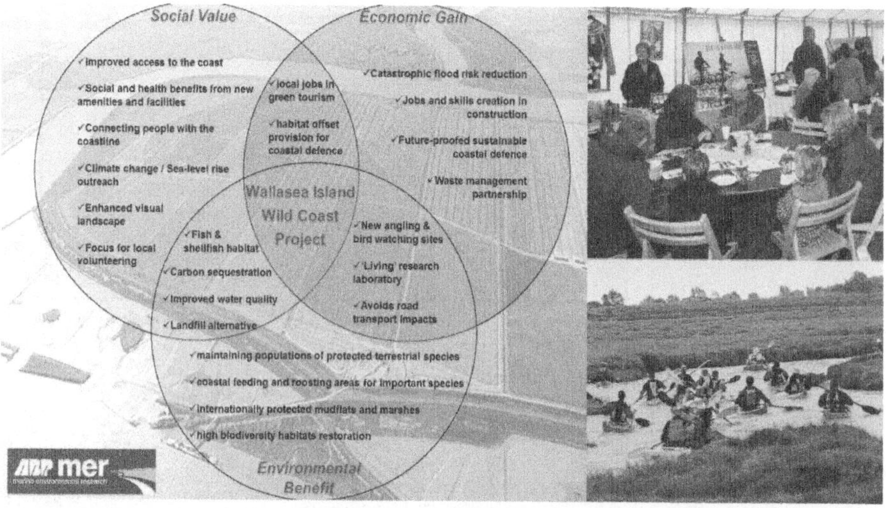


Figure 4: Multiple benefits from the project with photographs of community events

It is hoped that these lessons will now help inform further projects (including the work needed to deliver Cells 2 and 4 of the Wallasea Project) and lead to greater collaborations between major infrastructure projects to achieve multiple and mutual gains. It is also hoped that this project improves public understanding about the coastal environment and inspire decision makers to plan for future sea level rise. To describe the scale of the project, Figure 5 illustrates the work done on island up to August 2014.

## References

ABPmer (2008). Wallasea Island Wild Coast Project – creating a new coastal nature reserve. Environmental Statement Report for RSPB. Report R.1474.

ABPmer (2014). Wallasea Island Wild Coast Project: Environmental Impact Assessment for a Section 73 Application 31 December 2025 Report for RSPB Report R.2202

Bam Nuttall/Van Oord (2014) Submission for Green Apple Award commerce & Industry

Doyle A. and McNeill R (2014) Special Report: Flirting with retreat along Britain's battered shores

Environment Agency (2011). Adapting to Climate Change: Advice for Flood and Coastal Erosion Risk Management Authorities.

Environment Agency (2010) Essex and South Suffolk Shoreline Management Plan 2. Oct 2010

Temmerman, S., Meire, P., Bouma, T. J., Herman, P. M. J Ysebaert, T and De Vriend, H. J. (2013) Ecosystem-based coastal defence in the face of global change Nature 504, 79–83 (December 2013)

Townend I.H., Scott C.R. & Dixon M., (2010). Managed Realignment Chapter within Flood Risk, Science and Management Handbook. Pender G. Faulkner H Eds. Wiley-Blackwell SBN 9781405186575 Published 17 December 2010

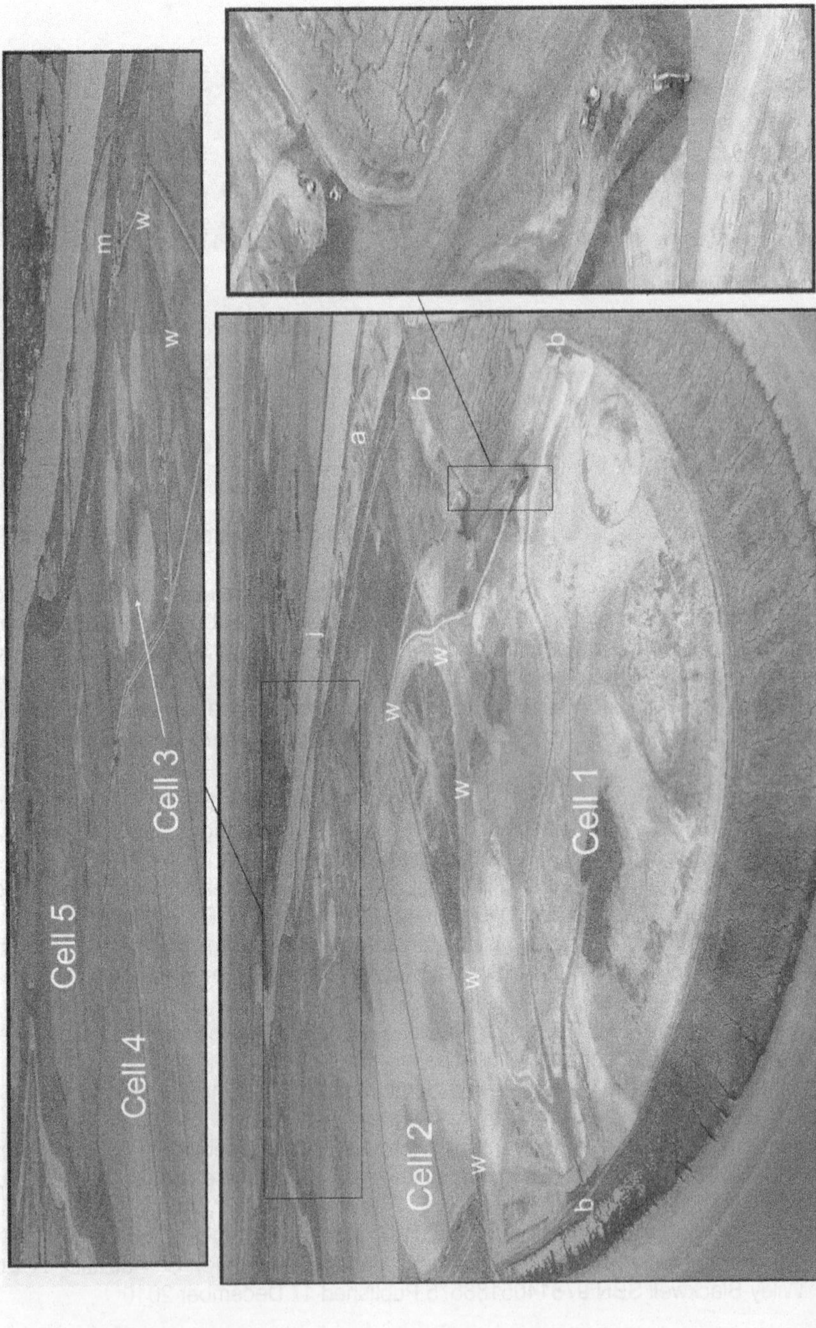


Figure 5: Wild Coast Project under development in August 2014 (showing plant on site and locations of: Cells 1 to 5, Cell 1 breach locations (b); new Cell 1 sea wall (w); unloading jetty (j) and material handling area (m) Allfleet's Marsh (a))

**Coastal Management: Changing coast, changing climate, changing minds**
ISBN 978-0-7277-6114-9

ICE Publishing: All rights reserved
doi: 10.1680/cm.61149.375

# Adapting to Climate Change on the Suffolk Coastline through the Minsmere Flood Risk Management Scheme

**Andrew E. Rouse**, Environment Agency, Peterborough, UK
**Victoria Stockwell,** Environment Agency, Peterborough, UK
**Stuart Barbrook**, Environment Agency, Peterborough, UK
**Alexandra Schofield,** ENBE, Crowland, UK, formerly Black & Veatch, Redhill, UK
**Comer Mead**, Royal HaskoningDHV, Peterborough, UK
**Frank Leonforte**, Interserve Construction Ltd, Uxbridge, UK

## Introduction

Minsmere is located on the Suffolk coast in the south east of England (see Figure 1). The low-lying historic valley of the Minsmere River is today blocked off from the sea by coastal defences with a tidal sluice providing gravity drainage of the hinterland to the sea. Should the defences fail then 926ha of predominantly rural land will potentially be inundated by the sea during an extreme coastal surge event along with 12 properties and the RSPB reserve's new visitor centre.

Much of the study area comprises of sites of international nature conservation significance, designated a Special Protection Area (SPA), Special Area of Conservation (SAC), and Ramsar wetland (collectively referred to as Natura 2000). Under international law, the UK Government is expected to conserve the overall integrity of the Natura 2000 complex. National designations are also present, including Site of Special Scientific Interest (SSSI) and Area of Outstanding Natural beauty (AONB) status. Much of the site is the RSPB's flagship Minsmere reserve, attracting over 90,000 visitors and thereby helping to support the local economy.

The existing coastal defences are under pressure from the sea as a result of natural processes, which is expected to increase due to climate change. The project investigated how best to manage the increase in flood (coastal and fluvial) and erosion risk, given the statutory conservation requirements and the regional significance of the nature reserve, resulting in the implementation of the Minsmere Flood Risk Management scheme in 2011/12. The Environment Agency procured technical, construction and contract project management services through their National Frameworks, embracing the overarching principle of working in partnership. The project roles were as follows:

- Designer (including environmental services) – Black & Veatch Ltd. (B&V)
- Early contactor involvement – Team Van Oord (TVO)
- Construction contractor - Interserve Construction Ltd. (ICL)
- ECC Project Manager - Royal HaskoningDHV
- CDM Coordinator - Callsafe Services Limited.

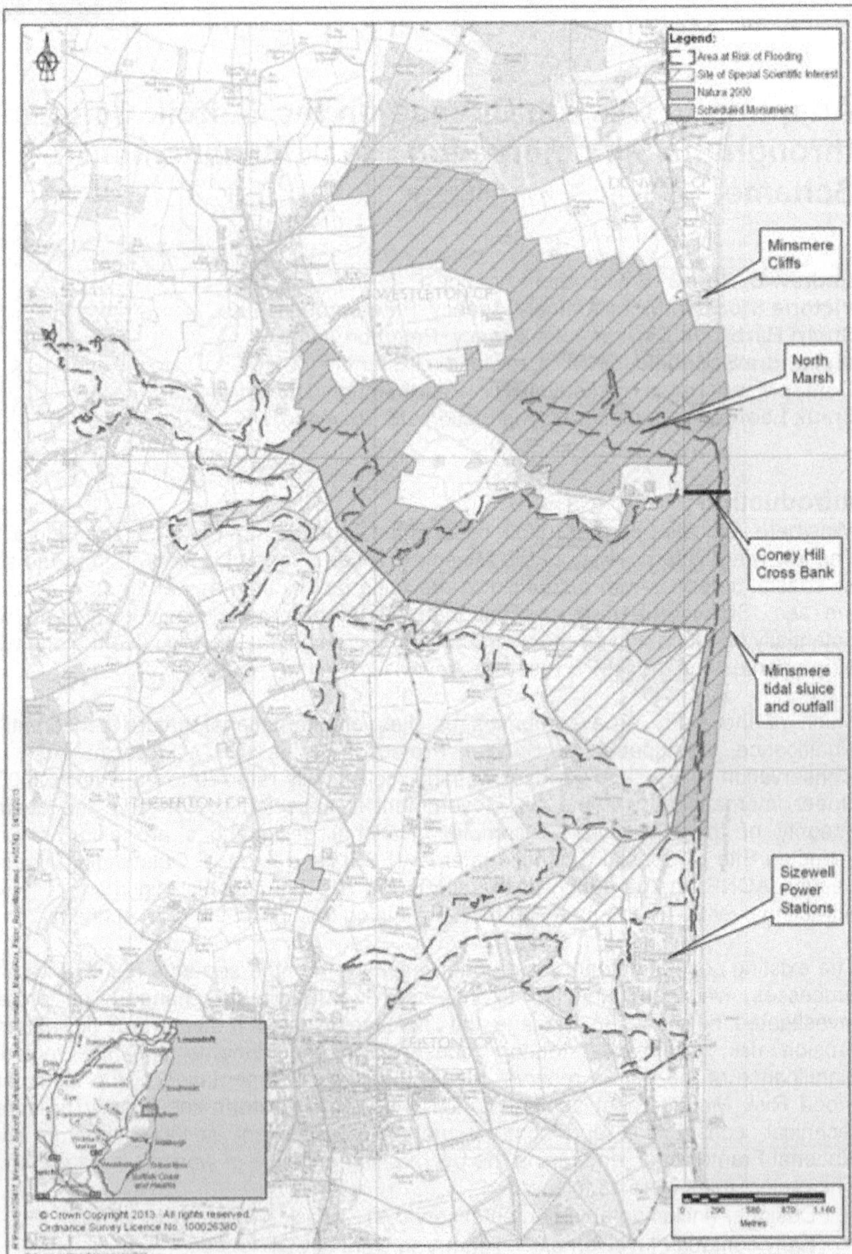


Figure 1: The Minsmere flood risk area

## The Problem

### Flood and Coastal Erosion Risk

The Minsmere River historic valley is protected from the sea by a modified dune embankment system than runs parallel to the coast for 3.1km from the Minsmere Cliffs in the north to Sizewell power station in the south. The 926ha hinterland drains to the Minsmere Tidal Sluice via gravity. North of the sluice, two embankments run parallel to the coast with the secondary (landward) embankment tying into the primary (seaward) embankment just north of the sluice. The primary embankment continues south developing into a wider dune system (see Figure 1).

Historically, the Environment Agency maintained the primary defence with reactive works to repair storm damage. The November 2006 surge event caused extensive damage to the primary bank north of the Coney Hill Cross Bank - the primary bank was breached, allowing the sea to flow into the dip between the primary and secondary banks. Rather than transport material offshore, the event pushed beach material through the breach site, creating a berm in front of the secondary bank. The following year a similar surge event occurred, further modifying the defence. Figure 2 shows the change to the defences that arose from these surge events.

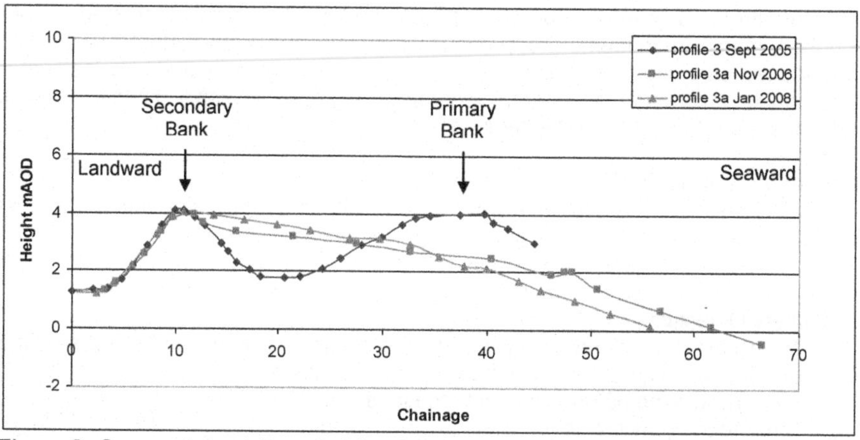


Figure 2: Survey data at Breach Site showing impact on defence cross section from surge events

The legislation relating to the conservation of the Natura 2000 site requires the UK Government to maintain existing sites in-situ, where it is sustainable to do so, providing compensatory measures (usually replacement habitat) where this is not possible. In the case of Minsmere, the area behind the defence is important in terms of its freshwater features, whilst that in front of the defence is important for its coastal features. However, by maintaining the existing line of coastal defence, as sea levels rise this strip of coastal habitat will reduce in width – referred to as coastal squeeze.

Without intervention, the existing defences were expected to be increasingly and more frequently overtopped, with each overtopping event likely reducing the crest level as wave action continues to mobilise the remaining bank landward. Climate change would further pressurise the coastal defences as sea levels rise and storminess increases. The existing defences were expected to be fully breached (at the location of the existing primary bank breach) within 20 years, exposing 246ha of internationally designated freshwater habitat to saline conditions. The development of an approach to provide continued protection of this internationally important environmental asset from the sea was required and needed to take a long term view as well as addressing immediate coastal pressures.

## Option Appraisal

### Environmental and Social Considerations

The team worked closely with Natural England to develop a long term sustainable approach to manage the designated habitats at Minsmere. Having considered the conflicting requirements of the freshwater and coastal designated features, Natural England's conclusion was the coastal habitat was to take priority as it was easier to create replacement freshwater habitat.

The local community and those who rely on the tourism economy regionally, were very concerned about change to Minsmere, which is a key recreational and tourism asset. Those managing the existing nuclear sites at Sizewell and planning for the possible future nuclear new build programme were also very interested in our proposed coastal management approach. Other environmental features of the site also required careful consideration such as the presence of protected species (water voles, bats, invertebrates, otters, vegetated shingle), invasive species, visual landscape, public access and heritage.

### Sustainability Appraisal

The option selection process was undertaken with a project steering group that included representatives from Natural England, English Heritage, the RSPB, the Suffolk Coast and Heaths AONB team, and the Sizewell Shoreline Management Steering Group. The following primary and secondary objectives were established:

- Primary Objectives:
    - To protect the internationally designated nature conservation features of the site and to mitigate the impacts of climate change.
    - To work with the natural coastal processes.
    - To be technically feasible.
    - To not entail excessive capital and / or maintenance costs disproportionate to the importance of the features under threat.
- Secondary objectives:
    - To protect heritage and archaeological assets.
    - To maintain the socio-economic environment.
    - To ensure the continued safe operation of the Sizewell power stations.

A list of all possible options was developed and a sustainability appraisal undertaken against the above objectives over the short (next 20 years), medium (20 to 50 years) and long term (50 to 100 years).

### Preferred Option

The preferred option in the short term was to construct a new defence landward of the existing defence at the northern end of the site by improving the existing Coney

Hill Cross Bank to provide a 1 in 10 (10%) annual exceedance probability (AEP) standard of protection against coastal flooding to the majority of the Natura 2000 site (218ha) for the next 50 years. The existing defences north of the Coney Hill Cross Bank would be allowed to evolve under natural coastal processes, exposing the 28ha of freshwater habitat to the north (an area known as North Marsh) to increasing risk of saline inundation. Replacement habitat will be required for this area via the Environment Agency's Regional Habitat Creation Programme.

It was recognised that longer term (post 50 years), as sea levels rose, the remaining coastal defences would start to be pressured by natural coastal processes and gravity drainage at Minsmere sluice would be ineffective. At this point it was considered that it would be unlikely to be sustainable to continue to protect the existing Natura 2000 site in situ, requiring the provision of replacement habitat for the remaining 218ha designated habitat.

In summary, the preferred option consisted of the following works (refer to Figure 3):

- Improving the existing 350m Coney Hill Cross Bank through widening and raising so that it could provide a new line of future coastal defence against coastal flood events with a AEP of 1 in 10 (10%) for the next 50 years.
- A new water control structure approximately halfway along the Coney Hill Cross Bank to enable the on-going management of water levels within North Marsh and to provide a complete shut off of drainage to the south in extreme coastal flood events.
- To construct two coastal bunds across the dip between the primary and secondary banks south of the Coney Hill Cross Bank to reduce the scour risk at the base of the secondary bank (a large scour hole occurred at the base of the secondary bank south of the Coney Hill Cross Bank during the 2006 surge event as sea water that entered through the breach flowed southwards with the naturally falling ground).

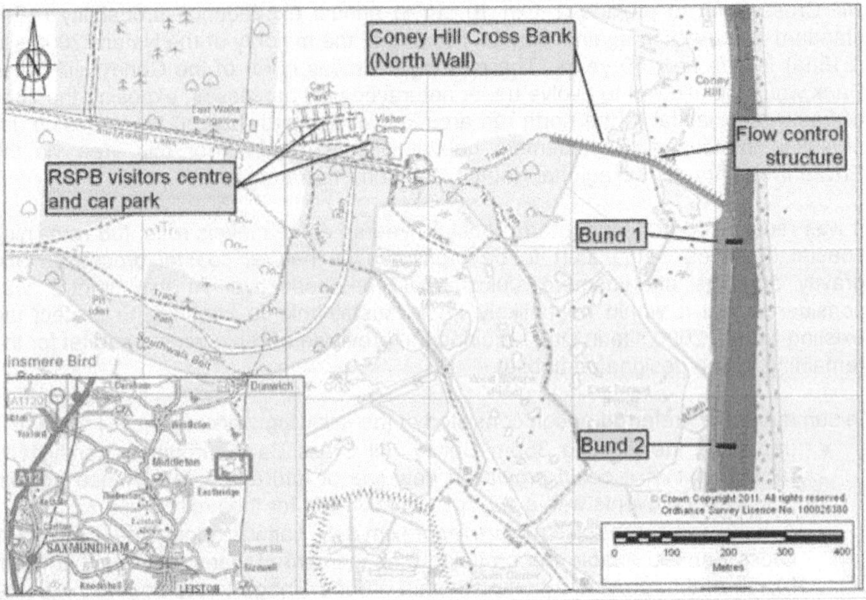


Figure 3: The preferred option

## Design and Construction of the Key Elements of Work

### Coney Hill Cross Bank

The bank traverses a reed bed and therefore ground conditions were far from ideal for construction work. In addition, the work had to take place between August and March to avoid the nesting birds season. In designing the final technical solution for the improvements to the Coney Hill Cross Bank, the following elements were incorporated:

- The alignment of the bank was chosen so that it made use of the better existing ground conditions. This meant predominantly widening the bank on its north side at the eastern end and on the south side at the western end.
- The crest level of the bank was designed for a 1 in 10 (10%) AEP coastal surge event, allowing for 50 years of sea level rise. At the seaward end the crest level ties in to the existing flood defence bank, whilst at the landward end we could afford a lower crest level as the effect of waves will be far less further inshore. The change in level was made over a 1:20 slope to aid easy public access along this RSPB footpath. Side slopes of 1:3 for ease of maintenance.
- Rather than key the toe of the bank into the existing ground, we opted to cut the grass short to make use of the natural "crust" to the ground and use geotextiles to provide additional strength and separation. Estimated seepage quantity was deemed to be acceptable at this location.
- The bank was designed so that the top of the existing bank would initially be cut off and placed adjacent to the bank to create a haul road for construction. The haul road was then incorporated into the final bank profile.

- The grass seeding of the embankment was designed in consultation with the RSPB and a landscape specialist to ensure it was in keeping with the nature conservation and landscape designations. Side slopes were initially blanketed with biodegradable coconut matting to allow grass seed to establish.

Topsoil was stripped from the existing embankment, with an ecologist carrying out a watching brief for reptiles. The arisings, for reuse on the new embankment, were stockpiled, away from the watercourses to prevent silt run off caused by rainwater. Testing was undertaken on the existing embankment material to ensure it was within the parameters set out in the specification.

Vegetation was cut back in the locations for widening the embankment prior to the placement of geotextile and geogrid. The imported fill material, sourced locally following CL:AIRE protocol, was then laid in layers not exceeding 225mm and compacted. The slopes were trimmed when the desired crest level was reached, using batter rails as a visual aide and engineering control. Stringent checks were placed over compaction methods, depths of layers placed, crest levels and slope gradients. Further testing was undertaken on the imported material at regular intervals to ensure it was at an acceptable standard.

As a section of the earthworks were completed, the stockpiled topsoil was placed over the bank and the timber edged, granular filled footpath was built as construction progressed from the coastal area, moving inland. Daily settlement monitoring was carried out throughout the construction stage and a means of future settlement monitoring was incorporated into the embankment construction.

Figure 4a: Before construction

Figure 4b: After construction

**The Water Control Structure**

For the water control structure, the team worked with the Environment Agency Operations team, the RSPB Minsmere team and eel pass specialists to develop the final bespoke design that incorporates the following elements:

- The existing informal water control structure, pipe and outfall were filled with grout and buried within the works, with a new arrangement being constructed alongside, removing uncertainly over the condition of the existing structure and maintaining flow through the bank during the works.
- Stop logs that can be safely removed and added from above, without entering a confined space, in order to provide day to day water level control for the RSPB.

- A manually operated penstock to provide a complete shut off between the North Marsh and the rest of the site. The penstock spindle was designed so that it can be operated in an emergency high surge tide scenario from the crest of the bank, accessed via an upper timber platform.
- Upper access platform to penstock spindle designed to be in timber to blend with the existing landscape. A GRP floor provided a robust, slip free surface.
- A ductile iron pipe was installed to connect the North Marsh to the main site, providing a more flexible material for future settlement. It was also laid with an inverted camber to mitigate for future settlement.
- Upstream and downstream of the structure new connecting water channels were dug out within the reed bed to connect to the existing drainage channels. At the upstream connecting channel, a bittern feeding ledge was incorporated in the final design, providing environmental enhancement.
- A bespoke eel pass was designed to enable continued eel migration when the penstock was open.
- A precast concrete outlet structure was selected for the outfall structure, with reno mattresses and gabion baskets downstream to reduce erosion.

A temporary works sheet pile cofferdam was installed in the North Marsh providing a safe and dry environment to construct the reinforced concrete flow control intake structure. Once the structure was complete and the concrete cured, the initial sections of pipe culvert were connected to the previously installed culvert beneath the embankment and the penstock, stoplogs, trash screen, GRP flooring and associated steelwork was then fitted and commissioned prior to the removal of the cofferdam.

A portable dam consisting of steel frames and heavy duty polythene sheeting was installed downstream of the outlet structure to facilitate dewatering of the area in which the downstream outlet structure and gabions were to be laid. An amphibious excavator was used for the excavation of the outlet and inlet channels, including a bittern feeding ledge. Using this specialist item of plant provided a number of benefits, including minimising construction time of the aforementioned activities, reducing the quantity and complexity of temporary works and providing a safe method for working in the reed bed. After the new culvert was pressure tested and commissioned, the existing culvert was filled with grout to render it redundant.

For landscape reasons, timber was selected as the preferred material for the visible operating platform for the penstock gate. In accordance with the Environment Agency's timber procurement policy, this timber (Oak) was procured from a sustainable timber source with a full chain of custody.

Figure 5a: Before construction

Figure 5b: After construction

**The Two Coastal Bunds**

For the two coastal bunds, the following details were incorporated into the final design:

- The crest level of both bunds was designed so that they would be overtopped rather than the adjacent primary and secondary defences. Hydraulic modelling was also undertaken to check the levels and score velocities.
- The slopes to the coastal bunds were designed at a 1:10 gradient to allow easy public access over the bunds as they were located across the Suffolk Coastal Path – a national trail.
- The specification of the profile to each bund was to create a smooth, natural looking profile so that it blended in with the existing landscape.
- A cellular geotextile was placed close to the surface of the bunds to provide protection against footfall, vehicular traffic (route is occasionally used by RSPB maintenance vehicles, the coastguard and Sizewell Police Patrol) and scour during overtopping.
- The bunds were designed so that material excavated from their footprint could be used to cover the clay bund, retaining their natural look.

Extensive measures were undertaken to divert the existing public footpath around the works whilst ensuring the public had uninterrupted access to the RSPB hides. Water suppression was also used to stop the dry coastal sand from becoming air-borne and causing discomfort to the public.

During construction the crest level of one of the bunds was reduced to allow more material for surface covering to be obtained. Once the bunds were blanketed with the excavated coastal material, they were fenced off to allow vegetation to become established prior to reopening the public footpath. This section of the work was further complicated by the presence of WWII anti-tank concrete blocks, which are a listed monument, and needed full recording prior to being buried/ partially buried.

Figure 6a: Before construction

Figure 6b: After construction

## Minimising Risk During Construction

At the early stages of the project it was identified that the importation of earth for the works would present several challenges associated with the interface between site traffic and other users of the transport route. The team investigated using on-site material from the reed bed area but testing of this material showed it to be too high in organic content to be a viable material, even with treatment. The material therefore needed to be imported. By incorporating the material from the existing bank into the works the volume of imported material was reduced by a third to 6,000m$^3$.

Pre-construction, the project team worked closely with the various partners to identify constraints, areas of high public safety risk, and possible mitigation measures. Early discussions with Suffolk County Council (SCC) Highways identified their preferred route from the main road, utilising roads most suitable for lorries and avoiding villages as much as possible. Traffic management was also discussed with the local community via the Parish Council and a public exhibition. We also discussed the impact on public footpaths with the local Public Rights of Way (PRoW) Officer (and RSPB) to develop practical and appropriate mitigation.

A detailed Traffic Management Plan (TMP) was submitted with the planning application to identify the constraints the team would work to and the proposed mitigation. This TMP was part of the construction contract, for the contractor to incorporate and develop for construction. During construction the TMP was regularly reviewed with stakeholders to ensure it continuously achieved its objectives to:

- provide a safe working environment for all site personnel
- ensure public safety
- maintain safe access to the site and reserve through the local road network by installing passing bays and diverting footpaths;
- minimise the impact on local communities and the surrounding environment;
- effectively manage the importation of 6,000m$^3$ of fill material;
- ensure deliveries were made outside of peak times;
- uphold relationships with the local community, Minsmere visitors and isolated property owners close to the reserve.

Letter drops in the local area were made in advance of construction works commencing and a leaflet explaining footpath closures and alternative routes was made available to visitors of the nature reserve. To keep the public and stakeholders informed on the progress of the scheme, large information boards were erected and a newsletter distributed to RSPB members and the local community. No complaints were received, and the team received positive feedback such as this "tweet" from a visitor to the RSPB reserve (a quantity surveyor by profession): *"if you want to see best practice traffic management, come to Minsmere".*

An Environmental Action Plan, a Site Waste Management Plan, a Materials Management Plan and a Carbon Calculator were prepared and updated regularly, with all significant environmental risks identified and mitigated. Opportunities for waste minimisation, material reuse, maximising the use of site won and recycled materials were identified.

## Lessons Learnt and Conclusions

The £1.04m construction contract was completed £74k under target and one month ahead of programme. Through working as an integrated team, whilst effectively communicating and importantly listening to local stakeholders and community, major challenges can be overcome to allow delivery of a complex flood risk management scheme. Although on the face of it the works involved were simple, it was the location along with its national and international nature conservation importance which raised numerous challenges that the team had to mitigate to the satisfaction of stakeholders. Furthermore, the technical challenge of building an earth embankment on what could be considered as water during the winter months gave the team much to think about that drove an innovative solution.

Following successful completion of the scheme, a lesson's learnt workshop was held to capture what went well and what could be improved for future similar schemes. For this scheme the key outputs from the workshop were:

- **Effective up front planning and stakeholder engagement** was show-cased by this project, resulting in minimal challenges during construction and limited cost increase through variations. This was only possible through good understanding of the constraints posed by the physical environment.
- **Lack of site communication** due to the remote rural location did hinder quick responses when challenges arose. Contract need be let earlier to enable communications to be put in place ahead of works commencing.

The scheme secures the short to medium term life of the frontage from coastal flooding and erosion whilst enhancing the area's important habitat. The long term future of this frontage is less uncertain as continuing to hold the line is unlikely to be sustainable and therefore replacement habitat is allowed for within the regional habitat creation plan.

**Coastal Management: Changing coast, changing climate, changing minds**
ISBN 978-0-7277-6114-9

ICE Publishing: All rights reserved
doi: 10.1680/cm.61149.387

# Dutch Strategy for Long-term Safety and Sustainable Use of the Coast

**Evelien van Eijsbergen**, Ministry of Infrastructure and the Environment, Rijkswaterstaat Water, Traffic and Environment Lelystad, The Netherlands
**Wout de Vries**, Ministry of Infrastructure and the Environment, Rijkswaterstaat Water, Traffic and Environment, Lelystad, The Netherlands
**Cees-Jan van Westen**, Ministry of Infrastructure and the Environment, Rijkswaterstaat Water, Traffic and Environment, Lelystad, The Netherlands

## Summary
Approximately 60% of the Netherlands is located below normal high tide level; the coast protects millions of Dutch inhabitants and their property in the hinterland against flooding from the sea. Owing to the anticipated effects of climate change, especially sea level rise, a timely adaptation is essential for safeguarding the current flood protection level. For this reason, the national Dutch Delta Program has created a coastal strategy, which combines the sustainable and long-term safety of the coast with spatial and economic developments in the coastal zone.

The Dutch coastline is maintained using sand replenishments. This practice is feasible owing to the large quantities of sand that's easily accessible for mining in the Dutch section of the North Sea. In the past decades, the coast has been appropriately managed and this has resulted in sufficient protection for the hinterland. The mortality rate related to flooding by the sea is practically remote. The coastal economy has however developed slower than that of the rest of the country. To a certain extent, the coastal zone is in danger of losing its attraction for leisure and tourism due to lack of investment. The quality of nature is also under pressure as a result of the lack of natural dynamics.

Within the framework of the national Dutch Delta Program (Delta Programma in Nederland, 2011), a joint effort of the Dutch government, regional and local authorities, and civil society organizations has resulted in a new National Strategy for the Coast (Deelprogramma Kust, 2010). The key objective is to create a safe, attractive and economic viable coast. To achieve this, it is strongly advised that flood safety measures are based on adaptive spatial concepts. This strategy provides the framework for future decisions and actions in this area.

The following challenges can be defined:
a) A sustainable safety level of the coastal zone requires that the sandy system remains in balance with rising sea levels;
b) Measures on quality improvement and spatial measures are required for realizing economic development and an attractive environment;

c) The interaction between flood defense, attractiveness and economic development can only be realized through an integrated approach.

## The Dutch Delta Program

Climate change is illustrated by rising sea levels, subsiding land levels, increased rainfall intensities, higher temperature and longer periods of drought in the summer. These developments result in changing demands on flood protection levels and fresh water supply.

In response and based on the advice of the so-called second national Delta Committee, the Dutch government started the national Dutch Delta Program in 2010. The goal of this program was to record the changing demands and to develop response strategies and solutions for the above mentioned developments. In the Delta Program, governments at all levels (national, provincial, municipal, water authorities), knowledge institutes and civil society organizations worked closely together to develop solutions and response strategies that are broadly accepted.
Important products of the Delta Program are:

- The four Delta Scenarios, which are used to identify opportunities and constraints of future developments on flood protection and fresh water supply. These scenarios are based on both the climate and socio-economic scenarios of respectively KNMI (the Dutch Met Office) and socio-economic planning institutes (Deltares en Planbureau voor de Leefomgeving, 2011);
- A systematic analyzing tool to weigh different response strategies and solutions systematically;
- The Dutch Delta Model, a computer simulation model, to allow for nationwide simulations of the Delta.

### Subprogram Coast

The Delta Program consisted of nine subprograms, including the Subprogram Coast. The activities of the Subprogram Coast lasted from 2010 to 2014 (Deelprogramma Kust, 2014a) and are presented in this paper. The goal of this subprogram was to identify strategies and measures aimed at realizing a safe, attractive and strong North Sea Coast, based on a broad and active participation of coastal actors and stakeholders. Two tasks were given (Deelprogramma Kust, 2010):

- Flood safety: analyze how to guarantee the required flood safety level and identify building blocks for an integral spatial vision on the coast;
- Coastal expansion: explore the possibilities and desirability of a large scale coastal expansion.

It was concluded that a large scale coastal expansion will not be needed in the near future to guarantee the safety level necessary to prevent flooding from the sea. Therefore, this subprogram aimed at the formulation of the so-called Decision Sand, being the adaptation agenda for the Dutch coastal sandy system for sustainable flood safety. Sandy solutions, like sand nourishment, are advocated to safeguard the Dutch coast against floods from the sea, while simultaneously safeguarding the other coastal functions in a sustainable and attractive manner.

## Description of the Dutch Coastal Zone

The length of Dutch coast is approximately 350 km. In the North, the coastal zone consists of the Wadden Islands with dunes on the North Sea side. Along the coast of the provinces of North and South Holland, the coastal zone mainly consists of dunes;

there are dikes over a limited length. The coast of the deltaic area of the province of Zeeland is protected by dunes, closure dams and dikes (Figure 1).

The coastal zone is habitat for a large variety of flora and fauna. The sea water is characterized by a nutritious sandy bottom environment, mainly caused by the nutrients originating from the rivers. The dune area has a rich flora and fauna. Some coastal stretches are managed by man in a dynamic way, meaning natural procedures are allowed and promoted, however between given safety boundaries. In this manner sea, wind, sun and salt can create a variety of coastal landscapes.

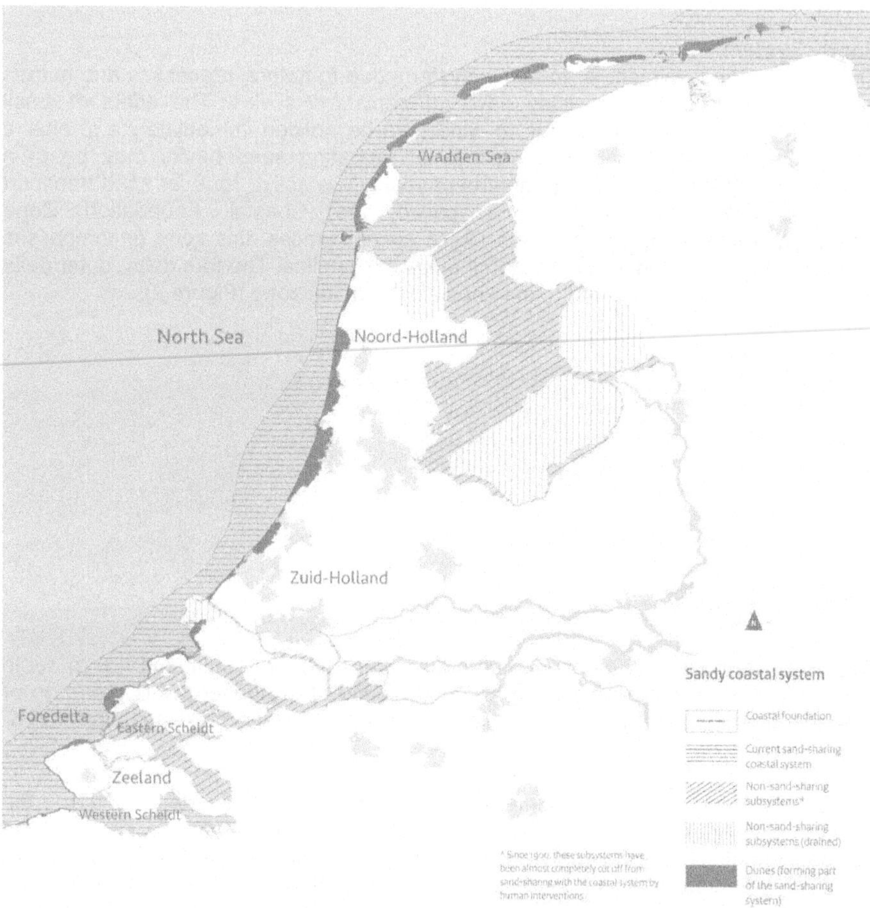


Figure 1: The Dutch sandy coastal system consists of sandy coastal waters, estuaries and tidal inlets, including sandy shores, e.g. beaches and dunes. The sandy system is classified into sand-sharing sections (coastal foundation, Wadden Sea and Western Scheldt) and non-sand-sharing sections (the semi-closed basins in the South-West Delta: Eastern Scheldt, Grevelingen, Haringvliet) and the basins along the Wadden Sea: IJsselmeer and Lauwersmeer. "Sand-sharing" means that an

active exchange of sand takes place between the sections of the sandy system as a result of currents and wind. (Source: DPK/Jelle Gulmans/Atelier 1:1)

The Dutch coastal zone is inhabited by some 1.2 million people, 7.5% of the Dutch population. Activities are mostly water related forms, including tourism, shipping, fisheries, agriculture, drinking water purification (Panteia, 2012).

The Dutch and many foreigners like the sandy coastal zone. Many people enjoy the water, wind, tides, seasons, gradients in the landscape, the contrast with the crowded hinterland, the emptiness, and darkness (Hisgen en Laane, 2007).

**Sandy System**

The coastal sandy system is an active system, due to natural processes and human activities. The coastal processes have different time scales. The effect of small changes, like coastal erosion due to wind, can be noticed immediately e.g. after a storm. On the other hand, the process of migrating sand banks may be of a significant longer time scale e.g. months or years. The active zone for sand transport and morphological processes is called the Coastal Foundation Zone (Deelprogramma Kust, 2013). In the Dutch circumstances, this zone encompasses the area between the inner dunes and the NAP -20 m line. The fore delta, outer delta and tidal channels also belong to the coastal foundation zone (Figure 2).

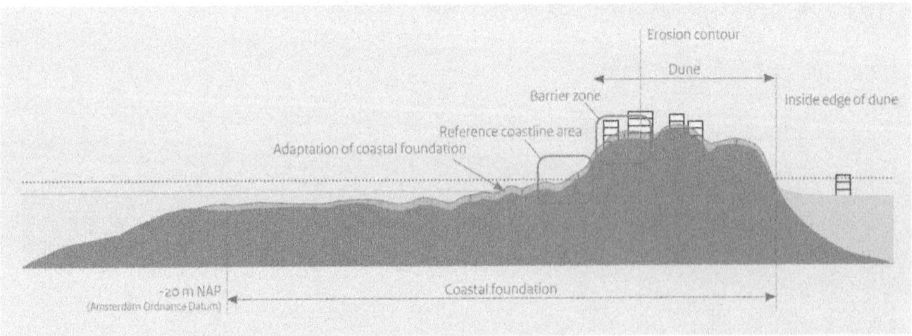


Figure 2: The coastal foundation covers the zone from the continuous -20 m depth contour at sea up to and including the inside edge of the dune and is made up of different management zones as shown in the cross-section (Illustration: DPK-DGW/Atelier 1:1/Studio Terp)

The development of a coast is determined by the available volume of sand and the required volume necessary to keep a coastal sandy system in balance. Eroded sand can be compensated with sedimentation. When the available volume is large, a coast will grow, whereas in the opposite case the coast will erode, banks will erode and channels will deepen.

For the last 1000 years, the sandy balance along the Dutch coast is negative; the continuing erosion leads to a retreating coast line. This is caused by a decreased sediment rate in the rivers, resulting from human activities in the catchment basins. In addition, the foreshore is deeper due to continuing sea level rise. Consequently, only limited volumes of sand reach the coast.

**Safety Norms**

Dunes, dikes and dams along the Dutch coast protect the hinterland against flooding from the sea. In The Netherlands, the safety level against flooding is set out in the national law, which is an unique situation in the world. This implies that safety norms were to be formulated. Current safety norms, expressed in probability of exceedance, date back to the analysis of the first Delta Committee in the 1950's. Since then, the population has doubled and the economic investment tripled. In addition, the meteorological and the technical knowledge on dikes and dunes has increased significantly. This is one of the main reasons why the second Delta Committee advised in 2008 to redo this analysis. In response, the Dutch government initiated the Delta Program. This was the begin point for the development of new flood safety norms (Deltaprogramma, 2015):

- It was decided to formulate a basic safety level: a nationwide maximum mortality rate as a result of flooding of 1/100.000 per annum;
- Extra protection is to be given at locations:
    - With a risk of large numbers of victims due to flooding;
    - Where flooding would lead to high economic and social damage;
    - Where vital infrastructural objects are located, like power stations.

The result of the analysis of the safety norms is set out in a proposal for new flood safety levels, expressed in probabilities of flooding.

**Basic Coastline**

Within the current policy, the coastline is held in its position. For this, a so-called Basic Coastline was defined. In addition, the coastal foundation zone should grow with the rising sea levels, so that is kept in its relative position to the sea level. This policy was defined in 1990, with the publication of the national Coastal Policy Document by the Dutch Government (Ministerie van Verkeer en Waterstaat, 1990). Two main goals of this policy were: maintaining the safety levels of the lower lying hinterland and protecting the dune area against structural erosion. The basic coastline can be considered the criterion of this policy. It is defined as the position of the coastline in 1990.

The maintenance of the coastline was to be realized including safeguarding the interests of nature, recreation and infrastructure. Sand nourishment was considered to be the best method to fulfill these requirements, because:

- It is the cheapest and most effective solution, because there is sufficient sand available in the North Sea. Mining areas are located between the 12 miles line and the NAP -20 m contour;
- It is in line with natural processes and contributes to the dynamic developments in the coastal zone;
- It can be planned in a flexible way and be applied at locations with the highest priority.

Since 2001, an extra volume of sand is being nourished annually to compensate the continued sea level rise. Currently, some 12 million $m^3$ is being applied annually, comprising a budget of some € 56 million (Figure 3).

## Problem Analysis of the Delta Program Sand

Based on the four Delta scenarios, a sea level rise along the Dutch coast of 0,15 – 0,35 m until 2050, and 0,35 – 0,85 m in the period until 2100 was predicted. The biggest uncertainty in this prediction is the time scale in which the effects of climate change will become evident. The problem analysis (Beslissing zand, 2014) covers the challenges in terms of safety, sand, spatial developments and related time scales.

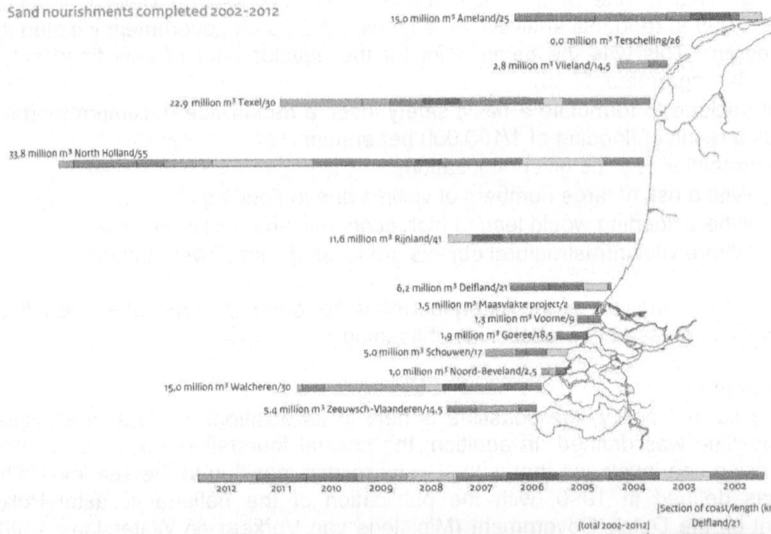


Figure 3: Total volume of sand nourishments carried out to maintain each section of the coast in the period 2002 to 2012. This does not include the sand motor or sand nourishments for coastal reinforcement purposes (Source: Rijkswaterstaat)

### Safety Challenges

Presently, flood defenses are being improved through a national flood protection enforcement program HWBP. After completion, these flood defenses fulfill the current safety norms. It is expected that, with the present regular sand nourishment program no significant challenges to the coastal defense system will develop before 2050. Further, is it anticipated that the new safety norms will not lead to big changes in the coastal flood safety challenge. In the longer term, the sea level rise and related changed wave characteristics may require new reinforcement measures to comply with the norms.

### Sand Challenges

The natural sediment recordkeeping is often influenced by human activities. The socio-economic and natural developments (esp. sea level rise) in the coming century will result in additional changes in the sediment fluxes, mostly in further reductions of the sediment volumes. For the Dutch coast, the following effects are foreseen:

- Increased coastal erosion. The Dutch coast is an eroding coast with an average rate of 10 m retreating coastline per decade during the past decades. This erosion threatens the flood safety level of the hinterland by attacking on dunes, dikes, dams. The other functions of the coastal zone e.g. drinking water supply, nature, housing, recreation will be affected too.

- Decreasing sand volumes of the outer deltas. The decrease and form of the reduction of the outer deltas of the Wadden area will lead to:
    - Decrease in wave reduction, leading to more wave attack at the Wadden islands;
    - Decrease sediment supply for the eastward positioned islands, thus increased erosion of the shores of these islands;
    - Decrease sediment supply for the Wadden area, leading to reduction of bank area and form.

  The outer delta of the Western and Eastern Scheldt in Zeeland will also reduce, leading to similar effects as for the Wadden Area. It is however not expected that these developments will be of such magnitude that they will result in serious flood safety problems in the coming 30 – 50 years.
- Tidal channels are positioned relatively close to the shore and thus to the coastal defense. The morphological dynamics of the channels is such that these will move closer to the shore with the continuing sea level rise. During storms, eroded sand from dunes will increasingly be transported through these channels out of the area. Further, steeper gradients of these channels will lead to instable foreshore with increased risk of sudden shear of these areas. The stability of the position of these channels is currently being researched
- Sediment shortages in the (semi-)closed estuaries in Zeeland. The closures of these former estuaries result in a decreased of flow in the former tidal channels, thus in higher sedimentation rates. This sediment comes from the eroding sandbanks in the delta. Erosion of sandbanks reduce their size and volume, and the importance for natural values. Also a reduction of the safety level of the flood defenses on the leeward side can be expected.

### Spatial Developments

The evaluation of the Third Coastal Policy Document (Twynstra Gudde and Alkyon, 2007) and research of the Water Governance Centre (2012) show that the required combination of sand nourishment/dike reinforcement projects with spatial developments is still not at the desired level. This is due to the large number of parties involved, each with different and sometimes conflicting interests, combined with a lack of financial resources. It is therefore the challenge to combine sand nourishment for safety with spatial ambitions. To achieve this, excellent communication with involved actors and stakeholders is required.

### Time Scales

The time scales of the spatial and safety challenges vary from short to very long term. The spatial developments are in the order of short to very short term (months, year), whereas the safety considerations cover long term (decades). Thus, these challenges have different times scales. To realize a sound combined development, much time is required to combine these two fields.

## Preferred Strategy

In the Netherlands, stakeholders play a pivotal role in the formulation and implementation of coastal policies. This is one of the specifications of the National Vision on Coastal Development. By following a strategic approach it was possible to develop a method to address the challenges of coastal safety, sand balance and spatial planning. This approach also included the uncertainties in causes and consequences, and the possibilities for new developments and interests. Special attention was paid to the governance aspects of combining coastal safety with spatial challenges.

The formulation of the Preferred Strategy (Deelprogramma Kust, 2014b) was the result of a joint effort of all relevant stakeholders. For this, ample attention was paid to this process to include all parties. This work also resulted in the formation of the 'Coastal Community', which are the new platform where coastal matters are addressed.

The Preferred Strategy was formulated based on the following principles:

- A safe, attractive and economic strong coast;
- A combination of coastal safety and spatial ambitions;
- Flexible solutions, which allow for inclusion of measures to realize future developments.

The different time scales of safety and spatial challenges can be dealt with as follows:

- No relation. In this case, challenges in both fields can be dealt with independently. However, good communication between actors is advised, so that possible combined actions can be identified;
- Both issues have a time scale in the order of 50 years. In this case, flexible solutions can be aimed at. For each of the 14 coastal types (e.g. young or older dunes, with or without buildings or harbor) basic flexible solutions were identified. These are characterized by sufficient space for future expansion of the flood defense and spatial use , including the consequences of future sea level rise and changed socio-economic requirements;
- Dealing with challenges that are foreseen for over 50 years are treated according to first procedure mentioned.

Given its general nature, the effects on environment and fisheries could be addressed only generally. These can be further specified when the solution is in a further, more detailed planning phase at local level.

## Formulation of Challenges

The Subprogram Coast resulted in the formulation of the above introduced preferred strategy. It has also given rise to the formulation of further challenges, namely:

- Knowledge challenges
  The Dutch knowledge of coastal processes and approaches is at a sound level. However, further research is needed, especially on the morphodynamics of the tidal channels located close to the shores and on the outer deltas. For this, the program Kustgenese 2.0 ('Coastal genesis 2.0') has been identified. The research and related monitoring aim at increasing the knowledge of the developments in the sandy systems and the measures that can contribute positively to these developments. In 2020 this program will be evaluated. Based on the outcome, it will be decided whether large scale pilots need to be identified and where these will be located. Another decision will refer to the need to reformulate the annual nourishment program ;

- Spatial development challenges
  The provinces are responsible for the spatial planning in their coastal areas. The four coastal provinces have jointly formulated the National Vision on Coastal Development, in which strategic agendas are presented. These agendas reflect the socio- economic and spatial ambitions along the North Sea coast, aiming at realizing an economic strong and spatial attractive coastal zone, whilst safeguarding the unique character of the coastal resorts. Within the boundary condition of coastal safety and spatial planning , these developments can be achieved using an integral approach. The strategic agendas are not explicitly based on the Delta scenarios, as the formulation began in an earlier stage. However, in general the agendas consider the effects of climate change and sea level rise on the possible socio-economic developments. Sand nourishment is considered a sound a sustainable measure, as it gives ample room for fulfilling the demands of most coastal functions.

The strategic agenda assumes that the current flood safety level will be maintained. It is expected that innovative sand nourishment techniques will develop, for instance aiming at stabilizing tidal channels with sand, realizing large scale coastal extension and higher dynamic conditions in the coastal zone. Further, the provinces aim at combining spatial quality with the socio-economic development and flood safety of the coastal zone. This can be realized through for instance a wider shore for recreation, dynamics in the dune zone or restructuring the old built up areas in the resorts.

**Dealing with Uncertainties in Space and Time**
Effective anticipation of the uncertainties related to socio-economic and spatial development over time will be achieved using a step wise approach. This approach is one of the building blocks of the Preferred Strategy. The main challenge is to bring the time scale of the desired spatial developments (20 – 30 years) in line with the time scale of flood protection (50 – 100 years).

## Conclusions

- The flood safety of the coastal zone of the Netherlands is realized through a system of dunes, dikes and dams;
- There are many functions in the coastal zone: recreation, nature, housing, agriculture, fisheries, drinking water purification and supply;
- The coastal zone offers a unique and highly appreciated environment in the densely populated country, given its space, quiet conditions, darkness at night and proximity to the natural elements;
- The so-called flexible grow concepts are an efficient means for connecting the spatial short term ambitions with the long term flood safety challenges;
- Sand nourishment is a sound and sustainable measure to guarantee the flood safety level in the long run;
- The new safety norms will not lead to significant changes in the flood safety level in the coastal zone;
- Sand nourishment is an attractive measure to realize a sustainable flood level. In the very long term there is sufficient sand available in the North Sea, which can be used in an economic manner;
- Sand nourishment offers optimal possibilities and flexibility for realizing other functions in the coastal zone, especially recreation, nature, housing;

- The coastal zone has many stakeholders and actors. To operate jointly with such diverse interests, the Subprogram Coast paid ample attention on the communication and integral measures. We can now speak of a coastal community, where the working atmosphere is such that integral approach and shared responsibility (including financing) is the standard. In this way, the Dutch coastal zone can be considered a sustainable one for many generations to come.

## Acknowledgements

The authors are very grateful to Mrs. Janneke van Bergen, Mrs. Moniek Löffler, Mrs. Josja Veraart, Mrs. Marja Wijnties and Mr. Stanford Wilson for their valuable contributions in writing and designing.

## References

Deelprogramma Kust (2010). Plan van Aanpak; (Plan of Approach); Den Haag.

Deelprogramma Kust, (2013). Nationale Visie Kust; (National Vision on Coastal Development), Den Haag.

Deelprogramma Kust, (2014a). Synthesedocument Kust (Synthesis report on Coastal development, background report of the Delta Program); Den Haag.

Deelprogramma Kust, (2014b). Voorkeursstrategie Kust (Preferred Strategy, background report of the Delta Program); Den Haag.

Deltaprogramma (2011). Delta Programma in Nederland; Den Haag.

Deltaprogramma (2014). Beslissing Zand (Decision on Sand, background report of the Delta Program); Den Haag.

Deltaprogramma (2015). De beslissingen om Nederland veilig en leefbaar te houden. (Delta decisions, resulting in a water safe Netherlands);

Deltares en Planbureau voor de Leefomgeving (2011). Deltascenario's: verkenning van mogelijke fysieke en sociaaleconomische ontwikkelingen in de 21ste eeuw op basis van KNMI'06 en WLO-scenario's, voor gebruik in het Deltaprogramma 2011-2012, (Deltascenario's) Delft.

Hisgen, R.G.W. en R.W.P.M. Laane (2007). De geheimen van de zee.(Secrets of the Sea). Rijkswaterstaat RIKZ. Den Haag.

Ministerie van Verkeer en Waterstaat (1990). Kustverdediging na 1990 (national Coastal Policy Document), Vergaderjaar 1989-1990, TK 21136. Den Haag.

Panteia (2012). De kracht van de kusteconomie; een foto van de economische motoren, een doorkijk naar hun ontwikkeling en een SWOT-analyse, (Strength of Coastal Economics). Zoetermeer;

Twynstra Gudde and Alkyon ( 2007). Evaluatie Derde Kustnota (Evaluation third national Coastal Policy Document), Amersfoort.

Water Governance Centre ( 2012). Verkenning en Advies Versterken Governance Kustontwikkeling.

**Coastal Management: Changing coast, changing climate, changing minds**
ISBN 978-0-7277-6114-9

ICE Publishing: All rights reserved
doi: 10.1680/cm.61149.397

ice
Institution of Civil Engineers
publishing

# Comparing the Morphological Behaviour of Dutch–Danish Shoreface Nourishments

**Quirijn J. Lodder**, Rijkswaterstaat, Lelystad, The Netherlands/ Hydraulic Engineering, Delft University of Technology, Delft, The Netherlands
**Per Sørensen**, Danish Coastal Authority, Lemvig, Denmark

## Introduction

In the Netherlands and Denmark, gradual structural coastal erosion is occurring due to sea level rise, natural redistribution of sand and human intervention. On longer time scales, this gradual erosion increases the risk of coastal flooding, threatens buildings on the dune, fresh water extraction, dune habitats and beach recreation.

Both countries have a coastal erosion policy aiming at compensating the sediment losses by means of sand nourishments. The nourished sand is being placed either at the beach (typically between +3 and -1 m MSL) or in deeper water on the shoreface (at depth around -4 to -5 m MSL). The effectiveness of beach nourishments in relation to coastal erosion is quite well known. However the effectiveness and morphological behaviour of shoreface nourishments is not fully understood (Spanhoff and van de Graaff, 2006). A better understanding of effective shoreface nourishment techniques/placement/management is necessary in order to be able to maintain our eroding sandy coastlines in an efficient way.

This paper focuses on the comparison of Dutch and Danish shoreface nourishments. The main objective is to indicate differences and similarities in the morphological evolution of two shoreface nourishment projects and trying to provide more insight in the physics governing the behavior of shoreface nourishments.

## Case Study Description

For this paper two shoreface nourishments projects have been assessed. One project near the Dutch coastal villages of Zandvoort and Bloemendaal and one near Skodbjerge in Denmark. In both cases the nourishment projects consisted of multiple shoreface nourishments. The characteristics and location of the nourishments are given in table 1 and figure 1.

| Country | Location | Year | Position (km-km) | Volume (Mm$^3$) | Volume per m (m$^3$/m) | D50 Nourished (µm) | Placement depth (m irt MSL) |
|---|---|---|---|---|---|---|---|
| NL | Zandvoort 1 | 2004 | 62.75 – 67.75 | 2.2 | 440 | 250-300 | -4 |
| NL | Bloemendaal | 2008 | 61 - 63 | 1.0 | 500 | 250-300 | -5 |
| NL | Zandvoort 2 | 2008 | 67.75 - 70.25 | 0.5 | 2 | 250-300 | -5 |

| Country | Location | Year | Position (km-km) | Volume ($Mm^3$) | Volume per m ($m^3/m$) | D50 Nourished (µm) | Placement depth (m irt MSL) |
|---|---|---|---|---|---|---|---|
| DK | Sdr. Holmsland Tange | 2010 | 11.6-21.0 | 7.7 | 57 | 300-400 | -5 |
| Dk | Skodbjerge North | 2011 | 17.2-18.0 | 0.3 | 400 | 300-400 | -5 |
| DK | Skodbjerge South | 2011 | 13.8-14.6 | 0.3 | 400 | 300-400 | -5 |

Table 1: Nourishment characteristics

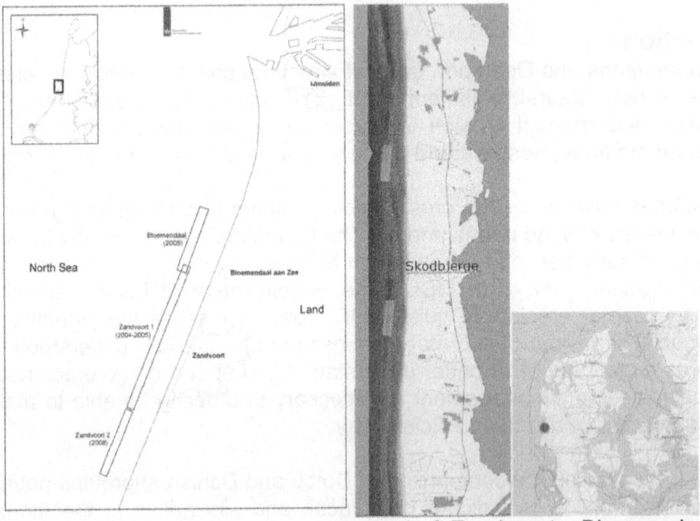


Figure 1: Location of the nourishment projects. Left Zandvoort – Bloemendaal (NL), Right Skodbjerge (DK) (yellow rectangles). The location in both countries is given in the upper left and lower right panel

**Zandvoort – Bloemendaal (NL)**

The Zandvoort – Bloemendaal nourishments are placed along the Dutch sandy beach-dune coast just a couple of kilom south of the IJmuiden harbour moles. The shoreface slope is around 1:300. The local grainsize is typically around 250 µm. Typically 2 to 3 breaker bars are present, based on annual bathymetrical data since 1965. Bar volumes amount up to 200 to 300 $m^3/m$. In the natural unnourished situation the breaker bars migrate in seaward direction. After reaching a seaward limit around 600 to 800 from the shore they fade away, while a new bar is formed at the shore (Wijnberg, 2002 and Spanhoff and van de Graaff, 2006). This process happens about every eight years on this site. The bar movement process is visualized in figure 2. The tidal range at IJmuiden just North of Zandvoort is 1.9 m. A storm surge with a return period of 1 year has a waterlevel of 2.35 m above MSL and the waterlevel for a storm with a return period of 100 years is 3.60 m. The wave climate is dominated by offshore waves in the range of 1.5 to 3.5 m (Van de Rest,

2004). The frequency of waves coming from the southerly directions is higher than the frequency of waves coming from the northerly directions. This results in a local net longshore sediment transport in northward direction of about 0.25 million $m^3$ per year (Van Rijn, 1997). The longshore sediment transport gradient is not precisely know, however it is estimated to be around 10.000 $m^3$ per year over the total nourished coastal stretch of about 8 km.

The Dutch shoreface nourishments have been placed at a distance of about 900 – 1100 m from the shore (figure 3). The nourishments where placed at the seaward side of the outer breaker bar. This position in the coastal profile was chosen because this position is reachable with a dredger with a medium draft and wave breaking occurs here providing the energy to disperse the nourished sand. The volume per m of coastline is about 2 times the volume of the natural breaker bars. The sand for the nourishments was dredged offshore at depth greater then 20 m below mean sea level.

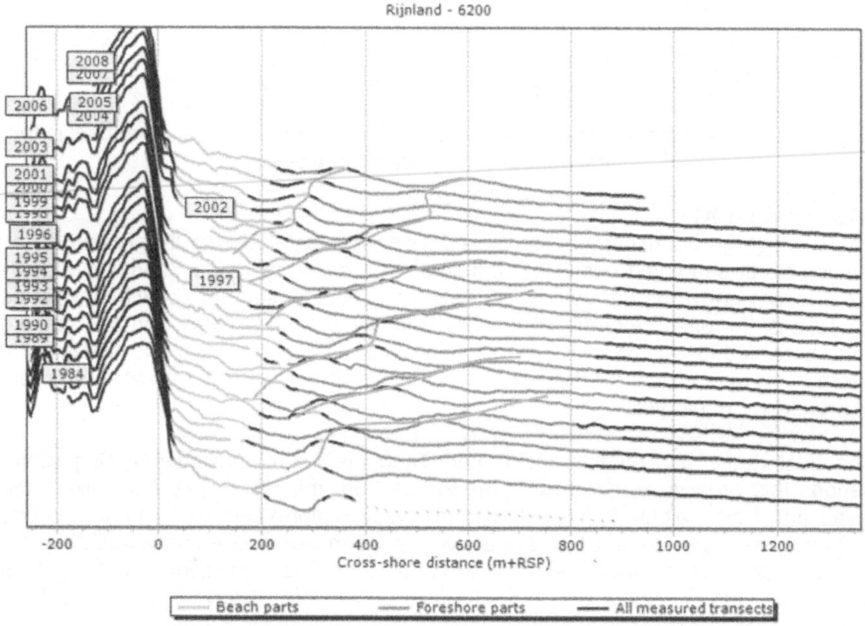


Figure 2: Timestack of yearly surveys of a coastal profile near Bloemendaal, visualizing the offshore migration of bars over time. The crest of the bar has manually been connected to visualize the bar movement. Note that this plot gives an overview of the morphological changes over time and not alongshore

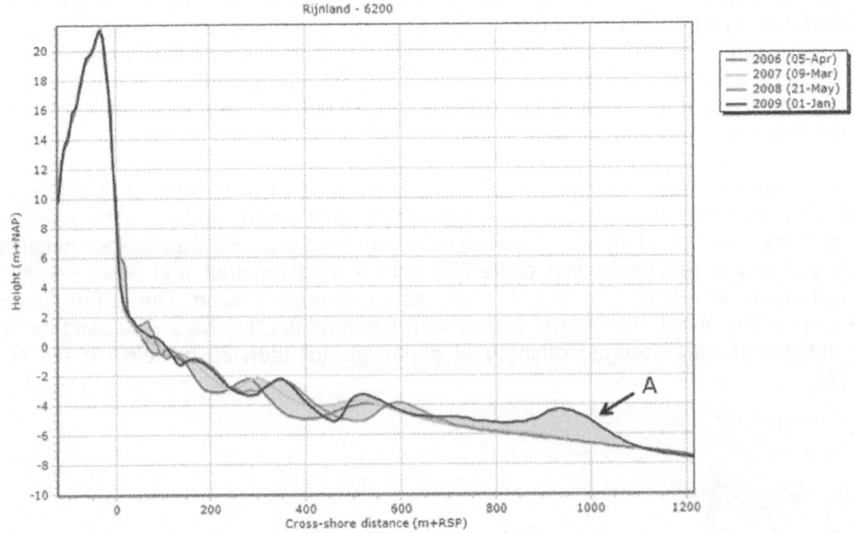


Figure 3: Cross shore position (A) of the 2008 nourishment near Bloemendaal in the first annual survey after completion

**Skodbjerge (DK)**

The Skodbjerge nourishments are placed along a coastal stretch that is part of the Danish North sea coast and is located south of the port of Hvide Sande. The study area is a narrow barrier which consists of a sandy seabed, beach and dune formations, the offshore profile also consists totally of sand typically with 1-3 offshore sand bars. A d50 sediment size of between 0.3 µm at the shoreline and 150 µm at a depth of 10 m is typical for this area. The coast in this location is retreating at a rate of around 1.7 m per year.

The tidal range at Southern Holmsland Tange is 1.1 m however, during winter periods this area is exposed to storm surges from the North Sea with raised sea levels and wave heights from west and northwest. A storm surge with a return period of 1 year has a waterlevel of 2.02 m above the mean sea level and the waterlevel for a storm with a return period of 100 years is 3.11 m. The significant wave height Hs measured north of Thorsminde is on average above 4 m 11 days pr. year based on the period 1992-2012. The 100-year significant wave height is 8.1 m

The wave direction forcing the dominant sediment transport is coming from approximately 300 deg north. The coast normal is approximately 265 deg, which results in oblique incoming waves with an angle of 35 deg, which is close to the maximum wave driven sediment transport rates. The net sediment transport rate is approximately 2.2 million $m^3$/year to the south (Sørensen, 2001). The net sediment transport gradient is estimated at 0.3 million $m^3$/year.

The coastal system consists of 1-2 bars which generally do not have a very clear behaviour unlike the bars near Zandvoort. Figure 7 shows a timestack of a profile in the middle of the northern nourishment. It can be seen that there is a trend that the outer bar moves shoreward, while the trend of the inner bar is not obvious.

The nourishment in 2011 was placed outside the outer bar as seen on fig 4. The nourishment in 2010 was placed outside the outer bar in the whole study area. This 11 km long area has been intensively monitored from 2005 to 2014 to identify the morphological changes of the area and the impact of the 3 shoreface nourishments. In total 19 surveys have been completed. Transects perpendicular to the coastline have been surveyed from dune top to seaward of the outer bar. The spacing between the survey lines is 200 m.

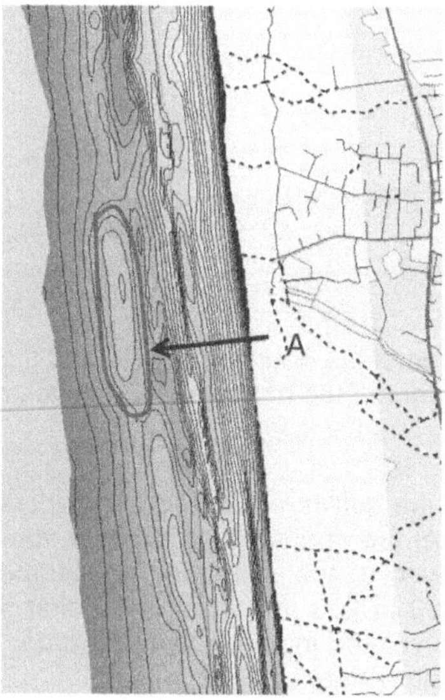


Figure 4: The 2011 shoreface nourishment Skodbjerge North encircled in red. A trough directly landward of the nourishment can be seen (A)

## Morphological Observations

### Zandvoort – Bloemendaal (NL)

Both the 2004 and 2008 nourishments where placed on a smooth profile seaward of the existing breaker bars. In the five months between the placement of the nourishment and the first annual survey, the nourishment transformed into bar and a trough formed directly landward of the nourishment. The nourishments and the existing breaker bar migrated in landward direction (figure 5). The behaviour of the nourishments is in agreement with the observations of Spanhoff and Van de Graaff, 2006.

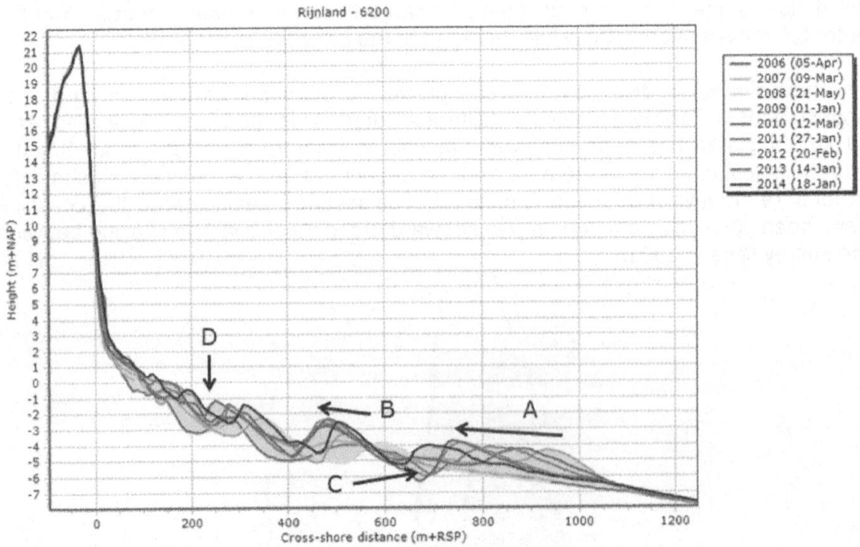


Figure 5: landward migration of the nourishment (A) and the inner bar (B), formation of trough (C) directly landward of the nourishment and volume increase in the upper parts of the profile (D).

Volume calculations with the software tool MorphAn (Lodder and Van Geer, 2012) show that the placement of the nourishment forces a volume increase in the upper parts of the profile landward of the shoreface nourishments. The increase per m coast is in the order of 150 to 200 $m^3$ per m in the first 5 months after placement. After the first 5 month the volume increase is in the order of 50 to 75 $m^3$ per m per year till 3 years after the nourishment.

The annual surveys show that the response of the morphological system is primarily in the cross-shore direction. Four years after placement the Bloemendaal 2008 nourishment has migrated about 300 m in landward direction and only about 50 m in northerly direction (figure 6). Further the nourishment became smaller in longshore direction and higher.

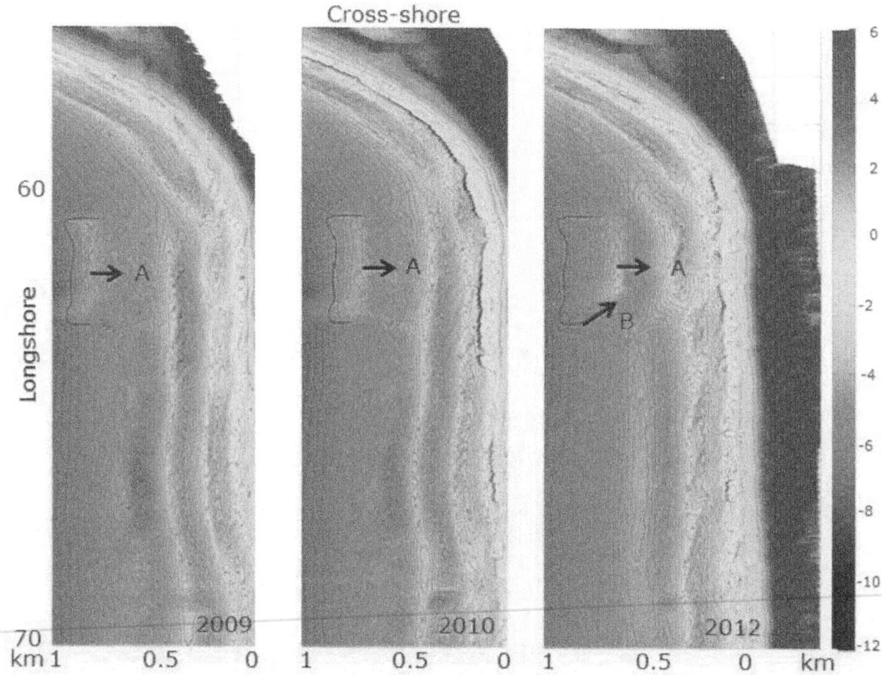


Figure 6: Bathymetry of the Zandvoort - Bloemendaal area. Indicated in red the Bloemendaal 2008 nourishment location. The nourishment migrated primarily in landward direction (A) and only partly alongshore (B). The Y-axis gives the longshore distance, the X-axis gives the cross-shore distance, both in km irt local coordinate system. The height is in m irt MSL.

**Skodbjerge (DK)**

After both nourishments it can be seen that the outer sandbar moves offshore, due to the placement of nourishment sand on the seaward side of the outer bar. This is most visible for the 2011 nourishment. The 2011nourishment continues to move offshore as it redistributes itself to the south and in some places creates an extra outer bar (figure 7). After both nourishments the outer sand bar begins to move towards the shore as was also seen in the stormy period before the nourishments and after the 2007.01 survey. The offshore movement of the bars, after a severe storm with a 20 year return period, between the 2013.01 and 2014.01 survey is clearly visible.

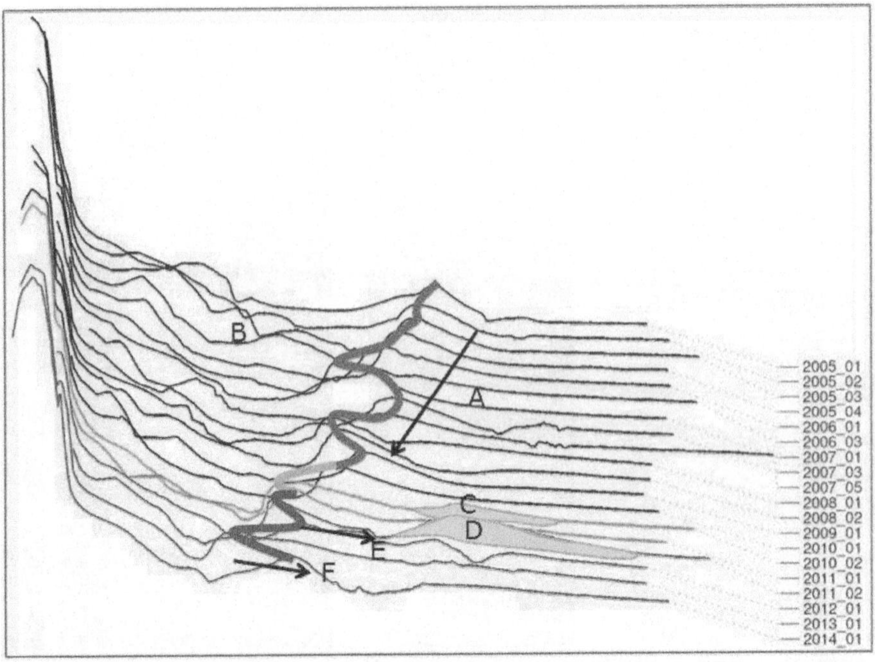


Figure 7: Timestack visualizing the bar migration at Skodbjege. The crest of the bar has manually been connected to visualize the bar movement. Before nourishment (blue), after nourishment (green and red). The location of the outer bar (A), inner bar (B), the nourishment 2010 and 2011 (C, D) are indicated. The migration of the outer bar after the placement of the nourishment and after the 2013 storm are also indicated (E, F). Note that the surveys are not collected at regular intervals. Hence intra annual morphological changes can be seen, unlike figure 2.

It is difficult to find an explanation for changes in coastal retreat based on the incoming wave energy. For instance coastal retreat occurred between 2011.01 and 2011.02 although wave energy during this period was not above normal. The highest wave energy recorded occurred between the 2011.02 and 2012.01 surveys. Despite this the coastline on average advanced during this period. There are clearly large natural variations in the position of the coastline that make the impact of the nourishments difficult to establish. However there is a clear coastal advance and morphological response in both surveys after the nourishment sand was placed. Volume and trend analysis suggest that the rate of coastal retreat decelerates from about 1.2 to 0.25 m/year after placement of the nourishments. However due to the large natural coastline variations it is not possible to validate this deceleration from a statistical point of view (Knudsen, S. B., Earnshaw M., 2014).

Spatial plots of the post nourishment surveys show that the morphological response of the morphological system is primarily in longshore direction. Within two years after placement both the Skodbjerge nourishments have migrated about 750 m alongshore in southerly direction. The cross shore position is more or less at the same position (figure 8).

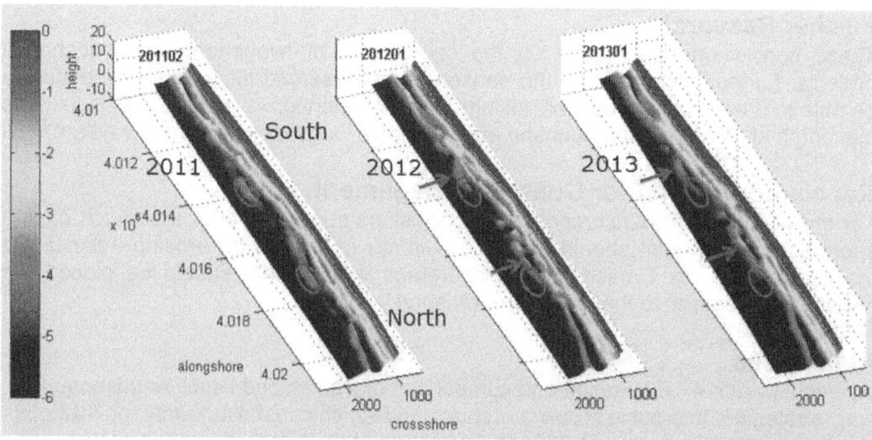


Figure 8: Bathymetry of the Skodbjerge area. Indicated with the red circles both the 2011 Skodbjerge nourishments. The red arrows indicate the position of the nourishment after one and two years. The nourishments migrated primarily alongshore. Note that the south is up in this figure. The Y-axis gives the longshore distance in km, the X-axis gives the cross-shore distance in m irt local coordinate system. The height is in m irt MSL.

## Comparison

When comparing the assessed Dutch and Danish shoreface nourishments it is clear that there are some striking differences in the morphological behaviour of both nourishment projects. Along the Zandvoort - Bloemendaal coast the response to a shoreface nourishment is primarily cross-shore. Along the Danish coast a primarily alongshore response was found.

We think that the opposite behaviour of the Dutch and Danish shoreface nourishments is caused by either or both:

- nourishment design parameters (e.g. length, height, volume and volume per m)
- the (gradient in) the net alongshore sediment transport rate.

The design parameters for both nourishment projects are quite the same except for the total volume and length of the nourishments. The 2011 Danish nourishments have around the same volume per m coast as the Dutch nourishments, however the total volume and hence the length of the nourishments is about 3 to 6 times smaller. The longshore sediment transport at the Danish nourishment site is approximately 10 times higher than at the Dutch site.

We think this significant difference in the longshore sediment transport rate is the most likely cause for the observed differences. The higher sediment transport rate along the Danish site will be able to transport the nourished sand along the coast more quickly than will be the case at the Dutch nourishment site. However from this assessment alone it is not possible to prove this hypothesis.

## Further Research

These conclusions are based on the comparison of two shoreface nourishment projects. Further research into the causes of the observed differences is needed to be able to draw more robust conclusions. Special attention is needed with regards to the length of the shoreface nourishments and local longshore transport (gradient).

## Recommendations for Coastal Management

For the planning of nourishments the conclusions suggest that in the Dutch case a shoreface nourishment should be placed directly in front of an erosion hotspot. On the contrary for the Danish case a shoreface nourishment should be placed "up stream" with regards to the longshore transport.

## References

Gawehn, M (2014), Evaluation and comparison of Danish and Dutch nourishment strategies, Internship Rijkswaterstaat and Danish coastal authority, pp 102

Knudsen, S. B., Earnshaw M. (2014), Evaluation of the effect of two shoreface Nourishments, Sdr Holmsland Tange, Denmark. Danish coastal authority

Lodder, Q.J. and Geer van, P.F.C. (2012), MorphAn: A new software tool to assess sandy coasts. In: NCK-days 2012 : Crossing borders in coastal research, 13 March 2012 - 16 March 2012, Enschede, the Netherlands. http://dx.doi.org/10.3990/2.189

Sørensen P., (2001), Sedimentbudget Vestkysten. Kystdirektoratet (in Danish)

Spanhoff, R., Van de Graaff, J. (2006), Towards a better understanding and design of shoreface nourishments. Coastal Engineering 2006, Proceedings of the 30th International Conference, pp. 4141-4153.

Van de Rest, P., (2004), Morfodynamica en hydrodynamica van de Hollandse kust, MSc Thesis, Delft University of Technology (in Dutch).

Van Rijn, L.C., (1997), Sediment transport and budget of the central coastal zone of Holland. Coastal Engineering 32, pp 61 – 90.

Wijnberg, K.M., (2002), Environmental controls on decadal morphologic bahaviour of the Holland Coast, Marine Geology, 189, pp. 227-247

**Coastal Management: Changing coast, changing climate, changing minds**
ISBN 978-0-7277-6114-9

ICE Publishing: All rights reserved
doi: 10.1680/cm.61149.407

# Learning from Flood Events on Critical Infrastructure: Relations and Consequences for Life and Environment (CIrcle)

**M.W.A. Hounjet**, Deltares, Delft, The Netherlands
**A.T.M.M. Kieftenburg**, Deltares, Delft, The Netherlands
**M. Altamirano**, Deltares, Delft, The Netherlands.

## Introduction

It becomes increasingly clear that when a natural disaster occurs in modern society, the direct effects to critical infrastructure networks are not the only effects. Dependencies and interdepencies between different networks, like the energy network (figure 1a), transport network (figure 1b) and communication network (figure 1c), can cause unexpected consequences. Furthermore, indirect effects and cascading effects may have a more significant impact over a longer period of time than that of the direct effects. In 2014 the Minister of Infrastructure and the Environment of The Netherlands wrote a letter to the Dutch parliament concerning awareness of critical infrastructure (CI) when dealing with climate change related topics, like floods and droughts (Rijksoverheid, 2014). She was not just directing her comment at the stakeholders concerned with the CI, but at the inhabitants of the Netherlands as well. The interactive CIrcle tool that has recently been developed will be able to contribute to these goals by analysing and visualising cascading effects of CI networks.

Figure 1: a, b and c: Electricity network, road transport network and communication network.

## Cascading Effects

A simple example of a cascading effect would be when during flooding at certain locations the water depth reaches 25 cm. In this case the substations of the electricity network will flood and there will be outages in the areas connected to these substations. Usually these areas will be flooded as well. These substations are connected to the main stations of the electricity network. When a certain percentage of substations is flooded, the entire main station and all the connected substations will cease working and electricity outages will be experienced in non-flooded areas. This is a cascading effect within one network. An example of a cascading effect between two different networks would be when communication substations are flooded with a water depth of about 50 cm. The consequence is that all communication within a certain area is affected. This includes the radio towers of the waterboards that are require to operate water pumps in the area. The pumps of the waterboards and pumps for tunnels in the area will stop functioning, which could worsen the effects of flooding in the area.

Until now connections between CI networks are poorly identified. CIs are dealt with separately from each other, even though different parties are aware of their interdependency and the possible cascading effects. Especially when recurrences are involved these effects become unclear. Furthermore, data is hardly available and dependencies are even less documented, which makes it difficult to determine the effects on a certain location. For that reason it may also hinder adequate coordination and disaster management.

In flood risk management there are concerns over indirect impacts. Because of the lack of quantitative research a certain factor per region multiplied by the total of the direct impact is used to calculate the indirect impact (Jonkhoff, 2012). There are qualitative studies indicating that the indirect impact could be much greater and other studies indicating that it is likely to be much less (Luijf, 2009). Provinces, waterboards and safety bodies are aware of these uncertainties and are in need of more data to support decisions. They need to know:

- what the weakest links are;
- how to protect the weakest links and minimise damage;
- where to evacuate people (non-flooded area but with working drinking water, energy and wastewater facilities);
- whether it is possible to take last-minute measures and what these measures could be.

## Open Data

It is not always possible to access detailed network information. This information is quite often consider to be vulnerable to terrorist attacks, especially that of the energy network. Because it is important to have easy access to network data and to analyse cascading effects quickly and in real-time in order to be helpful for stakeholders, open data is chosen instead of fictive network models. This open data needs to be combined with knowledge and experiences of the networks: at what water-depth will the network experience problems and what type of problems will they be?

Two studies in Groningen and West-Friesland in the Netherlands have shown that the availability of useful open data is sufficient for cascading effects analyses. Although the data was not as detailed as the data the network owners have themselves, they did indicate that the effects that were calculated were very close to reality and detailed enough for crisis management operations and decision making during events.

The abovementioned example of the electricity network is derived from these case studies. Contacts at the electricity network indicated that a water depth of 25 cm, although often 50 cm, is used (Royal HaskoningDHV, 2012).

## Causal Relationships

Although there are a lot of open data available and it is useful for cascading effects analyses, the data alone is not enough. How a network is affected, exactly when and how it interacts with other networks is not directly clear from the data alone. During meetings with different network owners it became clear that many outages and other effects can be described as causal relationships:

- When during a flood the water depth reaches 25 cm, the electricity substations stop functioning (see Figure 2).
- When electricity supply fails, our industry relies on temporary measures for 3 days.
- When water levels reach 30 cm, the gas network is damaged but is still repairable (also present in Figure 2).

With these types of relationships for different networks, it is possible to use open data for real-time cascading effect predictions. Figure 2 shows a screenshot of an automatic GIS-analysis of cascading effects for a part of Groningen, made entirely with open data and gathered causal relationships.

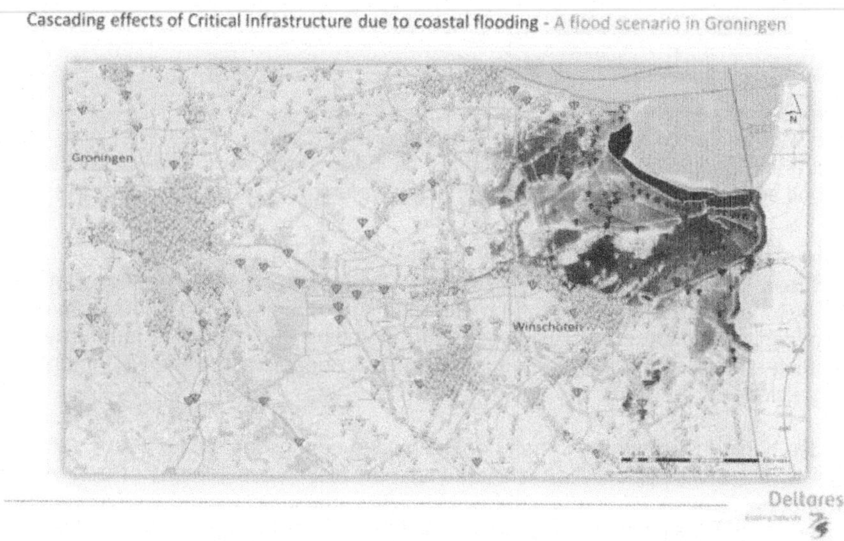


Figure 2: Screenshot of an automatic GIS-analysis of cascading effects for which only open data and causal relationships have been used

Figure 3 shows an overview of threatened pumps of the Waterboard in West-Friesland, based on open data. The flood scenario shown occurs after 30 hours of flooding based on a 3Di simulation, from a breach location at Wijdenes (see arrow in Figure 3). The effect of an outfall of the pumps is that the regulated water levels in the neighbouring canals are no longer regulated, probably causing more flooding. As a further result of the pump outfall in this area, drinking water distribution may be threatened since the pumps are fed with electricity from the Wervershoof area and it's pumps close to Andijk (i.e. red circle in Figure 3) is directly threatened as well. This will lead to an effect that is much larger than the shown area of West-Friesland, because the whole east site of the province of North Holland is dependant on the water supply from Andijk.

Figure 3: Threatened pumps of the responsible waterboard in West-Friesland, after 30 hours of flooding from a breach location as indicated; based on open data and a 3Di flood scenario. In the red circle the drinking water supply in Andijk is indicated

## CIrcle – Interactive Stakeholder Tool

With CIrcle (the tool for Critical Infrastructure: Relations and Consequences for Life and Environment), the knowledge and experience of the different network managers and experts are gathered, discussed and combined to visualise and analyse cascading effects. In this way the consequences for people and the environment will become clear.

The idea behind CIrcle is the sharing of knowledge during workshops. Participants of the workshops will typically be stakeholders in crisis management or owners of networks (such as pipes, roads, cables, etc.). All parties, including the coordinator of the workshop will sit round the CIrcle touch table. All of the parties can share their knowledge of the components of their own (part of the) CI. For example, the water pump of a waterboard which influences the water level of a watercourse depends on electricity supply to this particular pump. In case of emergency it may be dependent on the road leading to the pump as well when back-up power units need to be introduced. On the touch table application this will be drawn by the participants as an arrow between the pump and the substation and the road and the pump. By raising this dependency, the electricity distributor as well as the road owner will be able to show how they will safeguard the needs of the waterboard. The waterboard or province may know that a certain substation will be renewed and can ask for special measures. Since CIrcle is based on equality and is used as a circle of trust resulting in awareness it helps connecting knowledge on CI, by connecting people in the process and connecting technical knowledge.

Figure 4 shows CIrcle during a session where connections are drawn. CIrcle is used as a discussion tool. When knowledge is shared on direct effects and cascading effects, this can be entered into the system at each network block or at each connection. The data and information is gathered in an Excel-file. Some data is directly written down as a causal relationship, other data may be more qualitative but is important as well. The causal relationships are directly used in the automated GIS analysis to show results when, for instance, electricity substations can manage 50 cm of water instead of 25 cm.

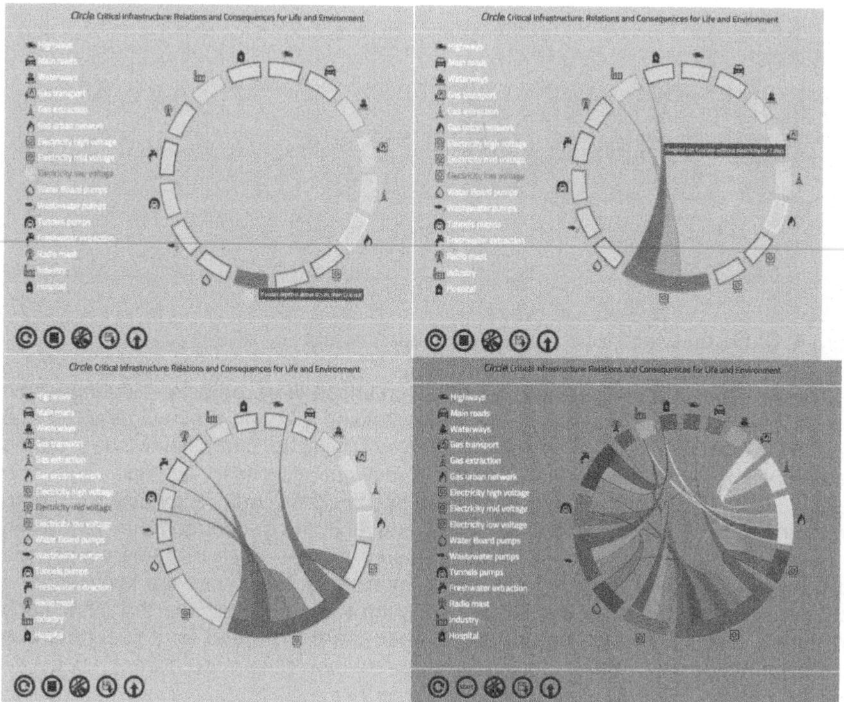


Figure 4a, b, c and d: CIrcle during a workshop session where in 4a a direct effect is added to the system and in 4b an indirect effect is entered into the system. 4C shows the gathering of different dependencies and 4d shows an end-result of a session where all discussed connections are shown

## Workshops

Sharing knowledge is crucial for the workshops. For CIs and cascading effects the combined knowledge tells us more than just having knowledge of one network. Because it is not necessary to have all the information of the different networks beforehand, network owners will not feel they have to share secret information and they can decide for themselves what to share. The visual and interactive techniques of CIrcle help to understand cascading effects and exchange important information on causal relationships, making stakeholders aware of their vulnerability, directly or through network interdependencies.

These causal relationships are collected in a database and shown in a time-dependent GIS analysis in order to visualise the shared knowledge immediately. Because the application has the ability to store the knowledge on interdependencies gathered in the workshops, it will build in each workshop and will become more powerful and more valuable to the different parties. Such a well-documented library of interdependencies and covariance between the risks affecting different infrastructure assets will be interesting to all participants, but also to other parties, such as insurance companies.

Figure 4: CIrcle on a mini touch table during a workshop

For the case area West-Friesland, several workshops were organised during which different stakeholders and network owners discussed CIs and their vulnerability and possible cascading effects. They were all very willing to share knowledge during discussions. All the participants were somewhat surprised to find out how interdependent they were within a small area like the case area. They also wondered how their networks could be best protected as everybody understood that increasing the height of dikes was not the ultimate solution. It is because of these workshops that CIrcle was created to be able to gather more data and to use the knowledge in such a way that every case and every workshop will get better and that after each workshop a better prediction can be given when there is a need for actual data and information during an event.

## Learning from Flood Events

There are many ways in which we can learn more from flood events. Good visualisation to enhance understanding of what actually happens during an event is one of the most important keys here. With CIrcle different stakeholders are able to learn from fictive or actual flood events in the following ways:

1. Participants of a workshop learn from each other's knowledge and the combined knowledge in cascading effects.
2. Flood experts learn from the combined knowledge gathered in CIrcle on direct and indirect flood impact and will be able better to quantify flood impact.
3. With CIrcle you can replay actual events and check whether the network knowledge can predict outages and dependencies properly.

4. Crisis management organisations can better predict where the safe areas are will be to which they can direct evacuees.
5. For other events other than flooding the knowledge gathered in these workshops provides insights in to the cascading effects that could be caused after an outage from a non-flooding another source. Some interdependencies are more universal than others and generically usable.

## Future Work

The coming months will be used to organise different workshops in order to gather as much data and knowledge as possible. After sufficient workshops have been carried out, further data analysis can be performed to find solutions for combining qualitative knowledge and causal relationships. At the moment they will be used separately: the causal relationships in the automated GIS-analysis and the qualitative knowledge to learn more about the impact for inhabitants and to find ways to quantify this qualitative knowledge. Over time it will become clear what causal relationships can be generally applied to other cases and what relationships are specific to a certain case or an event.

## References

JONKHOFF, W. (2009). *Flood risk assessment and policy in the Netherlands,* Green cities new approaches to confronting climate change

LUIJF, E. a.o. (2009). *Empirical findings on critical infrastructure dependencies in Europe*

RIJKSOVERHEID (2014). http://www.rijksoverheid.nl/nieuws/2014/03/17/schultz-oeso-rapport-bevestigt-noodzaak-wake-up-call-waterrisico-s.html?ns_campaign=nieuwsberichten-ministerie-van-infrastructuur-en-milieu&ns_channel=att

ROYAL HASKONINGDHV (2012). *Analyse Waterrobuuste inrichting voor nieuwbouw en vitale & kwetsbare functies*

www.3Di.nu

**Coastal Management: Changing coast, changing climate, changing minds**
ISBN 978-0-7277-6114-9

ICE Publishing: All rights reserved
doi: 10.1680/cm.61149.415

# Understanding the Fiscal and Regulatory Mechanisms Necessary to Achieve Climate Compatible Development (CCD) in the Coastal Zone

**Lesley King**, LTS International, Penicuik, UK
**Gerben Nap**, LTS International, Penicuik, UK
**Lucy Emerton**, Environment Management Group, Colombo, Sri Lanka
**Fiona Nunan**, International Development Department, Birmingham University, Birmingham, UK

## Introduction

With understanding of global climate impacts improving, it is becoming clear that climate change will have significant negative impact on the functioning of many coastal ecosystems (Stern 2007). These ecosystems support a large proportion of the world's poor (Food and Agricultural Organisation (FAO) 2005; Coulthard et al. 2011). Of these coastal populations, it is often the poorest communities that are in the worst position to adapt to changes in their environment and livelihoods (IPCC, 2007).

Climate compatible development is defined as development that minimises the harm caused by climate change impacts, while maximising the many human development opportunities presented by a low emissions, more resilient, future (Mitchell & Maxwell 2010).

Incorporating climate change into development planning is challenging. Policy makers are often hampered by incompatible or conflicting policies, low organisational capacity, restricted financial resources and poor understanding of the predicted impacts of climate change at the local level (Mirza 2003). These constraints can lead to policies that may have negative impacts for the very people that they are intended to benefit.

Typically, resources (financial and human) are finite in implementing development policy. Tackling issues of poverty and climate change requires planning so that interventions and investments are coherent, coordinated and therefore able to reach those that need it most (Celliers et al. 2013). This often means multiple levels of planning. National strategies frequently feed into sectoral plans which feeds into regional planning. At each level there are a myriad of decisions taken that can affect how a policy is applied and who may win/lose from its application (Wever et al. 2012). Sectoral planning can often be seen as hampering integrated planning such as Integrated Coastal Zone Management (ICZM) (Taljaard et al. 2012).

Of course, planning is not the only challenge in terms of balancing needs and priorities. There are numerous examples where policies may have been developed but government departments lack the capacity and resources to implement such policies leading to poorly integrated investments. Developing countries depend considerably on the financial support of aid agencies – either through loans or concessions – thus the implementation of policies affecting the coastal zone are strongly correlated with the policy and priorities of aid agencies such as the World Bank.

Whilst climate change is not a new or emerging issue, developing countries have been slow to respond to the issue. This has meant its integration in national planning, sector planning and regional planning has often been poor. Knowledge of climate change and its predicted risks is often very low at the sub-national level – in both developed and developing countries.

The aim of this paper is to describe participatory methods that can be used by government policy makers for the design of a CCD policy. It involves comparing a business as usual (BAU) scenario to a climate compatible scenario using evidence from multiple sources (social, economic and political) to develop an understanding of the potential for a climate compatible future and what the costs and benefits of achieving this future would be.

## Case Study Description

Fieldwork was conducted in 2 countries, Kenya and Sri Lanka and focused on the following sites:

- Kwale county, south of Mombasa
- Puttalam District, north west Sri Lanka

## Step 1: Understanding the Policy Environment

A review was undertaken of the existing legislation and policies pertaining to coastal development to understand whether there were existing opportunities for supporting climate compatible development. The areas of policy and legislation that were reviewed are as follows: Coastal zone management; Climate change; Forestry; Fisheries; National development framework; Planning; Environment; Land; Tourism.

An additional component of this review was to identify potential mechanisms (regulatory and fiscal) that could support climate compatible development in the coast.

### Kenya

From the review it appears as if there is a strong enabling environment for climate compatible development in Kenya. The legislation and policies clearly state an intention to achieve sustainable management of the environment that provides benefits to the human population. The references to climate change in the more recent Acts and policies also highlight the intention to incorporate issues of climate change into future development of Kenya's coast.

Kenya introduced performance contracts as part of its public sector reform which means each level of government has set itself measurable targets. The performance contracts should provide great insight into the capacity for government departments and Ministries to achieve the intended climate objectives. There are some programmes and projects planned that will support the intended objectives of climate compatible development and should be followed up. The lack of indicators specific to climate compatible development does mean it will be difficult for the Kenyan government to understand its contribution to the issue in future years. With the launch of the National Climate Change Action Plan in March 2013 this may improve with departments expected to report specifically on the issue of climate change. This will require substantial support however since even at the national level there is limited capacity and understanding of climate change issues.

**Sri Lanka**

Sri Lanka too has an enabling policy and legislative environment for climate compatible development, but this is largely on paper, with much scope for increased implementation and coordination between sectors.

While the review suggests there is a strong policy environment, implementation is not always supported by adequate resources, including financial, and coordination for effective policy implementation between sections and levels of government is needed.

## Step 2: Identifying Mechanisms for CCD

From the review of policy and legislation in Kenya and Sri Lanka of relevance to coastal areas, the following mechanisms have been identified that could contribute to a shift towards CCD:

- Coastal zone management
- Collaborative management/co-management
- Climate change mainstreaming
- Climate funds/grants
- Protected area status
- Regulations, including licensing and permits
- Extension support
- Ecosystem-based approaches
- Land use planning/ local development plans
- Environmental Impact Assessment (EIA) /Strategic Environmental Assessment (SEA)
- Land tenure arrangements
- Payment for ecosystem services

Most of these mechanisms are already supported by policy and legislation in Kenya and Sri Lanka, apart from Payment for Ecosystem Services (PES) being explicitly recognised as an instrument for natural resource governance.

## Step 3: Identification of Supporting Scenarios for CCD

Focus groups were used to develop BAU and CCD future scenarios for both Kenya and Sri Lanka for a 10-year horizon. The focus groups comprised of local stakeholders from a variety of economic sectors, Non-Governmental Organisations (NGO), Community-Based Organisations (CBO) and government agencies. The groups were led through a series of visioning exercises to narrate the eventual scenarios, which ensured a multi-disciplinary view of pressures on the area and potential for change. The focus group scenarios were further validated with expert interviews of stakeholders in Kenya (n=63) and Sri Lanka (n=65). These interviews sought to determine trends, and information on current conditions and existing mechanisms used to tackle pressing issues related to climate impacts and development.

The outputs of these focus groups were a 'Business as Usual' scenario and a 'Climate Compatible Development' scenario. These were story-based narratives that identified what would change, what the drivers of this change was and what the effect these changes would have on the political, social and environmental conditions of the area (Puttalam in Sri Lanka or Kwale in Kenya).

A shortened example of the Kenyan Business as Usual scenario and Climate Compatible Development scenario is given below.

**Kenya: Business as Usual Scenario**

By the year 2023, the population of the Kwale coastal area has nearly doubled. This has put pressure on adequate drinking water provision, sanitation services, electricity provision, and traffic density. The road from the airport has finally been completed, which has given a boost to the local economy as more tourists are making their way to Diani and other coastal resorts.

An urban expansion has resulted in the last remaining unprotected forests in the surrounding areas being cut down; even the protected *Kayas* (sacred forests) have suffered from illegal encroachment. The loss of ecosystem services has been substantial, and the appeal of this natural environment has dropped for tourists who come for a holiday.

The Diani Resort City has been built, but with poor marketing of the resort internationally it has not managed to attract the levels of tourists to reach a break-even point for the additional hotels it has built in Diani.

The drastic increase in population has also negatively impacted the labour market even more than expected. Unemployed and landless local residents and migrants seek refuge on unoccupied land near Diani with makeshift homes, thereby broadening the gap between the rich (e.g. tourists, hotel owners) and the poor (e.g. local residents, migrants). The all-inclusive hotels are still the norm, and they continue to advise tourists to remain within the premises, worried that petty violence – such as muggings – will make international headlines and therefore further disseminate negative publicity.

Despite the demographic challenges described, primary and secondary school attendance has continued to rise, and literacy rates have been higher than ever. However, those with a high school diploma seeking employment in rural areas are struggling because schools and institutions that teach specific trades and professions are too far away and thus too expensive to attend. The tourism sector provides few well-paid jobs for locals due to this same lack of opportunities for training and skills development.

**Kenya: Climate Compatible Development Scenario**
Strategically placed rural education centres have provided the necessary skill-focused training to generate better employment opportunities for local residents of the Kwale coastal area.

Greater access to financial and business education means there are more locally owned businesses. This has led to a wider market reach. As a result, local residents have slowly been able to improve their livelihoods by not spending all of their savings during the low season. This has significantly reduced the amount of disgruntled people that could start riots similar to the 1997 Likoni clashes. The more stable flow of capital into the local tourist economy has created a more stable work environment for many that are employed in the tourism value chain.

Opportunities for local businesses to cater to tourists have grown since the all-inclusive hotel model has proven unsustainable, and therefore unsuccessful. Self-catered accommodation served by small local businesses trading locally grown foods has become popular with Europeans looking for a different, local experience. An upgrade of several local roads and an improvement in infrastructure maintenance has helped facilitate all this.

There is a marked increase in tourists (50% on 2012 figures) to the Kwale coast due to effective tourism marketing with international tourists coming from Europe, the USA and East Asia. A focused campaign within Kenya and neighbouring countries has also increased the proportion of domestic and regional tourists (50% of tourist numbers). After several incidents with terrorist attacks on the North Coast, it is these tourists that have kept the South Coast tourism sector alive until international tourists started returning again. The contrast between high and low seasons has simultaneously become less sharp, due to quick coastal getaways for affordable prices becoming popular with the growing Kenyan middle-class during the rainy season.

Demand for natural resources including wood and fish has increased due to urbanisation and population growth. However this demand is not proportionate to population growth since increased electrification and access to international markets means fewer people are living a subsistence lifestyle.

Community control of natural resources, especially forests and fisheries, has helped address poverty and promote greater equality. The growth of community organisations has also led to improved transparency and governance at a local level.

**Plausibility of Scenarios Produced**
For these scenarios to be of value to policy makers in Kenya and Sri Lanka, it is essential that they are seen as plausible – that is, they are not ideas with no grounding in reality. Scenarios that require substantive changes to policy, legislation and institutions are seen as less plausible than those that do not require major changes. This is because the former may require greater investment (when resources are limited), more time (when time is pressing) and may rely on increasing capacity of existing institutions (when capacity may be limited).

The participatory approach used ensures that knowledge and experience of the local situation are built into these scenarios. In all the scenarios developed, the focus groups and interviews identified *how* it would be possible to get from BAU to CCD – that is, they identified the mechanisms necessary. For example, in both Kenya and Sri Lanka investment in localised vocational education was seen as a mechanism to increase employment prospects for local communities. In this way, tourism expansion could have a more equitable outcome and require less coastal migration.

Findings from focus groups and interviews were supported by the review of the existing legislation and policy in both Kenya and Sri Lanka. The review showed that for every mechanism identified by the stakeholders there is at least one existing piece of legislation, policy or government strategy that supports the application of this mechanism. Given the potentially supportive policy and regulatory environment in both countries, we believe that the CCD scenarios are plausible.

## Step 4: Providing Evidence of Economics of CCD Choices

Using the scenarios outlined by the focus groups we attempted to assess how the economic value of ecosystem services would change under alternative coastal development futures. The aim was to demonstrate how ecosystem valuation can be used to strengthen coastal development planning and implementation: by making the economic case for green CCD options, and highlighting needs and opportunities to capture ecosystem values as incentives and financing for CCD. The results for Sri Lanka are presented below (summarised from Emerton 2014, Emerton et. al. in prep.). The economic study assessed trade-offs associated with alternative land use development options in Puttalam Lagoon, with a view to demonstrating the opportunity costs of mangrove conversion in terms of climate compatible development benefits foregone. It built on work carried out under the iCoast project on the changes in land use and carbon storage that occurred in Puttalam Lagoon between 1992/4 and 2012 (Bournazel et al., in prep.) and on the relative profitability of alternative uses of mangrove land (Kumara 2014).

### Findings on the Economic Value of Mangrove Ecosystem Services

Six categories of mangrove ecosystem services were identified as being of key economic and livelihood importance in Puttalam: fuelwood and timber, non-wood/non-fish products, provision of breeding and nursery habitat for fisheries, water quality regulation, protection against saline intrusion, and carbon sequestration and avoided emissions[1]. The economic study indicated that the 731 hectares of mangroves in Puttalam Lagoon are currently providing these services to a value of some USD 2.7-2.8 million a year, or between USD 3,800-4,100 per hectare. These values are however much lower than they have been in the past. A net loss of just over 800 hectares of mangroves was registered over the period 1992-2012, most of which were converted to shrimp farms, agriculture, salt pans and other land uses such as settlements, buildings and infrastructure. This resulted in a progressive decline in the value of mangrove ecosystem services. The annual value of mangrove services today is almost USD 3.5 million lower than it was in 1992 (Figure 1).

[1]It should be noted that two 'classic' mangrove ecosystem services do not appear in Puttalam Lagoon: protection against shoreline erosion and extreme weather events. This is due to the fact that the sheltered lagoon/estuary system is not exposed directly to the sea, and mangroves are found only on the inside shores, not on the coastline abutting the Indian Ocean.

Figure 1: Puttalam Lagoon: change in mangrove ecosystem service values 1992-2012

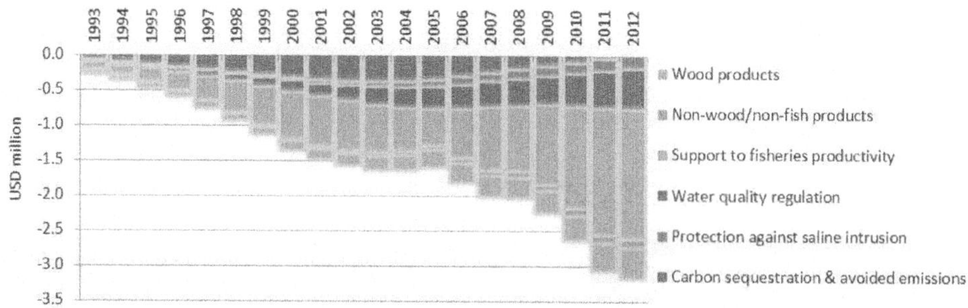


In total, mangrove ecosystem services worth almost USD 32 million (with a net present value (NPV) of USD 9.6 million) were lost between 1992 and 2012 (Table 1). This figure is around two and a half times more than the net income earned from the shrimp farms, coconut and salterns that were established on cleared mangrove land. Overall, the conversion and degradation of mangrove habitats has incurred economic losses of USD 18 million (a net present cost of just under USD 6 million), even taking account of the value of production from new land uses .

Table 1: Financial and economic impacts of land use change in Puttalam Lagoon 1992–2012

| | Total value (USD Mill) | NPV (USD Mill) |
|---|---|---|
| Added income from shrimp farms | 13.29 | 3.77 |
| Added income from coconut plantations | 0.04 | 0.01 |
| Added income from salterns | 0.13 | 0.00 |
| Total added income from mangroves | 13.46 | 3.78 |
| Foregone value of mangrove ecosystem services | -31.54 | -9.61 |
| Net economic gain/loss from land use change | -18.07 | -5.84 |

**Capturing Ecosystem Values as incentives and Finance for CCD**

On the one hand, the findings from the Puttalam Lagoon study demonstrate the costs and losses that can be incurred from failing to factor ecosystem values into coastal development planning. Yet, while this kind of information can help to make a convincing (and usually much-needed) economic case for CCD, it is important to underline that the economic analysis of CCD options extends beyond just valuation. However much coastal ecosystem services are demonstrated to be worth in theory, this has little meaning unless it translates into real-world changes in the way in which policies are formulated and decisions are made, and is reflected in the prices and profits that people face as they choose between alternative land, resource and investment options (Emerton, 2013).

The economic analysis of coastal development scenarios in Puttalam makes it clear that what might be the most profitable or desirable or beneficial land, resource or investment choice from the perspective of the wider economy is not necessarily the one which has the most immediate appeal to coastal landholders. Despite the high economic losses that arise from the conversion and degradation of mangrove habitats, shrimp aquaculture, coconut farming and salt production all generate better cash returns and more immediate sources of earnings for landholders at their full production value[2] (Table 2). This is because the bulk of mangrove ecosystem service values accrue as indirect support to other production, as non-cash benefits, or are received by beneficiaries other than the landholder.

Table 2: Puttalam Lagoon: comparison of the returns to alternative land uses in mangrove areas

| | Full production value (USD /ha) |
|---|---|
| Direct income and physical products | 669 |
| Off-site and non-market benefits | 3,663 |
| Total mangroves | 3,830 - 4,115* |
| Shrimp farms | 5,846 |
| Coconut | 2,825 |
| Salterns | 1,911 |

*A range of value is given, as ecosystem service values are not additive. Some are mutually exclusive: for example extraction of wood products and carbon sequestration / avoided emissions.

In the light of these figures, it is hardly surprising that the large-scale conversion and degradation or mangrove habitats took place over the 1990s and 2000s, or that shrimp farms accounted for the majority of this land use change. The bottom line is that landholders can gain higher profits from converting mangroves to other land uses – there remain few financial or economic incentives to maintain mangroves on their land. The key challenge in Puttalam becomes one of moving beyond merely articulating the value of coastal ecosystem services, to identifying how the costs, benefits and trade-offs associated with alternative land use and CCD options are distributed between different groups – and where (and for whom) the gaps and imbalances lie. The aim is to identify the kinds of policy and management instruments that can be used to capture and rebalance these values, change the economic conditions and circumstances that cause landholders to convert or degrade ecosystems in the course of their economic activities, and instead set in place the economic opportunities and rewards which will stimulate the investments which are required to encourage, enable and motivate ecosystem-based CCD.

## Discussion

The visioning exercise we carried out in Kenya and Sri Lanka has produced scenarios based on substantive stakeholder perspectives. In each case, our stakeholders developed a CCD scenario that was compatible with existing policy and legislation. The method proposed enables stakeholders to identify *mechanisms* that could be promoted to achieve CCD.

The benefit of the method is that it avoids unrealistic thinking in the development of future scenarios that require significant shift in policies, regulations and investment in

[2] In other words when operating at peak output levels: once the enterprise has been established and, in the case of shrimp farms, before productivity starts to taper off.

governance structures. For Kenya and Sri Lanka we have provided policy makers with a series of options for investment that will improve local conditions and meet the principles of climate compatible development.

Furthermore, we have highlighted to policy makers the mechanisms that can be used to achieve this equitable future. As opportunities to apply these mechanisms already exist in policy and legislation it suggests that CCD is not just a pipe dream in Kenya and Sri Lanka – it is a plausible outcome.

The economic analysis provides strong economic argument for green CCD options. They articulate the gains from investing in ecosystems as a key component of coastal development, adaptation and mitigation infrastructure. On the one hand, the findings from the Puttalam Lagoon study demonstrate clearly the losses that can arise from failing to factor ecosystem values into coastal development strategies. Since 1992/4, mangrove conversion has cost the local economy more than USD 30 million in foregone ecosystem service values. These losses are considerably greater than the additional income which was generated by shrimp farming and the other land uses that replaced mangroves (just under USD 13 million).

What is not yet clear from this work is whether the existing policies and legislation are sufficient for tackling the barriers to climate compatible development. That is, is there the necessary capacity and resources dedicated to these areas of policy and legislation? There is an opportunity here for authorities in Kenya and Sri Lanka to commit resources to existing legislation and policy in a way that will achieve a CCD future rather than looking to develop new, costly legislation to cope with the pressures of climate change.

Hopefully by highlighting that CCD is a real prospect and not yet another hurdle for policy makers, we may see improvements in regional CCD-related planning in Kenya and Sri Lanka.

## References

Bournazel, J., Kumara, P., Jayatissa, L., Viergever, K., Morel, V. and Huxham, M. (in prep.) The impacts of Shrimp farming on land-use and carbon storage around Puttalam Lagoon, Sri Lanka.

Celliers, L. et al., 2013. Pathways of integrated coastal management from national policy to local implementation: Enabling climate change adaptation. *Marine Policy*, 39, pp.72–86. Available at: http://linkinghub.elsevier.com/retrieve/pii/S0308597X12002035 [Accessed April 16, 2014].

Coulthard, S., Johnson, D. & McGregor, J.A., 2011. Poverty, sustainability and human wellbeing: A social wellbeing approach to the global fisheries crisis. *Global Environmental Change*, 21(2), pp.453–463. Available at: http://linkinghub.elsevier.com/retrieve/pii/S0959378011000045 [Accessed May 7, 2014].

Food and Agricultural Organisation (FAO), 2005. *Increasing the contribution of small-scale fisheries to poverty alleviation and food security*,

Gasper, R., Blohm, A. & Ruth, M., 2011. Social and economic impacts of climate change on the urban environment. *Current Opinion in Environmental Sustainability*, 3(3), pp.150–157. Available at:

http://linkinghub.elsevier.com/retrieve/pii/S187734351000148X [Accessed May 1, 2014].

IPCC, 2007. *ClimateChange2007:Impacts ,Adaptation and Vulnerability* P. J. van der L. and C. E. M.L. Parry, O.F. Canziani, J.P. Palutikof, ed., Cambridge University Press, Cambridge.

Mirza, M.M.Q., 2003. Climate change and extreme weather events: can developing countries adapt? *Climate Policy*, 3(3), pp.233–248. Available at: http://www.tandfonline.com/doi/abs/10.3763/cpol.2003.0330 [Accessed May 15, 2014].

Mitchell, T. & Maxwell, S., 2010. Defining climate compatible development. , (November).

Taljaard, S. et al., 2012. Implementing integrated coastal management in a sector-based governance system. *Ocean & Coastal Management*, 67, pp.39–53. Available at: http://linkinghub.elsevier.com/retrieve/pii/S0964569112001482 [Accessed March 23, 2014].

Wever, L. et al., 2012. Decentralization and participation in integrated coastal management: Policy lessons from Brazil and Indonesia. *Ocean & Coastal Management*, 66, pp.63–72. Available at: http://linkinghub.elsevier.com/retrieve/pii/S0964569112001044 [Accessed April 11, 2014].

**Coastal Management: Changing coast, changing climate, changing minds**
ISBN 978-0-7277-6114-9

ICE Publishing: All rights reserved
doi: 10.1680/cm.61149.425

ice
Institution of Civil Engineers
publishing

# Planning for Change on the Lincolnshire Coast

**Kevin Burgess**, CH2M HILL, Swindon, UK
**Paul Hunt**, Environment Agency, Peterborough, UK
**Marcello Cali**, CH2M HILL, Peterborough, UK
**Helen Jay**, CH2M HILL, Swindon, UK
**Deborah Campbell**, Environment Agency, Peterborough, UK

## Introduction

### Background

Around 20 years ago, the Environment Agency's predecessors embarked upon a strategy of intensive sand nourishment to prevent failure of the seawall over a 26km long frontage, thereby providing protection from tidal inundation to over 20,000 properties and up to 30,000 caravans, as well as key infrastructure, tourism assets, recreational amenities and agricultural land within a 35,000 hectare floodplain extending up to 15km inland. The present approach has been very successful and, up to now, has been considered the most appropriate long term strategy. However, changes to how schemes are appraised and choices made, have meant that a fundamental review of the sustainability of this approach has been required. This paper explores the implications of those changes for long term flood defence management along this coastline, highlighting how decisions on the most sustainable solutions are compounded by the present economic assessment approach, and why the current rules are restrictive in supporting a change in approach.

### Context

Coastal flood risk to the area is currently managed by a system comprising seawalls and a managed beach. In particular, beach nourishment has been carried out annually to address the lowering of beach levels and prevent undermining of the existing seawall structure, which could lead to a breach and resultant flooding of the low-lying hinterland. Since 1994, 15million $m^3$ of sand has been dredged from offshore licensed areas and placed along the shoreline, at a cost of approximately £130million. Current operations involve recharging approximately 500,000$m^3$ annually, spread over some 11km of coastline, at a cost of around £7million per year.

Historically the coastline has been prone to catastrophic flooding, with records of damage dating back to the medieval period. A change in attitudes towards defence of this shoreline followed development and expansion of the coastal towns and villages as tourist destinations in the late 1800s. This prompted the construction of seawalls and groynes along much of the shoreline. There was a major breach of the defences on the night of 31 January 1953 when a surge tide broke through in numerous places and over 40 people died in Lincolnshire alone, as a result of the flooding. Many defences were rebuilt in the aftermath; these have subsequently been improved and rebuilt over time, necessitated in part by depleted beaches and the potential for undermining of the seawalls. There is virtually no contemporary natural feed of sediment onto the beaches along this frontage; their current state relies on the annual re-nourishment to address continual movement and beach lowering in critical areas, without which the foreshore would return to a depleted state.

Although approx. 11million $m^3$ of re-nourished sand currently remains on the beaches, it has been established that if we cease re-nourishment, then most of the placed sand accumulated over the last 20 years will disappear, returning the foreshore to a state similar to that prior to the implementation of the current strategy. This will lead to a significant increase to wave exposure at the seawalls, higher levels of overtopping and a much increased probability of structural failure and breaching. Development lies directly behind the seawall in many places, and the consequences of this are unacceptable with uncontrolled flooding of property and considerable risk to life.

## Options for Future Coastal Defence

This can be considered an artificial coastline; it has been created and continues to be managed by man. Without any intervention the coastline would not revert to a 'natural, self-sustaining state' that would be compatible with the current use of the hinterland. Continued management of the coastal frontage is therefore necessary to maintain these existing land uses and to protect the social and environmental assets.

There are a range of approaches available to providing flood defence, which can be broadly grouped into the following high level concepts:

1) Protecting through holding the present line *with* beach nourishment;
2) Protecting through holding the present line *without* beach nourishment.
3) Compartmentalising the coast and addressing the management of each self-contained cells on the basis of local requirements.

These concepts are outlined below and summarised in figure 1.

### Concept 1 – Holding the Present Line *with* Beach Nourishment

This concept is to maintain the present defence line (the seawall) with a beach along the full frontage to provide a single and continuous littoral system, i.e. similar to present. The beach and the seawall together will provide continued protection to the local communities while also maintaining an amenity beach for recreation and tourism.

This assumes that there will be ongoing beach re-nourishment, either regular or occasional. The nourishment may be similar to that at present, or modified in some way in terms of volume, frequency, sediment size or placement approach. Those alternatives included single larger 'one-off' campaigns every few years and/or placement approaches such as sub-tidal stockpiles, combinations of recycling and re-nourishment, or using different (coarser) beach material. The idea of the Sand Motor, as currently being trialled in the Netherlands, was also considered amongst these.

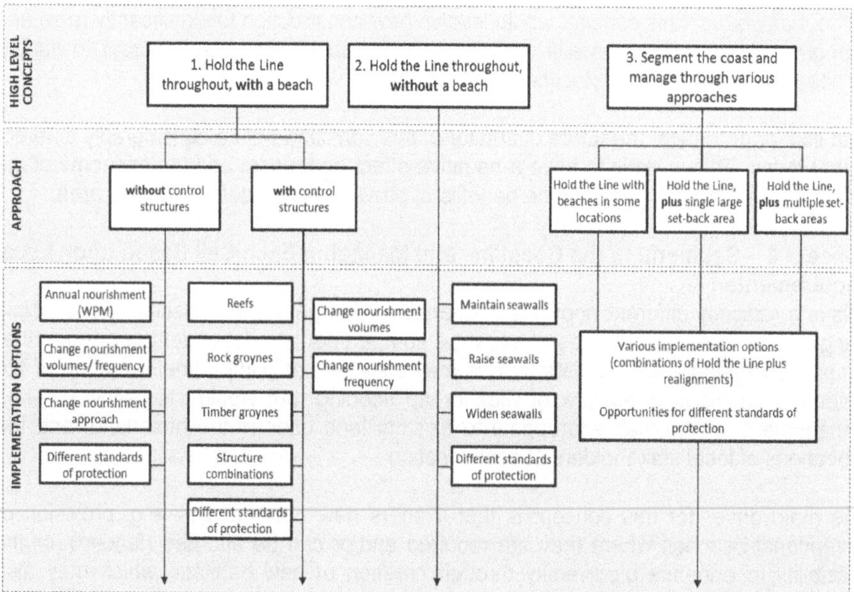


Figure 1: High level concepts, approaches, and implementation delivery options

A key issue with the re-nourishment approach is the increasing quantities of sand that are required over time to keep pace with predicted climate change impacts; up to twice as much will be required annually, i.e. 1million $m^3$/year, to nourish to a similar standard of protection in the future. Even with alternative nourishments, the total quantity of material to be placed over the full 100 year period would be very similar to that required with annual nourishment. Therefore, the sourcing of material and thus the sustainability of these approaches may also be an issue in the long term.

There are two broad approaches to delivering this concept; with or without control structures. The intention of the control structures, for example rock groynes, would be to partially interrupt or slow the movement of material along the shoreline, but not halt it altogether. This would provide greater stability in the areas needed, retaining a greater sand volume for longer and thus require less frequent and/or lower quantities of re-nourishment. Indeed, it is estimated that this approach could reduce the total re-nourishment volume over the course of 100 years to just 25-33% of that required without control structures.

**Concept 2 – Holding the Present Line *without* Beach Nourishment**

This concept is to maintain the present defence line (i.e. the seawall) throughout the frontage, but to cease nourishing the beach with sand. This will result in a gradual narrowing and eventual loss of the beach, and the seawall structure will therefore form the primary defence.

Without any beach this concept would involve new construction to significantly raise and strengthen the existing seawalls, or re-engineering them altogether in order to ensure an adequate standard of protection is still provided.

With little or no beach, the future character of this coastline will be considerably different to that today. This is likely to have a negative effect on tourism and the economy of the strategy area, thereby reducing the benefits of providing flood defence to this area.

**Concept 3 – Segmenting the Coastline and Managing Each Cell Based upon Local Requirements**

This is a radically different approach to managing this particular coastline, the principle being to consider it as a series of separate compartments independent of one another rather than a single littoral system. This may also require compartmentalisation of the hinterland, to manage the risk of widespread flooding, but would also allow differing management approaches appropriate to existing land uses in the hinterland, and the aspirations of local stakeholders at each location.

One main driver for this concept is that it offers new opportunities: e.g. provision of recreational beaches where they are required and/or can be afforded (funded), or the possibility to enhance biodiversity through creation of new habitats, which may also present different opportunities for the local economy. Alternatives for any individual cell may therefore include:

- creation of discrete tourist/amenity beaches for private developers or for specific uses
- realignment to a setback position to create a cove with new naturally functioning shoreline, backed by a flood embankment, offering new habitat creation opportunities
- other entrepreneurial options which may attract additional revenue and funding, e.g. a waterfront development or a marina.

Without a reliance on actions elsewhere, this has the potential to allow the local stakeholders to influence the strategy to best suit their needs/ financial ability and also to incorporate local planning and development aspirations. This approach may accommodate situations where defending front line lengths of the coast is become increasingly problematic, and expensive to do, and may no longer be justified. An example would be to accept a lower standard of front line protection and construct secondary defences to prevent wider flooding if the defence be overtopped or breached.

There are many permutations with such an approach and some scenarios have been considered to provide comparisons with other options. For example, a scheme which simply provides discrete self-contained bays and (larger) beaches throughout the entire frontage would still require less than half the total quantity of sand to be placed than would be required if the present re-nourishment approach continued. Various hybrid approaches can be developed, combining frontages with different characteristics. For the purposes of illustrating this an example has been generated incorporating a large managed realignment along an undeveloped part of the frontage, headland controlled bays to the north, and combinations of seawalls and further bays to the south.

## Economics and Funding

### Cash Costs

This study has raised some interesting issues with regard to the industry approach to evaluating schemes and informing choice on the preferred way forward.

To illustrate this, cost comparisons have been made for five approaches:

- Nourishment, as present practice (Concept 1a)
- Beach control structures, e.g. groynes, enabling reduced nourishment (Concept 1b)
- Seawall only providing defence, ceasing nourishment altogether (Concept 2)
- Headlands and embayments throughout, with no littoral exchange (Concept 3a)
- Hybrid of embayments, set back and walls, with no littoral exchange (Concept 3b)

The five approaches have both different total cost outlays and different expenditure profiles relating to their implementation, as illustrated in figure 2.

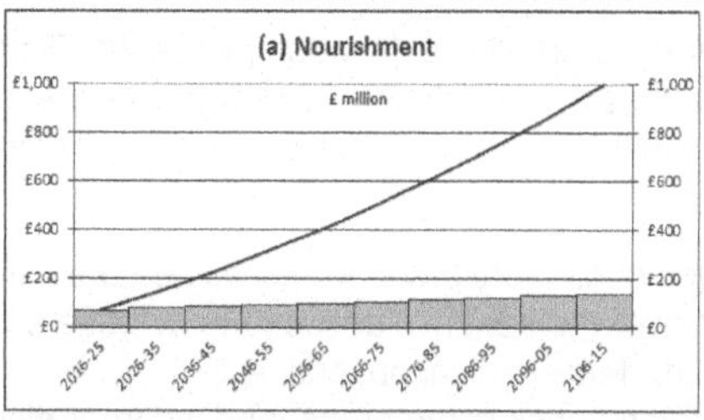


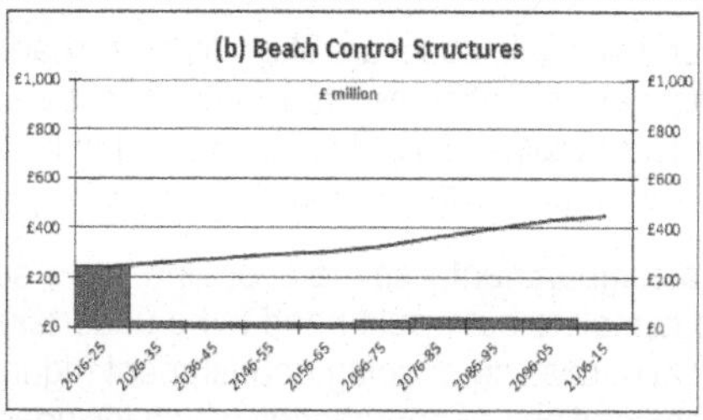


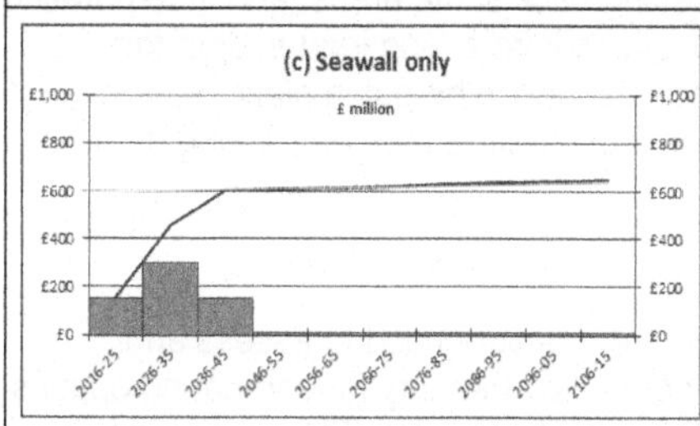


Figure 2:
Investment requirements by decade and cumulatively for each approach for:
(a) Nourishment;
(b) Beach Control Structures;
(c) Seawall only;
(d) Embayments;
(e) Hybrid

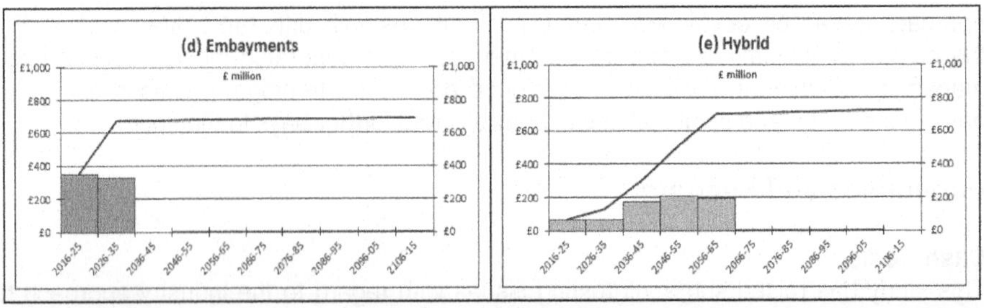


Although there are various options for implementation within these approaches, the most typical order of cost for that approach has been adopted. Allowances for predicted sea level rise over the next 100 years have been included in calculations of sediment quantities and defence structures.

Figure 2a shows that the cost of continuing with the present practice of regular re-nourishment is constant and relatively evenly distributed decade upon decade. There is a small increase in cost over time to continue to provide the same level of protection under rising sea levels.

Other options have quite different cost profiles however. To maintain the beach through the introduction of control structures would require a higher earlier investment, with only occasional recharge and modest maintenance thereafter over the following century (figure 2b).

For a seawall, costs are likely to be spread over the first three decades (figure 2c). This recognises that some areas would not lose the beach immediately, and in places the present wall is in good condition and able to withstand several years of exposure before major works to strengthen it would be required.

To compartmentalise the shoreline into a series of discrete bays, works would take time to complete but would need to be undertaken over the next two decades to achieve a goal of ceasing ongoing nourishment (figure 2d). More radical options, e.g. combinations of approaches which included setting back some areas, and incorporating walls without beaches in others, would realistically require a long lead-in time for acceptance, approvals, land acquisition and adaptations to be put in place, so are assumed to require a 20 year period to address those before then embarking on a 30 year implementation (figure 2e).

**Discounted Costs**

Straight cash cost comparison is not, however, the measure used to assess and determine a preferred scheme; the approach defined by treasury for comparing options has been to discount the costs to obtain the 'Present Value Cost' (PVc) over the scheme life, which are then compared with the Present Value Benefits provided by that scheme. Because of the diverse expenditure profiles of the various approaches, this generates a quite different picture, as presented later.

### Partnership Funding

Partnership Funding is a third component that now must be considered. Previous capital works to this frontage were entirely funded through central government with a 100% flood defence grant-in aid payment. Rates of payment for grant are now reduced and there is a general requirement for local contributions towards the cost of any works. Schemes are prioritised according to their Partnership Funding Score. This can be increased with higher local contributions, which improve a scheme's chance of a grant allocation in the desired year. The level of contributions necessary for a scheme also varies depending upon the costs and relative benefit to cost ratio of each option.

Local rates/council taxes may not be a permissible way to generate these additional contributions; so this may need to come from local businesses (such as holiday park/resort owners), investors, landowners, agriculture etc. who are willing to agree to contribute freely. These may need to be based on agreed parameters such as valuations against business rates and land area. The nature of, and proportions of, these may also influence option choice. For example:

- Continuing to provide a beach may mean tourism revenue remains constant
- If we don't provide a beach, there is potential for tourism revenue to drop significantly
- Providing bays or realignments/segmenting the coast may provide opportunities to bring in fresh additional revenue through attracting new businesses to the area

## Discussion

### Cost Comparisons

It is usual to adopt a 100 year period for evaluating schemes. Figure 3 summarises the cumulative costs of the five options: figure 3a shows cash costs, 3b shows the discounted present value costs[1] (PVc), and 3c illustrates the partnership funding contributions required.

Figure 3a clearly illustrates that the present practice of regular re-nourishment is in the long term by far the most expensive approach in cash terms. All other options would cost measurably less, and holding a beach with control structures substantially less, at no more than half the total cost of continuing with the present approach.

When we compare options on a discounted basis (figure 3b) a very different picture emerges; the highest cash cost option (present nourishment) has the lowest PVc. This is because the expenditure profile of this is spread evenly year on year, whereas the other options require a larger single outlay in the earlier decades. It is only as we consider the impact of discounting costs over a longer term that another option (control structures) becomes comparable.

The cumulative partnership contributions required are then compared in figure 3c. This illustrates the difference between options; some requiring a large 'one-off' (or relatively

[1] Strictly speaking, the horizontal axis in figure 3b shows the PVc depending on scheme life rather than this showing an expenditure profile, but it is useful to present this way to discuss the comparative costs.

short term) investment, whilst others, notably where there is an ongoing commitment to nourishment, requiring less up front but a continual commitment into the longer term. To this latter point, it is clear that continuing with the present management practice does not require such large sums to be contributed at any point, with approx. £2Million per year initially, but taken over the scheme life will ultimately require a higher total level of private investment than any other scheme.

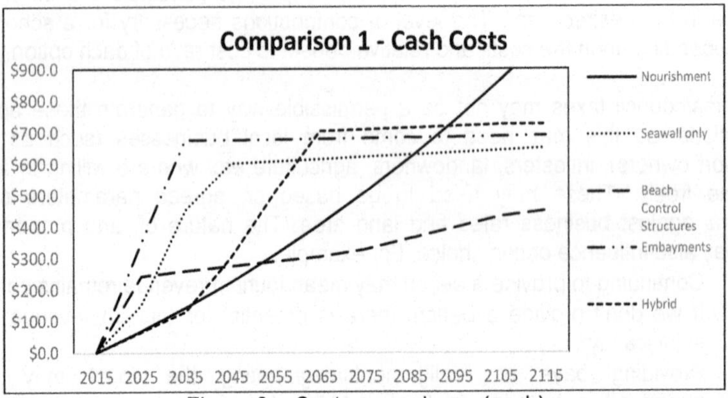


Figure 3a: Cost comparisons (cash)

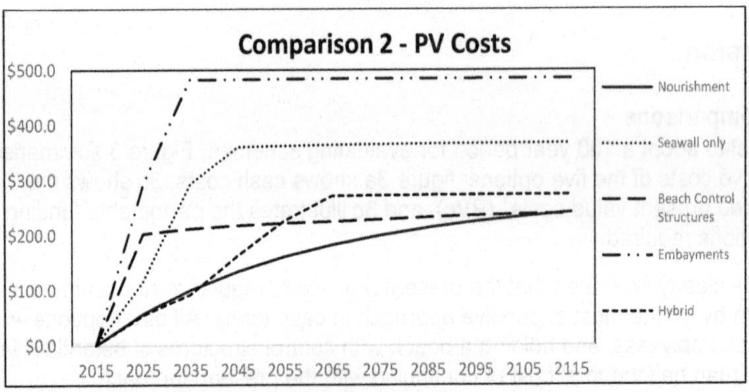


Figure 3b: Costs comparisons (PV)

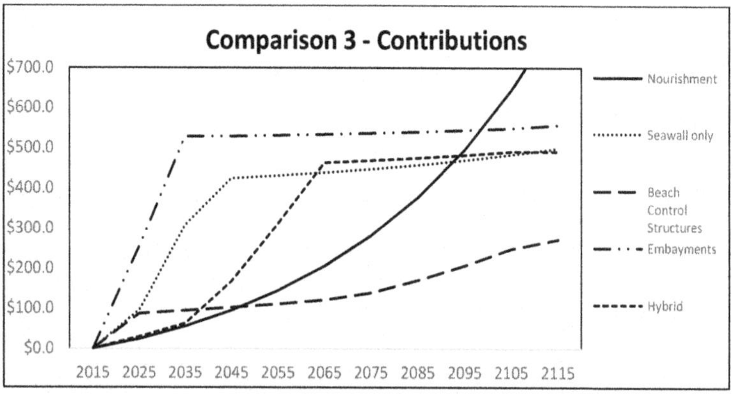


Figure 3c: Cost comparisons (partnership funding contributions)

**Implications**
Under normal circumstances we might simply take forward the scheme having the most favourable PVc, which in this case would mean carrying on with the present annual nourishment campaigns. However, this has other implications which are economic-related but not captured by the metrics described above.

The current approach has little legacy benefit in terms of risk avoidance. The nourishment has to be repeated each and every year to ensure that the required standard of protection is provided along the full 26km length of defences. In essence, the annual campaigns only provide the required protection for a single year; it is only if it is repeated annually that protection over a 100 year economic assessment period is provided. A different outcome is achieved from the other approaches; a scheme with control structures can be designed such that the required standard of protection is provided for several years between nourishment campaigns; whilst the compartmentalisation options would be designed such that long term protection is provided without further nourishment being required.

Critically, however, the cost of addressing this risk issue is strongly influenced by the commitment of third party funders to partnership contributions. Continuing with the present annual nourishment approach requires an in perpetuity commitment from third parties. What assurances can be provided that these contributions will continue to be forthcoming in a few decades time? If for example tourism of the type enjoyed here ceases to be fashionable, then that source of funding may cease almost overnight. Although it is not unreasonable to expect that those no longer receiving any revenue from a scheme would therefore no longer contribute, the bigger issue is the reliance on the same defences to provide on-going protection to life and property for thousands of people within the hinterland.

This residual risk would not exist with the other approaches, as already outlined, but why would businesses want to spend more now than they have too? Although other approaches will mean third parties need to spend less money in total,(as figure 3c shows for example even after just 30 years a scheme with control structures is cheaper), the present approach of annual nourishment would require a much lower *annual* contribution from them. At the present point in time this is likely to be more affordable/ acceptable to such parties.

This does though highlight another potential issue with respect to legacy risk. The funders' choice on how to invest their money may not actually be the best in the longer term for the wider community who could be affected by the consequences of that choice. For example, if the source for a significant amount of the third party funding is going to be local businesses, their ability and willingness to contribute is most likely and understandably going to be driven by immediate benefits to themselves rather than coastal defence to the wider population several decades in the future. Alternatively though, it may be that the option selected, for example compartmentalisation, actually inspires confidence in the future and acts as a catalyst for increased wider investment. We just cannot tell at the moment.

In summary if we are simply considering sustainability of finite resources (sand for nourishment), total cash expenditure, and the long term avoidance of risk to people and property, the annual nourishment approach would be least favourable. However, the current way in which we consider the economics of a scheme such as this, does not support that conclusion.

So what is the answer? In the past other approaches to evaluate schemes such as multi-criteria analysis have been developed and promoted, but are still to be pursued and introduced as primary decision making tools. Economic viability and affordability will always need to be at the core of the decision, but this project example demonstrates how we need to also be able to include in those calculations: the comparative sustainability of the options; the benefits of legacy protection provided by different approaches; the consequential risks arising from uncertain futures, of which long term third party funding commitments is a considerable one; and the robustness of the options to different circumstances, such as their resilience to 'above design' events.

## Conclusions

### Economic Conundrum

This paper has presented some of the options available for long term flood defence management along this coastline, highlighting how decisions on the most appropriate solution are compounded by the current economic assessment approach, which means that:

- The most expensive whole-life cost (and arguably least sustainable) solution returns the lowest discounted PVc. This also requires the highest total partnership contributions from local business/communities, albeit a limited (but not inconsiderable) amount year-on-year.
- Alternative schemes return a much less favourable PVc, due to them requiring higher upfront investment over a shorter time period (20 to 30 years), despite the fact that they may provide more sustainable solutions in terms of resources and

long-term investment commitments, and ultimately require only half as much contribution from both government and local partnering.

Along this coastline, continuing present practice not only requires the highest level of investment (over £1billion), it yields no improvement in the situation over the present day; we are simply maintaining the status quo. If for whatever reason nourishment ceases with the present approach, whatever monies have been spent over the previous decades is lost; each year's works effectively just 'buys' another year of protection. That is not the same with other options, where structures have been built and would remain if spending stops. These have lower levels of expenditure overall, providing the same outcome in terms of protection but with a longer lasting legacy, more certainty over risk avoidance, and with a lower demand on a finite resource.

**Partnership Funding**

The requirements for partnership funding are especially difficult on this particular scheme. Unlike many schemes which require 'one-off' partnership contributions over a short period of time, the size and duration of contributions required here for the leading option in PVc terms are both significant and continuing over indefinite timescales. Those who may need to contribute are moving from a situation of paying nothing at all to these considerable sums, and it is not immediately obvious how to generate that revenue. Although other options exist which require less total investment, these also require contributions of several hundreds of millions of pounds.

An important consequence of partnership funding is that parties have a stronger voice than before. Demonstrating this point, it was agreed that the option of a seawall only without a beach would not be considered further. We have a responsibility to give those funding the scheme different options to make their own choices on the matter, but subject to that funding being obtainable, the nature of that funding and the manner in which it can be secured, *these* will be the primary factors in deciding which option to take forward.

In short, the concept of partnership funding has obvious merits, but it is contended here that the implications of this have not yet been fully tested, particularly its application to schemes, such as this one, that involve significant financial commitments over very long time periods.

**Next Steps**

Although a partnership funding group has been convened along this coastline, at the time of writing this paper it is almost 2 years since the concepts above were first presented, but agreement on an acceptable funding mechanism is still to be achieved.

Further work is underway to explore possible solutions to progress the strategy. These are not discussions on the technical aspect of different approaches; indeed all parties are agreed that a move towards solutions that can provide greater certainty of outcome than the current practice are advantageous, and some of the opportunities arising from a degree of compartmentalisation of the coast are of great interest. The focus is all about how to obtain the third party funding, how much can be generate and when this can be generated, as this will define the approach that will be adopted for the next several decades.

This is a new issue for those with properties and businesses along this coast to have to deal with, and will take considerable time to resolve. So whilst the long term strategy cannot yet be agreed with partners, there is a general presumption by them that the current method of protection will be continued for at least another five years, to 2020. In the interim, partnership contributions will be sought to help fund continued nourishment for the period 2015 to 2020. It is currently assumed that agreement will be reached within 2 years and that the overall 5 year contribution will be paid in the final 3 years. Thereafter, from 2020/21 it is assumed (or hoped) that works arising from the agreed approach will commence with the required contributions. The nature of that approach, however, remains to be decided.

**Coastal Management: Changing coast, changing climate, changing minds**
ISBN 978-0-7277-6114-9

ICE Publishing: All rights reserved
doi: 10.1680/cm.61149.437

# Realignment of Coastal Defences by Land Owners

**Robert Caudwell**, Land Owner/Farmer, Lincolnshire, UK

## Introduction

For hundreds of years land around the "Wash" in South Lincolnshire has been reclaimed from the sea using traditional methods of raising sea defences using material accreted on the saltmarsh by tidal flows bringing silt into the "Wash" which has been eroded from land to the north.

This north-south drift is a continuous process that has enabled 1000's of hectares of the most versatile and productive land in the U.K to be reclaimed for agriculture and horticulture.

In Roman times the foreshore around the "Wash" was used for salt production using shallow pools which evaporated leaving salt deposits, the draining of the fens in the 17$^{th}$ century required the keeping out of tidal inundation and building of sea defences using the local material on the shore line.

This has continued with further reclamations by building banks further out on the saltmarsh, often in times of need by the U.K. of additional food production or high prices of commodities in times of shortage.

In our part of the coastline midway from Boston northwards towards Skegness reclamations have taken place in the 19$^{th}$ century during the Napoleonic wars often referred to as the 1810 defences. (Local folklore says that French Prisoners of War were used to construct the sea defences). The 20$^{th}$ century defences were constructed in 1947 and most importantly for the purposes of this paper in 1977 (known as the Jubilee Bank or enclosure).

It is the 1977 enclosure or "Jubilee Bank" that I will be referring to in this paper and where the issue of realignment of defences is current.

## The Jubilee Bank

The "Jubilee Bank" or enclosure was constructed using traditional methods of using material from the saltmarsh to construct the bank, no material was imported so the bank consists of silt soil relying on the vegetation growing on the bank to stabilise the soil material.

To a modern engineer this would not be the way you would look to build sea defences designed to withstand the waves from the North Sea.

This method of construction is known as using "borrow pits" this refers to soil being borrowed from the saltmarsh and being replenished by fresh silt brought in by tidal inundation. The rates of replenishment are contentious with conservation groups claiming that this process takes many years and land owners claiming that the process on takes a few seasons. The reality is that there is no definitive answer, silt deposition rates vary widely across the "Wash" and lifetime studies have not answered this question.

The driver of the construction of the "Jubilee Bank" was the high prices and shortages of horticultural crops during the 2 preceding years that were very dry and warm. Potato prices were particularly high reaching peaks of £250/ tonne. In today's prices and costs, prices would need to be in excess of £ 1500/tonne to compete. As this is written potato prices are less than £40/tonne across Europe.

In order to build the "Jubilee Bank" a reclamation Company was formed by a group of landowners whose land adjoined the "Wash" these are known as frontagers and have long established rights to graze the saltmarshes and to reclaim the land in front of the land currently farmed. The land below mean high water mark belongs to the Crown (or Duchy's) and a conveyance of the bottom of the creeks was necessary before work could commence.

The construction during 1977 was undertaken by contractors using "Draglines" to strip the vegetation of the area where the bank was constructed to enable soil from the "borrow pits" to be placed there and the material with some vegetation to be replaced on the front of the new bank, a 1 in 3 slope was achieved on the front face but the back was very steep with almost no slope at all, this became critical in December 2013.

Following construction in 1977 there was a storm tide in January 1978 which breached the bank causing realignment of the bank in sections during 1978.

Maintenance of the bank by the landowner has continued since construction and has involved sand bags, replacing eroded material and reseeding from 1978 to 2013.

## December 5th 2013

On the 5th December 2013 we saw the most serious storm surge since the 1953 floods.

The combination of high spring tides and the coastal surge peaking at approximately 2 meters caused flooding across the East Coast of the U.K. Over 1000 properties were flooded across the South Humber bank and Boston.

In 1953 Boston measured 5.3AOD. In December 2013 Boston measured 5.91 AOD overtopping defences in the town.

Agricultural land was flooded with 2,454 hectares flooded on the South Humber bank in several places.

The "Jubilee Bank" breached in 2 locations, 1 large breach of 50-60 meters width and a smaller breach of 20-30 meters. There was also some damage to other parts of the Jubilee Bank which required immediate repair, before following high tides could widen or breach the bank in other places.

This picture was taken on the 6th December, showing the breach points and the area of flooding of over 200 hectares.

Figure 1: Farmland flooded showing breached defences

The photograph clearly shows that the sea water was enclosed by the 1947 bank to the rear preventing further flooding of agricultural land or properties. It can also be seen that the sea water has remained on the site even when the tide has receeded.This is due to the continued accretion of the saltmarsh after reclamation making it higher than the land inside the "Jubilee Bank". This effect can be seen more clearly on this next photograph.

The next step was to try to prevent further sea water coming into the site by placing 1 tonne bags filled with soil as a temporary barrier; this could only be achieved by travelling on the marsh at low tide.

The water could then be pumped out using diesel pumps, pumping over the 1947 bank into the drainage system and out to sea via the Witham 4th Internal Drainage Boards electric pumps.

This was all done with agreement with the Environment Agency, Natural England and the Internal Drainage Board, but at landowner's cost.

Figure 2: Seawater flowing onto farmland via breaches

## Recovery

Having pumped the sea water off the site it was clear that the breaches were going to be difficult if not impossible to repair as the water had scoured a hole where the bank had breached to a depth of 7-8 meters below ground level, to repair the breaches would have required the filling in of these holes to provide a firm base to build a new section of bank.

Discussions with engineers made it clear that this would be expensive and need a large amount of imported material to fill and build the new bank and the risk of failure was great.

Alternative options were to build cross banks to the north of the breaches and abandon an area of land, or to attempt building around the breaches to minimise the loss of land.

The decision was taken to realign the defences, sacrificing some arable land but providing the best long term prospect of defence at a cost which the landowners could reasonably afford as there was no government support at this stage.

The material was to be taken from the land to be sacrificed to reduce cost and haulage of new material. This picture shows the first stage of building the new bank joining the "Jubilee Bank" to the 1947 bank.

Figure 3: Start of new bank construction using local material

## Realignment

A local contractor was used for this work to design and build the new bank with a 1 meter higher crest and a very long slope on the front and a 1 in 3 on the back of the new bank to provide greater stability in the future.

All the material was taken from the area in between the "Jubilee Bank" and the 1947 bank, whist the breaches were "bunded" using material from the site as well.

This was done so the new bank will have a full season of seeded growth in 2015 before it is exposed to sea water and wave action (possibly 2016).

Also the astronomic tides are at their peak and we wish to protect the new bank as much as possible

Experience had taught us that establishing vegetation and growth on the sea defences is key to withstanding high tides. This picture was taken in August 2014 showing the new bank nearing completion, but also the long area of foreshore that will hopefully tie the new bank in the future. The construction method is still traditional and would be understood by all the generations of bank builders in the "Wash".

Figure 4: Final works on new bank construction

## Saltmarsh Habitat

The building of the new section of defence between the "Jubilee Bank " and the 1947 bank has left an area of approximately 10 hectares of what was until 5th December 2013 grade 1 arable land , lying behind the old defences with temporary repairs, and in front of the new defences built during 2014.

The opportunity to create some habitat gain on this area is plain to see in this picture, but what and how has been more problematic. The land still belongs to the landowner and it will be his decision on how to proceed.

Saltmarsh habitat is a vital part of coastal defence in the "Wash" and a valued habitat for many species of birds and wildlife. Creation of new saltmarsh habitat is the obvious choice as it will protect the new embankment and add to a high value conservation asset.

There are government targets for creation of saltmarsh habitat to be delivered by the Environment Agency and other government bodies, but support for private landowners to register saltmarsh habitat as compensatory habitat for coastal squeeze to allow improvement of defences in the "Wash" has proved to be more of a problem.

Figure 5: Area of potential habitat creation

## Flood Defence Policy

The U.K. government's policy on flood defence focuses on protection of people and property. The Cost benefit analysis used to justify the building or improvement of defences uses the number of houses as a key indicator.

Agricultural land does not carry the same priority level and often struggles to justify expensive coastal defences where there are few houses but large areas of agricultural land.

To progress capital schemes under the Flood Defence Grant in Aid (FDGIA) calculator large contributions from beneficiaries would be required. The cost of applications to the FDGIA process would itself require thousands of pounds to be raised before any work was done.

I believe that this project of realignment on private defences, where the landowners faced with the prospect of no support had to deliver this project at minimal cost, may give an answer. And although it may not be engineered to modern standards for coastal defences, where people and homes are at risk, it is equal if not better than all the "Wash" coastal defences built using similar methods over the last 2 hundred years.

The cost of this new section of defence is extremely low compared to coastal defences being constructed or repaired elsewhere in the U.K. Whilst I would not want to give confidential economic data away; the cost per meter of construction of the new bank was less than £300.

Following the floods of 5th December 2013 and the winter flooding in the U.K. of 2013/14 the U.K. government launched a grant scheme for farmers to apply to help in flood recovery in order to get land back into production as soon as possible. This was administered by DEFRA and with good support from DEFRA officers a grant application was made for the land affected by the surge tide of 5th December 2013.

This covered some of the work done to recover the land, but not the loss of crops in the land at the time, or 2014 cropping which has been affected .But this support has been welcome at a time when due to falling commodity prices many agricultural businesses are under pressure.

## Conclusions

Is this type of grant aid process a more efficient way of supporting flood defence projects in rural areas than traditional FDGIA? For a small amount of money farmers and landowners could be encouraged to invest in flood defence, and using local contractors who can design and build defences with locally sourced materials minimising environmental impact, whilst allow National Bodies such as the Environment Agency to concentrate on protecting communities.

The need to improve defences around the "Wash" was brought into focus by the surge tide on the 5th December. All frontagers around the "Wash "have been contacted by a lobbying group known as the "Wash Frontagers Group" who are lobbying politicians for investment in the coastal defences. But waiting for others to take action may be too late, we cannot wait till another surge tide prompts action, improvements need to be done before the next peak of astronomic tides. This gives us a 15 year window at best.

Local action with cost effective support from government could be the answer, and that should be the focus of lobbying at the present. The role of the Environment Agency and Natural England as government bodies should be to enable this to take place as they did on the "Jubilee Bank" and allowing private landowners to register realignments to compensate for coastal squeeze.

This does not need time-consuming new legislation or the creation of new departments, just a flexible view and encouragement for local action. The cost benefit for the taxpayer, of flood defence funding from central government in this way could be improved further still, and the risk of flood damage which impacts on us all avoided.

The "frontagers" thank DEFRA, The Environment Agency, Natural England and Witham 4 IDB for their help and support during the recovery and realignment on the "Jubilee Bank".

I see no reason for not taking a proactive response to flooding, not just reactive, with small investments in maintenance and defence improvement working in partnership with Government agencies, Internal Drainage Boards and private land owners. Better standards of defence can be delivered at a lower cost in many rural areas.

Pictures: Mr H.C. Drake
Flood data: Environment Agency Lincolnshire and Northamptonshire Area.

**Coastal Management: Changing coast, changing climate, changing minds**
ISBN 978-0-7277-6114-9

ICE Publishing: All rights reserved
doi: 10.1680/cm.61149.445

# Wash-ing up an Adaptable Coastal Strategy

**Pippa van Kuijk MCIWEM C.WEM** Royal HaskoningDHV, Edinburgh, UK

## The Scene

A coastal management strategy is a non-statutory document that sets out the long term management approach of the coastline; it bridges the gap between a Shoreline Management Plan (SMP) and implementation of a scheme. The primary aim of a coastal strategy is to implement the higher level SMP policies. The coastal strategy develops a route to deliver technically feasible, environmentally acceptable and economically viable solutions to manage the risk to people and property from coastal erosion and flooding. The 2010 Flood and Water Management Act specifies the need for risk management of flood and coastal erosion and the introduction of Flood and Coastal Resilience Partnership Funding: a funding approach that provides transparency over funding levels for flood and coastal defence projects and allows local communities to have a greater say over what their coastal and flood defence management looks like.

These changes to legislation, a shifting economy and the uncertainties surrounding climate change into the future have meant a radical change in the way we conduct coastal management. Flood and coastal risk management approaches need to provide a flexible and adaptive response to uncertainties through setting a proactive management approach.

## Introduction

The Wash SMP, completed in 2010 identified a number of issues in relation to the management of the coastal frontage from Hunstanton Cliffs to Wolferton Creek in Norfolk, United Kingdom (Figure **1**).

The Wash East Coastal Management Strategy (WECMS) sets out to identify the most suitable way of managing flood and erosion risk to local communities, whilst protecting the internationally important natural and historic environments along this stretch of coastline. The strategy facilitates the wider delivery of a workable sustainable strategy for a changing coastline.

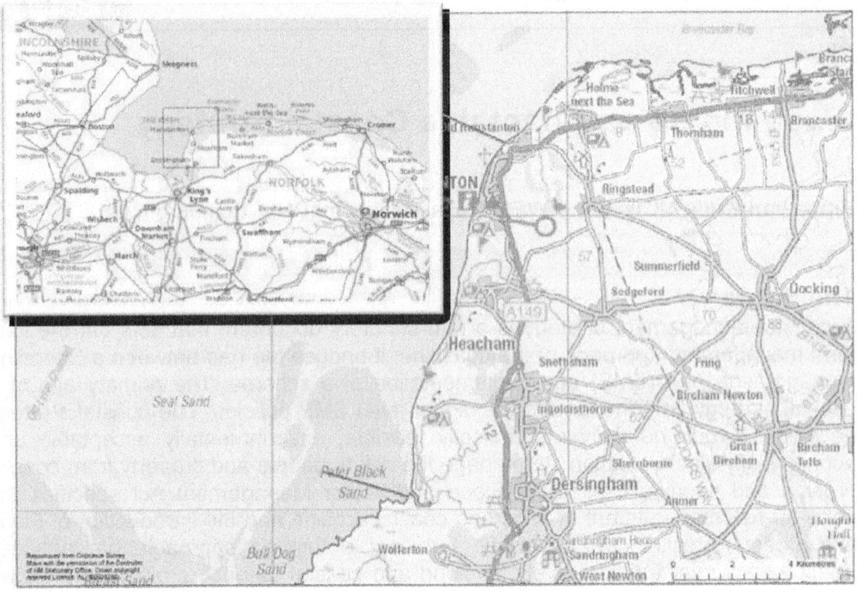


Figure 1: Wash East Location Map

## The Wash East story

### Working in Partnership

Working together with the Borough Council of King's Lynn & West Norfolk (BCKLWN), the Environment Agency has involved representatives from local communities and businesses to inform the Strategy. The Strategy makes a number of management recommendations for the short term (next 25 years) and an approach to future management over the long term (next 100 years) building on the work completed by the SMP.

A Key Stakeholder Group was set up during the development of the SMP to represent the interests of those who live and work in the flood risk frontage. The group comprises representatives from landowners, caravan park owners, holiday home owners, Local Councillors and Parish Council representatives. This group continued to guide the development of the Strategy. Each representative agreed to a set of terms and conditions which ensured that the local community was kept informed throughout the development of the Strategy. The Key Stakeholder Group ensured that the Strategy delivered the outcomes as initially guided by the SMP policies.

An Advisory Group was subsequently created to further support the Key Stakeholder Group; their interest additionally captured the views and included representatives from Hunstanton Town and the cliffs as well as the flood risk frontage.

Throughout the project lifecycle, these groups were engaged and a number of meetings and workshops were held at key stages:

- Objectives setting
- Baseline processes and scenarios
- Longlist optioneering
- Shortlist optioneering
- Preferred options; and
- Seeking funding.

This open and honest approach to the Strategy enabled decisions to be ratified by the stakeholders and project work rapidly progressed.

**The Setting**

The Strategy was divided into three distinct units; chosen because of the legal responsibility and their distinctive characteristics.

Unit A is an erosion frontage characterised by the infamous Hunstanton cliffs with tri-coloured cliff sediment, rock pools on the shore, a popular green space on the cliff top and locally important heritage assets. This is an undefended unit.

Unit B is an erosion frontage characterised by Hunstanton town, the local economy is drawn here by the variety of shops and amenities. Existing defences comprise a sea wall, promenade and a number of groynes of mixed function. The BCKLWN is responsible for both Units A and B.

Unit C is the flood risk frontage. The Environment Agency is responsible for the defences along Unit C. Unit C comprise the villages of Heacham, Shepard's Port and Snettisham, numerous caravan parks, holiday homes and low grade agricultural land. The existing flood defences are characterised by a shingle ridge interspersed with hard defences. Behind there is a steep-sided, narrow ridged earth embankment. The shingle ridge provides a standard of protection currently estimated to be between 10% Annual Exceedance Probability (AEP – the chance of flooding in any one year) and 2% AEP. A high proportion of the caravan parks and holiday homes are located in between the two defences. The settlements of Heacham, Shepard's Port and Snettisham are located behind the existing defences. The shingle ridge is currently managed through annual recycling with nourishment every ten years, or as required (Figure **2**). The hard defences are a mix of sea wall and concrete revetment, a number of them unsupported at the rear. An existing flood warning system is in place, supported by a Precautionary Evacuation Notice (PEN). This has been activated four times in the past 20 years.

Figure 2: Unit C Shingle Recycling

In 1953 there was a catastrophic defence failure and 65 lives were lost. The high tides in 1978 caused significant damage to the caravans and infrastructure between the two defence lines. In the winter of 2013, the highest water levels ever recorded on the frontage were seen resulting in localised breaching of the defences and damage to caravans and property (Figure **3**).

Figure 3: Unit C Storm Damage Winter 2013

**The Science**

This coastline is dynamic and diverse. The unique coastal processes have been investigated by using the latest survey data and engaging with local people about their experiences of the coast. Assumptions made as part of the SMP were built upon and informed the need to understand how this coastline works to help us best decide how to manage it in the future.

Coastal processes along the coastline are influenced by the constantly shifting sediment patterns in The Wash basin; a highly variable environment. The natural sediment drift along The Wash East frontage is from the north east to the south west. The Strategy confirmed that there is little interaction between the coastal erosion at Hunstanton cliffs to the north and the deposition of beach material to the south; this was a question raised by the SMP and answered early on in the Strategy and as a result focusing the option development.

Erosion at Hunstanton cliffs is estimated to be between 0.1m and 0.4m per year. At Hunstanton town, the frontage has been defended since 1885 to halt erosion. However should the defences fail, there would be a significant rapid response to the exposure to wind and waves and it is likely that the cliffs would recede to realign with those in Unit A. Units A and B are shown in Figure **4**.

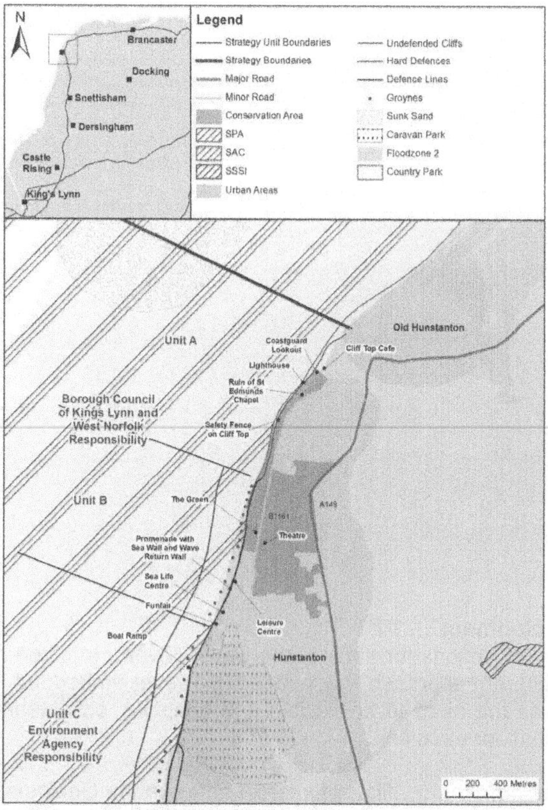


Figure 4: Units A and B

Unit C is a low lying vulnerable area (Figure **5**). The vulnerability is magnified by the low standard of protection provided by the existing defences. To the southern end of the frontage under Do Nothing the shingle ridge defences may only remain for another 3 to 5 years, with a reduction in standard in this time. To the north of Unit C the shingle ridge may remain for between 5 and 10 years. The hard defences would be expected to fail soon after the shingle ridge, not specifically to structural instability, though a solid back wall is missing on many of the defence lengths, but due to wash out from the back of the defences when the shingle ridge breaches.

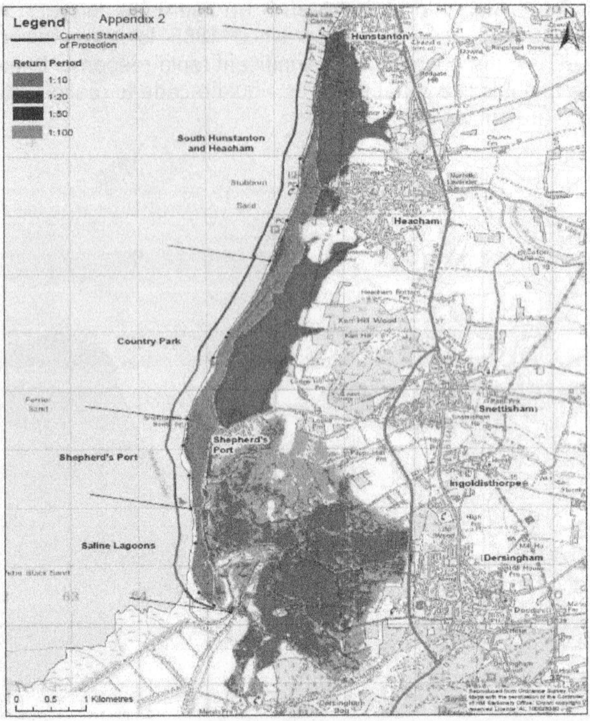


Figure 5: Unit C Flood Risk

**The Principles of Option Development**

The Strategy looked at a range of options for each of the Units. Given the different characteristics of each unit a different approach was needed. However embedded in the principle of the Strategy was the need to be flexible and adaptive, considering the variables and the uncertainties for each unit.

Unit A's complexities emanated from the conflicts of interest between the Geological SSSI for the erosion and colourful front of Hunstanton cliffs and protection of the locally valuable green space and assets, namely the lighthouse and the ancient St Edmunds Chapel. The SMP set out the management approach with a policy of No Active Intervention until much later when the road and properties were at risk of loss.

The BCKLWN was keen that the agreed regeneration plan for Hunstanton was supported by the Strategy outcomes. The regeneration plan focused on continuing the economic and social value of Hunstanton as it is now with no plans for further development in the area. The SMP mirrored this with a policy of Hold the Line across the unit.

The understanding of how to manage Unit C was drawn from the complex interactions between the environmental designations, coastal processes, the uncertainty of the impacts of climate change and the funding challenges that lay ahead. As a result, flexible, adaptable and affordable solutions are necessitated. The SMP unfastened the ability for a flexible approach through a policy of Hold the Line in the short term and an optional Hold the Line or Realignment policy in the future, subject to review of uncertain impacts of climate change.

**The Recommendations**

The Strategy recommends an adaptable approach to the future management of this coastline which does not preclude any future opportunities to further improve the standard of protection, or to take a different approach. Early on during the SMP development it was identified that minimal national funding would be available to fund future management of the flood defences. A Coastal Pathfinder study in 2011 supported this further through working with the local community to understand the willingness to pay for flood defences.

The team has developed an innovative approach to managing future decisions. This is a "decision pathways" approach and is based on the method first developed for the Thames Estuary 2100 Strategy. The approach supports the Defra policy statement on appraisal and coastal change leading to flexible future management allowing for future uncertainties such as climate change. This is a good example of bringing high level concepts down to the practicalities of a local scheme. The approach manages the interface between the coastal processes, economic assessment, environmental sensitivities and funding requirements. The process uses triggers that encourage a decision point. The decision could be to continue or change the existing approach. The decision point accounts for a buffer zone to manage the uncertainties that include time for:

- The process of decision making on the management approach;
- Planning and implementation of a preferred option; and
- A safety margin to manage public health and safety risks.

An example of the development of a decision-pathway for the cliffs at Hunstanton is shown in Figure 6. This figure illustrates the need for a flexible approach in managing risk. Figure 6 uses the example of the lighthouse for illustration purposes. The adaptive approach includes a margin for safety, in this case, a set distance from the cliffs to the lighthouse that is considered acceptable for risk management. at the edge of this margin is the implementation point; here the selected action should be implemented. This action could be, in this instance, erosion protection for the lighthouse, relocation of the lighthouse or to do nothing. The decision point is a trigger and a physical location at which there is a call to review and consider alternative management options for the lighthouse. The time needed allows actions to be developed, engagement with the public and the necessary authorities occurs and approvals are sought.

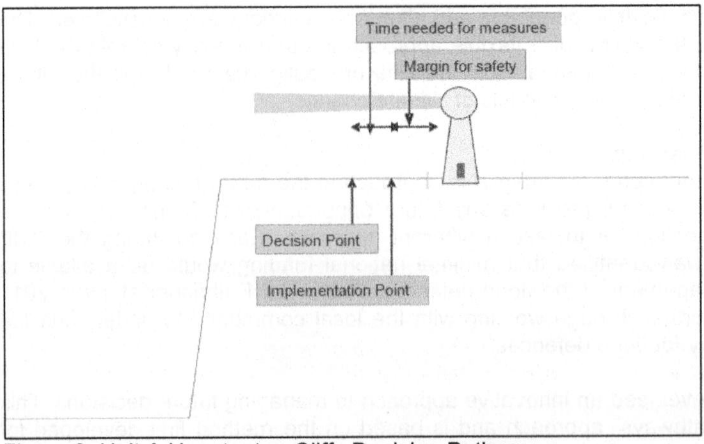


Figure 6: Unit A Hunstanton Cliffs Decision Pathway

In Unit C, there are 3 principal triggers identified; environmental conditions, risk to life and funding availability. Each of these triggers is very sensitive to change. If at any point the existing management regime becomes unacceptable environmentally, a change in approach is required. If at any point the risk to life is perceived to be too high, a change in approach is required. If the funding changes and the available finances cannot cover the costs, a change in approach is required.

## The Funding

It is unlikely that any works arising as a result of the Strategy would be fully funded by national government. This is due to the limited number of properties and infrastructure at risk of flooding and erosion resulting in a relatively low benefit to the nation. However, the changes to government funding, introducing a partnership funding approach, makes it possible for government to part-fund works as long as local funding can be secured. This allows for communities to contribute towards managing their coastline helping to reduce the risk of flooding and erosion under a Partnership Funding approach.

The Pathfinder Study, supported by the BCKL&WN and the Environment Agency, ran a number of workshops and engagement sessions to help understand the local perspective. This assisted the local community in drawing their own conclusions on flood risk, to help determine how change should be implemented and to identify how much they might be willing or prepared to pay for defences into the future.

Within the strategy project, the local community led the way in suggesting alternative funding mechanisms through discussing the amount payable, the distribution of the expenditure over the local area and if those in higher risk areas should be paying a premium. Realistic processes were identified for putting these funding mechanisms in place within the local financial system.

## The Future

The Strategy requires that a partnership approach to funding is established between public sector bodies and the local community - primarily caravan site owners, other local business and landowners who are most likely to directly benefit from the continued protection of this coastline.

Approximately £260k is needed per year to maintain the shingle ridge and the existing standard of protection with a spike of investment in approximately 40 years' time. Approximately £6.7m is required upfront to improve the defences to a 1.3% AEP. The majority of these funds are required from local contributions.

The BCKLWN support the Strategy and are jointly committed to developing a partnership funded approach to the future management of the coastline to protect people, property and the natural environment. The BCKLWN, Norfolk County Council and the Environment Agency are willing to make a contribution to the up-front capital costs required to sustain the current standard of protection and manage future climate risks.

Stakeholder engagement has been critical to the drive and the success of the strategy. The local community and businesses have been involved in the development of the strategy from definition of the problem, objective setting through to the preferred strategic approach. This has presented a number of challenges but ensured a thorough understanding of the strategic decision-making process and funding options.

From the start it was felt important to ensure the funding mechanism is both equitable and workable. Only two mechanisms of a long list proved to be workable. A charitable trust or Community Interest Company (CIC) but these have their own limitations.

The CIC was investigated which demonstrated that it provides a mechanism for capturing voluntary funding contributions which would then be managed for the purpose of maintaining or improving the coastal defences. It is anticipated that those local businesses and land owners that would benefit directly from maintained or improved standards of protection would contribute to and manage the CIC.

## Closing Remarks

The WECMS and The Wash SMP2, supported by the Pathfinder study have generated an innovative method that supports Defra policy on appraisal, funding and coastal change using growing knowledge and understanding. The WECMS has worked to take the funding framework one step further, working alongside the option development and assessment process, including Strategic Environmental Assessment, to identify available national funding, costs of potential management options and how the local community can work to bring the preferred option into practice by contributing money to fill the shortfall in funding. The WECMS promotes local community ownership and accountability in developing future coastal management through an approach that recognises the future uncertainties.

There is not one easy answer for Wash East; delivering the strategy can only happen the hard way.

**Coastal Management: Changing coast, changing climate, changing minds**
ISBN 978-0-7277-6114-9

ICE Publishing: All rights reserved
doi: 10.1680/cm.61149.455

# The St Jude's Storm – Just the Start

**Travis Mason**, Channel Coastal Observatory, Southampton, UK
**Andrew Bradbury**, Channel Coastal Observatory, Southampton, UK *(deceased)*

## Introduction

The storm which hit southern England on 28 October 2013 (the St Jude's Storm) turned out to be a harbinger of one of the stormiest winters for at least 2 decades. By the time the worst was over in March 2014, beaches along the south coast had experienced up to 7 storms exceeding the 1:1 year return period, with several storms representing the 1:50 year return period wave height. The stormy conditions featured widely in the national press and television, but the effects of the storms at individual sites varied considerably. Although no individual storm was considered particularly exceptional, the clustering and persistence was highly unusual (Met Office 2014).

Furthermore, attention during the winter was also focussed on the lengthy and extensive pluvial flooding in the Somerset Levels, and more briefly on the largest tidal surge along the North Sea coast (04/05 December 2013) since 1953. The storm surge also affected parts of the eastern English Channel coast *e.g.* at Shoreham, but for the majority of southern Britain, it was the effects of waves that was most significant for coastal management. The twin problems were (i) the magnitude and (ii) the frequency of the storms.

The six Regional Coastal Monitoring Programmes of England maintain an extensive network of coastal Waverider buoys and captured wave measurements throughout the storm season. The buoys were of significant operational benefit for incident management since the real-time data could demonstrate whether the storm conditions were rising or past their peak, or data from more westerly buoys (in most cases) could be used to help validate very short-term forecasts of what was likely to come. For example, long period swell arriving at the Porthleven buoy in Cornwall was used to estimate when it would reach the Chesil buoy along the central Channel coast, and hence whether the swell might coincide with High Water with potential beach over-topping.

In addition to the operational considerations, the buoy data measured during this unusual period of storms is invaluable for longer-term statistics for both design and beach management purposes and to put the winter of 2013/14 into context.

## Methods

**Measured Wave Data**

The wave buoys are Datawell Directional Waveriders (DWR) Mk III and are located in about 10-12m CD water depth from the shoreline since 10m is the depth used typically for design wave conditions for beach defences. As a result, the distance of the buoy site from the shoreline varies from typically 1 to 9 km; the example shown in Figure 1 is the DWR approximately 0.4 km off Chesil beach.

Figure 1: Datawell Directional Waverider Mk III off Chesil Beach (photo courtesy of Fugro EMU Limited)

During periods of very high waves, with accompanying longer mean wave periods, the wave heights can become depth-limited as the buoy is in the region of wave breaking. In this situation, some potential contamination of the wave signal is possible due to the buoy "surfing" down a breaking wave front or, possibly, hitting the seabed. In most cases, automated quality control of the half-hourly processed parameters of significant wave height ($H_s$) and periods, based around standard "jump" or out-of-range tests, were sufficient to highlight data for flagging but, on occasions, the automated processes were unable to detect all contamination due to wave breaking. Accordingly, the 1.28 Hz time series (the high-frequency data from which the 30-minute parameters *e.g.* $H_s$ are derived) was examined for all storm periods in order to assess the potential for contamination, for example, the recording period from around 26 –29 minutes of the 30 minute time series shown in Figure 2. If contamination of the high frequency signal was observed, the half-hourly processed parameters were flagged and removed from subsequent analysis.

Quality-controlled wave parameters were down-sampled to 3-hourly time series. Traditionally, return periods are calculated with 3-hourly data because early measurements and Met Office hindcasts are at this sampling interval. This produces lower return period significant wave heights than data at hourly or half-hourly intervals since statistically there is more chance of capturing the peak of the storm conditions using near-continuous (30 minute) sampling, but it allows a better comparison for beach or structure behaviour since in the majority of cases 3-hourly data is likely to have been used for their design.

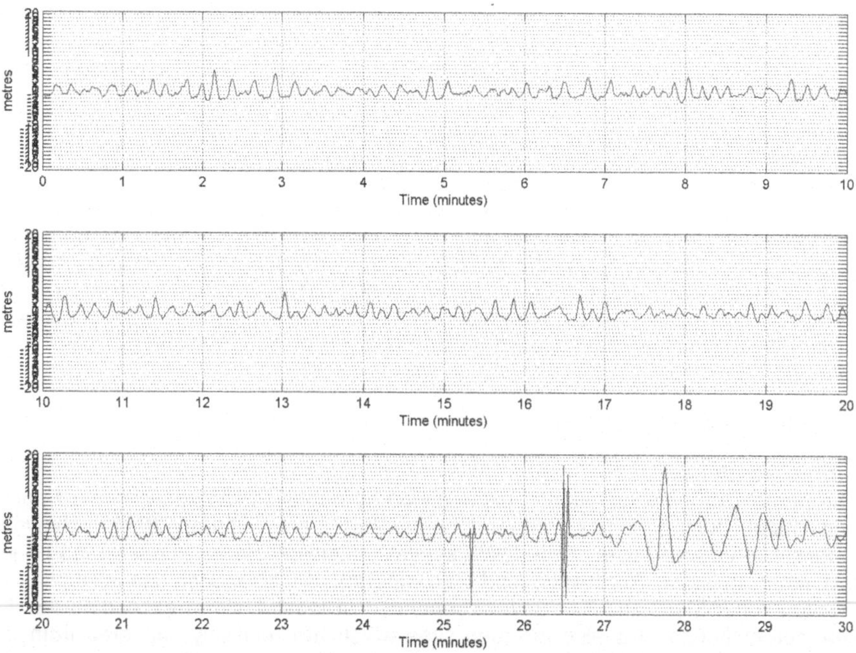


Figure 2: 30-minute time series, Bideford Bay 03 January 2014 at 12:30Z, showing contamination probably due to breaking waves (the "shock" event at around 26.5 minutes)

A "storm" is here defined as having measured significant wave height equal to or above that which would be expected, on average, once a year *i.e.* the 1:1 year return period. Return period wave heights were calculated in SANDS (CH2MHill) using the Weibul distribution, for periods up to 10 times the record length *e.g.* the Chesil DWR was deployed in January 2007 so return periods of up to 50 years could be derived. Potential depth-limited conditions were estimated using Goda (2000). The 1:1 year Return Period for each buoy was set as a storm threshold to identify individual storms using the Peaks-over-Threshold period (Figure 3), as recommended by the CIRIA Beach Management Manual (Rogers *et al.*, 2010).

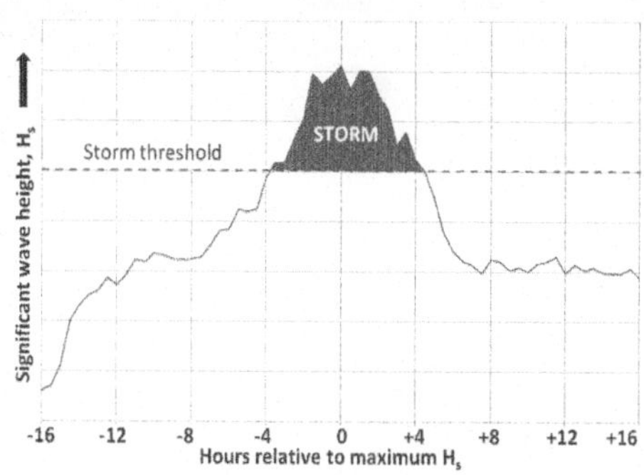


Figure 3: Derivation of storm

### Survey Data
By the late spring 2014, more than 30 post-storm beach surveys had been carried out by the Southeast and Southwest Regional Coastal Monitoring Programmes (Bradbury and Mason, 2104; Siggery and Wiggins, 2014). In most cases, the surveys were profiles along defined transects but at some sites a full beach survey was conducted using a laser scanner. Subsequent analysis compared the post-storm survey with the latest pre-storm survey to assess the change in beach volume. For laser scan surveys, a difference model was produced directly from the pre- and post-storm digital terrain models, and for profile data, the volume difference was calculated by assuming that the change in cross-sectional area between surveys was representative of the adjacent beach; whilst not as quantitatively accurate as using full DTM surveys, the method allowed an assessment of sediment losses over many more beaches than would have been possible if full surveys had been required.

## Results and Discussion

### Storm Frequency
The number of individual storms experienced by deployed DWRs along the English and Bristol Channel coasts between October 23013 and February 2014 is shown in Figure 4 (note that some of the DWR's were unserviceable for some period during this time, including Porthleven and Perranporth). All sites, with the exception of Weston-super-Mare and Tor Bay, experienced at least 1 storm exceeding the 1:1 year return period; this also included sites which are relatively sheltered from the prevailing southwesterly Atlantic storms, such as Weymouth and Dawlish.

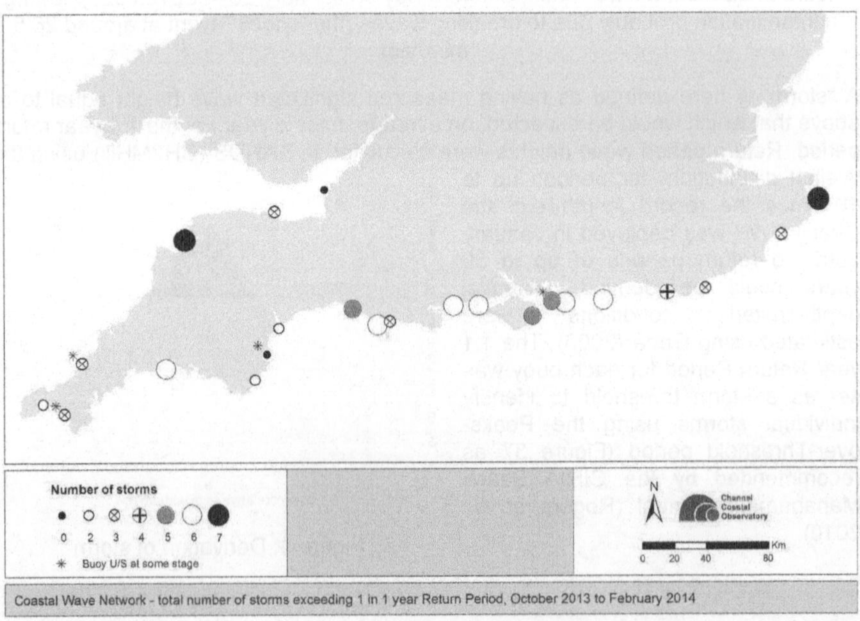


Figure 4: Number of storms measured by DWR's between Oct 2013 and Feb 2014

Particularly extreme, in terms of its typical wave climate, was Goodwin Sands (Kent). Although the highest measured significant wave height (3.69m) was considerably lower than along much of the southeast coast *e.g.* Seaford (5.34m), Goodwin Sands received 7 storms exceeding its 1:1 year return period, compared to only 2 or 3 1:1 year storms between Dover and Beachy Head. In terms of its own wave climate, therefore, Goodwin Sands was as "extreme" as Bideford Bay in north Cornwall (7.36m maximum Hs). Similarly, in terms of perturbation from typical local wave climate, the whole western and central English Channel as far eastwards as Shoreham (Sussex), was as severely affected as the west coast.

Analysis of data from 23 DWRs found that 50% of all storms measured during the last 10 years occurred during the 2013/14 winter period.

**Storm Magnitude**

The "storm calendar" for Chesil DWR is given in Figure 5; each point on the graph represents the highest Hs of each individual storm and also demonstrates clearly the clustering in addition to how extreme each storm was (note that the storm calendars are produced from the 30 minute time series of measured wave parameters, not the 3-hourly down-sampled data). Since the Chesil DWR's deployment in 2007, 12 individual storms exceeded the 1:1 year return period, of which 6 occurred during last winter, but also included the 3 highest storms. The 7.70m $H_s$ measured at Chesil on 14 February 2014 is the highest measured by the entire coastal wave network; waves measured during the Christmas Eve and 05 February 2014 storms would be depth-limited in normal conditions and are hence an indication of a significant storm surge spanning High Water during those storms.

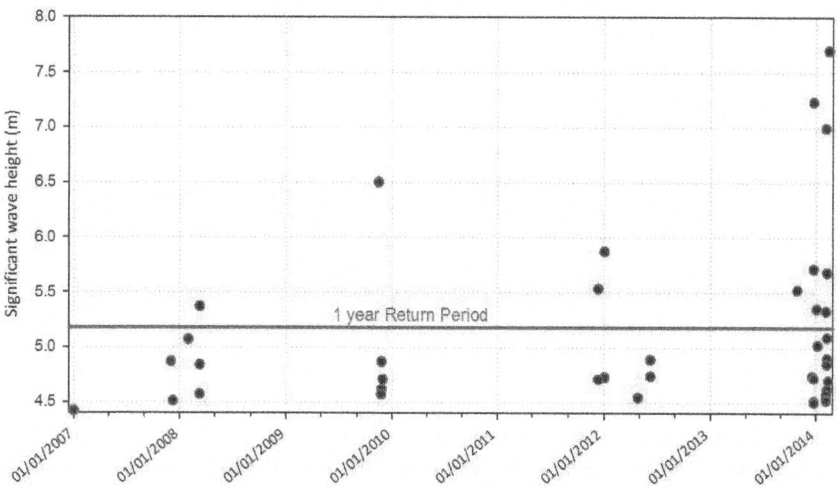


Figure 5: Storm calendar for Chesil DWR

At most operational DWR sites along the English and Bristol Channels, this pattern was repeated, with the highest measured conditions since deployment occurring during winter 2013/14, including 6.06m at Penzance (04 Feb), 7.32m at Looe Bay (14 Feb) and 7.28m at Perranporth (01 Feb, became unserviceable shortly afterwards). Tidal ranges at these last-mentioned sites are greater than 4m, but even with an anticipated tidal surge, total water depths are unlikely to have exceeded around 16-18m even around High Water, which highlights the magnitude of such waves in the nearshore.

However, between Beachy Head and Dover, the highest storm was only marginally higher than previously recorded whilst waves heights at southeasterly-facing Sandown Bay DWR (Isle of Wight) were lower than previously measured in 2005 and 2008, although the storm clustering was still evident at these sites.

The relative "storminess" of a year can also be reflected in an analysis of measured data in terms of wave height exceedance - notably in the 0.05% analysis which represents the average wave height of about 12 hours-worth of the highest waves - although the effect of storm sequencing within the year is missed. The example in Figure 6 suggests that 2014 was broadly similar to 2000 in terms of wave climate, yet the impacts of the storms on the beaches was quite different; of course, the condition of the beach may also have been quite different.

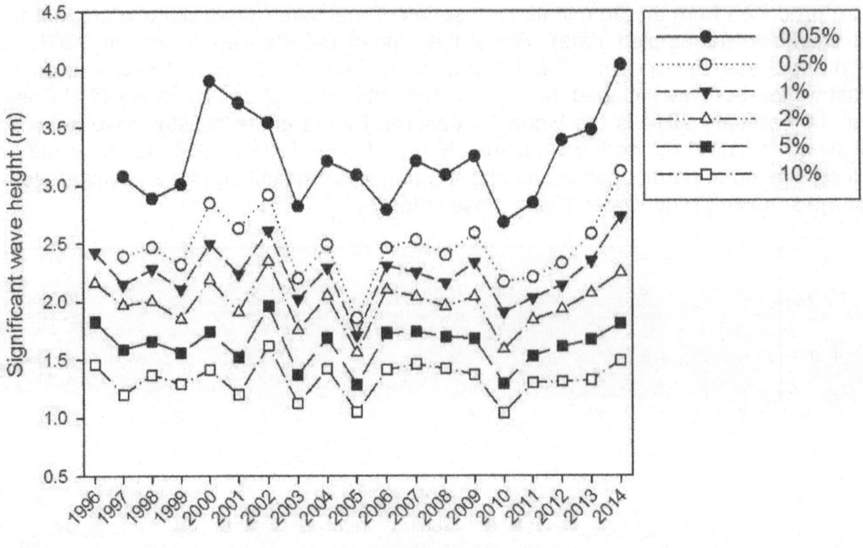


Figure 6: Wave exceedance at Milford-on-Sea

**Variation in Return Periods**

The assessment of return periods was based on values last calculated in 2013 for all measured data to 31 December 2012, when the majority of DWR's had been deployed for at least 5 years. Following lengthy and detailed quality-control of the storm data, return periods were re-calculated taking into account the new data. Table 1 shows the change in $H_s$ return periods for the Chesil DWR, and illustrates the sensitivity to the length of record used and the heavy influence of a relatively short period of intense storms (storm clustering). This is notable even at return periods as short as 2 or 5 years. As a result, the St Valentine's storm at Chesil, originally calculated as a 1:50 year storm, is closer to a 1:20 year storm. Even using the re-worked return period values, Chesil still experienced 3 *x* 1:20 year storms from 24 Dec to 14 Feb, with a further 4 storms in between.

Table 1: Hs return periods for Chesil DWR (note that some of the higher values are likely to be depth-limited)

| Return Period (years) | Jan 2007 to Dec 2012 | Jan 2007 to Feb 2014 |
|---|---|---|
| 0.5 | 4.84 | 5.45 |
| 1 | 5.18 | 5.95 |
| 2 | 5.52 | 6.45 |
| 5 | 5.95 | 6.74 |
| 10 | 6.26 | 7.10 |
| 20 | 6.58 | 7.60 |
| 30 | 6.76 | 8.09 |
| 50 | 6.98 | 8.38 |

The variation in return period reduces with longer record length shown by return periods for Milford-on-Sea, which has an unusually long record of measurements (Table 2). Based on 6 or 7 years period, the addition of the 2013/14 winter data leads to a 7% increase in 1:50 year $H_s$, whilst the same addition of storm data to the 16 year record results in only a 2% increase. In most cases of beach design, 5 years of measured nearshore waves on which to base management decisions would be considered particularly useful information, but the results must be considered in the light of the time span when the measurements were made.

Table 2: Variation of return period with record length, Milford-on-Sea DWR

| Record length | 6 years | 7 years | 9 years | 10 years | 16 years | 17 years |
|---|---|---|---|---|---|---|
| RP (years) | Jan 07 - Dec 12 | Jan 07 - Feb 14 | Jul 03 - Dec 12 | Jul 03 - Feb 14 | May 96 - Dec 12 | May 96 - Feb 14 |
| 0.1 | 2.63 | 2.91 | 2.68 | 2.86 | 2.82 | 2.88 |
| 0.2 | 2.88 | 3.17 | 2.91 | 3.11 | 3.06 | 3.13 |
| 0.3 | 3.02 | 3.33 | 3.04 | 3.25 | 3.20 | 3.27 |
| 0.5 | 3.20 | 3.51 | 3.20 | 3.42 | 3.37 | 3.45 |
| 1 | 3.43 | 3.76 | 3.42 | 3.66 | 3.60 | 3.68 |
| 2 | 3.66 | 4.00 | 3.63 | 3.88 | 3.82 | 3.90 |
| 3 | 3.80 | 4.14 | 3.75 | 4.01 | 3.95 | 4.03 |
| 5 | 3.96 | 4.31 | 3.90 | 4.18 | 4.10 | 4.19 |
| 10 | 4.19 | 4.55 | 4.10 | 4.39 | 4.32 | 4.41 |
| 20 | 4.41 | 4.77 | 4.29 | 4.61 | 4.52 | 4.62 |
| 30 | 4.53 | 4.91 | 4.40 | 4.73 | 4.64 | 4.74 |
| 50 | 4.69 | 5.07 | 4.54 | 4.88 | 4.79 | 4.89 |
| 100 | - | - | - | - | 4.99 | 5.09 |

**Beach Erosion**

Post-storm surveys were concentrated at sites where beaches themselves form a primary role in sea defence, many of which have a defined Beach Management Plan. In the southeast of England alone, erosion rates were found to be up to 25 times the annual average loss with average beach losses during the winter storms in the order of 25,000 $m^3$ per km length of coastline. This amounted to some 470,000$m^3$ across just 25km of frontage. In the Southwest, beach volumes were lowered typically to levels last seen in 2007.

At some sites, the beach damage was cumulative, and it was the frequency of storms which was perhaps more significant that their absolute magnitude. For example, at Portsmouth, a sea wall was undamaged by earlier storms but beach levels were progressively reduced with no time to recover naturally (or by management operations) before damage finally occurred (Figure 7a). Similarly, at Hurst Spit, Hampshire, repeated storms in January and early February 2014 prevented full emergency works to restore beach levels, and a combination of the highest measured waves since 1996 and a significant storm surge led to severe overtopping and breaching during the St Valentine's storm (Figure 7b).

Figure 7a: Seawall damage at Portsmouth (photo courtesy of East Solent Coastal Partnership)

Figure 7b: Hurst Spit, 15 February 2014 (photo courtesy of New Forest District Council)

**Beach Recovery**

Some estimate of the recovery of beaches since the winter storms has been made following surveys during the summer and early autumn of 2014. In this case, the comparison is for surveys taken ideally a year apart to assess the comparative volume of beaches *e.g.* summer 2013 to winter 2013/2014 to summer 2014. Results are expressed as the % of winter losses recovered by summer/autumn 2014, for example +100% recovery means that the beach has recovered all the volume of sediment lost over the winter, -50% means that the beach has lost half as much sediment again. Table 3 lists some results from the southeast, but by default incorporates the not-inconsiderable beach management operations so is not necessarily due to natural processes alone.

Table 3: Beach recovery by late summer/early autumn 2014

| Site | Sediment volume above MLWS recovered since winter 2013/14 |
|---|---|
| Seaford | 61% |
| Elmer | 31% |
| Bournemouth | 56% |
| Sandbanks | 19% |
| Ringstead | 39% |
| Weymouth | 66% |

**Forcing and Response Together**

By measuring both the forcing (nearshore wave climate) and response (beach surveys), the data collected by the Regional Coastal Monitoring Programmes during the winter of 2013/14 is a hugely valuable data source for future beach management and design. Already, steps are underway to refine beach alarm and crisis levels in terms of their cross-sectional area.

Further work is needed to examine in detail the characteristics of each storm so that the particular threshold parameters for beach response can be assessed. Wave height, although the representative parameter generally used in flood forecasts, is not necessarily the most decisive feature of the wave climate for beach behaviour at a given site. It is increasingly appreciated that wave period can be a crucial factor for enhanced wave overtopping or beach failure, and that a combination of long period waves with moderate wave heights can lead to more damage than high waves without underlying swell. This became evident at Chesil where the extended period of moderate-to-high waves (~4m $H_s$) in early January were accompanied by swell, around 22s $T_p$, and caused significant beach losses, gabion damage and undercutting of the seawall; more damage than was sustained during the higher waves of the St Valentine's storm. Since the threshold forcing parameters and critical beach response are inherently site-specific, the challenge is to re-visit existing warning factors in the light of experience of winter 2013/14.

## Conclusions

- The effect of storm clustering can be of equal or more significance than the absolute magnitude of a storm, by progressively lowering beach levels, if the sequence of storms is sufficiently close to prevent adequate natural or engineered beach recovery.
- Significant wave height remains the main parameter to describe the potential severity of a storm, but consideration must also be given to peak wave period ($T_p$) if the storm is accompanied by long period swell energy. Measurements of $T_p$ from the westerly-sited DWRs can be used to gauge propagation times to sites along the Chanel coast.
- Wave direction is particularly important at sites which are considered relatively sheltered from prevailing SW/SSW waves, when the effects of a large single storm from *e.g.* S or SSE can still outweigh a sequence of moderate storms.

- Post-storm surveys, whilst difficult to undertake, are an invaluable source of information both for immediate assessment of beach damage but also as a baseline for understanding natural recovery rates.
- The value of real-time wave measurements is of vital operational significance during severe storms. The importance of long-term wave measurements in the coastal zone cannot be over-stated, along with highlighting difficulties with statistical calculation of return periods even over decadal measurements, when faced with storm clustering.

## Acknowledgements

Tragically, Professor Andy Bradbury died in August 2014.

## References

Bradbury A.P. and Mason T.E. (2014). *Review of south coast beach response to wave conditions in the winter of 2013-2014,* Channel Coastal Observatory Report SR01, April 2014.

Goda Y. (2000) "Random seas and design of maritime structures (2nd edition)", World Scientific Publishing Company, 443p.

Met Office (2014). *The recent storms and floods in the UK,* Met Office Briefing Report February 2014.

Rogers J., Hamer B and Brampton A. (2010). *Beach Management Manual (2nd edition),* CIRIA

Siggery E.A. and Wiggins M.A. (2014). *Review of south west coast beach response to wave conditions in the winter of 2013-2014,* Plymouth Coastal Observatory Report SWSR01, June 2014.

**Coastal Management: Changing coast, changing climate, changing minds**
ISBN 978-0-7277-6114-9

ICE Publishing: All rights reserved
doi: 10.1680/cm.61149.465

# Sandwich Town Tidal Defences: Preserving the Past, Protecting the Future

**Jon Holland**, CH2M Hill, London, UK
**Richard Townson**, Environment Agency, Worthing, UK

## Introduction: Social, environmental and climatic change - flood risk

The Sandwich Town Tidal Defence scheme forms part of the Pegwell Bay to Kingsdown coastal flood and erosion risk management strategy for East Kent. This strategy identified innovative solutions to address the various flood and erosion risk problems. In particular, the solution developed for the tidal river promotes sustainable construction techniques, reduces environmental risk and creates the potential for habitat creation.

The medieval town of Sandwich in Kent, and extensive commercial areas nearby, are at risk from tidal flooding from the sea. Sandwich is one of the five Cinque Ports and is considered to be one of the best preserved medieval towns in England. Much of the historic interest lies along the frontage of the tidal River Stour. Working in partnership with Kent County Council and Pfizer (private sector land owner), the Environment Agency has successfully promoted a comprehensive scheme, using an unusual approach for managing extreme tidal events to minimise the scale of hard flood defences. The outcome is a cost effective and environmentally sensitive scheme providing a 0.5% annual exceedance probability (aep) (previously defined as 1 in 200 year) standard of protection to residential and business properties. The scheme provides a high level of protection to the uniquely important heritage of the area, having due consideration of potential social and environmental impacts. It takes account of long term climate change impacts both through future-proofing (precautionary) and taking an adaptive approach which re-uses and improves existing flood defences as far as possible, and allows for future adaptation when needed.

## Options Considered and Scheme Concept

With the deteriorating condition of defences, it was estimated that a pre-scheme 5% aep or 1 in 20 year tidal surge event would cause breaches in the existing defences, flooding 409 residential properties and 94 commercial properties, including "Discovery Park" – a large Research and Development science park formerly occupied by Pfizer. With climate change and rising sea levels, this flood risk increases to 864 residential and 185 commercial properties over the next 100 years.

During the preparation of the strategy, a range of options was considered, with a focus on minimising or avoiding impacts on the heritage of the town. A further constraint was the Special Protection Area (SPA) status of Pegwell Bay, internationally important for wading birds. A flood barrier option, for instance, was ruled out not just on cost, but also due to the likely impacts on the river habitats and ecology. Raised flood defences alone were considered unacceptable due to the visual impact on the town, as well as the high cost of implementing a scheme that would accommodate future climate change.

The chosen scheme is based upon reducing the peak level of surge tides in the River Stour by storing more than three million cubic metres of water on farmland adjacent to the coast in a tidal flood relief area. The area is formed using some six kilometres of new tidal defence embankments, incorporating a two kilometre section of spillway to control the peak of level of tidal surges as the rising tide approaches the town. This reduces the peak level of the tide by as much as one metre, thereby requiring only a modest raising of existing flood defences.

**Options**

Tidal Barrier
Studies undertaken some years before had identified a preferred option of a tidal barrier located close to the open estuary – across the 30m wide main channel and 100m of saltmarsh and inter-tidal habitat. Although straightforward in concept, this would have resulted in construction within the internationally designated estuary, requiring the provision of compensatory habitat. In addition to being an expensive option, the barrier would be a significant future liability in terms of operation and maintenance, needing substantial ongoing investment. To avoid the need for frequent closures of the barrier with the resulting impact on the tidal ecology and navigation this option would still require new flood walls and raised embankments upstream towards Sandwich.

Raised Defences
New walls and raised embankments were considered, without the tidal relief area. However, the height required at the quayside in Sandwich was considered unacceptable due to the landscape and heritage impact of a flood wall up to two metres tall. As well as the visual impact, constraining the river on both banks funnels the tidal surge, raising river levels further upstream, and requiring a far greater extent of flood defence structures to defend upstream properties from an increased flood risk.

Tidal Flood Relief Area
In reviewing the habitat designations and 1953 flood maps, the project team conceived a new and innovative approach to flood risk management. The team noted that the international environmental designations in the area literally surround an area known as Broad Salts: 220ha of agricultural land reclaimed from the sea as the port of Sandwich silted up in the medieval period. The records of the January 1953 flood event indicated the embankments protecting the Broad Salts site were breached, causing extensive flooding to this area, whereas the flood impact in Sandwich upstream was relatively minor.

**Hydraulic Model**

The Broad Salts site is conveniently located approximately one kilometre downstream of Sandwich town, and approximately two kilometres upstream of the estuary at Pegwell Bay. This fortunate arrangement meant that the area could be used to control river levels during the peak of a tidal surge event, much as had occurred in 1953, but engineered to be most effective for the chosen standard of protection. Tidal flows in excess of a 5% aep surge event would be directed by a reinforced grass spillway onto the agricultural land. Using a Tuflow Estry 1D-2D hydraulic model the river and Broad Salts floodplain system was 'tuned' to maximise the river level control. The tidal river embankment fronting the Broad Salts area along the River Stour would function as a spillway at the critical threshold, with appropriate strengthening and reinforcement to avoid a breach failure. A peer review of the hydraulic model was undertaken to improve confidence in the suitability of this option.

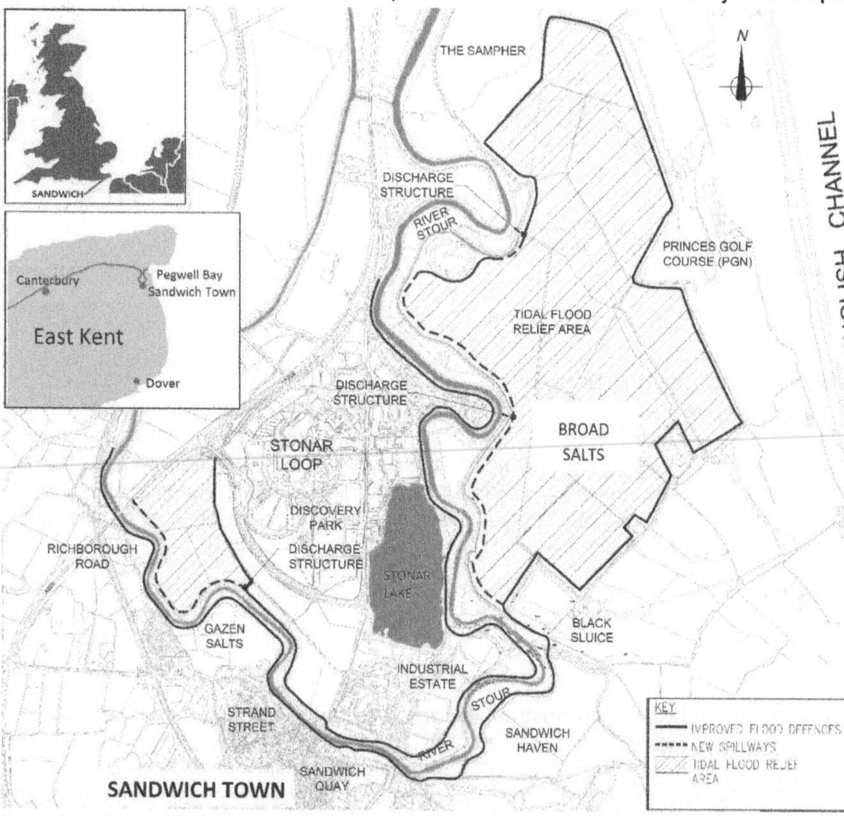


Location Plan and Arrangement of Sandwich Town Tidal Defence Scheme

**Public consultation**

The three flood risk management options shortlisted in the Strategy were presented to the local communities and councils through presentations, exhibitions and a consultation booklet. This gave the public an opportunity to comment on the options. A simple animation was prepared to help explain the location, size and timing of the tidal flood relief area, which helped to gain the support of the council. The economic and environmental case for the tidal relief area was clearly understood, it provided the least whole life cost scheme with the shortest delivery programme. As well as delivering the scheme benefits as early as possible this option helped to secure substantial financial contributions from Kent County Council and Pfizer, safeguarding the future of Discovery Park as a commercially viable business area, providing a vital source of employment in the region and promoting the regeneration of the wider East Kent economy. The Sandwich project has been an exemplar in realising the recommendations of the Pitt Review[1] in relation to generating partnership funding in both public and private sectors.

Public Drop-in Sessions

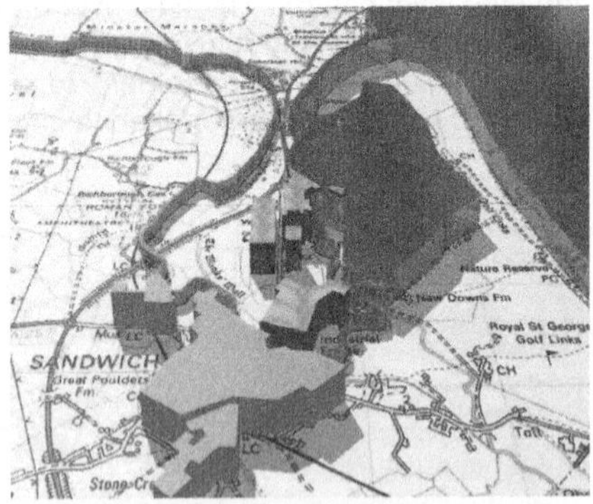


Animation shows scheme concept

## Environmental Considerations

### Interaction with Existing Habitats

The Sandwich and Pegwell Bay National Nature Reserve, at the mouth of the River Stour where it meets the English Channel is of international importance for its bird habitat: waders and wildfowl. It is designated as a Special Protection Area (SPA). The reserve is made up of a complex mosaic of habitats: inter-tidal mudflats, saltmarsh, shingle beach, sand dunes, ancient dune pastures, chalk cliffs, wave cut platform and coastal scrubland.

The important SPA designation significantly constrained the options available for providing improved flood defence to the Sandwich town area. Further, the existing coastal floodplain area, which forms a fundamental element of the scheme, is criss-crossed with drainage ditches – a haven for a large population of water voles.

The proposed earthworks to construct the spillway and bunds required approximately 150,000m$^3$ of fill material, sourced from within the storage area where there are former alluvial deposits. Local winning of material eliminated the need to import materials to the site through the narrow streets of Sandwich, significantly reducing the cost and carbon footprint of the scheme.

For the chosen scheme design the potential ecological constraints became a unique opportunity for returning tidal floodplain farmland to new and important habitat areas. All new embankments were constructed on arable farmland, with the existing embankments retained and breached, allowing the entire existing ditch system to be retained and thereby avoiding any impact on existing habitats. Material for the new embankments was won from the arable farmland area, with the creation of more than 2km of new ditches, enhancing the reserve area with new water vole habitat, and approximately 14ha of new wetland habitat. The win areas were carefully designed with islands and gentle side slopes formed for wading birds and wet grazing land to form a transition into the adjacent farmland. An early indication of the success of this approach was the presence of breeding Avocets during construction, unfazed by the heavy plant operating around them – this bird species had not been recorded in the reserve for many years.

New flood embankment, with existing embankment and drainage ditch retained (on left) and new wetland habitat for birds created in win area (on right)

Public access to the nature reserve area has also been greatly improved by the scheme, along with a wider public amenity. Discussions with Natural England have enabled part of the National Coastal Footpath to be incorporated with the flood embankments at Broad Salts.

**Landscape and Heritage**

Much of the historic interest in the mediaeval town of Sandwich lies along the frontage of the tidal River Stour. Through the town area the focus of the scheme design was to avoid or minimise any adverse impact on the historic town setting. As with the ecology benefits, the scheme yielded significant opportunities for enhancing the landscape and spatial context of the defences for the town. Where existing flood walls have been replaced, the existing reinforced concrete core has been re-used, reducing the cost of the new works, with the replacement wall approximately 400mm higher than the original. Footpath surfaces have been raised by a corresponding amount to give the impression that there is no change in wall height.

Sandwich Town Quay pre-scheme

Sandwich Town Quay post-scheme

At the main town quay, an area used simply as a car park for a number of decades, has been transformed with a new amenity area incorporated into the new flood walls. This has provided a new focal point for the town, bringing this part of the town back to life with a greatly improved environment for the local community and visiting tourists. The careful design blends well with other historic buildings and features, including the old toll bridge, fronting the riverside scene. The award winning scheme has been recognised as a significant enhancement to the town, providing a vital flood defence function whilst complementing the heritage setting. This has been repeatedly affirmed by local residents and business owners expressing their gratitude for the improvements to the town brought by the new defences.

## Climate Change Implications

### History of Coastal Flooding in East Kent

Major inundation in the Sandwich and Deal areas of east Kent, England, occurred in January 1953 when a North Sea surge (4.7m above Ordnance Datum (aOD) at Pegwell Bay) caused the banks of the River Stour to overtop and breach, resulting in extensive flooding within the study area. The breaches led to flooding of approximately 12km$^2$ of primarily agricultural land east of Sandwich, and within parts of Sandwich town. There are no records of the number of properties flooded although inspection of the flood record map indicates that there would have been approximately 20. In addition there are reliable reports of a number of properties flooding in Strand Street area of the town and people rescued from their houses by rowing boats. Fortunately, no lives were lost here, although along the east coast of the UK elsewhere 307 people died and 24,000 properties were damaged.

Examination of the 1953 event as it impacted East Kent revealed that flooding to the populated areas was limited because inundation of the mainly agricultural land nearer the sea acted as a buffer, by storing flood water as the surge advanced up the river Stour, reducing the tidal river level further upstream. This effect was apparently not understood at the time since significant bank raising works were undertaken during the second half of the last century following the 1953 event, which would serve to retain surge tides within the river channel, increasing the risk of flooding to the upstream urban areas.

A surge event in 1976 (4.1m aOD at Pegwell Bay, estimated 4% aep) occurred but there are no records of property flooding. In 1983 a surge event caused flooding at Sandwich Town Quay with 16 properties suffering direct flooding. The water level on the quayside was recorded at 3.7m aOD resulting in flooding approximately 0.2m deep across the quayside area. This is estimated to be equivalent to a 10% to 20% aep event. Water levels just below 3.7m aOD have been experienced since 1983, but not caused direct flooding.

In December 2013 the east coast of England and Scotland experienced the worst tidal surge event since 1953, hitting the north Norfolk coast on the evening of Thursday 5$^{th}$ December 2013, before moving south overnight along the Essex and Kent coasts. Thousands of homes were evacuated, including many properties in Sandwich. Fortunately, as a result of good planning, and greatly improved flood defences – lessons learnt from 1953 in particular – no lives were lost as a result of the tidal surge, although many homes were flooded, and some lost altogether on the north east coast. In Sandwich the tidal defence project had been under construction for nine months and, although not complete, was already sufficiently advanced to successfully prevent flooding to an estimated 111 homes and commercial properties, avoiding flood damage estimated at £2m. Modelling indicated that the 2013 event in the Sandwich area was between 1.33% and 1.0% aep and, once again, the inundation of the coastal floodplain areas served to reduce the height of the peak level further upstream in the populated areas of the town.

December 2013 tidal flood event with flood water stored on coastal floodplain

**Predicted Future Flood Risk**

Whilst the town and adjacent commercial areas previously benefitted from existing flood defence, these were of a relatively low and variable standard of protection, and generally of poor construction or in poor condition. Modelling indicated that without intervention these flood defences had a 50% probability of failure by 2020, and 80% probability of failure by 2030. Detailed hydraulic and breach modelling at appraisal stage for the new scheme indicated that overtopping damage would lead to breaching and rapid inundation to the low lying land behind. The likely breach locations are Stonar Lake, Gazen Salts and Broad Salts. The Stonar Lake and Gazen Salts areas contain significant numbers of residential and commercial properties and thus breaches of defences would have a significant economic impact and would pose a risk to life. Table 1 shows the pre-scheme properties at risk from tidal flood at appraisal stage in 2010, and in the future as a result of predicted climate change.

Table 1: Properties at risk from flooding

| Event | 2010 (Do Nothing) | | | 2110 (Do Nothing) | | |
|---|---|---|---|---|---|---|
| | Residential Properties | Commercial Properties | Total | Residential Properties | Commercial Properties | Total |
| 10% aep (10 yr) | 19 | 2 | 21 | 711 | 154 | 865 |
| 5% aep (20 yr) | 409 | 94 | 503 | 864 | 185 | 1,049 |
| 1.33% aep (75 yr) | 468 | 112 | 580 | 1,046 | 231 | 1,277 |
| 0.5% aep (200 yr) | 486 | 123 | 609 | 1,350 | 287 | 1,637 |

The Sandwich TTD scheme concept was developed in the light of the river and coastal floodplain response observed from the 1953 event. The concept is to use tidal floodplain areas, downstream of the town, to reduce peak surge tide levels in the river channel. Thus, as the surge tide advances up the river the peak level is reduced, with a corresponding reduction in the required height of flood defences in the town.

The proposed tidal flood relief area functions using a reinforced grass spillway set at approximately 5% aep level, controlling water levels further upstream to not significantly exceed the spillway level. The option was designed to mitigate climate change (sea level rise) as a precautionary system, with allowance for 50 years, but adaptation intervention in year 50 to raise defences as required for future sea level rise beyond. The coast adjacent to the tidal relief area has been accreting for many centuries, indicating a minimal risk of erosion of the tidal relief area over the appraisal period. Without future interventions the scheme will continue to provide a standard of protection, but reducing in line with climate change, to an estimated 1% aep by 2100 under current predictions.

The Supplementary Note to Operating Authorities – Climate Change Impacts[2] was used to model climate change effects over the next 100 years. Under this guidance the rate of sea level rise in the east Kent area increases from 4mm/year to 15mm/year over the 100 year period to 2110 – resulting in a total of 1,045mm increase by 2110 compared to the level datum (Extreme Sea Levels for Kent to Isle of Wight[3]). Updated climate change guidance was issued during the appraisal stage of the scheme. Sensitivity testing at that stage indicated that the climate change effects on the scheme following the new guidance would be similar to those already modelled.

A combination of precautionary and adaptive approaches have been used within the design of the adopted scheme option to mitigate the effect of climate change, with 50 years of sea level rise incorporated within the design (i.e. +360mm). The initial standard of protection on completion of the scheme is therefore considerably higher than the 0.5% aep standard which formed the basis of the scheme justification. The scheme generally allows for further improvement works to adapt to future climate change impact, to sustain the design standard beyond the design year of 2057.

The tidal flood relief area has a total capacity of approximately three million cubic metres. Under the design event this is allowed to fill over three successive extreme high tides, optimised to manage a 0.5% aep storm event in 2057. The area is drained by two new large outfall structures, discharging stored water to the River Stour on low tides during and following the surge event over a period of up to four days.

To provide the required volume capacity for the design event (0.5% aep in 2057), the tidal flood relief area must not be outflanked and filled by water from the sea overtopping the lower sand dunes at Princes Golf Course. The scheme incorporates some 4km of new embankment along its eastern flank, linking in with the sand dunes of Royal St Georges Golf Course to provide a fully contained tidal flood relief area.

Pre-scheme 0.5% aep event (2057)

Post-scheme 0.5% aep event (2057)

## Climate Change – Adaptive Approach

The Sandwich TTD scheme is a current day example of adaptation to climate change. Embankments existed in the 1950s and are believed to have been raised following the 1953 surge event. A series of new embankments and walls were constructed in the 1980s providing improved defence to many parts of the town and outlying areas. These were constructed with the best information available at the time, but to differing standards of protection, driven, to some extent, by local geographical constraints. So, for example, flood defence walls were constructed to defend some parts of the historic town, whilst other areas remained undefended, and some areas downstream of the town, of less historic importance, were afforded a greater standard of protection.

A fundamental requirement of the Sandwich TTD scheme was that it should re-use existing flood defence structures where possible. Approximately 80% of the scheme area of interest had existing structures at the time of scheme design, either embankments, sheet piled or gravity walls. The process for determining which of these structures could and should be re-used involved the following assessment process:

- Is the current route of flood defence still appropriate?
- Are the immediate environmental impacts (including heritage) acceptable or can they be mitigated?
- Are the future environmental impacts acceptable (eg from set-back defences)?
- Can the structure be adapted – what are the environmental and technical implications?
- What consents and land agreements are needed – what consultations are required with landowners (normally the owner of the existing flood defence asset), planning authority, statutory consultees and the public?

Good examples of re-using and adapting existing structures were:

- Flood embankments raised where local space constraints could accommodate the increased footprint;
- Flood embankments with constrained footprint, for instance by high value trees, raised and improved with waterproof geotextile membranes;
- Ground raising at boatyards, maintaining connectivity to the river to enable these businesses to continue to operate as before;

Waterproof membrane incorporated into narrow flood embankments

Ground raising at boatyards to avoid impact on crane movements

- Sheet piled walls, adapted with a new, raised flood wall, following detailed condition investigations of existing structure and in the light of as-built details;
- Existing reinforced masonry gravity walls re-used by removing brick facing, and incorporating the concrete base and cores, as part of the restraining weight, into the new brick clad reinforced concrete walls;
- Low gravity walls added to the top of existing embankments where constrained by roads.

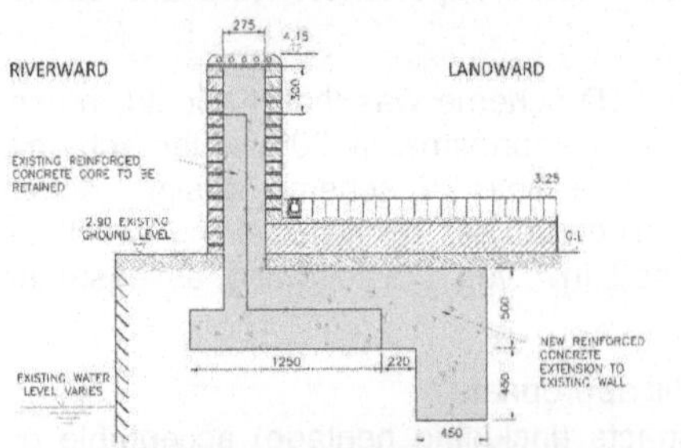


Detail of adaptation of flood wall, re-using existing flood wall core

Construction of new flood wall re-using existing flood wall core

Good examples of replacing existing structures were:

- New flood embankments constructed to replace existing where the immediate impact on ecology would be unacceptable and difficult to mitigate within the programme constraints of the scheme. For instance, impacted water vole habitat in ditch systems immediately adjacent to existing embankments.
- New sheet piled walls constructed where all other options had been explored, and all factors had been investigated to ensure a fully informed and balanced

view was taken, including archaeology and heritage, landscape and land ecology (trees), river ecology, geomorphology and navigation.

The need for greater and more cost effective adaptation in the future underlines the great importance of preparing and storing, in a readily retrievable way, the design basis and as-built details of the constructed works.

## Conclusion

Although it had devastating effects to coastal areas of the UK and mainland Europe the 1953 surge provided the empirical evidence that pointed towards a scheme for Sandwich town that "makes space for water". The surge tide sixty years later in 2013 occurred at just the right time to help demonstrate the validity of the approach.

The availability of coastal floodplain, to manage tidal flood water, has enabled this innovative and adaptive approach for flood risk management for Sandwich town. The resulting flood defence scheme has yielded improved amenity for all, with significant environmental benefits, avoiding unacceptable impacts, preserving our heritage for the future.

## References

1. Pitt, M. (2008) Learning Lessons from the 2007 Floods, Cabinet Office
2. DEFRA (2006) Supplementary Note to Operating Authorities, DEFRA
3. JBA (2004) Extreme Sea Levels for Kent to Isle of Wight, v10, JBA

Coastal Management: Changing coast, changing climate, changing minds
ISBN 978-0-7277-6114-9

ICE Publishing: All rights reserved
doi: 10.1680/cm.61149.477

# The Reality of Change: Involving Communities in Planning in Northwest Wales

**Gregor Guthrie**, Royal HaskoningDHV, Peterborough, UK
**Huw Williams**, Gwynedd Council, Gwynedd, UK
**Huw Davies**, Gwynedd Council, Gwynedd, UK
**Lisa Marshall**, Gwynedd Council, Gwynedd, UK
**Rachel Smith**, Royal HaskoningDHV, Birmingham, UK
**Victoria Clipsham**, Royal HaskoningDHV, Birmingham, UK

## Moving on

### To Start at the Beginning

A Shoreline Management Plan (SMP) *"does not set policy for anything other than coastal defence management"*, but within these simple words, within the Defra Guidance for preparation of SMPs (Defra 2006), can be hidden the start of a much broader process that may well outlive those involved with the SMP's development; especially when that policy involves change.

The MCCIP Arc Science Review (K. Horsburgh et al 2010) reported that:

- *Defra has estimated that approximately one third of the coastal defences in England and Wales could not be justifiably improved under current economic rules on investment (Defra, 2001).*
- *At many coastal sites (including major industrial and infrastructure facilities) there is limited scope to retreat inland without major economic and/or social implications.*
- *Shoreline Management Plans (SMPs) are the backbone of coastal management in England and Wales. The new SMPs will provide a "route map" for local authorities and other decision makers to move from the present situation towards meeting future needs.*

Are SMPs a route map and do they encourage change, or do they just sit on the shelf gathering dust? In many SMPs the need for change has been deferred to epoch 3 (50 to 100 years). Some would argue that this allows time for change to happen; some that this allows us to forget the need for change so as not to disrupt the present, paying lip service to the future, without the need to face the challenges now. Comment has been made that SMPs have not gone far enough in promoting change. For some, SMPs have gone too far, setting out policy for nothing else *"than coastal defence management"* without examining the broader implications: "*A decision to allow the sea to flood 50 coastal villages in Wales, in a bid to cut costs in sea defences, could cost the country billions of pounds in lost tourism revenue, an expert has claimed.*" (Daily Mail 2014)

While messages at a strategic level remain mixed:
*"Rising sea levels mean that waves and storm surges could cause greater coastal erosion. This could expose new risks from coastal flooding, lead to a greater risk of coastal defences failing and increase the need for maintenance work on defences and more extensive warning systems."* (Understanding the risks, empowering communities, building resilience. National FCERM Strategy, Defra 2010.)

Where in this is the challenge presented in the Guidance for Shoreline Management Plans (Defra 2006)? *"providing a long-term 'sustainable' plan which may require significant changes from present management practices – policies which avoid, as far as possible, committing future generations to inflexible and expensive options for defence."*

*"It's always interesting, taking the long view." ('Shrinking Britain' will force land to be abandoned to the sea.* (Guardian May 2011).

In reality, it is more than interesting, it is essential but it is challenging.

**Picking Up the Challenge**

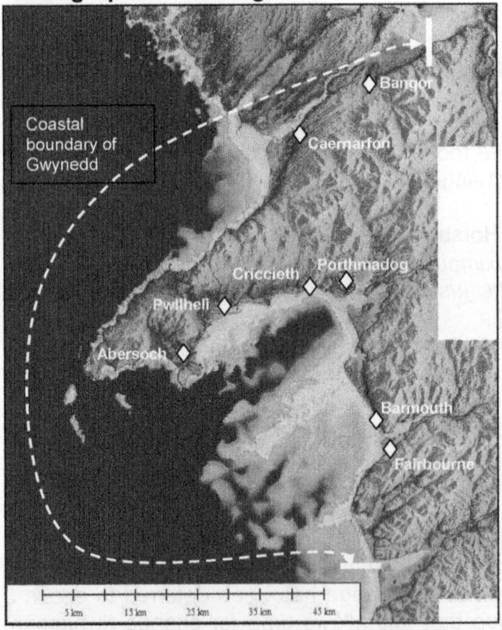

Figure 1: Gwynedd

The county of Gwynedd (Figure 1) is located to the northwest Wales. Much of the hinterland is mountainous and the majority of people live and work within the coastal area. There are over 30 coastal communities, including the main towns such as Abersoch, Fairbourne, Pwllheli, Caernarfon, Bangor, Porthmadog, Barmouth and Criccieth all situated at the shoreline, all with some degree risk (Table 1 sets out the number of properties at risk within the towns and city listed above). Many of the transport links run along the coast, including the A55 and the main railway line through north Wales to the Isle of Anglesey and the Cambrian railway running along the western coastline. Despite its large land area, the coast lies at the heart of Gwynedd.

Table 1: Number of Properties at risk, as defined by the SMP over 100 years

| Location | Flooding | Erosion | Location | Flooding | Erosion |
|---|---|---|---|---|---|
| Abersoch | 3 | 6 | Porthmadog | 1650 | 110 |
| Pwllehli | 1200 | 6 | Barmouth | 250 | 40 |
| Caernarfon | 176 | - | Criccieth | - | 100 |
| Bangor | 330 | - | Fairbourne | 400 | - |

Planning for change, within Gwynedd, has been recognised as a priority.

Coastal communities interact with their shoreline in many different ways. Some have been built around traditional fishing communities or developing in association with major ports and harbours, or linked to historic sites such as castles or traditional centres of communication and transport. Increasingly over the last century, many have developed as residential areas making use of available land, or land reclaimed from the sea, along the coastline, developing as tourism centres, making use of their beaches and the sea for recreational activities. Throughout this development, coastal communities develop and adapted over time. Change has happened. It has been a change driven by changing circumstance and attitudes, by changes in society and by socio-economic factors; it is a change that has typically, but not always, taken time.

This process of change has generally been accepted in hindsight, if not always welcomed at the time by those most affected; if not without some regret. The process has always brought into relief the balance of benefit and harm in relation to society, community and the individual and how this balance plays out over time.

**The Need for and Nature of Change**
Coastal communities exist within a changing environment. The coast is in a constant state of flux. In many cases, coastal communities have come to rely upon defences; either as communities have developed closer to the sea or as the sea has come closer to the community. As the risk of flooding or erosion has become more apparent, so defences have been constructed to prevent erosion or reduce flood risk to an acceptable level. Can existing defences be sustained in the future?

At Pwllheli (Figure 2), as part of the land enclosures in the 1700s, as the town grew, defences were constructed to defend the new land. These areas, won from the sea, now form the core features of the town. Six main rivers run through the area. The area is protected by a narrow and increasingly fragile coastal ridge, which even now is under threat and is being reinforced. Businesses have suffered recent flooding and there is a need to address these issues in a sustainable way.

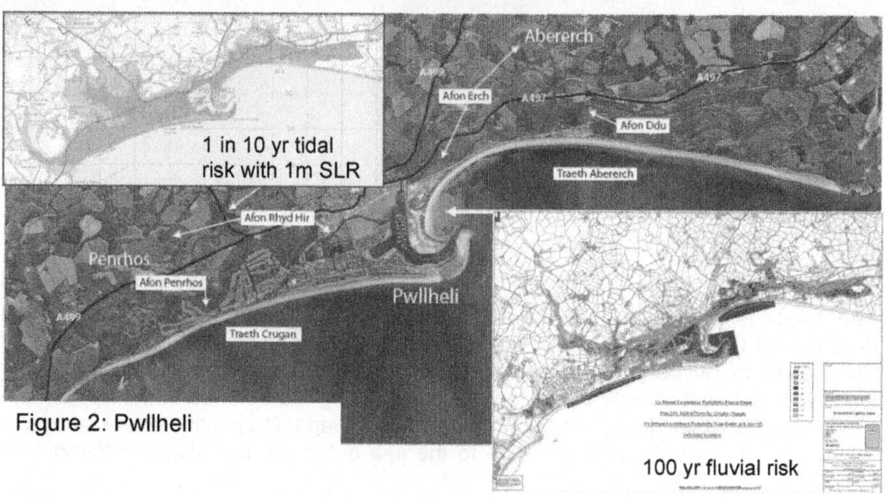

Figure 2: Pwllheli

At Abersoch, the dune system suffered significant damage during the storms of 2014, flooding occurred within the harbour, threatening important local businesses supporting the regionally important water sport centre. The natural reaction has been to call for strengthening and raising defences, with the potential of private investment contributing to hardening the coast, although discussions with some businesses recognise the need for change and the threats of climate change. At Aberdaron, the community face the prospect of choosing a beach or a village, as coastal squeeze occurs.

Figure 3: Porth Dinllaen

In other areas such as Porth Dinllaen, the iconic Welsh hamlet is under threat as sea levels rise (Figure 3). This small community, owned largely by the National Trust, faces the risk of increased flooding, and nowhere to go, or the prospect of higher defences destroying the setting of the village.

From the few properties of Aberdesach, to the 330 properties at risk within the city of Bangor, similar and different issues are being faced. Every community is different but each faces issues related to coastal change and sea level rise.

Many of these differences are brought together at Fairbourne (Figure 4). At present, there is a community with a population of some 800 individuals and around 400 properties. The community supports and is supported by thriving local businesses, including shops, the miniature railway which runs along the coast to the ferry service to Barmouth, local builders and other trade services.

Figure 4: Fairbourne

The village was developed during the late 1800s on reclaimed land, increasing in size over the early part of the 20th century.

The village is located to the south side of the Mawddach estuary, within reclaimed land behind the Ro Wen shingle ridge. Defences on the seaward side of the village were raised in 1981 and the embankment on the estuary side was recently improved, with an investment of over £6 million in strengthening the defence and improvement to the water courses flowing through the low lying village. During the recent storms of 2014, while there was some damage to the sea defences, the village suffered only minor flooding due to local overtopping. The main Cambrian railway line runs to the rear of the village and the land then rises steeply.

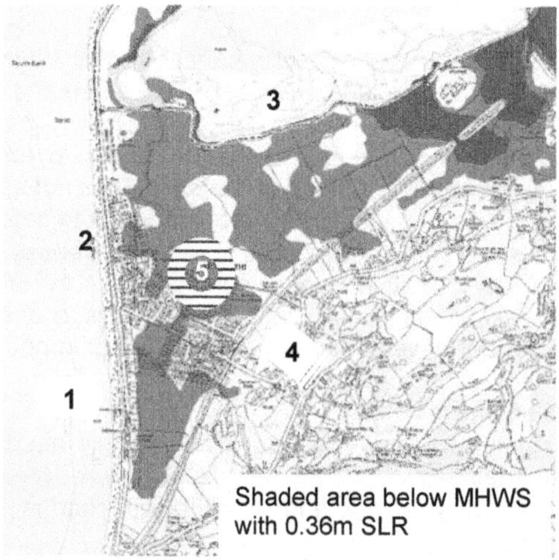


Figure 5: Fairbourne at Risk

There is one road into and out of the village. The village is developed on flat land, at present only marginally above mean high water spring tides (MHWS) (Figure 5).

With sea level rise, the SMP highlights the risks due to erosion (1), wave overtopping (2), water levels within the estuary (3), fluvial and surface water flows (4) and ground water (5).

It is of course possible to defend, it is always technically possible to defend, but the SMP questions whether it is sustainable to do so.

## Moving Forward

### Taking the Initiative

The West of Wales SMP was effectively finalised in 2012, with the agreement of the steering group comprising all six operating authorities. In January 2013 Gwynedd Council adopted the findings of the SMP and approved the SMP Action Plan, which included actions necessary to address the issues raised by the SMP. This was critical to the subsequent process. The Council recognised that, to use the work undertaken through the SMP, the SMP had to be used as a starting point.

Equally essential, Gwynedd Council has an approach to all their business to increase the involvement of those they represent. While ownership of decisions needs to be shared, the Council took ownership of the process by which that can be made to happen, supported by Welsh Government.

### Getting it Right Internally

Three parallel initiatives were taken forward on the basis of the above decisions. In the first, it was recognised that, internally, officers within the Council needed to gain a better understanding of the issues involved with coastal change. While this process had to be ongoing, an initial workshop was set up to discuss the different views and background between Planners and Engineers. This was a two way process, building on work undertaken previously and jointly on the Pwllheli Pilot by both officer groups. This looked at the specific areas and engendered discussion of broader principles. Key issues are summarised below.

The balance between individual, county and national response and ability to respond, remained unclear. Specifically:

- The ability of planning policy to directly determine the future use of existing property within a community is limited. This possibly limits the way in which communities can be influenced to adapt.
- Some regional policies, meeting broader objectives, may constrain relocation; for example of caravan parks. Relocation of people impacts on housing demand at a regional level.
- Socio-economic background has a significant impact on the ability to adapt. Merely including this as a factor in assessing the need for defence does not assist in providing the support needed in areas where decisions are made not to defend.
- There is a need for on-going discussion with communities to raise awareness and to encourage ideas to emerge from within those communities. In effect, communities are being asked to take on a spatial planning role. This is quite a major step in function and needs further development of skills or further support.

In an integrated process, the aspect of time is critical:

- Time is required to develop long term planning and to a degree this needs to remain flexible in dealing with uncertainties.
- There is a need for basic understanding of risk and how this will change over time. This requires better communication.
- Issues can to a degree be defined in terms of immediate impact, medium term adjustment and long-term adaptation. These each need to be developed now with equal priority, accepting that the medium to long term issues need to focus on what needs to be put in place in the present to support the ability to change in the future.

Some issues require further consideration at the national level. One of the key issues, in this regard, is that of no compensation. The existing policy, under permissive powers, is argued to apply in the case of individual properties. This, in relation to managing change at a community scale needs review, alongside:

- Social justice; the decision made to defend in some areas while deciding not to defend or actually preventing private defence in other areas.
- The expectation of continued defence and the actual ability to walk away from defending, and what this means in the case of a village.
- An upward informing process where local plans or regional spatial planning, developed in conjunction with consideration of local shoreline management issues, feeds back into and influences national spatial planning and policy.

The SMP had been developed with an open, forward looking manner, with permission "to think the unthinkable" (E. Williams 2011). This was matched from a planning perspective in seriously considering the possible need for a "New Pwllheli", an alternative town to meet the needs of the future.

**Walking and Talking**

Notwithstanding the consultation process undertaken during the development of the SMP, it was recognised that messages had not been absorbed at the local level. The Council's coastal management team, committed to meetings with all 35 of Gwynedd's coastal communities over the spring and summer of 2013, supported by locally specific pamphlets.

The emphasis was on involvement, recognising that this could not be achieved without greater awareness and understanding. The key questions asked by the Council were:

- What does the SMP 2 mean to you?
- How will the SMP 2 policy affect individuals and the community?
- What are your views about how the community needs to plan for the future?
- What is needed by the community to manage its future in the face of what the SMP 2 is saying?

Every community was different. Possibly naively, there was an expectation that community representatives would understand the process. In some locations the response was positive “yup, thanks, leave it with us”. In other cases it was far more “tell us when you have decided what needs to happen”. The gulf between consultation and engagement has been discussed and addressed within coastal management for years. The gulf between engagement and involvement and ownership is only really coming to the fore.

Such involvement cannot be “scheme” based. One of the key messages coming from the process, initiated by Gwynedd Council, is that involvement and ownership is an ongoing process. The conventional attitude of “raise the SMP awareness of local communities, full stop” has to be replaced by an ongoing discussion and exchange of information on coastal management. Traditional risk funding mechanisms tend to be based on a project and an outcome. In this case, the process is ongoing.

Two years on, further meetings in the bag, the understanding of this process is still developing. The SMP is only the start of the process.

The other key message is the practical and pragmatic question “what constitutes a community?”. Typically, community councils are representatives of a community, they are not of necessity elected as community leaders. When we discuss community involvement, especially in discussing change, communities can rapidly diffuse into individuals. This does not imply that the community no longer exists, merely that change impacts at an individual level as well as that of the community, the county and the nation. This has been brought home in considering the issues around Fairbourne.

## The Future of Fairbourne

The third initiative coming from the SMP has been to examine and plan for future change at Fairbourne.

The SMP has stated that in the future Fairbourne would not be defended. While the need to maintain existing defences and reduce flood risk to the area in the short term is recognised, any major increase in protection in response to climate change and sea level rise leads down a path for an unsustainable approach to management. This would lead to significant residual risk and the possibility of a catastrophic failure and flooding should the standard of defence be exceeded or any part of the defence system fail.

It is argued that eventually, at some point in time, a decision will be made to stop defending. There is, therefore, a need to plan for this now and a need to decide who decides.

In developing this intent to management, a meeting was held in the village in 2011, attended by over 30 members of the community, as part of the SMP process. Despite open discussion and debate over the SMP policies, the lack of any formal response from people in Fairbourne suggested that the key issue remained, that people did not fully recognise the significance of the SMP policy. Addressing this and understanding how the SMP message can be developed has been the primary purpose behind Gwynedd's initiative in Fairbourne. The SMP has been the start of the process.

**Planning the Journey**

While many SMPs have identified areas where in the future major loss may occur, the Gwynedd team found themselves in relatively uncharted territory in trying to take such change forward. The SMP had identified a change in policy during epoch 2 (2025 to 2055) but, based on the initial estimate of critical sea level rise of 0.5m, the point at which the village might no longer be defended occurs typically between 2042 and 2072. This uncertainty created its own problems.

Despite setting up a multi-agency approach in 2013, following the adoption of the SMP by the Council, and despite involvement of "the community", in effect it was the BBC's "Week in Week out" programme broadcast in late 2013 that suddenly raised the profile overnight, highlighting the policy change within 10 years and raising the fear that defences would fail in 2025. The initiative has had to respond to these needs and concerns of the individual; not least the concern that they would lose their property tomorrow. Until people are ready to listen, it is hard to explain the complexity of issues. Even with planning, there comes a point at which people feel they should have been told earlier.

Bad news is never popular. Fairbourne is the face of the future, whether we like it or not, it is not unique and it highlights the hard decisions will need to be addressed. Our present structures and procedures are not set up well to face this challenge. The process has had to inform upwards as well as down to where people are affected.

In facing these complex problems it was recognised that the project had to be broken down by time and by process; essentially the project needed structure and clarity. Core principles have been developed and while it has been accepted that not all members of the community accept the basic premise of abandoning the community, these principles have been accepted in taking the initiative forward.

*It is intended, through the project, to manage this process of change so that:*

- *The community remains a viable community over the period of sustainability, providing a secure place to live.*
- *Individuals within the community are not disadvantaged and are treated equitably.*
- *Businesses within the community continue to thrive, without constraining future change.*
- *The basic infrastructure and services are maintained.*
- *Fairbourne continues to contribute to the sustainable economic development of Gwynedd.*

*But that these principles are taken forward in the expectation and anticipation of longer term change.*

The project will be developed over time recognising that there are issues that need to be addressed now to address those problems thrown up by the findings of the SMP and to address the uncertainties raised by the SMP. The structure of the project is shown in Figure 6.

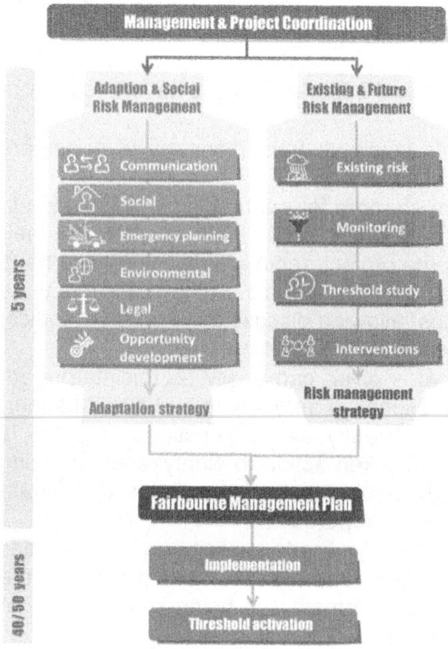


Figure 6: Project Plan

The project is split into three phases:

- Phase 1 (five years). Understanding, communicating and explaining where we are now. Developing the evidence base.
- Phase 2 (year five to year 39(?)). Monitoring and adapting the plan.
- Phase 3 (year 39(?)). Implementing the plan.

This is a very different process to that normally considered in flood and erosion risk management.

In the initial phase, the plan breaks this down into examining the flood and erosion risk. This includes continuing monitoring of the beaches, the integrity of defences and critically understanding the relationship between sea level and ground water levels. The aim of this work is to develop a medium term risk management strategy.

Alongside this, the project aims to develop communication, examine social issues; now and in the future, emergency planning, the environmental issues and the legal aspects of change.

The two principle interrelated themes (risk management and adaptation) will be brought together in the Fairbourne Management Plan that will give all involved and all those who are affected a clear vison for the future.

**Progress**

Progression along this path has not been easy or direct. The initial year, 2013, was one very much of exploration. Progress was made in many areas. Many of the issues, now encapsulated in the above planning process, developed from discussion with different agencies and with representatives of the community. However, while the Week in Week out programme was alarming, it did also provide an important test of progress. It highlighted flaws in the way in which the project was being taken forward. Not least, the programme resulted in the formation of a local action group, Fairbourne Facing Change (FFC). This group does not claim to be fully representative of the community but has provided an additional and important route for communication.

It would be wrong to suggest that prior to the Television coverage, or the formation of FFC, the project had not made significant progress but it has refocused efforts and has drawn out issues that might not otherwise have been understood. The Fairbourne Project Board, which includes Welsh Government, Natural Resources Wales alongside Gwynedd Council, now works closely with FFC, while recognising their very different perspective.

Important lessons have been learnt from this and important issues have arisen from the continuing work on the project:

- The range of agencies and departments involved, from planning, to engineers and social services.
- Be blunt and address the issues, there is no room for euphemisms. Adaptation in the case of Fairbourne means the loss of peoples' (individual's) livelihoods, their property and their investments. Through the project, regular public meetings have been held. Attendance at the first reached over 300 people. Subsequent meetings have adapted their format from workshops, allowing discussion of different topics, to surgery-style sessions to deal with specific individual issues.
- Misinformation is rife and information that is given can easily be misinterpreted. Gwynedd has committed to answering questions in an ongoing and open manner. This list to date is covered in the 42 frequently asked questions, but this is growing all the time. The project group has taken action to clarify issues relating to insurance and has written to all estate agents in the area so as to mitigate concerns over the future and current value of properties.
- Uncertainty is inherent in the process. It has been made explicit how uncertainties are being addressed. Some issues will take time to resolve, not least that of the rate of sea level rise. Monitoring data and information is being provided as an ongoing process, to allow greater understanding.

And for now and into the future:

- Who is at risk and who is vulnerable. In developing the emergency planning now, those at greatest risk and most vulnerable are well defined. However, in terms of vulnerability, we now have to include those who cannot sell their properties, whose options for moving for work can no longer do so, those whose savings are tied up in their property who may not be able to afford care in the future and local builders who have lost business due to people no longer looking to improve their properties.
- What is actually meant by "no longer defended" (and who actually decides?). Is this a case of bulldozers on day "X" in the future, a case of "well that's it, it's up to you now" or a gradual attrition creating a ghost town as standards of defence reduce "the last one out, please switch off the lights".

## The SMP Is Just the Beginning

Gwynedd Council has understood the purpose of the SMP is to raise issues that now need to be addressed. By approving the SMP, its findings and the action plan, the Council has accepted a challenge to take forward change. But ownership of that change and the implications of change needs the involvement of those most affected, involvement of other agencies, service providers and all levels of government.

Critically, we need to recognise that properties are really at risk, not just generically, and this needs to be identified in property searches.

The paper attempts to demonstrate the need for change both in a practical manner and in attitudes and understanding. Change has to be examined and has to be planned for now, to reduce uncertainty, to explore and address issues from now into the future and to support those who are at risk.

This initiative is not providing a one off solution; it is a continuing process of management. Headlined at Fairbourne but also, possibly down the line, at other locations, this is a different approach to flood and risk management; a sustainable ongoing approach, that sits uncomfortably within our deterministic processes and sectoral funding streams.

The project does not set out to learn lessons but rather to provide a planned approach to change. However, important lessons have been learnt:

- Plain speaking, translating strategy speak into the practical real implications for the individual. People understand more if given the opportunity.
- Involving all, especially those who disagree.
- Change at the coast goes well beyond management of the risk of getting wet.

You have to start somewhere and that means you have to learn as you go along. That can be challenging but you cannot plan for change, once change has happened.

## References

DEFRA *Shoreline management plan guidance. Volumes 1 and 2. London, HMSO. 2006*

DEFRA. *Understanding the risks, empowering communities, building resilience. National FCERM Strategy,* London, HMSO 2010.)

Horsburgh, K., Ball, T., Donovan, B., Westbrook, G. *MCCIP ARC Science Review 2010-11 Coastal Flooding*. 2010.

Williams, E., Guthrie, J.G.L. *Scenarios in Shoreline Management. Coastal Management. ICE Belfast* 2011

**Coastal Management: Changing coast, changing climate, changing minds**
ISBN 978-0-7277-6114-9

ICE Publishing: All rights reserved
doi: 10.1680/cm.61149.489

# How Do Civil Engineers See the Coast They Manage? A Comparative Approach between the Netherlands, England and France

**Servane Gueben-Venière**, University of Paris 1 Panthéon-Sorbonne, UMR 8586 PRODIG, Paris, France

## Introduction

Civil engineers are major actors of the European northwest coastal management. Their powerful influence in decision making for territorial planning has never been challenged since the birth of engineers' corps. It is only during the second half of the 20$^{th}$ century that a growing criticism towards certain practices gradually raised concerns against their authority. Rows of groynes, raised like gigantic scars along the coast, have been largely slandered from the 1980s, denying the essence of the coast: the mobility. However we observe today new coastal management practices using the dynamic of the coast. The evolution of the environmental and social context during the last decades may explain part of these new practices. On the one hand, an increasingly binding environmental legislation has made environmental issues inescapable in all coastal management projects. On the other hand, multi-disciplinary reflection is also widely encouraged by the European recommendation *Integrated Coastal Zone Management*. This double evolution raised two main questions: 1/ How do civil engineers see the coast they manage today? 2/ Is there a link between their vision of the coast and how they manage it?

This paper results from a thesis in coastal geography, completed in July 2014, using geographical but also sociological research methods to answer these questions, and exploring a comparison between the situation in the Netherlands, the UK and France.

## Research methods implemented and case studies description

### A Double Social and Geographical Comparison

To answer these questions, a double comparison has been conducted. A social one, as the vision of the coast and its management (discourses and practices) given by engineers and non-engineers has been compared to possibly highlight an "engineer's way of thinking". By "engineers", people trained as civil engineers were considered and as "non-engineers", all the people who received another specialized training in hard or social sciences whichever degree level they have. The underlying question was to assess the abilities of the first ones against the second ones, the competition between them or their complementarity to better highlight the characteristics of the "civil engineers' way of thinking" and to understand to which extent belonging to this professional group has an influence that would overtake the borders or not. This leads to the second comparison that was conducted, a geographical one. Three main reasons justify the comparison between the Netherlands, UK and France. Firstly, Dutch, British and French civil engineers have shown a continuous presence in coastal management for nearly one millennium. Whether it is sea dikes and drainage techniques widely exported by the Dutch since the 17$^{th}$ century, the invention of the steam engine by the British which increased the strengths of drainage since the 18$^{th}$ century, or the centralised organisation, inherited from the Napoleonic era, the mutual influence for centuries of Dutch, British and French engineers is still a reality. Secondly, the north west of Europe show a climatic coherence illustrated by the historical presence of violent storms as well as common predictions regarding climate change. Thirdly, the three countries belong to the European Community and so are subject to the same politic recommendations like ICZM (Integrated Coastal Zone Management).

**The Analysis of the Articulation Representations/Practices of Coastal Management**

To analyse the link between interviewees' vision of the coast and their practices, the implemented research method focussed on 3 objectives. Firstly, 13 management projects were studied to understand engineers' practices and 3 environments have been compared: estuaries, embanked coasts, shingle and sandy beaches (figure 1).

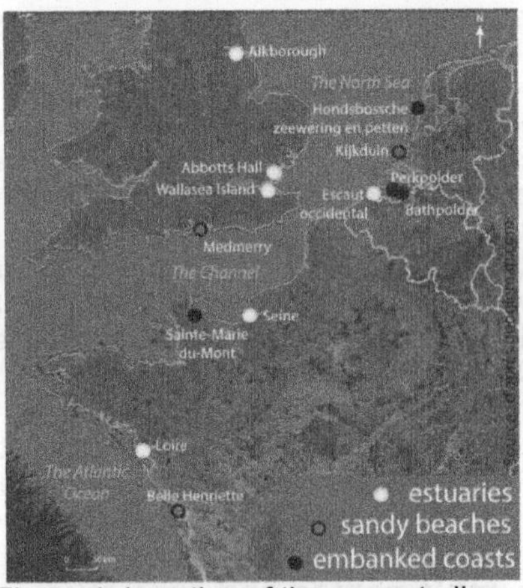


Figure 1: Location of the case studies

Secondly, the discourses of engineers and non-engineers were analysed through 102 interviews of which 46 with engineers, conducted between 2010 and 2014 in the three countries. Then graphic techniques have completed the interviews: associative graphics helped to highlight points of reflection (figures 2 and 4), whereas mental maps helped to give a spatial character to the representations collected[1] (figure 3).

[1] For more information on how mental maps have been used and analyzed, please refer to the article online: http://echogeo.revues.org/12573

Lastly, the training of engineers met was compared through consultation of archives in l'École Nationale des Ponts et Chaussées in France, Delft University of Technology in the Netherlands and Institution of Civil Engineers in England. In addition, interviews were conducted with people in charge of engineers' education in France and in the Netherlands.

## Three Main Results

Three main results have been highlighted in this study: 1/ a renewal of the practices of coastal management, 2/ a strong homogeneity between engineers and non-engineers in their speeches dedicated to the coast and its management, 3/ however showing few nuances mainly linked to cultural differences.

### A Renewal of Coastal Engineering, Giving More and More Space to Environment

Environment is considered in its geographical definition: a social and hybrid object that comes under complex interactions between the coastal environment and the possible uses made of it. Y. Veyret named it the geo-environment that must be analyzed through temporal and spatial scales (Veyret, 2007). In other words, taking environmental issues into account implies the implementation of a dynamic management of the coast, which has been particularly noticed in the Netherlands and in the UK. Dutch, for instance, have developed a dynamic coastal management by directing their reflection "seaside". In front of greater risk of marine erosion of the dunes, the Dutch government registered indeed by law in 1990 the obligation to maintain the coastline such as it was then positioned, without ever allowing any backward movement of the coastline. A massive annual beach nourishment policy was put into place. In order to save money but also to reduce the impacts of the disturbances on the marine environment provoked by these annual sand deposits, the idea was raised in 2008 to use the force of the littoral drift to scatter the sand for many years and over several kilometres (Mulder, Tonnon, 2010). In 2011, the project ZandMotor was implemented and some 22 million cubic meters of sand have been spread in less than 2 months. The use of the dynamic nature of the coast helped manage the recurrent problem of erosion at the very local but mostly regional scale, taking into account the long term results, as defensive benefits are expected until 2030. This also helped meet several socio-economics and environmental objectives.

On their side, British engineers have also implemented a dynamic coastal management, by developing, since the early 1990's, a reflection "landside" this time, through the technique of managed realignment.

The case study of Selsey Medmerry is one of the most recent examples. On the South coast of England, the headland of Selsey was protected by a shingle beach, regularly recharged, but nevertheless submerged with each large storm. During the winter 2013/2014, a managed realignment operation was put into place: a sort of de-polderisation consisting in allowing the sea to enter the land through a breach opened in the shingle beach.

Managed realignment indeed, gives the possibility to implement a sustainable solution of coastal protection: the creation of a minimum 10m strip of dense saltmarsh mitigates the strength of the wave before it hits the seawall and helps to stabilize the coastline on the long term (Shepard et al., 2011). Lastly, this technique reaches environmental objectives as the re-creation of a coastal area is very beneficial for coastal biodiversity but in also economical objectives, which partly explains why South and East coasts of England count today around 25 managed realignment operations, created since the 90's (Goeldner-Gianella, 2013 ; Gueben-Venière, 2014).

Two complementary ways of looking at a dynamic and more environmentally friendly coastal management were exposed, in which the involvement of engineers is increasing. Considering this, we examined in more details the engineers' vision of the coast they manage to understand the reasons of these practices.

### A Strong Homogeneity between the Visions of the Coast and its Management Given by Engineers and Other Scientists

The coast is seen as a complex territory and a dynamic system.

Interview reports show a general tendency of engineers to give a larger place to environmental issues, which implies complex interactions between the shore environment and its practices. The complexity of the coast was revealed at first by the difficulty to define it, showing the importance of the subjective factors linked to it. The coast has been depicted like "an interface", "a fuzzy thing" and also "the feeling of being near the sea". Also, several interviewees, engineers and others, have spontaneously approached the notion of complex systems to define the coast. A French engineer explained, for instance, that "the coast is an archetypal complex system. [...] We need to take into account multiple interactions regarding legal but also economical, technical, sociological and ecological matters. Generally speaking, that also means that there are retro-actions [...] that haven't been enough taken into account". This notion of complexity also appeared in associative graphics related to the coast (figure 2).

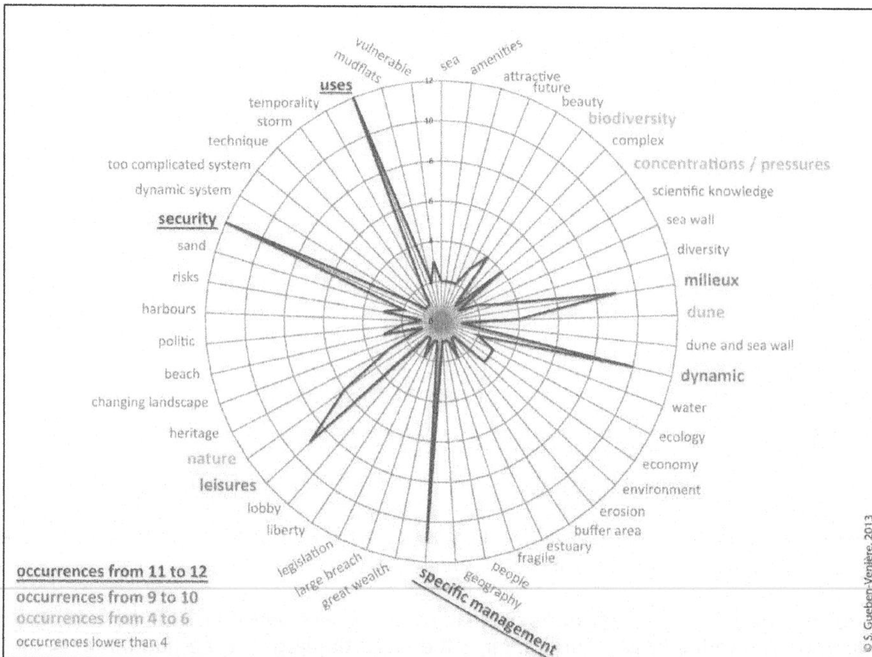


Figure 2: Words used by engineers and non-engineers to describe the coast

The outcome from all actors, engineers or non-engineers, across the 3 countries, highlights the words: *uses* and *security* (mentioned 12 times) and *specific management* (11 times). These words describe the complexity linked to the competition of multiple uses on a restricted territory and so confirm the attitudes drawn from interviews: the coast allows numerous uses and consequently requires a specific management to ensure a secure development. Besides, the word *dynamic* (mentioned 10 times) completes this first social vision of the coast, driven by security aspects and a dynamic vision. Yet, the distinction based on initial training shows that this complex and dynamic vision of the coast is largely shared by engineers and non-engineers, which also emerges in the interviews.

Although its perception varies from an interviewee to the other, all of them take into account the mobility of the coast, considering tidal movements and describing the dynamic currents of the coast. An English civil engineer explained about it: "I don't like the word coastline because the line implies that it's fixed here. But it moves, the coast is dynamic. And I do think that this idea is now shared by everybody". The results given by mental maps also confirm this dynamic vision. Here is the example for the Netherlands (figure 3)

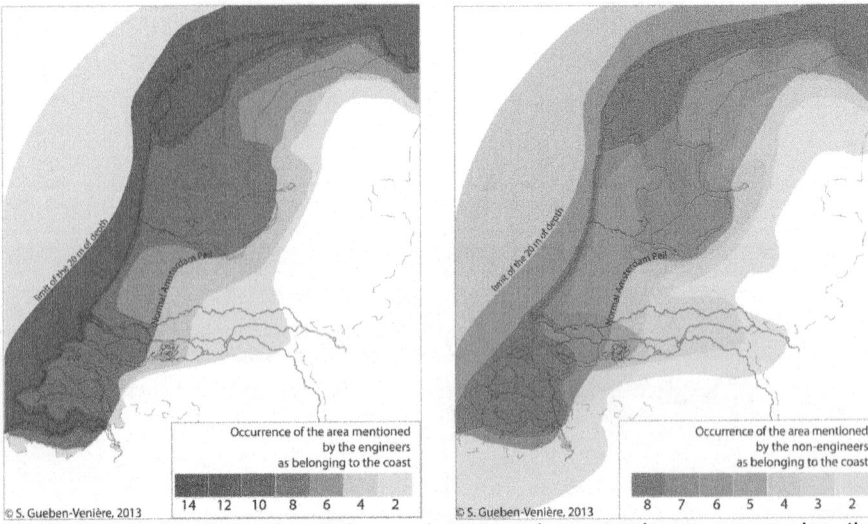


Figure 3: Mental maps drawn by engineers and non-engineers, answering the question: "What limit would you give to the coast?"

The maps above describe areas that Dutch engineers and non-engineers consider as belonging to the coast. The darker is the area, the larger is the number of people having defined it as "coastal zone". In total, 14 engineers and 8 non-engineers produced a mental map. The example is then relatively poor and non-equivalent for engineers and non-engineers, so the comparison between these maps must be done cautiously.

On the maps drawn by engineers (on the left side), the darkest area - a little larger than the simple coastline - has been described as belonging to the coast by 12 engineers over 14. That means that only 2 engineers reduced the notion of *coast* to the one of *coastline*. Also the area corresponding to the number 10 in the legend shows that 10 engineers over 14 widen the coast up to -20m depth which is associated to the closure-depth. Therefore, two thirds of the engineers attributed to the coast a submarine volume and considered it as a dynamic system. On the maps drawn by non-engineers (on the right), the coast is straightaway delimited widely: none for instance would reduce it to the coastline.

These 2 "average mental maps" show similarities: the characteristics of the coast given by engineers get closer to the ones of non-engineers. The opposition between a vision long imputed to engineers of a linear and static coast and the one attributed to other scientists who view the coast as something larger and more systemic, is now reduced. Yet the dynamic vision of the coast is no longer the monopoly of non-engineers and these two categories of actors seem from now on to share a common vision of the coast. The link between discourses and practices seem now evident: the actors of a project, and particularly engineers, envisage the coast as a dynamic system and therefore can imagine new solutions precisely using these natural coastal dynamics. Certainly, without a comprehension of the coast as a dynamic system, Dutch actors couldn't have conceived a massive sand nourishment project developing a "seaside reflection" or the English actors, a "landside reflection" with managed realignment.

Towards a more environmental management of the "coast of tomorrow".

During the interviews, the answers to the question: "According to you what would be an ideal coastal management? How do you envisage the coast of tomorrow?", were particularly interesting.

The results describe a will to reinforce the integration of environmental issues but also show that this will is shared by engineers and non-engineers. An interesting evolution has been noticed between the description of "today's coast" which certainly included several issues but in a sectored way, and the approach to implement in order to reach an optimized management of the "coast of tomorrow". This offers a systemic vision making the link between several issues to be taken into account in a dynamic approach.

In other words, we are no longer in the simple equipment of the coastline and not yet completely in the more complex approach of an integrated coastal management - the objective to pursue for all the actors interviewed.

The associative graphics confirm this tendency (figure 4). For instance, the word *adapted* was mentioned 4 times by 17 engineers and 4 times by 20 non-engineers. As well, the expression *working with nature* was mentioned as much by both categories of actors. Engineers quoted 3 times the notion of *research and development*, and the non-engineers 3 times *improve the knowledge*. *Coordination*, *security* and *planning* have also been mentioned as much. The environmental vision of coastal management is then no more the privilege of one or the other of the two distinguished categories.

The historical influence and the technical training of civil engineers could have possibly let thought that a distinction would be quickly established between the engineers' technical point of view and the one of non-engineers. However non-engineers as well as engineers have both raised an integrated approach of coastal management.

To the original question we can then answer positively: the notion of environment, in its geographical definition, has a growing focus in the discourses and the practices dedicated to the coast and a strong homogeneity has been highlighted between the vision given by engineers and non-engineers.

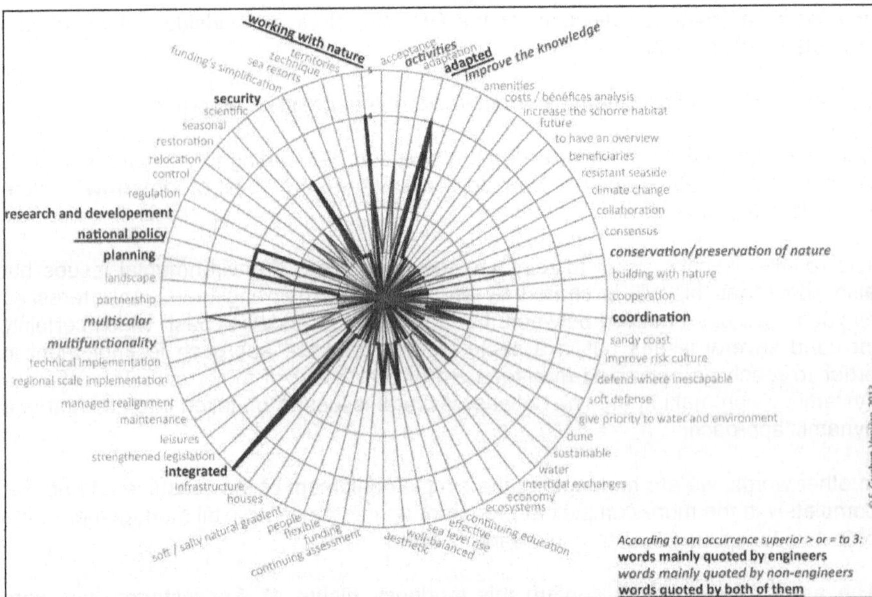

Figure 4: Words used by engineers and non-engineers to describe how they would manage the "coast of tomorrow".

**Nevertheless Nuances Appear...**

... mainly linked to cultural influences.

A second analysis shows however several nuances. Nuances linked to strong national speeches. The coast is for instance much more described by the Dutch as a territory where you feel secure, whereas the notion of risk is very present in the French speeches. These national nuances become much more pronounced in the description of what could be an ideal management of "tomorrow's coast".

In the Netherlands, several actors, engineers or not, explained that the national strategy, as well as the management methods currently implemented, are matching an optimized coastal management and could nearly be qualified of ideal: "For me, dialogue and multi-disciplinary involvement are both keys for an optimized coastal management. These two aspects exist well and truly in the Netherlands"[2]. Therefore quite a strong and positive discourse emerges from the Dutch actors. The few remarks concerning improvements are more a matter of flexibility, and according to the people interviewed, the organisation of coastal management in the Netherlands seem to answer the Dutch issues, security ones, but also environmental ones, to reach an integrated management of the coast.

Without going so far as asserting that their present organisation is already ideal, English interviewees expressed a kind of satisfaction with the latest evolution of coastal management. The concept of sedimentary cell adopted through Shoreline

[2] Extract of the interview of a Dutch non-engineer

Management Plans at the beginning of the 90's has been mentioned several times and in a very positive way. The idea is to no longer take into account administrative boundaries but a coastal section containing a complete cycle of sedimentation (production area, transport and deposit). It seems to be for the English actors the starting point for a smart coastal management, using the dynamics of the coast and giving the opportunity to consider the long term. To improve this way of doing, English interviewees would reinforce the influence of the Coastal Groups, the working groups based on sedimentary cells.

On the French side, the outcome of the interviews has been much more critical and the path to reach a form of ideal coastal management seems to be some way off: "Even if I caricature a little, I don't think to be far from the truth by saying that there isn't two people in France who understand in the same way how the current organisation all plays out".

Two main reasons can explain this disorganisation. Firstly, a lack of clear understanding of the role of the State. Without questioning the political decision of decentralisation taken in 1982, engineers and non-engineers explain the difficulties they still encounter today in terms of responsibility and implementation: "At present there is no clear communication between the State and local authorities. The State is responsible for civil security, but has been gradually opting out for the benefit of local authorities, to which the State imposes a certain number of boundaries and constraints to lead the mission that was initially the responsibility of the State. It's a vicious circle...". Secondly, there still is a partitioning of the central services of the State: the risk of erosion, which is under the responsibility of the DGALN[3] is not considered as a priority risk, while the risk of flooding is under the responsibility of the DGPR[4]. Yet, if the erosion of the low coasts is not taken into account early enough, it can increase the risk of flooding. The present organisation of the central services dedicated to coastal management is therefore not totally efficient.

The importance of the year of graduation.

For the majority of the interviewees, the generational factor is more important to take into account than the initial training to explain possible differences of points of view. This has been particularly true in France and in England: "Even if we come from different backgrounds, we now have almost everywhere the same training about the coast [...]: we all have in mind the errors to avoid. I think in particular in the Racine mission[5]". According to this young civil engineer, the search for sustainable coastal management solutions is now accepted by everybody, engineer or not. On the other side of the Channel, the question of generations has been openly formulated by several people, like this geo-morphologist: "Since I have joined the Environment Agency, the issue of not necessarily defending, not necessarily building something, has become much more common subject. When I joined in 1999, it was very new, and their (engineers') surprise was there: what are you doing that for? Put nice walls up and that will be okay!" In other words, the more recently engineers graduated, the

[3] Direction Générale de l'Aménagement, du Logement et de la Nature
[4] Direction Générale de la Prévention des Risques
[5] The « Mission Racine » consisted in an intense and planned urbanisation of the South coast of France launched in 1963 to catch the tourist flow before it reached Spain. This plan had a lot of negative consequences, particularly in terms of erosion and risks of flooding.

more easily they consider environmental issues as new opportunities to catch in coastal management, whereas engineers who graduated before 2000's, still can envisage them as an additional constraint.

## Between the Fear of Storms and the Improvement of the Scientific Knowledge: A Range of Explanatory Factors

Several factors were analysed to explain these results showing a general tendency, practices and speeches giving a larger space to environmental aims, but also particularities such as cultural and politic differences. Here are few examples.

### Two Main General Explicative Factors

From the 1950's, general knowledge on the functioning of the coast improved: engineers played a fundamental role in technical advances, in particular in numerical modelling of waves, whereas geo-morphologists played another fundamental role in the dynamic understanding of the coastal system, working with very large time scales. It is only since the 1990's that a multidisciplinary work has taken place, encouraged by the European Community through European research program like MAST (Marine Science and Technology) or INTERREG IIIB on the North Sea. Also, from the late 1980's, engineers' education evolved in the same direction, giving an increasingly place to the environment. As a result, it became possible for engineers, whether Dutch, English or French, to share a common vocabulary with other scientists like environmentalists, geographers or ecologists (Gueben-Venière, 2014).

### Specific Explicative Factors

Beyond these general explicative factors, common to the three countries, two other factors, specific to each country, must be taken into account. Firstly, the way the ecological and social movement of the 1970's has been experienced by Dutch, English and French engineers can explain a large part of the advances observed in the way coastal management may be implemented.

In 1973, a real turning point had turned Dutch engineers of the Rijkswaterstaat's authority upside down, up to the point of making them abandon the construction of the last dam of the Delta Plan (Disco, 2002; Meijerink, 2005; van den Brink, 2009). Consequently, these engineers experimented very early the multidisciplinarity by working together with other scientists to imagine another movable dam, satisfying security aims but also environmental ones. In the UK, this transition materialized, for coastal management, into the constitution of Coastal Groups and Shoreline Management Plans during the late 1980's. British engineers, whose social recognition is lesser than on the continent, have then spontaneously been integrated in these new working groups, becoming inescapable according to themselves. In France this movement shows a more complex development with vague outlines: an early institutionalization of the "green" movement with the creation of the ministry of environment in 1971 but a difficult adaptation for engineers who conserved a fragmented vision of the notion of environment. It is only during late 2000's that people without a specific engineering background have been able to occupy a key position in coastal management, whereas this transition appeared in the 1980's in the Netherlands and in the late 1990's in the UK (Disco, 2006 ; Gueben-Venière, 2014).

Secondly, the engineers' education models, well differentiated in the three countries, may explain the choices made. The education of English engineers has always put the accent on practice. Four years of professional experience are necessary to obtain the title of Chartered Engineer. Thus, an English engineer, when graduating, has already worked with other scientists with not only a technical education. In the Netherlands, research and innovation have always been the key of civil engineers' education, whereas the French model cherishes the excellence of mathematics and hard sciences, very selective during the 2 first years (Laffont, 2006). These education models can explain part of the advanced developments, in particular in the Netherlands and in England.

## Conclusion: A Double Consequence for Engineers

Despite an evolution of their larger and more systemic vision of the coast and of its management, engineers seem to show a certain withdrawal in the decision making process: they do not occupy the first-rank position that they used to hold systematically before. By opening key coastal management positions to other scientists, central authorities like Rijkswaterstaat in the Netherlands, Environment Agency in the UK and the Ministry of Environment in France, allowed them to manage engineers: mainly biologists and ecologists managed civil engineers from the beginning of the 1980's in the Netherlands (Disco, 2006), geo-morphologists did at the beginning of the 1990's in England and mainly jurists from the second half of the 2000's in France (Gueben-Venière, 2014). This could have led thinking of a slowing down for civil engineers, but that would have been without taking into accounting on their ability to innovate. This is why this relative downturn comes along with new perspectives for engineers' corps.

This downturn comes along with new perspectives for engineers' corps. Without ever forgetting their first role – to ensure the safety of people against the risk of marine flood - civil engineers seem to overcome the new outlines of the environmental and social context and take ownership of them. By now on taking into account the complexity of the coastal system at the starting point of their reflection, civil engineers are now able to renew the management of coastal risks by reducing the vulnerability of inhabitants. This means accepting but also using uncertainty, multiple interactions and retroactions of a phenomena, contradictions put forward, articulation of the knowledge in hard and human sciences, existing and future sciences.

For instance, few Dutch civil engineers work together with architects to imagine a new type of architecture more adapted to the coast, based on the fact that classical houses, with static structures, could not give a satisfying answer to dynamic environments in essence. In Ijburg, suburb of Amsterdam, floating houses, then floating buildings have been existing for ten years now. For this type of construction, the paving stones of houses slide along two pillars anchored in the ground, so they can adapt to a progressive rise of the level of water. Other techniques were since born: the Floating House of Rotterdam for instance, uses a paving stone entirely floating thanks to an alternate layers of concrete and expanded polystyrene. The relevance of solutions brought by floating houses to the sea level rise as well as scientific uncertainties, opens a very interesting line of thoughts and invites to question the vision of other influential actors: the architects.

The last 30 years can be described as a real upheaval for the corps of civil engineers. By denouncing their coastal management practices, it is also their way of thinking that was questioned and consequently their role within society and the reasons for their training. This upheaval, still ongoing, paradoxically led engineers to renew their reflection: imposed in the first place, it can, now on, help them to revive with the essence of their job: technique and innovation.

**References**

Bijker W. E. (2002) *The Oosterschelde Storm Surge Barrier. A Test Case for Dutch Water Technology, Management, and Politics*, in Society for the History of Technology, pp. 569-584.

Brink (van den) M. (2009) *Rijkswaterstaat on the horns of a dilemma*, Eburon, Delft, 335 p.

Didier C. & Talin K. (2014) *Engineer's ecoskepticism as an ethical problem*, in Christensen S. H., Newberry B., Meganck M., Didier C., (eds.), Engineering Identities, Values, Contexts and Epistemologies, Springer, p. nc.

Disco C. (2006) *De natuur herboren. De ecologische wending in het Nederlandse waterbeheer*, in Tijdschrift voor Waterstaatsgeschiedenis, n°9, pp. 1-18.

Disco C. (2002) *Remaking 'Nature': The Ecological Turn in Dutch Water Management*, in Science, Technology & Human Values, pp. 206-235.

Goeldner-Gianella L. (2013) *Dépoldériser en Europe occidentale. Pour une géographie et une gestion intégrées du littoral*, PUS, collection Territoires en mouvements – 1, Paris, p.350 p.

Goeldner-Gianella, L. (2007) *Perceptions and Attitudes towards Depolderisation in Europe: A comparison of Five Opinion Surveys (F., U.K.)*, Journal of coastal Research, vol. 23, n° 5, pp. 1218-1230.

Gueben-Venière S. (2014) *Vers une gestion renouvelée du littoral nord-ouest européen: des ingénieurs néerlandais, anglais et français de plus en plus « verts » ?*, Thèse pour l'obtention du doctorat de géographie, Université Paris 1 Panthéon-Sorbonne, 466 p.

Gueben-Venière S. (2011) *How can mental maps, applied to the coast environment, help in collecting and analyzing spatial representations?*, EchoGéo, n°17, 7 p.

Laffont H. (2006) *Un ingénieur est-il bien an engineer? Contribution à l'analyse contrastive du milieu professionnel de l'ingénieur en France, au Royaume-Uni et aux Etats-Unis*, ASp, n° 49-50, varia, pp. 1-9.

Meijerink S. (2005) *Understanding Policy stability and change. The interplay of advocacy coalitions and epistemic communities, Windows of opportunity, and Dutch coastal flooding Policy 1945-2003*, in Journal of European Public Policy, pp. 1060-1077.

Mulder J. P. M. & Tonnon, P. K. (2010) *'Sand Engine': background and design of a mega-nourishment pilot in the Netherlands*, Coastal Engineering Proceedings, n°32, 10 p.

Nicholls R. J., Larson M., Capobianco M. & Birkemeier W. A. (1998) *Depth of Closure: Improving Understanding and Prediction*, Coastal Engineer, pp. 2888 – 2901.

Shepard C. C., Crain C. M. & Beck M. W. (2011) *The Protective Role of Coastal Marshes: A Systematic Review and Meta-analysis*, Plos ONE, vol. 6, issue 11, pp. 1-11.

Veyret Y. (2007) *Dictionnaire de l'Environnement*, Paris, Armand Colin, 403 p.

**Coastal Management: Changing coast, changing climate, changing minds**
ISBN 978-0-7277-6114-9

ICE Publishing: All rights reserved
doi: 10.1680/cm.61149.501

# Coastal Protection for Ada, Ghana: a Case Study

**Annelies Bolle**, International Marine and Dredging Consultants (IMDC NV), Antwerp, Belgium
**Luciana das Neves**, IMDC NV, Antwerp, Belgium
**Joris Rooseleer**, IMDC NV, Antwerp, Belgium

## Introduction

Coastal erosion is a major threat to the entire West-African coast. Along the northwest coast of Africa average rates of coastal retreat are between one and two meters per year (UNESCO, 2012). At the coast in Ada, Ghana, the average rate of coastal retreat is locally, close to the Volta River mouth, more than 6 m/year (IMDC, 2010).

In 2010, the Ghanaian government decided to protect the coast in Ada along a distance of about 16km. The coastal protection scheme consists of beach nourishment and groyne structures for its structural stabilisation, and is split into two phases: first the groynes in the most critical stretch near the Volta river mouth are built (phase 1, finished in the summer of 2013), then the remaining groynes and the beach nourishment are executed in a second phase (phase 2, currently ongoing: 2014-2015). The purpose of the project is twofold, to stabilise the coastline, and to limit the amount of overwash towards the villages.

This paper discusses the aims and objectives of the project, the challenges, and the methods used to overcome these. Success and lessons learned are included in the concluding remarks.

## Aims and Objectives of the Project

### Site-specific conditions

The project site covers the length of eroded beach from the retreating delta of the Volta River, where the coastline suffers an approximate 6m erosion per year (see, e.g., IMDC, 2010), in the east to the village of Totope in the west. The project areas for phases 1 and 2 are indicated in the Figure 1 below.

The wave climate off the coast of Ghana is swell dominated with occasionally a minor wind sea component.

Extreme wave conditions along the coast of Ghana (and most of Western Africa) are completely determined by extreme swell waves and not by extreme wind waves, since no long local wind storms occur (IMDC, 2013). Peak wave directions of the present swell systems near the Ada coast are between 170°N and 210°N and induce locally a net West to East longshore sediment transport.

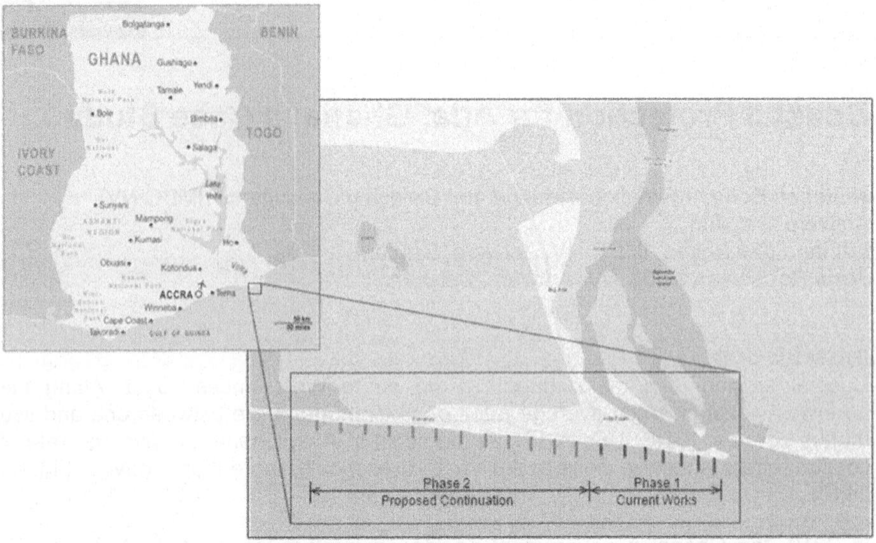


Figure 1: Location of phases 1 (current works ) and 2 project areas with schematic indication of the groyne system under construction (proposed continuation)

The nearshore significant wave height (at -10m LAT) is between 1 and 2m for 75% of the time, with wave periods higher than 10s for 80% of the time (IMDC, 2013).

The water level variation along the coast of Ghana is dominated by the astronomical tide, since no long local wind storms occur and wind induced (storm) surge is limited. At Ada the average tidal range is about 1m, mean sea level is about 1.15m LAT (Ministry of Water Resources, Works and Housing, 2010).

Tidal and wind-induced currents are small compared to the ocean currents (Guinea current and counter-current). Offshore Ada (-10m LAT) currents are mainly oriented to the East and vary between 0.15 and 0.60m/s (IMDC, 2013). Nearshore the wave-induced longshore currents are dominant.

The net longshore transport on the coast of Ghana is mostly directed from West to East. Estimations by IMDC (2010), based on the growth of the Volta and Keta sand spits, suggest net longshore transport rates near Ada in the order of 0.4 to $0.8Mm^3$/year.

Relevant physical features with potential implications on the coastline morphology and the beach erosion are the Volta delta (with impact on larger temporal and spatial scales), bathymetric features (such as nearshore depressions or troughs in the bottom almost perpendicular to the coast or parallel with the main wave direction), rip currents, the alignment and shape of coastline (the coastline alignment rotation from the west-east direction increases in the phase 1 project area towards the Volta river mouth), low-lying areas (where overwash occurs), and the presence of unconsolidated sediments (silt and clay layers that may influence locally the morphological response).

**Coastal Erosion and Flooding**

Ada Foah and other surrounding villages have been suffering from ongoing erosion for a very long period of time. As a result of this various properties and existing infrastructure including schools, post offices, public offices, graveyards, factories, private residential homes and roads are being damaged and destroyed (Figure 2).

Figure 2: Aspects of the damage or destruction of property and infrastructure (left: January 2010, right: May 2009) © IMDC

This coast stretch is also prone to flooding, due to the poor condition of the beaches and the presence of extensive low lying areas. In the advent of storms or other extreme weather conditions, overtopping, severe flooding and even breaching of the sand barrier (which acts as a natural flood protection) are likely to occur with possible loss of human lives, property and livelihood.

**Overview of the Project**

The design and construction were both carried in two phases, of which the first phase was accomplished by the summer of 2013 and the second is anticipated to be completed by 2015.

During project design, specialised studies and investigations, including measurement campaigns and modelling efforts, were combined to study and understand the hydro- and morphodynamics in the project reach and to optimise the design.

Final design is thus informed by site investigations, physical and numerical modelling and value engineering undertaken by the designer and the contractor appointed to undertake construction. It consists of the placement of 5,000,000m$^3$ of sand (measured in the hopper) as beach nourishment and dune heightening over the entire project reach (i.e. the phase 1 and the phase 2 project areas), and the construction of 21 groynes, of which 7 in the phase 1 project area (accomplished by the summer of 2013), and another 15 groynes along the entire phase 2 project area (currently under construction: 2014-2015). For the nourishment works, a uniform design template, with fixed alignments and fixed elevations, is used over the entire project reach.

The design concept has been developed to best realise project requirements within the contractual Bill of Quantities.

## Challenges

The challenges of this project include, amongst others, the high-energetic environment with long swell waves, which is not only being responsible for the highly dynamic beaches, but also for pushing designers and contractors beyond their comfort zone. This meant that they were forced to adapt to the local conditions and to be creative during the design and the execution of the works in order to find alternative, workable solutions that still fit into the tight budget and the project requirements. Along the following paragraphs three main topics are discussed in that respect: data availability and collection, modelling studies in support of the design process, and selection of execution methods.

### Data Availability and Collection

The amount of data available for the project site, and its surroundings, has a direct impact on the level of insight in the system's behaviour and the detail possible for the design. Typically for remote and rather undeveloped areas, a limited amount of data is available from literature. This is then complimented by field measurements at the start of the project (e.g. bathymetry and topography, hydrodynamic and geotechnical data), and therefore becomes a well-documented case by the end of the project (large data set of beach morphology measurements). The situation was no different for the Ada case. A timeline of the measurements, the design and the construction is shown in Figure 3.

In 2010, well before the commencement of the design for the first phase, a long-term Metocean measurements campaign (1 year) has been performed near Ada to collect site-specific data as basis for the design. A fist topographic and bathymetric campaign was performed beginning 2010 (as baseline for the preliminary studies for the design of phase 1), and repeated beginning 2011 (as in-survey for the design of phase 1). However, since at that time, no project team nor offices were on site, access to the right material on site was not evident.

A sudden increase in the amount and type of data measured is visible from mid-2012 onwards. This coincides with the presence of a project team on site for the execution and supervision of the construction works for phase 1. All the in- and out-surveys during the construction works of phase 1, together with the frequently monitored beach profiles and continuous wave measurements provided already a better insight into the behaviour in the local hydro- and morphodynamics.

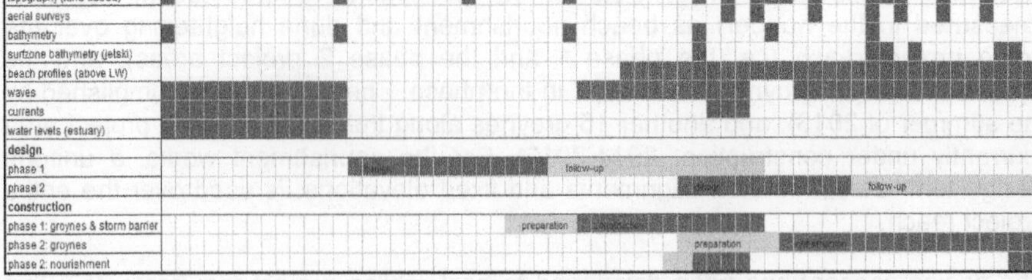


Figure 3: Timeline of the measurements, the design and the construction works for the Ada coastal protection project

At the start of the second phase, even more effort was put into measurements, by for example organising jet-ski surveys and aerial surveys. By deploying a jet-ski the surf zone could be covered, which could not be surveyed with a vessel due to the heavy swell conditions and (in fair weather conditions). The gap in between the lowest beach surveys and the shallowest jet-ski bathymetry could be reduced to 10 to 15m.

The aerial surveys performed every 3 months with an X100 mapping system by Gatewing (a lightweight Unmanned Aerial System, UAS) provided detailed aerial pictures and topography of the beach and surf zone over the entire project area. These frequent measurements, not only allow to monitor the natural processes (coastline evolution, surf zone processes, e.g. rip currents), but also to follow-up the progress and effect of the construction works.

As part of the Ada Coastal protection Works, morphological monitoring is also undertaken in both the phase 1 and phase 2 project area to evaluate the effect of the coastal protection works in place (i.e. the effect of the groynes and the nourishment). This is done on two levels: a general evolution of the coastline and morphology and a detailed evolution of beach profiles. The general evolution is based on analysing the changes in elevation, the volumetric changes and the coastline evolution, including the sub-aerial beach. The profile analysis is based on analysing the topographic evolution of a set of relevant parameters for the 120 beach profiles, monitored on a frequent basis.

The results of such extensive monitoring campaigns provide understanding of the coastal processes at specific locations and for the whole project area, with and without protection measures, namely it provides information about:

1. the natural behaviour of the coastal system (before the works);
2. the influence of the works on the coastal system (during the works); and
3. the effectiveness of the works on the coastline and beach profile evolution (after the works).

**Specialised Studies in Support of the Design Process**

The aim of the specialised studies and investigations is to determine the physical conditions which are generally the primary determinant for the design and construction of coastal protection structures. The physical site conditions of principal interest include the bathymetry, topography, morphology, geotechnical conditions and hydraulic conditions (such as water levels, winds, currents and waves). Measurements and modelling efforts are combined to study and understand the hydro- and morphodynamics in the project area and to optimise the design of the groynes and nourishment works.

To get insight on the coastal system and to determine the design conditions, wave and hydrodynamic modelling was undertaken (with SWAN and Delft3D). The hydrodynamic and morphological modelling focused on physical processes on the local and short/medium term time scale. The cross-shore processes such as wave run-up and overtopping causing beach erosion and overwash were investigated; and also longshore processes such as wave induced longshore currents causing longshore sand transport. It is clear that these morphological processes are a direct consequence of the hydrodynamic processes and that a numerical model which incorporates both was needed for simulation. All numerical models were calibrated and validated with results obtained from the various measurement campaigns.

A 2DH XBeach (Roelvink *et al.*, 2010) forecast model was implemented to help determine the best choice for the beach nourishment design profile shape, to qualitatively predict the coastal morphological evolution, and to understand the combination of coastal processes that alter the physical features of importance, and conversely.

1DH and 2DH XBeach models were also used to simulate wave conditions, currents and sediment transport near the groynes.

To study the long-term coastal evolution (over a decade) and to understand the effect and influence of the coastal protection works on the coastline evolution, a LITPACK model was set-up.

Apart from satellite images being available immediately for the calibration of long-term coastal evolution, the gathering of datasets for the more detailed morphological models only begun after the start of the project. This means that for instance LITPACK modelling could be performed for both project phases, whereas the detailed XBeach models, including the beach and surf zone area, could only be developed and calibrated once also the surf zone had been surveyed. This detailed topography and bathymetry, together with (at least) monthly beach profiles for about a year, allowed us to perform a Hindcast to obtain a calibrated model (Figure 4) (Gruwez *et al.*, 2014). These settings could then be used to do some predictive modelling for the phase 2 area (Verheyen *et al.*, 2014).

These detailed XBeach models provided additional insight in the coastal processes: e.g. the beach erosion and recovery, the wave run-up levels, the presence of rip currents and the impact of the nearshore depressions in the bathymetry on the coastline evolution. From in situ observations it was already established that rip currents are highly likely to occur along the beach of the phase 2 area (IMDC, 2012). Aerial pictures have indeed verified that this phenomenon occurs. Also in the results of the 2DH XBeach forecast model rip currents seem to have been identified for some time steps.

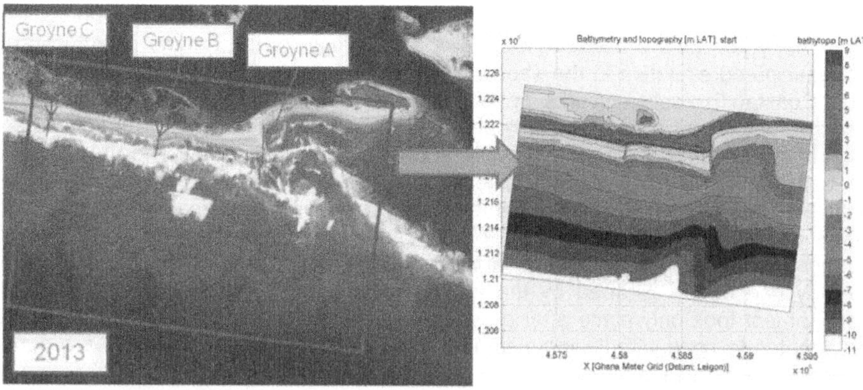


Figure 4: Hindcast model area (red rectangle) on background of © Google Earth (2013). Right: interpolated measured bathymetry in March 2013. Note the nearshore depression reaching towards the tip of groyne A

As the design studies for the groynes took place in parallel (e.g. physical modelling for the groyne stability), several feedback rounds occurred, both in phase 1 and phase 2, once more data became available and a better insight in the coastal system was obtained (Mercelis *et al.*, 2013).

Based on the morphological modelling it was decided to construct groynes with an average length of 180m. Physical modelling confirmed the dimensions of the groynes: a crest width of 3.5m and side slopes of 1V:3H. The 1m thick filter consists of a 0.35m thick layer at the bottom of a mix (30/70) between 5-40mm gravel with 1-300kg rock and on top of that a layer of 0.65m of 1-300kg rock. On top of the filter layers the groyne will be constructed from 2000-4000kg rock. A perpendicular core layer of 1-300kg rock is placed inside the groyne to reduce the permeability and increase the sand trapping efficiency. In the phase 2 project area, the groynes are built on top of the nourishment.

The longitudinal profile of the groynes consist of an onshore horizontal part, an onshore sloping part, an offshore horizontal part, and a seaward sloping part (Figure 5). The offshore horizontal part is placed at the level of the observed berm or platform in some of the beach profiles, and improves not only the hydraulic stability of the groyne, but also supports the shallow part of the beach profile, where wave energy is dissipated by wave breaking.

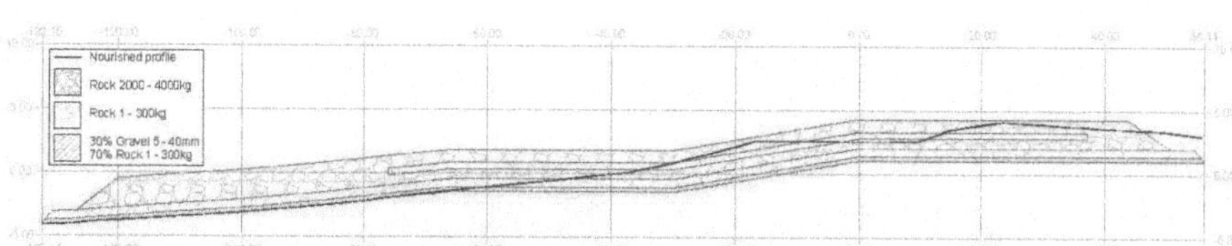


Figure 5: Scheme of a typical longitudinal profile of a groyne in phase 2 (different vertical and horizontal axis)

Each part of the groyne is defined before construction, based on the nourished bathymetry just before construction (i.e. the in-survey for the groyne construction), the anticipated evolution in the short- to long-term, and the overall objectives of the Ada Coastal Protection Works project. This means that the crest elevation and length of each part of the longitudinal profile of each individual groyne is evaluated as part of the whole protection scheme. A further complicating factor for this groyne lay-out and the adjustment of the different groynes with respect to each other is the different state of the beach nourishment at the start of the groyne construction: for some groynes the nourishment had only recently been performed, and not a lot of changes occurred yet, for others a bit more time occurred so the beach profile already had been reshaped by the natural processes. For a few groynes, beach nourishment took only place after groyne construction, so the groyne design had to be adapted to cope with the nourishment afterwards. These optimisations are done during the construction phase.

The beach nourishment and the dune heightening together will provide a buffer against beach erosion during storm events. The total nourishment consists of 0.6Mm³ in the phase 1 and 3.4Mm³ in the phase 2 project area, and a uniform nourished profile with a slope of 1V:25H. Based on the 1DH XBeach simulations it was concluded that for an adequate level of protection against wave run-up and overwash events the dune should be heightened up to +6m LAT with a seaside slope of 1V:6H (IMDC, 2014).

Sand for the nourishment has been searched and sourced offshore, in the vicinity of the project area. Sufficiently coarse sand (corresponding with the sand already present on the beach) has been found towards the west, and dredged for the nourishment works.

**Selection of Execution Methods**

Due to the severe wave climate, a water-based construction within the surf zone was not an option, so a land-based construction was considered. This classically results in a cross sectional design with a high and wide core to be able to install the armour layer elements by means of equipment driving on top of the core. However, in order to limit the amount of rock materials and to obtain a hydraulically stable design within the armour stone sizes available from the nearby quarries, a concept based on low-crested groyne structure was designed.

The land-based construction method developed makes use of temporary steel jetties accessible from the beach with excavators and providing direct access to the construction materials (Figure 6).

Figure 6: Top panel: construction of the groynes from land by means of jetties and a small cofferdam in the surf zone (October 2014); bottom panel: change in construction method between phase 1, one main jetty and two auxiliary ones (left: August 2012) and phase 2, two main jetties (right: August 2014) © IMDC

The groynes in phase 1 have been constructed around and below the main jetty, flanked by two auxiliary jetties to allow for construction and removal of the temporary cofferdam in the surf zone (top panel left in Figure 6). To facilitate and speed up construction pace however, the contractor modified and optimised the jetty lay-out for the second phase. Instead of one long and two short jetties, two long jetties were constructed this time, so that both the cofferdam and the groyne itself could be constructed from these temporary structures (top panel and right bottom panel in Figure 6). The main advantage was that the armour stones could now be placed in between the two jetties, instead of around and under the main jetty.

Since sand for the beach nourishment was dredged offshore, the method to hydraulically pump sand onshore with sinker lines also had to be adapted to the high-energetic environment (Figure 7). Most especially the movement from the sinker line from one spot to another had to be done with care, since the alongshore and wave-driven currents in the surf zone, together with the heavy swell and breaking waves made this a vulnerable operation.

Figure 7: Installation of the pipelines onshore (December 2013) and the sinking line in the surf zone (October 2014) © IMDC

## Concluding Remarks

As for many construction projects in remote areas, challenges had to be faced during the studies, the design and the execution of the Ada Coastal Protection project. In this case challenges included, amongst others, an initially limited data set and more important the high-energetic environment with long swell waves, not only being responsible for the highly dynamic beaches, but also pushing designers and contractors beyond their comfort zone. This meant that during the project designers and contractors were forced to adapt to the local conditions and to be creative during the design and the execution of the works in order to find alternative, workable solutions that still fit into the tight budget and the project requirements.

Survey data analysis denote, by December 2014 (almost about one year after the first nourishment works' campaign has been completed), that:

1. Profiles in the nourished areas along the phase 1 (Figure 8) and along the phase 2 area were generally above the reference situation after the completion of works (that is the situation at the moment of the nourishment works out-survey, in March-May 2014), despite transients of erosion during the multi-day sequence of storms with maximum wave heights reaching 3m and maximum wave-periods over 16s, which occurred in August 2014. Note that aerial pictures in Figure 8 are both from after the completion of nourishment works.
2. Profiles in the non-nourished areas in the phase 2 were generally eroding and receded with respect to the reference situation before the commencement of the phase 2 (that is the situation at the moment of the in-survey for the design of phase 2, in February 2013).

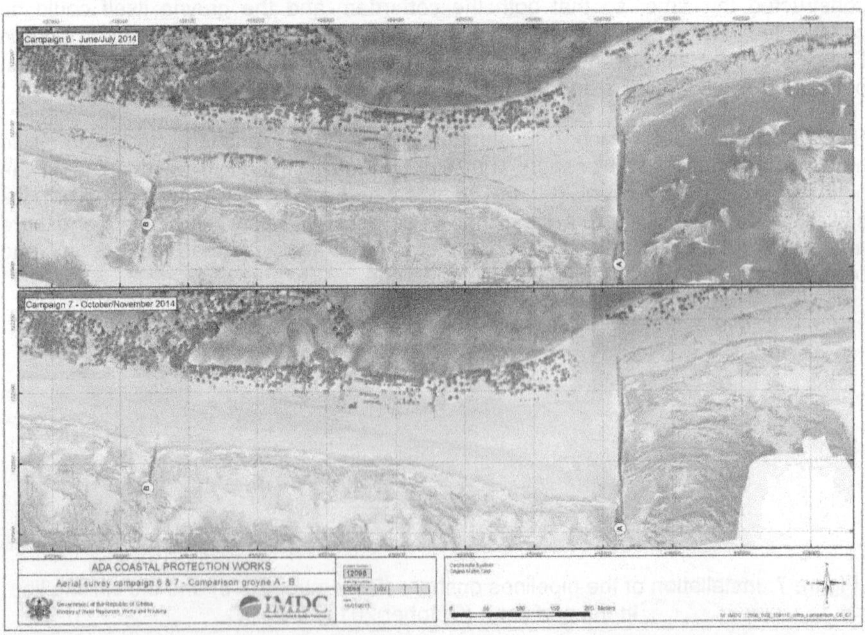


Figure 8: Comparison of aerial campaigns 6 and 7 (both after nourishment works), compartment between groynes B and A, phase 1 project area © IMDC

Despite all the challenges, it is possible to conclude that a technically sound and robust design responding to Client requirements and within the pre-established budget has been delivered. Further an efficient construction method, which makes use of temporary steel jetties, allows land based-construction of the submerged rubble-mound groynes, and warrants a precise and concise execution of the works, was applied.

## References

Gruwez V., Verheyen B., Wauters P. & Bolle A. (2014) *2DH morphodynamic time-dependent Hindcast modelling of a groyne system in Ghana*, 11th International Conference on Hydroscience & Engineering, 28 September – 2 October 2014, Hamburg, Germany.

IMDC (2010) *Ada coastal protection works: urgent measures. Coastal protection schemes: morphological studies*, I/RA/12062/10.138/NZI, v4.0, Antwerp, Belgium.

IMDC (2013) *Ada Coastal Protection Works Phase II: Wave Modelling*, I/RA/12062/13159/SUD, v1.0, Antwerp, Belgium.

Mercelis P., Bolle A., Olivier A., Audenaert S. & Smits J. (2013). Hydraulic Design of a Groyne System in Extreme Swell Conditions: a Case Study. Coasts, Marine Structures and Breakwaters 2013, 18 - 20 September 2013 - Edinburgh, UK.

Ministry of Water Resources, Works and Housing (2010) *Ghana water level datum.*

Roelvink J.A., Reniers A., van Dongeren A., van Thiel De Vries J., Lescinski J. & McCall R. (2010) *XBeach Model Description and Manual*, Delft, The Netherlands.

UNESCO (2012) *Coastal erosion major threat to West-Africa*, article on website (access date: January 2015): http://www.unesco.org

Verheyen B., Gruwez V., Zimmermann N., Wauters P. & Bolle, A. (2014) *Medium term time-dependent morphodynamic modelling of beach profile evolution in Ada, Ghana*, 11th International Conference on Hydroscience & Engineering, 28 September – 2 October 2014, Hamburg, Germany.

**Coastal Management: Changing coast, changing climate, changing minds**
ISBN 978-0-7277-6114-9

ICE Publishing: All rights reserved
doi: 10.1680/cm.61149.513

# Developing an Approach to Implement Adaptive Coastal Management: a Case Study from North Norfolk, UK

**Stephanie Hampshire**, Mott MacDonald, Croydon, UK

## Introduction

Adaptation of coastal communities to the impacts of climate change is becoming increasingly important, as it is unlikely that it will be possible in the future to maintain all areas of the coast as they are today. Coastal management needs to take into account the dynamic nature of coastal processes and how to adapt communities, services and infrastructure. As highlighted in the recent UK Climate Change Adaptation Sub- Committee Progress Report (2013), adaptation must look beyond the short term, to a long term vision, considering what the coastline could look like in the future. This long term vision needs to sit within the current guidance and processes for coastal management to ensure it is practically achievable.

There are abundant definitions regarding adaptation with regards to coastal management. For example a study undertaken by RPA (2008) defines it as '*the process for managing the impacts of coastal change on communities and individuals, in advance of erosion and or realignment, with the aim of reducing the risk and mitigating the adverse effects*'. Potential adaptation options include: the rebuilding of key assets/infrastructure as they are eroded; relocation/roll back of properties; and assisting property owners, businesses and communities with adaptation.

### Adaptation in UK coastal management policy

Many responsible authorities in the UK are increasingly coming under pressure to save money and to allow more natural coastal processes to occur, whilst also protecting local communities against coastal erosion. One of the largest, most recent studies of coastal adaptation in the UK was the Coastal Pathfinder Study. The aim of the £11m Government funded study (2009-2011) was to road test approaches to planning for and managing coastal change across several parts of the UK. The North Norfolk Coastal Pathfinder was the largest and most complex. As part of the North Norfolk Study, a small scale property acquisition and roll back programme enabled the purchase of 9 out of 12 properties identified as being imminently at risk of coastal erosion (within the first 20 years) in one key settlement. One of the key outcomes of the Study was that it achieved community buy-in and successfully trialled a methodology for residential property rollback. Further information on the steps/lessons learned for implementing rollback can be found in the Coastal Pathfinder Evaluation report (DEFRA, 2011).The property acquisition scheme enabled communities to think about planning for their futures and delivery of wider coastal adaptation projects. It also provided homeowners access to finances which would otherwise be unobtainable via property sale on the open market.

Now that the Pathfinder Study has been completed a key issue for rollback of communities on a wider scale includes defining the mechanism for funding adaptation measures and who should pay. There is therefore a real opportunity now to explore potential adaptation measures to provide an economic analysis on the reality of implementing such measures. By undertaking this analysis, the potential viability of implementing adaptation measures under the current UK framework could be realised.

When considering coastal management within the UK, it is typical to consider a long-term (100 year) plan and business case for managing coastal erosion or flooding in the form of a Project Appraisal Report (PAR). This is in accordance with the 100 year policies recommended under the Shoreline Management Plans. A PAR is a business case application for the UK Government's Flood Defence Grant in Aid (FDGiA) funding stream. This includes requirements under the Flood and Coastal Resilience Partnership Funding[1] (FCRPF) regime.

Historically the production of PARs has tended to focus on the scoping and funding application for capital works that meet 'Hold The Line' (HTL) policies. However, as increasingly highlighted in coastal management plans (such as Shoreline Management Plans (SMPs) in the UK), it is unlikely to be economically viable under the current criteria, nor environmentally sustainable to continue to 'Hold The Line' in some locations. Therefore alternative options to manage a changing coast including the potential for adaptation need to be considered. The policy of Managed Realignment (MR) which involves the realignment of defences landwards of the existing defences could be facilitated by considering adaptation measures.

In the long term it is not considered viable to continue to protect some communities. As such it is more sustainable to relocate communities with the most vulnerable properties rather than defending the properties through coastal defences. However no known PARs produced in the UK to date have actually considered how adaptation measures could be implemented, how much they might cost and how they compare and contrast to implementing capital works.

**Paper Aims**

This paper aims to use a case study in North Norfolk to:

- Contrast the economic viability of adaptation options against capital works options; and
- Consider whether this can be captured in the current UK coastal management framework, guidance and funding processes;

If adaptation options and an associated business case could be considered feasible and accepted by the UK Government then North Norfolk District Council would be well positioned to develop an innovative and industry first adaptation focussed PAR. For clarity the paper does not suggest a change to existing HTL policies.

[1] Partnership Funding is a process which aims to allow more flood and coastal protection schemes to go ahead and to give communities more say in what is done to protect them. A calculation is used to determine the amount of funding available depending on the project benefits. See www.gov.uk/government/publications/flood-defence-partnership-funding for more information.

## Case Study from North Norfolk, UK

### Coastal Management in North Norfolk, UK

The Cromer to Winterton Ness Coastal Management Strategy Study (Figure 1) was completed in 2012 (hereafter referred to as the Strategy Study (Mott MacDonald, 2012)).

**Figure 1: Cromer to Winterton Ness Coastline with SMP Policy Units (Crown Copyright : LA 079707 2003).** *Location in UK indicated by inset map.*

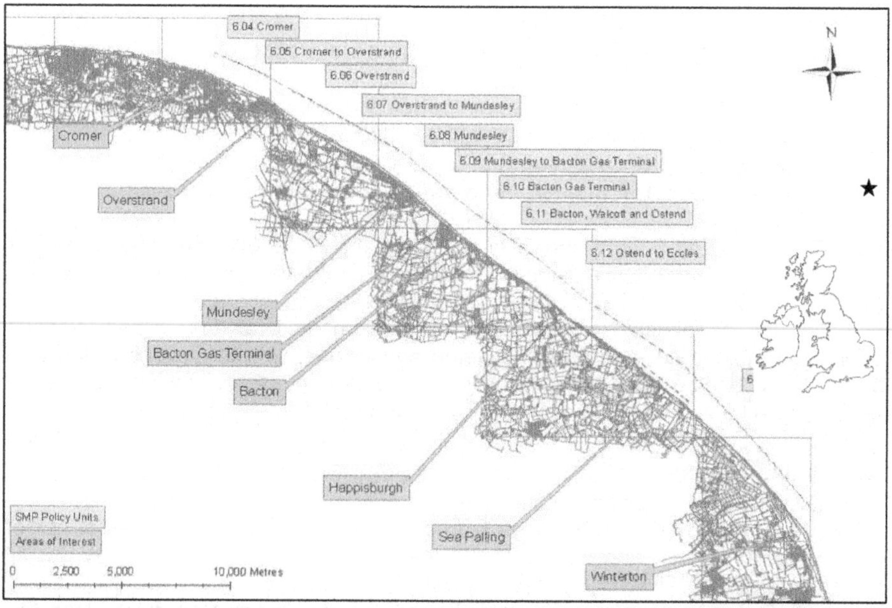


The Strategy Study aimed to assess the assumptions and impacts of the 2005 coastal management policy recommendations from the Kelling to Lowestoft Ness Shoreline Management Plan (SMP) (AECOM, 2010). Shoreline Management Plans present a large-scale assessment of future coastal evolution risks and provide a policy framework to address these risks to the developed, historic and natural environment.

The Strategy Study modelled potential changes in the coastline between Cromer and Winterton over the next 100 years to support a more holistic understanding of inter-linked coastal processes across the frontage. This included consideration of current and future coastal defence condition, future options and an economic assessment under the recommended management policies. The findings from the Strategy Study indicated that holding the line in the long term at various points could have an overall detrimental effect on the coast down drift by starving the frontages from sediment; consequently there could be accelerated erosion of the coastlines down drift whilst the updrift coastlines are held in position, causing an unstable coastline to form in the long term.

**Case Study of Adaptation in North Norfolk**

A typical village situated on soft cliffs and vulnerable to coastal erosion along the North Norfolk coastline, UK has been selected to provide a baseline case study for assessment in this paper. A specific area has not been named for sensitivity reasons. For the purpose of this paper the typical village sits on top of a cliff which is at risk of coastal erosion; timber groynes manage sediment movement by longshore drift; however the defences are not currently identified as viable for investment in the long term. The village has been considered to have a recommended 'HTL' policy for 0- 50 years and a 'MR' policy for 50-100 years from the current day.

**Coastal Protection Options (for 0-50 Years Only)**

Three coastal protection options for this village were proposed and investigated for the first 50 years:

- **Option 1: Do Nothing baseline option for comparison** (assumes that no maintenance or capital works would be undertaken);
- **Option 2: Monitor and maintain** (Do Something);
- **Option 3: Monitor, maintain and rock placement** (Do Something).

Both Options 2 and 3 were shown to be financially viable over 50 years. Option 1 was only considered for economic comparison and assessment of the relative benefits of 'Doing Nothing' against 'Doing Something'. In order to consider the potential for adaptation measures to be implemented at the village, various adaptation options need to be investigated and compared with Options 2 and 3 to see if a sustainable business case utilising adaptive measures (both economically and technically) can be made over a 100 year timeframe.

**Adaptation Options**

To assess the potential for adaptation, several options were selected:

**A) Buy and lease properties from year 50-** buy all the properties at risk of coastal erosion and lease back to their owners (or others) until they are all eroded.

**B) Outright purchase, demolish and rebuild properties from year 50-** demolish all the properties and relocate and rebuild the properties further inland (away from the risk of erosion).

**C) Buy and lease properties from year 1-** as above but based on predicted erosion risk bands.

**D) Outright purchase and demolish properties from year 1-** as above but based on predicted erosion risk bands.

These have been combined with capital works to HTL for the first 50 years (Table 1). This is to allow time for adaptation of communities.

**Table 1: Combined coastal protection and adaptation options for implementing the HTL and MR policies at the village case study**

| Coast protection option | Adaptation option | Combined option description |
|---|---|---|
| 2 Monitor and maintain | None | 0 to 50 years: HTL. Monitoring and maintenance only. |
| | | 50 to 100 years: MR. Reduction in the coastal defence monitoring and maintenance costs. |
| | A from year 50 | 0 to 50 years: HTL. Monitoring and maintenance only. |
| | | 50-100 years: All the properties bought at year 50 and rented until erosion takes place. No additional monitoring and maintenance. |
| | B from year 50 | 0 to 50 years: HTL. Monitoring and maintenance only. |
| | | 45-50 years: Over a 5 year period all properties are demolished, relocated and re-built. Maintenance and monitoring to continue until year 50. |
| | | 50-100 years: No further actions are undertaken. |
| 3 Monitor and maintain and rock placement | None | 0 to 50 years: HTL with rock placement, monitoring and maintenance. |
| | | 50 to 100 years: MR. Reduction in the coastal defence monitoring and maintenance costs. |
| | A from year 50 | 0 to 50 years: HTL with rock placement, monitoring and maintenance. |
| | | 50-100 years: All the properties bought at year 50 and rented until erosion takes place. No additional monitoring and maintenance. |
| | B from year 50 | 0 to 50 years: HTL with rock placement, monitoring and maintaining the coastal defences |
| | | 45-50 years: Over a 5 year period all the properties are demolished, relocated and re-built. Maintenance and monitoring of the defences will continue until year 50. |
| | | 50-100 years: No further actions are undertaken. |
| 1 'Do Nothing' | C from year 1 | Properties are bought in year 1, year 20 and year 50 (depending on the predicted risk of erosion) and rented until erosion occurs. No additional monitoring and maintenance. |
| | D from year 1 | All properties are demolished in in year 1, year 20 and year 50 (depending on the predicted risk of erosion), relocated and re-built. No additional monitoring and maintenance. |

The costs of the coastal protection options (Options 2 and 3) were determined by combining the capital and maintenance costs correct as of January 2013. These values were then discounted back over 100 years from current prices to calculate Present Values (PV). In the same way the PV of the adaptation options (options A,B, C, and D) were determined by combining capital costs (initial upfront costs of purchasing properties/land/rebuilding etc.) and maintenance costs (maintaining properties being rented out and potential income from renting). In this case 'maintenance costs' could be negative as it would include potential income from rent.

A description and key assumptions for each of the adaptation options is provided below. All costs were estimated as realistically as possible using contractor information and recent costs of construction of similar works. A risk allowance of 60%, as typical in the UK guidance (Environment Agency, 2010), was added.

**Options 2 and 3: Monitoring and Maintenance (and Rock Placement)**
Monitoring and maintenance of existing defences were based on existing maintenance spend from experience along the North Norfolk coastline. The premise was to 'sweat the existing assets' until the MR policy is implemented in year 50. Defence spend is then assumed to be reduced in the long term (50-100 years) as defences fail, become a health and safety risk, and are ultimately removed. In addition rock placement was considered (Option 3) at particularly vulnerable sections along the village frontage. Further details can be found in the Strategy Study (Mott MacDonald, 2013). These options alone do not provide a solution for protecting or allowing adaptation of communities in the long term (50-100 years).

**Option A and C: Buy and Lease Properties**
For Option A the assessment has assumed that all the properties will be bought at year 50 and rented until they are lost to erosion. Realistically it would be more likely that properties would be bought in batches, with those at most risk bought first; therefore this assessment takes a conservative estimate of upfront costs.

For Option C it is assumed that all properties are bought from year 1 (i.e. year 1, 20 and 50) and rented according to the erosion band in which they are situated i.e. 0-20 year, 20-50 years and 50-100 years. In year 1 all properties that will erode in the next 10 years are bought and rented until year 9. It is assumed that a property cannot be rented the year that it will erode. Similarly, in year 20 all properties from the 20-50 erosion band are bought and rented until year 35, and in year 50 all properties from the 70-100 erosion band are purchased and rented until their assumed erosion in year 75.

Erosion zones for this high level study of 0-20 years, 20-50 years and 50-100 years mean that the assessment has assumed that properties will be lost in the midpoint between the two years i.e. year 10, 35 and 75. This means that the year of loss for some properties may be overestimated and some may be underestimated; overall this should not make a significant difference to the benefit estimates. It is recommended further detailed erosion assessments should be undertaken at PAR stage. There may be a case for purchasing properties earlier to maximise the income to cover costs; further sensitivity analysis is required.

The full market value of each of the properties at risk of erosion over 100 years was considered. The market value was obtained from the North Norfolk Coastal Strategy (HR Wallingford, 2004) and amended according to the Retail Price Index (RPI) (a measure of inflation published monthly by the UK Office for National Statistics). An average value of the annual rent per property type was obtained using Zoopla.co.uk. In this assessment it has been assumed that 100% of the properties will be rented 100% of the time from the purchase until erosion takes place. Realistically some properties may be utilised as holiday lets, vacant periods will occur and some properties may require excessive investment in order for them to be let.

**Options B and D: Outright purchase, Demolish and Rebuild**
Options B and D assume that the purchase, demolition and rebuilding of the properties will be implemented according to the erosion band in which are located, e.g. Option B all properties will be bought, demolished and rebuilt over a 5 year period (from year 45 to 50). For Option D properties located in the predicted 0-20 erosion band will be demolished, relocated and re-built in year 1. Similar, in year 20

all properties from the 20-50 erosion band will be demolished, relocated and re-built; and, in year 50 all properties from the 50-100 erosion band will be demolished, relocated and re-built.

Demolition costs were based on SPON'S Civil Engineering and Highway Works Price Book (2014). The estimation of the relocation of each residential property cost was based on the purchase of an area of land per property. The calculations were based on the North Norfolk Management Plan (2008), where an average of 333m$^2$ of land was assumed to be purchased for the relocation of each residential property. Considering an average floor area of 100m$^2$, 333m$^2$ of new land per property will include a large garden, similar to the current properties in the village. This is a generous assumption which has been included in the assessment. An estimated rebuild cost was obtained from the North Norfolk Management Plan (RPA, 2008) which was also updated utilising the RPI.

The land purchase prices for relocation were based on agricultural land surrounding the village; prices were also obtained from RPA (2008) and also adjusted using the RPI. At this stage, this assessment has not considered potential compensation for farmers, but this would need to be considered at the PAR stage.

## Calculating the Benefits

The potential damages from erosion were calculated from the value of the residential properties that are at risk from erosion over 100 years for the village (base date is January 2013) under both a 'Do Nothing' Scenario and 'SMP Policy' Scenario. The projected erosion rates were calculated using the SCAPE model (Royal Haskoning, 2013). This is based on the residual life of the defences (Mott MacDonald, 2013) for the short term (0-20 years), medium term (21-50 years) and long term (51-100 years). The predicted erosion extents were then mapped within GIS and used to calculate the number of properties at risk of erosion within each zone over the next 100 years. The year in which an asset is considered to be at risk from erosion is dependent on both the location of the property and/or when services or infrastructure to the property are lost. The total number of predicted properties lost under a 'Do Nothing' scenario was 465.

Due to the high level nature of this Study only residential properties were considered during this assessment; commercial properties were not valued. Residential properties not only outnumber commercial properties at the typical village but they also are considered to attract more UK Government funding. Therefore the small number of commercial properties is unlikely to have a significant impact in the outcome of the assessment.

The values of properties that were affected by erosion under the 'Do Nothing' option were entered into the appraisal spreadsheets (Environment Agency, 2012) used in the UK to calculate benefits of a scheme. The discount rate was then applied to each year to determine the Present Value (PV) of the properties lost to erosion. The Present Value of damages avoided (benefits) were then calculated to provide a measure of the benefit of 'Do Something' rather than 'Do Nothing'.

### Other Benefits

Under adaptation options the 'benefits' or 'damages avoided' are different to traditional coastal protection works. In this case there are additional benefits to households who have new properties. Therefore in this assessment it has been assumed that the benefit value is assumed to be the market value of the properties plus an uplift per property (discounted for the year of erosion/rebuild).

Similarly, it was assumed for this assessment, that, where properties may be bought and rented until the time of erosion (Options B and D), households will be able to move elsewhere (beyond the village) with the income from their sale. Therefore whilst houses may be still present in the area at risk of erosion, no damages have been considered as it is perceived that the 'households' have been given the means to move out of the area of risk. The overall benefits are lower compared to when houses are demolished and rebuilt, as there is no uplift.

Other benefits which could be captured for adaptation options are the social benefits which were considered briefly as part of the Pathfinder Study (DEFRA,2012). In cases where households are moved to a new area close to the village this could have the benefit that local schools/roads etc. can be rebuilt in a nearby area. Therefore whilst the overall characteristics of the village may change, the overall community spirit can be maintained. Conversely if the houses and infrastructure were left to erode, those properties remaining (even beyond the 100 year erosion boundary) would still suffer. This is because their amenities and infrastructure will be lost but also it is likely that house values will fall as the whole area suffers decline and people move away.

The social value of these benefits is very difficult to ascertain and has not been included in this high level study. Also despite the potentially important influence of tourism for this area of coastline, it was considered that the current data available is not sufficient to include within the economic assessment. However, it is likely that the benefits from tourism may increase the economic viability across all the options.

## Economic Analysis

Economic analysis has been undertaken in accordance with the latest UK Guidance (Environment Agency, 2010), which provides the methodology to undertake effective appraisals. The economic assessment utilises the spreadsheet templates provided by the Environment Agency (2012), which is the basis on which the UK Government approves coastal defence schemes and grants funding. The economic assessment includes information from Her Majesty's Treasury Green Book (2011) and the Multi-Coloured Manual (Middlesex University, 2014).

The economic assessment provides a framework for assessing the advantages and disadvantages of the options by expressing all of the potential effects and benefits of an option in terms of its monetary cost. This considers the value (cost) of the options and whether investment in any option is worthwhile against the benefits. In order to compare the different options it is useful to consider the benefit cost ratios for each option. The benefit cost ratio compares the cost of each option over the next 100 years against the benefits over the same period.

Under recent changes in guidance, the approach of Partnership Funding was brought in to allocate funding to flood and coastal schemes based on the numbers of households protected in England and Wales, the damages being prevented and other benefits a project would deliver. From this approach a percentage score is achieved; all schemes must achieve at least 100% scores in order to be considered further for funding. Partnership Funding calculations for the options are presented in Table 2.

**Table 2: Benefit cost ratios and Partnership Funding for each option**

| Coastal protection option | Adaptation option | PV Costs (£k) inc 60% bias | PV Benefits (£k) | Average Benefit Cost Ratio | Partnership Funding score % |
|---|---|---|---|---|---|
| 2- Monitor and maintain | None | 2,402 | 15,566 | 6.5 | NA |
| | A from year 50 | -179 | 20,393 | ? | ? |
| | B from year 50 | 9,722 | 23,067 | 2.1 | 54% |
| 3 - Monitor and maintain and rock placement | None | 2,901 | 15,566 | 5.4 | 169% |
| | A from year 50 | -58 | 20,393 | ? | ? |
| | B from year 50 | 11,878 | 23,067 | 2.0 | 45% |
| None | C from year 1 | 17,669 | 20,393 | 0.9 | 29% |
| None | D from year 1 | 28,691 | 27,714 | 1.0 | 19% |

*? indicates negative costs which means BCRs and PF scores can't be calculated.*

Overall the results suggest that:

- The options that involve HTL until year 50 (Option 2 and 3) and then adaptation (Options A and B) have overall higher benefit cost ratios than those options which involve adaptation measures from year 1 (Options C and D).
- In addition the benefit cost ratio is higher if adaptation is not included at all (Options 2 and 3) – however this still leaves the critical question of what would happen to the community after year 50 when the defences are not maintained.
- Options C and D are sensitive to small changes within the economics and assumptions made. Within the Study three erosion bands have been assessed. However, considering smaller erosion bands (i.e. 5 year intervals) will result in changes to the number of properties eroded in a specific year. This will have an impact on the properties PV cost and also the number of years the rent can be charged (for Option C). Furthermore, this option is sensitive to the fluctuations in rent value and the cost of purchase; i.e. a decrease in the properties purchase cost or an increase in the rent income per year will result in change in the benefits and costs for this option.
- Options B and D are generally more expensive options but may be considered more socially advantageous as the community could be relocated and maintained. The benefits of this haven't been captured within the economic analysis but would be important to investigate in a more detailed Study.

## Conclusions

This paper has considered the economic feasibility of implementing adaptation options in contrast and in conjunction with traditional coastal protection options for a case study village in North Norfolk. Adaptation is potentially an economically viable and sustainable option for coastal management in this area, but depends on a costly upfront injection of funding. Under the current UK coastal management framework there are issues regarding the calculation of Partnership Funding scores for the adaptation schemes. Where the costs are negative this indicates that the rental of the properties would actually provide an income which would cancel out the costs of implementing adaptation. In this case it raises the question of who should pay for the adaptation. Should the UK Government provide the initial funding if other Authorities could financially gain from it?

All adaptation options considered in this Study are open to variation and other approaches are possible. All options should not only be viewed as solutions in their own right but with other advantages and disadvantages which are beyond consideration of this Study. However the investigations undertaken in this Study do form a basis on which to build a case for adaptation; ultimately a mix of options may be best over different timeframes.

Several assumptions have been made in using the current UK coastal management framework to consider adaptation options. Ultimately further discussion is required to establish whether the assumptions and manipulations undertaken are valid; consultation is needed between Authorities and the UK Government to consider a realistic approach to implement adaptation measures in current framework. If not then could an alternative financing mechanism be arranged? This discussion is required before a formal business case submission in the form of a PAR for coastal adaptation can be prepared.

## References


- AECOM, 2010. Kelling to Lowestoft Ness Shoreline Management Plan.
- Climate Change Adaptation Sub-Committee, 2013. Managing the land in a changing climate. Progress Report.
- DEFRA, 2011. Coastal Pathfinder Evaluation: An Assessment of the Five Largest Pathfinder Projects. A Final Report by Regeneris Consulting.
- Environment Agency, 2010. Flood and Coastal Erosion Risk Management appraisal guidance. Available at: https://www.gov.uk/government/publications
- Environment Agency, 2012.FCERM economic appraisal spreadsheet. https://www.gov.uk/government/publications
- HM Treasury, 2011. The Green Book: Appraisal and Evaluation in Central Government. Available at: https://www.gov.uk/government/publications
- HR Wallingford, 2004. Overstrand to Walcott Strategy Study. Report EX 4692.
- Middlesex University, 2014. Flood and Coastal Erosion Risk Management: A Manual for Economic Appraisal (Multi-Coloured Manual). Available at: http://www.mcm-online.co.uk/
- Mott MacDonald, 2013. Cromer to Winterton Ness Coastal Strategy Study.
- Risk & Policy Analysts (RPA) Limited, 2008. North Norfolk Coastal Management Plan Evidence Gathering Study 01.
- Royal Haskoning, 2013. Appendix B: SCAPE Modelling of Shore Evolution: Cromer to Cart Gap. In Mott MacDonald, 2013. Cromer to Winterton Ness Coastal Strategy Study.

**Coastal Management: Changing coast, changing climate, changing minds**
ISBN 978-0-7277-6114-9

ICE Publishing: All rights reserved
doi: 10.1680/cm.61149.523

# Coastal Catch-up Following Defence Removal at Happisburgh

**Mike Walkden**, WSP Group, Exeter, UK
**Gary Watson**, Environment Agency, Ipswich, Suffolk, UK
**Alice Johnson**, Royal HaskoningDHV, Exeter, UK
**Eleanor Heron**, Environment Agency, Solihull, UK
**Owen Tarrant**, Environment Agency Horizon House, Bristol, UK

## Introduction

Understanding coastal 'catch-up' is central to effective coastal management decision making. This is the behaviour exhibited by certain coasts, where the shoreline retreats rapidly after the removal of defensive structures, before returning to a more stable recession rate. Understanding the scale of such response (and therefore the governing physical processes) is necessary to gauge the benefits and costs of decisions to remove or maintain defences. The uncertainties around these responses are quite large and this is a significant issue for the prediction of shore retreat (as identified by the Risk Assessment of Coastal Erosion project, Halcrow, 2007). This problem will grow as coastal management in the UK increasingly seeks sustainable policies, including managed realignment and decisions not to actively intervene in natural processes.

Two key reasons for the current uncertainty around the scale of catch-up response are a shortage of well recorded examples (due to historic policies of defence maintenance), and a lack of modelling tools able to represent the governing processes. Given this uncertainty it might be tempting to assume that, following defence loss, the coast will 'catch-up' to the position it would have had if the defence had never been constructed. This assumption carries the risk of neglecting residual benefits (the coast only partially catching-up) or residual losses (the coast overtaking the position it would have had if structures had never been built).
This problem will be the subject of an upcoming Environment Agency project, described elsewhere in these proceedings (Hardiman, 2015). Technical elements of this issue have also been studied within the Environment Agency's project SC20017 Cliff and Shore Sensitivity to Accelerated Sea Level Rise (from which this paper has emerged). This project aims to improve understanding of long term coastal recession for the purposes of coastal management. It is focussed mainly on the sensitivity of coastal recession rates to sea level rise, but has also addressed questions around the relative importance of rainfall and catch-up processes. In this paper we describe some of the results of that project (described more fully in WSP & RoyalHaskoningDHV, 2014). These findings arose from a study of a site of coastal catch-up, and involved the use of a numerical modelling tool to explore the governing processes.

Due to limitations of resource this is not an in-depth study of the involved processes. Instead this paper presents relatively simple two-dimensional simulations aimed at understanding the key process governing coastal catch-up. The model is also used to explore potential future recession rates and their dependence on beach volume, and therefore neighbouring coast protection.

In presenting these results we use the concept of a shoreline 'trajectory', which is simply the position of a chosen point on the shore as it retreats through time. A 'natural' trajectory refers to coastal recession in the absence of any defences. Example of such trajectories can be seen in figures below.

The simulations and the site observations reported here show both rapid recession following structure loss and greater overall recession than might be expected from simple extrapolation of historic rates. It follows that the economic case for continued investment/ maintenance of such situations is likely to be increased (by accounting for the behaviours reported here) as significant losses are 'nearer-term' and therefore not discounted so heavily, and the overall losses are greater.

Balanced against this is the need to consider, when planning new defences, the whole-life cost of the scheme, which should include the costs of decommissioning. If this decommissioning coincides with an aspirational policy of 'no further active intervention' then accounting for the behaviours reported here could adversely affect the business case for the initial investment.

**Happisburgh**

Happisburgh (pronounced 'Haisborough') is a village situated in North Norfolk, in the south of the east coast of England. The geology here is soft till and (net southerly) sediment transport rates increase with distance south. Sea levels in this location have been rising for millennia, and this means that, under natural conditions, the coast of Norfolk is erosional. A continuous line of defences were constructed in the 1960s to protect the village, and these comprised sheet piles, crested with a sloping timber palisade, fronted by groynes. Defences of this type were constructed along a considerable length of coast, extending north to Trimingham, a distance of almost 15km. The design was intended to reduce cliff recession rather than entirely prevent it, to allow some sediment to feed the beaches. A short distance to the south of Happisburgh, at Sea Palling, the Environment Agency have undertaken a programme of beach nourishment (around 150,000 $m^3$/year on average) since 1996, to offset the reduction in sediment supply from cliff erosion (due to coast protection), and maintain sea defence.

Deterioration of the Happisburgh structures led to their progressive failure in the early to mid-1990s. Sheet piles buckled, palisades and groynes were broken and destroyed by wave action. This led, very rapidly, to the formation of an embayment, partially stabilised at its northern end by the placement of a rock armour revetment.

The recession that followed structure removal, and the resulting embayment formation have been recorded in various ways.

Figure 1: Aerial images of Happisburgh (recorded in 1992 and 2012) showing the formation of an embayment (from Environment Agency 2013); red lines indicate the location of profile monitoring surveys, the magenta line shows the approximate cliff toe position in 1992

The natural retreat rate of this coast was used in this study to understand the scale of coastal catch-up, and so is a key input to this work. Dickson *et al,* (2007) shows the results of historic map analysis for the region (in their Figure 6). They show that recession at Happisburgh was approximately 0.8 m/y between 1885 and 1907 (22 years) and approximately 0.4 m/y between 1907 and 1950 (43 years). This provides an average rate of approximately 0.54 m/y. It may be noted that coast protection works updrift of Happisburgh predate historic mapping, and so may well influence recession rates derived from these maps. Dickson *et al*, (2007) present numerical simulations of this coast under a scenario in which no structures are ever built (their Figure 8). Their 'low sea level rise' scenario is comparable with historic conditions, and shows a recession rate at Happisburgh very similar to the value derived from the historic mapping. 0.54 m/y has therefore been adopted to represent the historic natural rate of recession.

The Environment Agency (2013) report that, following structure failure, recession within the Happisburgh embayment between 1992 and 2012 ranged up to 140 metres. This is a maximum, but the coast is fairly straight within the embayment (see Figure 1). If it is assumed that zero recession occurred from 1960 to 1992 (whilst the structure was in place) then 140 metres represents the total recession from 1960 to 2012. If the structure had never been built then the total recession that might be expected over the same time period would be around 30 metres (52 years at a rate of 0.54 m/y).

These data strongly suggest that, in response to structure loss, the recession trajectory of this coast has overtaken its natural trajectory. The recession that has occurred appears to be around 4.7 times that which would have occurred under natural conditions (i.e. 140 / 30).

It is clearly important to understand the conditions and processes behind this response, and this was explored using the Soft Cliff And Platform Erosion (SCAPE) modelling tool.

**The SCAPE Model**

SCAPE was developed at the University of Bristol (Walkden & Hall, 2005) to describe medium to long term processes governing the emergence of shore profiles comprised of a platform, cliff and beach. It was designed from the outset to allow the representation of structures, like those present along the Norfolk coast. It has been used in a series of applied coastal management studies and research projects. These include regional scale Coastal Strategy Studies for the Norfolk coast in 2002 (University of Bristol 2002) and 2013 (Mott Macdonald & RoyalHaskoningDHV, 2013). Various stages of the regional Norfolk SCAPE model are described in Walkden & Hall, (2011), Dickson *et al*, (2007), Dawson *et al*, (2009) and Walkden *et al*, (in press). Those models demonstrated the suitability of SCAPE for coasts like that of Norfolk, but were too spatially coarse to resolve the Happisburgh embayment. A new model was therefore constructed for this study.

The inputs to this new model were synthetic wave data generated by the UK Met Office (WaveWatchIII model), tide levels derived from the Cromer gauge (and transformed to be suitable for Happisburgh) and a historic sea level rise trajectory derived from Bradley *et al*, (2011), and Woodworth *et al*, (2009). A fuller description of these inputs, and model parameters, is given in WSP & RoyalHaskoningDHV (2014).

The main calibration parameter in a SCAPE model is a term representing rock resistance to wave erosion. This was adjusted to give a similar recession rate to that observed in reality. However, precise representation of Happisburgh was outside the scope of the study, and so a precise calibration was not sought. The study instead explored relative scales of change.

For reasons of resource, a 2D (i.e. shore profile) model was constructed, and this brought certain limitations. The most significant was that changes in beach volume, driven by 'alongshore' processes and management decisions had, necessarily to be represented using scenarios (see below). Despite this, the model provided valuable insights into catch-up processes.

Figure 2 shows shore profiles output from an early simulation made with the SCAPE model. The cliff toe and surface of the shore platform are shown at various years. Although a beach was represented in the simulation, this is not shown to avoid over-complicating the image. The results are taken from the last forty years of a long simulation that represented many hundreds of years of profile erosion. The model was run with site specific wave, tide and sea level rise conditions and a pre-defined beach volume. The simulation period was very long to allow the shore profile (platform, beach and cliff) to emerge and reach dynamic equilibrium.

A structure (sheet pile and palisade) was then assumed to be built close to the cliff toe in 1960. This has the effect of reducing erosion of the profile behind the structure, but not in front. Over the following years the profile in front of the structure erodes down. Behind the structure the profile is largely dormant. This results in a step change in platform elevation across the structure.

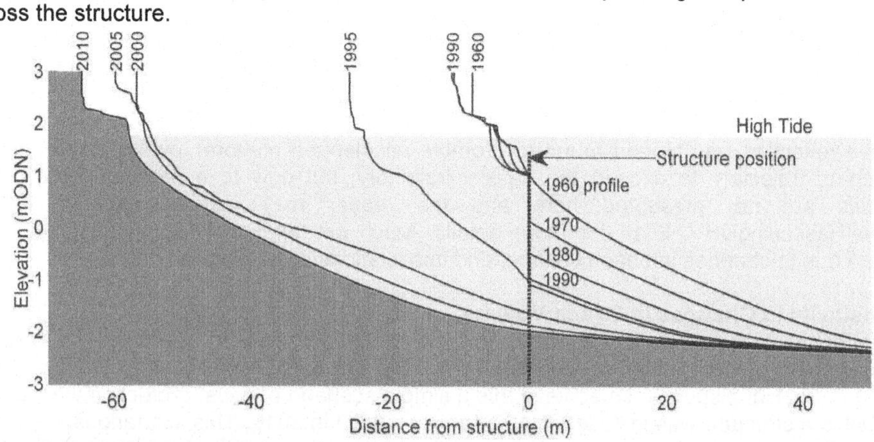


Figure 2: SCAPE simulation output showing the development and loss of a step in the shore platform due to the construction and removal of a vertical coast protection structure; note that (for clarity) the beach is not shown

The profiles in Figure 2 span forty years, beginning just before the structure is constructed in 1960. They progress from right to left, through erosion. The step change can be seen developing until the early 1990s when the structure is assumed to be lost. At its maximum the vertical step extends between around -1m and +1 m.

By 1995 the structure is absent, and a period of accelerated recession has begun. By 2000 a more stable profile form has emerged and recession rates reduce. This, in its simplest form, represents coastal 'catch-up'.

**Processes Governing Catch-up**

The vertical step seen in Figure 2 develops simply because the structure reduces platform lowering on its landward side (by reducing wave energy), but not its seaward side. In the model, coastal catch-up is driven by the (re)emergence of a more natural profile form, during the removal of this 'step'. Such steps are known to have existed at Happisburgh, although they tend to be hidden by beach deposits, and cannot be seen clearly in shore profile observations. The size of the step north of Happisburgh (where structures survive) is estimated to be greater than 1.5m and less than 4m (WSP & Royal HaskoningDHV, 2014). At other times site observations have recorded drops of up to 2m in front of the structures. It seems clear that (as in the model) coastal catch-up at Happisburgh has been driven by the emergence of a more natural profile form during the removal of this step by wave action.

As noted above the recession trajectory at Happisburgh seems to have exceeded the natural trajectory. This behaviour cannot be explained simply by the removal of a step in the platform. It suggests that some process has increased the size of the step, by accelerating the lowering of the platform on the seaward side. The following possible causes of accelerated platform lowering were explored:

1 Wave reflection at the face of the structure; and

2 Reductions in beach volume diminishing the protection that the beach provides to the platform against wave attack.

Wave reflection was indeed found to promote accelerated platform lowering and cause the catch-up trajectory to exceed the natural trajectory, but only to a modest degree. Those results are not presented here and the reader may instead refer to WSP & RoyalHaskoningDHV, 2014 for more details. Much greater sensitivity of cliff position was found due to changes in beach volume, and this relationship is explored below.

**Sensitivity to Changes in Beach Volume**

A set of simulations were designed to explore whether reductions in beach volume associated with coast protection might be responsible for the large scale of coastal catch-up that has occurred at Happisburgh. To achieve this a simple scenario of coast protection was assumed in which a structure was installed in 1960 and removed in 2010. This scenario is not intended to represent Happisburgh directly, but is similar to conditions north of the embayment, where defences continue to fail.

Changes in beach volume as a consequence of coast protection would be relatively straightforward to represent explicitly with a three-dimensional model (see for example Dickson *et al*, 2007 and Walkden & Hall, 2011), but not however with two-dimensional simulations of the sort undertaken in this study. Instead it was necessary to proscribe scenarios of beach volume change. Four scenarios of beach change were defined, and these are described in Table 1 and illustrated in Figure 3.

Table 1: Scenarios of beach change

| Number | Name | Description of beach response | Line style (Figure 3 & Figure 4) |
|---|---|---|---|
| 1 | Zero | Falling to zero when the structure is installed | Thick black |
| 2 | Low | Falling to one-quarter of its original value | Grey |
| 3 | Recovery | Falling but then recovering once the structure is lost | Thin black |
| 4 | Constant | Remaining constant throughout the simulation | Dashed |

These scenarios generally comprised an initial 'natural' average beach volume, which was adopted to represent the pre-construction period (i.e. before 1960), followed by a reduced volume, representing the effect of structures reducing cliff recession (and therefore sediment supply). The last stage of the scenarios represented how the beach responded to the loss of the local structures.

Under Scenario 2 (the grey line) the beach volume remains low despite structure failure in 2010, and this is intended to represent the effect of ongoing presence of structures along neighbouring (updrift) coasts, despite the loss of local defences. Under Scenario 3 (the thin black line) a transient peak in beach volume occurs as the structure fails, followed by a return to pre-structure levels. This is intended to represent a large influx of sediment due to the failure of updrift structures.

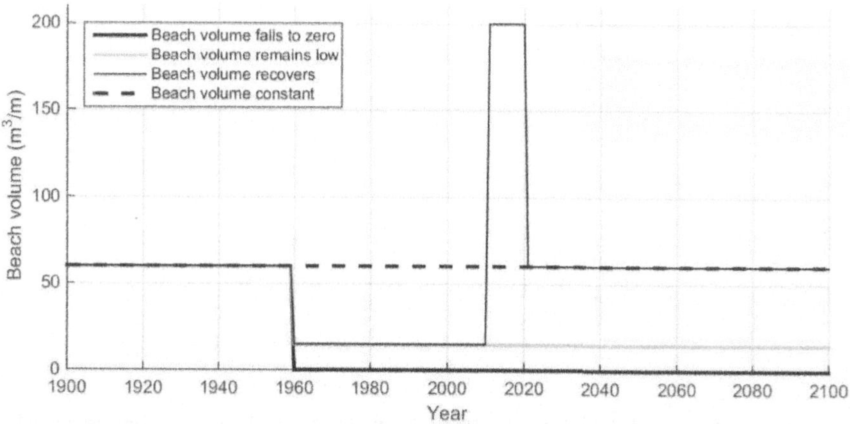


Figure 3: Alternative scenarios of beach volume change resulting from structure installation in 1960, and structure failure in 2010

The other two scenarios represent the beach volume falling to zero (Scenario 1) and remaining unchanged by the construction (Scenario 4). These are not intended to be realistic but to help bound the total range of response.

The actual beach volumes at Happisburgh, and the way they have changed through time, are unknown; no records are available. For this reason this modelling can only explore whether relative changes in beach volume *may* be the driver of the behaviour seen at Happisburgh; it is not presented as a precise representation of Happisburgh itself. Model responses were assessed in *relative* terms; i.e. the recession trajectories were assessed in terms of whether they exceeded the natural trajectory of the model, and if so by what factor.

The simulation trajectories resulting from these beach change scenarios are shown in Figure 4, along with a simulation in which no structure is built (to allow relative changes to be quantified). As noted above, the coast protection structure is assumed to be present between 1960 and 2010. Between these dates the shoreline position is relatively constant. After structure loss the shoreline retreat is very rapid for a small number of years (seen as a step increase in shoreline retreat at 2010) and then becomes steadier. It can be useful to think of the outputs after 2010 as 'decommission trajectories', i.e. change resulting from structure decommissioning.

These results demonstrate the great sensitivity of the recession trajectory to beach volume. The low beach scenario and zero beach scenario result in very high recession trajectories, and the sensitivity appears to be non-linear, in that the response increases as the beach volumes become smaller. It is worth noting that twenty years after the loss of the structure the recession of the 'low' beach scenario is 3.2 times the recession that would be expected in the absence of any construction (the ratio of the distances *b/a* in Figure 4). The equivalent ratio of the 'zero' beach scenario is 4.7. These relative losses are clearly comparable to the ratio of 4.7 derived above for Happisburgh.

Figure 4: Shoreline trajectories under alternative scenarios of beach volume change; the dotted line represents the natural trajectory, i.e. that which would be expected in the absence of any structures

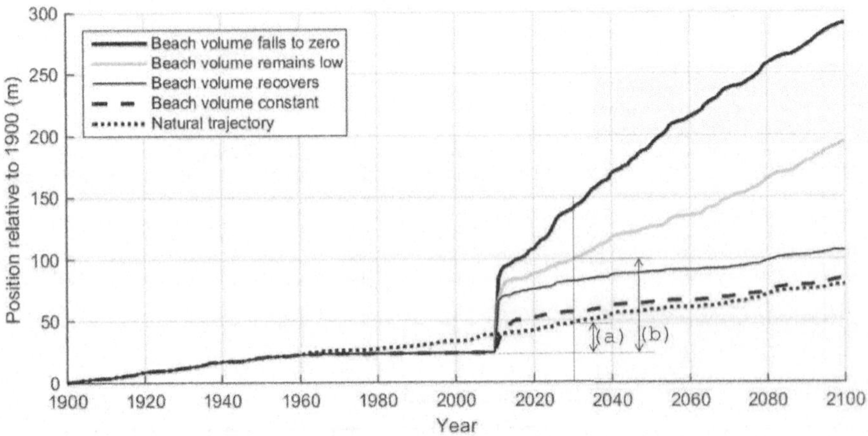


The similarity of the 'constant' beach recession trajectory to the 'no structure' recession trajectory helps to illustrate the importance of the beach. If the beach remains constant (perhaps representing the situation of a short length of local defence, held for a short time) then the degree to which the recession trajectory overtakes the natural trajectory is much reduced, and also diminishes with time.

Interestingly, the 'recovering' beach scenario converges very little on the 'no structure' scenario before 2100. Even though the beach has recovered immediately after the structure is lost, the temporary presence of the coast protection has resulted in a net loss of land. In other words, the defence of this location has ultimately (at least by 2100) resulted in more loss of land than would otherwise have occurred.

**Summary and Implications for Coastal Management**

This study has explored an example of coastal 'catch-up' behaviour, observed at Happisburgh, in North Norfolk. The concept of a shoreline 'trajectory' has been used to represent its change in position through time. Shoreline retreat in the absence of any coastal structures is termed here the 'natural' trajectory (e.g. the dotted line in Figure 4). Similarly retreat following structure loss is referred to as a 'decommission' trajectory (e.g. the other lines in in Figure 4 *after* 2010).

At Happisburgh, and probably at many other cliffed sites, the tendency to retreat rapidly ('catch-up') during the decommission phase is caused by lowering of the shore platform during the structure lifetime. At many places, including Happisburgh, such lowering is obscured by the presence of a beach, so may not be immediately recognised. The observed decommission trajectory at Happisburgh has strongly overtaken the natural trajectory. This suggests that some process has accelerated the lowering of the shore platform to a significant degree. In model simulations, reductions in beach volume were found to have this effect, by reducing the protection given to the shore platform against wave attack.

Simulations of future change under alternative beach volume scenarios indicate that the associated losses (of land) are permanent (i.e. the natural trajectory and the decommission trajectory do not converge over coastal management timeframes). Moreover, if beach volumes remain low at a site after coast protection is lost (for example due to neighbouring 'updrift' defences) then the ensuring shoreline retreat will continue to outpace the natural trajectory (i.e. net losses will grow).

This implies that net losses at Happisburgh may still be increasing, due to the presence of extensive updrift defences. These losses appear to be suspended by the current embayment, but this situation is strongly dependent on presence of the temporary rock revetment at its northern limit. Ultimately the losses here may be expected to increase because of the maintenance of updrift structures (i.e. those structures make the 'low' trajectory of Figure 4 more realistic than the 'recovery' trajectory).

More generally, catch-up should be expected at cliffed sites where coast protection is removed, unless that protection extended to the shore platform. At sites where coast protection is accompanied by a reduction in beach volume, the resulting decommission trajectory should be expected to overtake the natural trajectory. The scale of the increased loss will vary from place to place, but at Happisburgh it is estimated to be up to around 4.7 times the natural loss (after around twenty years).

The changes seen at Happisburgh, and simulated in the model, include an early stage of very rapid retreat, as the discontinuity in the shore platform is removed. The economic losses arising from such change should, of course, be accounted for when determining the economic benefit of maintaining a defence. It may be noted that such early losses would not be subject to strong discounting in the assessment of the economics.

Taking a broader perspective, these results suggest that decommissioning should be accounted for when assessing the value of a potential defence. Furthermore such assessment should extend beyond the local to encompass neighbouring (particularly updrift) defences.

Assuming the conclusions of this study are generally correct, and that decommission losses are accounted for by extending the timescales over which appraisal are undertaken, then it follows that:

1. A decision to construct a new defence may become less likely, because of the loss of land during decommissioning; but
2. Once built, a decision to maintain a local defence may seem more attractive, because of the large losses that would otherwise occur.

**Acknowledgements**
The Environment Agency funded this work and provided much support through advice and data (including waves, shore profile data and LiDAR topography). Jonathan Tinker, Jason Low and Andy Saulter (UK Met Office) provided advice on the availability of wave data, future and historic sea level change and upcoming developments in climate change science. Sarah Bradley (British Antarctic Survey) kindly provided Holocene sea level rise trajectories. Tide data was supplied by the British Oceanic Data Centre. The SCAPE model was used by permission of Mike Walkden (WSP Group) and Jim Hall (Oxford University).

**References**

Bradley, S.L., Milne, G., Shennan, I., and Edwards, R., (2011), An improved glacial isostatic adjustment model for the British Isles. Journal of Quaternary Science 2011 26(5) 541–552

Dawson, R., Dickson, M., Nicholls, R., Hall, J., Walkden, M., Stansby, P., Mokrech, M., Richards, J., Zhou, J., Milligan, J., Jordan, A., Pearson, S., Rees, S., Bates, P., Koukoulas, S., Watkinson, A., (2009) Integrated analysis of risks of coastal flooding and cliff erosion under scenarios of long term change. Climatic Change, Climatic Change (2009) 95:249-288. DOI 10.1007/s10584-008-9532-8. Springer

Dickson, M, Walkden, M, and Hall, J. (2007), Systemic impacts of climate change on an eroding coastal region over the twenty-first century, In: Climatic Change. Springer

Environment Agency (2013), Coastal trends report: north east Norfolk and North Suffolk (Kelling Hard to Lowestoft Ness), RP033/N/2013 June

Hardiman, N. (2015) A Short Guide to Doing Nothing at the Seaside, proceedings of Coastal Management 2015, the Netherlands

RoyalHaskoningDHV (2013), SCAPE Modelling of Shore Evolution: Cromer to Cart Gap, Appendix C of the Cromer to Winterton Ness Coastal Study, (Mott MacDonald, for North Norfolk District Council), http://www.northnorfolk.org/files/C2WS_Appendix_C_SCAPE_Report.pdf

University of Bristol (2002) for HR Wallingford and North Norfolk District Council SCAPE modelling of the Norfolk Coast, report for the Overstrand to Walcott Strategy Study

Walkden, M.J.A. and Hall, J.W., (2005), A predictive Mesoscale model of the erosion and profile development of soft rock shores. Coastal Engineering, 52(6): 535-563

Walkden, M. and Hall, J. (2011), A Mesoscale Predictive Model of the Evolution and Management of a Soft-Rock Coast. Journal of Coastal Research 27 (3) 529–543

Walkden, M., Dickson, M., Thomas, J., and Hall, J. Simulating the Shore of North Norfolk (In press), Chapter 7 of Broad scale coastal simulation: new techniques to understand and manage shorelines in the third millennium, Springer

Woodworth, P.L.Teferle, F.N., Bingley, R.M., Shennan, I., Williams, S.D.P. (2009), Trends in UK mean sea level revisited, Geophys. J. Int. 176, 19–30 doi: 10.1111/j.1365-246X.2008.03942.x

WSP and RoyalHaskoningDHV (2014). Cliff and Shore Erosion under Accelerating Sea Level Rise: Results of Case Study Analysis (Draft). Environment Agency Research and Development Programme project SC120017

**Coastal Management: Changing coast, changing climate, changing minds**
ISBN 978-0-7277-6114-9

ICE Publishing: All rights reserved
doi: 10.1680/cm.61149.533

# Holes Bay Rail Crossing – Maintaining Critical Infrastructure in a Highly Sensitive Environment

**Simon Burchett**, WSP UK Ltd., Basingstoke, UK
**Hamish Hall**, WSP UK Ltd., Exeter, UK
**Stuart Terry**, Poole Borough Council, UK
**Rebecca Wells**, Network Rail, London, UK

## Introduction

Holes Bay is located within Poole Harbour (Dorset, UK) which forms the largest natural harbour in Europe. A large area is drained at low water, revealing salt-marsh and mud-flat ecosystems which support wintering wading birds and wildfowl leading to the area being designated as a Wetland of International Importance (under the Ramsar convention), Special Protection Area (SPA) and Site of Special Scientific Interest (SSSI). A vital rail link crosses the bay, consisting of a causeway, which includes two viaduct crossings to allow tidal exchange in the upper reaches of the bay. Two tracks run on top of the causeway, which serve the London Waterloo to Weymouth main line which is a vital link in both regional and national terms.

This paper presents the project from inception (condition assessments, planning and coastal modelling) through option assessment (stakeholder engagement, multi-criteria analysis and environmental assessment) and into design development and construction. The project is considered of wider interest due to the complexity of the existing asset condition, the heavily protected local environment and unique safety and access challenges, all of which required an unconventional design approach.

A sensitive design philosophy and collaborative approach amongst project stakeholders has led to wider benefits through the identification of an opportunity for beneficial re-use of dredged material. The approach taken also meant that environmental protection was a key driver in the formulation of engineering approaches, to the extent that impacts were deemed sufficiently minimal and hence an Environmental Impact Assessment was not required. This approach ultimately meant that remedial works to preserve the integrity of this vital piece of infrastructure could take place during 2015, without the need to risk further deterioration of the asset, which could lead to rail service disruptions or costly emergency works with potentially adverse environmental impacts.

## Local environment

The structure consists of a causeway crossing Holes Bay between Poole and Hamworthy (as shown in Figure 1). The causeway crosses to the south of Pergins Island (a local nature reserve) and the surrounding land is relatively low-lying and flat, with shallow foreshore slopes typical of a brackish tidal basin.

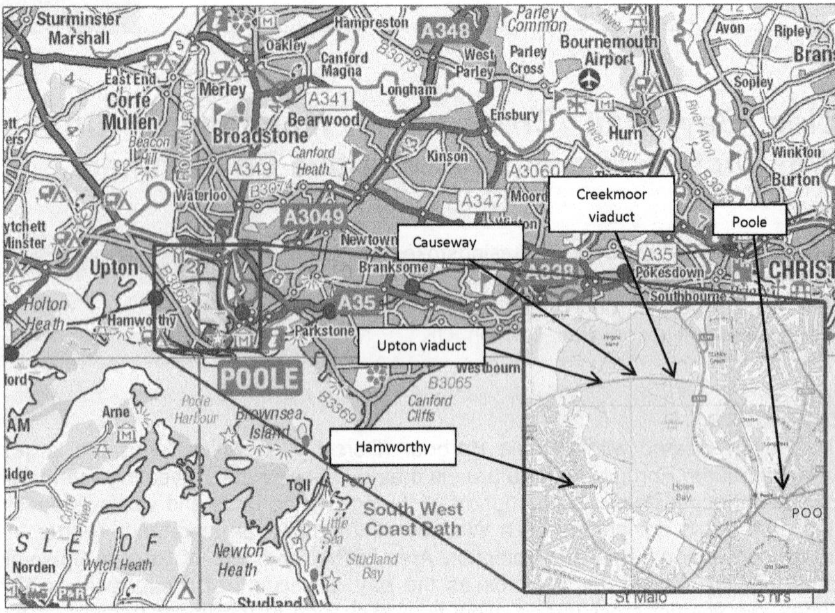


Figure 1: Location plan

Poole Harbour is the largest natural harbour in Europe, and whilst the overall tidal range is relatively small by UK standards (1.6m spring tide range compared to 3.9m at Portsmouth, approximately 60km east), a large area is drained at low water, revealing salt-marsh and mud-flat ecosystems. Tides are influenced by the nearby Isle of Wight and a double tide is notable on the ebb, giving rise to a unique tidal regime. This means that high-tide periods last longer than low-tide periods; placing considerable time constraints on accessing and constructing work during low-tide conditions.

Salt-marsh dominates the western margin of the bay and some of the centre of Holes Bay, close to the rail causeway. This feature provides a degree of passive protection to the causeway from all but the highest water levels. Tidal mud-flats are present across much of the main central area of the bay and are characterised by dendritic meandering drainage channels, typical of a relatively low energy depositional environment, although monitoring reports, Poole Harbour Commissioners (2011/ 2012), have indicated a trend toward reducing saltmarsh coverage.

Holes Bay is used for fishing, bait digging/ dragging and is also popular for bird watching. Upton Country Park lies in the north-western corner of Holes Bay which covers over 100 acres and is popular for informal recreation.

The southern half of Holes Bay contains Cobb's Quay marina and a number of boat moorings. The marina has approximately 850 berths (up to length 25m) and an additional 240 dry-stack berths (up to length 10m) and undertakes annual maintenance dredging (using a backhoe excavator on a floating barge) the arisings from which are then disposed offshore. Due to its relatively high degree of shelter and ease of access, the bay is a popular location for small leisure boats and watercraft use with access for larger vessels (greater than 2m draft) restricted by the opening of the lifting Twin Sails Bridge, and further restricted due to limited water depths within the bay. The Royal National Lifeboat Institution (RNLI) have their headquarters and training school located on the southern periphery of the bay.

Navigational access around the causeway itself is highly restricted due to the presence of extensive mud-flats and salt-marsh and also the two railway viaducts. These are of a relatively low height meaning that access for boats is restricted beneath them and limited to kayaks/ small craft (<5m length and <0.5m draft).

## Structure Condition

The approximately 1.1km rail causeway dates back to 1872 and comprises a chalk core and combination of stone pitched slope and unprotected slope, with more recent toe protection in some areas. Both sides of the embankment are exposed to tidal inundation and a series of repairs have been conducted over the years including localised placement of rock armour, sprayed concrete render and concrete wave-walls (at the crest in more exposed locations). Recent condition assessments conducted in January 2014 revealed a wide variety of structure forms and condition, some of which are in need of attention to prevent further deterioration which could pose a risk to the integrity of the rail crossing.

Typical exposed revetment section (no fronting salt-marsh)

Certain lengths of the structure are exposed to locally generated wind-waves which have led to scour/ lowering of the foreshore and exposure of the structure toe and damage to the stone pitching. These areas have typically been repaired with sprayed concrete, although this is broken-up and missing over substantial areas.

Typical protected revetment section (with fronting salt-marsh)

Other sections, with reduced wave exposure due to high foreshore levels are typically heavily vegetated and in places large trees and bushes are present. These sections benefit from natural wave attenuation offered by a combination of salt-marsh, mud-flat and shingle foreshore. Where shingle foreshore is present, this is primarily due to fronting salt-marsh which offers a degree of protection (allowing shingle to accrete). Reinstatement of the original condition would require vegetation removal and re-pointing of the masonry blocks. A cautious approach has been adopted however, as the high foreshore levels and established vegetation are currently acting to protect the causeway, and removal of vegetation may compromise the integrity of the defence system.

Following a structural inspection/ walkover survey conducted in January 2014, a comprehensive condition record was established and used as the basis for an asset management tool/ database using a Geographic Information System (GIS) platform. This tool has provided the basis for subsequent investment decisions, allowing prioritisation of capital and maintenance spend to maximise the asset life.

## Environmental Character

**Environmental Designations**

The project site lies within the following designations (see Figure 2):

- o Poole Harbour Ramsar site
- o Poole Harbour Special Protection Area (SPA)

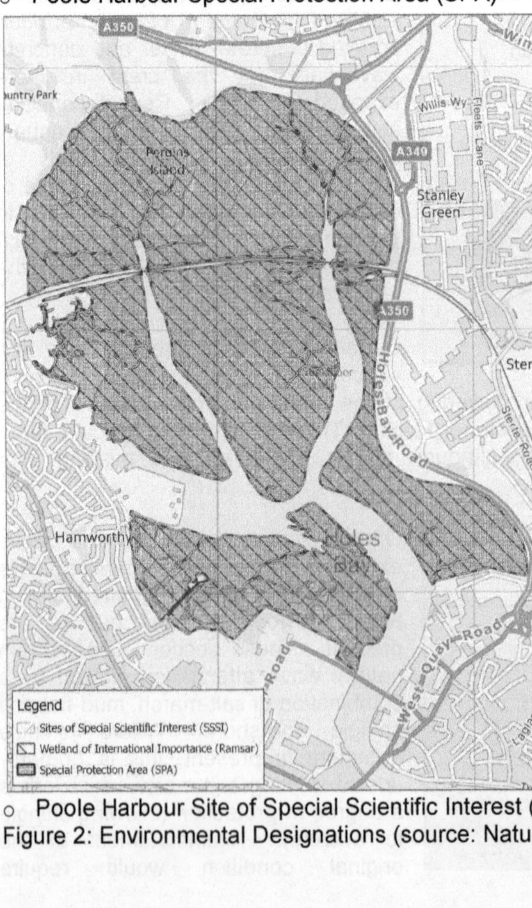


- o Poole Harbour Site of Special Scientific Interest (SSSI)

Figure 2: Environmental Designations (source: Natural England)

These designations result from the large areas of intertidal mud-flat and salt-marsh, and the wintering waders and wildfowl that these support. Species for which the site is important include shelduck, black-tailed godwit, avocet, common tern and Mediterranean gull. Holes Bay is covered by three SSSI management units and in 2010, two of these were assessed as being in unfavourable but recovering condition. This was due to high percentage cover (>10%) of macro-algal mats which can have a detrimental effect on benthic invertebrate populations. The remaining unit is in favourable condition, Natural England (2010), but is at risk of algal mats.

**Marine and Coastal Ecology**

The slopes of the causeway are colonized by a mixture of coastal scrub plants, including brambles, grasses, sea beet and several large pine trees. The shoreline either side of the causeway comprises a mixture of shingle, mud-flats and salt-marsh. The southern side of the seawall is exposed to wave action from across the extent of Holes Bay. In places along the seawall, a band of shingle is present on the upper shore immediately seawards of the wall. The band of shingle is approximately 1-2m wide, with grasses colonizing it in places. Large accumulations of dead seaweed and plastic rubbish have built up on the strandline in some places.

A large area of salt-marsh extends southwards from the central causeway section for approximately 600m and is roughly 200m wide and protects the seawall from waves. The salt-marsh is species poor, and dominated by *Spartina* spp. and is eroding in places along its seaward edge. Some strands of sea purslane are also present along the upper edges of the marsh. Large extents of mud-flat are exposed at low tide making this a valuable bird feeding habitat. The northern side of the causeway is very sheltered from wave action with very stable and sheltered mud-flats.

**Bird Interest**

Holes Bay is very important for waterfowl with the pattern of usage being extremely complex and varying during the year, O'Rouke (2005). Winter typically contains the highest number of birds with the bay supporting wigeon and teal in the north-east corner and other species such as pintail, shoveler, goldeneye, tufted duck, red breasted merganser and great-crested grebe on the larger water bodies. Holes Bay stays unusually warm during extreme cold weather due to an outlet in the north-east. This attracts a large number of birds to the bay during cold periods.

In spring, smaller flocks of teal, wigeon, shelduck, pintail and shoveler still use the bay. Waders such as black-tailed godwit and redshank also feed on the mudflats. Summer is normally the quietest time of the year for birds but common and sandwich tern use the bay as a feeding area. In the autumn, many waders pass through the bay such as redshank and greenshank. Osprey have also been recorded feeding in the bay from the end of August through September. Holes Bay has been designated in the Poole Harbour Aquatic Management Plan as a Bird Sensitive Area. This is because Holes Bay is an important roosting site including for avocet and black-tailed godwit (Poole Harbour Steering Group, 2011).

**Historic Environment**
Poole Harbour has been identified by English Heritage as one of the most important areas for coastal archaeology in the country. This is because it has a long period of human occupation combined with excellent preservation of coastal and marine structures and artefacts. Available Historic Environment Records (HER) indicate the causeway itself and the railway line as being opened in 1872. Other records in the bay are the wrecked remains of vessels, peat beds, a Roman road, Upton House, limekilns and a jug found in the mud-flats.

**Landscape**
The site does not lie within an Area of Outstanding Natural Beauty or other landscape designation. The causeway in general is not particularly visible. It cannot be seen by train passengers very well because of the speed of the trains as well as the steep slope of the seawall. The middle section of the causeway lies at least 200m from the nearest road and cannot therefore be viewed at close proximity and the seawall does not form a particularly prominent feature of the landscape. This has meant that deterioration of the asset can only be assessed through carefully planned site visits, requiring line possessions (and service disruptions) or via boat access, restricted by tides.

## Dredging Activities

Initial stakeholder workshops identified the dredging activities taking place within Holes Bay, the current practise of disposing of this material offshore, as well as the need to view dredging and disposal activities in the wider context of the sediment budget of the entire Poole Harbour and explore opportunities for beneficial re-use.

The volumes of sediment removed from Poole Harbour through maintenance dredging activities are reported in detail by PHC (2012) and can be broadly described as follows:

- Bi-annual dredging of the main navigation channel using a trailer suction hopper dredger, with volumes of approximately 110,000m$^3$ of sand and silt.
- Smaller scale dredging of marinas and boatyards using backhoe or grab dredging plant and barges, with volumes varying each year but over the last 20 years being in the range of 3,000m$^3$ to 38,000m$^3$ per year, with an average of 18,000m$^3$ per year, the majority of which is dredged from Holes Bay.

Historically, sediment dredged from within Poole Harbour has been deposited offshore at the dedicated spoil ground in Swanage Bay, but in recent years opportunities have been sought for beneficial re-use of this dredged material within the harbour. Examples of re-use of dredged sediment have included beach recharge schemes and disposal of fine silty material to the east of Brownsea island (approximately 28,000t per year since 1991) for which a comprehensive monitoring programme has been established. Several other small disposal sites exist and several agreements are in place with small boatyard and marina operators to dispose in-harbour, with further agreements being continuously explored.

The primary dredging site within Holes Bay is the Cobbs Quay marina, which has approximately 850 marina berths and a further 240 dry-stack berths and the adjacent Davis Boatyard. Annual maintenance dredging is carried out with volumes varying between approximately 5,000m$^3$ and 22,000m$^3$ per year since 2002, all of which is disposed offshore. A small amount of contaminated sediment was previously removed and disposed to landfill with the sediment extracted as part of the ongoing maintenance dredging programme expected to be suitable for re-use within the harbour, but to date the quantity and frequency of the operation has not been compatible with the controls in place for in-harbour disposal.

Bait dragging and illegal fishing, particularly for clams is known to take place and has been observed during site visits. Although a byelaw is in place to prohibit use of a dredge within sensitive areas of the harbour, there are limited resources to enforce this.

## Coastal Process Understanding

The project involved hydrodynamic modelling of tidal flows to estimate current velocities and tidal exchange during the planning and option selection stage of the project. As previously noted, high-tide periods last longer than low-tide periods leading to considerable time constraints for constructing the works during low-tide conditions.

The tidal currents through the two viaducts were modelled for normal and extreme tidal conditions. The modelling indicated current velocities that could be expected to mobilise sand and small gravel fractions, indicating the potential for scour in these areas. The bed-forms around the structures are expected to have reached dynamic equilibrium, although there is no clear evidence available to allow a firm conclusion to be drawn. A key recommendation has therefore been to conduct periodic bathymetric surveys.

Following selection of the preferred option, project team workshops led to the subsequent application of the hydrodynamic model to support construction access planning for the works due to the limited tidal working windows.

From a review of the structural condition and observations made on site, it was apparent that observed undercutting of the earth face of the north side of the causeway is due to the action of small waves at or around the high water level. During periods of sustained high wind speeds, this wave action is fairly persistent due to the extended high-tide period.

The south face of the causeway is exposed to larger waves with a longer period and therefore appears to have been originally built (circa 1872) with a more robust form of construction, although this has subsequently required repairs on several occasions. It was apparent during project team site visits that the more frequently repaired areas were those most exposed to wave action, notably where the foreshore was lower and the toe of the structure was exposed and that these lengths tended to be closer to the viaduct abutments.

The higher areas of foreshore act to reduce the wave energy that impinges on the structure and hence more of the original masonry construction was found to be present here (and in better condition). Significantly more vegetation was also found on the causeway structure in the areas of lower exposure, i.e. with fronting shingle foreshore or salt-marsh.

Further modelling was performed to estimate overtopping rates due to wind-generated waves and the potential impacts on the structure (in terms of damage) and rail operations (in terms of safety). For the purpose of overtopping discharge estimates and outline design of remedial measures, the following scenarios were assumed:

- 'Operational' scenario, i.e. relatively frequently occurring, assuming Mean High Water Spring (MHWS) tide conditions combined with 1 in 1 year return period wave conditions (100% Annual Event Probability, AEP)
- 'Extreme' scenario, i.e. infrequent, assuming 1 in 200 year return period water-level conditions (0.5% AEP) combined with 1 in 100 year RP wave conditions (1% AEP).

These scenarios were assessed based on current wave and water-level values and also extended to take account of increasing wave and water-level values due to projected climate change influences over a 60 year time horizon/ design working life. This assessment involved application of the latest Intergovernmental Panel on Climate Change 5$^{th}$ Assessment Report, in advance of formal guidance being issued by the Environment Agency. Overtopping estimates produced for these scenarios were found to be close to tolerable limits.

## Options Assessment

The assessment of options followed a standard approach, intended to initially create an unconstrained and wide range of potential elements (the 'long-list' of elements), which were then screened through Multi Criteria Analysis (MCA) to derive a 'short-list' for further consideration.

A wide range of options were derived during a stakeholder workshop held at the RNLI headquarters in January 2014. During the workshop the 'long-list' options were qualitatively appraised for their environmental, engineering and economic merits, and subsequent assessments involved short-listed intervention techniques being used (in combination) to present a series of possible options ranging from 'Do Nothing', through 'Do Minimum' and three differing 'Do Something' options. The Do Something options were estimated to have similar whole-life costs and it was therefore felt that environmental aspects would dictate the preferred option, with particular constraints identified during workshops already factored into the outline designs.

A high-level review of temporary access and safety considerations was performed during feasibility stage, and subsequently extended to a more comprehensive review in collaboration with Network Rail and an experienced contractor team. The causeway is in an isolated location, with only limited rail access and intermittent boat access and the design development of the preferred option considered the safe movement of materials and labour to undertake the works in a cost effective and environmentally acceptable manner.

Proposed works will repair existing damage as identified from structural condition surveys and extend the life of the structure by giving additional protection against waves and erosion.

**Key issues Identified Through Collaborative Workshops**
The following issues were identified as being pivotal to the future success of the project.

- Delivery to programme (high level of confidence/ certainty on programming for the Network Rail contractor and service operator).
- Collaborative working across the Network Rail, main contractor and sub-contractor teams.
- Accurate predictions of tide times and levels to aid construction planning and minimise delays.
- Sufficiently detailed and clearly presented design/ construction drawings and specifications.
- Environmentally conscious design development to ensure stakeholders remain supportive and wider benefits are realised.
- Efficiency in construction – maximising use of prefabricated elements or otherwise lightweight materials that can be assembled by hand in-situ between tide restrictions and delivered during night rail possessions.
- Provision of floating barges to provide welfare facilities and material storage areas during construction.
- Engagement with key stakeholders including Network Rail teams, Poole Borough Council (PBC), Poole Harbour Commissioners (PHC), Natural England (NE), the Marine Management Organisation (MMO) and Environment Agency (EA).
- Attendance by key people at Early Contractor Involvement (ECI) and stakeholder workshops.
- Cross project team support for a potential related project to establish areas of salt-marsh within Holes Bay using dredged material from Cobbs Quay marina.
- Flexibility to change the design approach as ground conditions, consents, access or condition require (considering the potential for further structural deterioration over winter).

## Environmentally Driven Engineering Solutions

The SPA, Ramsar and SSSI designations have been accounted for in the planning and implementation of works to the structure. The works do not permanently encroach on the foreshore beyond the line of the existing structures, thus reducing long-term impacts and minimising the risk of provision of compensatory habitat, due to requirements of the EU Habitats Directive.

The works have been planned to minimise disturbance to birds, primarily by minimising noise impacts (such as piling) and the presence of humans (and dogs) on the foreshore. Such disturbance has been minimised by avoiding work during the winter, when the bay is most heavily populated by birds (November through March are typically used to define the winter bird season). Furthermore, construction plant and personnel are to be kept within a small working area and high noise impact activities have been avoided where practicable.

So far as possible, vegetation clearance works have been conducted outside the bird breeding season (March through July). Repair works have also been designed to minimise disruption to the foreshore by avoiding plant crossing salt-marsh and mud-flats with the railway used to bring materials to site so far as possible.

The archaeological significance of Poole Harbour has also been accounted for when planning the works, minimising excavation so far as possible.

## Collaborative Working and Wider Benefits

From the outset of the project, a collaborative approach was established, involving and engaging with stakeholders not only to take their views into account, but to seek wider opportunities and benefits. The WSP and Network Rail project team worked alongside Natural England, Poole Harbour Commissioners, Poole Borough Council, the Environment Agency and Marina Developments Ltd (a local marina owner/ operator).

From this approach, a pilot project to establish new areas of salt-marsh has been initiated. The pilot project will involve beneficial use of dredged spoil arising from maintenance dredging of Cobbs Quay marina, which if successful, could lead to a sustainable long-term habitat that counteracts gradual loss of salt-marsh within Holes Bay, and provides a natural form of wave attenuation to protect the causeway.

The approach taken to stakeholder engagement and environmentally conscious design, led through workshops and continuing open dialogue has meant that the Marine Management Organisation and Local Planning Authority have agreed that an Environmental Impact Assessment (EIA) is not required. This has been achieved as the design and construction planning took stakeholder views into account from the outset and hence managed to minimise negative environmental impacts to acceptable levels, meaning that the urgent repair works can proceed on a fast-track basis.

## Conclusions and Recommendations

This paper seeks to demonstrate the wide variety of features, activities, and varying interests of a large group of stakeholders, within the relatively small but highly sensitive coastal environment of Holes Bay, and the high level of interdependency between them.

The lessons learned and recommendations for similar projects relate to active promotion of a collaborative approach from the outset of a project, sometimes requiring skilled leadership. This should allow the key challenges and constraints to be identified at an early stage, which then require further investment to consider and arrive at (through consensus) sufficiently detailed solutions at an early stage in the project, but maintaining flexibility to adapt to changes as the project progresses through subsequent stages.

## References

Natural England (2010) *SSSI unit information* www.sssi.naturalengland.org.uk
O'Rourke (2005) *Poole Bridge Regeneration Initiative Appropriate Assessment*
Poole Harbour Commissioners (2011) *Report on Harbour Monitoring 2012*
Poole Harbour Commissioners (2012) *Report on Harbour Monitoring 2011*
Poole Harbour Commissioners (June 2012) *Poole Harbour Baseline Document for Maintenance Dredging Protocol*
Poole Harbour Steering Group (2011) *Poole Harbour Aquatic Management Plan*

## Acknowledgements

The authors would like to acknowledge the following persons and organisations, all of whom are playing a vital role in working together to deliver sustainable coastal management practises at Holes Bay:

**James Buckley,** One Team Wessex, London, UK
**Sue Burton,** Natural England, Dorset, UK
**Laura Jade Calvert,** The Marine Management Organisation, Newcastle, UK
**Mark Quilliam,** Marina Developments Ltd, Poole, UK
**Andy Ramsbottom,** Poole Harbour Commissioners, Dorset, UK
**Nicola Solly,** Independent Environmental Consultant, UK
**Royal National Lifeboat Institution,** Poole, Dorset, UK

**Coastal Management: Changing coast, changing climate, changing minds**
ISBN 978-0-7277-6114-9

ICE Publishing: All rights reserved
doi: 10.1680/cm.61149.545

# Monitoring Dune Response to the East Coast Storm Surge of 2013 Using Laser Scanning and SfM Photogrammetry

**David Ackerley**, Dept of Geography, Loughborough University, UK.
**Jim Chandler**, School of Civil and Building Engineering, Loughborough University, UK.
**Joanna Bullard**, Dept of Geography, Loughborough University, UK.

## Introduction

On 5 December 2013, the UK coastline experienced one of the biggest storm surges on record. This storm caused substantial coastal erosion, particularly along the North Sea coast where the southwards travelling surge was heightened by funnelling between the European mainland and the East Coast of England. This paper focuses on a field site located north of Mablethorpe on the Lincolnshire coast (53°21'43"N 0°15'03"E) which has undergone long-term (1891-2010) accretion of, on average, 2m $yr^{-1}$ (Montreuil and Bullard, 2012). Prior to the 5 December storm event, well-developed coastal foredunes were present with primary foredunes up to 3.5m ODN and secondary (inland) dunes up to 7m high. The storm surge eroded the dunes laterally by up to 20m, creating a vertical scarp face up to 2m high. Provided there are no further storm surges, is expected that the dune system will gradually recover and the scarp will eventually be infilled. The overall aim of this research project is to combine high resolution topographic and environmental data to quantify the rate and nature of post-storm coastal recovery at this site. This paper draws upon this project to highlight how modern geomatic methods can be used for detailed monitoring of coastal phenomena at close range. These methods include both terrestrial laser scanning (TLS) and close range digital photogrammetry, which incorporates the very latest structure from motion (SfM) techniques.

## Fieldwork and Routine Data Capture

Since February 2014 monthly surveys have been conducted at the Mablethorpe site using a Leica terrestrial laser scanner or TLS (ScanStation 2 or P20) to capture approximately 10 million XYZ coordinates to represent the beach, sand cliff and evolving embryonic dunes (area: 50x40m, approx.: 1 $ppcm^2$). The scanner is mounted on a surveying tripod, approximately 1.6 m above the beach surface. This low elevation combined with vegetation and small morphological features necessitates moving the scanner to between 3 and 4 scanning locations to capture most of the terrain surface. However, capturing complete data proves challenging because of the shadowing effect or "dead ground" created by one feature obscuring another. Data captured from each of the scanner locations are registered into the Ordnance Survey (OS) National Grid coordinate system, using four high-definition surveying (HDS) Leica scanner targets. The position of these targets is determined using a Trimble R6 survey grade differential GPS system (dGPS). Initially a complete survey of the site took a full day using the rather slow ScanStation 2 TLS. The more transportable and efficient Leica P20 scanner makes it possible to survey the site in less than 90 minutes, however the equipment remains expensive and heavy to transport, requiring the use of a four-wheel-drive vehicle on the beach.

Figure 1 Leica P20 terrestrial laser scanner with 1 HDS target used for registration and geo-referencing.

## TLS Data-processing

Post-processing the data gathered by the TLS systems involved four steps: i) registration, ii) geo-referencing, iii) filtering and iv) digital elevation model (DEM) interpretation. The first two of these stages has been achieved using Leica Geosystems' Cyclone software, capable of handling and manipulating high-resolution, dense point-cloud datasets. Registration involved transforming individual scan surveys into one congruent local coordinate system, utilising the relative locations of the HDS targets and a 3D bundle adjustment algorithm. The mean absolute error for this local registration procedure was < 4mm for all surveys.

Ordnance Survey (OS) Grid Coordinates of each control point was determined using the dGPS system to allow geo-referencing of the registered point-cloud dataset. Initial target coordinates were transformed to a local OS related system by assigning OS Grid Coordinates to the centrally located control point. Significantly, the remaining coordinates were rescaled using the reciprocal of the local Ordnance Survey Projection Scale Factor, to achieve a local scaling factor of unity. This matched the 1:1 scale factor of the TLS and improves the accuracy of the alignment procedure; 6 mm being the mean absolute error between the TLS and local OS Grid coordinate systems. Following geo-referencing, erroneous outliers were manually removed from the point-cloud and objects (targets, tripods) cleaned semi-automatically using a plane region-grow algorithm. A representative 40 x 50m subset area (Figure 2) was clipped from the original full scan view to moderate file size (reduce processing time) and avoid regions of low point density.

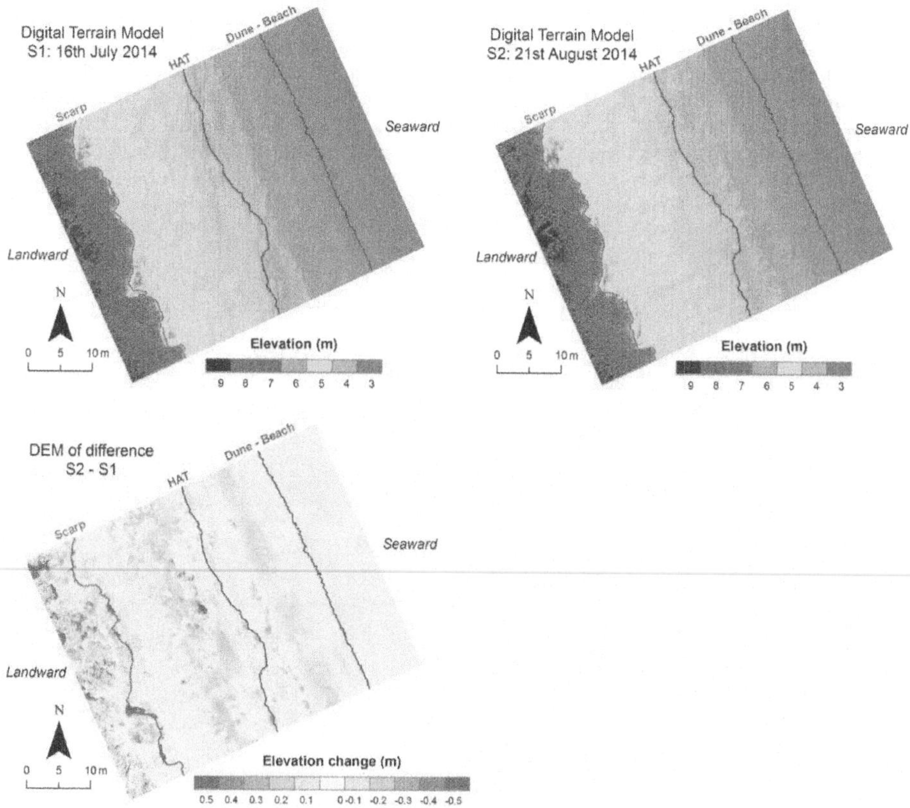


Figure 2 Filtered digital terrain models and surface difference captured by TLS (July/August 2014). Elevations are given in metres ODN.

Filtering the point-cloud subset was achieved within CloudCompare (version 2.6.0) and served to segment the original scene into points representing either bare-earth or vegetation. A number of related studies have employed a local lowest-points window filter to differentiate between 'ground' and 'vegetation' elevations (Guarnieri et al., 2009; Montreuil et al., 2013a; Fan et al., 2014; Feagin et al., 2014). This approach has been found particularly effective in minimizing vegetation-induced error in low-relief saltmarsh environments, the magnitude of which can be far larger than other survey-related errors (Guarnieri et al., 2009; Fan et al., 2014). The size of this filter was determined through a series of sensitivity tests whereby over 100 dGPS-recorded ground measurements were compared to local minimum elevations at varying window dimensions. For small moving window filters the likelihood of one of the laser pulses reaching the ground surface is low. As the size of the moving window is increased this chance increases. As such, a tradeoff has to be made between the accuracy of elevations - 'vegetation error' in Fan et al. (2014) - and the resolution of the filtered raster, the 'quality' of which may have implications for the ability to identify landforms from the resulting DEM and determine morphological change between repeat surveys (Coveney and Fotheringham, 2011). For this study, the best compromise was achieved using a 0.05m x 0.05m block window with mean absolute differences of 0.0218m ($\sigma$ = 0.01289m). The filter was also used to calculate local highest values (Figure 3) which, for low-relief grassland, have been found to provide a good approximation for vegetation height (Fan et al., 2014).

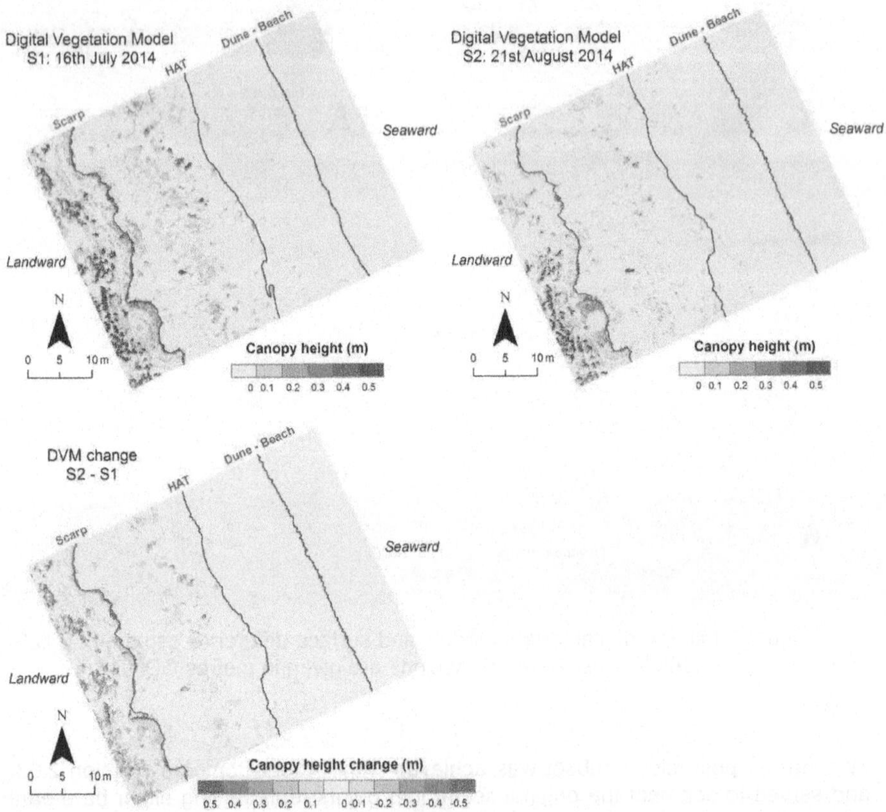


Figure 3 Canopy height and canopy height changes, captured by TLS (July/August 2014)

## TLS Results and Analysis

Cloud-to-cloud differences between the original dataset and the filtered local highest and lowest elevations were assessed within CloudCompare. The nature of the differences between the filtered clouds and how this compares to the original dataset defines whether there are systematic changes between the techniques - i.e. are there components within the scene (areas of bare-ground or vegetation) that are better described using one or other of the datasets?

After filtering, the local lowest and highest point-clouds were further processed and classified to generate **digital terrain models** (DTMs- Figure 2) and digital surface models. The elevation difference between these two models was taken to represent canopy height (Fan et al., 2014) as illustrated by **digital vegetation models** (DVMs– Figure 3). The grid spacing used to produce the 'ground', 'surface' and 'vegetation' models was assigned to the grid step of the moving window filter.

DEMs of difference (DoD- Figure 2c) provided a visual comparison of monthly changes in elevation. These changes were quantified through the calculation of monthly changes in volume of sediment in accordance with Montreuil et al. (2013). The locations of the scarp (6m) and contours representing highest astronomical tide (HAT) (4.09 m) and the boundary between 'Primary foredune' and 'Upper beach' active aeolian sub-units (3.5 m) have been included in DEMs and DoD for reference (Montreuil and Bullard, 2012a). Pronounced elevation changes below high water are assumed to be influenced by the tide, whereas changes above this contour can be interpreted as aeolian-driven.

Comparisons between sequential 2½d digital terrain models derived using the TLS provides some insight into the processes allowing the dune systems to recover (Figure 2). It is apparent that appropriate wind conditions facilitate the formation and development of small dune features which develop at varying rates through time. These embryonic dune features are particularly active in the upper beach area close to the scarp feature, showing rates of accretion that exceed 100mm per month. It is also noticeable how distinctive "troughs" develop several months following the storm surge event and are maintained behind the embryonic "echo dune" thereafter. These appear to be particularly prominent adjacent to small planimetric indentations in the scarp. It can be envisaged that these embayments cause localised increases in wind velocity which help to maintain a clear channel stripped of vegetation.

While these landforms require several months to establish, there are features within the DTMs and DoDs that develop rather more rapidly and which may be crucial in the trapping of material seen in subsequent surveys. For example, a number of parallel dune ridges form between the highest astronomical and 'dune - beach' tidal marks (20-30m from the scarp face; Figure 5). These features stabilised relatively quickly in the months following the storm surge, showing little deviation throughout all subsequent DTMs.

Changes in canopy height (DSM-DTM) help to understand the interrelationship between patterns of sediment change and vegetation dynamics. For the July to August 2014 survey period, canopy height decreased significantly toward what could be interpreted as a near un-vegetated state (Figure 3). The associated DEM of difference would seem to suggest, however, that the clumps of loose, semi-vegetated material deposited following the storm surge have been subsumed within this repeat survey period. As a consequence, terrain and vegetation models must be used in tandem to differentiate morphological change from either vegetation growth or dieback.

## Alternative Monitoring Approach Using Structure from Motion (SfM) Photogrammetry

In July 2014, the topography of the site was also captured using a wholly image-based approach involving "structure from motion" (SfM), the latest technical innovation in close range digital photogrammetry (Fonstad et al, 2013; Micheletti et al, 2014). A Nikon D7000 digital SLR camera (14 megapixel) was used to capture 55 images rapidly (15 minutes) from a range of locations, all field equipment being portable and cheap. The camera was mounted on top of a 3 m length pole (Figure 4) to try and reduce the well-known "shadowing" or "dead ground" effect, which also affected the TLS. This was important as the low beach vegetation would obscure features if a hand held camera vantage point had been used. Six photogrammetric targets had been distributed around the site (3 visible in Figure 5) and their positions were coordinated using the Trimble dGPS system. Back in the Office, the 55 images (Figure 5) were processed fully automatically using SfM and "multiview stereo" (MVS) techniques provided by PhotoModeler, a commercial software package.

Figure 4 Digital SLR located on a 3m pole

Figure 5 One of 55 images used for data-extraction using SfM photogrammetry

Figure 6 High resolution point cloud derived using SfM photogrammetry (similar perspective as Figure 5)

Ten million 3D points were recovered within hours, each colour-coded using the RGB values from the camera (Figure 6). The level of detail is exceptional with many vegetative elements being recovered. The SfM data was compared with the TLS (raw digital surface model) to quantify SfM accuracy (Table 1 and Figure 7). This demonstrates the effectiveness of SfM photogrammetry to capture surface detail accurately but also shows how localised effects degrade point density and subsequent surface modelling. Specifically, smooth areas free of vegetation exhibit the highest accuracies whilst variability increases in vegetated areas.

Table 1 Accuracies of SfM, related to cover type

| Area/vegetation cover | Absolute Mean 3D error (m) | Standard deviation of error (m) |
|---|---|---|
| Sloping sand, no plants | 0.0108 | 0.0063 |
| Whole area | 0.0194 | 0.0193 |

## Discussion

The data captured has again demonstrated the suitability of a terrestrial laser scanner for monitoring change at a coastal site. This is not entirely novel, a variety of terrestrial laser scanners has been used in the past to measure coastal change at a range of scales and applications including cliffs (Lim et al, 2005; Dewez et al, 2013) and sand dunes (Montreuil et al, 2013a; Montreuil et al, 2013b; Feagin et al, 2014). However, with a one monthly cycle of repeat observation achieved here, the temporal frequency is high and the beach's response to a major storm surge event has been captured effectively.

It is estimated that the sensitivity in detecting change is approximately 15 mm and this remains challenging to achieve during fieldwork and data processing. Specifically, errors are introduced during registration, both when registering individual scan sets to each other (4mm) and to an external coordinate system (6mm) and considerable care needs to be taken in targeting. The low scanner position relative to the beach surface has also created difficulties, both in terms of geometry (Soudarissanane et al., 2011) and object measurement. As Coveney and Fotheringham (2012) found, classification of scanned points into "terrain" and "vegetation" also requires judgement and some differences due to interpretation are inevitably introduced.

Specifically, the monthly TLS scans have been able to detect and quantify the early beach recovery processes. Embryonic dunes have developed, which are critical in smoothing the scars left by the tidal surge. Sand accretion rates have varied

Figure 7 Surface of difference (absolute) between laser scanner and SfM photogrammetry (same perspective as Figure 6)

spatially, but have exceeded 100 mm per month, rates that Montreuil et al. (2013a) have recorded at a nearby site during notable periods of dune formation. Unsurprisingly, vegetation plays an important and localised role in influencing sediment change, encouraging the formation of a variety of landforms (parallel dune ridges, shadow dunes, echo dunes) which appear to be crucial to the post-event recovery process.

Although effective, a TLS remains expensive and is time-consuming to use in the field. SfM photogrammetry appears to offer an alternative, being both cheaper and more efficient (Micheletti et al, 2014), particularly in terms of time spent on the beach. Both technologies require extensive data-processing to generate point clouds, particularly if registered to some form of external coordinate system; essential when true morphological change needs to be detected. SfM photogrammetry does require "image texture" and may not be effective for all surfaces. However, the resolution of data extracted can be decided post image acquisition, unlike TLS which needs to be defined in the field and is often a function of time available!

## Concluding Remarks

Through the ongoing monitoring work described on the East Coast, this paper has demonstrated how modern geomatics methods can provide very detailed morphological data suitable for monitoring and measuring coastal change at close range. Surface differences detected by terrestrial laser scanner have quantified exactly how aeolian processes are smoothing the sharp features created by the storm surge. Vegetation is becoming re-established and although welcome from a dune regeneration perspective, does create difficulties in representing the actual terrain surface digitally. Vegetation can be filtered effectively, provided the laser pulse can penetrate the vegetation and a reflective pulse can be obtained from the actual terrain surface. The study also demonstrates how rapidly modern geomatics methods are evolving and are becoming increasingly cost effective for a range of coastal applications. Specifically, Structure from Motion photogrammetry appears to have great potential of generating 3-D models, which have hitherto only been practicable using a terrestrial laser scanner. Although a cheap camera can replace the expensive scanner and fieldwork efficiency can be improved, there remains the necessity to define and utilise a consistent coordinate system during any monitoring project. Only then, can true change be detected and quantified.

## Acknowledgements

The study is supported by a NERC "Urgent Grant" project, funded between February 2014 and February 2015 (NE/M000052/1). The authors would like to thank Tom Mockford and Annie Ockelford for field assistance and Natural England for site access.

## References

Coveney, S., Fotheringham, A.S. (2011), Terrestrial laser scan error in the presence of dense ground vegetation. The Photogrammetric Record, 26, 307-324. doi: 10.1111/j.1477-9730.2011.00647.x

Dewez, T. J. B., Rohmer, J., Regard, V., Cnudde, C. (2013), Probabilistic coastal cliff collapse hazard from repeated terrestrial laser surveys: case study from Mesnil Val (Normandy, northern France). Journal of Coastal Research, Special Issue 65, 702-707.

Fan, L., Powrie, W., Smethurst, J., Atkinson, P.M., Einstein, H. (2014), The effect of short ground vegetation on terrestrial laser scans at a local scale. ISPRS Journal of Photogrammetry and Remote Sensing, 95, 42-52. doi: 10.1016/j.isprsjprs.2014.06.003.

Feagin, R.A, Williams, A.M., Popescu, S., Stukey, J., Washington-Allen, R.A. (2014), The use of terrestrial laser scanning (TLS) in dune ecosystems: The lessons learned. Journal of Coastal Research, 30, 111-119. doi: 10.2112/JCOASTRES-D-11-00223.1.

Fonstad, M.A., Dietrich, J.T., Courville, B.C., Carbonneau, P.E. (2013), Topographic structure from motion: a new development in photogrammetric measurements. Earth Surface Processes and Landforms 20: 817-827.

Guarnieri, A., Vettore, A., Pirotti, F., Menenti, M. Marani, M. (2009), Retrieval of small-relief marsh morphology from Terrestrial Laser Scanner, optimal spatial filtering and laser return intensity. Geomorphology, 113, 12-20. doi: 10.1016/j.geomorph.2009.06.005.

Lim, M., Petley, D. N., Rosser, N. J., Allison, R. J., Long, A. J., Pybus, D. (2005), Combined Digital Photogrammetry and Time-of-Flight Laser Scanning for Monitoring Cliff Evolution. The Photogrammetric Record, 20: 109–129. doi: 10.1111/j.1477-9730.2005.00315.x

Micheletti, N., Chandler J.H., Lane S.N. (2014), Investigating the geomorphological potential of freely available and accessible structure-from-motion photogrammetry using a smartphone. Earth Surface Processes and Landforms, DOI: 10.1002/esp.3648.

Montreuil, A-L., Bullard, J.E. (2012), A 150-year record of coastline dynamics within a sediment cell: Eastern England. Geomorphology, 179, 168-185. Doi: 10.1016/j.geomorph.2012.08.008.

Montreuil, A-L., Bullard, J.E.,Chandler, J.H., Millett, J. (2013a), Decadal and seasonal development of embryo dunes on an accreting macrotidal beach: North Lincolnshire, UK. Earth Surface Processes and Landforms. doi: 10.1002/esp.3432.

Montreuil, A-L, Bullard, J.E., Chandler, J.H. (2013b), Detecting seasonal variations in embryo dune morphology using a Terrestrial Laser Scanner. In: Conley, D.C., Masselink, G., Russell, P.E., O'Hare, T.J. (eds), Proceedings 12th International Coastal Symposium (Plymouth, England), Journal of Coastal Research, SI 65,1313-1318.

Soudarissanane S, Lindenbergh R, Menenti M, Teunissen P. (2011), Scanning geometry: Influencing factor on the quality of terrestrial laser scanning points. ISPRS Journal of Photogrammetry and Remote Sensing: 66(4):389-399. doi:10.1016/j.isprsjprs.2011.01.005.

**Coastal Management: Changing coast, changing climate, changing minds**
ISBN 978-0-7277-6114-9

ICE Publishing: All rights reserved
doi: 10.1680/cm.61149.555

# Mandela, Sinterklaas, Xaver, Bodil: Impacts and Lessons from the December 2013 Storm Around the North Sea

**Jaap Flikweert**, Royal HaskoningDHV, Peterborough, UK
**Michael Cranston**, Scottish Environment Protection Agency (Scotland), Perth, UK
**Craig Woolhouse**, Environment Agency (England), Bristol, UK
**Patrik Peeters**, Ministry for Mobility and Public Works (Flanders), Antwerp, Belgium
**Bart Vonk**, Ministry of Infrastructure and Environment, Lelystad, The Netherlands
**Jan-Moritz Müller**, Agency of Roads, Bridges and Water (City of Hamburg), Hamburg, Germany
**Per Sørensen**, Kystdirektoratet, Lemvig, Denmark

## Introduction

The December 2013 tidal storm had a similar intensity across the North Sea. This makes it an excellent case study to compare and contrast how different countries manage coastal flood risk and to identify opportunities to learn from each other and support adaptation to climate change. The fact that the Storm was named in most countries emphasises its importance and impact. It was named Bodil in Denmark and Xaver in Germany, following storm naming systems in those two countries. In The Netherlands and Flanders it is referred to as the 'Sinterklaas Storm' as it occurred on the date of the annual Santa Claus celebrations in those countries. In England it is called the Mandela Storm as it coincided with Nelson Mandela's death.

This paper first summarises the general meteorological and hydraulic aspects of the storm. This is followed by country-specific sections that describe how the storm affected each country and was perceived there, within its own flood and coastal management context. The paper then concludes with comparisons and lessons learned that the authors developed on the basis of a joint workshop in September 2014's Deltas in Times of Climate Change Conference, and highlights how this event has triggered improved communication between experts in different countries during storm events.

## December 2013: Meteorology and Hydraulics

On 4 December a depression moved south from Iceland, and then eastward across Scotland. The resulting depression above the North Sea was very deep and large and remained in place for much of 5 and 6 December, see Figure 1. It caused north westerly winds ranging from 9 to 11 Beaufort. On 6 December a high pressure area approached to the west of Ireland, which sustained the strength of the storm, even when the depression moved further east toward the Baltic region. Only in the course of Friday 6 December did the storm gradually reduce.

The resulting storm surge moved south west into the North Sea toward the Channel and then rebounded The storm coincided with spring tide, and in some places around the North Sea the peak of the storm surge occurred around high tide. Similarly, in some places the storm also caused high waves around the same time, but in other places the waves were unremarkable. This is discussed in more detail per country in the following sections.

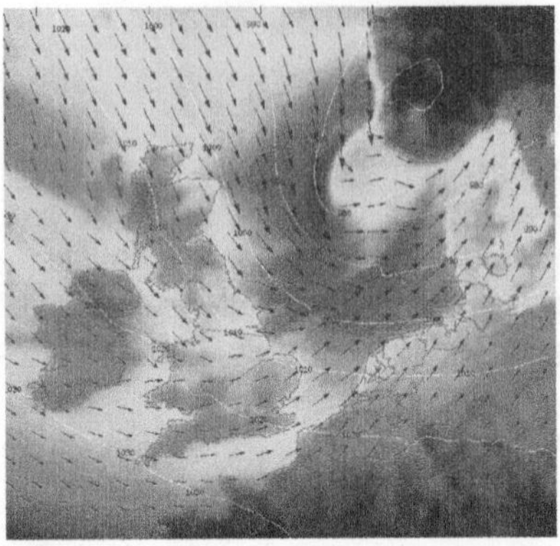

Figure 1: Wind speeds on 5 December at 12:00 GMT, showing 25 m/s on various coasts and offshore maximums above 30 m/s

## Experiences per Country

This section describes how each of the countries around the North Sea experienced the December 2013 storm, broadly following the course of the surge around the North Sea coasts. Each section starts with a concise coastal flooding history and how this has influenced governance and policy of flood risk management. This is followed by a description of the technical and policy experiences with the event, covering key aspects such as understanding of the risk mitigation, preparedness and response. Finally, lessons learned are highlighted, and how these will impact future coastal management.

### Scotland

The Scottish coastline extends to over 18,000 km with a diverse range of islands, Firths and estuaries. As a consequence, Scotland is prone to severe weather which can cause coastal flooding with over 21,000 properties estimated to be at risk (SEPA, 2011). According to historical records, coastal flooding occurs on average once or twice a year (records from 1849), with these events typically occurring in the winter months of January and February. Understanding the history of flooding over the past decade has brought coastal flood risk into the political focus with a focused effort in this sector following the storm of 2005 which resulted in the tragic deaths on the island of Uist (Ball *et al.*, 2008).

This 2005 storm resulted in a step change in coastal flood forecasting capabilities in Scotland with the extension of the Storm Tide Forecasting Service to Scottish coastal areas and the development of early warning systems. Since then a series of new systems have been developed to better prepare and warn for coastal flooding (SEPA, 2012) – many of which were tested during the December 2013 storm.

The earliest flood impacts of the December 2013 storm were reported in the north west of the country with flooding to the Highland town of Oban in the early hours of the 5th December. With strong to gale force winds combined with a 1 metre surge and high tides, the east coast of Scotland experienced the worst of the flooding impacts. Flooding to infrastructure and properties was recorded in Aberdeen and Eyemouth; however the greatest co-ordinated flood response centred on Grangemouth with a predicted combined tide and surge expected to inundate an area of the petrochemical plant. There was a high degree of preparedness to the flood and emergency response, largely due to the introduced new early warning systems along the east coast (Cranston *et al.* 2013) and over this period over 9,000 members of the public were directly warned of the potential for flooding. In addition the five-day flood guidance supported by surge forecasting information at the same timescale has supported the early response. Response included the establishment of multi-agency resilience forums and local scale flood risk management measures such as the deployment of flood barriers.

Looking ahead, with 17% of all flooding impacts in Scotland from the sea, coastal flood risk is an important component of flood risk management. Key legislation for the management of coastal risk is the Flood Risk Management (Scotland) Act 2009 and the Marine (Scotland) Act 2010 and in 2015 a series of Flood Management Strategies will be published outlining the planned objectives for reducing future coastal flood risk which will include further enhancements to early warning systems (Werritty *et al.*, 2012).

Key learning points for Scotland were as follows:

- The value of improvements since 2005 was demonstrated.
- It is important to have a detailed understanding of the receptors at risk.

**England**

England has experienced a series of floods during the last 15 years, the most serious of which were in Easter 1998, winter 2000/2001, summer 2007 and winter 2013/2014. In combination with the recent remembrance of the 1953 floods, this has raised flood awareness with the public. As a result of the regular floods in recent years, and subsequent evaluations, England's flood incident management arrangements have improved over time. They have also been the driver for a risk-based approach to flooding: using a broad portfolio of measures (prevent, protect, prepare, respond) to achieve the best possible risk reduction with the budget available.

During the December storm the highest sea levels ever recorded were experienced on most of the east coast of England and in Liverpool. Calculated return periods were in excess of 800 years in some places (annual exceedance probability less than 1:800 per year). In many places the storm coincided with high tide which explains the extreme water levels, but fortunately there were limited waves. Models were underpredicting peak surge levels in northern England, due to the very rapid deepening of the low pressure system, but this was recognised during the event and manually corrected successfully through expert judgement.

Six major breaches and a number of smaller ones occurred, most of them at places that were already identified as requiring improvement. The storm caused flooding to some 2,800 properties, significant cliff erosion in East Yorkshire and Norfolk, inundation of 6,800 ha of agricultural land and wildlife areas and led to the evacuation of 18,000 people. The early forecasting and warning was very effective, supported through social media and direct messages. The Hull and Thames Flood Barriers were successfully operated against the highest sea levels since they were built and prevented flooding in Hull and London. Though a lot of homes were flooded, due to the investment in flood defences in the last decades around 800,000 homes were protected from flooding.

For England the December storm was the start of three months of extreme coastal, fluvial and groundwater flood events throughout the south of the country in particular. The winter of 2013/14 turned out to be the wettest in 250 years and the stormiest in over 30 years. Sea level records were broken at over half of the primary sites and the 30, 60 and 120 day total river discharge from rivers in England was either the highest or second highest on record. Damage to coastal flood defences was particularly significant and over 800 repair projects were completed at a cost of £100 m between March and November 2014.

Key learning points for England were as follows:

- Forecasting lead times were excellent and ensured an effective multi-agency response; rapid changes in atmospheric pressure in the last 24 hours were a challenge and some local estuarial forecasting was relatively poor which means expert validation remains essential;
- The response benefitted from recent flood incident experience, multi-agency exercises and preparatory work as part of the 60 year remembrance of 1953. Specific lessons at Central Government level revolved around wide area coordination and prioritisation of national assets. For the Local Resilience Forums community engagement is important. At both levels, coordination of evacuations was flagged up.
- The 'opt-out' Flood Warnings Direct system was very effective, in combination with much enhanced use of social media. The event showed that it is essential to communicate through a narrative, giving appropriate context. Very clear infographics successfully communicated often complex and rapidly changing situations.

**Flanders (Belgium)**

In Flanders the open coast defences experienced wave overtopping which led to evacuation and placing of temporary defences. The protection measures realized within the frame work of the coastal safety master plan in the past few years proved to be effective. The main protection measure is beach nourishment.

The safety plan for the tidal river Schelde, called Sigmaplan (referring to the Greek letter 'S' of Schelde, with reference to the Dutch Deltaplan), dates from 1977 and aimed at dike heightening and strengthening as well as the construction of flood control areas in combination with a storm surge barrier downstream of Antwerp. The Sigmaplan was recently updated to take account of changing physical circumstances and new insights into water management. The so-called Updated Sigmaplan aims at satisfying safety and ecological needs along the river Schelde in a sustainable way. Therefore different restoration techniques are elaborated which combine safety with estuarine restoration, e.g. dike strengthening together with more space for the river (e.g. managed realignment), flood control areas with or without a controlled reduced tide, non-tidal wetlands, etc.

The predicted water level of the December 2013 storm corresponded with a 1:50 per year storm flood; the highest water level recorded since 1953. This included 1.5m of surge. The maximum surge at Antwerp was just below 2.5m, but this did not coincide with high tide.

Real time predictions turned out to be good and reliable. Three days in advance the water height was known. A range of actions were taken before and during the storm. Along the Schelde, Flood Control Areas (FCAs) came into action, performed well and proved their necessity. For the coastal area, the weakest points of the coastline are the port areas because of their many temporary defences that need closing, but no major problems occurred. The main damage was essentially loss of sand and sand cliffs on the beaches.

Key learning points for Flanders were as follows:

- The value of the protection strategies for the coast and estuary was demonstrated.
- Forecasting and response during the storm was very successful.

**The Netherlands**

The Dutch have a love-hate relationship with water. Over many centuries they reclaimed land, but there have also been many serious flood events. The last serious coastal flood was in 1953. 1836 people died, 100,000 lost their houses and belongings, ten thousands of cattle drowned, 200,000 ha flooded. After this traumatic event, the Dutch Government stated that this should never happen again. The Dutch started building the Delta Works (in particular shortening the coastline by closing estuaries). They also established legal flood defence safety standards, as laid down in the Water Act. This divides the country into dike rings that each have a safety standard, ranging from 1:250 to 1:10,000 per year.

It is recognized that the country must be prepared for the small chance that the design standards are exceeded. The water boards that manage the flood defences are very well prepared, but national coordination is also necessary. The National Coordination Centre on Flood Threat (LCO) gives early warnings and real-time information, gives advice on how to act at a national level and on mobilization of nationally available emergency measures like pumps, F16s, big bags. It also provides adequate information for external communication.

The LCO first warned of a possible storm surge one week before the event, on Friday 30 November. On 1 December it predicted a 50% chance of exceedance of a 1:5 per year event, and on 3 December a 30% chance of exceeding a 1:40 per year event. This led to upscaling to a national crisis team on 4 December, while the focus shifted to Delfzijl in the north. In the course of 5 December, it became clear that the evening tide would be the most critical, nearly reaching design levels. In the end, the resulting water levels in the south were the second highest ever, although more than

0.5m lower than those that caused the 1953 disaster. In Delfzijl the water level equaled the record level and has been assessed as a 1:50 per year event.

The event led to closure of some coastal barriers, but not all of them. There was no flooding and only limited damage to the defences. The authorities communicated extensively. This revealed clearly that there was no anxiety among the people, other than: will the train still go and will I be home before Sinterklaas arrives with all his presents? However, if the storm had reached the Dutch coast 6 hours later, the maximum surge and high tide would have coincided, causing water levels close to the defences' design loading. This would have triggered the highest alert and put the safety regions in charge.

Key learning points for The Netherlands were as follows:

- The the coastal defence system, much improved since 1953, performed well during this storm.
- The development of the meteorological and hydraulic models in the last decades has led to reliable early warning (a week ahead!), providing frequent prognoses and adequate information for the Waterboards and Rijkswaterstaat.
- There is however a potential downside: the event showed that, because of the excellent flood defence, the public perception is now that there is no danger of flooding. People are not prepared for the situation that their houses will be flooded. In recent years there have been various campaigns to make people aware of this (very small) chance that they will be flooded and what they can do to prepare. This includes awareness of nearby higher ground, storing drinkwater, food, clothes on the higher floors of their houses, etc. There is even an application at playstore available that gives information on how high the water can come around your house (www.overstroomik.nl).

**Hamburg (Germany)**
Significant parts of North West Germany are at risk of coastal flooding and are protected by sea dikes. The focus of this section of the paper is on the City of Hamburg, an independent authority whose centre is located in the floodzone, housing more than 326,000 people. The river Elbe is influenced by tides and storm surges and these can cause high water levels in Hamburg. The city has water level records since 1750, which show that some flooding has always occurred, but high water levels have become more regular in recent years due to construction of flood defences elsewhere. After floods in 1962 and 1975 the sea dikes of the Elbe were raised to 8m above mean sea level.

The December storm surge was named 'Orkan Xaver'. The public warnings, based on the forecasts, indicated storm gusts up to 150 km/h, a long lasting storm, a heavy storm surge and snow. The resulting water level was the 2nd highest since 1750 and water levels of +6.08m above mean sea level were measured, including a surge of more than 4m.

Hamburg provided extensive risk communication. They have developed a system where it is possible to register for an sms texting warning service. Several warnings were given, including also the use of gun shots, radio warnings and even sub-titles on TV channels. The highest alert level, which would trigger siren signals and local tannoy announcements, was not reached.

The flood protection facilities were in a good state and performed well. The flooding that did occur on the seaward side of the dikes did not cause any permanent damage. The only financial impact concerned the removal of pollution that had accumulated on the dikes and flooded areas after the storm.

Key learning points for Hamburg were as follows:

- The dike inspectors were well trained;
- The communication between the different parts of the dike defence organisation will be improved in future;
- Flood protection facilities were in a good state;
- Openings in dikes and sea walls are the weakest points in the system → a second safety is always necessary

**Denmark**

Denmark has over 7,000km of coastline, but much of this is sparsely developed, so its vulnerability is limited. Historic storms, such as the one in 1872, did cause damage on some of the islands near Copenhagen. Storms in 2006 and 2007 have caused surges up to 2m, but did not cause significant damage. Coastal protection is coordinated by the government's Kystdirektoratet (approvals, advice, monitoring and warning), but protection works are initiated and funded by local beneficiaries. Protection works are present on approximately 1,800km of the coastline.

In Denmark, the December 2013 storm was named Bodil (following Denmark's alphabetical storm naming system). It caused record water levels around Denmark: higher than 1:100 per year in many areas, and even more than 1:1,000 per year in the northern part of Zealand at Hornbæk, which has a more than 100 year long water level record. This extreme value was caused by the change in wind direction from a westerly to a more northerly wind direction.

In the southern Wadden Sea area, the storm triggered the highest level in Denmark's flood alert system, nearly leading to evacuation of people. However, the sea dikes withstood the storm. In the inner Danish waters 100 to 200 people had to be evacuated because their homes were flooded, and there was damage to properties due to minor flooding. Coastal erosion caused significant damage to coastal protection structures, and one house was lost. On the North Sea coast the soft defences withstood the storm which had a 1:600 per year intensity around Thorsminde. The dune erosion in that part of Denmark was less than expected, probably due to the rapid change in wind direction.

There was significant media interest including some pressure on politicians, and this has had positive effects on their willingness to cooperate. Since the storm, people have taken the initiative to carry out recovery works, although many of them without the required permit.

Key learning points for Denmark were as follows:

- Soft defences performed well.
- Bodil being the first recorded major coastal flood event, it caused a high level of attention with public, politicians and media.
- Extreme events trigger improvements to risk management systems.

## Comparison and Lessons Learned

A joint workshop at September 2014's Deltas in Times of Climate Change Conference identified differences and similarities between how each country experienced the December 2013 storm and what lessons were learned in each country. This then led to the identification of good practice: which country is currently leading the way in which topic. This can then be a starting point for real cross-North Sea learning. This comparison is structured per aspect of flood risk management.

### Understanding the Risk (Sources, Pathways, Receptors)

| Differences and similarities | Lessons learned and good practice |
|---|---|
| All countries around the North Sea have high quality water level and wave monitoring systems, have a good overview of the existing defences and know which areas are at risk of flooding and erosion.<br><br>The focus of each country reflects which element plays the dominant role in its coastal management approach, and also the overall level of flood risk.<br><br>Public understanding of risk is strongly influenced by the occurrence of recent flood events. The Dutch case is different: the public is aware they live below sea level, but they trust the thorough defence system so much that they don't really believe flooding could happen. | Investment in forecasting and warning is worthwhile, as illustrated by all seven countries.<br><br>The examples of Scotland and Hamburg show that it's important to have a detailed understanding of the potential flood area and the receptors at risk (supported by the Flood Maps required by the European Union's Floods Directive).<br><br>The English example shows the importance of using expert judgement to verify and correct model predictions.<br><br>England and Scotland already have a process for expert level cross-border telecoms to discuss forecast surge uncertainty (UKCMF).<br><br>England is developing evacuation modelling for its East Coast, which will help understanding of vulnerability and improve preparedness.<br><br>The Netherlands have a very thorough knowledge of their flood defences. England benefitted from a good knowledge of where the weak spots in the defences were. |

**Mitigation and preparedness**

| Differences and similarities | Lessons learned and good practice |
|---|---|
| Historically in each country there was first a flood and then a plan to prevent future flooding. Every country learned from experience.<br><br>All countries have well-developed coastal flood forecasting systems and emergency response procedures, in most cases developed after previous flood events. These systems performed well during the event.<br><br>Most countries have centrally planned flood protection systems, while for others the construction of defences is locally driven.<br><br>All countries use temporary defences to supplement the static defences. | All countries work with a combination of different types of risk reduction measures, in line with the principles of the EU Floods Directive. The emphasis varies; for example England's response processes have improved over time because they have been tested regularly in recent years.<br><br>England and Hamburg have very extensive flood warning and incident communication systems, including the use of direct messages.<br><br>The high level of flood protection in Belgium, the Netherlands and Hamburg meant that the storm did not cause significant damage, even though the event was very extreme.<br><br>Flanders' structured use of flood control areas worked as planned.<br><br>Soft defences in England, Flanders, the Netherlands and Denmark performed well. |

**Response**

| Differences and similarities | Lessons learned and good practice |
|---|---|
| Most countries appear to have had multi-day forecasts which were very important in supporting the response.<br><br>Each country implemented its response plans and these worked well.<br><br>For England and Denmark, and to some extent Scotland, the response involved dealing with actual flooding (e.g. evacuation). In Flanders, the Netherlands and Hamburg there was no flooding, and the response focused on flood defence inspection, closure of temporary defences and awareness raising.<br><br>The awareness of the public, media and politicians was very different in each country. In England the flooding was part of a long series, but it was the first coastal flood in many years. In Denmark it was the first recorded major coastal flood event, which explains the high level of attention. Hamburg made a big effort to raise awareness. In the Netherlands the public was not concerned at all, even though the storm was the most extreme in many years. | Long term forecasts help to streamline response.<br><br>Temporary defences and barriers in all countries worked well, supported by the multi-day forecasts.<br><br>England and Hamburg are good examples of raising awareness. For England this included targeted flood warnings (opt-out system) and social media; for Hamburg social media and television sub-titles.<br><br>Hamburg's experience emphasised it is important to also be aware of flooding due to storm water and to have a plan in case flooding occurs at night<br><br>The Netherlands made sure to inspect the flood defences directly after the storm to avoid that a second storm would aggravate any damage. |

**Summary**
Key similarities are the presence of good systems for forecasting and warning, and of risk mitigation measures that reflect the nature of the risk. Improvements have usually been driven by historic flood events. All countries make an effort to raise public awareness; the difference in public response reflects both the level of risk that this storm posed but also the existing awareness and perception of coastal risk in each country.

The main lessons learned concern the importance of expert judgement in the interpretation of forecasts of such extreme storms and the use of a wide range of methods for public communication, for which different countries have each developed their own useful approaches.

## The 2013 Storm as a Trigger to Improve Communication Between Eexperts

International exchange of information is already happening during storms, but what can be improved is exchange of expert judgement about interpretation during and after the event, building on the existing collaboration between Scotland and England. This can support and improve flood risk management. The cooperation between the authors in the September 2014 workshop already initiated this type of communication during the October 2014 North Sea storm, and an initiative was started to explore enhancement of such a process, possibly with European Interreg funding.

**References**

Ball, T, Werritty, A., Duck, R., Edwards, A., Booth, L, and Black, A. (2008) *Coastal Flooding in Scotland: A scoping study*. SNIFFER, Edinburgh, Scotland, UK, 119pp.

Cranston, M., Naysmith, L., Tavendale, A., McLean, L., Hart, P., Whitlow, C., Powis, C. and Hu, K. (2013) The Firths of Forth and Tay flood warning scheme and the storm of December 2012,

Scottish Environment Protection Agency (2011) *The National Flood Risk Assessment*, SEPA, Stirling, Scotland, UK, 14pp.

Scottish Environment Protection Agency (2012) *Flood Warning Strategy (2012-2016)*, SEPA, Stirling, Scotland, UK, 12pp.

Werritty, A, Duck, R., Dawson, S., Powell, V., Ball, T., Dawson, A. and Muir, D. (2012) *Coastal Flooding in Scotland: A guidance document for coastal practitioners*. Centre of Expertise for Waters, The James Hutton Institute, Aberdeen, Scotland, UK, 74pp.

**Coastal Management: Changing coast, changing climate, changing minds**
ISBN 978-0-7277-6114-9

ICE Publishing: All rights reserved
doi: 10.1680/cm.61149.565

# Interpretation of High Resolution Coastal and Marine Monitoring Data for Shoreline Management and Geohazard Risk Mapping

**Andrew Colenutt**, Channel Coastal Observatory, Southampton, UK
**Keith Westhead**, British Geological Survey, Edinburgh, UK
**Jon Evans**, Channel Coastal Observatory, Southampton, UK
**Stuart McVey**, Channel Coastal Observatory, Southampton, UK
**Tim Le Bas**, National Oceanography Centre, Southampton, UK

## Introduction

The demand for high quality coastal and marine-related spatial information has become increasingly apparent in recent years at a European and national (UK) scale, due to a wide range of increasing pressures. The Defra-funded National Network of Regional Coastal Monitoring Programmes of England (Bradbury et al, 2002) provides important quantitative long-term coastal and nearshore marine monitoring data to improve understanding of coastal processes, rates and trends of coastal change, and enables robust shoreline management and engineering decisions to be identified with confidence.

Temporal and spatial analysis of such datasets enables assessment of: sediment transport studies; coastal slope stability and landslide risk; coastal erosion and sea flooding risk; the performance of beach management plans and coastal engineering works; and the impact and recovery of managed beaches from storm events. Integrating monitoring of physical processes with a range of marine surveys also enables modelling of biological, geological and oceanographic interactions.

## Data Collection Programmes

The National Network of Regional Coastal Monitoring Programmes of England comprises six regional programmes and is coordinated and delivered under the Open Government Licence by the Channel Coastal Observatory (CCO) (Bradbury, et al. 2002). The CCO data management centre is hosted by New Forest District Council, in partnership with the University of Southampton and is based in the National Oceanography Centre Southampton. The network of monitoring programmes is managed through a series of lead authorities on behalf of the Coastal Groups and provides a regionally consistent and integrated approach to evidence gathering, delivers strategic actions, maximises the use of data and provides best value.

The principal monitoring and analytical tasks of the regional coastal monitoring programmes include topographic surveys (cross-section profiles, baselines, control network, post-storm and emergency incident recovery surveys of beaches, cliffs and landslides); bathymetric surveys (multibeam and single beam surveys); network of coastal wave buoys, tide gauges and meteorological stations); ecological mapping; aerial photography and photogrammetry; airborne Lidar (Light Detection And Ranging). Data are freely available via CCO website and free web service delivery under Open Government Licence (NFDC, 2010). The regional coastal monitoring programmes have embraced recent advances in land-based topographic monitoring and marine acoustic survey technologies, which has resulted in increasingly cost-effective data acquisition in coastal zone areas.

In order to maximise efficiencies under the 'gather once, use many times' principle, effective pan-governmental agreements have also been successfully established. The UK Marine Environmental Mapping Programme (MAREMAP) was originally launched in 2010 by the Natural Environment Research Council centres of British Geological Survey (BGS), the National Oceanography Centre Southampton (NOCS), and the Scottish Association for Marine Science (SAMS), and aimed to improve seafloor and shallow geological mapping to achieve national objectives such as habitat mapping, Quaternary science, coastal and shelf sediment dynamics and the assessment of human impacts and geohazards in the marine environment. The programme expanded rapidly with a broadening of the partnership to include the CCO, Universities of Southampton and Plymouth, the Maritime and Coastguard Agency (MCA) principally through the Civil Hydrography Programme, the Centre for Environment, Fisheries and Aquaculture Science (CEFAS), and the Marine Scotland Science (MAREMAP, 2013). These collaborations are resulting in efficient integration of swath bathymetry survey planning, analysis and delivery of marine environmental data, and have resulted in seabed survey data of unprecedented detail being brought together for approximately 40% of UK Waters.

## Data Types

This paper will focus primarily on swath bathymetry, topographic laserscan and aerial photography data and present examples of integrated datasets from these data collection and mapping programmes, taken from southeast, southwest and northeast England.

### Swath Bathymetry

Until relatively recently, the most effective method for capturing bathymetric data within shallow coastal waters was with single beam echo sounders; this involved gathering one profile line at a time, but even when the line spacing is fairly close (~50m) such surveys provide poor spatial resolution and may fail to identify large and potentially important seabed features (Colenutt et al, 2013). High resolution multibeam echosounder systems collect a swath of georeferenced bathymetric soundings and can efficiently map large areas of the seafloor at metre-scale horizontal and centimetre-scale vertical resolution. The widespread application of such systems has meant that it is increasingly economically-viable to achieve 100% sea floor coverage in the nearshore zone, which has opened up new opportunities to understand and address a range of coastal engineering and shoreline management issues as well as contributing towards wider scientific initiatives (Mason et al, 2012). The regional monitoring programmes are increasingly utilising swath bathymetry technology to survey the nearshore zone from approximately Mean Low Water Springs (MLWS) to about 1km offshore, ensuring an overlap with aerial photography, Lidar, and beach topographic surveys (Figure 1). Data have been converted from Chart Datum to Ordnance Datum using the most appropriate vertical offsets and projected to British National Grid on OSGB36 datum in order that the monitoring programme can provide a seamless Digital Terrain Model of the coastal region (Mason, 2012).

Figure 1. Swath bathymetry and aerial photography data integrated to produce comprehensive coverage of the coastal zone. Seaford, East Sussex

Due to the collaborative partnerships and pan-government agreements that have been developed through the publicly-funded hydrographic survey community, there is an increasing consensus for the gathering of swath bathymetry data to comply with the MCA's Civil Hydrography Programme's Survey Specification, which ensures data collection and processing achieves the International Hydrographic Organisation Order 1a Standard (IHO S44 Edition 5), which is applicable for depths shallower than 100m. It requires detection of all features larger than 2m in depths up to 40m, and of features larger than 10% of the depth in deeper water. Sediment samples are also obtained for seafloor characterisation and sediment texture analysis. The Specification also requires backscatter data to be collected. Backscatter is a measure of how much power is lost in a sonar beam during its passage from the survey vessel to the seabed and back; the intensity and reflectivity of the returning acoustic signal provides information on the nature and composition of the seabed e.g. a flat rock seabed will reflect the sound beam well whilst, in contrast, mud substrate tends to absorb sound energy (Le Bas and Huvenne, 2008).

Using this Specification ensures that the data is truly multi-purpose since surveys conducted for safety of navigation are automatically suitable for a wealth of other purposes, whilst the converse is not necessarily true.

**Terrestrial Laserscanning**

The Southeast regional coastal monitoring programme have been using terrestrial laserscanning techniques, in addition to existing Real Time Kinematic (RTK) Global Positioning System (GPS) techniques, for monitoring beaches, cliffs and landslides for a number of years. Advances in technology and data processing have made the terrestrial laser scanning technique a viable technique in terms of cost, mobility of equipment on site and advances in computing power. The systems employed have relatively long range data acquisition capabilities, approximately 200-300m, with typical individual data point accuracy of 10-15mm. Digital colour photography can also be captured simultaneously with the laserscan data thus providing red-green-blue values for each point measurement, improving visualisation and thus interpretation of the data. Ground-based laserscanning technologies, combined with Lidar and aerial photography, are being used increasingly to focus on specific issues in the coastal zone, such as onshore-offshore bedrock mapping, and slope instability in cliff-dominated areas. Consistent and repeatable monitoring techniques allow improved understanding and temporal analyses of active landslide and coastal retreat mechanisms. Figure 2 shows a continuous elevation surface generated from integration of Lidar and swath bathymetric data, overlain in part with a higher resolution point cloud generated from a ground-based laser-scanning survey. The latter adds important surveying detail in the lower, near-vertical cliff section, that may not necessarily be adequately defined by Lidar alone.

Figure 2. Features of coastal landslides in Monmouth Ledges, Lyme Regis, West Dorset. Inset: Laserscanner mounted on tripod.

The coastal slope in this area is formed by interbedded limestones and mudstones and is actively landsliding and retreating through a combination of mudflows, rotational slides and cliff falls, enhanced by wave action at the cliff base. Narrow beach deposits can be seen at the base of the cliff and widening towards further to the east where they form a significant storm beach.

**Digital Aerial Photography**
Digital aerial photographs are images acquired of the ground surface using a nadir-pointing large format digital camera mounted on an airborne platform. The camera calibration properties are used in conjunction with a detailed elevation model to perform geometric correction in photogrammetry, generating a series of orthophotos that are commonly used in geographic information systems (GIS) as contextual overlays draped onto digital elevation models to simulate 3D landscapes.

## Applications

**Substrate Mapping**
Interpretation of the bathymetry and backscatter data, integrated with ground-truthing data provided by a range of other partners, has produced a series of detailed thematic maps including surficial substrate, broad-scale marine habitats, anthropogenic features and marine geology. These are derived through analysis and interpretation of a series of GIS layers. Generating a Hillshade layer, which is a form of artificial sun-illumination applied to the bathymetry, highlights and enhances depth changes and features in the bathymetry. This layer is particularly useful for displaying areas of bedforms and rocky seabed where there are numerous depth changes across relatively short distances. The hillshade and bathymetry layers can be displayed together so that the depth values can be viewed with the desired colour map. The seabed slope layer calculates the angle of slope of the seabed by using a central cell and comparing its value to those around it.

The processed backscatter data can aid the definition of the general distribution of coarse and fine sediment textures and differentiate between surficial sediment and exposed bedrock. Figure 3 clearly illustrates the boundaries in backscatter denoting variability in seabed sediments that are not detectable solely from bathymetry data. There are many environmental factors that affect the backscatter values, and it requires technical expertise to process and interpret the mosaic of backscatter tiles

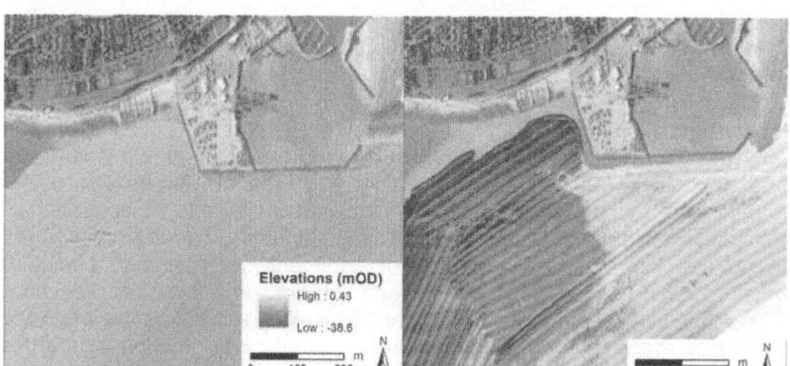

Figure 3. Seabed substrate differentiation and detection from [a] bathymetry and [b] backscatter datasets (Ramsgate, Kent, UK).

The relative importance of these seabed layers varies with seabed terrain and substrate. For example in areas with rocky seabed, rock platforms can be readily identified from the bathymetry, with the aid of the slope layer. In contrast, on a featureless seabed, careful interpretation of the backscatter is needed to infer a change in sediment composition. The type and extent of indicative seabed environments will depend on a range of factors, such as the depth of water, the underlying rocks, the thickness and type of substrate material, the direction and velocity of tidal and seabed currents, and exposure and influence of waves.

The identification of zones of exposed bedrock and surficial deposits of coarse, fine and mixed sediments as shown in Figure 4, provide spatial detail of seabed features, such as sand waves, rock outcrops, pinnacles and scour, and informs future monitoring programme requirements.

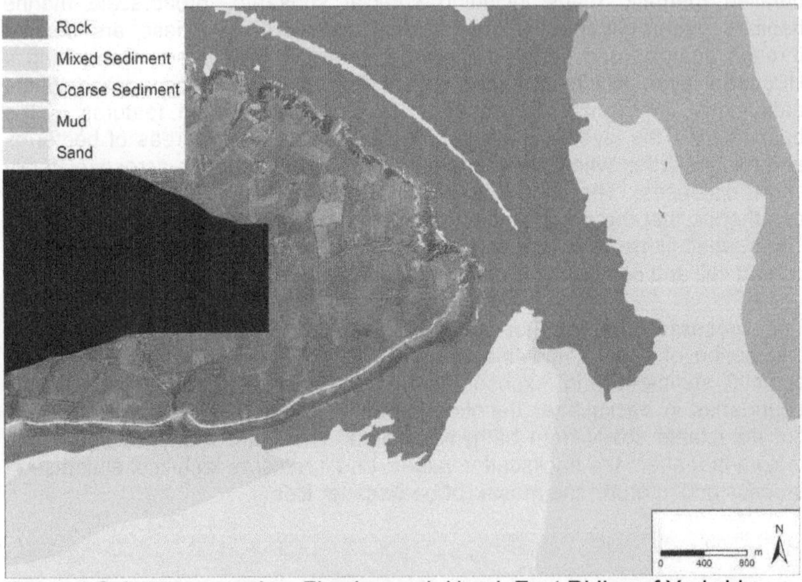


Figure 4 Substrate mapping, Flamborough Head, East Riding of Yorkshire.

**Anthropogenic Features**

Identifying location, type and condition of anthropogenic features such as pipelines, cables and trenches, wrecks, or trawl marks may provide useful information for marine planning, and indicate levels, type and concentration of marine activities and condition of marine assets. Identification of such features can provide information on mobility of sediment, for example, sediment plumes downstream of obstructions and wrecks, and rates of sediment transport by channel infilling and scouring. Assessing environmental impact or recovery of a site may contribute towards scientific evidence base, and for example, may support pragmatic conservation measures to be implemented or enable offshore developments to progress.

**Coastal Stability and Landslide Risk Mapping**

A further application of integrating swath bathymetry, terrestrial laserscanning, aerial photography, and seabed mapping data is the improvement in geomorphological and geological surveys (Westhead et al, in press). When combined with innovative 3D mapping techniques and visualisation technologies. These advances are allowing data collected during earlier mapping programs to be reinterpreted in much greater detail. Seamless high resolution surveys of the coast and nearshore zone have been achieved, and permit offshore extensions of onshore geological maps to be produced. This new generation of geological maps and science outputs provide important information about seafloor composition and structure, physical properties and geohazards.

In addition, the correlation of offshore relict landslide deposits with active onshore slopes, as illustrated by the example in Figure 5, has the potential to improve prediction and modelling of the most active areas of coastal retreat, and informs the mechanisms of failure and mass movement, geomorphological mapping and shoreline evolution studies, sediment inputs, geotechnical assessments. At this example site, 7 km to the east of Lyme Regis in Dorset, the cliffs are topped by more resistant sandstones of the Upper Greensand Formation, underlain by weaker mudstone-dominated formations (mainly Lias Group).

This combination of geological layers produces an actively landsliding and retreating coastal slope. The nearshore platform provides a record of this retreat through preserved 'boulder arcs' (mapped here with black lines) formed at the toes of former landslides. These are interpreted as recording former positions of the coastal cliff line and sea-level, and extend to at least 700m offshore and to 10 m below current sea-level. Such information will underpin future research into coastal evolution, which in turn could contribute to better predictive coastal management, and hazard and risk mapping.

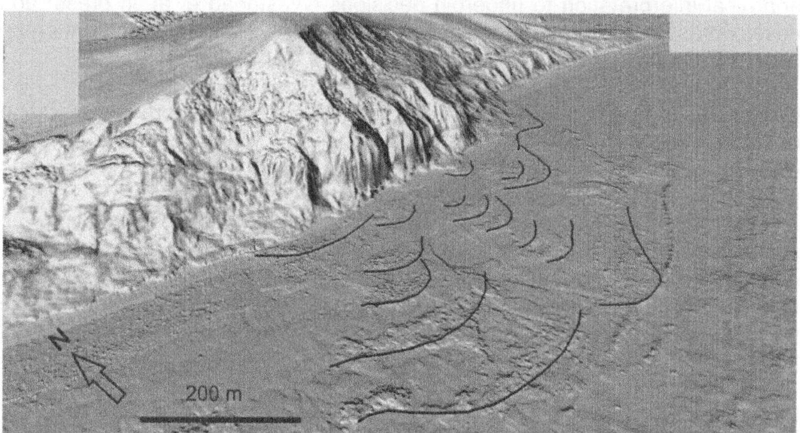


Figure 5. Former coastal positions, landslide features, 3D Golden Cap, Charmouth, Dorset

**Sediment Transport Studies**

Providing information on seabed texture and micro-scale nearshore and seafloor topography can be useful for management of nearshore zones. Such information, considered with other sources of evidence indicated from other seabed and anthropogenic features, can then be integrated with coastal monitoring data to further enhance the modelling of sediment transport within the coastal zone, which may indicate the mobility or stability of the seabed sediments. The regional coastal monitoring programme's robust, consistent and repeatable data acquisition methodologies enables rates, volumes and directions of shoreline and sub-tidal sediment transport to be assessed and quantified. Sediment budget analysis informs the design, operational performance and post-project appraisal of managed beach frontages, the management of coastal erosion and coastal instability risks, assessment of interaction between coastal processes and defence and offshore structures, and identification of potential sources of sediment that could be available for beach renourishment and recycling operations.

## Conclusions

Advances in data capture technologies and techniques, and improvements in consistency of data quality, coverage, availability and interpretation of coastal and marine-related spatial data directly contribute to multi-disciplinary baseline conditions, and have the potential to better inform shoreline management planning and coastal engineering, by providing a more robust scientific evidence base.

To enable planning authorities to consider and assess the impacts of developments at the coast or that contain connections between offshore development and shore-based works there is also an urgent requirement to integrate the overlapping terrestrial and marine planning regimes that cover the inter-tidal zone. It is clearly beneficial to regulators, developers and land and marine-area users to have access to the improved scientific evidence and analytical interpretation to underpin decisions (Westhead et al, in press) and development of pragmatic policies that aim to improve the management of the coast and marine resources.

Coastal management authorities could benefit significantly from interpretation of the nearshore bathymetry datasets to identify sediment types, direction and rates of alongshore and off-onshore sediment transport pathways, and potential nearshore stores and sinks of suitable sediment for future beach management operations. In the longer-term, with further surveys it may be possible to quantify sediment transport budgets in dynamic sub-tidal and nearshore zones. The availability of such marine spatial information may enable more-efficient and cost-effective management measures to be implemented and be of value in terms of evaluating scheme design and performance of structures and beach management operations.

High-resolution substrate mapping through innovative use of bathymetric and backscatter data can better define baseline conditions for monitoring of marine ecosystems. Integration of new nearshore seabed topographic and texture interpretations with nearshore monitoring data can provide improved definition of sediment mobility and stability mechanisms important in subtidal engineering and beach management activities.

Improved mapping of seabed anthropogenic features and their interaction with the environment (e.g. scouring) can support more pragmatic installation of conservation measures and engineering structures at the coast or further offshore. New techniques for repeated, high-resolution measurement of coastal morphology (e.g. repeated ground-based laserscanning) can improve prediction of coastal retreat rates and mechanisms, enabling more effective coastal management works. Innovative geological bedrock and landslide mapping using seamless elevation surfaces through the coastline have the potential to contribute to better predictive coastal management, and hazard and risk mapping.

Within the UK the availability of such high quality data informs the management of tidal flooding and coastal erosion risk and geohazards (Westhead et al, in press), enhances the integration of terrestrial and marine planning regimes, and sustainable use and conservation of coastal and marine resources (Colenutt et al 2013). Such work will have multiple benefits, including: better-informed environmental impact and economic assessments for coastal and marine developments; more effective coastal defence and offshore installations; improved identification of coastal hazards; improved identification and condition assessment of marine heritage features; and improved assessment of seafloor environments and management of resources.

The repeatable, consistent and standardized approach to data collection, analysis and interpretation, coupled with making the publicly-funded data and resulting GIS layers freely available, are significantly contributing towards meeting the requirements of EU Directives, and the UK's legal environmental obligations, and contributes towards improving the scientific evidence base for policy development.

## Acknowledgements

The team members of the data collection and management teams of the various regional coastal monitoring programmes are thanked, as are colleagues from MCA, UKHO, and BGS. The technical guidance of Dr Justin Dix (University of Southampton) is acknowledged. The provision of additional groundtruthing data of Kent and Sussex Wildlife Trusts and Sussex Inshore Fisheries and Conservation Authority has been invaluable.

The authors would also like to acknowledge the inspirational visionary Director of the CCO, Professor Andy Bradbury, who passed away in August 2014. His wisdom and enthusiasm is greatly missed.

## References

Bradbury, A. P., McFarland, S, Horne, J. And Eastick, C., (2002). Development of a Strategic Coastal Monitoring Programme for Southeast England. ICCE Conference, Cardiff, ASCE.

Cocuccio, A. (2010). The UK Civil Hydrography Programme. Hydro International. http://www.hydro-international.com/issues/articles/id1215-the_UK_Civil_Hydrography_Programme.html

Colenutt, A., Mason, T., Cocuccio, A., Kinnear, R. and Parker, D. (2013). Nearshore substrate and marine habitat mapping to inform marine policy and coastal management.International Coastal Symposium, Plymouth. Journal of Coastal Research Special Issue 65, 2013.

IHO Standards for Hydrographic Surveys, 5th Edition, February 2008, Special Publication No 44.

Le Bas, T. & Huvenne, V. (2008). Acquisition and processing of backscatter data for habitat mapping – comparison of multibeam and sidescan systems. Applied Acoustics 70 (2009) p1248-1257. Elsevier Ltd.

MAREMAP. (2013). Marine Environmental Mapping Programme. Website: http://www.maremap.ac.uk/index.html

Mason, T., Kinnear, R., Colenutt, A., Cocuccio, A. & Parker, D., (2012). Beyond bathymetry – coastal marine mapping. Proceedings of Hydro12 Conference, Rotterdam.

Mason, T. (2012). Regional Coastal Monitoring along England's coastline. Hydro International.

New Forest District Council (2010). National Network of Regional Coastal Monitoring Programmes. Strategy Appraisal Report.

Westhead, K., Smith, K., Campbell, E., Colenutt, A., & McVey, S. (in press). Pushing the boundaries: Integration of multi-source digital elevation model data for seamless geological mapping of the UK's coastal zone. Earth and Environmental Transaction of the Royal Society of Edinburgh, Conference Volume: Mapping the Seabed. Edinburgh 2013.

**Coastal Management: Changing coast, changing climate, changing minds**
ISBN 978-0-7277-6114-9

ICE Publishing: All rights reserved
doi: 10.1680/cm.61149.575

# Schematisation and Implementation of the Moray Firth Flood Forecasting System

**Darroch Kaye**, Scottish Environment Protection Agency, Perth, UK
**Daniel Rodger**, JBA Consulting, Skipton, UK
**Nicola Buckley**, JBA Consulting, Edinburgh, UK
**Michael D. Cranston**, Scottish Environment Protection Agency, Perth, UK
**Laurence D. Cload**, Mott MacDonald, Inverness, UK

## Introduction

The Moray Firth Forecasting System is the largest coastal flood forecasting project the Scottish Environment Protection Agency (SEPA) has ever undertaken and a priority area outlined in SEPA's Flood Warning Strategy (SEPA, 2012). The system covers approximately 500 miles of coastline from Fraserburgh adjacent to the North Sea to Wick near the Pentland Firth (Figure 1), encompassing three local authority areas and 44 coastal communities. The fully integrated forecasting system will provide forecasts of still water level, near-shore wave height and wave overtopping rate. In addition, a fluvial forecast model was developed for the River Ness catchment, feeding the tidal model to create a more accurate representation of the fluvial-tidal interaction in the Inverness area. The forecasting system assumes the Inverness Flood Protection Scheme currently being developed and due for completion in April 2015 is in place and in full operation. These forecasts will allow the Scottish Flood Forecasting Service (SFFS) and SEPA's Flood Warning Duty Officers (FWDOs) to issue regional Flood Alerts and targeted Flood Warnings to specific reaches of the coastline up to 48 hours in advance, allowing local authorities, emergency responders, businesses and local residents to prepare for the forecast conditions. The forecasting system will also provide input to the Flood Guidance Statement, a 5-day flood forecast issued daily by the SFFS for Professional Partners and Category 1 and 2 Responders as defined in the Civil Contingencies Act (Scotland) 2004.

The approach adopted for the development of the flood forecasting system included the following steps:

1. Define and assess which communities to forecast for and whether wave overtopping is a key factor in each location
2. Develop a wave model for the Moray Firth, covering the entire basin
3. Perform wave overtopping modelling at specific number of locations
4. Create flood inundation maps for locations that have wave overtopping modelling
5. Create fluvial modelling component for the River Ness catchment to accurately simulate the fluvial-tidal interaction in Inverness

6. Use JBA Consulting's ForeCoast software to perform long-term performance testing utilising hindcast meteorological data

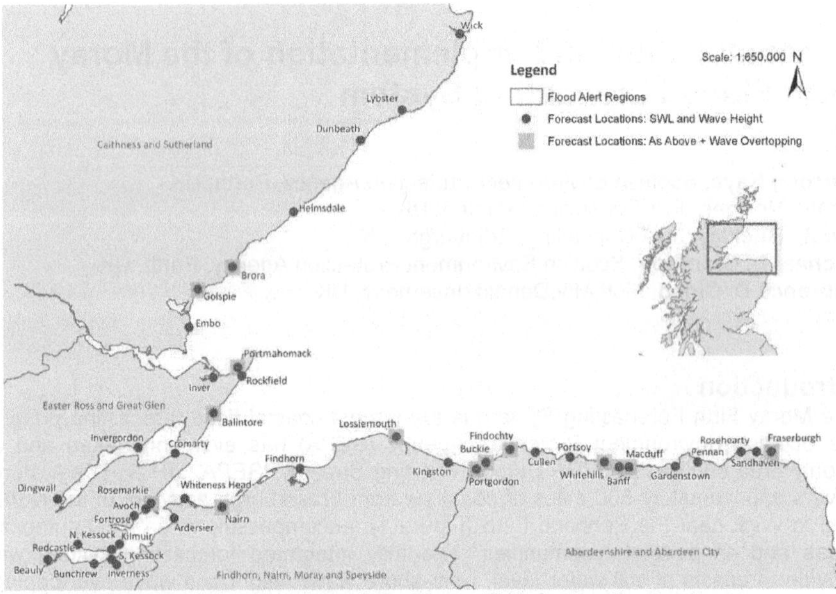


Figure 1. An overview of the Moray Firth forecasting system.

By adopting this above approach, the forecasting system should deliver the most appropriate forecasts for each location while remaining robust, comprehensive and flexible to further developments and/or adaptations.

## Historical Flooding

The winters of 2012-13 and 2013-14 saw several major coastal flood events in the Moray Firth area. In December 2012, a complex weather system pivoted over the region, allowing large waves to form in conjunction with a large storm surge, causing a provisional £2.55m worth of damage to coastal defences and property in the Highland Council region alone (The Highland Council, 2013 and Cranston *et al.*, 2013). In December 2013, a large storm surge event affected much of the UK's eastern seaboard with Aberdeenshire and Moray's exposed northern coastline suffering major damage once again as the surge propagated around the UK and Northern Europe (Flikweert *et al.*, 2015).

The two recent events from differing sources highlighted the need for a forecasting system in the area to provide vital targeted information with sufficient lead time to warn residents, emergency responders and businesses on what is at risk and when. In addition to these recent events, a wider historic review identified a number of potential dates where coastal flooding may have occurred. These dates were collated from data sources including SEPA's historical flooding database, local authority reports, storm damage reports and local media coverage available on the internet. This included over 50 events that are thought to have occurred since the early 1800s, equivalent to a frequency of around one event every 3 years. This information has been used for further validation of the forecasting system through long-term performance testing.

## System Development

The flood forecasting system is built using a suite of coupled oceanographic, nearshore and onshore models such as regional wave transformation and wave overtopping modelling. These models have been used to pre-compute the flooding consequences associated with 38,880 combinations of the key storm driving variables expected for the region, most notably tidal, storm surge, wind and wave conditions. These simulations have then been used to develop automated look-up matrices that relate offshore forecasts of sea-level and wind conditions (derived from Met Office models) to the expected coastal risk on land. These matrices are then queried by SEPA's in-house flood early warning system, FEWS Scotland, via an additional module called TRITON. This methodology was preferred over running a 'live' hydrodynamic model primarily due to the computational effort that would be required to run these in real-time, but also due to stability issues of running a model covering such a large area.

To calculate still water levels, the project team utilised the Coastal Flood Boundaries (CFB) dataset, a UK-wide project aiming "to provide values on design sea levels at a regular spacing around the coastline" (Environment Agency, 2011). Each forecast location has a parent tide gauge assigned to it and a relationship has been derived for each location between the nearest CFB points to the forecast location and the parent gauge. The average difference in sea level for the 1 year and 200 year return periods from the CFB dataset points are quantified, giving a correction factor relative to the parent gauge. This is applied to the observed and forecast sea level time series from the tide gauge to give a forecast time series at the forecast location.

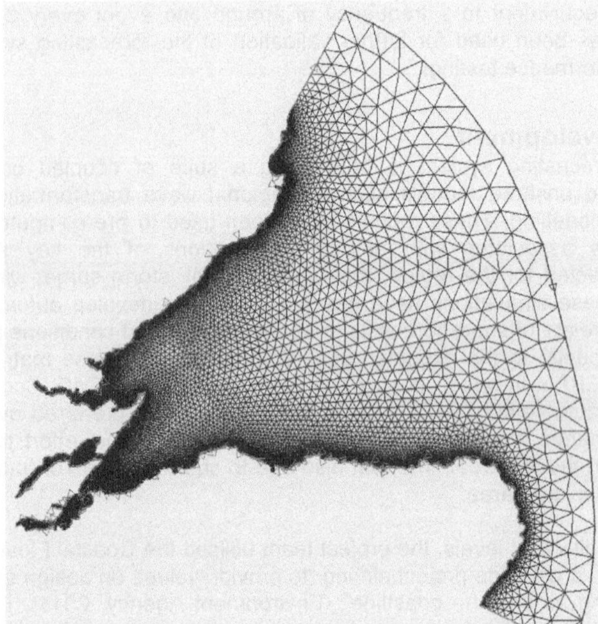

Figure 2. Triangular mesh used for wave transformation modelling.

A wave transformation model was constructed using the SWAN (Simulating Waves Nearshore) modelling software (The SWAN Team, 2014) and a flexible mesh grid (Figure 2). The flexible mesh approach allows for large spacing between grid points in the offshore area where detailed representation of the sea floor is unnecessary due to the deep water environment, and closer spacing to provide more detailed information in nearshore areas. This results in an offshore resolution of 10km and a nearshore resolution of 10m at forecast locations, supported by the latest TruDepth Grid bathymetry data available as well as survey data, LiDAR (if available) or NEXTMap topographic data. Several storm forcing variables are required to run the model, notably sea level, wind speed and direction, and offshore wave height, direction and period. A detailed list of the input parameters that make up the operational look-up matrices are listed in Table 1.

The wave transformation model was tested and calibrated using the WaveNet (Cefas, 2015) hindcast dataset, a 33 year record of hindcast wave data for UK waters from 1 January 1980 to 31 December 2012, hosted by the Centre for Ecology, Fisheries and Aquaculture Science (Cefas). Based on the WaveWatch III model (Tolman, 1989, 2009), the hindcast has been created by running the current Met Office operational wave forecast model with a wind field generated from re-analysis of historic observations. The dataset is provided on an 8km grid and contains around 6000 individual points, each of which has 48 parameters. This allowed the project team to test the wave model against more than 30 years of data, assessing its performance as if it had been in operation for this length of time. Utilising the dataset allows for greater confidence in the operational model and any thresholds set using ForeCoast for frequency analysis.

Three parameters are extracted from the wave model matrices for operational use: significant wave height ($H_s$, m), spectral wave period ($T_{m-1}$, secs), and peak wave direction (dir, degrees/N). These parameters have been collated with other forcing variables to construct the look-up matrices as described below.

Table 1: Model simulations performed to create the 38,880 matrices.

| Sea Level (mAOD) | Offshore Wave Height (m) | Offshore Wave Period (sec) | Offshore Wave Direction (deg, °) | Wind Speed ($ms^{-1}$) | Wind Direction (deg, °) |
|---|---|---|---|---|---|
| 1.50 | 0.00 | 4 | 0 | 0 | 0 |
| 1.75 | 0.25 | 8 | 45 | 5 | 45 |
| 2.00 | 0.50 | 12 | 90 | 11 | 90 |
| 2.25 | 0.75 | 16 | - | 17 | 135 |
| 2.50 | 1.00 | - | - | 25 | 180 |
| 2.75 | 2.00 | - | - | 33 | 315 |
| 3.00 | 4.00 | - | - | - | - |
| 3.25 | 6.00 | - | - | - | - |
| 3.50 | 8.00 | - | - | - | - |
| 4.00 | - | - | - | - | - |

Wave overtopping modelling has been undertaken at 14 of the 44 forecast locations, which were selected in consultation with local authorities based on historic events. The forecast location and applicable Flood Alert area are listed in Table 2. SEPA commissioned a survey project to capture beach and defence profiles to support the wave overtopping modelling, which was enhanced by some additional surveying performed by SEPA staff (JBA Consulting, 2013). Without this survey data, the project team would have had to rely on LiDAR or NEXTMap topographic data. The north of Scotland has yet to be fully mapped with LiDAR data and the NEXTMap dataset grid (5m) with a vertical accuracy of +/- 1m are too large for the required level of detailed analysis, so the survey project was key to the success and reliability of the wave overtopping element of the forecasting system.

Table 2: Locations where wave overtopping is forecast and their associated Flood Alert and local authority areas.

| Aberdeenshire Council | Moray Council | The Highland Council | |
|---|---|---|---|
| **Aberdeenshire and Aberdeen City** | **Findhorn, Nairn, Moray and Speyside** | **Easter Ross and Great Glen** | **Caithness and Sutherland** |
| Fraserburgh | Lossiemouth | Rosemarkie | Golspie |
| Macduff | Nairn | Balintore | Brora |
| Banff | - | Portmahomack | - |
| Whitehills | - | - | - |
| Findochty | - | - | - |
| Buckie | - | - | - |
| Portgordon | - | - | - |

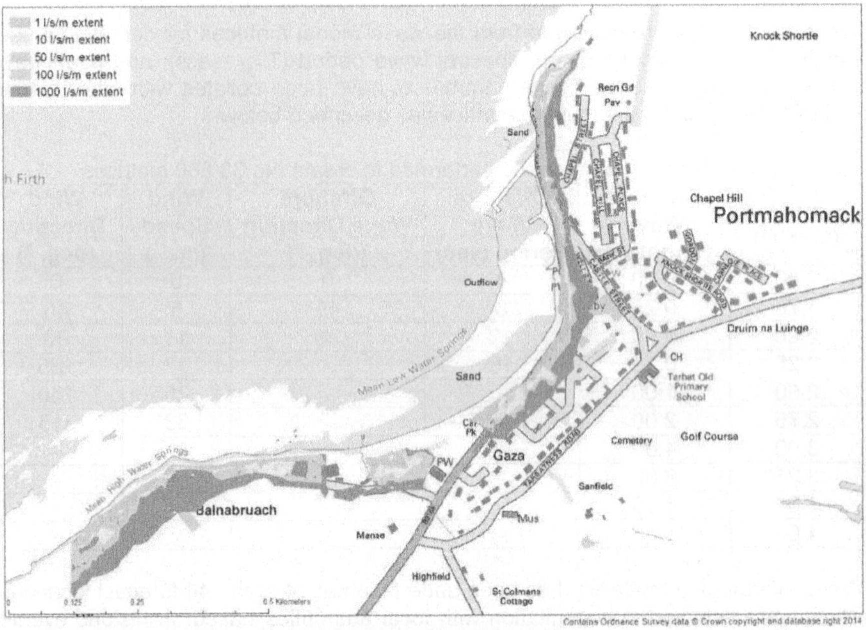


Figure 3. Inundation map of Portmahomack for a range of wave overtopping simulated events.

The European Wave Overtopping Manual (EurOtop, 2010) methodology was employed and a Neural Network (NN) approach was preferred as it provides the most suitable methodology for assessing complex multi-component defence structures (EU Commission, 2005). The NN model uses nearshore wave characteristics at the toe of a defence structure, defence geometry and sea level data to quantify a mean overtopping discharge rate, expressed in litres per second per meter of defence structure (l/s/m).

Inundation mapping at the 14 wave overtopping forecast locations has also been analysed. JFlow+ modelling software was utilised to create flood extents for a range of wave overtopping rates based on a 200 year sea level. This is a first for SEPA in a wave overtopping context and plans are in place for SEPA to perform some in-house modelling, adding more sea level return periods to this suite of maps (Figure 3) will give Duty Forecasters a far larger range of modelled extents that can be utilised for reference during a live flood event.

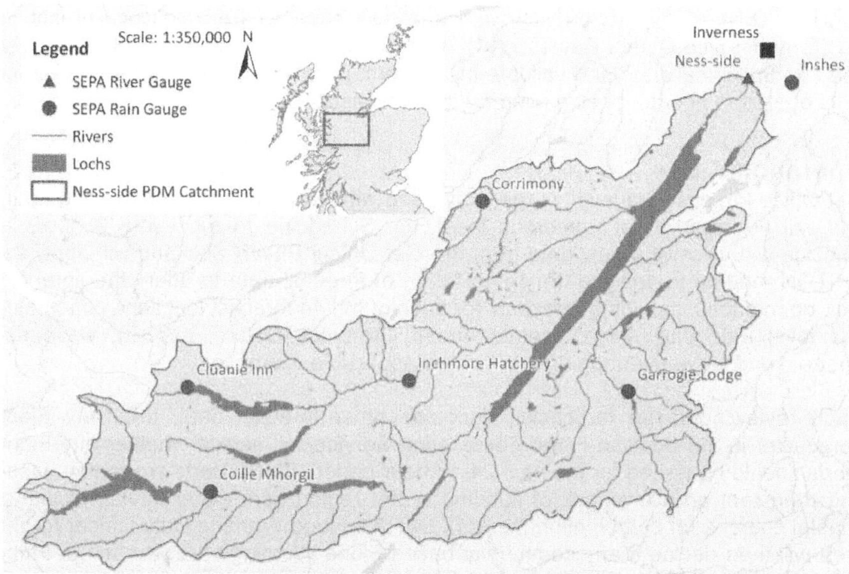


Figure 4. Schematic diagram of the River Ness catchment PDM to SEPA's Ness-side river level gauge.

There is a fluvial input to the coastal model from the River Ness, generated by a broad-scale rainfall-runoff model (Probability Distributed Model or PDM (Moore, 2007); Figure 4) coupled with a 1D ISIS routing model. The fluvial input will allow for a more accurate and representative coastal forecasting system, allowing greater confidence to inform, for example, the closure of flood gates on the River Ness Flood Alleviation Scheme (Mott MacDonald, 2013) and general incident management in and around Inverness. The River Ness Flood Alleviation Scheme (tidal section) is currently being constructed to alleviate flooding to the city of Inverness from both tidal and fluvial sources from the River Ness downstream of Ness Bridge (expected completion date April 2015). This multimillion pound scheme includes 4km of walls and embankments built through the centre of Inverness on permeable sands and gravel layers. The scheme contains 12 manually operated flood gates which will require time to put into place and close. Some gates will prevent access to the harbour and as such, any closure would likely affect local businesses and industry in this area. The forecasting system compliments the successful operation of the scheme and subsequent protection of the city from fluvial and tidal flooding.

Whilst this general approach for developing coastal forecasting systems has been in practice for some time, these systems have normally been implemented with little or no historical testing, limiting confidence in the predictions. In particular, many systems are known to perform poorly from the perspective of wave overtopping, the key source of coastal flood risk for many of the Moray communities. To optimise the performance of the system, a new approach will be applied using specialty software called developed by JBA and PlanB called ForeCoast. Using hindcast model data from the Met Office and the look-up tables generated from the ensemble modelling, ForeCoast will simulate how the system would have performed if it had been in operation historically, performing such analyses such as Peaks Over Thresholds (POT) and False Alarm Rates (FAR) amongst others. This information has been used to further calibrate and validate the system, providing much greater confidence in its operation prior to it being used for live forecasts.

## Operational Use

Met Office forecast sea level, surge, wave and wind data are received every 6 hours and will be sampled at numerous locations. Scheduled to run every 6 hours to coincide with receiving new data from the Met Office, FEWS Scotland will apply the TRITON module to analyse the combination of forecast data to query the library of look-up matrices, providing forecasts for each of the 44 forecast locations concerning sea level (tide plus surge; metres above Ordnance Datum, mAOD), wave risk (metres) and wave overtopping rate (where applicable; l/s/m).

Upon reviewing these forecasts, alongside other forecast data, the Duty Flood Forecaster in the Scottish Flood Forecasting Service will decide whether any Flood Alerts should be issued for the next 24-48 hour period. Flood Alerts are region based and represent an overall risk of flooding to the region from either fluvial, pluvial or coastal sources (or combination thereof). SEPA's Flood Warning Duty Officer for the area will then decide if any community-based Flood Warnings are warranted. Flood Warnings are site specific and targeted to each community however in this project these were grouped geographically due to the number of forecast locations. Flood Warnings can be triggered by either a forecast sea level or wave overtopping rate threshold exceedance.

Figure 5 gives an example of output presented to the Duty Flood Forecaster for Lossiemouth, a wave overtopping location, in FEWS Scotland. The time series of water level (including tide and surge), wave height and wave overtopping rate are clearly visible against their associated thresholds created during the ForeCoast analysis. This pre-set display gives the Duty Forecaster all the key information required to make an initial flood forecast assessment for this location, up to four tidal cycles in advance. In reality, the duty forecaster may wait to see if subsequent model runs continue along the same trend thereby increasing their confidence in the forecast before issuing a warning. Other data, such as wave period and wind direction, is available elsewhere in FEWS Scotland which can be used to help inform decision making. For non-wave overtopping sites, only the top two plots are shown.

Since it was launched in October 2014, the system has been activated over the winter of 2014-15, predicting impacts at various times including on the day of its launch (Press and Journal, 2014). In the first three months of the system being in operation, SEPA have issued 12 Flood Warnings based on forecasts from the Moray Firth forecasting system.

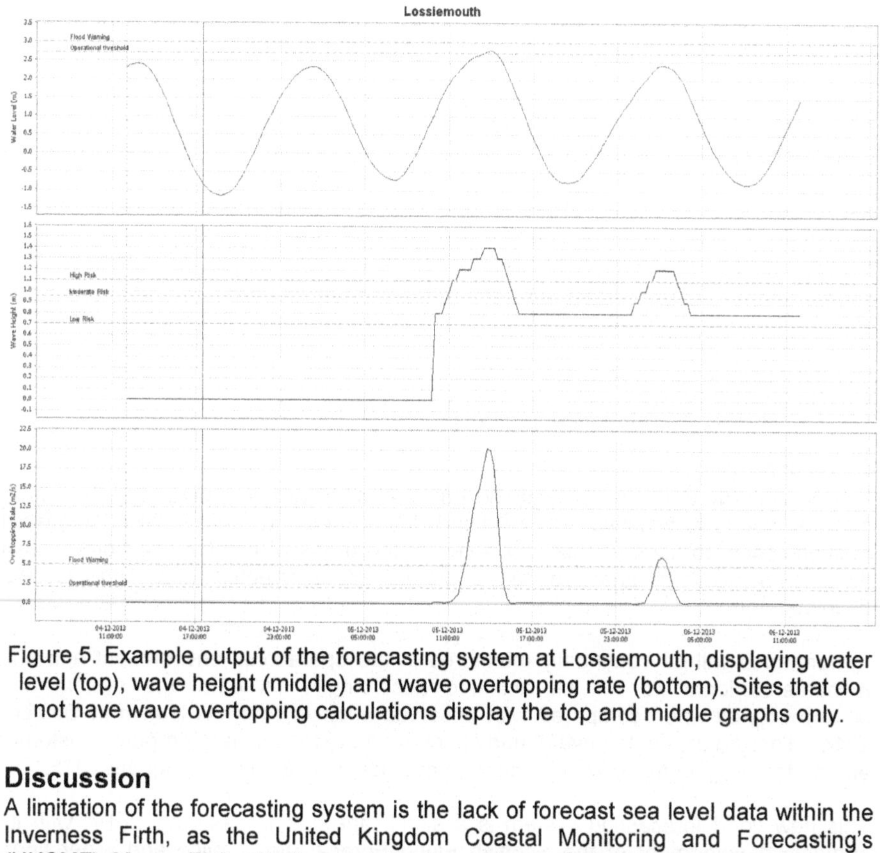


Figure 5. Example output of the forecasting system at Lossiemouth, displaying water level (top), wave height (middle) and wave overtopping rate (bottom). Sites that do not have wave overtopping calculations display the top and middle graphs only.

## Discussion

A limitation of the forecasting system is the lack of forecast sea level data within the Inverness Firth, as the United Kingdom Coastal Monitoring and Forecasting's (UKCMF) Moray Firth site is used for the astronomical time series in this area. SEPA, in collaboration with UKCMF and the National Oceanographic Centre (NOC), have recently derived an astronomical time series for the Clachnaharry tide gauge, using its 22 year record which is due to be incorporated into the forecasting suite in early 2015.

As with any flood forecasting project, periodic review of the forecasting system will be required to assess accuracy and performance. This may include continual assessment of thresholds compared with on the ground observations and post-event analysis, as well as identifying potential improvements to the forecasting system in response to new data collection methods, adjustments to existing or installation of new flood defence schemes, or resurveying of beach profiles for wave overtopping modelling.

The geometry of a beach and how it couples with any defence structure is pivotal to how wave overtopping occurs and behaves. Beaches are in a constant state of flux and although the degree of change varies with location, a periodic resurveying of beach and/or defence profiles should be commissioned at regular intervals. SEPA will need to work closely with the associated councils, the responsible authorities for the maintenance of flood defence structures in Scotland, to ensure the forecasting system is modelling what is currently in situ, as far as technology will permit.

Some further enhancements to the system approach include the use of ensemble forecasts and combined risk assessment. At present the system does not integrate surge ensembles which could allow for the quantification of any uncertainty in surge forecasts. SEPA is currently investigating its digital network structure and any associated resource implications, with the aim of including this data into the system by mid-2016. This would give the Duty Forecaster a vastly improved suite of time series, promoting increased comprehension of the uncertainty in decision making while temporally increasing this decision making process. SEPA are also investigating the merits of a combined risk based approach to triggering warnings. This is envisaged to account for a joint still water level and wave risk, rather than thresholds on the individual parameters. This method will allow FWDOs to assess the combined flood impacts for decision support purposes.

## Conclusion

The new Moray Firth Forecasting System is a major progression in approaches to coastal flood forecasting in Scotland. The new scheme offers improved methodologies to flood warning for some of the areas of highest risk from coastal flooding in Scotland and are based on the latest scientific and computational developments.

Performance will be continually monitored in the operational environment and any refinements identified will be actioned accordingly. To date, the system has performed well when compared with observed impacts since its launch in October 2014. Minor adjustments have already been made in response to on-going customer and partner engagement, with further improvements scheduled throughout 2015.

This new flood forecasting system is a significant step in assisting SEPA with its strategic development of the Scottish Flood Forecasting Service and wider flood warning operations.

## Acknowledgements

The authors would like to thank the support of the Scottish Government, The Highland Council, Moray Council and Aberdeenshire Council for their financial support in this project.

## References

Cefas (2015) "WaveNet", http://www.cefas.defra.gov.uk/our-science/observing-and-modelling/monitoring-programmes/wavenet.aspx (cited 18 January 2015).

Cranston, M., Naysmith, L., Tavendale, A., McLean, L., Hart, P., Whitlow, C., Powis, C. and Hu, K. (2013) "The Firths of Forth and Tay flood warning scheme and the storm of December 2012."

Environment Agency (2011) "Coastal flood boundary conditions for UK mainland and islands", https://www.gov.uk/government/uploads/system/uploads/attachment_data/file/291216/scho0111btki-e-e.pdf (cited 18 January 2015).

EU Commission (2005) "Crest Level Assessment of Coastal Structures by Full-Scale Monitoring, Neural Network Prediction and Hazard Analysis on Permissible Wave Overtopping – Neural Network Manual NN_Overtopping 2", WL Delft Hydraulics, June 2005.

EurOtop (2010) "Wave Overtopping of Sea Defence and Related Structures: Assessment Manual", Overtopping Edition, HR Wallingford, November 2010.

Flikweert, J., Cranston, M., Woolhouse, C., Sorenson, P., Muller, J., Vonk, B. and Peeters, P. (2015) "Mandela, Sinterklass, Xavier, Bodil – Impacts and lessons from the December 2013 Storm around the North Sea." 2015 ICE Coastal Management Conference, Amsterdam, 9-11 September 2015.

Highland Council (2013) "Storm Damage – 14/15 December 2012", http://www.highland.gov.uk/download/meetings/id/1182/item_8_storm_damage_%E2%80%93_1415_december_2013 (cited 18 January 2015).

JBA Consulting (2013) "Moray Coastal Survey Report", undertaken by JBA Consulting on behalf of SEPA.

Moore, R.J., (2007) "The PDM rainfall-runoff model", Hydrology and Earth System Sciences, 11 (1) 483-499.

Mott MacDonald (2015), "River Ness Flood Alleviation Scheme – Operation and Maintenance Manual (Rev. B)", Mott MacDonald for The Highland Council.

Press and Journal (2014) "Moray Firth flood scheme launched", https://www.pressandjournal.co.uk/fp/news/aberdeenshire/366501/moray-firth-flood-scheme-launched (cited 18 January 2015).

Scottish Government (2005) "The Civil Contingencies Act 2004 (Contingency Planning) (Scotland) Regulations 2005", http://www.legislation.gov.uk/ssi/2005/494/pdfs/ssi_20050494_en.pdf (cited 18 January 2015).

SEPA (2011) "Flood Warning Strategy 2012-16", http://www.sepa.org.uk/flooding/idoc.ashx?docid=694c7071-519d-4967-835d-9bd087c18d64&version=-1 (cited 18 January 2015).

SWAN (2014), "Scientific and Technical Documentation – SWAN Cycle III version 41.01", http://swanmodel.sourceforge.net/online_doc/swanuse/swanuse.html, Delft University of Technology, (cited 18 January 2015).

Tolman, H. L., (1989) "The numerical model WAVEWATCH: a third generation model for the hindcasting of wind waves on tides in shelf seas." Communications on Hydraulic and Geotechnical Engineering, Delft Univ. of Techn., ISSN 0169-6548, Rep. no.89-2, 72 pp.

Tolman, H. L., (2009) "User manual and system documentation of WAVEWATCH III version 3.14 (MMAB Contribution No. 276)" http://polar.ncep.noaa.gov/mmab/papers/tn276/MMAB_276.pdf (cited 18 January 2015).

**Coastal Management: Changing coast, changing climate, changing minds**
ISBN 978-0-7277-6114-9

ICE Publishing: All rights reserved
doi: 10.1680/cm.61149.587

# Innovative Approach in the Stabilisation and Erosion Protection of Coastal Slopes in Scotland

**Slobodan B. Mickovski**, School of Engineering and Built Environment, Glasgow Caledonian University, Glasgow, Scotland, UK
**Craig S. Thomson**, School of Engineering and Built Environment, Glasgow Caledonian University, Glasgow, Scotland, UK

## Introduction

The significant and lasting change in the statistical distribution of weather patterns over long- and short-term manifesting as a climate change (IPCC, 2013) affects all structures especially the ones on the interface between air, soil, and water such as the coastal slopes (e.g. natural or engineered; Meyer et al 2010). The design and management of these structures is traditionally based on less likely, detrimental, assumed climate characteristics that, if they happen, will bring relatively insignificant change in terms of affecting the stability of the structure.

The stability of coastal slopes, in general, depends on the spatial and temporal distribution of the weather patterns, especially rainfall and wave action which are often triggering soil mass wasting in terms of erosion and landslips (The Scottish Executive, 2005; Minder et al 2009). In order to minimise the risk to life and property on and near coastal slopes, the potential long and short term impacts of climate change should be taken into account when developing the design for stability and long-term management. Despite this, only a few theoretical frameworks have been reported in the literature (e.g. Iverson 2000; Wong et al 2004) that account for erosion and landslides as the most frequent forms of mass wasting on slopes. These are based on the analysis of historic events and systematic data collection in order to predict future behaviours, but significantly lack focus on coastal slopes.

In the current financial climate when the engineers and managers are tasked with providing 'more for less' or 'value for money', the main concern encompasses not only the traditional concern for slope stability (Transport Scotland, 2008) but also the sustainability and resilience of the design and management plan (Mickovski, 2014b). Apart from the financial aspects such as the available budget, projects are required to increasingly consider both social aspects such as public participation, risk, and social resilience and environmental aspects closely connected to climate change such as whole life costing and carbon footprint. Current local, national, and trans-national strategic planning strategies seek to promote a cultural change and target investment through a proactive and inclusive approach which pulls stakeholders together to promote dialogue between the strategic level (policy makers, regulators and funders), project teams and local communities (McFadden et al., 2009). Such approach to coastal slope management has the potential to promote double loop learning (Voss and Wanger, 2010) and could widen the evidence base for design and management (Astley-Reid, 2013; Thomson et al. 2014), promote mediation, transparency and permit the local system to question and share their learnt experiences and ultimately enhance the ability to respond to complex briefs within constrained budgets. This culture is advocated to promote innovation and avoid over-engineered projects which are costly and disproportionate and instead support solutions which are accepted and sensitive to local needs (Schmidt-Thome and Schmidt-Thome, 2007).

This study aims, through case study analysis, to demonstrate the current coastal slope engineering and management strategies in Scotland by investigating the past interventions and, through qualitative research, to explore the change in attitude of the involved parties with development of the sustainability paradigm. The innovative approach of using eco-engineering techniques for stabilisation and management of coastal slopes is dissected in the light of risk, sustainability, and resilience, as well as stakeholder engagement, innovation, and double-loop learning. A stabilisation and management framework in the context of risk and learning is proposed as the main focus of this paper is on using double loop learning in the development of innovative strategies for design and management of coastal slopes in a changing climate where ownership, community engagement and funding availability are the most important factors which shape the potential for innovation.

## Methodology

To outline the types of problem associated with the management of the coastal slopes in Scotland, we used a case study approach to establish an understanding of current practice. Comparing and critically analysing two case studies within a similar locality, they experience consistent instability problems, planning regulations, local culture and attitudes, many of the same stakeholders, and similar climate and topography. This approach allowed us to make an empirical inquiry analysing engineering, managerial, and social concepts and phenomena in their "real-life context" (Yin, 1994). As our research is problem-based, this descriptive and explanatory approach provides a pragmatic analytic framework around which to explore the complexity and challenge of delivering coastal slope management analysing the design, decisions, policies, institutions and stakeholder engagement.

Two study areas in the North-east Scotland (Fig.1 and Fig.2) were chosen to illustrate the problems associated with the instability of coastal slopes and the management of remediation both in the short- and long-term providing stability and resilience. Published information was extensively studied and a number of site visits were carried out within a four-year period to monitor the processes occurring on the slope and the management strategies applied. Semi-structured interviews with a range of stakeholders such as the Local Authority (LA), engineers, and members of the communities adjacent to the study sites were also conducted within this period with key themes emerging from the voices of the parties involved.

## Case Studies

**Bervie Braes (Bervie)** comprises a 850 m long, 55 m high vegetated coastal slope west of the old harbour in Stonehaven, Aberdeenshire. The former trunk road connecting Stonehaven to Dunnottar Castle and A94/A957 road runs sidelong and generally NW-SE across the slope (Fig.1). Around 60 residential properties lie towards the toe of the slope with the residents using the Braes for pedestrian access to the road and the nearby tourist attractions. The slope profile above the road is made up of a series of hollows, mounds, and bulges which represent washouts from ephemeral springs, material failed during one or more historical instability events (Table 1) and areas of ongoing creep respectively. On the slope below the road a visible seepage line is visible around the locations of former public wells used to provide drinking water for Stonehaven. The overlapping of old, apparently inactive, and active areas of slope instability gives the slope with its hummocky appearance (Currie et al. 2009). A watercourse runs from the agricultural land above the Braes down a gulley in the slope and was culverted beneath the road near the north-western end of the study area leading to the adjacent burn which, in turn, partly causes periodic flooding to the old town of Stonehaven.

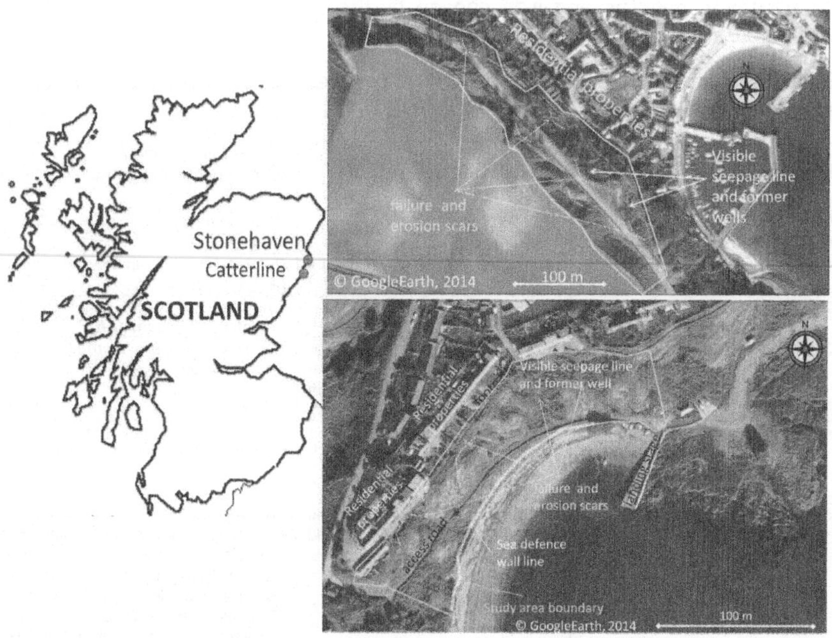


Figure 1: Location of the study sites (left). General arrangement and instability characteristics of the coastal slope in Stonehaven (Bervie, upper right) and Catterline (lower right). Source: Google Earth, 2014.

**History of Problem and Management**

Previous instability and erosion events on the slope (Table 1) were indicated to be the consequence of shallow groundwater, oversteepened ground profiles (Currie et al 2009) and extreme rainfall events. The Local Authority (LA) historically used traditional engineering measures to repair failures of the slopes adjacent to the road (Table 1) that they own albeit the Braes is owned by a charitable trust that has no direct funding for upkeep and maintenance of the location. The LA declined adopting the Braes from the trust due to potential risks associated with instability that would impact the residential properties at the toe of the slope. In 2010 the LA together with the Scottish Government raised limited funds to repair the damage to the road embankment following the most recent instability events in 2009 and decrease the risk of slope instability for the future. They appointed an Engineer and involved the residents and the Community Council in the planning and design process (Mickovski et al 2013). Innovative stabilisation works including structural and eco-engineering stabilisation (Mickovski, 2014b) were completed in 2013 and the road partially reopened the same year for cars (one way) and pedestrians/cyclists. The eco-engineering methods employed included engineered planting and seeding of selected native vegetation in slope areas of lower failure risk which was supplemented by soil nails recessed into the slope and covered by bio-degradable pre-seeded soil mix in the areas of higher risk of failure. The design took into cognisance the potential climate change effects in terms of changes in temperatures and precipitation patterns and was chosen by the community at risk of instability (Mickovski, 2014a and 2014b).

Table 1. Timeline of instability events at Bervie Braes. (LA) – Local Authority, (R)- Residents

| Year | Event | Consequence | Action (carried out by) |
|---|---|---|---|
| 1958 | Major slip above the south pier of the harbour | Collapse of a road section | Anchored sheet pile wall installed (LA) |
| 1950s – 1970s | Drains and water supply channels blocked | Vegetation overgrowth | Burning vegetation on the slope at the end of the summer (R) |
| 1980s | Slope failure at far western end | Major damage to residential property | Debris cleared, damage reinstated (R, LA) |
| 1992 | Distortion of the carriageway surface | Cracked pavement | Temporary single carriageway operation introduced (LA) |
| 1993 | | Localised bulging of the slope and development of longitudinal cracking along the footpath on the embankment | Permanent closure of road in one direction (uphill) (LA) |
| 1994 | Slope failures above and below the road due to heavy continuous rainfall over a period of 4 days | Road blockage | Debris cleared; full road closure between February and July (LA) |
| 2009 | Surface washouts and mud flows at the south-east end; major slip at the far western end | Undermined road embankment; major damage and abandonment of residential property | Full road closure; commissioning of a forensic study to determine the relationship between the local groundwater regime, including perched water levels, the locations of weak, sensitive soils and the movement and progressive failure of the road (LA) |

**Catterline Braes (Catterline)** is a 300 m long, 30 m high coastal slope forming the NE end of Catterline Bay in Aberdeenshire, Scotland (Fig. 1) located along the coast 10 km south of Bervie. Catterline comprises a series of exposed, mainly vegetated, slope sections interspersed with gully-outcrop forms ranging in height between 2 and 10 m. The slope is vegetated with a range of vegetation types including grasses, herbs, shrubs and planted trees covering the slope to a degree varying seasonally (approximately 20% in winter to approximately 80% during the vegetative period). A small tarmaced access road traverses the slope from SW to NE dropping in level towards the concrete landing stage on the NE end of the bay. A gravel drain runs along the road discharging into the sea near the landing stage. Private residential properties are located along the crest of the slope with a grassed, non-fenced public footpath offsetting them from the slope. A branch of the footpath leads to a midslope structure – former pumping station used in the past for fresh water supply to the adjacent properties. A seepage line is visible adjacent to the pumping house, as well as 25 m north of the pumping house, indicating a series of springs with seeping water flowing along naturally formed gullies towards the gravel drain along the access road which, in turn, discharges into the sea at the landing stage which is used by the local diving club. Remnants of a sea wall defence comprising sections of rounded boulders, mass-poured concrete, anchored gabions, and concrete cubes exists at the toe of the slope and delineates the slope from the beach. The line of the sea wall is damaged to various degrees due to the action of the waves and seepage from the slope.

**History of Problem and Management**

Landslide activity along the slope (Table 2) has been historically linked to the springs, surface water flow, and areas of ongoing creep (Fig.2a) and the LA reactively employed traditional engineering strategies for slope stability together with the residents albeit the slope has effectively no owner. Infrastructure belonging to water utility stakeholder is located on the slope, while no ownership of the seawall defence could be established. In the 1980s, the slope was owned by a trust founded by a resident but after the trust went into liquidation in the early 1990, the ownership status has not been resolved.

Table 2. Timeline of instability events at Catterline Braes. (LA) – Local Authority, (R) - Residents

| Year | Event | Consequence | Action (carried out by) |
|---|---|---|---|
| 1940 | Slips at slope toe | Coastal erosion | Mass concrete and concrete block sea defence wall built along a section of the beach (LA) |
| 1950s – 1970s | Drains (ditches) and water supply channels blocked | Vegetation overgrowth | Burning vegetation on the slope at the end of the summer by the residents; cleaning of the drain channels by the Local Authority (R) |
| 1965 | Slip at the crest of the slope | Mass wasting, buried existing well and pump on the slope | Manual clear up of the failed soil; extended sea defence wall by anchored gabions (R; LA) |
| 1980 | Sheet erosion of the slope | Surface water flow on the slope and access road | Drains trench built along the access road (LA) |
| 1983 | Debris flow | Mass wasting on the access road | Manual clean up; willows planted in a small section of the slope (R) |
| 1994 | Major slip at slope crest | Subsidence, soil mass wasting | Sheet piles installed to support ground in front of the house; attempt at ground investigation and monitoring (LA) |
| 1995 | Major slip at slope crest at the same location as in 1994 | Subsidence, soil mass wasting; blocked drains | Excavation into the slope and installation of gabion baskets (LA) |
| 2000 | Slip at slope crest near the entry to the access road | Blocked access road | Slope re-graded and anchored geotextile installed (R) |
| 2005 – 2012 | Slip along the embankment of the access road | Subsidence of a section of access road; cracking in the pavement; drains dysfunctional | Subsided areas 'topped up' with tarmac and surface water directed over the pavement and into the embankment (LA) |
| 2013 | Damaged sea wall defence and slips of the slope behind the wall | Subsidence of the access road, mass wasting and erosion of soil behind wall | Reinstatement of the damaged concrete blocks; attempt at ground investigation on the slope (LA) |

During a storm in 2013, high waves topped the seawall, eroded the lower slope and overturned and dislocated sections of seawall defences. This collapse allowed the movement of the soil on the slope (Fig.2b) which resulted in subsidence of the access road embankment. It is possible that this failure can migrate upslope and lead to further instability, potentially affecting the properties at the crest. After this event, the local community asked for help from the LA who due to limited budget and perception of low risk, commissioned an earthworks contractor to repair the road and reinstate the seawall which got damaged by the sea action later the same year, leaving large portion of the slope toe exposed to the waves. The local community then started to look for help and assistance elsewhere to minimise the risks to their property and life. An innovative approach based on the lessons learned from Bervie to see at least mitigation of the current instability is currently sought by the community. Such approach would be aimed at employing eco-engineering strategies such as the sustainable use of vegetation to provide erosion protection and resistance against shallow landslides which can be combined with structural stability measures in the areas of highest risk of instability (Mickovski, 2014a).

Figure 2. (a) Shallow translational slides and mudflow and (b) deeper slides behind the seawall defence.

## Analysis and Discussion

### Innovative Design and Management

The analysis of the history of problem and management showed that the case study sites have similar geology, climate, and environment, which allows the innovative design and management of stabilisation measures and experiences at Bervie to be used at Catterline to minimise the risks and provide resilience. The risk of shallow and deep failures as well as erosion and flooding associated with the effects of climate change (Mickovski 2014b) exist at both sites and affect the residential properties and infrastructure alike. The design applied at Bervie (Figure 3a), where the resistance to catastrophic landslides is provided by soil nails and a reduction in erosion and increased resilience with easier recovery from any potential shallow slips is provided through sustainable use of vegetation, could be modified at Catterline where a new or repaired and enhanced seawall defence (Figure 3b) designed to adapt to impacts of changing climate is combined with eco-engineering strategy that will mitigate the effects of historical instability on the slope and provide resilience in the long term. This innovative solution has an added sustainability value in terms of minimal maintenance in the long term and increasing the recreational value of the coastal slope such as pedestrian access to tourist attractions at Bervie and local fishermen and diving club at Catterline.

The interviews with the relevant parties at both locations showed that for innovative solutions to be delivered significant management support is required to capture the risk complexities, local needs and financial constraints, in order to foster creativity. The approach of integration of a slope stabilisation project within a risk management framework could be implemented at Catterline based on the promotion of experiential learning from Bervie (Figure 4) which is a change from the way these problems have been managed in the past (see Table 1 and 2) and in line with the current (trans-)national planning strategies.

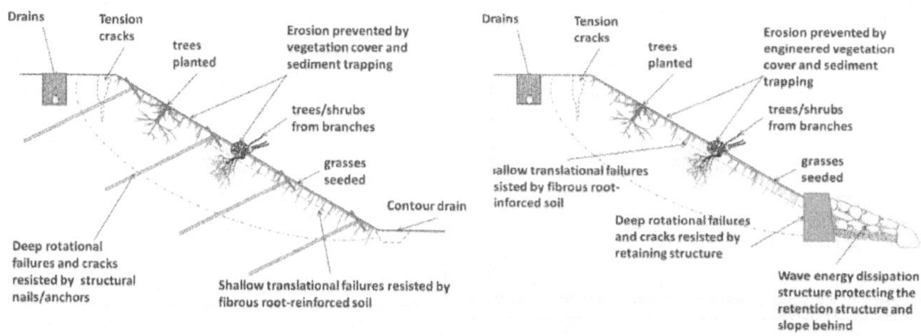


Figure 3. Integrated engineering solutions that a) was applied at Bervie b) could be applied at Catterline.

**Sustainability and Risk**

Reflecting the current state of management of unstable coastal slopes in Scotland, the two case studies showed that risks to life and property were the main drivers for actions leading to stabilisation but identified site ownership and funding for works as constraints in the process. The ownership status of the sites presented limitations which without community leadership could stall the project due to uncertainty. At Bervie the LA took ownership of the slope failure risks due to historical ownership of the road on the slope and the stabilisation works were procured. At Catterline, only the water authority owns assets but has not engaged in consultation to date and, in absence of a risk owner the community is forced to explore sources of funding for stabilisation works as a community project. In developing a community project, ownership problems can be overcome if a case for building resilience and aiding the common good can be established as this is basis for funding applications (Astley-Reid et al. 2013). Supporting communities to self-organise and represent themselves as they see fit is important, and evidence suggests it permits discourse and representation which is appropriate and meaningful (Manzo and Perkins 2006).

The limited funding raised at Bervie precluded stabilisation of the whole slope below and above the road. The Engineer had to further assess the risks of instability and prioritise areas for stabilisation, resulting in stabilisation of 60% of the slope below the road. The rest of the slope is remotely monitored and stability re-assessed during/after any critical rainfall event as part of the Emergency Plan developed by the LA and the Engineer in cooperation with the residents. At Catterline, the community managed to cover ca. 40% of the slope with planting more than 420 trees in priority areas as specified by the Engineer through a grant from Woodland Trust showing that community leadership plays a significant role in promoting involvement in the design process, aiding the local needs reflection and arriving at an appropriate solution which reflected the perceived risk and funding context. However, long-term monitoring and emergency planning at Catterline will depend on research interest from the academia and the motivation of the residents due to lack of funding.

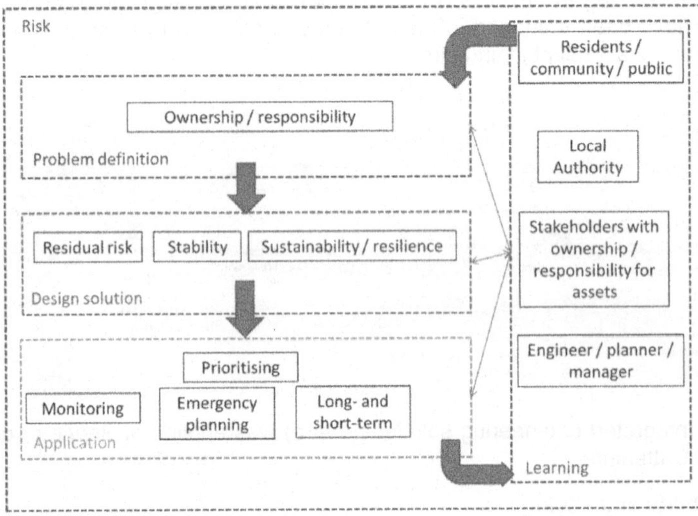


Figure 4 Proposed stabilisation and management framework in the context of risk and learning as used at the case study sites.

**Stakeholder Engagement**

Both case studies demonstrated that projects which are inclusive of stakeholders at a strategic, project, and community levels enhance evidence provision aiding the awareness, identification and understanding of the risk presented. This facilitated the consideration by engineers of suitable design solutions, and when shared with the community this promoted discourse, mediation of views, their input to help evolve the design and ultimately acceptance (Mickovski 2014a,b). The discourse between the stakeholders at Bervie ensured that they were receptive to ideas, informed about the potential risks, and responsive to the construction and maintenance plans for the slope which could not be achieved at Catterline due to a non-responsive stakeholder. The inclusive nature of the Bervie project promoted a culture which was capable of supporting a creative design process, but also aided to achieve mediation, learning and a trust in the engineer's technical judgement.

The varied nature of stakeholder engagement approaches at both study sites were important with a mix of face to face meetings, informal discussions, community public meetings and presentations to visually represent the design to the community. Levels of engagement were observed in the interviews with parties from both sites to be significant compared to similar projects due mainly to strong self organisation, interest and a willingness to be involved. On the other hand, building relationships, establishing trust and enabling openness during discussions helped the engineers to reflect local knowledge but also facilitated community understanding and acceptance of the proposals and their short- and long-term implications. In the early stages of a project this not only informs its development but aids the creativity process required for innovation in design, which otherwise may not have been possible. At Bervie, the Engineer and LA demonstrated motivation and desire to innovate and develop new stabilisation and management solutions to accommodate the budgetary constraints. **Double loop learning:** The stakeholder engagement resulted in a set of experiences structured in such a way that lessons learned were an opportunity to improve further similar experiences. Similarly, the residents at Catterline reported the benefits of the learning process based on the experiences at Bervie, the involvement of the same Engineer and the LA while being directly involved in the procurement and design processes. On the other hand, the Engineer and LA, together with other stakeholders involved in these projects benefitted from the local experiences and knowledge which helped in identifying the problem and potential solutions and creating possibilities for translating these to similar circumstances elsewhere (e.g. Catterline) as experiential learning (Yin, 1994) (Figure 4).

The approach to managing the coastal slopes in both of the case studies (Figure 4) highlights the need for a dynamic learning environment which ensures that the lessons learnt are carried forward by all stakeholders as they reflect on the project. A key characteristic to promoting this environment is for a project team, stakeholders and the community to be willing to explore ideas and solutions in an open and transparent manner, thus leading to solutions which are tailored and accepted for the local context. The Bervie case study reflects an environment which gets close to this, with the evolution of the design aided by an inclusive culture which promotes an exchange of knowledge and opportunities for social and mutual learning. The Catterline case study demonstrated a team who recognised the benefits of providing space for creative thinking and worked with stakeholders (project and community) to achieve implementation. Interviews revealed that realising double loop learning within the design stage and management of stabilisation projects is not reflective of wider practice. Its value within the Catterline project was highlighted by the delivery of an innovative solution within a project which lacks the same level of public attention or pressure for a solution as at Bervie.

In conclusion, our research showed that the awareness of the risks, potential solutions and mechanisms of funding is critical in the initial stages of a community project. Evidence shows that an engaged community has a better potential to find solutions within limited funding regimes based on an understanding of the risks and willingness to accept change once the risk ownership is clear. This can promote innovation in design but it can also, if the community or stakeholders are not engaged, stifle innovation. Knowledge sharing between academia, industry and community in regional, national, and international context should be encouraged in the management of coastal slopes because the particular technical solutions often can be site-specific but the generic design and management principles can have a wider application and trans-boundary significance.

## References

Astley-Reid, T., Burch, J. and Thomas, K (2013) Community-led Flood and Coastal Management Projects Supporting Adaptation and Localism in Suffolk and Essex, [available online at] http://www.suffolkcoastandheaths.org/assets/Projects--Partnerships/Stour--Orwell/Community-led-approaches-July-13.pdf (accessed 30/01/15)Schmidt-Thome´ P and Schmidt-Thome´ K (2007) Natural hazards and climate change:

stakeholder communication and decision-making processes. "Management of Environmental Quality: An International Journal", 18(3), 329-339.

Currie, C. Murdoch, W. Sallis, D. Smith, M. (2009) Forensic investigation of a failed road at Bervie Braes, Stonehaven - models and mechanisms. In: Forensic Engineering - From Failure to Understanding, B. Neale (ed.), Thomas Telford , London , 325-334.

IPCC (2013) Climate Change 2013 - The Physical Science Basis. Intergovernmental Panel on Climate Change, Switzerland.

Iverson RM (2000) Landslide triggering by rainfall infiltration. Water Resources Research, 36(7), 1897-1910.

Journal of Planning Literature, 20(4), 335-350.

Manzo, L. and Perkins, D. (2006) Planning Common Ground: The importance of place attachment to community participation and planning,

McFadden, L., Penning-Rowsell, E., Tapsell, S. (2009) Strategic coastal flood-risk management in practice: Actors' perspectives on the integration of flood risk management in London and the Thames Estuary, "Ocean & Coastal Management", 52, 636–645.

Meyer M.D., Amekudzi, A., O'Hare, J.P. (2010) Transportation Asset Management Systems and Climate Change, Transportation Research Record 2160,12-20.

Mickovski, S.B. (2014) Resilient design of landslip prevention measures: a case study. Forensic Engineering [DOI: 10.1680/feng.14.00001]

Mickovski, S.B. (2014) Why is the future ready for environmental geotechnics?, Environmental Geotechnics. [DOI: 10.1680/envgeo.13.00107].

Mickovski, S.B. Black, J.D, & Smith, M.J. (2013) Innovative use of ECC (NEC3) for procurement and management of infrastructure projects with limited funding: Bervie Braes case study In: Procs 29th Annual ARCOM Conference, 2-4 September 2013, Reading, UK, eds. S.D. Smith & D.D. Ahiaga-Dagbui, Association of Researchers in Construction Management, 799-808.

Minder JR, Roe GH, Montgomery DR (2009) Spatial patterns of rainfall and shallow landslide susceptibility, Water Resour. Res., 45, W04419.

Norris, J.E. Stokes, A. Mickovski, S.B. Cammeraat, E. van Beek, R. Nicoll, B.C. Achim, A. (Eds.) (2008) Slope Stability and Erosion Control: Ecotechnological Solutions, Springer, pp 290.

The Scottish Executive: Scottish Road Network Climate Change Study (2005) (eds). R.M. Galbraith, D.J. Price & L. Shackman. The Scottish Executive, Edinburgh.

Thomson, C, Mickovski, S and Orr, C (2014) Promoting double loop learning in flood risk management in the Scottish context In: Raiden, A B and Aboagye-Nimo, E (Eds) Procs 30th Annual ARCOM Conference, 1-3 September 2014, Portsmouth, UK, Association of Researchers in Construction Management, 1185-1194.

Transport Scotland: Scottish Road Network Landslides Study – Implementation (2008) (eds.) M. Winter, F. Macgregor & L. Shackman. TS, Edinburgh.

Voss M and Wagner K (2010) Learning from (small) disasters, Natural Hazards 55, 657-66.

Wong HN, Ko FWY, Hui THH, (2004). Assessment of Landslide Risk of Natural Hillsides in Hong Kong, Special Project Report SPR 5/2004, Geotechnical Engineering Office, Hong Kong.

Yin R. (1994) Case study research: design and methods. Thousand Oaks, CA: Sage Publishing.

**Coastal Management: Changing coast, changing climate, changing minds**
ISBN 978-0-7277-6114-9

ICE Publishing: All rights reserved
doi: 10.1680/cm.61149.597

# The Sand Motor: Building with Nature in Progress

**Carola van Gelder-Maas**, Rijkswaterstaat, Lelystad, The Netherlands
**Carrie de Wilde**, Rijkswaterstaat, Lelystad, The Netherlands
**Sarah Marx**, Rijkswaterstaat, Lelystad, The Netherlands
**Matthieu de Schipper**, Delft University of Technology, Delft, The Netherlands

## Introduction

Countries like the Netherlands, which are located partly below sea level, must defend themselves against coastal erosion and flooding. With projected climate change causing sea-level rise, and the increase in economic investments near the coast, this is becoming an increasingly pressing issue. More sustainable strategies are needed, especially given the prospect of sea-level rise.

The Sand Motor is a pilot project implemented by Rijkswaterstaat, the provincial authority of South Holland, research institutes and the private sector. It is a new multi-disciplinary approach to sand nourishment that provides natural coastal maintenance while, at the same time, creating an area where leisure activities and nature can thrive. The pilot project was organized to determine whether such a sustainable approach to coastal maintenance actually works in the short and long terms.

The Sand Motor nourishment serves a threefold purpose:

1. the enhancement of coastal protection in the long term;
2. knowledge development and innovation in the field of coastal management and protection;
3. nature and leisure development by widening beaches and dunes.

The Sand Motor is a peninsula comprising approximately 20 million cubic metres of sand (fig 1). It is located on the Dutch coast to the south of The Hague. The thinking behind the project was that natural elements such as the wind, waves, tides and currents will spread the sand along the coast of South Holland, replenishing beaches and dunes without the need for more frequent dredging and pumping operations. This approach of using the forces of nature to one's advantage is aptly referred to as "Building with Nature". The peninsula was initially hook-shaped. The sand is being transported in the direction of the dominant currents and so, over time, the peninsula will acquire a natural look, resembling an intertidal flat. In coastal regions where there is structural erosion, as seen on some sections of the Dutch coasts, regular sand replenishment is needed to maintain the coastline. One of the planned benefits of the new strategy – the deposition of a large amount of sand in a single operation – is that the vulnerable seabed ecology will not suffer repeated disruption because, if the Sand Motor works as expected, no further beach nourishment will be required on this stretch of coast for the next 20 years.

Figure 1: The Sand Motor shortly after construction in July 2011 (Rijkswaterstaat, Joop van Houdt)

To monitor whether the Sand Motor is developing in accordance with predictions and fulfilling the three objectives listed here, it is being extensively monitored. The first firm conclusions about progress will be possible in 2016 and 2021. Nevertheless, initial observations indicate that positive results are being achieved in the areas of coastal defence, knowledge development, nature and leisure activities. This knowledge can be applied to new "sandy strategies" like the Sand Motor in the Netherlands and other countries (Stive et al., 2013).

**Close Collaboration between Government, Research Institutes and the Private Sector**

The Sand Motor is a major innovation project bringing together the government, research institutes and the private sector. They have joined forces to assess the feasibility of this new form of coastal maintenance based on building with nature.

This is a unique partnership. The parties are looking beyond their own backyards, taking each other's interests, and those of the end users, into account. A steering group made up of representatives from the universities involved, the provincial authority of South Holland, EcoShape and Rijkswaterstaat was established to ensure the project runs smoothly.

*Government authorities*

The Ministry of Infrastructure and the Environment (working through Rijkswaterstaat) and the provincial authority of South Holland have jointly funded and executed the preparation and implementation of the project. The provincial authority is responsible for the day-to-day management of the Sand Motor, while Rijkswaterstaat will lead the monitoring and evaluation programme.

*Universities and research institutes*
New knowledge and insights will emerge from both fundamental and applied research. Researchers from Delft University of Technology, VU University Amsterdam, the University of Twente, Utrecht University, Wageningen University and the research consortium comprising Deltares, Imares and Witteveen en Bos are all conducting research in their own fields. There are two scientific research programmes in which the universities mentioned here are working together: NatureCoast is entirely devoted to the Sand Motor; NEMO includes activities that cover the Sand Motor. The research addresses several areas, including morphology, hydrology, ecology, governance and recreational use. The monitoring activities cover the dunes, the Sand Motor itself, and the surrounding area. The STW research programme NatureCoast has already resulted in a range of presentations and publications. A total of twelve doctorate students, three post-doctorate researchers and four graduates are involved. The NatureCoast and NEMO research programs complement the Rijkswaterstaat Monitoring and Evaluation programme.

*Private sector*
EcoShape is a consortium of private sector parties, government authorities, non-profit organisations and research institutes. The marine contractors Van Oord and Boskalis were behind the initiative for the consortium. They believe it is important for hydraulic engineering to become more sustainable where possible, and they are using the opportunity afforded by this project to expand their expertise in this field. (Ecoshape, 2013)

*European Regional Development Fund*
Knowledge development through the Sand Motor project is co-financed by the European Union's Regional Development Fund through the 'Kansen voor West programme' (a joint initiative for propelling the west of the Netherlands into the top five of Europe's metropolitan areas).

Rijkswaterstaat, Ecoshape and a range of universities and research institutes are hard at work collecting and analysing data. This research is crucial for the development of other Sand Motors around the world.

## First Results after 2.5 Years of Monitoring – Initial Evaluation of Objectives

The following objectives have been formulated for the Sand Motor:

**Objective 1:** Stimulating natural dune growth in the coastal zone between the Hook of Holland and Scheveningen. This dune growth serves various purposes: flood defence, nature and leisure activities.

**Objective 2:** Generating knowledge development and innovation to answer the question of the extent to which this form of coastal maintenance can deliver added value for leisure and nature.

**Objective 3:** Creating an appealing leisure and nature area on the Delfland Coast.

In 2013, on the basis of the data obtained during two and a half years of monitoring, an interim evaluation was conducted as part of the Sand Motor Monitoring and Evaluation Plan (MEP) to determine the extent to which the Sand Motor was fulfilling objectives at that point in time. The objectives were evaluated using a number of secondary objectives and hypotheses (Rijkswaterstaat, 2014). This article discusses the principal findings and some of the additional findings since the interim evaluation.

The results of the evaluation will be set out below for each objective.

**Results of the Evaluation of Objective 1**

The first objective is: "Encouraging natural dune growth in the coastal zone between the Hook of Holland and Scheveningen. This dune growth serves various purposes: flood defence, nature and leisure activities."

Three secondary objectives were formulated as part of this objective: coastal defence, nature and leisure.

The secondary objective of **coastal defence** was evaluated using the following hypotheses (from the *Sand Motor MEP Implementation Programme*(2011)):

- The Sand Motor and supplementary nourishment operations on either side of the Sand Motor will safeguard the defences in the coastal zone between the Hook of Holland and Scheveningen for 50 years and, as a result of dune growth, improve defences by comparison with the standard nourishment programme between 1990 and 2010.
- With the Sand Motor, less sand will be required in total by comparison with the standard nourishment programme for the maintenance of the base coastline in the coastal zone between the Hook of Holland and Scheveningen during a period of 20 years.
- With the Sand Motor and supplementary nourishment operations, the sand balance for the coastal foundation will be maintained in the coastal zone between the Hook of Holland and Scheveningen for at least 50 years assuming sea-level rise of 3 mm a year.

The evaluation of the three hypotheses above results in the following conclusions relating to coastal defence:

- The hypotheses relating to coastal defence relate to the long term (20 to 50 years) and cannot yet therefore be fully tested.
- On the basis of expert judgement, it is already possible to argue that coastal defence in the vicinity of the Sand Motor in the short and medium terms has been improved since:
  - the coastline at the location of the Sand Motor has shifted seawards;
  - the mega-nourishment operation is contributing to a positive sand balance.

However, the data currently available are still inadequate as a basis for the analyses required for an evaluation of coastal defences. For the time being, it is not possible to arrive at a definitive conclusion.

**Sand Movement**

Sand movement transformed the appearance of the Sand Motor between August 2011 and October 2014 (fig. 2):

- 21.5 million $m^3$ of sand was used to create the Sand Motor.
- Relative to the situation immediately after construction, a total of 3.0 million m3 of sand has been moved.
- Most of that sand, 1.38 million $m^3$, was transported northward.
- A smaller proportion, approximately 748,000 $m^3$, has been deposited to the south of the original peninsula.
- 820,000 $m^3$ of sand was moved outside the monitoring area: into deeper water, to places further along the coast, and into the dunes.
- Relatively large amounts of sand were moved in the early days of the Sand Motor while the Sand Motor still had its artificial shape. This artificial, blunt shape resulted in faster sand movement.
- Storms accelerate sand movement. In the first winter of the Sand Motor's existence (2011-2012), there were several storms, and large amounts of sand were also moved during the storm in early December of 2013. (Data from Shore Monitoring & Research)

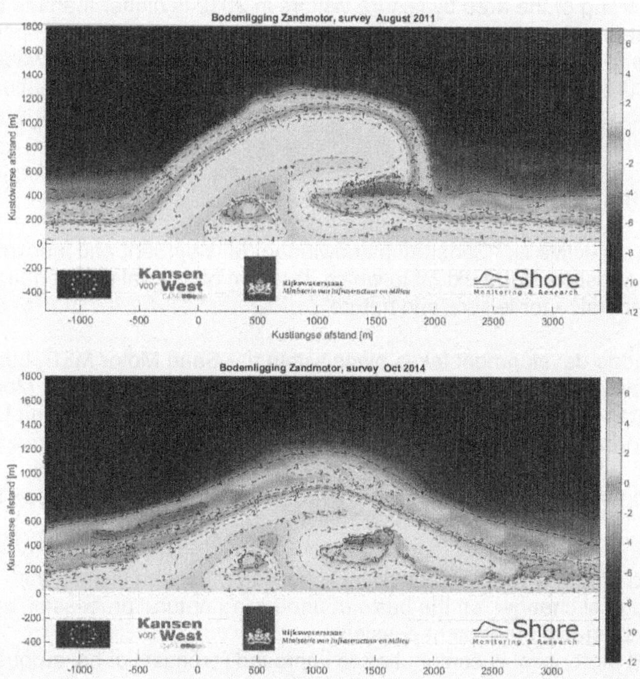


Figure 2: Topography surveys of the Sand Motor in August 2011 and October 2014. Colours indicate the bed level with respect to mean sea level. (Shore Monitoring & Research)

The secondary objective of **nature** (in relation to dune growth) was evaluated using the following hypothesis:

- Between the Hook of Holland and Scheveningen, wider, more natural and dynamic dunes will develop; leisure activities will not have a negative impact on the development of natural values.

Although the data has not yet been analysed, there are indications that young, embryonic dunes are developing. This development would appear to be negatively affected by car traffic, for example involving the rescue services, beach pavilion operators and researchers. The management agency Zuid-Hollands Landschap is now monitoring this area more strictly.

The secondary objective of **leisure activities** (in relation to dune growth) was evaluated using the following hypothesis:

- As a result of dune growth, more space will be created for leisure activities.

Data collection of leisure activities is in progress. It has already been found that the constantly changing landscape has a major appeal for leisure visitors. There is considerable demand for excursions organised by the management agency Zuid-Hollands Landschap and a foundation for nature development: Stichting Ark (Ark, 2013). The rating of the area by leisure visitors in 2012 is higher than the rating measured before the creation of the Sand Motor. Leisure visitors also spend quite long periods of time on the adjoining beach and the dunes (between two and four hours). Leisure activities and nature seem to be a harmonious combination. The area is robust enough. Most leisure visitors are very aware of the natural qualities of the area.

**Results of the Evaluation of Objective 2**

The second objective is: “Generating knowledge development and innovation to answer the question of the extent to which this form of coastal maintenance can deliver added value for leisure and nature”

This knowledge development takes place within the Sand Motor MEP, but also in knowledge programmes such as NatureCoast and NEMO (Nearshore Monitoring and Modelling), for which research institutes use the Sand Motor as a ‘Living Lab’. These research programmes study a wide range of subjects, including Governance.

The following secondary objectives and hypotheses were formulated as part of this objective:

- The Sand Motor is delivering physical knowledge that can explain the morphological changes on the basis of underlying natural processes, making more efficient coastal management possible.
- The Sand Motor pilot is contributing to improved forecasts of hazardous bathing situations, allowing for more effective supervision of swimmer safety.
- The Sand Motor is delivering ecological knowledge that allows us to determine the extent to which a Sand Motor has added value in terms of natural values by comparison with standard nourishment operations.

The evaluation of this second objective will provide the following:

*Knowledge development relating to morphology*
In accordance with the MEP, large amounts of data are being collected and analysed by Deltares (Rijkswaterstaat 2014). Within the NatureCoast programme a total of eight doctorate students, one postdoctorate researcher and multiple Master's students are working on the morphology of the Sand Motor.

*Knowledge development relating to bather safety*
An application has been developed for mobile phones to improve forecasting relating to hazardous bathing conditions. The assessment by the Regional Emergency Services Coordination Authority ("Veiligheidsregio") of initial experience with the application proved positive. Rescue services find the mobile application to be very practical. The application's strengths are its ease of use and the reliability of the forecasts. In early 2015, there will be an evaluation by the *Veiligheidsregio* to determine the number of rescue operations by comparison with the baseline.

*Knowledge development relating to ecology*
Knowledge development is proceeding apace. Data is being collected and analysed in accordance with the MEP (see the evaluation of objective 3). Two doctorate students started their research looking at both marine and land-based ecology in 2013. The NatureCoast research programme has invested in software to identify birds using the radar facility in Kijkduin. Samples have been taken on Ameland to serve as a point of reference for the Sand Motor. An experiment with vegetation was conducted in spring of this year. Doctorate students are starting now on the analysis of the data that has already been collected relating to the Sand Motor, but also at the Sand Motor itself a field campaign with marram grass has started in the spring of 2014.

*Knowledge development relating to research programmes*
The Rijkswaterstaat programme is used to evaluate the coastal response on the timescales of months to years, whereas the shorter intensive field campaigns of NatureCoast and NEMO are used to link this observed response to the detailed forces and processes shaping the coast.

One of these short intensive campaigns, *MegaPEX2014*, was conducted three years after the construction of the Sand Motor in the summer of 2014. During the six-week experiment, a group of approximately 40 international researchers met at the site. This international scientific interest is driven by the fact that the Sand Motor is a radical intervention, generating essential data about the coastal response to both extreme conditions and human interventions.
During the MegaPEX weeks, the researchers measured the coastal shapes, flows, waves and biota in detail. At about twenty locations, instruments were installed that measured waves and flow patterns around the peninsula (fig 3). Furthermore, several vessels and jet-skis were used to sample the different sediment compositions and the effect of the fresh water from the River Rhine nearby. Unmanned aerial vehicles ('drones'), kites and camera images from the Argus Station were used to obtain synopic information about the flow and the elevation of the Sand Motor. The first results of MegaPEX showed that energetic events with high winds or waves result in large amounts of sediment transport, and that they could therefore have a major

impact on the success of the nourishment project. This is a multi-disciplinary project, and the field experiments at the Sand Motor have a correspondingly unique multidisciplinary character. Simultaneous research looking at the interaction between biota and the seabed, the sediment composition, bather safety and the coastal currents at MegaPEX made it possible to bridge gaps between the various fields of research and to join forces where possible.

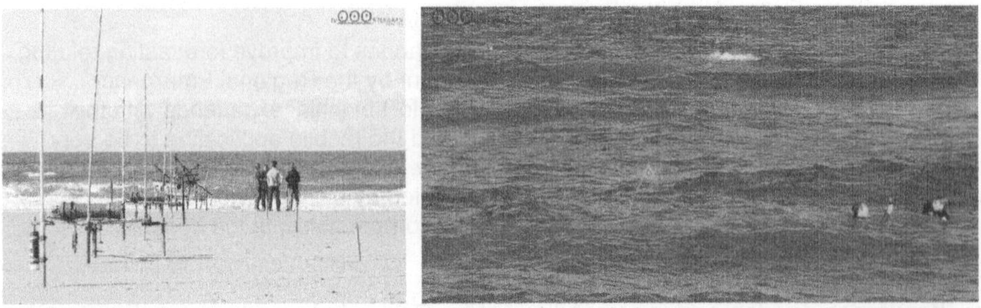

Figure 3: Photos of the MegaPEX2014 field experiment. Left: instruments measuring waves and currents near the waterline (Utrecht University); right: fluorescent dye and surface floats used to track coastal currents and mixing (Delft University). Photos: Ivo Oosterbaan

**Results of the Evaluation of Objective 3**

The third objective is: "Creating an appealing leisure and nature area on the Delfland Coast."

Two secondary objectives were formulated as part of this objective: an additional nature area and an additional leisure area.

The secondary objective of *an additional nature area* was evaluated on the basis of the following hypotheses:

- As a result of the presence of sheltered areas (essentially: the lagoon) and exposed areas (on the seaward side), the composition of the sediment of the sand hook varies.
- The creation of the Sand Motor will result in enhanced ecological values in the intertidal area and the shallow coastal zone as a result of new habitats and variation in habitats.
- The steep gradients (exposed beach and sheltered lagoon) resulting from the creation of the Sand Motor will be seen in a change, and more variation, in the benthic fauna.
- The relatively sheltered location of the lagoon and the high levels of available food will further enhance the nursery function of the area.
- The lagoon will lead to a rise in the wader and seabird populations in the area.
- The sandy hook will result in a feeding and resting area for sea mammals.

The secondary objective of *an additional leisure area* was evaluated on the basis of the following hypotheses:

- There will be room on the hook of the Sand Motor for leisure activities immediately upon completion.
- The dynamics of the Sand Motor will be seen in a changing pattern of leisure activities.
- The new space of the Sand Motor will encourage extensive leisure activities.

The evaluation of objective 3 results in the following insights:

- Most analyses have begun, but they have not yet been completed. However, the initial signals relating to nature development are predominantly positive (Ark, 2013):
  - Numbers of ecotopes and numbers of habitats have risen.
  - Sheltered areas house denser populations of benthic fauna.
  - The benthic fauna are more diverse.
  - The number of birds and species has increased, certainly in the lagoon.
  - Seals and a harbour porpoise have been seen on occasion.
- The range of leisure visitors has increased with the arrival of, among others, kite surfers. The other groups (walkers, dog-owners, bathers) have continued to visit the area. (RoyalHaskoningDHV, 2013)
- Appreciation from leisure visitors is high, not only from the existing groups of visitors, but also, in particular, from the kite surfers.

## Conclusions of the First Interim Evaluation

The natural processes take several years and so the initial conclusions can be drawn no sooner than in 2016. It is possible to state some initial indications on the basis of the monitoring in place:

- The quality and quantity of the data needed to evaluate the objectives generally match expectations.
- In a number of areas, including ecology, longer series of measurements will be needed as a basis for meaningful analyses.
- The initial signals on the basis of expert judgement are positive in terms of the achievement of the objectives of the Sand Motor. Initial positive developments can be seen in the areas of coastal defence, nature, leisure and knowledge development. The observed developments conform to the prediction models used.

### Innovation in Water Management

The pilot project Sand Motor is the first test of a new concept: large concentrated nourishment operations as a potential solution for the major coastal problems in delta regions in the 21$^{st}$ century. The project has therefore generated high levels of interest and there have been regular visits from foreign policymakers and coastal engineers.

First observations show that the Sand Motor is indeed spreading as hypothesized, but a first comprehensive evaluation is planned in 2016, i.e. 5 years after completion. If the Sand Motor fulfils expectations, the concept can be applied elsewhere in the Netherlands and the rest of the world. With the Sand Motor as its latest innovation, the Netherlands is once again setting the tone in water management.

## References

Deltares, Imares (2011), Uitvoeringsprogramma Monitoring en Evaluatie pilot Zandmotor, Rijkswaterstaat.

Rijkswaterstaat (2014), Beleidsevaluatie Zandmotor 2013.

Rijkswaterstaat (2014), Verkenning Bruikbaarheid Zandmotor 2013.

Stichting ARK (2013), Het tweede jaar Zandmotor, Natuurontwikkelingen op een dynamisch stukje Nederland.

RoyalHaskoning DHV (2013), Evaluatie Strand- en zwemveiligheid pilot Zandmotor.

Ecoshape (2013), Quick scan considerations and cases for sandy strategies.

Stive, Marcel J.F., de Schipper, Matthieu A., Luijendijk, Arjen P., Aarninkhof, Stefan G.J., van Gelder-Maas, Carola, van Thiel de Vries, Jaap S.M., de Vries, Sierd, Henriquez, Martijn, Marx, Sarah, and Ranasinghe, Roshanka (2013) A New Alternative to Saving Our Beaches from Sea-Level Rise: The Sand Engine, Journal of Coastal Research, Volume 29, Issue 5.

## Acknowledgement

Matthieu de Schipper was supported by the ERC-Advanced Grant 291206-NEMO

**Coastal Management: Changing coast, changing climate, changing minds**
ISBN 978-0-7277-6114-9

ICE Publishing: All rights reserved
doi: 10.1680/cm.61149.607

# Assessing Coastal Vulnerability in Yucatan (Mexico)

**E. Tonatiuh Mendoza**, Instituto de Ingeniería, Universidad Nacional Autónoma de México, Puerto de Abrigo s/n, 92718, Sisal, México
**Elena Ojeda**, Instituto de Ingeniería, Universidad Nacional Autónoma de México, Puerto de Abrigo s/n, 92718, Sisal, México
**Klaus J. Meyer-Arendt,** Dept. of Environmental Studies, University of West Florida, Pensacola, FL 32514, USA
**Paulo Salles**, Instituto de Ingeniería, Universidad Nacional Autónoma de México, Puerto de Abrigo s/n, 92718, Sisal, México.
**Christian M. Appendini**, Instituto de Ingeniería, Universidad Nacional Autónoma de México, Puerto de Abrigo s/n, 92718, Sisal, México.

## Introduction

Shorelines are very dynamic environments and respond to the net amount of sediments supplied, sea level variations, the frequency and intensity of storms, as well as human interventions in the system. For many centuries, human settlements developed more in lower fluvial regions that in coastal areas; however in the late nineteenth century people began to build along the coastlines. Subsequent economic booms since the start of the 20th century led to an increase in coastal urbanization worldwide. As shorelines became urbanized, problems began to arise. The dynamic nature of the coast, in which shoreline positions alternate between moving seaward and moving landward, was not appreciated by coastal landowners who claimed fixed areas of property on which they build permanent structures. This resulted in coastal infrastructure vulnerable to shoreline erosion and inundation. The present climate change only worsen this condition, as suggested by studies related to possible implications of sea level rise (IPCC, 1992, Nicholls, 2002) and a more immediate effect due to increased storm frequency and intensity (Allan and Komar, 2006, Choi and Moon, 2013, Emanuel, 2005). Although the frequency and intensity of cyclones is site specific, the general consensus is that there is certainty in the increased intensity in the most energetic tropical cyclones in the North Atlantic since the 70s (Knutson et al., 2010, Wong et al., 2014), with the inherent increase of wave height and storm surge, which has caused heavy economic loss on coastal areas.

With this scenario of increase in sea level and severe storms it is important to identify which stretches of coastline are most vulnerable to natural and human-induced hazards. This is even more critical in low-lying coasts, with a more vulnerable morphology to extreme events. In that sense, the northern coast of Yucatan, a carbonate ramp with a very low-lying morphology – both submerged (slope < 500) and emerged (slope < 1:2000), has experienced increased shorefront urbanization since the 60s (Meyer-Arendt, 2004), and coastal erosion is a common feature. This, together with an increased wave energy scenario demands the creation of a vulnerability index (or vulnerability assessment) to provide policy-makers a mechanism to make sound decisions regarding their jurisdictional shorelines. Such decisions may include where to implement erosion control, where best to provide infrastructure such as roads, and where best to prevent coastal areas from urbanization.

## Study Area

The north coast of Yucatan covers about 350 km which extends from the municipality of Celestun to the municipality of Tizimin (figure 1) with a wide continental shelf and a 1/1000 slope (Enriquez et al., 2010). It is a low lying area with a mean altitude of 2m, mainly characterized by coastal lagoons and wetlands separated from the sea by barrier islands, and sandy beaches (CINVESTAV, 2007), with a mixed tidal regime with diurnal dominance, and a range of 0.1 m for neap tides and 0.8 m for spring tides (Cuevas-Jiménez and Euán-Ávila, 2009). The grain size distribution of the zone corresponds to medium and fine sands and the majority of the beaches are dissipative (Mendoza et al., 2013) The Yucatan Peninsula is mainly affected by mid latitude anticyclonic and cyclonic systems that include tropical depressions, tropical storms and hurricanes. The longshore sediment transport predominantly to the west (Appendini et al., 2012) The State of Yucatan is divided in administrative municipalities of which 13 are present in the coastal stretch, Progreso being the most economically important (figure 1).

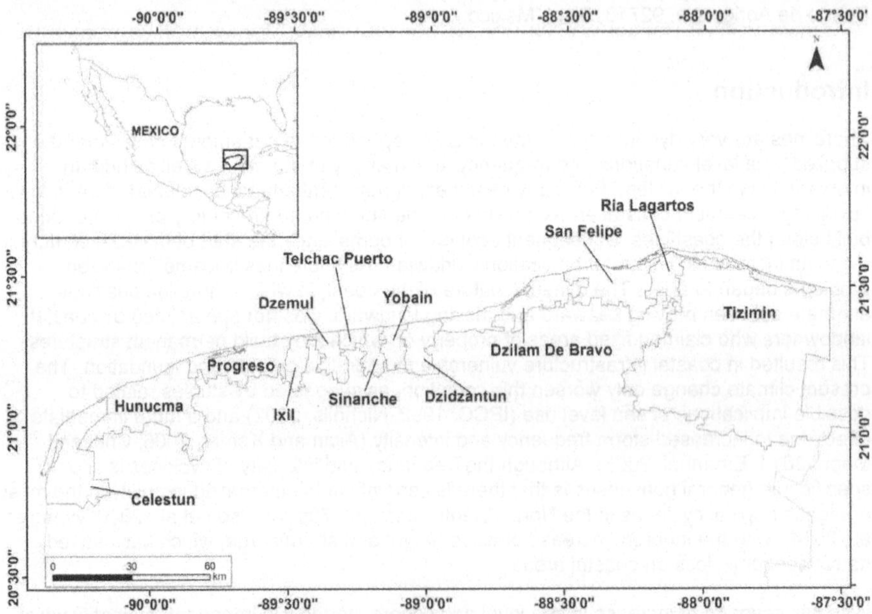


Figure 1. Study area, and location of the administrative municipalities in the coastal stretch.

## Methods

The vulnerability index was evaluated using transects separated every 200m for the total coastal stretch, except for a mangrove dominated section between Dzilam de Bravo and San Felipe (figure 1). The index is composed of three indicators, which integrate the physical, biological and socio-economic features. The physical indicator relates the physical characteristics of the beach with the potential storm impacts in terms of flood and erosion. In terms of erosion, this work only considers the transversal dynamics induced by extreme storms given its greater magnitude than longshore transport during an event, additionally this work has not taken on account the chronic erosion resulting from the cross-shore transport. The biological indicator takes into account the vegetation coverage – the presence of a vegetated dune –, and the anthropization indicator reports the presence or absence of human settlements.

### Physical Indicator

Physical Characterization

The beach characterization was based on a series of beach profiles obtained through a digital terrain model (DTM) generated from a LIDAR survey carried out in June 2011 with a spatial resolution of 1m. A total of 1276 beach profiles were obtained through transects spaced at 200m intervals. From these beach profiles, the beach width was extracted as defined from z=0 (i.e., mean water level) with the horizontal distance to the highest point of the dune location or the foot of infrastructure. Likewise, the maximum beach height was obtained as the vertical distance from z = 0 to the highest point of the dune system or the foot of infrastructure.

Potential Storm Impacts

This work has used the potential erosion and flood values for the most extreme storms (Tr 30-yr) previously found in the area for 26 coastal sections which cover the Yucatan coast (Mendoza et al., 2013) and are presented in Table 1.

For each section these values were used as a threshold compared to the beach width (for erosion) or maximum height (for flooding) for every transect within each section. Figure 2. gives an example of erosion assessment for the Yobain section which presented a maximum erosion of 13.8 m (Table 1). This erosion potential value was subtracted for each of the beach width (on top in italics) of the 47 transects (grey lines). The negative values (on bottom in bold) were considered as -1 and the positive values were considered as 1. The same methodology was applied for the flood index in which the maximum berm height was subtracted by the flood reach (runup + tide values) where negative values were considered -1 and positive values were considered 1.

### Vegetation Indicator

Dune systems act as buffers, diminishing the impact of storms, and contribute to the protection of low-lying coastal area communities. For this reason, this indicator takes on account the absence (value of -1) or presence (value of 1) of dune fields for each transect mentioned earlier using the LIDAR Digital Elevation Model and confirmed with Google Earth (©) images from 2011 with a 1 meter resolution.

Table 1. Erosion and flood values obtained for each section for the Yucatan coast after Mendoza et al. 2013.

| Section | Class V erosion values (m) | Class V flood values (m) | Section | Class V erosion values (m) | Class V flood values (m) |
|---|---|---|---|---|---|
| Tizimin A | 18.5 | 1.3 | Dzemul/Ixil | 13.6 | 1.2 |
| Tizimin B | 16 | 1.2 | Progreso A | 2 | 1.7 |
| Ria Lag A | 14 | 2.0 | Progreso B | 10 | 1.4 |
| Ria lag B | 6.8 | 1.9 | Progreso C | 0 | 1.2 |
| Ria Lag C | 18 | 1.0 | Playon | 0 | 1.4 |
| Ria Lag D | 15.4 | 1.5 | Progreso D | 12.4 | 2.4 |
| San Felipe | 16 | 2.4 | Hunucmá A | 13.6 | 4.9 |
| Dzilam A | 11.8 | 3.6 | Hunucma B | 0 | 1.6 |
| Dzidzantun | 5.6 | 2.2 | Hunucmá C | 12.4 | 2.5 |
| Yobain | 13.8 | 1.2 | Celestun A | 0 | 1.2 |
| Sinanche | 16.8 | 1.4 | Celestun B | 4.6 | 1.5 |
| Telchac A | 18.2 | 0.8 | Celestun C | 13.6 | 1.2 |
| Telchac B | 1.2 | 1.3 | Celestun D | 0 | 2.1 |

**Anthropization Indicator**

Human settlements interact with sediment transport processes and thus can intensify erosion, leading to increased vulnerability to extreme storms. Therefore, this indicator takes into account the absence (value of 1) or presence (value of -1) of human settlements for each transect using the LIDAR Digital Elevation Model and confirmed with Google Earth (©) images from 2011 with a 1 meter resolution.

**Coastal Vulnerability Index**

The determination of a coastal vulnerability index uses a straightforward aggregation method simply considering each element as the same and the sum of the involved elements.

**Erosion Vulnerability Index**

In the case of erosion, all three indicators were taken into account and the sum of all was divided into three different classes: negative values were classified with high vulnerability, values equal to 1 with medium vulnerability, and values equal to 3 with low vulnerability.

**Flood Vulnerability Index**

For the flood index, only the physical and the anthropization indicators were taken into account, given that the presence of vegetation does not play a major role in the inundation schemes. Therefore the sum of the two indicators was divided in three different classes: -2 values were considered as high vulnerability, 0 values as medium vulnerability and 2 values as low vulnerability.

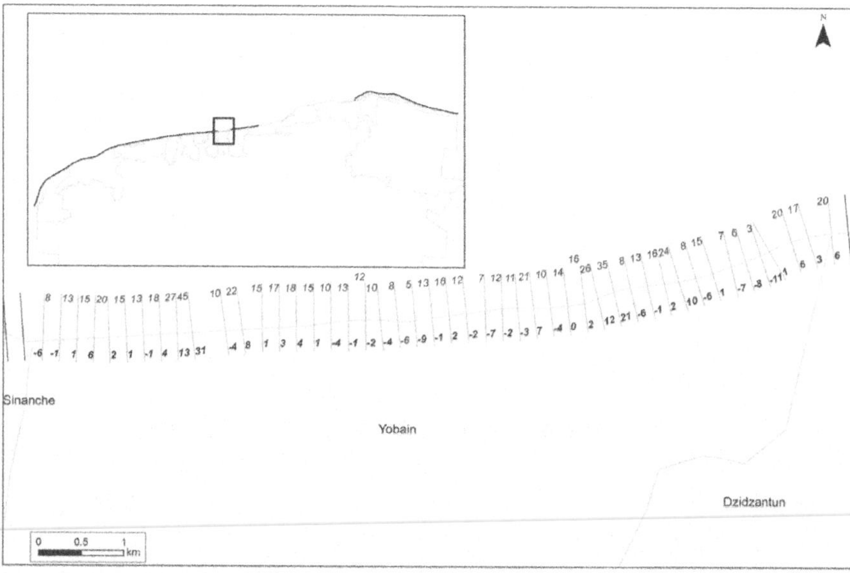


Figure 2. Example of Yobain section with 47 transects (grey) with beach width data (top values next to transect in italics) and erosion values (lower values next to transects in bold) after the 13.8 m subtraction.

## Results and Discussions

Erosion Vulnerability Index

The regional analysis identifies quite clearly the central coast of Yucatan as critical – high erosion vulnerability values -- along a sector extending from Progreso to Dzemul, and large portions of Telchac, Sinanche, Yobain, Dzinzantun (Figure 3). These values are due to a combination of narrow beaches, the absence of vegetated dunes and high urban density. Although the natural sedimentary balance suggests stability in the coast (Appendini et al., 2012), the implication on the degradation of the natural beach due to anthropogenic impacts is clear.

Celestun, Hunucma and Tizimin municipalities presented low erosion vulnerability values mainly due to wider beach values, absence or very small human settlements and healthy vegetated dune systems. Information for each administrative section –municipality- is very important for coastal management schemes, in this sense each beach length per municipality (Table 3) was considered as a 100% and the amount of transects with a given vulnerability value were considered for this evaluation. This assessment found that Progreso, Telchac and Dzemul presented the highest percentages (93%, 67% and 62% respectively), of high erosion vulnerability. In contrast with Ria Lagartos, whose total beach length is similar to Progreso, presented 87% of the transects with low erosion vulnerability and only 3% of the transects with high erosion vulnerability (Table 3).

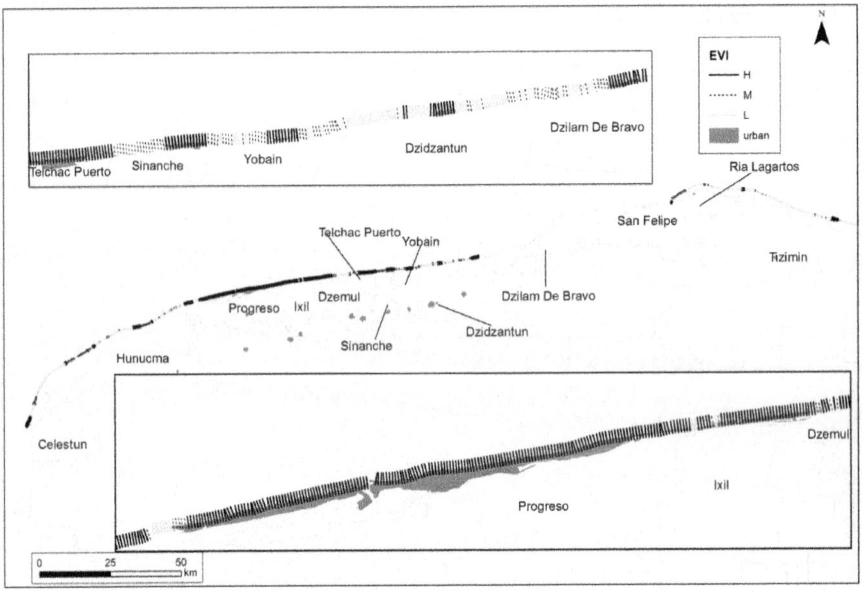


Figure 3. Erosion Vulnerability Index results for the study area.

Flood Vulnerability Index

The regional flood analysis suggests small sections of high flood vulnerability in Celestun, the western part of Progreso, Dzinzantun, Dzilam and San Felipe (Figure 4). These values are mainly caused by a lack of dune systems and the presence of infrastructure on the upper dune forefront in the back part of the beach. Furthermore, it must be taken into account that again the central part of the coast from Progreso to Dzemul, which is heavily urbanized, presented medium values for flood vulnerability. On the other hand Celestun, Ria Lagartos and Tizimin municipalities presented a wide majority of transects with low vulnerability, mainly due to the presence of only punctual small communities together with large extensions of natural areas and a healthy dune systems.

Table 4 presents the results in terms of percentage per beach length per municipality, where the most threatened is Dzilam (40%) followed by Progreso (33)% and Dzinzantun (22%). It must be stressed out that the flood threshold has not taken into account the storm surge, which has been reported to be over 2m (for Hurricane Isidore in Progreso in 2002), However, although no actual measurements have been made, this threshold can be used for to assess the erosion and flood risks due to the strong cold fronts (nortes) that affect the area quite frequently during the winter season (approximately 40 per year) and have small storm surges.

Table 3. Erosion Vulnerability Index results with the number of transects for each class beach length and percentage of beach with corresponding values per municipality.

| Municipality | Erosion Vulnerability Index | | | Beach Length | Percentage of beach with corresponding EVI values | | |
|---|---|---|---|---|---|---|---|
| | H | L | M | (km) | H | M | L |
| Celestun | 25 | 51 | 101 | 35.4 | 14 | 29 | 57 |
| Hunucma | 23 | 73 | 101 | 39.4 | 12 | 37 | 51 |
| Progreso | 184 | 6 | 8 | 39.6 | 93 | 3 | 4 |
| Dzemul/ixil | 52 | 14 | 18 | 16.8 | 62 | 17 | 21 |
| Telchac | 37 | 5 | 13 | 11 | 67 | 9 | 24 |
| Sinanche | 14 | 19 | 0 | 6.6 | 42 | 58 | 0 |
| Yobain | 11 | 27 | 9 | 9.4 | 23 | 57 | 19 |
| Dzinzantun | 11 | 4 | 39 | 10.8 | 20 | 7 | 72 |
| Dzilam | 13 | 23 | 12 | 9.6 | 27 | 48 | 25 |
| San Felipe | 3 | 28 | 6 | 7.4 | 8 | 76 | 16 |
| Ria Lagartos | 5 | 21 | 174 | 40 | 3 | 11 | 87 |
| Tizimin | 13 | 9 | 124 | 29.2 | 9 | 6 | 85 |

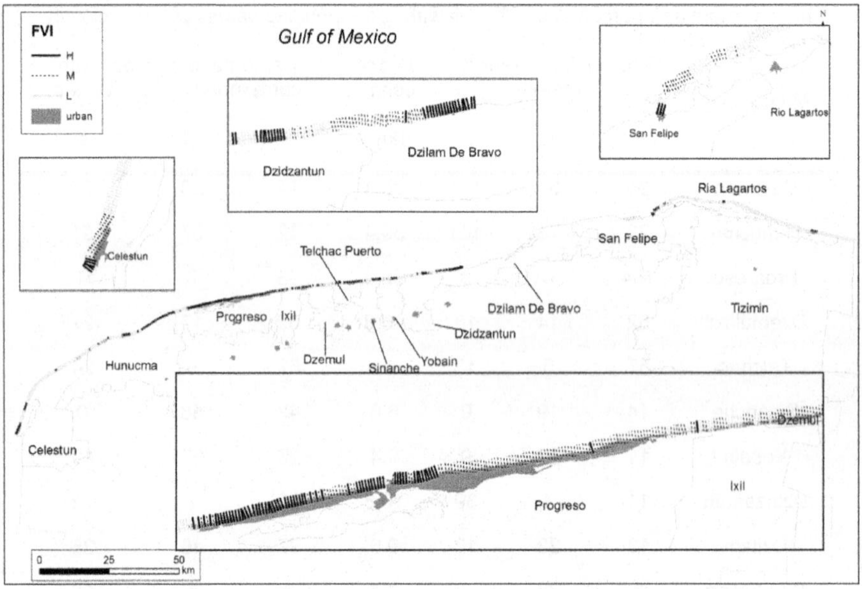


Figure 4. Flood Vulnerability Index results for the study area.

Table 4. Flood Vulnerability Index results with the number of transects for each class beach length and percentage of beach with corresponding values per municipality.

| Municipality | Flood Vulnerability Index | | | Beach Length | Percentage of beach with corresponding to FVI values | | |
|---|---|---|---|---|---|---|---|
| | H | M | L | (km) | H | M | L |
| Celestun | 5 | 34 | 138 | 35.4 | 3 | 19 | 78 |
| Hunucma | 1 | 100 | 96 | 39.4 | 1 | 51 | 49 |
| Progreso | 66 | 108 | 24 | 39.6 | 33 | 55 | 12 |
| Dzemul/ixil | 0 | 33 | 51 | 16.8 | 0 | 39 | 61 |
| Telchac | 2 | 41 | 12 | 11 | 4 | 75 | 22 |
| Sinanche | 4 | 16 | 13 | 6.6 | 12 | 48 | 39 |
| Yobain | 0 | 16 | 31 | 9.4 | 0 | 34 | 66 |
| Dzinzantun | 12 | 9 | 33 | 10.8 | 22 | 17 | 61 |
| Dzilam | 19 | 29 | 0 | 9.6 | 40 | 60 | 0 |
| San Felipe | 3 | 16 | 18 | 7.4 | 8 | 43 | 49 |

| | | | | | | | |
|---|---|---|---|---|---|---|---|
| Ria Lagartos | 1 | 5 | 194 | 40 | 1 | 3 | 97 |
| Tizimin | 1 | 12 | 133 | 29.2 | 1 | 8 | 91 |

## Concluding Remarks

This work presents an erosion and flood vulnerability assessment for the coast of Yucatan, Mexico. In terms of erosion, the central section of the coast which extends from the municipalities of Progreso to Dxemul/Ixil and Telchac (approx. 70 km.) was identified as critical, with high erosion values covering 93%, 62% and 67% of the beaches respectively. In terms of flooding, Progreso and Dzilam municipalities were identified as critical with 66 and 19 transects, respectively, exhibiting high vulnerability values. Therefore the Progreso, municipality should be prepared for coastal management problems involving erosion and flooding.

Although this research has shown that the Yucatan coast has more transects with higher vulnerability in terms of erosion than flooding, it must be noted that this work did not consider the storm surges, and therefore the results should be interpreted with caution.

## References

Allan, J. C. & Komar, P. D. 2006. Climate Controls On US West Coast Erosion Processes. *Journal Of Coastal Research, 22, 3,* 511-529.

Appendini, C. M., Salles, P., Mendoza, E. T., López, J. & Torres-Freyermuth, A. 2012. Longshore Sediment Transport On The Northern Coast Of The Yucatan Peninsula. *Journal Of Coastal Research,* 28, 1404-1417.

Choi, K.-S. & Moon, I.-J. 2013. Relationship Between The Frequency Of Tropical Cyclones In Taiwan And The Pacific/North American Pattern. *Dynamics Of Atmospheres And Oceans,* 63, 131-141.

Cinvestav 2007. Programa De Ordenamiento Ecológico Del Territorio Costero Del Estado De Yucatán (Poetcy). *In:* Yucatán, S. D. E. D. G. D. E. D. (Ed.).

Cuevas-Jiménez, A. & Euán-Ávila, J. 2009. Morphodynamics Of Carbonate Beaches In The Yucatán Peninsula. *Ciencias Marinas,* 35, 307-320.

Emanuel, K. 2005. Increasing Destructiveness Of Tropical Cyclones Over The Past 30 Years. *Nature,* 686-688.

Enriquez, C., Mariño-Tapia, I. J. & Herrera-Silveira, J. A. 2010. Dispersion In The Yucatan Coastal Zone: Implications For Red Tide Events. *Continental Shelf Research,* 30, 127-137.

IPCC 1992. Global Climate Change And The Rising Challenge Of The Sea. *Ipcc Response Strategies Working Group.* The Hague: Rijkswaterstaat.

Knutson, T. R., Mcbride, J. L., Chan, J., Emanuel, K., Holland, G., Landsea, C., Held, I., Kossin, J. P., Srivastava, A. K. & Sugi, M. 2010. Tropical Cyclones And Climate Change. *Nature Geosci,* 3, 157-163.

Mendoza, E. T., Trejo-Rangel, M. A., Salles, P., Appendini, C. M., Gonzalez, J. L. & Torres-Freyermuth, A. 2013. Storm Characterization And Coastal Hazards In The Yucatan Peninsula. *Journal Of Coastal Research,* 790-795.

Meyer-Arendt, K. J., 2004, Human Alteration of the North Yucatán Coast, *In: The Coastal Zone: Papers in Honor of H. Jesse Walker.* Davis, D.W. & Richardson, M., (eds.) Geoscience and Man, vol. 38, Dept. of Geography & Anthropology, Louisiana State University, Baton Rouge, LA, 65-80.

Nicholls, R. J. 2002. Analysis Of Global Impacts Of Sea-Level Rise: A Case Study Of Flooding. *Physics And Chemistry Of The Earth, Parts A/B/C,* 27, 1455-1466.

Wong, P. P., Losada, I. J., Gattuso, J. P., Hinkel, J., Khattabi, A., Mcinnes, K. L., Saito, Y. & Sallenger, A. 2014. Coastal Systems And Low-Lying Areas. *In:* Field, C. B., Barros, V. R., Dokken, D. J., Mach, K. J., Mastrandrea, M. D., Bilir, T. E., Chatterjee, M., Ebi, K. L., Estrada, Y. O., Genova, R. C., Girma, B., Kissel, E. S., Levy, A. N., Maccracken, S., Mastrandrea, P. R. & White, L. L. (Eds.) *Climate Change 2014: Impacts, Adaptation, And Vulnerability. Part A: Global And Sectoral Aspects. Contribution Of Working Group Ii To The Fifth Assessment Report Of The Intergovernmental Panel Of Climate Change.* Cambridge, United Kingdom and New York, NY, USA: Cambridge University Press.

**Coastal Management: Changing coast, changing climate, changing minds**
ISBN 978-0-7277-6114-9

ICE Publishing: All rights reserved
doi: 10.1680/cm.61149.617

# Colwyn Bay Waterfront Regeneration Project

**G.B. Edwards**, Conwy County Borough Council, Colwyn Bay, UK
**A.J. Williams,** Coastal Engineering UK Ltd, Wirral, UK

## Introduction

This paper provides the background to the development and progress to date (2014) of a coastal regeneration scheme in Colwyn Bay, North Wales that integrates improved coastal defence and amenity with regeneration of the waterfront and the town as a whole, from the perspective of the Council Officers charged with delivering the project.

The development of proposals for regeneration of the Colwyn Bay Waterfront arose primarily from the amalgamation of two separate initiatives:

- The Colwyn Bay Waterfront Strategy in 2002
- The Colwyn Bay Coastal Defence Strategy, produced in 2007

The Waterfront Strategy provided recommendations for regenerating the Waterfront area of Colwyn Bay. The ability to regenerate the Waterfront however was identified as being reliant on carrying out improvements to the ageing coastal defences that date from the early 1900s, and which were in need of refurbishment, set against a background of lowering beach levels across the frontage that had been on-going for most of the 20th century. Identification of how best to carry this out was developed through production of a Coastal Defence Strategy.

The paper deals with how these two initiatives were brought together to develop a vision for regeneration of the Colwyn Bay waterfront as part of a wider regeneration package for the whole town. It provides a summary of the development of the resort from the mid-1800s to the present day, the contemporary background to the regeneration project, how the vision for the frontage was developed and ultimately delivered and how that vision will be managed in the future.

## History

Colwyn Bay is located on the north facing coast of Wales, as shown in Figure 1. The town has a relatively short history having developed rapidly in the late 19th and early 20th century. Development of the town was directly linked to the construction of the north wales coast railway main line from Chester to Holyhead in the mid-19th century.

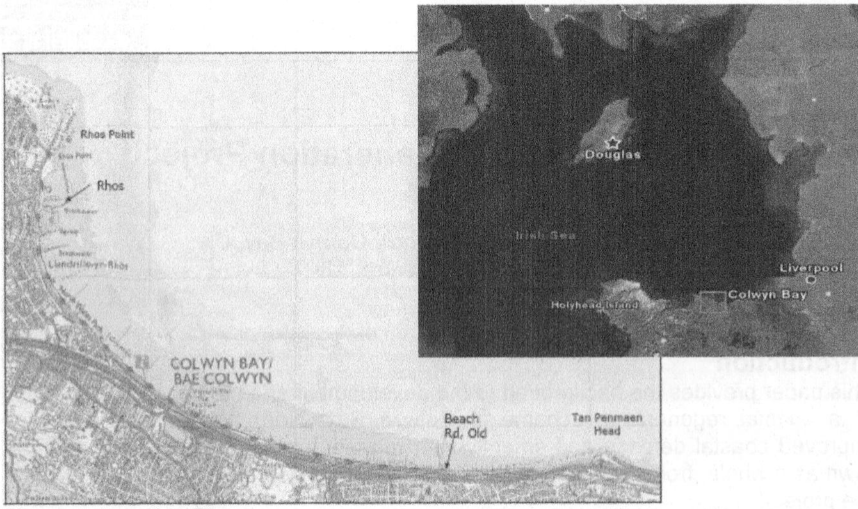

Figure 1: Location Plans

This railway had great significance offering easy access from the industrial urban areas of NW England and the Midlands, and its construction is attributed to the rise of coastal resorts along the North Wales Coast including Colwyn Bay, Rhyl, Prestatyn and Llandudno, through the development of tourism.

Colwyn Bay began to develop properly after the establishment of its railway station in 1849 and expansion of the town radiated outwards from this focal point, with land which had been farm fields becoming the scene of enterprise. Within 30 years the new town changed from New Colwyn to Colwyn Bay and was subsequently recognised as an urban district in 1894. Rapid growth and pristine countryside quickly attracted persons to the seaside resort and the town of Colwyn Bay grew apace doubling its population from 1891-1901 and again between 1901 and 1911 (Harvey, 1931). The new council built the Victoria pier and pavilion as well as completing a mile of promenade and associated coastal defences at a then cost of £15,000 in the early 1890s. The pier pavilion was rebuilt after a fire in 1920 which in turn was also destroyed by a fire being replaced by a third pavilion in 1934. The promenade was extended further eastwards to old Colwyn in 1897 and westwards to Rhos on Sea in 1905.

The coastal defences across the Colwyn Bay frontage primarily comprised vertical or near vertical masonry blockwork sea walls. The construction of these defences whilst providing the improved amenity function overall were detrimental to the beach conditions applying and throughout the 20th century beach levels gradually lowered necessitating the construction of toe aprons to protect the foundations to the walls. Evidence provided by the levels of natural river outfalls that have been culverted through the existing walls confirm the beach records that upper beach levels dropped by up to 1.5 metres during the 20th century. Additional local observations and reference to historic photographs (ref figure 2) suggests that in places against the walls levels may have locally dropped by up to 3 metres.

Figure 2: Changes in Conditions (1930s left, 2000s right)

The construction of a road bridge across the Clwyd Estuary in 1953 connected Rhyl to the east to many of the seafront towns in the west encouraging further growth and prosperity within the area of Colwyn Bay and the town continued to grow. However by the 1960's, as air travel became more affordable and the British public's holidaying habits began to change to travelling, Colwyn Bay, in keeping with many other British seaside towns, went into a steady decline as a tourist destination.

The 1980's saw renewed investment in Colwyn Bay with the construction of a new shopping centre at Colwyn Bay and the North Wales A55 Expressway to provide improved road links along the North Wales coast to Holyhead. However construction of this new highway infrastructure did nothing to reverse Colwyn Bay's decline, as it not only reduced the amount of traffic stopping in the town but it effectively severed the main part of the town from its waterfront.

## Contemporary Background

Gradual lowering of the beach led to increased exposure of the defences and as a consequence increased risk to the structural integrity of the walls as well as an increased frequency of overtopping.

At the beginning of the 1990s low level rock groynes were introduced to take the place of the dilapidated timber groynes across the frontage in a bid to stabilise beach levels. Whilst some benefit was observed it was becoming increasingly clear that without major investment, there was an increasing risk of failure of the walls which would have significant impacts on the road, rail infrastructure and property located immediately behind the defences.

In 2005 the Council commissioned development of a Coastal Defence Strategy (Conwy County Borough Council, 2007) for the frontage to examine the justification for and definition of appropriate forms of coastal defence for the Bay.

It was identified at an early stage of the Coastal Defence Strategy Development that merely improving the ageing linear defences alongside improved promenade facilities would not attract people back to the resort, as such an approach would not fully address the issue of low beach levels. Such defences had been constructed across the frontage to the east of Rhos Point during the 1990s. Whilst this approach was functional in terms of providing a structurally stable and hydraulically efficient coastal defence and, due to the close proximity of suitable sources of rock, would probably be the most cost effective form of defence, Council officers felt that continuation of this approach across the Colwyn Bay frontage would accelerate rather than reverse the decline in the fortunes of the town. There was therefore a local determination to investigate more innovative solutions.

The benefits of this approach were confirmed during early public consultation on potential options, when a strong desire to implement proposals that provided improved beach conditions in parts of the bay resonated with the public. Subsequently an assessment of options, including beach recharge, linear defences and other potential solutions, was carried out during the second phase of Strategy development in 2007.

The Colwyn Bay Waterfront Strategy had been produced in 2002 and the coastal defence strategy dovetailed with proposals to provide recommendations for a way forward that would both improve coastal defence and regenerate the Colwyn Bay waterfront.

The Waterfront strategy had suggested splitting the shoreline into 4 zones with specific proposals for primary usage and activities identified within each zone. The zones, as shown on figure 3, where:

- Zone 3 – Beach Zone;
- Zone 4 – Pier and Watersports Zone;
- Zone 5 – Old Colwyn Zone; and
- Zone 6 – Links to Old Colwyn Zone

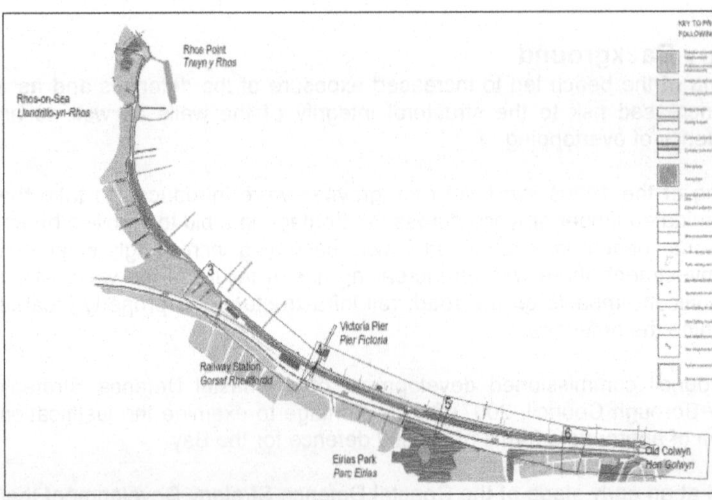


Figure 3: Waterfront Strategy Zones

## Strategic Development

Amalgamating these two initiatives underpinned the Council's approach to developing a vision for Colwyn Bay. This was initially driven by Council Officers within the Environmental directorate, however buy-in from other Council departments and most importantly Council members and the local community was required if it was ultimately to be delivered.

Specific challenges faced in developing this vision, included:

- How to get local "buy-in";
- How to develop suitable proposals integrating regeneration and coastal defence; and
- How to obtain funding.

'Communities First' is a Welsh Assembly Government programme to help tackle those areas of Wales experiencing the most poverty and disadvantage. Colwyn Bay was identified in the 2005 Welsh Index of Multiple Deprivation with parts of the town being in the 10% most deprived wards in Wales.

In October 2008, the Welsh Government designated the North Wales coast as a Strategic Regeneration Area (SRA). In Conwy, the designated area extends from Kinmel Bay, located approximately 15km to the east of Colwyn Bay in the east to Mochdre, just to the west. This area consists primarily of coastal communities which are linked by the A55 trunk route, the A547 coastal road and the north Wales mainline railway.

In order to give focus and direction to regeneration in Colwyn Bay, Conwy County Borough Council established the Bay Life+ partnership in early 2009, run by a board made up of key local business people and council officials that works in close partnership with organisations and community groups throughout the local area.

The Bay Life+ Programme is more than just physical regeneration - it aims to improve and bring together all the things that make up a community. The vision for Colwyn Bay is a thriving, attractive and vibrant town that is welcoming, safe and friendly; a place with unique character that people are proud to live in, based on:

- Improvements in space for living & housing;
- Better community cohesion;
- People enjoying better health & greater wellbeing;
- More integrated transport systems and an improved environment;
- Greater opportunities for people to improve their education and develop their skills; and
- A sustainable economy.

Bay Life+ laid the foundations for the regeneration of the town and provided the opportunity to build an identity for Colwyn Bay that moved on from the traditional seaside town into a modern, vibrant centre for commerce, tourism, employment and leisure, making Colwyn Bay an attractive and satisfying place to live.
In order to deliver sustainable regeneration a plan for 2007-2014 was developed with three key initiatives:

- a major programme of coastal defence, to enhance the beach and the promenade, linked inextricably to the need for regeneration;
- creation of a centre of sporting excellence at Parc Eirias; and
- the relocation of Council offices to the town centre;

In 2012 'The Bay Life+ Programme' was selected from projects across Europe as an example of good practice in sustainable development, and was showcased at the Committee of the Regions 2012 Summit in Copenhagen.

## Development of the Waterfront Project

The Bay Life+ programme provided the framework for development of the Colwyn Bay Waterfront Project to combine the renewal of coastal defences with regeneration improvements to the promenade, creating a modern, sustainable and attractive waterfront, which would be an asset to the local community and an attraction for visitors - "*Integrating coastal protection with regeneration delivering innovative solutions and value for money*".

Development of the Coastal Defence Strategy for Colwyn Bay had followed the then Welsh Government guidelines for Strategy development and justification of the use of public funds for coastal defence improvements, identifying the risks of defence failure and overtopping and quantifying the damages that would occur if investment in defence provision did not take place.

Optioneering examined a range of options and although linear defence provision through bolstering of the sea wall with rock armour offered a cheaper capital investment solution, preliminary examination of the technical feasibility of improving beach levels suggested that a scheme could be implemented. Furthermore associated economic assessment indicated that the additional economic benefits that provision of a sustainable recharged beach would provide, would outweigh the initial increased construction costs of such a form of defence, as well as ensuring that the coastal defence provision did not detract from the initiatives and investment of others tasked with regeneration of the town.

Fundamental to this process was Welsh Government acceptance of the technical and business cases put forward, allowing for detailed development of the proposals.

Concurrently, Conwy County Borough Council commissioned landscape architecture and urban design consultants to develop visions for the development of regeneration proposals for the promenade and public realm works through a design completion. Crucially, the preferred bidder showed a high degree of appreciation of the need to integrate the waterfront design and coastal defence elements of the project, reinforcing the need to develop the regeneration ideas alongside coastal defence scheme proposals.

Through development of the Colwyn Bay Waterfront Project the two processes were amalgamated (see figure 4) into a single, coordinated, deliverable entity, the overall objectives of which were to provide:

- "robust and resilient coastal defences to protect the town, its residents, properties, businesses and infrastructure for years to come"; and
- "an attractive, high quality and sustainable public realm along the promenade to attract visitors to the town and drive its economic redevelopment".

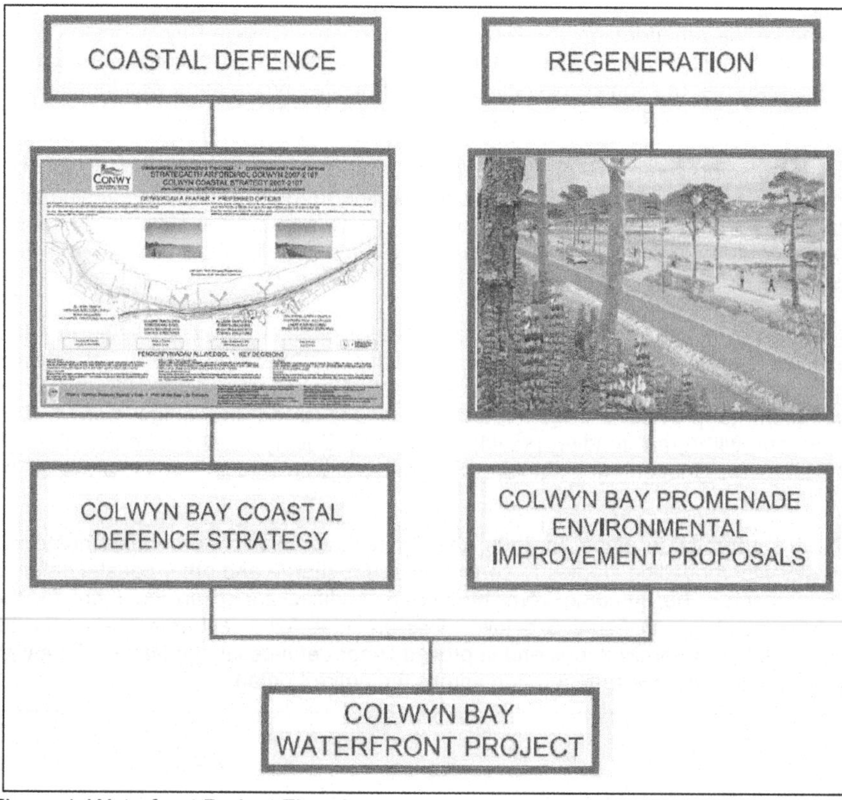


Figure 4: Waterfront Project Flowchart

## Delivering the Vision

Developing the project required an adaptive approach, building upon but not constrained by earlier thinking. The original Waterfront Strategy (see above) had split the shoreline down into a number of activity zones. During development of the Coastal Defence Strategy public consultation identified that solutions which provided improved beach amenity and water sports provision were preferred across the western part of the frontage (zones 3 and 4), whilst modification of linear defences and the inclusion of facilities for other amenity interests e.g. angling was considered to be more appropriate across the easterly parts of the frontage (zones 5 and 6).

An integrated approach required integration of different funding streams. It was only through developing a clear vision for the area that this was possible. The vision created the environment and framework that allowed the Council to draw upon different funding in a consistent and convincing manner.

Grant Aid funding for coastal defence elements of the Waterfront project was available, subject to availability, through Welsh Government's Flood and Coastal Defence budget for up to 75% of the capital cost. In addition during Strategy development European Structural Funding was identified as a potential source of funding for the regeneration elements of the proposals. Colwyn Bay is located within the 'West Wales and the Valleys' region, which qualified for Convergence Funding, the highest level of support available and the Bay life+ programme was identified as one of the major projects proposed to bid this funding.

Pursuant to presentation of the Strategy and development of the Waterfront proposals, the Council was able to secure funding from the European Convergence Funding Programme with match funding provided by the Welsh Assembly Government, for the first phase of the coastal defence element of the Colwyn Bay Waterfront Project.

Concurrently, a Promenade Environmental Improvements Concept Design that had been developed by the appointed landscape architecture and urban design consultants enabled Conwy County Borough Council to secure £8 million of investment to provide a regenerated promenade along the entire Colwyn Bay Waterfront, with grant funding provided to Conwy County Borough Council through the European Convergence Funding Programme, with match funding provided by WG DE&T.

With this approval in place, in 2009, the Council commissioned detailed numerical and physical modelling studies to confirm beach recharge and other coastal defence arrangements. The outcome from this was a modified zoned arrangement for the frontage as shown in figure 5 below. This provides for a recharged beach in the Watersports and Family zones and improved linear defences in the Nature Zone with associated promenade regeneration across the entire frontage.

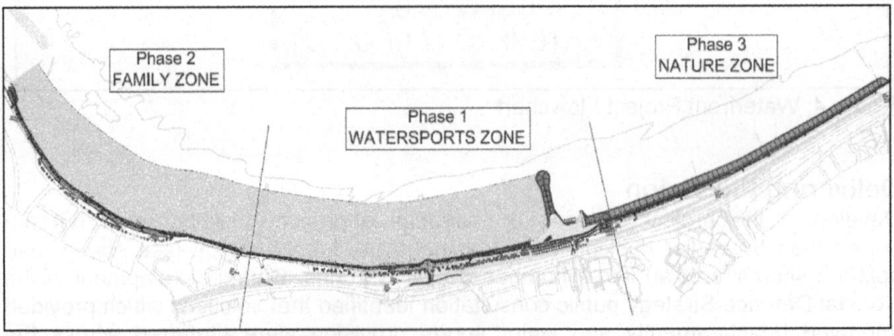


Figure 5: Waterfront Project Phases

## Phasing

The first phases of the Waterfront Project comprised approximately £5.0 million of investment in a terminal beach retention structure and localised adjacent linear defence improvements at Porth Eirias, at the eastern end of the Watersports Zone, approximately 300 metres west of the Victoria Pier. In addition £5.0 million of funds was available for the first phase of promenade regeneration enabling works.

Again, as the project developed, there was a need to continually review progress and opportunities, building on the benefits of an integrated approach. Essentially, as detailed design was being undertaken it became clear that delivery of the works as a single package would provide significant cost savings. The initial ideas had been for a terminal groyne, required as part of the coastal works, with a promenade focal point being constructed elsewhere. Integration of coastal works with regeneration proposals identified opportunities to create a focal point at the groyne location with associated land reclamation. This offered potential savings of £2-£3 million.

The challenge for Conwy CBC was to convince the funders of the benefits of this joined up approach, linking funding streams in a manner that delivered such benefits.

This was successful and a Phase 1a contract to construct the "joined up" works was let in January 2011. These works were completed in March 2012 and following on from this a £3.5 million contract was let to construct a new promenade building on the completed promontory. This comprised a single storey steel frame building with accessible roof area to provide a restaurant / bistro, classroom training facilities, changing room facilities, a watersports venture and retail space.

In 2012, following confirmation from Welsh Government that a further £6.0million of capital investment would be available to provide coastal defence improvements, development of Phase 1b of the works commenced comprising the importation of approximately 300,000m$^3$ of sand to recharge the beach. Available sources of sea dredged material were available near to the site in areas that were already licensed for extraction. This element of the scheme was carried out during the first half of 2013, by contractor Volker Stevin.

This phase of the Works did not provide full level of protection across the Phase 1 frontage as further phases of promenade enhancement works were required before this could be completed. In February 2014, a Contract for the third phase of works across the Phase 1 frontage, comprising £2.5 million of investment in promenade enhancement works and a further £2.0 million in beach recharge and associated sea wall improvement works was let to contractor Galliford Try. This work was completed in Autumn 2014.

The promenade improvements in this phase as well as improving the physical environment of the promenade also included for better linkage between the town centre, which lies on the landward side of the railway, and the Waterfront, through improved signage and environmental enhancement of existing accesses.

## The Future

Future management of the beach is an integral part of the Waterfront Project. In 2013 the Council commissioned production of beach maintenance and management plan to develop arrangements and procedures to ensure that beach is managed and maintained to provide its coastal defence and amenity function. This plan covers specific requirements including:

- Monitoring;
- Beach recycling and future topping up;
- Wind blown sand management;
- Beach Cleansing and tidying requirements; and
- Environmental Impact Management.

Furthermore the project is only partially complete and there are a further two phases of works yet to be carried out.

Phase 2 at the western end of the bay in the Family Zone (ref Figure 5) comprises extension of the beach recharge and sea wall improvements, at an estimated cost of the order of £10-12 million. Allied to this will be further environmental improvements to the promenade. Additional investigations and modelling and being undertaken to develop specific proposals for this area.

Phase 3 comprises the modification of linear defence arrangements and provision of improved amenity at the eastern end of the bay in the Nature Zone, where current defence standards are low and there is a residual risk of promenade and highway damage and of damage to railway infrastructure.

The Council are now looking to secure funding for the remaining elements of the scheme and whilst national and European funding has been available to date it is likely that there will need to be additional third party funding to supplement public funds in order to complete the remaining phases.

## Conclusions

This overall project has highlighted the difficulties in delivering an integrated approach to coastal management but also the significant value and benefits that such an approach can achieve. In reality, this project started over a decade ago through recognising that a joined up, visionary attitude had to be taken.

In progressing this, that vision, linking the needs of coastal risk management with ideas for regeneration, has driven the project and sustained its momentum. The journey has not been easy. There has been the need to work with people at both a local level and within funding organisations.

The project has shown the need for clear leadership and on-going commitment to an idea. However, it has also shown the need for flexibility, a willingness to adapt ideas and not be constrained by past thinking. It has been a journey of many twists and turns, with a continuing need for review and a willingness to take opportunities as they emerge. Almost certainly, in hindsight, different approaches might have been taken to get to where we are now. However, without the enthusiasm and drive of Conwy Council officers and the unswerving vision that without doing something different, Colwyn Bay would face a difficult and uncertain future, none of what has been described would have been possible.

The journey has not ended. There is still considerable energy and passion needed to complete the project. However, what has been developed so far and what is envisaged for the future stays faithful to the overall aim of changing the waterfront of Colwyn Bay into an area that supports and inspires the regeneration of the area and the whole town for the current and future generations.

## References

Conwy County Borough Council/Coastal Engineering UK Ltd, November 2007. *Colwyn Bay Coastal Defence Strategy Plan – Strategic Assessment & Proposals;*

Mott McDonald, August 2002. *Colwyn Bay Waterfront Strategy – Phase 2 Report No. 61938/02/C.*

LDA Design, March 2009. Colwyn Bay Promenade. Design Report Stage D.

## Acknowledgements

The authors wish to acknowledge the role of a number of institutions and firms without whose support and professional inputs the Colwyn Bay Waterfront Project would not have come to fruition, including but not limited to - Conwy County Borough Council members and officers; Welsh Government; LDA Design; Laboratório Nacional de Engenharia Civil (LNEC); Royal Haskoning; ABPmer; Volker Stevin BV; Galliford Try; Mott McDonald Consulting Engineers.

**Coastal Management: Changing coast, changing climate, changing minds**
ISBN 978-0-7277-6114-9

ICE Publishing: All rights reserved
doi: 10.1680/cm.61149.629

# USACE Adaptation Approach for Future Coastal Climate Conditions and Total Water Level Impacts

**Heidi Moritz**, P.E., USACE, Portland, USA
**Kate White**, PhD, P.E., USACE, Washington, D.C., USA
**Ben Gouldby**, HR Wallingford, Oxfordshire, United Kingdom
**William Sweet,** PhD, NOAA, Silver Spring, USA
**Peter Ruggiero**, P.E, OSU, Corvallis, USA
**Mark Gravens**, P.E., USACE, Vicksburg, USA
**Patrick O'Brien**, P.E., USACE, Norwich, USA
**Hans Moritz**, P.E., USACE, Portland, USA
**Thomas Wahl**, University of Siegen, Siegen, Germany
**Norberto Nadal-Caraballo**, USACE, Vicksburg, USA
**Will Veatch**, USACE, New Orleans, USA

## Introduction

The U.S. Army Corps of Engineers (USACE) plays a major role in the effectiveness of coastal risk management within the United States. Based on lessons learned from Hurricane Katrina in 2005 and subsequent events, including the Mississippi River flood of 2011, the 2012-2014 droughts, and Hurricane Sandy in 2012, USACE has worked to improve our understanding of climate change and its impacts to our missions and operations within the coastal zone. USACE is mainstreaming climate preparedness and resilience with a focus on priority areas. We engage in external collaboration to improve our understanding of climate change impacts and vulnerabilities so that we can develop new policy and guidance to support adaptation implementation based on the best available and actionable science.

Climate change has provided us with an opportunity to re-think how we analyze and design our projects. Past approaches do not always consider the full combinations of water level components, their level of uncertainty, and how those might change over time. Since our projects are typically designed for a minimum of 50 years of service life, both performance and maintenance over the project life are important considerations in coastal risk reduction planning and design. Variable temporal and geographic scales for total water level and event loading projections, including storm description and characterization are considered in the planning and design of coastal risk reduction projects. Capturing and articulating the appropriate level of uncertainty is important to developing a realistic understanding of potential residual risks. At the project level, this requires us to define specific components of total water level and their varying impacts on project stability and performance. The goal is to adequately and cost-effectively project future climate conditions that may impact performance in a manner that will support a robust analysis of alternatives, effective long term planning and a realistic evaluation of project expenditures.

The USACE approach to adaptation to changing sea levels (USACE 2014) includes a carefully planned adaptive management process that permits transitions from one adaptation approach to another, depending on the magnitude and rate of changing sea levels. Identifying thresholds beyond which stability or performance are adversely affected is an important way to understand current and future vulnerability with respect to water levels. This process requires that future management options be developed to be robust to a range of future climate change conditions, as well as projected changes in demographics and coastal development. Adaptation implementation allows for lead time so that measures are in place prior to reaching identified trigger points and thresholds. This paper provides an introduction to technical guidance now being developed by an interdisciplinary and international expert team to address total water levels at coastal projects.

## Background Information on Total Water Levels

Coastal total water levels (TWLs) are the result of complex interactions between multiple oceanographic, hydrological, geological and meteorological forcings that act over a wide range of scales. Important components of TWL include astronomical tide, wave set-up, wind set-up, large-scale storm surge, precipitation, fluvial discharges, monthly mean sea level, and land subsidence/uplift. The contribution of each component of a TWL has the potential to significantly alter spatially and temporally varying flood hazards, resulting in a complex and non-linear problem. Exposure and shoreline type influences the relative contribution of components. However, under certain conditions, some TWL factors can be neglected, and linear superposition of various terms is appropriate, resulting in a relatively manageable problem from a hydrodynamic and statistical point of view. Figures 1 and 2 contrast a fairly generic combination of components for an open coast application to a more complex application of possible components on a coastline with variable geomorphology as well as inter-connected water bodies.

TWLs in coastal environments change over an extremely wide range of time scales from seconds to hours or days (e.g., atmospherically- driven wind waves and storm surges) to long-term changes in sea level occurring over hundreds and thousands of years. Intermediate temporal scales important in the planning and design of coastal risk reduction measures include decadal, interannual, and seasonal variability. Irregular phenomena including tsunamis and abrupt land level changes generated by earthquakes can produce dramatic local changes in water levels. Depending on the geomorphic setting, seasonal riverine flooding can also be a significant contributor to water level fluctuations.

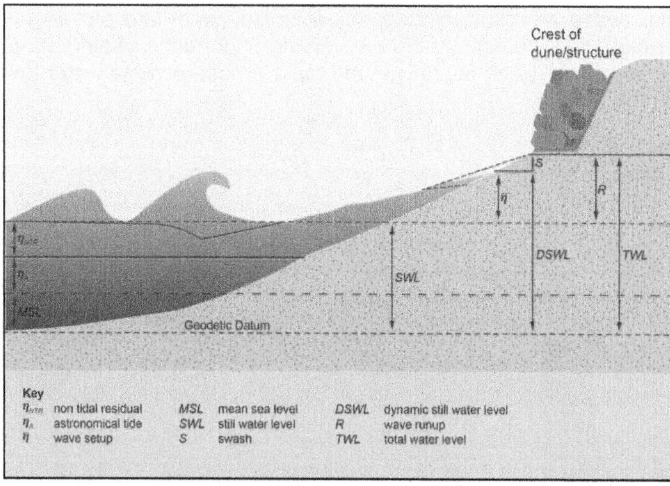


Figure 1: Generic profile view schematic of total water level components

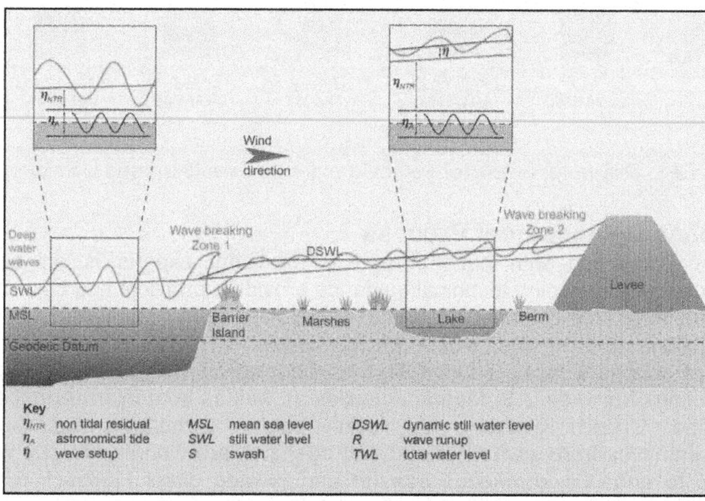


Figure 2: Profile view schematic of 'representative' U.S. Gulf Coast geometry and associated influence on total water levels

The local geomorphic setting can significantly influence the relative magnitude of the water level changes that are observed at any given location, with extreme events at some locations much more dependent upon the range of the tide and phasing during an extreme event. There are also regional scale characteristics that produce variability in the individual TWL components. The tide range for example, depends heavily on physical factors in the area of interest including topography, water depth, shoreline configuration, size of the ocean basin etc., whereas storm surge reflects both the intensity of the storm event as well as the bathymetric constraint of surge set-up imparted by the width of the adjacent continental shelf. Figure 3 illustrates variability of water level component contributions for select events on different coastlines of the U.S. A representative extreme event is shown for a tide gauge on each of five US coast lines: Alaska, Hawaii, West Coast, Gulf Coast, and East Coast. The components of the still water level (wave effects excluded) for a single event are plotted in terms of absolute magnitude, demonstrating the varying importance of different processes in different locations. Each event is in the top 10 most extreme water levels at the particular tide gauge.

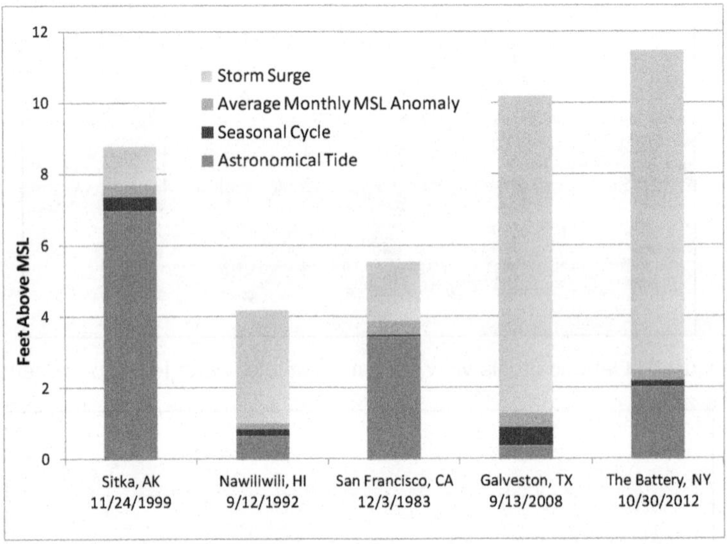


Figure 3: Still water levels for selected extreme events around US coastline

## Guidance Development Process

Close collaboration with national and international experts is an essential component of developing technical guidance providing practical and cost-effective measures to reduce potential vulnerabilities resulting from coastal climate change. Contributions and perspectives from academia, science and implementation agencies provide a broader foundation for guidance. The guidance incorporates contributions from key U.S. federal agencies as well as from international experts from the United Kingdom, the Netherlands, Germany, and Australia. Representatives from several US federal agencies are working directly with the USACE to enhance knowledge transfer and provide direct feedback regarding data and tools. The National Oceanic and Atmospheric Administration (NOAA) provides real-time and historical water level data (NOAA 2014), wave data (NOAA 2014b), and near term oceanographic forecasts (NOAA 2014c) along and within the waters of the U.S. Coast and Great Lakes (NOAA 2014d). The Federal Emergency Management Association (FEMA) maps coastal and riverine one-percent exceedance flood hazard areas to administer the National Flood Insurance Program. United States Geological Survey (USGS) basic and applied science that address extreme water levels generally falls under two main categories: direct measurements and models of extreme water levels, and indirect measurements and models of parameters such as nearshore bathymetry and beach topography that are primary controls on extreme water levels at the land-water interface on exposed coasts.

The guidance development team considered USACE primary coastal mission areas that are impacted by TWLs. In doing so, they considered framing questions that guide the level of analysis, topics to be included in the study documentation, and performance and functional risk analyses in support of the project. Example questions include the following:

- What assumptions have we made and do we know if those assumptions might change over time? (With climate change does "recurrence interval" end up on a sliding scale? Do we need to be using another term that does not imply stationarity?)
- Are the loading parameters related or independent and which of these might change with climate change?
- How certain are we of the loading and the responses?
- Have we projected what could happen if the project fails or does not perform?

A thorough physical understanding of the project area and project purpose is required in order to effectively assess the project's sensitivity to TWL. Projects can be assessed in terms of both their stability against the design loading and their ability to perform their function under these loadings. Table 1 summarizes a range of project stability and performance processes that may be impacted by total water level changes. Each process can be described using a dominant frequency of interest.

Table 1: Project processes of interest impacted by total water levels

| Performance Category | Process of Interest | Frequency of Interest | Typical Return Period Range of Interest |
|---|---|---|---|
| **Inundation** | short-term flooding<br>long-term flooding | high frequency<br>low frequency | 5 to 20 years<br>20 to 1000 years |
| **Erosion** | short-term erosion<br>long-term erosion | high frequency<br>low and high frequency | 5 to 20 years<br>10 to 50 yrs |
| **Wave Damage** | structure damage / stability | low frequency | 50 to 100 yrs |
| **Hydrostatic Loading** | differential loading | low and high frequency | 5 to 50 yrs |
| **Hydrodynamic Efficiency** | system drainage | low and high frequency | 5 to 50 yrs |
| **Operating Conditions** | wave run-up, overtopping, transmission, resonance | high frequency | 2 to 20 yrs |
| **Water Quality** | water quality | high frequency | 2 to 20 yrs |

Performance of coastal risk reduction factors under a variety of loading conditions is an important aspect that can have implications in the type and level of detail of analyses conducted for planning and engineering. Levels of performance and failure can range from full performance, stable with non-performance, damaged with non-performance, and complete failure with non-performance.

Projecting future conditions over the project life recognizes that there will be both stationary (e.g., storm surge frequency; Zhang et al., 2000) and nonstationary contributions to the TWL over time necessitating the consideration of scenarios in project alternative development. The most obvious potential non-stationary contributor to TWL is sea level change, which can increase the likelihood for low-probability (e.g., Tebaldi et al., 2012) and high-probability (Sweet and Park, 2014) TWL extreme event components. Just as different types of projects have different performance requirements, there are also differing levels of analysis required, depending on the project purpose, consequences of failure, and the likelihood of nonstationary vs. stationary conditions. As a result, the collaborative team expanded the concept of a tiered approach to adaptation (USACE 2014) with additional considerations for TWLs.

## Proposed Outline for Adaptation to Future Total Water Levels

The USACE approach to adaptation to changing sea levels (USACE 2014) provides a foundation for further guidance considering TWLs. Key elements of the sea level change guidance include understanding the project's vulnerability and thresholds. A proposed outline of steps to follow for adaptation with respect to TWLs follows:

1. **Define the forcing and exposure climate:**
   a. Define the potential envelope of total water level components over the project life cycle in terms of magnitude, range, duration, and frequency.
   b. Identify the parameters that have the potential for nonstationarity or wide confidence bands.
   c. Understand the range and combinations of water level components that control the project's stability and performance. (If the water level components result in a sliding recurrence interval over time, does the project stability and performance also have a variable definition with time?)
2. **Identify the project's stability and performance response to the forcing climate:**
   a. Outline the design functions' sensitivity to total water level. (low, high, or combinations of relevant frequencies)
   b. Articulate how the project responds to individual and combinations of parameters.
   c. Identify thresholds relative to non-performance, instability, and tipping points in project function.
   d. Define critical cross-over points that will be primary drivers of required design or maintenance changes.
3. **Outline expected project maintenance and performance over time:**
   a. What stability or performance issues would result in a design, management, or maintenance change?
   b. How much lead time would be required to respond in a cost-effective manner?
   c. What type of monitoring activities can be used to identify a required decision point?
   d. What level of risk are the users and stakeholders willing to accept? This will vary with project size and project type.

For example, one of the primary mission areas that will be impacted by changing water levels is Coastal Storm Damage Reduction (CSDR). The purpose of CSDR projects is to reduce the amount and frequency of damage to homes, businesses, and infrastructure caused by coastal storms due to the associated processes of erosion, inundation and direct wave impact. Damage reduction can be achieved either through measures to reduce the hazard (structural) or strategies to reduce the consequences of the event (non-structural). Below a summary is provided of

TWL-relevant processes that may impact either the loading or the performance of the CSDR project with revetments, seawalls, beach nourishment, and groins.

**Forcing and exposure climate:**

- Extreme high water level
  - Increased storm water elevation overtopping seawall, revetment, breakwater, levee, or dune resulting in degradation of protective system, erosion of structure volume, loss of structure crest elevation, landward-side erosion.
  - Water flow over protective system resulting in seawall failure (loss of backfill/compromise of tie-backs/toe failure), de-stabilization of revetment/breakwater armor, dune crest height reduction or dune feature removal. (threshold – structure crest elevation)
  - Increased duration of elevated storm water elevation with associated storm waves contributing to erosion of dune feature, de-stabilization of revetment/breakwater armor, scour of toe protection at base of seawall, revetment, breakwater, or levee.
- Extreme high energy wave conditions
  - Increased wave energy dissipation leading to erosion of beach berm, reduction of protective berm width, scarping of dune, loss of dune volume, loss of dune crest width, and loss of dune crest elevation.
  - Increased wave energy dissipation leading to displacement of revetment armor, scour of seawall, revetment, or breakwater toe protection, failure of structure foundation.
- Extended storm duration or multiple storms in succession.
  - Extent of beach berm and dune erosion is directly proportional to the duration of elevated water levels and high energy wave conditions leading to loss of berm and dune volume, loss of dune crest width, and reduction of dune crest elevation.
  - Multiple successive storm events increase the extent of beach berm and dune erosion leading to loss of berm and dune volume, loss of dune crest width and reduction of dune crest elevation.

**Stability and Performance Response:**

- Extreme high water level
  - Increased storm water elevation overtopping seawall, revetment, breakwater, levee, or dune resulting in upland inundation and wave propagation into protected infrastructure.
  - Flooding of berm resulting in ponding on back-berm resulting in aesthetic and health concerns, reduction in recreation area.
- Extreme high energy wave conditions
  - Contributes to total water level elevations increasing extent and elevation of inundation.
  - Direct wave impact on protected infrastructure resulting in extensive economic damages.

- Wave energy dissipation resulting in erosion of upland and compromise of upland infrastructure foundations, roadbeds, leading to structure damage, washout of road, etc.

- Extended storm duration or multiple storms in succession.
  - Long duration storms and multiple storms in succession lead to increased landward penetration of inundation, direct wave impacts, and erosion processes.

**CSDR Project Maintenance and Performance over Time:**

- Since CSDR projects typically form the first, or the second, line of defense against open coast forces, elevations and thresholds can be critical to performance. Exceedance of those thresholds at a pre-determined frequency would shift the project to a change in design or an alternative pathway.

- CSDR projects almost by definition are typically constructed in heavily developed areas. Any adaptive alternative pathway which may include shifting measures inland or retreat options will most likely have significant coordination steps to effectively complete. Lead time in this situation may take longer than just to design and execute the alternative.

- In some cases reaching a decision point for a given project area may be related to the level of risk and conditions that the stakeholders are willing to accept. Combinations of water level components including nonstationary elements can gradually transition to more frequent and unacceptable impacts.

## Conclusions

USACE is building climate change adaptation guidance that is intended to be used by field offices in the execution of planning, design, and operations of the wide inventory of USACE projects to ensure resilience to climate change. Project performance-related total water levels vary widely from region to region and between project types. Adequately preparing for performance and stability risks requires the ability to de-construct total water levels into time-varying components and projecting relative uncertainty. Concepts such as recurrence interval which imply stationarity may require new analysis and presentation techniques. USACE total water level guidance will discuss available data, graduated levels of analysis, and a range of tools necessary to produce study products which articulate residual risk by project type. The goal for USACE projects is to establish designs and project operations that are robust and adaptable to current and future total water levels.

## References

NOAA 2014 URL: http://tidesandcurrents.noaa.gov
NOAA 2014b URL: http://www.ndbc.noaa.gov/
NOAA 2014c URL: http://www.stormsurge.noaa.gov/models_obs_modeling.html
NOAA 2014d URL: http://www.nowcoast.noaa.gov

Tebaldi, C., B. H. Strauss, and C. E. Zervas (2012), Modeling sea level rise impacts on storm surges along US coasts, *Environ. Res. Lett.*, 7, 014032, doi:10.1088/1748-9327/7/1014032.

Sweet, W. V., and J. Park (2014), From the extreme to the mean: Acceleration and tipping points of coastal inundation from sea level rise, *Earth's Future*, 2 579-600, doi:10.1002/2014EF00272.

USACE, 2010. Standards and Procedures for Referencing Project Elevation Grades to Nationwide Vertical Datums, EM 1110-2-6056, Engineering Manual, 2010.

USACE, 2014. Climate Change Adaptation Plan, June 2014.

USACE, 2014. Procedures to Evaluate Sea Level Change: Impacts, Responses, and Adaptation, ETL 1100-2-1, Engineering Technical Letter, June 2014.

USACE, 2013. Incorporating Sea Level Change in Civil Works Programs, ER 1100-2-8162, Engineering Regulation, December 2013.

Zhang, K., B. C. Douglas, and S. P. Leatherman (2000), Twentieth century storm activity along the U.S. East coast, J. Clim., 13, 1748-1760, doi:10.1175/1520-0442(2000)013<1748:TCSAAT>2.0.CO;2.

**Coastal Management: Changing coast, changing climate, changing minds**
ISBN 978-0-7277-6114-9

ICE Publishing: All rights reserved
doi: 10.1680/cm.61149.639

# US North Atlantic Coast Comprehensive Study: Resilient Adaptation to Increasing Risk

**Jonathan Simm**, HR Wallingford, Wallingford, United Kingdom
**Amy Guise**[1], US Army Corps of Engineers, Baltimore MD, USA
**David Robbins**[1], US Army Corps of Engineers, Baltimore MD, USA
**Jason Engle**[1], US Army Corps of Engineers, Jacksonville FL, USA

## Introduction

Hurricane Sandy originated as a late season hurricane in the south-western Caribbean Sea on October 22, 2012. On October 29, 2012, the remnants of Hurricane Sandy in the form of a post-tropical cyclone made landfall near Brigantine, NJ. Because of its tremendous size, the storm drove a catastrophic storm surge into the densely developed New Jersey and New York coastlines, accentuating the vulnerability of the North Atlantic Coast to coastal flood hazard. As a result, there was considerable loss of life, extensive damage to development (Figure 1), and massive disruption to communities.

Figure 1: Hurricane Sandy impacts New Jersey coast at Atlantic City

On January 29, 2013, the Disaster Relief Appropriations Act, 2013 (Public Law 113-2), was enacted to assist in the recovery in the aftermath of Hurricane Sandy. Public Law 113-2, Chapter 4, authorized U.S. Army Corps of Engineers (USACE) to conduct a US$ 15-20 million comprehensive study to address the flood risks of vulnerable coastal populations and infrastructure at risk (Figure 2) throughout the

[1] The views and opinions expressed or implied in this article are those of the authors and are not officially sanctioned by an agency or department of the US government.

Figure 2: Metropolitan Transit Authority employees working around the clock to remove seawater from a flooded subway tunnel in Manhattan, NY, November 5, 2012 (Source: http://www.theatlantic.com/infocus/2012/11/hurricane-sandy-the-long-recovery/100405/ )

31,200 miles of the North Atlantic coastal region in areas that were affected by Hurricane Sandy. The goals of the NACCS were to:

- Provide a risk management framework, consistent with the NOAA-USACE Infrastructure Systems Rebuilding Principles; and
- Support resilient coastal communities and robust, sustainable coastal landscape systems, considering future sea level and climate change scenarios, to manage risk to vulnerable populations, property, ecosystems, and infrastructure.

The resulting North Atlantic Coast Comprehensive Study (NACCS) outputs provide a step-by-step approach, with advancements in the state of the science and tools to conduct or facilitate three levels of analysis (see http://www.nad.usace.army.mil/CompStudy). Tier 1 (completed as part of the NACCS) is a regional scale analysis, Tier 2 would be conducted at a State or catchment scale (conceptual Tier 2 evaluations were completed for each State and the District of Columbia), and Tier 3 would be a local-scale analysis that incorporates benefit-cost evaluations of coastal storm risk management plans. Using the tiered analyses is expected to enable communities to understand and manage their short-term and long-term coastal risk in a systems context. The geographical scope of the NACCS includes all the coastal areas significantly affected by Hurricane Sandy's storm surge, including in the District of Columbia and the States of New Hampshire, Connecticut, Rhode Island, New York, New Jersey, Delaware, and Maryland, and the Commonwealths of Massachusetts and Virginia (see Figure 3).

Managing short-term and long-term risk across local, regional, Tribal, State, and Federal agencies, nongovernmental organizations (NGOs), academia, business, and industry requires collaboration, data sharing, overcoming barriers, advancing the state of the science, and developing new partnerships and incentives for a renewed era of coastal storm risk management and action. Comprehensive action and monitoring to assess system responses are imperative to increase resilience and manage risk from future storms and a changing climate, including sea level change. Resilience as defined by the Infrastructure Systems Rebuilding Principles (established by NOAA and USACE) is the ability to adapt to changing conditions and withstand and rapidly recover from disruptions due to disasters. The Infrastructure Systems Rebuilding Principles (NOAA and USACE 2013), which were developed

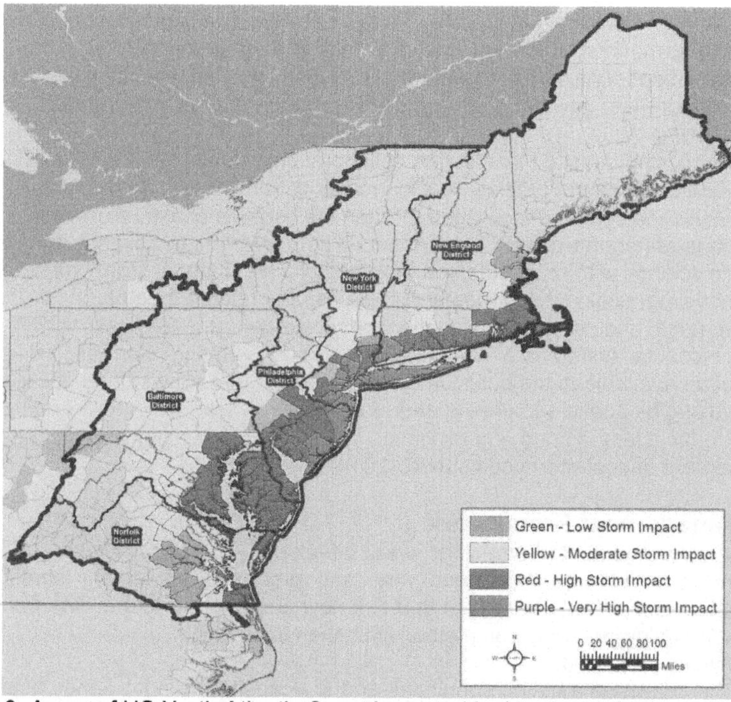


Figure 3: Areas of US North Atlantic Coast impacted by Hurricane Sandy

following Hurricane Sandy, provided the foundation for a framework to address flood risk to vulnerable coastal populations. These included:

- Working together in a collaborative manner across multiple levels of governance (including Federal, Tribal, State, and local) and with relevant entities outside of the government to develop long-term strategies that promote public safety, protect and restore natural resources and functions of the coast, and enhance coastal resilience;
- Improving coastal resilience by pursuing a systems approach that incorporates natural, social, and built systems as a whole; and
- Promoting increased recognition and awareness of risks and consequences among decision-makers, stakeholders, and the public.

As the NACCS is not a major Federal action, it does not include designs, evaluations for specific projects, or National Environmental Policy Act (NEPA) documentation.

## Findings Outcomes and Opportunities

Key findings, outcomes, and opportunities of the NACCS included the following:

- Flood risk is increasing for coastal populations and supporting infrastructure.
- Improved land use, wise use of floodplains, responsible evacuation planning, and strategic retreat are important and cost-effective actions.
- Communities should adopt combinations of solutions, including non-structural, structural, natural and nature-based, and governance measures to manage risk, where avoidance is not possible.

- Communities must identify their acceptable level of residual risk to plan for long-term, comprehensive, and resilient risk management.
- Many opportunities exist to improve risk management, including enhancing collaboration, building new partnerships, and strengthening pre-storm planning.
- Addressing coastal risk requires collaboration among local, regional, Tribal, State and Federal entities, NGOs, academia, business, and industry.
- Resilience can be encouraged through the use of a coastal storm risk management framework and continued commitments to advance the state of the science with respect to sea level and climate change, storm surge modeling, ecosystem goods and services, and related themes.
- Strategic and comprehensive monitoring is required to fully assess and adapt the coastal system to avoid future damages. Monitoring information must be made available to the public in a timely manner that allows rapid decision-making by public and private partners.
- Pre-disaster planning and mitigation can save communities approximately 75 percent of post-storm costs (NRC 2014).

## Risk Management Framework

The risk management framework was developed to allow States and local communities to identify their flood risk, and plan and implement strategies in collaboration with others, to reduce that risk now and into the future. The framework includes the following steps completed within the NACCS:

1. Initiating analysis.
2. Characterising existing conditions and forecasting the post-Sandy landscape. This included delineating the 31,200 linear miles of coastal shoreline into 39 planning reaches based on State boundaries, similar shoreline types, geomorphic features, and extent of existing or planned risk management projects. Future inundation conditions considered USACE and NOAA sea level rise scenarios over a 100-year time horizon.
3. Analysing the risk and identifying problems and opportunities. Exposure indices used included population density and infrastructure, social vulnerability, and environmental and cultural resources together with an overall exposure index including the probability of inundation.
4. Identifying possible solutions by shoreline type (Figure 4). Structural, non-structural, natural and nature-based features (NNBF), and policy/programmatic risk management measures (including floodplain and evacuation planning, managed retreat and buyouts) were gathered from a wide range of stakeholders and coastal experts and evaluated to identify those most appropriate or likely to succeed for different shoreline types.
5. Evaluating and comparing solutions using a systems approach. Potential reductions in risk were compared to the relative costs of the strategies/measures and the combinations of strategies and measures that provided the greatest risk reduction for the lowest cost.

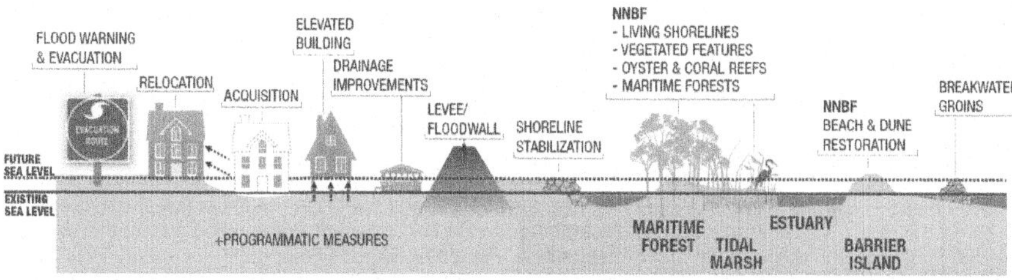


Figure 4: Layers of protection options for consideration for rehabilitation

Specific products generated by the NACCS to support these stages of the framework are set out in Table 1. Further steps which will need to be completed at a more local level include:

6. Selecting plan
7. Developing implementation plan (including completing pre-construction engineering and design, considering operation and maintenance issues, establishing adaptation thresholds and developing strategic monitoring plan)
8. Execute plan
9. Monitor and adapt (including measuring performance and benefit generation, assessing resilience and managing adaptively)

## Overcoming Institutional and Other Barriers

The study identified a number of institutional and other barriers to improving the situation but also some significant opportunities for action. These are summarised in Table 2.

## Example NACCS Product: Barrier Island and Back Bay Example Analysis

Based on the documented impacts from Hurricane Sandy and the NACCS sea level change evaluation, barrier islands, back bay areas, and embayments along the North Atlantic Coast, including the coastlines of New York, New Jersey, Delaware, Maryland, and Virginia are at a significant exposure to the impacts of sea level change and corresponding increases in flood risk. The exposure is probably increased by the urban development of areas on the barrier islands as well as along the back bay mainland shorelines which tends to preclude or inhibit the natural migration of wetlands in the back bays. Additionally, the back bays of barrier islands to the bay side of beaches and dunes as well as other areas of the North Atlantic coast including embayments and harbours are at risk of storm surge and tidal flooding via barrier island inlets. Long Beach Island, New Jersey was therefore identified as an illustrative example of a barrier island that might be impacted under sea level change inundation scenarios. The study therefore examined for this island both the performance of the Atlantic freeboard beach and dune system as well as coastal flooding from back bay areas. The example was not intended to evaluate the detail of actual coastal storm risk and consequences; rather it was intended to be an example of how such an evaluation would be conducted and to illustrate the relatively higher flooding risks associated with sea level change that exist on interior coastal shorelines.

Table 1: NACCS Products Supporting Steps of Risk Management Framework

| Framework Step | Technical Product[1] | Value Added |
|---|---|---|
| Initiate Analysis | Visioning Sessions Report & Focus Area Analyses | Identifies specific problems, needs, and opportunities for each focus area. |
| | Institutional & Other Barriers Report | Identifies six themes and each theme's institutional and other barriers, successes, and opportunities for action. Results are documented in the NACCS Main Report and also in the NACCS Institutional and Other Barriers Report. |
| | Collaboration Report | Documents outreach conducted throughout the course of the study. |
| Characterise Conditions | GIS Geodatabase | Includes data layers derived from the NACCS in a central location that can be used for additional analyses. |
| | Environmental & Cultural Resources Conditions Report | Provides a comprehensive report of environmental and cultural conditions for the North Atlantic Region. This information can be used in future National Environmental Policy Act (NEPA) documentation or environmental studies. |
| | US Fish and Wildlife Service Planning Aid Report | Provides a report of environmental conditions, including species and habitat considerations for the North Atlantic Region |
| Analyse Risk and Vulnerability | Storm Surge Modelling | Provides information about future storms and climate change, which will inform future studies and analyses. |
| | Barrier Island Sea Level Rise Inundation Assessment Report | Provides an example to complete an assessment of flood risk to a barrier island and back bay and vulnerability to the impacts of sea level change. |
| | Extreme Water Levels Report | Provides current and future extreme water levels for each of the NACCS sea level change scenarios for the 1, 0.1, 0.04, 0.02, 0.01, and 0.002 percent events for all 23 tide gages along the North Atlantic coastline of sufficient record length. |
| Identify Possible Solutions | NNBF Report and Brochures | Advances the science on NNBF (Natural and Nature-Based Features) strategic placement, how these features can be applied, and the benefits they provide. Includes a technical report (Bridges et al. 2014), as well as user-friendly consolidated brochures. |
| | Conceptual Regional Sediment Budget | Identifies the sources and sinks for sediment. Also identifies opportunities for the strategic placement of dredged material for NNBF. |
| | State and District of Columbia Analyses Appendix | Provides State by State chapters that discuss each State and District's post Hurricane Sandy landscape, sea level change considerations, and vulnerability assessment. |
| | Vulnerability Decision Tree | Provides a question tree that guides local users through the exposure and vulnerability assessment criteria and weightings. |
| | Coastal Program Guide | Provides detailed information on coastal storm risk management programs, including their authority, types of grants available, point of contact information, goals, etc. |
| | Measures Infographics | Displays various risk management measures discussed in the NACCS and depicts graphically how the measures work. |
| Evaluate and Compare Solutions | Enhanced Depth-Damage Functions for Coastal Storms | Provides generic coastal depth-damage curves for the region, as well as a report documenting the relationship of secondary and tertiary impacts. |

[1]Products are available at http://www.nad.usace.army.mil/CompStudy

Table 2: Summary of Institutional and Other Barriers to Achieving Study Goals and Opportunities for Action

| Barrier Theme | Opportunities for Action |
|---|---|
| 1. Risk/Resilience Standards | Develop consistent definitions for risk, vulnerability, resilience, and related terms and conduct research, as necessary, to develop design standards for resilience, performance metrics, a resilience scorecard, and other standards. |
| | Conduct a national vulnerability study of constructed USACE coastal storm risk management projects. |
| | Develop a national strategy for coastal storm risk management |
| | Develop regional and catchment-based plans, including a broad base of benefits, benefit quantification, and multi-objective approaches |
| 2. Communication and Outreach | Conduct coastal storm risk management visioning sessions with the public |
| | Continue to develop information and programs to educate the public about flood vulnerabilities, flood risk, residual risk, blended solutions, and pre-disaster and evacuation planning |
| | Develop a community of practice for Natural and Nature-Based Features (NNBF) |
| 3. Risk Management | Strengthen and enforce floodplain management policies |
| | Simplify the complicated network of coastal programs for communities |
| 4. Science Engineering and Technology | Improve research, coordination, and collection of pre- and post-storm data collection (e.g., climate and sea level change), including more rigorous instrumentation and monitoring for adaptive management. |
| | Develop better design and implementation guidance for NNBF for use in coastal storm risk management, including effects on long-term maintenance |
| | Compile information on ecosystem goods and services provided by NNBF |
| 5. Leadership and Institutional Coordination | Re-evaluate and complete authorized or planned projects in a comprehensive systems approach |
| | Increase coordination between Federal, State, local, and Tribal governmental agencies with responsibility for coastal management to foster mutual understanding of roles and responsibilities and to foster consistency between Federal programs affecting coastal management |
| | Support national adaptation planning |

## Submergence Assessment

An initial simple submergence assessment was applied to bands of sea level against the elevation of the island to identify the area that would be lost at varying levels without levees or flood walls. The assessment was based on a digital terrain model with a resolution of 6 feet and analyzed into 1 foot bands of ground height. The analysis assumes full hydraulic connectivity across the flood plain and therefore should be considered a conservatively high estimate. Figure 2 shows the resulting percentage loss in land area compared with the four sea level change scenarios considered in the NACCS. As previous studies have shown, this kind of analysis indicates significant loss of land for just 1 to 2 feet of sea level rise, with only the large beach berm and dune systems on the Atlantic Ocean escaping much of the inundation. The analysis was completed for several barrier islands and the results were broadly similar for each. The results confirm the vulnerability of back bays and the lack of a comprehensive coastal storm risk management solution.

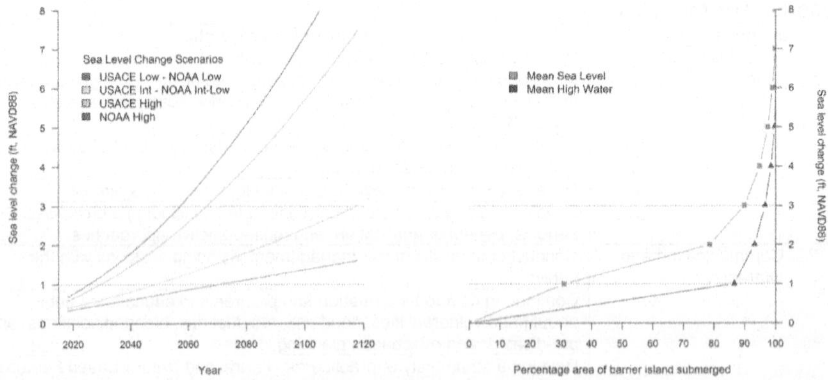


Figure 2: Submergence due to sea level change

**Storm Inundation Analysis**

Table 3 presents a description of the various analyses completed as part of the NACCS barrier island and back bay example. The storm inundation analysis emphasizes the sensitivity of the island to a relatively modest rise in sea level. In particular, risk of the island road network to back bay flooding, even during relatively modest storms, could affect access. Although less than 20 percent of the road network is flooded during such annual storms and only to an average depth of about a foot, just a foot increase in relative sea level increases the percentage to about 70 percent, with some roads flooded up to 4 feet. With an increase in 3 feet of relative sea level, the road network becomes unusable.

Flood risk increases with sea level change. As part of the storm inundation analysis, annual damages were estimated based on market valuations of structures in the different zones of the island, including ocean front, ocean block, ocean side, bay side, and bay front. The costs were normalized to present the concept of increasing risk and corresponding damages associated with future storm events (excluding wave attack and erosion because the analysis used only depth-damage relationships) coupled with sea level change. Table 4 explains the future expected annual damages as a fraction of total current (sea level rise 0 feet) damages, with a quadrupling of damages for just 1 foot of sea level rise.

**Implications of Barrier Island Analysis for Adaptation**

The analysis suggests that the beach berm and dune system can be maintained in a relatively robust condition even with 6 feet of sea level rise. Coastal storm risk management efforts should instead focus on back bay flooding. Combinations of the following measures could be considered instead:

- Comprehensive back bay risk management. Maintenance of leisure access for boating will be important where bulkheads already exist; new (or elevated) bulkheads could be considered. Stepped features may be possible. For locations where natural beaches exist, terraced features incorporating ecological components could be considered. In both cases, paths for access and viewing are possible.

Table 3: Storm analysis stages

| Storm analysis stage | Description |
|---|---|
| Scenario assumptions | Because the issues with barrier islands are primarily associated with the amount rather than the rate of sea level change, future scenarios were examined following 1foot, 3foot and 6foot increases rather than selecting a specific sea level change rate scenario.. The increased water levels also affect the wave heights, both to a limited extent in conditions offshore (e.g. in 30 feet of water) but more significantly in the near-shore conditions where the wave breaking occurs. Here, as nearshore wave heights become strongly dependent on the available water depth, the generation of greater water depths by increases in mean sea levels lead to correspondingly increased nearshore wave heights. For each scenario, extreme wave and water level conditions for the 100, 10, 3, 1, and 0.1 percent flood events were examined. Calculations were also performed for two response scenarios: one where defences and dune systems were raised in line with sea level and the other where no such improvements were included. |
| Extreme waves and water levels | The analysis uses offshore waves based on data from the National Data Buoy Center at the nearest available offshore location. A coincident wave and water level record length of 24.9 years was available which included 139 events (including Hurricane Sandy) where the significant wave height exceeded 13 feet. Equivalent coincident water levels were based on recorded sea level data, using Monte Carlo simulation to fill any data gaps and a joint probability distribution of waves and water levels obtained. Wave heights were transformed to the nearshore taking account of wave refraction and breaking and then the data was analysed to obtain estimates of extreme wave heights. Because there is a relatively low tidal range on the North Atlantic coast, a strong correlation exists between the most extreme waves and hurricane or other significant storm surges. |
| Beach/Dune Profile response | A DUROS+ empirical dune model (van Rijn, 2013) validated by good field data obtained before and after Hurricane Sandy (Stockton, 2012), was used to predict beach-dune profile response, using a representative uniform sediment size of 0.152 mm. The predicted run-up (exceeded by 33 per cent of the waves) necessary for the empirical model was calculated using van Rijn (2008) with results ranging from 6.5 feet for the 100 percent storm event to 8.7 feet for the 0.1 percent storm event. In the sea level rise scenarios with nourishment, these results barely changed. In the sea level rise scenarios without nourishment, greater cut-back of the dunes occurs, but as sea levels continue to rise, the cut back of the dune crest seems to reach a threshold beyond which further erosion does not occur. Instead, under extreme conditions, further erosion is focussed on the submerged part of the beach profile. |
| Dune overtopping calculations | The EurOtop manual (Pullen *et al*, 2007), was used to assess overtopping rates based on the modelled beach-dune profile and wave heights at the toe of the beach and taking account of the crest height, toe level and a simplified structure slope. Large overtopping rates can be maintained over a large part of the tidal cycle and hence overtopping rates were calculated over a full tidal cycle. Because of the significant variation in beach-dune profiles, a sensitivity analysis was conducted and found several orders of magnitude difference in overtopping rates depending on which beach-dune profile was selected. This issue could have been explored further, but in practice the inundation of the island was dominated by inflows from the back bay. |
| Inundation modelling (ocean & bay shorelines) | The computational mesh for the flood spreading model RFSM-EDA (Jamieson et al. 2012a and b) was made up of relatively small irregular polygon impact zones to capture the inundation spreading across the narrow barrier island. Discharge boundary conditions were applied using the dune overtopping rates on the ocean shore and a water level on the bayshore which represented the average sea level over the 24 hours that follow the peak of the event. Plots for different sea level rise scenarios were created indicating the proportions of the whole island, and of the road network, that would be inundated by different flood depths. The proportions of inundation are slightly higher for the case of the road network reflecting lower ground elevations for the roads than for the property parcels. |
| Flood risk analysis | The total impact of flooding for each scenario (return period, sea level rise) was calculated by combining the maximum flood depths with a depth-damage function in each grid cell based on those used in HAZUS/ HEC-FIA (Hydrologic Engineering Center Flood Impact Analysis) . Property values were based on average real-estate prices for different zones across the width of the island. |

Table 4. Future Expected Annual Damages as a Fraction of Total Current Damages

| | Relative sea level rise | | | |
|---|---|---|---|---|
| **Property Zone** | **0 feet** | **1 foot** | **3 feet** | **6 feet** |
| Bay front | 0.34 | 0.64 | 0.80 | 1.16 |
| Bay side | 0.46 | 1.59 | 2.07 | 3.03 |
| Ocean side | 0.10 | 1.05 | 2.09 | 3.17 |
| Ocean front | 0.00 | 0.03 | 0.29 | 0.56 |
| Ocean block | 0.10 | 0.58 | 0.78 | 1.15 |
| **TOTAL** | **1.00** | **3.88** | **6.04** | **9.07** |

- Modifications to drainage systems. Structural coastal storm risk management measures alone would not solve all the flooding problems because, during high water events, flooding can occur by water backing up the drainage systems. Flap valves or sluices would, therefore, need to be installed on all outfalls. Significant rainwater storage would also be required, possibly located within modified features.
- Elevation measures. Property elevation (Figure 3) remains a valuable tool to limit damage in the event of flooding. Elevation of the road network to improve access could be considered, but would require proper drainage and rainwater storage, for example located beneath such elevated roads.

Figure 3: Flooding on 11th Street Ocean City NJ on October 31, 2012 (Courtesy: USACE, see: https://www.flickr.com/photos/philadelphiausace)

## Conclusions

Hurricane Sandy revealed to the Northeast United States of America where the vulnerabilities to coastal storms exist, now and into the future. With projected population increases, climate change, and existing barriers to comprehensive coastal storm risk management, the risk to populations and infrastructure will continue to increase. Local municipalities were the first to feel the impacts of Hurricane Sandy and are also the first line of defence in hazard mitigation planning. The subsequent NACCS study revealed an opportunity for local governments in the NE USA to lead a new era of coastal storm risk management through intensive and proactive pre-storm initiatives. As stated by NRC (2014) "*Every dollar spent before an event saves four to five dollars in reconstruction costs after.*" The study identified that integration of a common coastal storm risk management framework, evaluations of blended solutions and adaptation, and collaborative, strategic investments in coastal storm risk management would facilitate resilient, thriving communities. It also concluded that local governments would need to commit to wise land use planning and zoning, use of floodplains, and evacuation planning so that future impacts from coastal storms and sea level rise could be reduced and resilience could be increased.

## References

Bridges et al (2015) Bridges, T., Wagner, P., Burks-Copes, K., Bates, M., Collier, Z., Fischenich, J, Gailani, J., Leuck, L., Rosati, J., Russo, E., Shafer, D., Suedel, B., Vuxton, E., and Wamsley, T. (2015). *Use of Natural and Nature-based Features for Coastal Resilience.* Vicksburg, MS: U.S. Army Engineer Research and Development Center.

Jamieson, S., Lhomme, J., Wright, G. and Gouldby, B. (2012a). Highly Efficient 2D Inundation Modelling with Enhanced Diffusion-wave and Sub-element Topography. *Proc. Inst. Civil Engineers – Water Management* 165(10), 581–595.

Jamieson, S., Wright, G., Lhomme, J., and Gouldby, B. (2012b). Validation of a Computationally Efficient 2D Inundation Model on Multiple Scales. *Floodrisk 2012 Proceedings.* Rotterdam: Taylor & Francis Group.

NOAA and USACE (2013). Infrastructure Systems Rebuilding Principles. Available at http://coastalmanagement.noaa.gov/resources/docs/infsysrebuildingprinciples.pdf Accessed September 13, 2014.

NRC (2014). *Reducing Coastal Risk on the East and Gulf Coasts.* Washington, DC: Committee on U.S. Army Corps of Engineers Water Resources Science, Engineering, and Planning: Coastal Risk Reduction; Water Science and Technology Board; Ocean Studies Board; Division on Earth and Life Studies; National Research Council. The National Academy Press.

Pullen, T., Allsop, N.W.H., Bruce, T., Kortenhaus, A., Schüttrumpf, H. and van der Meer, J.W. (2007). *EurOtop: Wave Overtopping of Sea Defence and Related Structures: Assessment Manual.*

Stockton (2012). *Beach Dune Performance Assessment of New Jersey Beach Profile Network (NJBPN) Sites at Long Beach Island, New Jersey After Hurricane Sandy.* The Richard Stockton College of NJ Coastal Research Centre.

van Rijn, L.C. (2008). *Beach and Dune Erosion Due to Storms.* Proc. Int. Conf. Coastal Engineering 2008.

van Rijn, L.C. (2013). *Erosion of Coastal Dunes Due to Storms.* Available at http://www.leovanrijnsediment.com/papers/Coastalduneerosion2013.pdf. Accessed September 15, 2014.

**Coastal Management: Changing coast, changing climate, changing minds**
ISBN 978-0-7277-6114-9

ICE Publishing: All rights reserved
doi: 10.1680/cm.61149.651

# Nature Based Solutions: Lessons from Across the World

**Siddharth Narayan**, NCEAS, UC Santa Barbara, USA
**Nigel Pontee**, CH2MHILL, Wiltshire, UK
**Michael W. Beck**, TNC, UC Santa Cruz, USA
**Adam H. Hosking**, CH2MHILL, Wiltshire, UK

## Introduction

This paper reviews the range of nature-based solutions for wave attenuation and erosion reduction on the coast. A whole suite of coastal protection approaches exist that incorporate natural or nature-based features in some form. We use the term "Nature – based Solutions (NBS)" to describe these approaches. The paper presents initial results of efforts to synthesise and understand existing knowledge on the role of natural coastal habitats in coastal protection. This is based on work carried out as part of a Science for Nature and People (SNAP) project (www.snap.is/coastaldefenses) as well as practical engineering experience. The paper first examines the terminology surrounding NBS. The following section describes key factors that need to be considered when choosing and implementing NBS, especially in comparison with more traditional hard engineered solutions. We then provide a brief overview of some NBS projects to further illustrate the range of approaches that exist. Finally, we conclude with a summary of the key ideas and considerations regarding NBS in coastal engineering today.

## Context and Terminology

Recent decades have seen a shift in coastal protection and risk reduction from traditional 'hard' or 'grey' engineering solutions that exclusively involve structural features (e.g. seawalls and breakwaters) to 'softer', more eco-friendly solutions. The term 'soft' engineering came into use in the 1980's to describe those solutions which attempt to have beneficial influence on coastal processes and in doing so improve the level of service provided by a sea defence or coast protection structure. In the last 5 to 10 years, a variety of new terms have started to be used to describe these type of solutions including "Building with Nature", "Living Shorelines", "Engineering with Nature", "Ecological Engineering", "Green Infrastructure", etc. (Borsje et al., 2011, Bridges et al., 2015, Chapman and Underwood, 2011). This variety of terms, coupled with a lack of specific details about the physical elements of the solution can cause confusion. In this paper we use the term Nature Based Solution (NBS) as the catch-all term for these approaches.

In our definition, NBS consist either wholly or partially of natural features that are designed to offer or improve coastal protection. This includes (1) fully natural solutions (e.g. naturally occurring marshes and mangroves), (2) managed natural solutions (e.g. creation of oyster reefs or beach/dune widening), (3) hybrid solutions that combine structural engineering with natural features (e.g. marsh-levee systems or dune-dyke systems) and (4) 'environment-friendly' structural engineering (e.g. vegetated engineering or bamboo sediment fences). In this discussion we exclude non-structural measures such as policies, building codes and emergency response (e.g. early warning and evacuation plans).

For fully natural NBS the coastal protection service (i.e. erosion control, wave reduction or flood storage) is provided entirely by a coastal habitat. In this case, the habitat occurs naturally and is not specifically managed for coastal protection (i.e. offshore coral reefs, nearshore mangrove belts or marshes). For managed NBS the coastal protection service is provided by a coastal habitat that is created and/or managed for this purpose (i.e. artificial coral/oyster reefs, planted salt marsh). Examples of managed NBS include the coastal dune management programme in the UK (Pye *et al.*, 2007) and Netherlands as well as oyster reefs in Louisiana (Kirkpatrick, 2013). For hybrid NBS the coastal protection is provided by a coastal habitat in combination with a 'hard' defence (i.e. salt marsh in front of a levee). Flood reduction schemes that combine hard flood defences such as seawalls with habitats, or erosion mitigation measures that combine concrete sills with salt-marshes are also examples of hybrid NBS (Wheeler *et al.*, 2008, Hardaway *et al.*, 2009).

As mentioned above, there is some confusion as to what constitutes NBS. Views differ between countries and between different professions. For instance, ecologists might consider NBS to only include solutions with a living element (e.g. marshes, mangroves, reefs). Engineers, on the other hand, commonly use the term NBS to describe solutions that use or mimic natural processes (e.g. beach nourishment projects that rely on coastal processes to redistribute the sediment). Offshore breakwaters are typically seen as hard engineering solutions though these are often used to protect or enhance habitats such as beaches, dunes or cliffs in order to provide coastal protection. Such a combination of breakwater and beach falls within the hybrid NBS category in our classification. Additionally, there are solutions that modify hard engineering solutions to make them more environment-friendly (or 'softer'). 'Vegetated engineering' or 'greenwashing' is the practice of modifying coastal protection structures to create or enhance ecological value. One example is the provision of ecological niches in breakwaters and seawalls (Chapman and Underwood, 2011). Another example of 'softer' hard engineering solutions is 'bio-engineering', such as the use of bamboo wood fences to trap sediment (Halide *et al.*, 2004). These examples illustrate the importance of specifying the physical elements of the solutions being considered (e.g. embankments fronted by saltmarsh, revetments fronted by beaches, offshore breakwater backed by beaches etc.) in order that different parties understand what is being proposed.

## Considerations for NBS Implementation

A number of countries have started to implement NBS as part of wider coastal risk reduction strategies in response to sea level rise and the ongoing loss and deterioration of coastal habitat. Since such solutions are comprised in some way of natural habitats they often also provide a number of additional benefits collectively termed ecosystem services. Ecosystem services are defined as services of a supporting (e.g. nutrient recycling), provisioning (e.g. raw materials such as wood, fodder), regulatory (e.g. coastal protection, carbon sequestration) or cultural nature (e.g. recreational, science).

In addition to the structural and hydraulic parameters analysed for a conventional engineering structure, NBS design requires consideration of a range of biological, chemical and physical parameters to ensure the successful establishment and functioning of habitats. For example, mangrove restoration is unlikely to be possible in environments that do not have the right range of soil moisture, salinity, nutrient and energy gradients required for mangrove establishment (Balke *et al.*, 2011). Where the correct conditions for the habitat do not exist, considerable anthropogenic interventions (and costs) are likely to be required.

Guidelines and manuals for NBS are less well established than for conventional engineering approaches although there are an increasing number of studies that can contribute to the development of NBS guidelines and manuals. A few design guidelines and manuals have been developed, such as the Building with Nature guidelines in the Netherlands, the Managed Realignment guidelines in the UK (CIRIA, 2004) that build on lessons learnt from multiple NBS projects. More recently, the US Army Corps of Engineers provide a framework and metrics for assessing and ranking NBS alternatives alongside other coastal protection measures for the Atlantic coast of the USA (Bridges *et al.*, 2015). The Living Shorelines projects on the east coast of the United States and 'greenbelt' mangrove and salt-marsh restoration projects in China (Chung, 2006) and Vietnam (see case-studies below), are some examples of large-scale NBS implementation. Here we examine a few case-studies from the US, UK, Netherlands and Vietnam that represent a range of NBS types, drivers and settings.

While specific design considerations may differ between NBS and grey engineering solutions, the design process is essentially the same. Existing frameworks for NBS design for coastal protection follow a process very similar to the design process for a conventional coastal engineering structure (e.g. the Dutch Building with Nature framework and the USACE Engineering with Nature framework).

The standard of protection provided by an NBS, is, in general, less well understood than an engineering structure. This arises due to the higher natural variability of a natural habitat and the relative lack of experience and knowledge regarding its sea defence performance. The coastal protection effectiveness of an NBS will be determined by several factors:

- site exposure:(e.g. mangroves and marshes tend to be located on low energy coasts; coastal habitats are more effective in lower water levels); and
- habitat characteristics (e.g. the higher the reef crest, or the height and density of a mangrove stand, the higher is its effectiveness).

It is often perceived that NBS provide a lower standard of service than grey engineering structures. However in some cases, such as the managed coastal dunes in The Netherlands, NBS may provide levels as high as traditional hard engineering structures (achieving a 1 in 10,000 return period water level in some places; Most and Wehrung, 2005). Coral reefs have also been found to be highly effective as off-shore breakwaters (Ferrario *et al.*, 2014). Hybrid NBS make use of natural habitats to complement an engineering structure either by protecting the structure or by increasing its standard of service. In some environments, oyster reefs may be a suitable alternative to traditional breakwaters where some form of seaward protection is required for an inter-tidal or coastal habitat (Kirkpatrick, 2013).

The costs of NBS can vary significantly depending on the habitat and site characteristics. For instance, while marsh creation in a naturally favourable environment is relatively inexpensive, restoration after damage in a storm or in unsuitable (e.g. polluted) environments can be difficult and expensive (Barbier, 2013). Once established, NBS are typically seen as low cost 'no regret' options that are cheaper and easier to maintain than grey engineering structures (Tinch & Ledoux, 2006). Examples also exist of hybrid solutions where a salt-marsh, willow or mangrove belt seaward of a dyke have been shown to greatly reduce the cost of dyke maintenance, and in some cases, have allowed a reduction in dyke crest height by complementing its protective function (Anthony and Gratiot, 2012).

A survey of expert opinions suggests that ecosystem-based options are more affordable and have positive additional consequences although they are not as effective as other options in reducing the impact of the hazard (Royal Society, 2014). However, we caution against making generalisations about the relative costs and standards of protection of NBS versus grey infrastructure due the variety of approaches that exist under the NBS banner.

## NBS Case Studies

The following section presents a number of case studies of NBS from around the world. These case studies illustrate the range of different approaches that can be adopted and include managed natural and hybrid.

### Howard Beach Flood Protection Study, New York, USA – Hybrid

Study Objective: This study examined strategies for Howard Beach (neighbourhood of New York City borough of Queens) in the aftermath of super-storm Sandy (The Nature Conservancy, 2013a). The main drivers of these mitigation strategies were to help coastal communities to reduce erosion and flood reduction risk and provide recreation benefits.

Reason for Consideration of NBS: While hard infrastructure was beneficial for storm events, NBS provided cost savings as well as significant aesthetic and environmental benefits to the community. NBS also improves water quality, provide new habitats and, under the right conditions, keep pace with sea level rise.

Alternative solutions considered: It was necessary to integrate NBS with more traditional coastal protection techniques to provide the necessary levels of protection to the densely populated coastal community and surrounding areas, which include the nearby JFK airport. The study identified a number of alternative hybrid approaches that would provide robust flood and erosion protection to the landward urban areas (Figure 1).

Project Description and Outcomes: The project included the definition of options ranging from barriers and gates across the entrance to Jamaica Bay (within which Howard Beach is located), through vegetated levees of various forms, to wetland restoration options as a storm surge buffer. The study undertook detailed hydrodynamic modelling of the various options to quantify their performance in reducing the flood risk exposure of the vulnerable assets and infrastructure. This analysis demonstrated that grey options are necessary to ensure the requisite level of protection under large storm surge conditions, and that the incorporation of NBS elements into a hybrid approach reduces maintenance burdens, protects against smaller events, and provides improved overall socio-environmental outcomes.

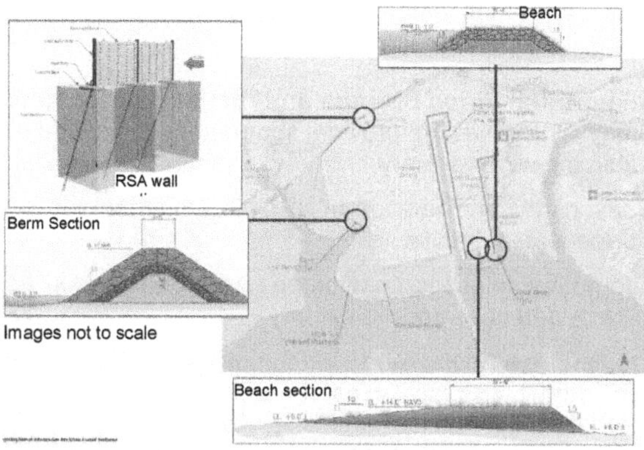

Figure 1: Design of grey and green coastal infrastructure

Lessons learnt: Where there is dense urban development with a high degree of flood or erosion exposure at the coast, it will likely be necessary to incorporate elements of 'grey' protection structures into a solution in order to provide sufficient certainty of protection under extreme weather and climate related events. Hybrid options are often more cost beneficial and environmentally preferable to a purely grey approach. Developing alternative strategies integrating natural infrastructure with more traditional coastal defense techniques can provide sustainable solutions that balance robust engineering and environmental benefits. This will be important in future rebuilding and resilience coastal projects.

**Ship Shoal Pipeline Erosion Control, Louisiana, USA – Managed Natural**

Study Objective: This study investigated alternative natural infrastructure designs that will enhance wetland shoreline stability where pipelines intersect waterways (The Nature Conservancy, 2013b).

Reason for Consideration of NBS: The traditional grey infrastructure, such as bulkheads and rip rap plugs, were being undermined by eroding tidal channels.

Alternative Solutions Considered: Given the dynamism of the wetland environment and its environmental value, NBS were considered to provide multiple benefits and flexibility to adapt over time. A wide range of fully natural and hybrid NBS were investigated (Figure 2) and various options proposed based on local context.

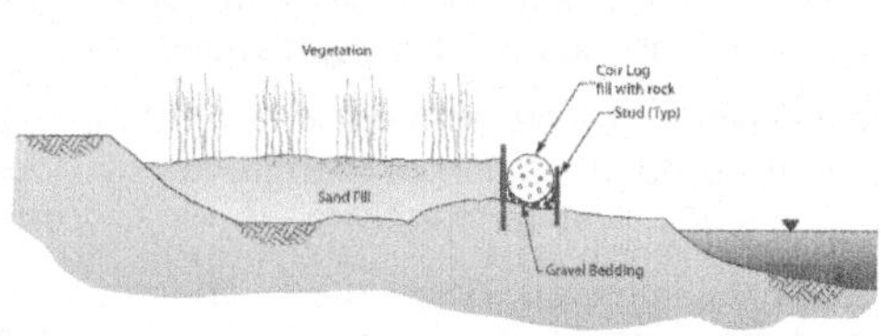

Figure 2: Schematic of NBS coastal protection options

Project Description and Outcomes: Shell filled gabions, permeable concrete mattresses with planted vegetation, and vegetated embankments are amongst the recommended solutions. Each of these options had the benefit of biological production and the ability to naturally adapt to changing conditions thereby providing sustainable solutions to the erosion problem.

Lessons learnt: Through the design review process for this project, a wide range of NBS approaches were defined and appraised. While each had various merits and opportunities in relation to its habitat value, the certainty of protection provided to the vulnerable infrastructure was ultimately the key factor in determining the preferred approach. Traditionally this might have pointed to simply expanding or rebuilding the existing 'grey' infrastructure, but with this assessment, the project's implementation, and future monitoring will help demonstrate how NBS performs to improve the longevity for infrastructure assets.

**Alkborough Flood Storage and Wetland Restoration Scheme, UK – Hybrid**

Study Objective: To provide compensatory intertidal habitat and reduce flood risk within the wider estuary.

Reason for Consideration of NBS: Requirement for compensatory habitat.

Alternative solutions considered: Additional works to raise defences around estuary.

Project Description and Outcomes: The project converted 1.4 square miles of arable farmland into new inter-tidal habitat (Figure 3). The also acts as a flood storage reservoir during extreme events helping to lower water levels in the inner and middle parts of the Humber Estuary. This reduces the amount of work needed to improve the existing hard defences around the estuary. The scheme incorporates new flood defences around the rear of the site. Further details can be found in Wheeler *et al.*, (2008).

Figure 3: Alkborough Flood Storage scheme in the Humber Estuary, UK

Lessons learnt: The erosion of exit channel leading from scheme to main estuary was extremely rapid with the majority of change occurring over the first week following opening.

**Vietnam Red Cross Mangrove Restoration Scheme – Hybrid**

Objective: The Vietnam chapter of the International Federation of Red Cross Societies spear-headed a mangrove reforestation effort from 1993 to 2010 (IFRC, 2011). The objective of the project was to provide coastal protection and other ecosystem services to local populations.

Main Drivers: The project was motivated by large-scale deforestation of mangroves by logging and shrimp farm activity.

Figure 3: IFRC mangrove afforestation project (Source: IFRC, 2011)

Alternative solutions considered: The majority of these efforts were intended as hybrid measures and integrated with artificial flood defences (sea dykes) from the start. No alternative coastal management measures were examined.

Project Description: The project comprised mangrove reforestation of nearly 9000 ha, to protect 100 km of sea-dykes and coastal assets (Figure 4).

Outcomes: The measures were found to be highly cost-effective, with Benefit-Cost Ratios (BCRs) varying from 3:1 to 28:1. Three communes that were hit by similar storms before and after the mangrove reforestation experienced much less flood and sea dyke damage in the presence of the mangroves, the avoided damage in these communes being estimated at £53,000 to £196,000.

Lessons Learnt: The project showed that large-scale mangrove restoration can be done cheaply and successfully and can provide significant coastal protection among other ecosystem services.

**MacDill AFB Oyster Reef– Managed Natural**

Figure 4: Oyster Reef Balls in Macdill Air Force Base (Source: Kirkpatrick, 2013)

Objective: The MacDill Air Force Base near Tampa, Florida in the USA conducted an oyster reef restoration program over five years to assist further salt-marsh restoration activities. The objective of this project was to restore and stabilise an eroding shoreline using NBS to protect valuable coastal assets and provide conditions suitable for further salt-marsh restoration activities (Kirkpatrick, 2013)

Alternative Solutions Considered: Two alternative NBS solutions were trialled – planting salt-marsh and mangroves, neither of which succeeded.

Project Description: The three phase programme was funded by multiple institutions including the Air Force, the US Fish and Wildlife Service and other institutions. The oyster reefs were built and installed with the help of volunteers using Reef Ball and Oyster Shell Bag construction techniques (Figure 5).

Outcomes: Over 300 m of shoreline were protected by the reefs and monitoring shows considerable shoreline accretion accompanied by salt-marsh colonisation.

Lessons Learnt: The project provided useful experience in the design and construction methodologies for oyster reefs, including details of reef profiles, positioning, spacing, etc. The project provided educational opportunities and opportunities to engage local communities that may not have been achieved with more conventional coastal protection measures.

## Lessons for NBS Implementation

NBS implementation is often driven by a combination of considerations such as coastal protection and habitat conservation. These include conservation or restoration of coastal habitats, environment – friendly and sustainable strategies for private industries and local or regional policies such as the Living Shorelines programme in the USA that incentivises NBS (Swann, 2008). NBS implementation can also be a result of legislation (e.g. The EU Habitats Directive; European Commission, 2009) that promotes the use of 'green infrastructure' for coastal protection and conservation. This is comparable to the European Commission's Green Infrastructure Strategy (Fritz, 2014) or projects like BRANCH in the UK, under the aegis of Natural England, the Environment Agency and other non-profit and government organisations (BRANCH partnership, 2007). A major motivation for NBS is in the immediate aftermath of coastal disasters when billions of dollars are spent on recovery and restoration. An example of this is the EU Floods Directive (European Commission, 2007) which arose as a response to a series of coastal and fluvial flood disasters and which encourages the use of NBS in coastal protection.

Solutions involving inter-tidal habitats typically require more space than a conventional engineering structure, and as such, land availability and costs can be major constraints to their implementation. On the other hand, these solutions offer an opportunity for creating new habitat which is a key motivation for projects in Europe (BRANCH partnership, 2007). NBS projects that combine multiple objectives, such as coastal protection and habitat provision may face additional constraints on location and design. Actual design and implementation are often a combination of experimental design and experience gained from previous projects. The 'Building with Nature' programme in the Netherlands has used extensive experience in this field to come up with a basic design process and framework for similar projects elsewhere (van Slobbe *et al.*, 2013). Engineering and management guidelines also exist for the implementation of Managed Realignment projects that build on experience with these schemes in the Europe. Design guidelines for NBS also vary by habitat type. Management of dune and beach systems for coastal protection (Pye *et al.*, 2007), for example, requires a different approach in comparison with salt-marsh or mangrove systems (IFRC, 2011).

Figure 5: Waves break over coral reefs (Source: http://upload.wikimedia.org/wikipedia/commons/7/70/Papeete_-_Lagon.JPG)

Other more direct constraints for NBS are site conditions such as exposure of the habitats to high wave energy and/or extreme events. NBS that involve inter-tidal habitats are typically located in low energy environments and as such, have a limited potential for coastal protection in terms of risk reduction. In environments where sediment availability is not a constraint however, they can be very effective in trapping sediment and reducing rates of erosion. Terrestrial NBS such as coastal dunes and beaches, or sub-tidal NBS such as coral reefs provide higher standards of protection and can often be found in higher energy environments (Figure 7), though the higher levels of exposure also make them more susceptible to damage.

The cost-effectiveness of an NBS will vary depending on the standard of protection desired. In some cases the NBS are measures that complement and reduce the cost of creation and/or maintenance of traditional engineering approaches, such as the Ship Shoal Pipeline erosion protection. In some situations, such as the Vietnam case-study, NBS are reported to provide relatively cheap coastal protection alternatives in favourable environments. In such cases they are commonly reported to have low maintenance costs relative to grey engineering structures on account of their biological production and self-organization. Sometimes NBS are chosen despite their greater creation costs due to the wider benefits it provides in terms of other eco-system services, such as carbon storage, increased ecosystem diversity and/or health, water treatment, tourism, aesthetic and recreational benefits, etc. It should be noted that tourism, aesthetic and recreational benefits can also be obtained with suitably designed grey structures. A comprehensive cost-benefit comparison of NBS and grey engineering options is therefore necessary when evaluating the viability of a particular solution. However, the consideration of NBS alongside other engineering options is not yet widespread. Equally, some projects are specifically designed with an NBS in mind, and therefore do not compare these with more conventional engineering options (for example see MacDill Bay case study).

## Conclusions

NBS make use of natural coastal habitats to provide coastal protection. We recommend that such options be assessed alongside other structural and non-structural risk reduction measures when considering risk reduction projects. At present, comparison of various alternatives is not common practice within NBS projects and the information available on these projects are often qualitative or are not detailed enough to perform ex-post evaluations. An online interactive database of NBS projects from across the globe is now being developed as part of the SNAP Coastal Defenses Working Group (www.maps.coastalresilience.org/global, "Natural Defenses Projects" tab). The aim of this database is to gather and improve access to information on NBS projects of various types and settings, to enable better evaluation of the feasibility of future projects.

NBS follow a design process similar to conventional coastal protection measures, with additional requirement to consider specific habitat development needs. Like any other coastal engineering measure, design guidelines, project-specific studies and cost benefit assessments are needed before decisions can be made on the applicability and viability of NBS at individual sites. Consideration of these solutions alongside other conventional, 'hard' engineering solutions will help improve understanding of their advantages and limitations.

Guidelines for considering NBS as coastal protection solutions are limited and implementation is still predominantly case-specific. As such, information from existing projects can provide valuable information on appropriate designs, implementation techniques and techniques for cost-benefit assessments. From a review of existing literature and a number of case studies the following lessons are identified:

- A key advantage of NBS over hard engineering is the provision of additional eco-system services.
- By their nature NBS are constrained by the suitability of the habitat to the environment at the site, as well as by the suitability of its coastal protection service.
- Given the range of NBS solutions, including hybrid solutions, evaluation of viability requires specific analyses of affordability and effectiveness for each case. General statements with regards to costs and standards of protection of NBS compared to grey structures can be misleading.
- We believe that that NBS approaches can be more widely considered as an integral part of a suite of long term coastal protection solutions. However, this will requires further guidance and tools to select, design, construct and monitor these solutions.

## Acknowledgements

This work was partly supported by the Science for Nature and People (SNAP) Coastal Defenses Working Group (http://www.snap.is/groups/coastal-defenses/). The comments of the anonymous reviewers have helped clarify the paper and are gratefully acknowledged.

## References

ANTHONY, E. J. & GRATIOT, N. 2012. Coastal engineering and large-scale mangrove destruction in Guyana, South America: Averting an environmental catastrophe in the making. *Ecological Engineering,* 47, 268-273.

BALKE, T., BOUMA, T. J., HORSTMAN, E. M., WEBB, E. L., ERFTEMEIJER, P. L. & HERMAN, P. M. 2011. Windows of opportunity: thresholds to mangrove seedling establishment on tidal flats. *Marine Ecology Progress Series,* 440.

BARBIER, E. B. 2013. Valuing Ecosystem Services for Coastal Wetland Protection and Restoration: Progress and Challenges. *Resources,* 2, 213-230.

BORSJE, B. W., VAN WESENBEECK, B. K., DEKKER, F., PAALVAST, P., BOUMA, T. J., VAN KATWIJK, M. M. & DE VRIES, M. B. 2011. How ecological engineering can serve in coastal protection. *Ecological Engineering,* 37, 113-122.

BRANCH PARTNERSHIP 2007. Planning for biodiversity in a changing climate - BRANCH project final report. Natural England, UK.

BRIDGES, T., WAGNER, P. W., BURKS-COPES, K. A., BATES, M. E., COLLIER, Z. A., FISCHENICH, C. J., GAILANI, J. Z., LEUCK, L. D., PIERCY, C. D., ROSATI, J. D., RUSSO, E. J., SHAFER, D. J., SUEDEL, B. C., VUXTON, E. A. & WAMSLEY, T. V. 2015. Use of Natural and Nature-Based Features (NNBF) for Coastal Resilience. North Atlantic Coast Comprehensive Study: Resilient Adaptation to Increasing Risk. Vicksburb, MS: US Army Corps of Engineers: Engineer Research and Development Center.

CHAPMAN, M. G. & UNDERWOOD, A. J. 2011. Evaluation of ecological engineering of "armoured" shorelines to improve their value as habitat. *Journal of Experimental Marine Biology and Ecology,* 400, 302-313.

CHUNG, C.-H. 2006. Forty years of ecological engineering with Spartina plantations in China. *Ecological Engineering,* 27, 49-57.

h, 2004. Coastal and estuarine managed realignment: design issues. Leggett, D.J., Cooper, N.J. and Harvey, R., CIRIA Publications, RP681.

EUROPEAN COMMISSION. 2007. *The Habitats Directive* [Online]. http://ec.europa.eu/environment/nature/legislation/habitatsdirective/index_en.htm. Available: http://ec.europa.eu/environment/nature/legislation/habitatsdirective/index_en.htm [Accessed August 2012].

EUROPEAN COMMISSION. 2009. *The Floods Directive* [Online]. Available: http://floods.jrc.ec.europa.eu/eu-floods-directive.html [Accessed October 2014].

FERRARIO, F., BECK, M. W., STORLAZZI, C. D., MICHELI, F., SHEPARD, C. C. & AIROLDI, L. 2014. The effectiveness of coral reefs for coastal hazard risk reduction and adaptation. *Nat Commun,* 5.

FRITZ, M. 2014. Green Infrastructure - Enhancing Europe's Natural Capital. Budapest, Hungary: The European Commission, CEEweb Academy.

HALIDE, H., BRINKMAN, R. & RIDD, P. 2004. Designing bamboo wave attenuators for mangrove plantations. *Indian Journal of Marine Sciences,* 33, 220-225.

HARDAWAY C.S., J., MILLIGAN, D. A., O'BRIEN, K. P., WILCOX, C. A., SHEN, J. & HOBBS.III, C. H. 2009. Encroachment of Sills onto State-Owned Bottom: Design Guidelines for Chesapeake Bay. Gloucester Point, Virginia, USA: Virginia Institute of Marine Science.

IFRC (INTERNATIONAL FEDERATION OF RED CROSS AND RED CRESCENT SOCIETIES) 2011. Breaking the waves: Impact analysis of coastal afforestation for disaster risk reduction in Vietnam. Geneva: IFRC,.

KIRKPATRICK, J. 2013. Constructing Oyster Reef for Shoreline Stabilization and Restoration. *www.oyster-restoration.org.* MacDill AFB.

MOST, H. & WEHRUNG, M. 2005. Dealing with Uncertainty in Flood Risk Assessment of Dike Rings in the Netherlands. *Natural Hazards,* 36, 191-206.

PYE, K., SAYE, S. & BLOTT, S. 2007. Part 3: The geomorphological and management status of coastal dunes in England and Wales. *In:* DEFRA/EA (ed.) *Sand dune processes and management for flood and coastal defence.* Bristol, UK: Environment Agency,.

ROYAL SOCIETY 2014. Resilience to extreme weather. London: The Royal Society Science Policy Centre.

SWANN, L. The use of living shorelines to mitigate the effects of storm events on Dauphin Island, Alabama, USA. American Fisheries Society Symposium, 2008. 000-000.

THE NATURE CONSERVANCY 2013a. Integrating Natural Infrastructure Into Urban Coastal Resilience: Howard Beach, Queens. New York: The Nature Conservancy.

THE NATURE CONSERVANCY 2013b. Shell and TNC: Coastal Pipeline Erosion Control Using Oyster Reefs, Louisiana, US. In: THE NATURE CONSERVANCY (ed.) Green Infrastructure Case Studies.

TINCH, R. & LEDOUX, L. 2006. Economics of Managed Realignment in the UK: Final Report to the Coastal Futures Project. Environmental Futures Limited.

VAN SLOBBE, E., DE VRIEND, H. J., AARNINKHOF, S., LULOFS, K., DE VRIES, M. & DIRCKE, P. 2013. Building with Nature: in search of resilient storm surge protection strategies. *Natural Hazards,* 66, 1461-1480.

WHEELER, D., TAN, S., PONTEE, N., PYGOTT, J.., 2008. Alkborough scheme reduces extreme water levels in the Humber Estuary and creates new habitat. FLOODrisk 2008 - The European conference on flood risk management research in to practice 30 September - 2 October 2008 Keble College, Oxford, UK.

**Coastal Management: Changing coast, changing climate, changing minds**
ISBN 978-0-7277-6114-9

ICE Publishing: All rights reserved
doi: 10.1680/cm.61149.663

# Delivering Large Habitat Restoration Schemes: Lessons from the Steart Coastal Management Project

**James Scott**, CH2M, Exeter, UK
**Nigel Pontee**, CH2M, Swindon, UK
**Tim McGrath**, Wildfowl and Wetlands Trust, Steart Marshes, UK
**Richard Cox,** Environment Agency, Exeter, UK
**Matt Philips,** Kier, Exeter, UK

## Summary
The Steart Coastal Management Project is one of the largest managed realignment and habitat creation projects ever carried out in the UK. As compensation for disappearing habitat elsewhere in the Severn Estuary the Environment Agency (EA) has transformed 477 hectares of the Steart Peninsula into wildlife rich habitats whilst improving the flood defences and providing greatly improved provision for public access and community involvement. This complex scheme has faced and overcome numerous challenges including: community engagement, which turned opposition into support; development of a mosaic of habitat types; the design and construction of earthworks, which worked within the constraints of the site; the complete replacement of an extensive water management network; and the construction of a breach and exit channel. This paper outlines the key aspects of the project and documents the lessons that were learnt during the planning, design and construction phases of the project.

### Background
The extensive mudflats and saltmarsh of the Severn Estuary are under threat due to coastal squeeze. In the UK, the combined effect of sea level rise and existing flood defences are causing approximately 100ha of saltmarsh to be lost annually (Environment Agency, 1999). The Steart scheme was undertaken to provide compensatory habitat to replace the habitat being lost whilst also improving flood defences to the local area. The scheme involved the construction of new improved set back defences, an artificial creek network, a new water management network, and breaching the existing defences.

The Steart Peninsula is located on the north Somerset coast in Bridgwater Bay at the confluence of the River Parrett and the Bristol Channel (Figure 1). The Stolford to Steart Sea Defences Pre-Feasibility Report (Environment Agency, 1998) and the Shoreline Management Plan (Halcrow, 2010) for the area identified that it was a suitable location to create new intertidal habitat in the Severn Estuary and Bristol Channel area. Prior to the scheme being undertaken the peninsula was characterised by low lying ground which had been reclaimed in historical times for farming through the construction of embankments.

The scheme area covers around 600ha most of which was formerly used for agriculture. The area is bisected by strategically important high voltage power lines to the nearby Hinkley nuclear power station and has a single track road, Stert Drove, which serves the village of Steart near the end of the peninsula.

The peninsula is important for wildlife and people. It is an historic 'anciently enclosed' landscape and archaeologically sensitive zone. It features several archaeological monuments, including a deserted medieval village, mill mounds, ditched enclosures and a ship wreck. Parts of the peninsula lie within the Severn Estuary Site of Special Scientific Interest (SSSI), Special Protection Area (SPA), Ramsar Site and Special Area of Conservation (SAC) all of which must be safeguarded. Numerous protected species exist on the peninsula including water voles, otters, bats, badgers, great crested newts, reptiles and breeding and wintering birds.

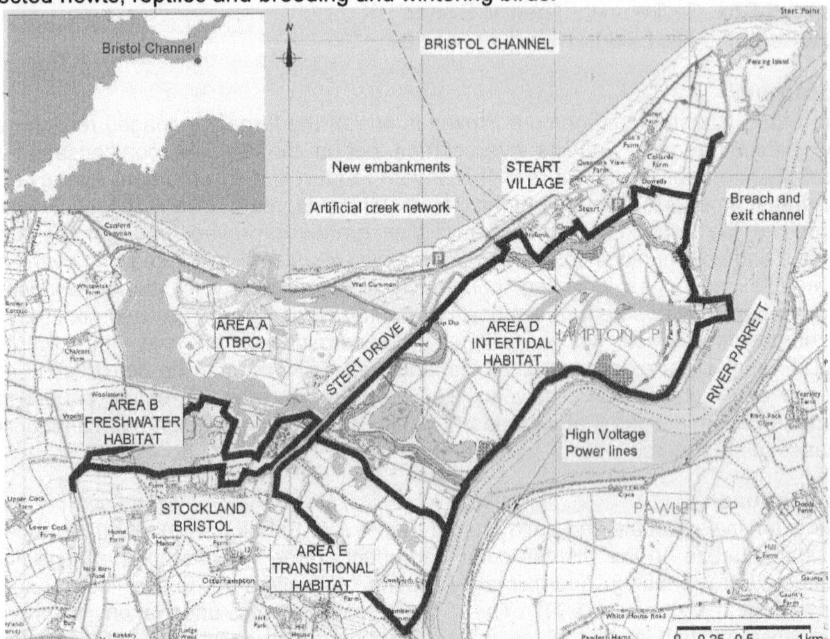


Figure 1: Study area

Prior to undertaking the scheme the existing earth embankment defences along the Parrett Estuary were in various states of repair. The project has transformed the peninsula by the creation of three distinct habitat types, (Figure 1). The project will generate environmental benefits in the form of major new wetlands, including almost 300 hectares of intertidal salt marsh and mudflats, 70 hectares of transitional brackish habitat, over 100 hectares of coastal grazing marsh and freshwater habitat.

## Planning and Management

### Master-planning

Throughout the development of the scheme, from 1998 to 2014, a key success was wining and retaining the trust of the local residents. Working in partnership with the community, their elected representatives, and various concerned organisations, the 'vision' for the future of Steart Peninsula, first drafted in 2002 (Babtie Brown & Root, 2002), formed the basis for the dialogue with the community and helped to educate and reassure the community of the EA's commitment to that vision.

In 2011, the creation of the Joint Environment Agency and Bristol Port Company Master Plan for the Steart Peninsula between the Environment Agency and the Bristol Port Company developed the original concepts to provide a vision for the after-use of the development helping to remove uncertainties for the local residents and local authorities (Environment Agency; The Bristol Port Company, 2011) (Figure 2).

Figure 2: Working with the local community to develop an economic, social and environmental vision for a sustainable future on the Steart Peninsula

This joint master plan helped the project team maintain focus and ensured effective communications with stakeholders and local community.

**Partnership**

The EA realised from the project's inception that there was an opportunity for a long-term site management partner to be integrated into the project team. After a competitive process the Wildfowl & Wetlands Trust (WWT) were selected in 2009 to deliver a shared vision for the site.

The addition of WWT completed a core project team which had a broad depth and breadth of knowledge to deliver the project's vision and guide the scheme through all the legal and regulatory challenges. Where necessary, specialist advice and input was sought to help secure over 20 different statutory consents and licences.

A strong team was formed by selecting key individuals for their strengths and by fostering a team approach throughout, whereby problems and credit were shared, issues were raised in no-blame culture and key decisions were supported by short 'decision papers' which examined the options and documented the rationale for the chosen approach.

### Community and Stakeholder Engagement

The project required extensive engagement with the community and stakeholders throughout the development of the scheme. In 2008 a small collaborative working group, Steart Environmental Advisory Group (SEAG), was set up between major stakeholders to assist the scheme's development. Additionally a wide range of interested organisations and individuals were invited to the Stakeholder Working Group to comment on the scheme's development. This was followed by public consultations to allow preferred decisions to be identified. The involvement of the community in the scheme and the formation of strong and open relations allowed any concerns or opposition to be turned into understanding and support, enabling the project team to quickly develop the most satisfactory proposals.

Early consultation allowed the local stakeholders to be kept informed, enabling them to feel engaged and take ownership of shaping the landscape on their doorstep. This culminated in the local community choosing the name of the new reserve 'Steart Marshes'. Significantly, by 2011, this process meant that there were no objections received during the public rights of way diversion order or at the planning application stage.

### Collaboration

High value ecological and environmental interests meant that a close working relationship with Natural England (NE) was essential. One important aspect was the high level collaboration between the EA and NE at the start of the process to secure common corporate objectives for the site, particularly concerning legally protected species and habitats. Close working relationships were also set up with Somerset County Council and Sedgemoor District Council. This collaborative approach resulted in the development of new and improved public rights of way, the creation of new job opportunities and the provision of an asset for local environmental tourism.

Furthermore, working with the Internal Drainage Board (IDB) the project team developed a joint agreement (EA, IDB, WWT) for the maintenance of the new water management network and structures. The close collaboration between the partners facilitated the design of a wholly new water level management system. This system is simple and safe to operate, is supported by neighbouring landowners and has clear maintenance obligations on all parties.

### Programme

From the start of detailed design to completion the project, numerous inter-dependant tasks were led by different parts of the project team, including: consenting, community design input, environmental work-embargo periods, inclement weather and tides. At some points some tasks related to design, construction, site investigation and planning consents were run in parallel in order to achieve the most efficient programme. A specialist programmer was brought in to manage a 'Total Programme' covering all project-related activities.

## Design

Detailed design commenced in 2010. The physical and topographical characteristics of the site created the opportunity to form a range of habitat types across the peninsula.

Within the intertidal area extensive modelling was undertaken to evaluate the tidal inundation characteristics and assess the impacts on the surrounding estuary. Multiple options were considered for the connections of the scheme with the wider environment. Seaward breaches connecting to Bridgwater Bay were rejected primarily due to the requirement for large bridges to maintain access to Steart village. Multiple estuary breach options were shown to produce unacceptable increases in tidal levels within the estuary. The chosen design was for a single northerly breach into the River Parrett. Further details of the options appraisal process and the modelling that was carried out are given in Burgess *et al.* (2013). The key elements of detailed design are described below.

**Embankment Design**

New set back defences consisting of earth embankments were needed to improve flood defences to the peninsula. Extensive ground investigations were undertaken as the new embankments were intended to be constructed from locally won site material. These showed that the ground profile typically comprised around 15m of generally low strength and highly compressible Quaternary alluvium overlying weak mudstone. The water table was generally less than 1m below ground surface. Historical evidence indicated the possible occurrence of peat filled palaeochannels in some areas of the site.

The project team chose to undertake field trials to confirm the design, assess the performance of the embankment and provide valuable data on the buildability of the earthworks. A 150m long, 4m high dual trapezoidal profile trial embankment was constructed in spring 2011. 5m wide berms consisting of additional earth buttressing the lower half were constructed on one section of the embankment's profile whilst berms were omitted for the other section. This successfully proved that the selective extraction of the uppermost 1m of the ground profile at the site would produce fills suitable for reuse within the embankments. Furthermore it showed that industry standard method compaction specifications could reasonably be employed during the scheme's construction, removing the need for intensive earthworks compliance testing.

The trial embankment was safely constructed within one calendar month proving that the proposed scheme embankments could be built within one earthworks season with the available material. The dual profile on the embankment proved that berms were not required until the 4m height threshold was exceeded. Confirmation of the embankments performance allowed stakeholders to reassess the generous allowances in the initial design. The removal of the berms and the reassessment of the allowance offered significant reduction, 20%, to the overall earthworks required.

In addition to the extensive geotechnical data the trial embankment provided, a number of unforeseen benefits were realised which included:

- A physical structure for local stakeholders to appraise.
- Grass sward trial to identify preferred seed mix during the latter stages of the scheme.

The benefit of field trials and its intensive monitoring cannot be overstated in assisting the project team to understand how the embankments will behave when constructed. This insight provided programme reassurance and allowed additional savings to be made by refining the design. Further benefits could have been realised by undertaking a grass sward trial earlier to understand grass growth during a whole season.

**Earthworks Design**

The design team recognised the benefits of using 3D computer modelling to determine accurate quantities, plan the earthworks efficiently and manage risk. An extensive site survey allowed the project team to understand the technical challenges and design the embankments and creeks accordingly to achieve a 'cut to fill' balance. The 3D model (Figure 3) ensured that all the material excavated was accounted for. Landscaped areas were designed to accommodate unsuitable engineering material. In some instances the landscape material was used to buttress the new engineered embankment, especially where it crossed palaeochannels. This provided increased stability where underlying strata was likely to be poorer. Other locations were selected for aesthetic reasons.

Careful planning ensured that haulage routes were minimised. To mitigate the potential for unforeseen ground conditions a number of reserve borrow pits were identified and checked for environmental constraints. Recognising the environmental constraints, new embankments were designed to:

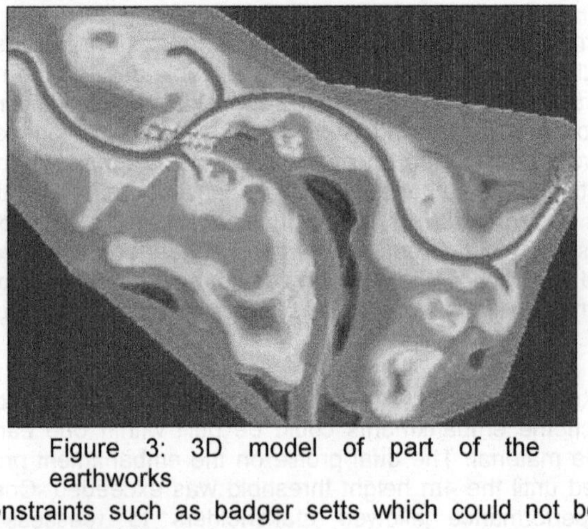

Figure 3: 3D model of part of the earthworks

- Avoid constraints such as badger setts which could not be closed due to programme constraints.
- Avoid important archaeological features which required preservation.
- Avoid community important sites.

**Creek Design**

An artificial creek system was designed for the site in order to distribute water into the farthest parts of the site, allow the site to drain, aid the development of creeks and consequently habitat to form.

The creek network had to fulfil a number of objectives:

- To be constructed safely.
- To maximise the area of functional and sustainable intertidal habitat created.
- To maximise the diversity of intertidal habitat created.

- To avoid, minimise and mitigate adverse environmental impacts.
- To minimise whole life costs.
- To achieve an aesthetically pleasing landscape.

Figure 4: Excavated artificial creek network © Sasha Dench/WWT

The initial creek design consisted of a dendritic form based loosely on the initial ditch network. However, consultation highlighted the environmental constraints of this approach, specifically with water voles within the existing ditch network which were unlikely to be relocated prior to the earthworks. A further consideration was imposed by the requirement to graze the site. This led to the design being replaced with a herring bone form which allowed cattle to safely migrate away from rising flood tides (Figure 4). The cross sectional form of the creeks was designed in conjunction with the contractor to maximise the excavation of suitable material, consisting of the uppermost 1m, for the embankments.

**Lagoons and Islands**
The project team created a number of lagoons to encourage birds and wildlife to them whilst providing focal points for visitors. These features were easily visible from visitor access points. The lagoons had the added benefit of minimising haulage distances by providing local engineering fill for the new embankment. The lagoons were designed to retain water and were carefully contoured to gradually expose mudflats as the tide receded which would attract a diverse range of species. Islands were created within the intertidal habitat from excess unsuitable material. These provided protected roosts and nesting location for birds during spring tides. Additional lagoons were also constructed within the transitional habitat area.

The holistic consideration of the various requirements of the project team during the design process allowed subtle changes to be made which enhance the opportunities and the habitat created with no impact on the cost or programme of the project.

**Water Management Network**
Prior to the scheme the existing farmland was managed by an extensive ditch network. The arrangement of the ditch network was complex in terms of its interconnectivity and any changes to this system needed to meet the different requirements of the local landowners and local community. Ensuring that the project did not impact on landowners or the critical drainage function for the local community

beyond the confines of the site presented the design team with considerable challenges. The project also needed to work with the available water resource to maximise and enhance the habitat that was created.

The project team designed a watercourse network which utilised the existing ditch network as much as possible. This enabled two watercourses to remain at the periphery of the site to ensure that the existing function was maintained. New larger channels were constructed, existing connections were sealed up and new control structures were designed to manage water levels between the internal and the periphery ditch network. This enabled new freshwater wetlands to be created and the input of freshwater into a regulated tidal exchange area to be controlled to create transitional habitat.

Modelling was carried out to ensure that the creation of freshwater wetlands and formation of transitional habitat did not adversely impact the drainage or flood storage. New deeper and wider ditches provided increased storage capacity and reduced the impact of vegetation growth on the conveyance. Over 100 structures were designed to manage water and access across the scheme.

## Construction

### Creek Construction

Construction commenced in 2012 with the artificial creek being constructed using a 3D model. The 3D model combined with GPS enabled earthmoving equipment allowed the earthworks to be accurately constructed whilst eliminating pedestrian plant contact. The accuracy that was achieved minimised waste, reduced plant movements and reduced the environmental impact of the project. This approach required an accurate survey at the outset as LiDAR was not sufficiently accurate. The excavation verified the presence of highly compressible Quaternary alluvium with a much higher moisture content below the uppermost 1m of the ground profile.

2012 was exceptionally wet which significantly hampered the earthworks. Rainfall was 125% of the long term average (Met Office, 2013) and consequently delayed the programme. The project team chose to persevere through the winter utilising tracked dumpers to construct landscape areas. This approach, whilst unconventional, allowed the team to recommence embankment construction 1 month ahead of programme.

Figure 5: (a) Small creeks forming, (b) cutting back and deepening of creeks.

Following the opening of the site in September 2014 the creek system has evolved. Some creeks are cutting back and deepening and new smaller creeks are forming (Figure 5(a)). The creeks are being cut back in a similar fashion to a river system eroding a hard substrate. A vertical face is formed with water cascading over the top gradually eroding the lower substrate. This causes the upper substrate to facture and collapse before being washed away whilst the vertical face moves further upstream (Figure 5(b)).

**Embankment Construction**

The embankment was constructed along standard engineering lines with the 3D model and GPS controlled plant removing the need for setting out. The success of the trial bank aided the construction of the embankment which involved moving 500,000m$^3$ of material.

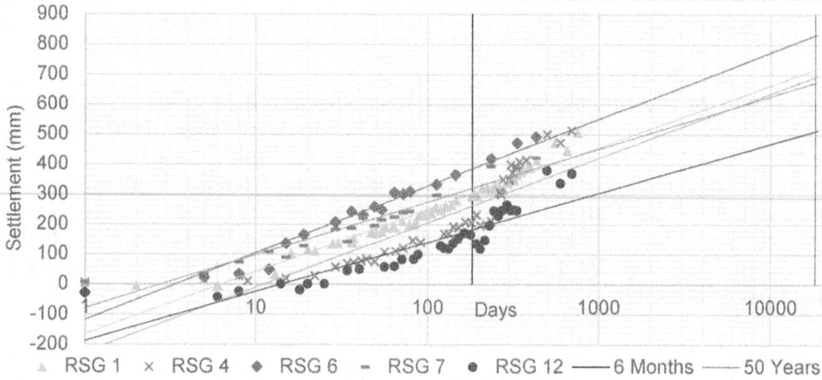


Figure 6: Logarithmic plot of settlement against time until 50 years after installation.

Regular monitoring was undertaken of the embankment with rod settlement gauges, (RSG) and lateral displacement markers placed every 250m. The monitoring of the instrumentation buried below and within the trial embankment indicated that 300mm of settlement occurred within the first 6 months following construction. This was corroborated by the results for the main embankment. The rate of pore water pressure dissipation below the trial embankment indicated that significant consolidation settlements would continue for a substantial period of time. Logarithmic plots of the settlement rates provided the necessary reassurance that the settlement rates were reducing and the overall settlement of the embankment was within acceptable tolerances (**Error! Reference source not found.**).

**Construction Constraints**

Unexploded ordnance (UXO) and archaeology posed a significant risk to the earthworks. The historic use of the site necessitated surveys and clearance to be undertaken. These were managed by the contractor allowing them to programme these works efficiently.

Protected species had the potential to significantly delay the earthworks programme. Species translocation was undertaken prior to the main works where appropriate.

The wide distribution of protected species created an ever-changing set of site constraints as wildlife had the potential to re-colonise previously cleared areas. These works followed the lean principles of just in time delivery with any displacement undertaken under the supervision of a dedicated Environmental Clerk of Work prior to the works. The project was delivered with no delays due to wildlife, undertook:

- Successful widescale displacement of water voles without the need for trapping,
- Innovative methods to mitigate for ground nesting birds ahead of the earthworks by surveying and monitoring, followed by mowing and cultivation, providing alternative habitat and applying deterrents
- Construction of several artificial badger setts and successful relocation of badger populations
- Displacement of reptiles away from working areas
- Translocation of over 3000 great crested newts ahead of the main works to prevent programme delays.

**Breach and Exit Channel Construction**
The design of the breach has already been discussed in detail (Burgess, *et al.*, 2013); however an exit channel was required due to the location of the breach which lay landward of a 100m wide existing salt marsh. The salt marsh was characterised by an existing small channel which carried freshwater from the Thirteen Acre Clyce sluice into the River Parrett. The development of the salt marsh had resulted in the topography of the salt marsh being higher than the land within the site. Unaltered, this would have reduced the number of inundations of the site (**Error! Reference source not found.**(a)).

The existing saltmarsh, which was inundated on high spring tides, had poor load bearing capacity. This limited the amount of excavation that could be safely carried out. The form of the exit channel had to balance the need to remove as much sediment as possible to limit future erosion whilst considering the safety implications of excavating on the existing salt marsh. The project team adopted a pragmatic approach whereby an exit channel was excavated up to the existing Thirteen Acre Clyce channel. The full creek profile was excavated 25m beyond the existing banks before returning at 45° to form a 30m wide deep central channel (**Error! Reference source not found.**(b)). The excavation was undertaken during dry weather using tracked excavators distributing their weight on 100mm deep bog mats due to the risk of loss of control during wet weather.

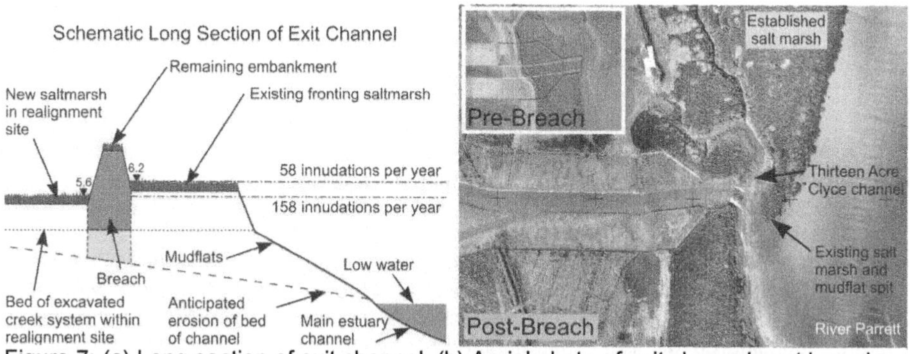

Figure 7: (a) Long section of exit channel. (b) Aerial photo of exit channel post breach

Following the opening of the site the rapid velocities, particularly on the ebb tide, led to a rapid deepening of the exit channel which quickly widened over success tides (). The enlargement of the channel has led to improved connectivity between the site and the wider estuary. The success of the exit channel excavation was dependent upon careful planning and the fortunate occurrence of good weather during the excavation.

Figure 8: Erosion at the breach (a) 8th September 2014 (b) 23rd October 2014

## Conclusions

The Steart Coastal Management Project has successfully created important new wetland habitats. Numerous challenges were overcome in the design and construction of the scheme to provide a multifunctional landscape that has exceeded the expectations of the client and local community. The main lesson learnt in the development of this scheme are:

- Early, continual and effective stakeholder engagement was needed to ensure that the community were kept informed, understood the changes and felt engaged. This helped to eliminate any potential objections.
- Early collaboration between key stakeholders enabled common objectives to be identified preventing costly changes during design of construction.
- The 'one team' approach was a key success. This allowed problems and credit to be shared and key decisions to be made swiftly.
- The construction of a trial embankment and the extensive monitoring was hugely beneficial. It improved the design and construction methods, providing

confidence in settlement rates, enabled quantities to be reduced and provided a structure for local stakeholders to appraise. A comprehensive grass sward trial would have added further benefit.

- The 3D model of the earthworks and the use of GPS enabled equipment offered numerous benefits. These included improvements in: safety, logistical planning, quantity certainty, cost savings as well as environmental benefits from reduced traffic. However such an approach means that an accurate survey of the area is essential as LiDAR data alone does not provide sufficient accuracy.
- A holistic understanding of all the constraints and how to avoid them with ownership given to parties best able to manage the risk was vital. Collaboration between the designer and contractor ensured that the design considered all the constraints and was optimised to suit the construction programme.

It is hoped that these lessons will assist other practitioners in creating sustainable wetlands habitats whilst maintaining a high level of tidal flood protection.

**Acknowledgements**

The authors would like to thank the contribution of the Environment Agency, Wildfowl and Wetlands Trust (site managers), CH2M (Consultants) and Team Van Oord (Contractor) in supporting this paper.

## References

Babtie Brown & Root, 2002. *Stolford to Combwich Strategy Study - Volume 1 Strategy Summary Plan (consultation draft),* South-West Region: Environment Agency.

Burgess, K. et al., 2013. Steart Costal Management Project: Engineering Challenges in a Hyper-tidal Environmnent. *ICE (Institution of Civil Engineers) 10th Coasts, Marine Structures & Breakwaters*, 18-19 September.

Environment Agency; The Bristol Port Company, 2011. *Joint Environment Agency and Bristol Port Company Master Plan for the Steart Peninsula,* s.l.: Environment Agency, The Bristol Port Company.

Environment Agency, 1998. *Pre Feasibility Report Stolford to Steart Sea Defences,* s.l.: Environment Agency.

Environment Agency, 1999. *The state of the environment of England and Wales: Coasts.* Bristol: Environment Agency.

Halcrow, 2010. *Shoreline Management Plan Review (SMP2) Hartland Point to Anchor Head,* s.l.: North Devon and Somerset Coastal Advisory Group.

Met Office, 2013. *Statistics for December and 2012.* [Online] Available at: http://www.metoffice.gov.uk/news/releases/archive/2013/2012-weather-statistics
[Accessed 18 Jan 2014].

**Coastal Management: Changing coast, changing climate, changing minds**
ISBN 978-0-7277-6114-9

ICE Publishing: All rights reserved
doi: 10.1680/cm.61149.675

# Tsunami Hazard and Risk Assessment and the Planning of Mitigation Measures: Case Study City of Galle

**S.S.L. Hettiarachchi**, Department of Civil Engineering, University of Moratuwa, Sri Lanka
**S.P. Samarawickrama**, Department of Civil Engineering, University of Moratuwa, Sri Lanka
**N. Wijeratne**, Department of Civil Engineering, University of Moratuwa, Sri Lanka

## Mechanics of Tsunami Wave Impact and Overall Impact on the City

Many coastal cities of Sri Lanka, particularly in the east and the south of the island were severely affected by the Indian Ocean Tsunami of 26$^{th}$ December 2004. The city of Galle in the southern province was subjected to the severe impact of tsunami waves, their magnitude having increased due to near-shore transformation processes. The persons killed, disappeared, injured and affected were of the order of 500, 90, 1000 and 8120 respectively. Houses and other buildings damaged were of the order of 1600 and 1300.

The tsunami waves, which reached the offshore waters of Galle were primarily diffracted waves around the southern coast of Sri Lanka. In the context of tsunamis the location of Galle is extremely exposed. It lies besides a wide bay and a natural headland on which is located the historic Galle Fort with vertical non-porous walls on all sides **(Figure 1)**. In the vicinity of the headland, the wave energy concentrates due to refraction. These waves then reflected from the vertical solid walls of the Fort and moved around the headland. On the west of the headland the waves moved ferociously into the Dutch Canal. On the east it moved along the bay. The wide bay in Galle further contributed to the increase in wave height by modifying the shoaling process via reduced wave crest width to accommodate the bay shape. The combined effect of this phenomenon and the wave coming around the eastern side of the Fort caused a massive wave of destruction along the Marine Drive located around the bay.

## Post Tsunami Field Investigations for Hazard Analysis

A two pronged approach was undertaken to assess the impact of tsunami on the Galle City. These studies primarily related to the assessment of inundation, flow directions and damage to buildings and infrastructure.

In the first approach the assessment of inundation was carried out in an organized manner by dividing the area under study into 250m x 250m grids. At least one location was selected for each grid and a total of 138 points used.

People living within the respective areas were interviewed for all grids. Information on inundation depth and flow direction was obtained together with other associated parameters of the hydraulic regime. The collected data were used to identify the

- Inundation profile comprising depth, length, run up and spatial distribution
- Inundation contours with wave direction and information that may help to estimate the flow speed
- Distance from sea along the tsunami flow path

The said data generated a tsunami hazard map for the City of Galle based on the field measurement arising from the Indian Ocean Tsunami.

Figure 1: Galle Bay and Headland

In the second approach studies were undertaken to study the damage to buildings and infrastructure and specific flow regimes which were generated due to the location and spacing among buildings. Those who survived the tsunami were able to describe devastating impact of flow regimes such as that of jetting effects which occurred along streets separating rows of buildings. These studies were more relevant to study the vulnerability of infrastructure.

Both studies focused attention on human and structural vulnerability and in particular the impact of strong currents at a given depth as witnessed by those exposed to the event.

**Deterministic Tsunami Hazard Modelling**
Deterministic tsunami hazard modelling comprising, deep water, near-shore and inundation modelling is primarily carried out with the objectives of studying the overall exposure of the country/island, region or city to a given hazard source, simulate tsunamis which have occurred and compare them with field measurements on height, inundation length and run up and finally simulate potential tsunamis based on 'credible scenarios' from geologic and seismic studies

Results from modelling of tsunamis which have taken place and 'different credible scenarios', provided a data base of the key parameters relating to inundation and the flow regime. Inundation height, length and its distribution, run up and velocity are some of those parameters. These parameters can then be used for the development of critical hazard scenario by relating to threshold values for security of people and infrastructure. Modeling studies generate a valuable data base which can be used for scenario based risk assessment.

Generation and deepwater propagation of the tsunami waves were modeled using the AVI-NAMI model. ANUGA fluid dynamics model based on a finite-volume method for solving Shallow Water Wave Equations was used for the inundation modeling. High resolution near shore bathymetric data were obtained from the Galle Port Development Project in 2007 and high resolution topographic data were obtained from the LIDAR surveys for near shore coastal areas implemented with the assistance of the Italian Government after the 2004 Tsunami.

The results of this exercise and those from field studies were used for the preparation of hazard maps for the City of Galle for different scenarios based on mathematical modelling. It was evident that the Indian Ocean Tsunami represented the worst case scenario.

**Probabilistic Tsunami Hazard Modelling**
Probabilistic tsunami hazard modelling seeks to assess the probabilities of certain wave heights being exceeded due to the arrival of a tsunami at locations under investigations. These probabilities are expressed in terms of expected return periods. A Probabilistic Tsunami Hazard Assessment of Indian Ocean Nations was carried out under the leadership of Geoscience Australia (Burbidge et al 2009).

Two views quickly emerged in discussions among the panel of developers. On one hand, the hazard assessment should avoid over-estimating the hazard by considering only those sources for which there is solid evidence for generation of large tsunami. On the other hand, the assessment should be careful not to miss source zones that may generate large tsunami even if they have not done so historically – as was the case for the 2004 Indian Ocean Tsunami (IOT). The panel decided that these two important views could best be accommodated by developing two assessments, referred to here as low-hazard and high hazard end member assessments. This affords an expression of uncertainty in the hazard as the difference between the two end members, while it was hoped that any additional confusion created in application of the two assessments to mitigation would be manageable.

The important observation from the outcome of probabilistic tsunami hazard modelling is that for Sri Lanka the low hazard and high hazard maps are very similar in character with maximum hazard along the east coast and the high hazard case being greater than the low hazard case by about 30%. Both low and high hazard cases for Sri Lanka are dominated by the events in North Sumatra and Nicobar Islands. Probably this seems that the Indian Ocean Tsunami was the worst case scenario for Sri Lanka.

**Tsunami Hazard Analysis for Risk Assessment**

Tsunami hazard analysis for risk assessment could be achieved either via a multiple scenario based or an event based approach. For risk assessment against the tsunami hazard it is important to assess scientifically and establish the basis and criteria on which such an exercise is undertaken. In developing hazard maps for risk assessment it is necessary to develop hazard levels and many approaches are available for defining such levels.

For the City of Galle it was decided to develop an initial risk assessment based on the Indian Ocean Tsunami which represents the worst case of the hazard scenarios. The event based approach is relevant to countries which are located at a considerable distance from the fault lines and therefore usually impacted by tsunamis arising from very large earthquakes. Even if planning is based on observations arising from a single extreme event it is important to analyse impacts of credible scenarios and operate within a framework of a scenario based approach. This provides a justification for the use of event based approach for selected scenarios.

For the city of Galle, hazard levels were developed based on inundation levels and flow speeds. The four hazard levels were classified as given below.
High- Inundation level above 0.5m with high flow/current speeds (>1.5 m/sec)
Medium- Inundation level between 1m to 2m with low flow speeds
Low- Inundation level less than 1m and low flow speeds
Very low (Zero) - Very small or no inundation to have any impact on humans
**Figure 2** represents the Hazard Map developed from this analysis

Figure 2: Hazard Map for City of Galle

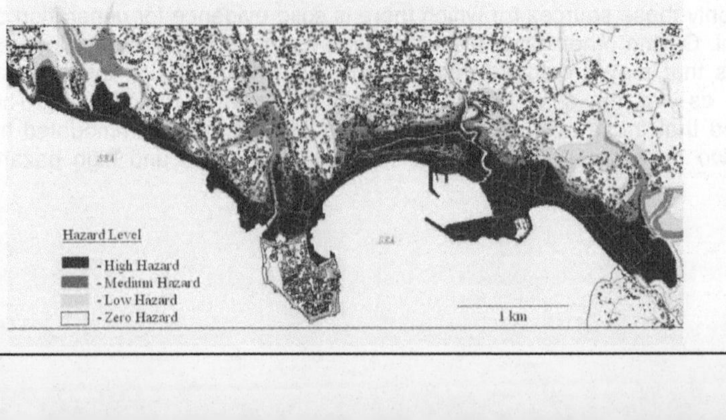

**Assessment of Vulnerability**

Vulnerability represents characteristics and circumstances of a community, a system or an asset that make it susceptible to the damaging effects of a hazard such as tsunami, an earthquake, a flood, etc. It arises as a consequence of conditions determined by physical, social, economic, political, institutional and environmental factors that characterize the framework of development employed in every society. The basic components of vulnerability can be broadly classified as human, physical, socio-economic, environmental, functional and administrative (Villagran 2006) and is therefore dependent on several factors related to the said components. These include, among others, population characteristics and density, degree of poverty, livelihoods, building types their strength (structural vulnerability), and a variety of other factors. This approach is primarily a one dimensional approach of susceptibility.

Detailed assessment of vulnerability remains a complex area in view of the widely varying parameters which have to be analysed and in view of the difficulties in defining and quantifying certain parameters. Assessment of vulnerability can therefore be implemented at different levels using varied approaches from simplified to highly sophisticated, commencing from very basic data bases to highly sophisticated data bases both referral and relational. Description of approaches and models on vulnerability analysis is presented in detail by Birkman (2006) and a summary is available in UNESCO 2009a.

A framework to breakdown vulnerability assessment into components by analyzing how disasters can impact the different sectors which comprise society was proposed by Villagran (2008) thereby moving from single dimension of susceptibility to three dimensional sector approach. A detailed sector approach for vulnerability was also applied in a benchmark project for the City of Galle, Sri Lanka by Villagran (2008) and its advantages were clearly observed. The application of this method requires vast amount of data and consumes time. However, one of the important strengths of the study was that it provided guidance in identifying the critical parameters which could be adopted for a simplified approach on vulnerability.

Having reviewed all relevant information including the detailed study of Villagran (2008) it was decided to adopt a simplified approach giving attention to the identification of relevant parameters via stakeholder consultation.

Systems exposed, primarily include the population (human), buildings and infrastructure (physical), socio economic fabric of society (socio economic), ecosystems/ natural environment (environment). Services and infrastructure essential for the functioning (functional) and administering (administrative) of society are the other two important aspects.

A community which is well informed of the hazards and prepared is less vulnerable because they are able to move away prior to the hazard sets in or in a better position to respond intelligently to emergency situations.

Based on the review of above parameters, the following broad groups were identified to accommodate the relevant parameters relating to the vulnerability of the City of Galle.

- Population, its profile and geographical distribution
- Socio-economic profile of residents including livelihood
- Buildings, infrastructure and its status,
- Ecosystems and areas of environmental importance.
- Capacity to evacuate, within the broader framework of awareness, preparedness, early warning, response and safe evacuation

In addition attention was focused on profile of the occupants, sources of income and household economic level, conditions of buildings and infrastructure and community knowledge.

In the final analysis it is important adopt weightage factors for relative importance after reaching agreement with stakeholders. **Figure 3** illustrates the Vulnerability Map developed on the above analysis.

Figure 3: Vulnerability Map for the City of Galle

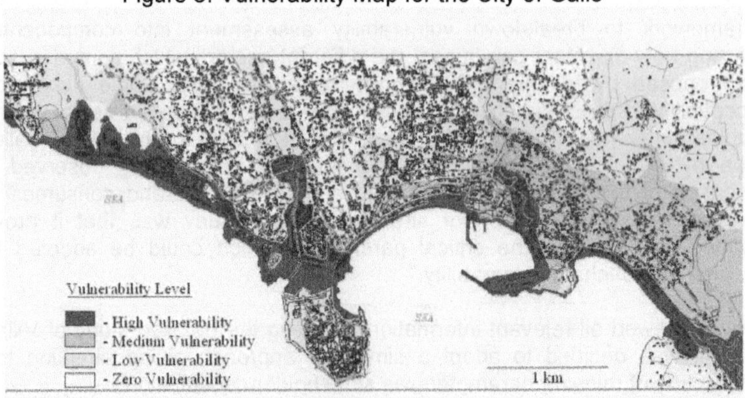


**Assessment of Risk**

Risk Assessment is an essential task that is conducted initially to determine the degree of risk that coastal communities face, and subsequently to identify the measures that are needed to reduce such risks through a variety of structural and non structural measures aiming at reduced exposure to such hazards, reduce vulnerability, increased preparedness and increased coping capacities.

While risk can be described using plain text and risk related data presented in a tabular form, it is advantageous in representing hazard, vulnerability and risk in the format of maps.

Risk maps identify areas of comparable risk. Typically these maps can be developed by the superimposition of hazard and vulnerability maps. Superimposition is based on criteria which should be developed for defining risk levels. It is also necessary to educate the user on the value of information presented and potential limitations in its applications. On most occasions risk maps provide a qualitative classification (similar to hazard and vulnerability maps). For some situations quantitative scales can be adopted (in the form of deaths, damage/economic losses per year). The approach to be adopted is the prerogative of those who develop the risk map but it is important to adopt a logical and consistent approach in developing hazard, vulnerability and risk maps as well as to clearly state the methods adopted.

In this study the hazard and vulnerability maps, each comprising four levels of classification, namely, high, medium, low and very low (zero) have been prepared. For superimposing them it was necessary to establish the criteria for risk levels. This meant the definition of risk levels corresponding to the superimposition of hazards and vulnerability of different levels (four levels in each case). **Figure 4** illustrates the Risk Map developed by superimposing hazard and vulnerability maps. For example, according to this approach, a level of high risk will be present for large populations, in particular comprising a high percentage of elderly and children, residing at low elevations, within short distances from the seafront, in an environment of easily damageable infrastructure, with easy disruption to livelihood and not having the capacity to evacuate quickly and not protected by coastal flood protection schemes, thus having direct exposure to the sea.

Figure 4: Risk Map for the City of Galle

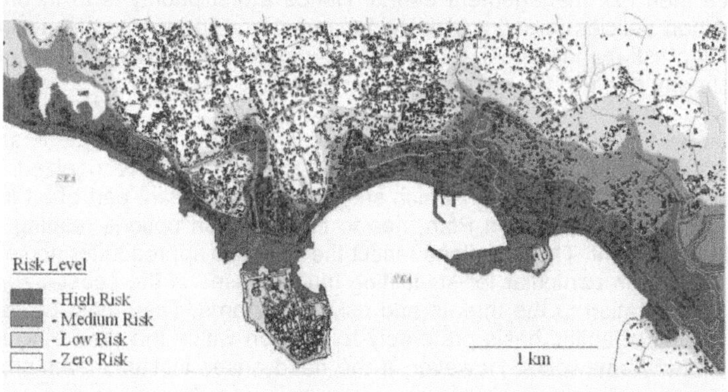

It is also noted that sufficient information was available from this investigative study to conduct a scenario based hazard analysis and levels of risk could have been developed based on such an analysis. This study focused on event based hazard analysis which can be used for strategic planning buildings and infrastructure and preparedness for safe evacuation. The Indian Ocean Tsunami certainly represents the worst case scenario but other scenarios can be used to identify the potential impacts from more likely events having a high frequency of occurrence. However in the case of Sri Lanka which is at considerable distance from the fault lines, other credible scenarios have had very little impact.

The development of Hazard, Vulnerability and Risk Assessment Maps was followed by the preparation of Disaster Management Maps. In this respect two maps were prepared, namely, Maps of safe places and buildings and Maps of Evacuation Routes and Sign Posts.

**Managing Risk: Planning and Classification and Risk Management Measures**

There are many measures that could be adopted for risk management in coastal zone management when planning for a tsunami and other coastal hazards that accompany high waves and high inundation. Measures for risk management can be broadly classified into four categories, namely, those which mitigate the impact of the hazard, mitigate exposure and vulnerability to the hazard, improve preparedness and response and risk transfer via insurance. These include physical interventions such as protection structures, utilizing the full potential of coastal ecosystems, regulatory interventions in the form of extending existing setback defense line and early warning systems for community preparedness. These have to be supplemented with public awareness on disaster preparedness, efficient evacuation procedures, incorporating planned evacuation routes, safe places and evacuation structures that effectively integrate with the overall planning process.

In many countries, prevailing policies target disaster response efforts to a larger degree than risk management efforts. Hence a first priority is to incorporate or strengthen policies targeting risk management and to incorporate accountability regarding risk management activities as a way to complement the accountability that already exists in the case of disaster response.

Subsequently, it is important that risk and disaster management should be undertaken within a framework of multiple hazards. It is recognized that Risk Management for a city/region which should be an important part of an Integrated Coastal Area Management Plan, has to be based on options relating to Policy and Management. These options reflect the strategic approach for achieving long term stability in particular for sustaining multiple uses of the coastal zone giving due consideration to the threats and risks of hazards. They must be formulated on a sound scientific basis preferably to function within the prevailing legal and institutional frameworks. However, if the need arises institutional improvements should be affected and new laws should be imposed. In this process high priority should be given to stakeholder participation.

**Risk Management: Hazard Mitigation and its Integration with Development Projects**

In the year 2000, Japanese Port Consultants (JPC) developed a Master Plan for the development of the Port of Galle **(Figure 5).** In view of critical environmental issues relating to the Galle Bay, including the presence of the historic Galle Fort (west side), buried archaeological sites (exceeding twenty) within the bay and the famous coral reef system (on the east side) it was recognized that the development should be restricted to a two berth medium size harbour. In order to maintain healthy exchange of tidal flow for the well being of the coral reef system, JPC in consultation with the environmental specialists incorporated an offshore detached breakwater, which coincidentally has all the characteristics of an effective tsunami breakwater. It must be admitted tsunamis were not part of the agenda of the engineering and environmental terms at that stage, long before the Indian Ocean Tsunami impacted Sri Lanka.

Figure 5: Master Plan for the development of the Port of Galle (Japanese Port Consultants)

On analyzing the damage aftermath of the Indian Ocean Tsunami, the proposed development of the Galle harbour and the use of the offshore breakwater were revisited. Through numerical modelling it was determined that by implementing this project with a slightly extended offshore breakwater in the direction of the Galle Fort, the City of Galle will have the benefit of a Tsunami Breakwater as part of a port development project. Considerable reduction of the inundation area could be achieved. Once this extension is designed it may be possible to reduce the length of the revetment protruding near the berth. The construction of a small offshore breakwater in front of the channel located west of the headland would provide additional protection.

The breakdown in reduction in inundation area which can be achieved by the construction of the offshore breakwater for the Indian Ocean Tsunami scenario was determined by numerical modelling. Although the reduction in the inundation area is around 13%, considerable energy is dissipated as wave travel over the offshore breakwater thereby the destructive wave velocities are considerably reduced thus limiting potential damage. Inundation with lower velocities of flow will cause considerably less damage.

A protection wall (tsunami dyke) of modest proportions, along the coastline, can supplement the tsunami breakwater. The design details and the structural configurations can only be determined after carrying out mathematical modeling and physical modeling to assess the performance of the proposed configuration. A hybrid approach using both artificial and natural methods seems very appropriate for this situation. These are some of the physical mitigation measures which can be successfully adopted as feasible options for the protection of the city of Galle. These measures will also be effective against potential coastal hazards that have a greater probability of occurrence than an extreme tsunami wave event.

**Concluding Remarks**

This paper has described in detail and presented the results of an investigative study on tsunami risk assessment for the port city of Galle in Sri Lanka. The paper focuses on hazard, vulnerability the principal components of risk assessment followed by risk management measures.

Hazard analysis comprised both probabilistic and deterministic hazard analysis. The probabilistic tsunami hazard analysis has revealed that for Sri Lanka the low hazard and high hazard maps developed for the two scenarios of the Indian Ocean Tsunami Hazard Map are very similar in character. Since both high and low hazard cases for Sri Lanka are dominated by the events in North Sumatra and Nicobar Islands, it seems that the Indian Ocean Tsunami (IOT) was the worst case scenario. The use of deterministic tsunami hazard modelling permitted a scenario based analysis. The IOT hazard scenario can therefore be used for long term strategic urban development planning, in particular to identify locations for critical infrastructure and installations. This hazard scenario was therefore selected for risk assessment in this study. However, there are no constraints for the use of multiple scenarios for preparation of an overall hazard map and the scenario modeling carried out for this study can be used for such an exercise with the adoption of weightage factors.

The report presents a simplified approach for vulnerability and capacity analysis whereby only the critical parameters have been used. These parameters were identified in consultation with stakeholders. The report also provides a cross comparison with a very detailed study on vulnerability assessment undertaken for the city of Galle using the Sector Approach. The Sector Approach demands a very detailed analysis based on an extensive database. It was evident that the simplified approach provides an effective vulnerability analysis on which risk assessment can be undertaken with confidence.

The risk assessment has been achieved via superimposition of hazard and vulnerability maps. It is recognized that the frequency of occurrence of tsunamis which may have severe affects on the Sri Lankan coast is small and this aspect has to be given due recognition in adopting measures for risk management. It is in this respect that that an extreme case scenario of IOT has been considered and the results of which can be used for strategic planning as identified earlier

The paper then focuses on tsunami risk management. It presents a four pronged strategic approach and identifies the importance of operating within a multi hazard coastal hazard protection plan. In the case of the City of Galle the report presents a case for integrating national port development with tsunami hazard protection. It recommends that the breakwaters of the proposed new port in Galle can be utilized to provide protection against the tsunami hazard and modeling will have to be carried out to estimate the desired crest height of the breakwater to achieve the expected level of protection

**References**

Burbidge, D.R. Cummins, P.R. Latief, H. Mleczko, R. Mokhtari, M. Natawidjaja, D. Rajendran, C.P. & Thomas, C. A Probabilistic Tsunami Hazard Assessment of the Indian Ocean Nations. Geoscience Australia (2009). Professional Opinion No. 2009/11, September 2009.

Birkman, J. (2006) Measuring Vulnerability to Natural Hazards: Towards Disaster Resilient Societies. Edited by Jorn Birkman, United Nations University Press.

UNESCO (2009 a). Hazard awareness and risk mitigation in ICAM. IOC Manuals and Guides No.50, ICAM Dossier No.5 Paris UNESCO 2009.

UNESCO (2009 b). Tsunami risk assessment and mitigation in the Indian Ocean; Knowing your tsunami risk-and what do about it. IOC Manual and Guides No 52, Paris: UNESCO 2009

Villagran de Leon, J.C. (2006). Vulnerability. A Conceptual and Methodological Review No.4, SOURCE Publication Series of UNU-EHS, Bonn. 64 pages.

Villagran de Leon, J.C. (2008). Rapid assessment of potential impacts of a tsunami: Lessons learnt from the port of Galle in Sri Lanka. No.9, SOURCE Publication Series of UNU-EHS, Bonn. 96 pages.

**Coastal Management: Changing coast, changing climate, changing minds**
ISBN 978-0-7277-6114-9

ICE Publishing: All rights reserved
doi: 10.1680/cm.61149.687

# Long Term Loss of Beach Material from the Eastern English Channel

**Ian Thomas**, Pevensey Coastal Defence Ltd, Pevensey Bay, UK

Pevensey Bay, on the English Channel coast of East Sussex is the site of a unique sea defence project. In 2000 Pevensey Coastal Defence Ltd (PCDL) entered into a 25 year contract with the Environment Agency (the Agency) to provide sea defence services along part of the East Sussex coast. The Agency is part of Defra, the United Kingdom government's Department for Environment, Food and Rural Affairs, and has a strategic overview role for all flood and coastal erosion risk management. It is also an operating authority in its own right, having permissive powers to undertake works to manage flood risk. Whilst having these powers, the Agency does not in general have a legal obligation to provide flood defence.

PCDL is an SPV (special purpose vehicle) company formed specifically to undertake the Pevensey Bay Sea Defence scheme. The SPV is formed of four independent companies that were already operating and well-known in the UK sea defence industry, all of whom had specific roles within the company;

Boskalis Westminster Ltd – provision of sea dredged material
Dean & Dyball Ltd – to undertake any rock work
Mackley Construction – supply of beach plant and groyne works
Mouchel – project design and management

Figure 1: The position of Pevensey Bay on the English Channel coast
(Image courtesy of Google earth)

Pevensey Bay's sea defences are formed by a 9km long mixed shingle embankment that runs from Eastbourne in the west to Bexhill-on-Sea in the East, as shown in Figure 1 above. Immediately inland there is a low-lying area of some 50km$^2$, which includes over 10,000 properties, several caravan parks and important road and rail links, in addition to an ecologically sensitive and important area. These would all be flooded at high tide if the defences were breached.

Historically the defences had been maintained by a field of 150 timber groynes. However, due to underinvestment during the 1990s, by 1999 they were reaching the end of their useful life. The Agency, as part of a strategic study for the area, identified a preferred option for the management of the defences to 2050. Its principal component was to thicken the shingle embankment to provide a 1 in 400 year standard of protection against breach. Groynes would be progressively removed as they failed, and the shoreline managed as an open beach. Cross-shore response and littoral drift would be managed by a combination of replenishment, reprofiling and recycling of shingle. (Hardacre & Chester, 2001)

Encouraged to do so by the UK government, in 1997 the Agency selected the Pevensey Bay Sea Defence scheme as one of two projects to be undertaken as Private Finance Initiative (PFI) or, as in this case, a Public Private Partnership (PPP) scheme. PPP contractors are engaged to deliver services for an extended period, typically, 25 years. This contrasts sharply with traditional public procurement, where the contractor delivers an asset, which the public sector then takes ownership of and manages. The PPP contractor must finance the design, construction, operation and maintenance of the asset necessary to deliver the required service. The public sector then pays the contractor a fixed periodic sum in return for the delivery of the service.This guaranteed income enables the contractor to borrow the required capital. The contractor of course expects to make a profit and the cost of borrowing will always be higher for the private sector. So, if PPP is to deliver improved value for money, these additional costs must be exceeded by savings made possible by the PPP type of contract. Savings can be generated because:-

- Contract risks are allocated to the party best able to manage them. The contractor therefore will assume appropriate identified risks hitherto borne by the public sector and through better management the cost of these risks will be reduced and absorbed into the overall payment system.
- The private sector is not subject to the same spending constraint as the public sector. This enables the contractor to make commercial decisions on the basis of whole life costings. This is frequently not possible for the public sector. This can also lead to quicker delivery of the service.
- Due to the long term nature of the contract the contractor is able to apply technically and managerially innovative methods of delivering the service and make long term investments. (Hardacre & Chester, 2001)

Given 'pathfinder project' status by the Treasury, detailed negotiations proved to be a much longer and more difficult exercise than either party expected, and financial closure was not finally achieved until April 2000, some 3 years after the procurement process started. Now 15 years into the contract, both parties agree the process was well worthwhile and the contract has been a great success on several levels.

## Allocation of Risk and Best Practice

As already stated, the allocation of risk is one of the most important aspects of a PPP contract. In this case, as the contractor, PCDL has accepted all risk associated with design, construction and maintenance of the improved sea defences. It is obvious why this is of benefit to the client, but at first glance it may not be quite so clear why this is the case for the contractor. It is important to recognise that the obligation accepted by PCDL is to provide a service, the service being that East Sussex hinterland will not flood. In itself this is an unusual concept to embrace – being paid for something not to happen.

In reality there are only a limited number of options available when it comes to maintaining a shingle embankment sea defence. As part of the Environmental Impact Assessment completed in 2002, each was considered so it could be used if, as and when any of them became necessary. In reality it was the duration of the contract that presented the greatest opportunity for PCDL to make efficiency savings. In real terms annual income for the full 25 years was known from the outset, as payments made by the Environment Agency are linked to government produced inflation indices, in this case RPIX – the Retail Price Index (RPI) excluding mortgage interest payments. Thus, from the outset PCDL was able to consider whole life costs as a matter of course, knowing that resources would continue to be available to meet any ongoing financial commitments.

Being responsible for the defences in their entirety means that variations to the management regime can be implemented as and when the contractor sees fit. As soon as it becomes clear that a process needs changing, or control measure amending, it can be immediately implemented without reference to the client, need for a variation order or having to make potential claim. This ethos actively encourages innovation and quickly leads to improved best practice.

Figure 2: Eroded beach crest widened by beach recycling

**Best Practice – Beach Recycling**

One instance of this is demonstrated by how periods of recycling are planned and initiated. Beach recycling involves excavating accreting sediment from the intertidal zone and returning it areas of erosion. Typically this has been carried out by contractors as an annual or biannual exercise, involving movement of fixed volumes of material over known distances for payment on a cubic metre rate. At Pevensey, this approach was very quickly discarded. From the outset, the site employed a rigorous monitoring regime, with a centimetre accurate GPS receiver mounted on an ATV (All-Terrain Vehicle) – in this case a four wheel drive quad bike. Monthly surveys have been completed on low water spring tides every month since January 2003 by driving along the beach and recording three dimensional coordinates every second. From the resulting digital model, comparisons to the design surface quickly identify not only where erosion has occurred, but also where the lost sediment has accumulated. Knowing the detail of what is required to reinstate damaged defences means that appropriate sizes and quantities of machine can be acquired to quickly complete necessary repairs.

Since the greatest erosion usually coincides with the highest spring tides, recycling is now deliberately targeted for the same periods. Thus, any additional erosion can be immediately addressed without increasing risk of breach by having a reduced width embankment for more than a single tide. Most erosion areas are the result of past failures to restrict development on the beach crest. In these locations it is often impossible to work continuously through a tidal cycle on the highest tides. Rather than be restricted by a fixed volume and distance contract, all recycling works are now carried out by hiring items of plant and then directing it's on-site activity on a day by day, or, if necessary, hour by hour basis. It should be pointed out that on a publically accessible beach, it is impossible to segregate the public from working machines. With no fixed daily volumetric targets, this methodology removes the requirement (or temptation) for operatives to drive faster and maximise output, which would increase the risk posed to other beach users in the vicinity of the works.

Figure 3: Dredger Sospan Dau discharging sea dredged aggregates

**Best Practice - Replenishment**

A second example came in the way handling of beach replenishment deliveries has evolved. Through a combination of natural processes and man-made intervention, there is an annual net loss of sediment from the frontage of approximately 20,000m$^3$. Although 5-10,000m$^3$ arrives from Eastbourne in the west, 25-30,000m$^3$ is lost to Bexhill-on-Sea in the east. Importantly, the contract recognises that movement of sediment along the coast must be maintained if frontages to the east are to be themselves maintained. Consequently PCDL has to allow at least 16,000m$^3$ of beach to pass beyond the terminal groyne every year. Precise quantities for these annual losses are calculated from the regular GPS surveys, with the shortfall being replaced by sea dredged aggregate delivered by Boskalis Westminster dredger Sospan Dau.

Sospan Dau's first ever job as a dredger was to deliver shingle to Pevensey in 2001. At that time it was common practice to recover material delivered at high tide over the following low water. This would be stockpiled on the beach crest until there was sufficient to allow sectional completion, whereupon it would be bladed into a 1:7

slope and surveyed for alignment and payment purposes. Whenever sediment is deposited below high water, there are inevitable 'losses' (the difference between hopper and beach measured volumes) before it can be recovered. Amounts vary according sea conditions at the time. Although the discharge procedure begins the process of re-sorting sediment on the foreshore, transporting it all to the beach crest causes it to become randomly mixed and introduces far more sand to the upper beach than would appear naturally. As no beach has a natural 1:7 slope, the formed gradient is quickly eroded by the tide, with the result that the beach 'cliffs'.

Subsequent monitoring led to PCDL realising that if sediment is left in situ it behaves far more naturally than when it is recovered and placed. The initial design process suggested that at Pevensey very little coarse sediment moves cross-shore, only moving longshore with littoral drift. Therefore, confident that it would not be lost, PCDL decided to accept this risk rather than leaving it with Boskalis Westminster. Delivered sediments were then left on the foreshore to be distributed both cross and longshore by natural processes. For this to work contractually, average 'losses' were agreed so payment is made as a percentage of the hopper measurement. This not only produces a more natural and safe beach, but has also led to an overall reduction in costs of 15%, as there is no longer a requirement for any resources on the beach. Moreover, it has significantly reduced the impact on the local community, as now the only disturbance comes from the delivery of the shingle itself.

Figure 4: Top – After delivery of 20,000m$^3$ of renourishment;
Bottom – natural sediment redistribution four weeks later

**Community Views**

Although the client for the scheme is the Environment Agency, on a practical level it is the local community that receives the benefit of the service provided by PCDL. When the project started in June 2000, it is fair to say there was a degree of scepticism within local interest groups and residents associations that formed the Key Stakeholder group. This was largely because the badly-needed scheme had failed to get government funding several times during the 1990s. From a contracting perspective, the most important moment was recognising that there needed to be full and regular engagement with local communities. Often, contractors generate profit by being efficient and completing projects ahead of schedule. On a 25 year scheme that has to run until May 2025, this was not an option. Add to the equation that PCDL was a new company, hence a new name with no track record, then the only sensible option was to be as open and accessible as possible. To that end, there has always been an 'open door' policy so that residents may visit and learn for themselves, not only how the beach is being managed, but also how coastal processes are the result of interaction between wind, sea, tide and local bathymetry, and how the storms that result affect the coastal community.

Reactions within the local community vary considerably. Long term residents are far more accepting of works that have to be done, mainly because they have direct experience of the worst that the sea has to offer. Those who have more recently moved to the area are less tolerant of any disturbance. By their very nature, soft defence shingle beaches are reshaped on a daily basis by wave action. A summer gradient approaching 1:7 will be replaced by a winter one closer to 1:10. This generally means a drawdown of sediment from the crest to the beach toe. The contract allows for up to 5m crest recession before intervention activities are triggered. On a day to day basis this means using heavy plant on the beach so that crest widths can be reinstated. All such machines generate noise, dust, fumes and vibration and at times have to operate within a few metres of property boundaries. There is no doubt that this can be inconvenient. However, without regular maintenance, many homes would be lost to the sea, so it is seen as a necessary evil. On the other hand, all works are confined to winter months, so homeowners are able to enjoy well over half the year with little or no disturbance.

Winter 2013-2014 was the worst seen in Pevensey for many years. Although there have been more significant individual storms, none have produced such continuously stormy weather, especially from early December 2013 to early March 2014. Because the contract requires that PCDL protect against breach for up to 1:400 year storms, it contains a series of Joint Return Probability (JRP) tables that define all weather events from 1:1 through to 1:400 year. Of measurable weather events, the contract anticipates that six or seven should be experienced in a typical year. Winter 2013-2014 recorded 29. And not just any; there were 1:250, 1:100, two 1:50s and several 1:20 year events. The net result was vastly exaggerated littoral drift rates. In a normal winter beach reprofiling occurs throughout, with short period recycling initiated as and when significant erosion occurs. That winter recycling became a constant activity, starting on 31$^{st}$ December 2013 and continuing until early April 2014. This effectively doubled machine costs, yet annual beach losses were still more than double those of a normal year.

Ironically, this elevated workload resulted in a huge increase in the numbers of residents thanking PCDL for all the effort made to protect their homes. Suddenly there was a wider understanding of how important it is to maintain the level of defence and that erosion does need to be repaired as soon as it occurs. Community appreciation has also contributed to the site receiving awards in the Contract Journal Construction Industry and Public Private Finance Awards, where letters of commendation from local organisations were an integral part of the submission. The site is also recognised by the Considerate Constructors Scheme as being in the top 7% of registered sites as it has received awards in each of the last eight years, culminating in a gold award for 2014.

Figure 5: Top – Normal summer beach profile;
Bottom – Erosion typical of winter 2013-2014

## Challenges of managing a shingle beach

Details provided above relating to winter 2013-2014 illustrate some of the challenges inherent in managing a shingle embankment sea defence. Equally, they show that even in particularly bad weather, the chosen management regime has been effective as no properties were flooded or damaged.

Those directly involved in the Pevensey scheme recognise that a major part of its success is due to the nature of the contract. Yes, it is a PPP contract, but with all capital elements completed prior to 2006, for almost ten years it has virtually operated as a term maintenance contract.

Far from being a drawback, the current contract means PCDL has a distinct advantage over more traditionally maintained frontages;

- It is the service that is specified not how it is to be achieved. Hence, there is great flexibility in how, when and where to initiate works
- There is budget certainty for 25 years
- There are no budget constraints. If necessary PCDL could borrow against future income to deliver services earlier and hence more effectively
- There is no underspend penalty. Any money not spent in a particular year is rolled forward to provide additional resources for the following year
- Continuity of public facing staff is comforting to local communities and helps develop trust and encourages healthy working relationships with stakeholders
- Local knowledge gained from beach monitoring is retained over the long term
- 25 year certainty is attractive to subcontractors, who become treated more as partners for the long term good of the project
- Machine operatives return regularly, developing their own understanding of how the beach behaves and how repairs need to be completed

## Coping with Climate Change

The last 15 years has demonstrated that the contract can cope with changing climate on a decadal scale. On the other hand, 25 years is a short period when looking at long term global climate change and in particular sea level rise. At the start of the project it was stated that a rise in sea level had to be included in the service level provided, and at that time the perceived rate was approximately 0.006m per year. Actual values are evaluated every five years by independent experts HR Wallingford, and so far reviews in 2005 and 2010 indicate that levels have risen by little over 50% of the supposed amount.

Of course sea level rise is only part of the picture, and increased storminess could be a more important factor. The concept of measuring and proving increased storminess is already complex, but in this case it could be even more difficult. We only need to consider whether there has been increased storminess at Pevensey, not the whole English Channel, nor the UK in general. If historic data do not exist at a sufficiently local level, how could an increase or decrease be adequately proven at a specific location? Whereas it is quite straightforward to examine local sea levels, evaluating Pevensey's climate over an appropriate time scale would be immensely difficult. Equally, it is not as straightforward as reviewing the number of JRP Storm Events in each category. The JRP tables are purely a statistical analysis and do not reflect a specific outcome on the beach. For instance a 1:1 year event can result from an unusually high sea level. On its own, this will not affect the defences. Alternatively, a change in the average wave direction of only a few degrees is likely to increase or decrease sediment drift rates, which would significantly affect levels of beach maintenance work.

The way the contract deals with climate change in general is twofold;

1. At the end of years 10 and 20, there is an option to "review the material increase (or decrease) in Storm Events". As stated above, this has to be specific to Pevensey, so may prove difficult. Neither party believed a material change could be demonstrated in 2010.

2. A basic premise of the contract is that specific weather risks have been accepted by both parties. Those risks were based on data and modelling available in 1999, which covered most of the 1980s and 1990s. Fifteen years later, there is almost double the amount of data available, as well as advances in modelling capability. Consequently, the definition of risk can, and should, change to reflect actual conditions experienced. A good example of this occurred in 2014 when HR Wallingford, acting as an independent expert, recalculated Joint Return Probability tables used to define Weather Events from 1:1 to 1:400 years. In going through this process, both parties accepted the importance of retaining the balance of risk and hence the integrity of the contract.

In reality, it is this last point that is the most important. Throughout, the contract has been managed in the spirit of its original intent, with provision of a suitable flood defence service being at the forefront of all decisions.

There is however one further element to be considered. Monthly GPS surveys have been completed for more than 10 years, primarily to monitor sea defence volumes. Contractually, these are measured above 0.0mOD, which is roughly mean sea level. It soon became clear that the beach above 0.0mOD, or indeed the embankment itself, is not isolated from the whole marine sediment system. Volume analysis indicates that there is regular movement of finer particles up and down the beach profile, including to and from the seabed below mean low water.

So, regardless of any contractual requirements, it is important to understand how the main embankment structure interacts with its surroundings in order to provide the most suitable and sustainable management. Thus, surveys are always on low water spring tides, always going to -2.5mOD but often beyond -3.0mOD.

Unfortunately, it is not possible to isolate the embankment from all the other intertidal sediment because its toe position moves in both time and space. In some areas it is as low as -2.2mOD, whilst in others at -0.5mOD or occasionally higher. Equally, sand terrace levels fluctuate by a metre or more over indeterminate periods, whilst the toe position can vary by up to 15m cross-shore.

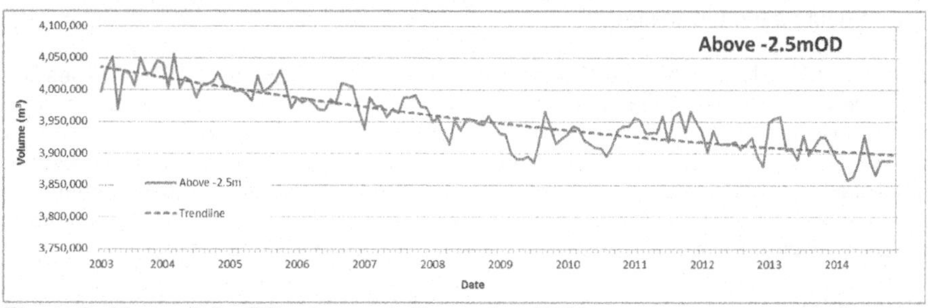


Figure 6: Above -2.5mOD volumes to December 2014

Following each GPS survey, total beach volumes are extracted at 0.5m intervals between 0.0mOD and -2.5mOD, which is the lowest extent possible on solstice spring tides. Individual volumes show high levels of variation partially because of storm activity and addition of beach replenishment, but also because it is difficult to accurately survey. Although the sand terrace as an area is greater than that of the embankment, there is much less time in the tidal cycle when it is accessible. It is a more irregular surface as the size, shape and number of exposed sand bars changes all the time, and usually retains pools of water lined with silt deposits that are impractical to access. Often it cannot be accurately represented in the time available. Consequently, it was only when 10 years of survey data were available that trends suggesting the long term lowering of the sand terrace became evident.

In Figure 6 above, wide variations between individual surveys are clear, but it is equally obvious that the overall trend is down. Given that contract volumes above 0.0mOD have been maintained, data to December 2014 indicates that about 10,000m$^3$ per year has been disappearing from the sand terrace.

If this trend is genuine and continues, in time its impact on our ability to 'hold-the-line' on existing coasts will rise. Water depth will increase, allowing waves to approach ever closer to the shore and increasing wave energy in breaking waves. Similarly, the foreshore will steepen, further increasing wave energy, making the defences less effective and less sustainable. Although this problem may be as a result of climate change, there may be additional factors involved.

## Maintenance Dredging

At the western end of the Pevensey frontage is Sovereign Harbour. It is not a natural harbour, and its approach channel has to be dredged from the seabed. About 2m lower than its surroundings, it requires an annual maintenance dredge that removes approximately 35,000 m$^3$ of sediment every year. Since the harbour was completed in 1993 this suggests that 700,000 m$^3$ has been taken from the harbour and its approaches. Currently it is deposited 2.5km offshore, into a hollow in the seabed, at about -16mOD. The material consists equally of silt from between the breakwaters and sand from the marina approaches. This further suggests that since Pevensey's sea defence scheme started in 2000, around 225,000 m$^3$ of sand has been removed from the intertidal zone. Given the location of the existing disposal site, it seems unlikely that much, if any, of that material will have returned to the swash zone It is hard to believe that the loss of this much material has not had some impact on surrounding sand terrace levels.

The CONSCIENCE project's "Concise report for policy makers", 2010, stresses the importance of coastal foundation in being able to guarantee residual dune strength along the Dutch coast. Essentially this means that the whole system to a depth of 20m below sea level has to be managed as one. Thus, since 1990 in Holland, shoreface nourishment has been as important as beach replenishment in maintaining the volume of sediment in the littoral system. Re-use of dredged sand in a similar way at Pevensey should help sustain important sand terraces.

## Summary

Few, if any, UK beaches have such a detailed survey record as Pevensey Bay. Therefore it is entirely possible that similar losses are being experienced elsewhere. According to the MMO (Project No. 1073, 2014) up to 50 million tonnes of dredged materials could be available in the South marine plan areas over the next 10 years, with only around 5 million tonnes currently envisaged to be re-used in planned projects. Clearly there is scope for increasing the amount of beneficial re-use of material that takes place, particularly if it can be demonstrated that the dredging itself is impacting on our ability to provide adequate sea defences.

There is no doubt that the management style fostered by the PPP project at Pevensey Bay is ideal for the coastal scenario found on that part of the Sussex coast. Lessons learnt and advances in best practice can – and are – being applied elsewhere. It is satisfying to think that data captured by the detailed monitoring programme can help to inform the coastal community about long term trends in coastal processes that in the long run could affect a much wider area. However there is little doubt that in the UK more attention needs to be paid to the coastal foundation as described by CONSCIENCE if we wish to continue protecting our eroding coastlines.

Sea levels are undoubtedly rising, and as a result seemingly changing near shore seabed levels. It remains to be seen whether increased storminess will also have a material impact on our ability to manage eroding coastlines. After 15 of the 25 years at Pevensey, it has been demonstrated that a contract of this nature is an appropriate way of managing shingle embankment sea defences. But, with a further decade still to go, no one knows what changes may be just round the corner.

## References

Chester, F., ICE Southern Association - Maritime Group Meeting - 16/11/00, Pevensey Bay Sea Defences - Public Private Partnership Contract

Chester, F. & Hardacre, G., MAFF Conference 2001, Pevensey Bay Sea Defences PPP – 'The Contractors Perspective'

Marchand M., Concepts and Science for Coastal erosion Management. Concise report for policy makers. Deltares, Delft. 2010

ABPMer, (2014), MMO Project No. 1073, "Use of Beneficial Dredged Materials in the South Inshore and South Offshore Marine Plan Areas", p36

**Coastal Management: Changing coast, changing climate, changing minds**
ISBN 978-0-7277-6114-9

ICE Publishing: All rights reserved
doi: 10.1680/cm.61149.699

# Changing Coast at Salthouse

**Mark Johnson**, Environment Agency, Ipswich, UK
**Steve Hayman**, Environment Agency, Norwich, UK
**Gary Watson**, Environment Agency, Ipswich, UK

## Introduction

The North Norfolk coast between Weybourne and Morston (*fig 1*) has seen considerable change in past centuries as a result of natural coastal processes and human intervention, most particularly large reclamations of saltmarsh. This coast is dominated by a 10km long barrier beach known as the Blakeney Ridge which shelters the villages of Morston, Blakeney, Wiveton, Cley and Salthouse from the worst ravages of the North Sea.

The outfall of the River Glaven runs northwards between Cley and Blakeney before being diverted westwards by the Blakeney Ridge. To both sides of the river channel there are freshwater marshes of high environmental and ecological value. Landward of these marshes the A149 main coast road and significant numbers of properties are within the area at risk from coastal flooding. Behind these, the land rises steeply into chalk escarpment.

The sea defences protecting the low-lying coastal strip comprise grassed clay embankments and the shingle bank which forms the eastern half of the Blakeney Ridge. There is a long history of these defences being breached by storm surge tides with the more notable flood events occurring in January 1953, January 1978 and, most recently, December 2013. Rising sea level will impose increasing strain on the defence system and, over the past 10 years, the Environment Agency has been working with its partners to introduce a management regime which will secure the unique features and character of this coast in the longer term.

This paper describes how the Environment Agency's relationship with the local community and stakeholders has evolved during the period covering the introduction of changed coastal management and the most severe East Coast tidal surge for 60 years.

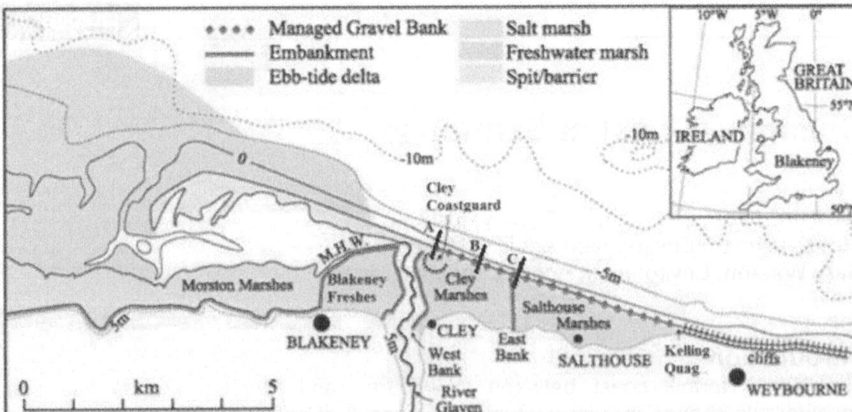


Figure 1: The Coast at Salthouse

## The Physical Setting

The modern landscape of North Norfolk reflects the legacy of the last ice age and the subsequent rise in sea level. The present coastline was the maximum extent of the advancing glacier and it left behind large quantities of tills, sands and gravels which have been reworked by the sea to form today's Blakeney Ridge. The fact that the shingle which comprises the ridge was derived from the sea bed in the early Holocene means that no significant contribution to the total volume is being made by present-day processes.

With sea level increasing during the post-glacial period, the sea penetrated into the Glaven valley and large deposits of sands and muds were laid down in the lee of the shingle ridge to form the marshlands, with a maximum elevation of +2metres AOD. Between 1522 and 1630, a series of embankments were built to enclose the salt marshes which have evolved into the sea banks which are maintained by the Environment Agency today.

A combination of deepwater close inshore and a relatively long fetch to the north results in large waves penetrating to this length of coast and the shelter offered by the shingle bank has, in the past, provided a safe haven for ships exporting Norfolk's farm produce. From the 13th to the 16th centuries the 'Glaven Ports' of Blakeney and Wiveton were among the most important in the country, but the 17th century reclamations exacerbated problems with siltation in the harbours and channels, bringing about their decline. Today, the small harbours are used for recreation, boat trips and fishing.

## The Shingle Ridge

The naturally occurring Blakeney Ridge is a relict post glacial feature. The shingle size in the ridge diminishes from east to west, with an increase in sand content west of Cley Coastguards. Sand forms the distal end of Blakeney Spit, which has been growing westwards at an estimated rate of 5metres per year as a result of the feed of fine material from both east and west. As well as extending westwards, the ridge has been moving landwards at approximately 1metre per year since at least 1630.

The structure of the ridge comprises a relatively impermeable core covered by well-sorted shingle which is capable of adapting to and dissipating wave energy. It has a wide flat natural profile which can be overtopped by wave run-up under the most severe combinations of high tide levels with heavy waves, lowering the crest and causing shingle washover fans to form on its landward side. Then, under more normal wave conditions, shingle is transported up the beach slope to gradually rebuild the crest elevation and maintain the integrity of the ridge.

The 5km length of the ridge between Kelling Quag and Cley Coastguards provides flood defence for properties in Salthouse and east Cley, the coast road and 300ha of environmentally sensitive reclaimed marshes. From the 1950s until 2005 this part of the ridge was managed by the Environment Agency and its predecessors to provide an over steepened profile with an artificially high crest. This annual bulldozing operation also had the effect of maintaining the ridge on a line seaward of its natural position.

To the west of Cley Coastguards the shingle ridge fronts 160ha of reclaimed coastal grazing marsh, known as Blakeney Freshes, bounded by rising ground to the south and clay embankments on the other three sides. The Glaven outfall channel runs between the most northern section of embankment and the shingle ridge. This unmanaged length of the ridge has a history of encroachment into the river channel, necessitating regular dredging operations in order to maintain a clear passage.

## Scheme Development 1997-2004

The first generation North Norfolk Shoreline Management Plan (SMP) was published in July 1996. This was only a short time after a surge tide event in February 1996 had breached the managed length of the ridge resulting in serious flood damage to both the marshes and coast road. The SMP called into question the benefits of continuing the practice of reprofiling the shingle ridge with bulldozers and recommended studies to investigate ways of implementing a managed realignment policy. It suggested a possible option whereby the shingle ridge was allowed to roll back with a new secondary embankment constructed approximately halfway between the ridge and the coast road.

The Environment Agency then commissioned the Coastal Research Unit at Cambridge University (CCRU) to advise on natural and human-modified processes which characterise the shingle ridge and its adjacent environment. CCRU reported that the regular bulldozing of the ridge to maintain an artificially high crest was damaging its internal structure and decreasing the stability of the seaward slope. They recommended discontinuation of the ridge maintenance programme in order to allow the ridge to restore a more robust lower and wider profile, with natural landward progression over time. This would result in increased overtopping frequency, but the overall stability would be enhanced with a reduced risk of breaching.

Halcrow were appointed as the Agency's consultants for preparation of a project appraisal report and Environmental Statement. Their brief included full public consultation and the University of East Anglia's Centre for Social and Economic Research in the Global Environment (CSERGE) were engaged to lead on this aspect. The appraisal confirmed a partial setback scheme with a new sea defence embankment on a retired line as the preferred option, although the economic justification was marginal, at best.

Detailed discussions with the Department for Environment, Food and Rural Affairs (Defra) during 2002 concluded that the partial setback scheme did not satisfy Government criteria for investment of public funds and could not be progressed. It was agreed that the emphasis would be placed on the works needed to maintain the freshwater and brackish environmental features of the Salthouse and Cley Marshes, with the anticipated increased quantity of seawater overtopping the ridge. With regard to people and property, local flood warning improvements were introduced and technical advice on self-help property protection options was provided on an individual basis.

In 2003/04, the Environment Agency worked closely with English Nature (now Natural England) to agree the flood management measures necessary to protect the ecological integrity of the marshes. A drainage improvement scheme was then designed to evacuate saline waters rapidly and deliver the required nature conservation outcomes. The two main elements of this scheme were the replacement of the tidal outfall sluice draining the marshes with a much larger structure and the landward realignment by 200 metres of the length of the River Glaven to the north of Blakeney Freshes where it was in imminent danger of becoming blocked by the southern migration of the ridge.

## Scheme Implementation 2005 to 2013

There has been no active maintenance of the shingle ridge since the 2004 annual bulldozing. A management plan was agreed with English Nature in 2005 with the objective to allow the ridge to develop naturally and move to a 'no intervention' policy once it has achieved favourable conservation status. It was recognised that there are uncertainties about the behaviour of the ridge during the transition phase, and some form of intervention was not completely ruled out in the event that recovery does not proceed as hoped and there is a risk of barrier breakdown. This approach has been incorporated in the 2$^{nd}$ generation North Norfolk SMP.

The drainage improvement scheme was completed as planned in 2006. The scheme provided match funding for the EU-funded Glaven Education and Environmental Sustainability Enterprise project (GEESE), a partnership between the Environment Agency, Norfolk Wildlife Trust and The National Trust. This project financed a new visitor and education centre for the Cley Marshes Nature Reserve which has proved extremely popular.

Changes to the geometry of the shingle ridge are monitored by a combination of LiDAR and topographical profile surveys. These show that a substantial proportion of the formerly-managed length of the ridge has gradually been reworked by wave action and has reverted to a more natural form. Overwashing events occurred on average once every two years, and none caused major change. The reworking has increased the permeability of the beach making it more effective at dissipating waves.

## Impacts on the Natural Environment

The importance of the natural habitats and species found within the reclaimed freshwater areas behind the shingle ridge is recognised by multiple national and international nature conservation designations including SPA, SAC, SSSI and Ramsar. The marshes and reedbeds of Blakeney Freshes and the Cley Nature Reserve, managed by The National Trust and Norfolk Wildlife Trust, attract many species of breeding and overwintering birds including marsh harrier, bittern, avocet, lapwing and redshank.

Experience has shown that coastal freshwater habitats such as these are mostly able to withstand occasional saltwater flooding provided the period of tidal inundation is not too prolonged. Even if the mosaic of habitats does gradually change as a result of natural processes, this can allow new nature conservation interests to evolve and does not necessarily mean the value of the site is diminished. If it is the case that changes in habitats constitute an adverse effect on the integrity of a designated site it may be that the most sustainable approach is to provide compensatory habitat in another more secure location.

With regard to the long term ecological health of the Salthouse and Cley Marshes, it was recognised that the environmental legislation has a presumption in favour of protecting designated features in-situ, but the need for realism in adapting to dynamic coastal change was fully appreciated. The decision-making process relied heavily on advice from English Nature. It was essential that human intervention on the shingle ridge stopped so as to allow it to recover favourable conservation status, and the scheme was considered to be necessary to conserve the features of the SAC and Ramsar site. Increased salinity will result from overtopping of the lowered ridge and there could be an adverse impact on the integrity of the SPA. After detailed assessment it was concluded that, with the improved drainage measures, there would not be a significant effect on the grazing marsh components of the SPA. There is, however, likely to be a long term adverse effect on the freshwater reedbed habitat of the bitterns. This has been addressed by providing compensatory habitat on an inland site, which is also being managed by Norfolk Wildlife Trust.

## The Consultation Process

Following breaching of the shingle ridge in 1996, the Environment Agency made a point of involving partners and the local community from the outset in an interactive debate about the best way to manage flood risk in the area. The development process for the SMP had already involved full public consultation but it was clear that the disadvantages of continuing the annual ridge re-profiling programme were not appreciated locally. The bulldozing operation at the start of each storm tide season was still perceived to be maintaining an effective flood defence for the villages.

Independently facilitated forums were arranged in the villages attended by parish councillors and other local representatives. Decision making needs to be underpinned by sound science and expert advice was provided on geomorphology, and other subjects. The participatory process had consensus as its goal, with stakeholders having a sense of shared ownership in a certain output, but appreciating the potential pitfalls on the way. The Environment Agency therefore agreed that the scheme based upon cessation of bulldozing would include constructing a new secondary flood embankment, as suggested in the SMP. However, it was made clear that previous attempts to promote similar improvement schemes had floundered because the reduction in flood risk did not justify the cost.

The conclusion, in 2002, that the secondary embankment was not deliverable came as a disappointment, but not a great surprise. The most positive aspect is that the relationships and trust which had developed between participants in the forums survived this set back, and have proved very helpful in communicating the reasons for subsequent decisions, which could otherwise have resulted in conflict. This shows there are potential long term benefits to be gained from involving the community in decision-making, irrespective of any specific outcome.

## The Challenge of Winning Hearts and Minds

It is human nature to be suspicious of change and the move, in 2005, from annual bulldozing to a non-interventionist approach was no exception. It would be fair to say that there was a mixed reception to this change in policy locally. Local residents were used to seeing a uniform height bank on the horizon which gave them a sense of security, and many felt uneasy as they witnessed the bank start to evolve to a lower broader shape. There have been many local meetings over the last 10 years and throughout this time there has never been a unanimous view on the best way to manage the ridge. The views of the local nature reserve warden, who has known the area for all of his life, are interesting. Initially he was in the sceptical majority but over the years has come to accept that the new approach is more sensible.

The management of the dynamic coastline of Norfolk and Suffolk brings with it many challenges around managing expectations of what is fundable, technically feasible and environmentally acceptable and Salthouse is one of many locations where matters have to be handled carefully. Our organisation's approach over the last ten years has been to work with local communities and wherever possible reach a consensus on future approaches to flood and coastal erosion risk management. We do have to accept that unanimous support for some projects is not achievable, but by being open, honest and available, we have managed to dissolve conflict.

We have consulted closely with the residents of Salthouse over many years, attending public exhibitions, parish council meetings; site meetings and informal meetings at residents' houses. As such, although there are some residents that would still like to see more active intervention, there is growing recognition that we are working in their best interests, with advice from nationally-respected experts. As with any situation trust and respect is hard earned and can be very easily lost. This is something we are acutely aware of.

## The Storm Surge of 5th December 2013

The surge tide which struck much of the east coast of England on 5th and 6th December 2013 was the most severe such event in more than 60 years. On the North Norfolk coast, peak tide levels exceeded those recorded for the great flood of 31st January 1953. The surge caused extensive damage to property and infrastructure but past investment in improving sea defences and the major advances in flood forecasting and warning prevented the devastating consequences experienced in 1953.

The surge seriously damaged the sea defence embankments which enclose the Blakeney Freshes with numerous breaches and extensive scour to the rear face along the remaining sections. The marshes were totally inundated and it took more than 3 weeks for the saltwater to drain through two outfall sluices. The very popular North Norfolk Coastal Path which runs along the top of these embankments had to be closed for several months.

The shingle ridge to the west of Cley Coastguards, which has not been managed previously, responded in a natural manner to the extreme hydrodynamic conditions. Although there was significant overtopping along this section, the net impact of the event was an increase in the ridge crest elevation. Overwashing onto the lee face initially occurred along most of its length temporarily lowering the crest. However, the latter part of the storm was constructive rather than destructive and the crest was rebuilt to an elevation typically 0.3metre above its original level.

Figure 2: The Eastern Storm Surge Breach at Salthouse Ridge (image courtesy of Mike Page)

By contrast, the beach response east of Cley Coastguards was far more variable, reflecting the fact that the ridge is in transition from a heavily managed system to a natural barrier. Significant changes to the ridge geometry resulted from the combined effects of extreme sea level and severe wave conditions. Waves exceeded the ridge height causing crest lowering and rollback of sediments with shingle fans extending more than 100metres across the marshes in places.

The widespread seawater overtopping of the lowered ridge crest resulted in deep flooding of the Cley and Salthouse marshes. The differential hydrostatic head between the back barrier area and the beach and foreshore at low water caused destabilisation of the weakest points of the barrier and two deep-seated breaches (*fig* 2) formed. The trapped seawater was then able to escape at high velocity carrying with it large volumes of sediment which temporarily formed ebb deltas to seawards of the barrier. The sites of the breaches appear to be linked to relic channels.

## The Post-Surge Recovery Period

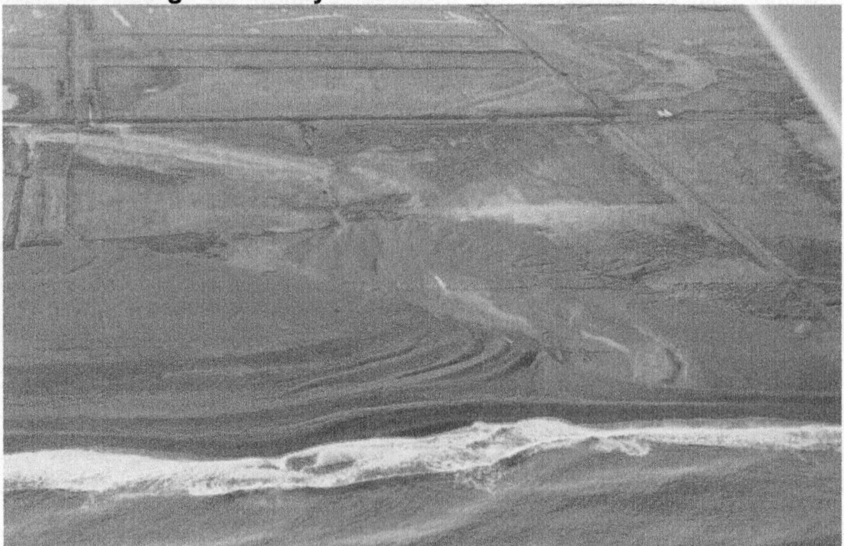

Figure 3: A Naturally Repaired Beach at Salthouse (image courtesy of Mike Page)

Both breaches in the ridge healed naturally within 7 weeks of the storm event (*fig 3*). The healing process, which was predicted, is a product of particular wave conditions in combination with tidal elevations. By this time it is estimated that approaching 50% of the previously managed length of the shingle ridge had evolved to a point where it was performing in a natural manner and moving towards favourable conservation condition.

The elevation of the healed ridge gradually increased but the cross section at both of the former breach sites remained small and it will take a considerable time to build resistance to further flooding. The continued growth of the ridge is dependent upon an adequate longshore supply of material and the wave and tidal conditions needed to deposit it on the crest. There is a thin line between the moderate wave action which is favourable for the beach building process and the more severe conditions which can result in significant overwashing and lowering of the ridge crest. This was demonstrated on 21 October 2014 when northerly winds associated with a deep area of low pressure (ex-hurricane Gonzalo) generated waves which overtopped the ridge and caused extensive flooding of the marshes. On this occasion, the majority of the saltwater had drained in 48 hours.

When adverse weather conditions pose a threat of flooding, the Agency provides landowners and residents with information on the level of risk through it's flood forecasting and warning service. These warnings take into account the current condition of the sea defences and at Salthouse a helpful 2-way exchange of information has developed with the local flood wardens.

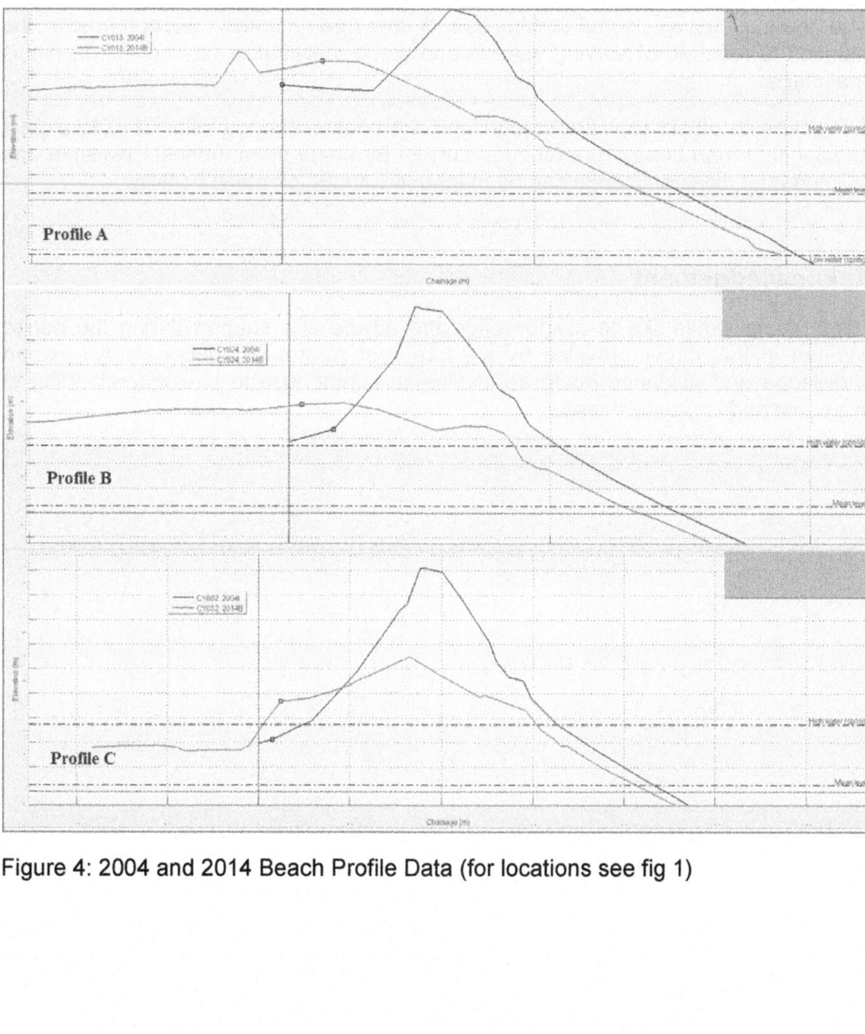


Figure 4: 2004 and 2014 Beach Profile Data (for locations see fig 1)

## Concluding Remarks

Prior to 2005, the annual appearance of heavy plant on the shingle ridge provided visible evidence of a commitment to minimise the coastal flood risk at Salthouse. It has been important to ensure that allowing the coast to function more naturally with less intervention in coastal processes does not give rise to a feeling of abandonment within the at-risk community. This has been addressed through a determined attempt to encourage public participation in decision-making and the ongoing monitoring of coastal change (*fig 4*). We are now used to engaging with communities during the development and implementation of specific projects but, in the case of Salthouse, this participatory process has been in progress for more than 10 years – and is ongoing.

Our experiences working with communities at both Salthouse and other locations on the Norfolk Suffolk and Essex Coast mirror reactions to change in the work place. For example there is initial shock, followed by occasional anger. After a period of time there is usually some recognition of the new situation. Acceptance of the situation is reached at varying speeds and some individuals may never reach this final place.

Our advice to others who are embarking on a coastal change initiative is to expect traditional human behaviours (change curve). By being open, honest, available and building trust, coastal change can be managed without too much hostility

## Acknowledgement

The authors would like to acknowledge the advice and support during the period covered in this paper provided by the late Prof Andrew Bradbury. Andy's expert knowledge and guidance made an invaluable contribution to our understanding of this coast and is greatly missed.

**Coastal Management: Changing coast, changing climate, changing minds**
ISBN 978-0-7277-6114-9

ICE Publishing: All rights reserved
doi: 10.1680/cm.61149.709

# A Short Guide to Doing Nothing at the Seaside

**Nick Hardiman,** Environment Agency, Bristol, UK

## Introduction

This paper builds on the ICE Coastal Management 2011 conference paper on 'Increasing the Dialogue Between Shoreline Management Plans and Spatial Planning'[1], which considered the extent of interaction between the second generation of SMPs and the spatial planning process delivered at the time through the Local Development Frameworks of Local Planning Authorities. It also considered the principles upon which SMPs and spatial planning documents might interact. Between 2011 and 2015, the coastal management 'narrative' has moved from one focussed on SMP planning, development and approval to one focussed on implementation of the many measures within each completed Plan. As such, this paper moves the conversation on to the practicalities of achieving the over-arching goals of SMPs as expressed through the management policy options assigned to each stretch of coast[2] - in particular, 'No Active Intervention', sometimes denoted as 'do nothing' in SMPs themselves. This is not to imply that the challenges associated with spatial planning have gone away, or that SMP development has ended – shoreline management planning is an iterative and living process – but they are now framed more in the language of approved action than potential decision.

### The developing framework for SMP delivery

Since 2011, many positive developments are discernible that support this implementation agenda. At the Environment Agency, information on risk is improving with updates to the National Flood Risk Assessment, more consistent asset condition recording and the publication of new flood maps targeted at both spatial planners and the general public. National Coastal Erosion Maps are also online, with SMP policies displayed[3]. Meanwhile the concept of 'Coastal Change Management Areas' – in which risks and opportunities associated with erosion and permanent inundation are mapped with specific planning provisions embedded – enjoyed a smooth transition to the National Planning Policy Framework[4], and are beginning to take shape in some Local Plans. Community engagement experiences and techniques for coastal adaptation have been documented through the Defra Coastal Change Pathfinders and projects such as *Coastal Communities 2150 and Beyond* on the south coast. The Defra Pathfinders have in particular explored adaption options in the spatial planning system, and techniques are being reviewed to ascertain their wider

[1] Hardiman, N, *et al*, 2011
[2] Hold the Line, No Active Intervention, Managed Re-alignment and Advance the Line: see Defra, 2006 vol 1 p13-14 for definitions
[3] National Coastal Erosion Risk Maps: note this URL may be subject to change in 2015 as the Environment Agency maps are hosted at www.gov.uk
[4] National Planning Policy Framework, p25 https://www.gov.uk/government/publications/national-planning-policy-framework--2

applicability. Above all, the approach to funding flood and coastal erosion risk management (FCRM) has become more flexible and inclusive in the Partnership Funding approach, with added confidence provided by the Government's six year investment programme.

These developments mean that England is moving from a position where flood and erosion risk, and plans for its management, was little understood to one where those at or potentially at risk are far more engaged and involved in shoreline planning, shaping and funding delivery, and incident response. It is a positive transition, but one which hangs substantially off rising to the challenge presented by those areas not earmarked for future investment. High profile events such as the 2014 east coast surge and subsequent storms in the south of England have raised the profile of the investment that has been made, and have tested the physical risk management measures in place. They have, however, also brought SMP No Active Intervention policies into sharp relief for many. Earlier-than-predicted damage or destruction of some defences tested the soundness of the timing or choice of some SMP management policies; it also tested the resilience of unpopular No Active Intervention policies to pressures from the public, media and political arenas. It was in effect the first real test of the 'mettle' of SMPs, and the preparedness of the coastal management community for the delivery of management policies in them. The precedent set will be important for ongoing confidence in the SMP system.

This paper will start by setting out the scale of No Active Intervention policies in SMPs in temporal and geographical terms. It will then explore what is meant by 'No Active Intervention' (NAI) in SMPs, and the caveats and subtexts associated with it as a management policy. It is such 'small print' behind the over-arching policy that may be missed by the general public, prompting misunderstanding about how, when and why NAI might be borne out. This is followed by a consideration of the key practicalities surrounding NAI that will usually merit consideration well in advance by Risk Management Authorities[5]. This subject will be the main focus of an Environment Agency R&D initiative beginning in February 2015, the results of which will inform the ICE conference presentation in September 2015. The paper therefore sets the scene without anticipating the outcomes of the study, and concludes with provisional recommendations for Risk Management Authorities, based on examples referred to throughout.

## No Active Intervention: Timing and Location

The first important point to note is that the designation of this policy throughout England barely changes across the next 100 years in SMPs as they currently stand. Within the first planning horizon of the SMPs (the first 20 years – roughly considered to approximate to c.2030 depending on when the SMP was approved), 38% of England's shoreline has a NAI policy, and this increases by only 1% - less than 100km - in total by the final planning 'epoch', which runs into the next century[6]. This does not necessarily mean that, just because the aggregate figure remains constant that the same places will have a NAI policy now and in 100 years: some will lose this policy, others will gain it. But in general the shift is towards NAI from Hold the Line (HTL) rather than vice versa and withdrawal of existing management interventions is

[5] As defined in the Flood and Water Management Act 2010. In this context, coastal local authorities and the Environment Agency.
[6] Environment Agency analysis, unpublished.

to happen substantially in the first 20 years of most SMPs.

**NAI and Managed Re-alignment**

In fact, the greater shift is from HTL to Managed Realignment. 52% of England's coast is to be defended in the first 20 years or so of the SMPs, of which 6% will, according to current policy intentions, be re-aligned in a managed way thereafter. Managed re-alignment – largely a policy attributed to low lying flood risk areas – is often controversial as it still prompts objections around 'giving up land to the sea', and perceptions that the risk of flooding is being brought closer to hinterland communities. Many of the community engagement or political challenges associated with NAI exist for managed re-alignment also.

However, the principles upon which Managed re-alignment is framed with landowners is different: options exist under Countryside Stewardship to receive payments for making the change, and in some cases there may be recourse to financial compensation[7]. It may also involve re-investment of capital and maintenance expenditure from the national FCRM budget into new defences that rural communities may not otherwise have benefitted from – an example being the experience of the Steart Peninsula re-alignment in the Bristol Channel, which ended up with strong community support after initial negative reactions to the scheme[8]. Most managed re-alignment in England is designed to create inter-tidal habitat as a compensatory measure for lost salt marsh or mudflat elsewhere resulting from 'squeeze' between rising sea levels and sea defences: locations are often chosen on ecological grounds and fewer properties are at immediate risk from managed re-alignment than in NAI areas[9].

**Spatial Distribution**

The distribution of NAI policies for the first planning horizon is shown in Figure 1 (overleaf)[10]. Much of the coast of Devon and Cornwall is NAI, in many cases reflecting a historic approach to shoreline management and the relatively resistant geology that dominates the south west. There are, however, significant pockets where a need for adaptation of access routes, tourism infrastructure and property is identified, and the storm climate of Devon and Cornwall presents important challenges surrounding residual risk behind sea defences and other structures. This need is amplified along the coasts of Somerset, Dorset, the Isle of Wight, where the natural resistance provided by geology, the interaction between flooding and erosion, vulnerability of coastal settlements and the management interventions to date, are collectively more variable.

Elsewhere, on the coasts of Norfolk and Yorkshire, the predominantly soft sedimentary geology coupled with strings of vulnerable open coast settlements of varying size with predominantly NAI policies makes for a series of high-profile examples of households and communities at risk. These areas have provided the

7 Environment Agency, 2013 Appendix 3
8 Burgess *et al*, 2013
9 Environment Agency, 2013b
10 Note that the map in Figure 1 also displays stretches of 'retreat the line' and 'natural defence management', extra policy descriptions included in some SMPs to denote controlled allowance for cliff retreat and limited intervention in natural systems respectively. These policies are often provisional dependent on monitoring, and are generally described as NAI or HTL respectively in most SMPs.

main focus for recent Defra-sponsored work to trial practical ways to adapt to imminent coastal change.

Figure 1: Distribution of adopted or provisionally adopted SMP management policies in England (Channel Coastal Observatory, 2013). Dark outlines show HTL, light outlines predominantly indicate NAI and small areas of managed realignment

Location is also important in determining the overarching NAI policy in a different way. Different SMPs, developed by different authorities and consultancy suppliers, took different approaches to the themes and principles set out in Defra's SMP development guidance[11]. 'Managed realignment', interpreted strictly in association with habitat creation schemes in some SMPs, was interpreted more loosely in others, including the monitored allowance of cliff retreat by a given extent. Elsewhere, allowance of inundation to gently rising ground is characterised as NAI. The interplay between limited managed intervention and the overall management intent of a given shoreline area may be complex and subject to underlying detail in the rationale and premise for policy decisions. The expression of this varies between plans, making

[11] Defra, 2006 (vols 1 & 2)

representation of SMP intent such as that shown in Figure 1 an approximation to the actual situation.

## Rationale and Premise for No Active Intervention Policies

NAI is defined simply in Defra's guidance on developing SMPs as "no investment in coastal defences or operations"[12]. This can be divided into two main situations:

1) NAI will be a continuation of the historical approach to a particular shoreline stretch;

2) NAI will involve a departure from previous investment, even though investment in coastal defences or operations may have only been relatively recent and/or short term.

It is the second of these situations where the main expectations of Government compensation or other intervention arise among the affected parties.

### The Detail Behind SMP Policies

There are important caveats associated with many SMP policies, whether HTL, NAI, managed re-alignment or the little-used 'advance the line'. Firstly, the policy option may have been chosen for a continuous stretch of shoreline – and displayed as such on a map for that Policy Unit[13] – but there may be localised exceptions within the Unit. Such exceptions might arise where a different approach is required to accommodate particular features, where this will do little to compromise the overall management intent within the Unit.

Secondly, how, when or whether the policy option is applied in practice may be conditional on further monitoring of coastal evolution – especially where changes are planned for the longer term. This allowance for flexibility within the SMP is common and is especially important when considering how NAI policies are to be implemented: a switch from HTL to NAI in the second SMP 'epoch' of 20-50 years after Plan adoption will hardly mean an abrupt withdrawal of investment on the SMP's '21st birthday'. In many respects, SMP policy options are all underpinned by an 'adaptive management' approach, where the possibility of future change should be accommodated. In some locations, such as the settlements adjacent to the dynamic spit at Pagham Harbour in West Sussex, this 'adaptive management' approach has been made explicit in the policy description[14].

### The Different Premises for No Active Intervention

Notwithstanding the generic definition of NAI by Defra, further expansion on policy option selection in Volume 2 of the SMP Guidance[15] on developing SMPs highlights that the caveats to SMP policies might extend beyond timing and method of intervention or withdrawal. Management Policies are arrived at through the process of scenario assessment, identification of preferred options and sensitivity testing against key uncertainties that might elicit a future change in direction, such as

[12] Defra, 2006 (Volume 1) p14
[13] A Policy Unit in an SMP denotes the discrete section of shoreline to which any SMP policy option applies. Maps of SMP policies by Unit are often included in SMP documents, and SMP policies are displayed by Unit on the Environment Agency National Coastal Erosion Risk Map.
[14] New Forest District Council, 2010a and b; Defra, 2015
[15] Defra, 2006 (Volume 2), pp52-58

variations in our understanding of sea level rise. The decision to designate a Policy Unit as NAI may therefore be taken on the basis of a number of different – often complementary, though sometimes potentially conflicting – factors, broadly:

1. Environmental sustainability: including

   i) statutory requirements to maintain the ecological integrity of the Natura 2000 network of internationally protected nature conservation sites and/or national Sites of Special Scientific Interest[16];

   ii) a broader aim to achieve greater working with natural processes to promote efficiencies in environmental management, improve or maintain landscape and wildlife more generally (including in the marine environment), and/or to enable or maintain sediment transport to other Policy Units currently or potentially at flood or erosion risk.

2. Technical feasibility, especially associated with the application of an isolated HTL policy on an otherwise dynamic coastline into the long term.

3. Financial sustainability, including

   i) a broader aim to achieve greater working with natural process to promote efficiencies in coastal flood and erosion risk management, within and beyond the Policy Unit;

   ii) likelihood of funding. The broad-brush assessment in an SMP may underestimate the benefits of a HTL policy which would be picked up in a more detailed assessment, potentially leading to a NAI policy being changed. More likely overall, however, is that some HTL policies adopted in SMPs are unlikely to attract sufficient funding from the FCRM budget.

The implication in 3. ii) above is a third situation in addition to the two at the beginning of this section, i.e. one where a HTL policy exists in principle but remains unfunded, leading to a changing standard of protection or a change to NAI over time through failure to implement the preferred option. Managing the change in public expectations will be especially difficult in these areas[17].

## Flexibility – Its Opportunities and Risks

However, Defra's Guidance highlights that "Where works are not having a detrimental effect [i.e. under 1. 2. Or 3 (i) above] but are simply not economically justified, this [situation] might be identified for private funding of works if this is the intention as they may be permitted under such circumstances."[18]

---

[16] i.e. where *not* maintaining or building a defence is necessary to ensure Favourable Condition or Favourable Conservation Status of a site, as opposed to Category 2 situations in the Environment Agency asset maintenance protocol (Environment Agency, 2013) concerning where intervention (defence) is necessary for this.

[17] Improving strategic understanding of where these areas are and the level of risk is the subject of ongoing Environment Agency assessment during 2015-16.

[18] Defra, 2006 (Volume 2) p58.

**Third-party Maintenance in NAI Areas**

This provision is an important caveat to the generic NAI definition: in some places, third party defences were already in existence in sections of Policy Unit where NAI was approved in the SMP – NAI was chosen purely to highlight that investment would not be forthcoming from H.M. Treasury. In other areas, the NAI policy makes explicit the potential for third party defences to be erected subject to planning permission from the Local Planning Authority and/or Flood Defence Consent from the Environment Agency (and now a Marine Licence from the Marine Management Organisation) being granted. The Partnership Funding mechanism, introduced after most SMPs had been substantially agreed, now extends this potential. Indeed, further caveats to NAI policies – or even policy changes to HTL – might now be expected to accommodate the potential for limited state funding coupled with local sources to sustain a level of intervention. These principles are fleshed out fully in the Environment Agency's 'asset maintenance protocol'[19].

An example of this is in the local partnership that has been formed in the Deben Estuary in Suffolk. One of many grass-roots local community initiatives at the coast that are generically badged under the 'Coastal Partnerships' banner, this group brings together various strands of estuary management but was essentially formed to refine the preferred FCRM options spanning the Estuary in the SMP and associated Suffolk Estuary Strategies. Set up as a charity, the Estuary Partnership has been able to fund further studies into adaptive management options with the Environment Agency and Suffolk Coasts and Heaths Area of Outstanding Natural Beauty. These studies have indicated options for targeted maintenance, salt marsh creation and the strengthening of defences with allowances for overtopping. Together, such options have led to a successful re-appraisal by the Environment Agency using the normal FCRM Appraisal Guidance, leading to the re-categorisation of NAI from being the preferred option in some cases. This is one of a number of similar initiatives with similar aims evolving across the country, particularly on the east coast[20].

**Maintaining the Strategic Intent of NAI**

Some such initiatives might beg the question: where does local self-determination, innovation and enterprise stop and procrastination begin? On the one hand, locally funded works can be a long-term sustainable option that effectively replaces (or substantially supplements) Government investment. Or they may help 'buy time' to allow a strategic change in management to no – or significantly reduced and better targeted - intervention alongside other measures such as property-level protection or resilience, building modification or removal, changes to land management or access, and improved warning and evacuation systems. This is the approach being taken at Cockersands on the Lancashire coast, where both HTL and managed re-alignment have been shown to be uneconomic as current defences begin to fail after an estimated 30 years. The direction of travel set by the SMP is maintained, with greater local 'ownership' of the SMP policy, its implementation and associated liabilities.

On the other hand, unilateral initiatives to maintain a defence might undermine the overall management intent of a Policy Unit, and set a precedent whereby others in a similar situation follow suit in isolated locations with potentially detrimental effects upon adjacent areas unable or unwilling to follow suit. The potential for this can be

---

19 Environment Agency, 2013a

20 www.coastalpartnershipsnetwork.org.uk for more information on nationwide initiatives

particularly borne out in responses to storm damage. Even where the intention is to withdraw or reduce intervention in the first 20 years, the immediacy and violence of storm damage is liable to prompt pressure to rebuild, thereby re-furbish, and (subject to permission) possibly improve and extend existing defences rather than using the opportunity to adapt the coastline as a system. Yet the infrastructure behind those defences - caravan parks in particular - might be highly suitable for the type of relocation or rollback initiatives that have been trialled through the Defra Coastal Change Adaptation Pathfinders.

Risk Management Authorities must clearly continue to weigh the immediate pressures to secure important features of the local economy with the longer term management intent of the SMP Policy Unit. This can be a difficult balance for them to strike. On the East Anglian coast, initial unilateral investments to bolster defences to protect movable tourism facilities have led to further adjacent investment and moves to change NAI policies entirely.

**Physical Responses to NAI**

Isolated exceptions to a NAI policy may therefore become less isolated and less exceptional within and across Policy Units, and cumulatively begin to impact upon other stretches of coast. A change in land ownership or change in the financial viability of unilateral schemes, if it led to the withdrawal of one or more of these defences, might further complicate the evolution of a dynamic coastline. In particular, people and property in the vicinity of the failing or removed defences could experience more rapid erosion in the short term, and a temporarily exaggerated recession of the coastal baseline position – phenomena colloquially known as coastal 'catch-up' and 'overshoot' respectively.

The Environment Agency has recently funded research into the processes governing an example of coastal 'overshoot', which occurred at Happisburgh, North Norfolk, after the implementation of a NAI policy through defence removal. That study, which is also reported in these proceedings (Walkden *et al*, 2015) found the scale of catch-up to be large (around five times 'natural' losses) and used a numerical model to explore the reasons for this response. The study concludes by arguing that structure decommissioning and 'catch-up' losses should be included in the whole-life costing of coast protection options.

**Maintaining SMP Integrity in the Face of Change**

Local authorities and other Risk Management Authorities will wish to guard against an unmanaged evolution of the NAI policies in their SMPs, where the strategic intent of the SMP could be compromised rather than moderated and adapted. National and local political pressures to maintain communities and infrastructure in place through local empowerment are frequently brought to bear on Local Planning Authorities. In many cases, such initiatives are positive and Risk Management Authorities will seek to facilitate and support them. In others, they may point to a return to more parochial approaches to shoreline management that SMPs are designed to counter.

As such, an agreed process now exists to ensure that proposed changes to an agreed management policy, or the terms associated with it, are scrutinised at the relevant Coastal Group, according to specified criteria that might justify such a change. A proportionate approach to consultation, legal consideration and environmental assessment is enabled, and amendments are advertised alongside the SMP and reported to the Environment Agency[21]. Maintaining this audit trail and transparency is crucial for the onward maintenance of the SMPs as 'living documents', improved by evidence and experience but still compliant with environmental, health and safety and other legislation.

## Principles Associated with Not Actively Intervening

Given the variety of reasons for and details associated with NAI, are there governing principles to applying this policy that Risk Management Authorities should use? The challenge of withdrawing FCRM activity is obviously not new, and is the subject of a range of protocols, guidance and papers. Currently disparate, there is an intention at the time of writing that the Environment Agency will draw the key elements together into a revised Coastal Handbook. They cover the full range of responses to NAI policies, including:

- establishment of Coastal Change Management Areas, in guidance produced by East Riding of Yorkshire Council on behalf of Defra[22];
- community engagement approaches, in guidance produced during the development of SMPs and in the initial appraisal of Defra's Coastal Change Adaptation Pathfinders[23];
- practical adaptation methods experienced during the Adaptation Pathfinders, in Defra's review of property rollback, relocation and buy-and-lease-back schemes[24];
- environmental and legal implications surrounding third party asset adoption, health and safety considerations and the approach to landowner engagement, in the Environment Agency's Maintenance Protocol and associated Landowner Pack[25];
- managing environmental impacts, in the Environment Agency-Natural England joint statement in relation to the winter surge and storm events of 2013-14[26];

The conference presentation associated with this paper draws upon these in the context of outputs from the Environment Agency project work in 2015. However, themes do emerge from the preceding discussion that should guide the approach of Risk Management Authorities to NAI.

1. **Anticipate:** Cost considerations suggest that NAI is likely to emerge as a management approach in some places currently set out as 'Hold the Line', thereby increasing risk. But equally there are likely to be more opportunities, through Partnership Funding and local initiative, to maintain smaller and more targeted interventions alongside other risk management measures where it is sustainable to do so. Risk Management Authorities need to map these risks and opportunities so that their cumulative effects can be anticipated, and NAI 'hotspots' can be identified.

---

[21] The governing process for making changes to SMPs, which details circumstances when consultation and assessments (such as revisions to the Habitats Regulations Assessment for the SMP) are required, is provided in Environment Agency, 2012.
[22] Halcrow, 2015
[23] Scott Wilson, 2009; and Defra, 2012
[24] Risk Policy Analysts *et al*, 2015
[25] Environment Agency 2010; and ibid, 2013
[26] Environment Agency/Natural England, 2015

2. **Prioritise:** NAI, particularly withdrawal of maintenance, is likely to entail a suite of environmental, health and safety and other legal considerations in some locations. Depending on the rationale for agreeing the NAI policy, there may be flexibility around the timing of or conditions attached to its implementation. The priority associated with communications, engagement and activity at NAI 'hotspots' should be kept under review so that effort can be aligned with the full suite of flood/erosion, communications and legal risk.

3. **Communicate:** The need for early clarity around timescales, responsibilities and future options among communities and landowners at risk is well-rehearsed. Where there is a change in policy, it is important that Risk Management Authorities and others are able to discuss and agree this at (and, if necessary, beyond) the Coastal Group according to the established SMP Change Process to ensure it is auditable and transparent. Changes to timing of a NAI policy implementation are less important than changes that alter the SMP policy intent, but both require clear communication.

In a changeable political climate where austerity wrestles with local determination, the application, proportion and distribution of NAI policies in SMPs is likely to evolve with the evidence and policy that underpins them. The challenge for the risk management community is to balance consistency, fairness and legal compliance with innovation and efficiency in SMP delivery.

**References**

Burgess, K. Pontee, N. Wilson, T. Lee, S.C. Cox, R. 2013, Steart Coastal Management Project: Engineering Challenges in a Hyper-tidal Environment

Defra 2006, 'Shoreline Management Plan Guidance, Volume 1: Aims and Requirements', Defra publications

Defra 2006, 'Shoreline Management Plan Guidance, Volume 2: Procedures', Defra publications

Defra 2011, 'Flood and Coastal Resilience Partnership Funding: Defra policy statement on an outcome-focused, partnership approach to funding flood and coastal erosion risk management', Defra publications

Defra 2012, Coastal Change Pathfinder Review, Defra publications

Defra 2015, Report on the Coastal Management Practices at Pagham, Government publications

Environment Agency 2010, 'Landowner Information Pack' produced by Environment Agency Anglian Region

Environment Agency 2011, Appendices to the Protocol for the Maintenance of Flood and Coastal Risk Management Assets

Environment Agency 2013, Protocol for the Maintenance of Flood and Coastal Risk Management Assets

Environment Agency, 2013b, National Annual Review of Habitat Creation Programmes, England

Environment Agency/Natural England 2015, Responding to impacts of storms and

Halcrow 2015 (on behalf of Defra and East Riding of Yorkshire Council), Coastal Change Adaptation Planning Guidance (unpublished at time of writing)

Hardiman, N. Baptiste, A. Guthrie, G. Flikweert, 2011 'Will, or should, the RTPI be

running the ICE Coastal Management Conference in four years time?'
Increasing the dialogue between Shoreline Management Plans and spatial planning
New Forest District Council 2010b Beach Head to Selsey Bill Shoreline Management Plan Appendix C (Additional) - Recommendations for a No Active Intervention Policy at Bracklesham Bay and Pagham Harbour
New Forest District Council, 2010a North Solent Shoreline Management Plan
North Norfolk District Council 2010 Kelling to Lowestoft Ness Shoreline Management Plan
Risk Policy Analysts, The Planning Co-operative, Royal Haskoning 2015 (on behalf of Defra), Adapting to Coastal Erosion: Evaluation of rollback and leaseback schemes in Coastal Change Pathfinder projects (unpublished at time of writing)
Scott Wilson 2009 (on behalf of Defra), Community Adaptation and Engagement Planning on the Coast
tidal surges on designated wildlife sites behind seawalls
Walkden, M. Watson, G. Johnson, A. Heron, E. Tarrant, O. 2015 Coastal catch-up following defence removal at Happisburgh

**Coastal Management: Changing coast, changing climate, changing minds**
ISBN 978-0-7277-6114-9

ICE Publishing: All rights reserved
doi: 10.1680/cm.61149.721

# Indigenous Coping Mechanisms to Coastal Perturbations in the Philippines

**Carmela G. Taguiam**, Ecosystems Research and Development Bureau, Department of Environment and Natural Resources, Philippines
**Honorato G. Palis**, Freelance Consultant, Philippines

## Introduction

Climate change is a phenomenon that greatly affects the world population living in low lying coastal zones. As more than half of the populace lives within the 60km of the shorelines (Turner et al., 1996), it is inevitable that people would be exposed to the threats of climate change such as sea level rise, storm surge, coastal flooding and erratic tidal movement. Despite these threats, the estimated 634 million people (Balk in Greenfieldboyce, 2007) gravitate in coastal zones due to the number of benefits that can be drawn by living in low lying zones (Darwin and Tol, 2001; Cruz et al., 2007;Lal et al., 2007) .

The Philippines is an archipelagic country with a total coastline of 36,289km (http://www.wepa-db.net/policies/state/philippines/seaareas.htm). It experiences an average of 20 typhoons in one year as it fronts the Pacific Ocean where extreme weather events emanate. Considering that 60% of its population lives in 832 coastal communities (http://ecology.org/downloads/publications/blue-environment.pdf), a large number of the people are vulnerable to various forms of coastal perturbations triggered by climate-related changes. Coastal perturbation being referred to in this paper includes coastal flooding, shoreline erosion and retreat, sea level rise, salt water intrusion, and storm surge among others.

Owing to the overwhelming impacts of climate change specifically coastal perturbations such as the chronic coastal flooding, shoreline erosion sea level rise and storm surge from which the Philippines is not an exemption, adaptation is inevitable (Ford et al, 2010). This paper presents the coastal communities' perception of climate change focusing on sea level rise to highlight their awareness of the changes in the coastal environment. Corollary, it showcases the indigenous coping mechanisms applied by coastal communities who continuously experience the impacts of coastal perturbations.

## Coastal Perturbations in the Study Sites

*Camiguin Island* is a part of the Calayan group of islands located between the Batanes Island and main land province of Cagayan. With the Pacific Ocean on its east side, the island is directly hit by strong wind, wave, and typhoon. This in effect was perceived to cause sea level rise, storm surge, coastal flooding and erosion.

*Casiguran* is a municipality in the northern part of Aurora province. Aurora faces the Pacific Ocean and has no barriers from typhoons coming from the east. Casiguran and other municipalities within the province experiences periodic tropical cyclone. The seasonal tropical cyclone coupled with strong wind and wave led to coastal erosion.

*Gubat* and *Prieto Diaz* are coastal communities of Sorsogon directly fronting the Pacific Ocean. Due to geographical location, coastal communities have been wont to occurrence of typhoon, coastal flooding and storm surge. Shoreline retreat has also been observed.

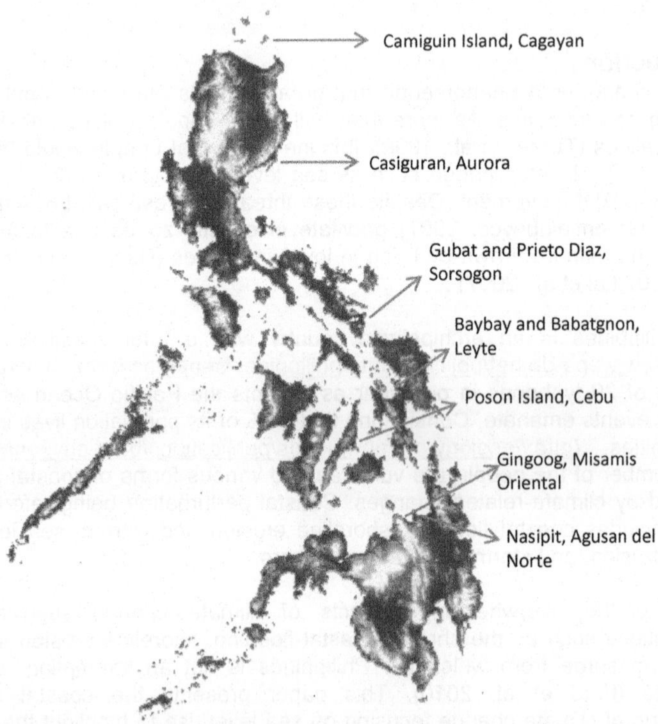


Figure 1: Map of the Philippines and location of study sites.

*Baybay* and *Babatngon* are municipalities in the island province of Leyte. The island has a history of being struck by super typhoons and storm surges. The most recent typhoons to hit the island were Super Typhoons Yolanda (Haiyan) and Ruby in 2013 and 2014, respectively.

*Ponson Island* is lies in the northeastern part of Cebu. Incidences of coastal perturbations in the island include flooding, sea level rise, shoreline erosion and retreat. A significant number of tropical cyclones and typhoons affected the province of Cebu. In 2013, Super Typhoon Haiyan nearly wiped-out the vegetation of Ponson particularly the mangroves found abundant in the island.

*Gingoog* is one of the cities in Misamis Oriental. Among the study sites, the perturbation encountered by Gingoog was limited to coastal flooding due to the presence of mangroves.

*Nasipit* is a coastal municipality in the province of Agusan del Norte. The coastal community experienced shoreline retreat, sea level rise and sedimentation.

## Methods

The research was carried out with the purpose to identify and document the coping mechanisms of communities to various forms of coastal perturbations that occur in the Philippines. Corollary, indicators of climate change were drawn to determine how aware and sensitive the coastal communities are to their environment.

In the identification of study sites, consultation and coordination with regional offices of the Ecosystems Research and Development Service (ERDS), local government units (LGUs) at municipal level were done to make sure the study sites indeed experience coastal perturbations.

To achieve the purpose of the research, the non-experimental field survey utilizing both sociological and anthropological approaches in the collection of primary data was applied. In the collection of primary data, a pre-tested participatory interview schedule was used in the interview of 400 household-respondents from the selected coastal provinces (Fig. 1). The interview schedule was pre-tested to determine the reliability of the research tool and information obtained from the household-respondents accurate. Household-respondents were selected through simple random sampling and based on the households' residence proximity along and within the shorelines. Respondent of the interview was the head of the family (male); in the absence of the male/husband, the female counterpart/spouse was interviewed. During the interview process, the researchers explained to household-respondents the merit of using the research tool in getting information from them.

Descriptive statistics particularly the degree of central tendency (frequency count) was used in the analysis of data such as the demographic characteristics, awareness, and indicators of climate change. To determine perception of sea level rise, 11 positive and negative constructed statements were prepared. To answer the perception statement, the Likert-type scale with 5-stage rating was adopted (Jackson, 2010). For the positive statement, the rating was: 1-strongly disagree, 2-disagree, 3-neutral/undecided, 4-agree, and 5-strongly agree. The scoring pattern was reversed to answer the negative statement. The summated scores obtained from the 11 statements were analyzed through factor reduction analysis. Factor analysis is the grouping of variables in the fewest possible wholes to form a new independent factors responsible for the groupings (Cattell, 1952). The software used in the analysis was the Statistical Package for Social Sciences (SPSS). Also, focus group discussion (FGD) and key informant interview (Rietbergen-McCracken and Narayan, 1998) were done to validate the information from the interview of household-respondents.

Identification of indigenous coping mechanisms to coastal perturbations was through recall method of the household-respondents. Documentation of these mechanisms was done during the field survey. Field survey was conducted from 2011-2012.

## Results and Discussion

### Demographic Profile of Coastal Household-Respondents

A total of 400 household-respondents composed of men (48.75%) and women (51.5%) were interviewed. Household-respondents age ranged from 18-91 years; average age was 46.3 years. Nearly all (98.75%) except for one (.30%) were literate. Over half of household-respondents were born-residents (51.5%) and the others as migrants (48.5%). Length of stay in the study sites was from two months to 91 years with the mean residence at 34.4 years.

### Awareness and Perceived Indicators of Climate Change

Household-respondents (82.25%) were aware of the climate change phenomenon (Table1). Of the 329 household-respondents affirming the existence of climate change, 68.38% said it occurred in recent years (less than five years). However, a number of household-respondents claimed that climate change begun 6 – 10 years (14.58%) and 11 -15 years (16.41%), respectively. The remaining two household-respondents said climate change was felt for more than 16 years (.6%).

Numerous indicators of climate change were identified by household-respondents. The list was quite long revealing that household-respondents have different perspectives of the changes in their environment. As shown in Table 1, unpredictable weather was the most felt indicator of climate change. Household-respondents elaborated the weather had become unusual - they could not tell when the dry and wet seasons start. Likewise, the coastal environment especially the sea and its tide was perceived to be different from what it used to be. It was followed by hotter temperature, long dry season, frequent occurrence of typhoon, and increase rainfall. Other given signs seemed to be the effect like frequent typhoon and increase rainfall cause frequent coastal flooding.

Table 1: Household-respondents awareness and indicators of climate change

| Particulars | Frequency | Percentage |
|---|---|---|
| Awareness of climate change | | |
| Aware of the climate change | 329 | 82.25 |
| Unaware of the climate change | 71 | 17.75 |
| Start of climate change (in years) | | |
| ≤ 5 | 225 | 68.38 |
| 6 – 10 | 48 | 14.58 |
| 11 – 15 | 54 | 16.41 |
| ≥ 16 | 2 | .60 |

| | | |
|---|---|---|
| Perceived indicators of climate change* | | |
| Unpredictable weather | 267 | |
| Extreme temperature | 51 | |
| Hotter temperature | 148 | |
| Cooler temperature | 43 | |
| Drought | 95 | |
| Long dry season | 108 | |
| Short dry season | 40 | |
| Early dry season | 12 | |
| Late dry season | 18 | |
| Long rainy season | 87 | |
| Short rainy season | 26 | |
| Early rainy season | 22 | |
| Late rainy season | 17 | |
| Frequent/more typhoon | 107 | |
| Increase rainfall | 104 | |
| Decrease rainfall | 17 | |
| Frequent coastal flooding | 65 | |
| *multiple response | | |

**Perception of Sea Level Rise**

Result of the factor reduction analysis revealed that sea level rise was perceived by household-respondents in three dimensions and these are shown in Table 2.

Table 2: Perception of sea level rise.

| Dimensions | Eigen value | Percent variance | Cumulative percent |
|---|---|---|---|
| Dimension 1 (shoreline erosion) | 2.692 | 24.474 | 24.474 |
| Dimension 2 (decrease of marine resources) | 2.206 | 20.051 | 44.526 |
| Dimension 3 (salt water intrusion) | 1.205 | 10.953 | 55.478 |

Household-respondents first perceptual dimension of sea level rise is attributed to shoreline erosion. Household-respondents supposed the continuing rise of sea level triggered shoreline erosion. Related to this, Cruz et al (2007) noted that erosion in Asia will occur due to the continuing rise of sea level. Corollary, household-respondents observed that shoreline erosion destroyed the root system of different plant species and eventually led to its disappearance. In the same way, a number of households were forced to change place of residence as the shoreline slowly eroded destroying and inundating their houses by the onslaught of sea level rise.

The second dimension highlights the household-respondents perception on the decrease of marine resources. Majority of household-respondents were engaged in fishing. In recent years, a reduction on the volume of fish catch was observed; the decrease of fish catch affected the economic activity of household-respondents and the blame was ascribed to sea level rise. Household-respondents reasoned that the increasing level of sea water push the school of fishery and other marine species in the midst of the ocean due to the disruption of their ecological niche (Dossou and Glehouenou-Dossou, 2007).

As regards to the third dimension, household-respondents perceived the advent of sea level rise resulted to intrusion of salt on surface water. Household-respondents primary sources of water for domestic purposes were either the deep well or hand-held water pump. Due to the proximity of the residential area to shoreline and rise of sea water, households observed that water drawn from their sources had a taste of salt. Scientists working on coastal ecosystem had foretold that salt water intrusion would be one of the impacts of sea level rise (Turner et al., 1996; Nicholls and Cazenave, 2010).

**Indigenous Coping Mechanisms to Coastal Perturbations**

Result of interview showed that household-respondents adopt various ways to cope with coastal perturbations brought by climate change. Through the years, household-respondents repeatedly experience the impacts of coastal flooding, tropical cyclone or typhoon, and storm surge. These coping mechanisms were handed down to household-respondents by their ancestors and parents, learned from experience and extension service-initiatives of local government units and private organizations. While household-respondents came from various geographical locations, the indigenous coping mechanisms manifested similarities and differences due to what was acceptable and needed in a particular location. In the end, the indigenous coping mechanisms observed by household-respondents with the exemption of retreat as a strategy are apt and in consonance to the adaptation responses being espoused by the IPCC-CZMS (1990) and UNFCCC (2006).

Among the sites surveyed, household-respondents from the provinces of Cagayan, Cebu, Leyte, and Agusan del Norte had the most number of indigenous coping mechanisms adopted to address coastal perturbations (Table 3). Planting of mangroves was the universal coping mechanisms in all sites. The provinces of Aurora and Sorsogon are the pathway or entry point of tropical cyclone and typhoon, as such, planting of mangroves was deemed the most effective way to downplay the impacts of coastal perturbations. In the seven provinces studied, rehabilitation of mangrove forests was a joint effort of people's organizations and national government in Aurora, Sorsogon, Leyte, and Misamis Oriental. Management of natural and rehabilitated mangrove plantations were the most acceptable and efficient strategy to deter the impacts of coastal perturbations. In Panay Island, it was observed that the strong and structurally complex above-ground root systems found in mature mangroves serve as an effective wind and wave break (Walters, 2004). Similarly, in a storm-related study due to super cyclone in India, villages with wide expanse of mangroves reported fewer incidence of deaths as compared to villages with absent or limited mangrove area (Vincent et al., 2009).

Another coping mechanism common to all household-respondents was the temporary transfer in evacuation center. Household-respondents houses were generally made of light materials and as a precautionary measure, the government infrastructures such as public school, covered court, and sports facilities are usually the designated evacuation centers. At times, the church also serves as a refuge of the affected people. Evacuation and transfer of residence is one of the conventional strategies that have been institutionalized in the Philippines to minimize the impacts of coastal perturbations (Palis and Taguiam, 2011).

Dredging of canal, sandbagging, construction of seawall and stilt house sandbagging were undertaken to deflect flooding of the settlement areas. Indigenous option such as placing tire on the roof of the house was done to prevent the roof from being blown by the wind.

As shown in Table 3, indigenous mechanisms have evolved in different places to cope with the impact of coastal perturbations. The repeated incidences of coastal perturbations have made the coastal populace resilient in times of adversity.

Table 3: Matrix of indigenous coping mechanisms to coastal perturbations by household-respondents.

| Particulars | Camiguin Island, Cagayan | Casiguran, Aurora | Gubat and Prieto Diaz, Sorsogon | Poson Island, Cebu | Baybay and Babatngon, Leyte | Gingoog, Misamis Oriental | Nasipit, Agusan del Norte |
|---|---|---|---|---|---|---|---|
| Plant coconut | √ | | | | | | |
| Maintenance and plant of the naturally-growing trees/beach-type species along the shore | √ | | | | | | √ |
| Pile rocks along the shore and house perimeter | √ | | | | | | |
| Dredge/Canal construction | √ | | | | √ | | |
| Plant mangroves | √ | √ | √ | √ | √ | √ | √ |
| Temporary transfer in evacuation center | √ | √ | √ | √ | √ | | |
| Construct sea wall | | √ | √ | | | | √ |
| Arrange sandbags | | | √ | √ | √ | | √ |
| Breakwater | | | √ | | | √ | |
| Construct dikes and drainage | | | | √ | √ | | √ |
| Indigenous options | | | | √ | √ | | √ |
| Recite a prayer | | | | | √ | | √ |
| Build stilt/two storey house | | | | √ | √ | | |

Figure 2: a) construction of canal to divert water; b) pile of rocks within house Perimeter; c) construction of riprap to prevent coastal erosion; d) arrange Sandbag; e) planting of mangroves; and f) build stilt house.

Figure 3: a) and b) stilt house made of indigenous materials; c) maintenance of mangrove plantation; d) and e) construction of seawall and canal; and, f) indigenous options with tires arranged on the roof.

## Conclusion

Household-respondents are exposed and incessantly experience the impacts of climate change particularly coastal perturbations. To safeguard their well-being, indigenous coping mechanisms evolved and being institutionalized to curb the impacts of coastal perturbations.

Coping mechanisms are diverse and continue to evolve. However, it must be remembered that coping mechanisms vary. What is effective in one place is not necessarily applicable in another place. This occurs due to geography and distinct features of the place including the cultural attributes of the people. Nonetheless, it is important that people be aware of the risks and dangers they are exposed to; that they arm themselves with knowledge and skills to limit if not overcome the detrimental impacts of coastal perturbations.

## Recommendations

- Information education campaign to highlight impacts of coastal perturbations to households vis-à-vis environmental awareness, protection and conservation
- Encourage participation of households to adopt mechanisms to ensure their safety and heed the calls or warnings of government during occurrence of natural calamities
- Institutionalize laws to limit establishment of settlement area along the coastline

## References

Berang-Ford,L.,Ford, J.D.,Paterson, J. (2010), "Are we adapting to climate change?", Global Environmental Changes (2010), doi:10.1016/j.gloenvcha.2010.09.012

Cattell, R.B. (1952), Factor analysis: An introduction and manual for psychological and social scientist, New York, Harper and Brothers

Cruz, R.V., Harasawa, H., Lal, M., Wu, S., Anokhin, Y., Punsalmaa, B., Honda, Y., Jafari, M., Li, C., Huu Ninh, N. (2007), In Climate Change 2007: Impacts, Adaptability and Vulnerability. Contribution of Working Group II to the Fourth Assessment Report of the Intergovernmental Panel on Climate Change, Cambridge, UK, pp 469-506

Darwin, R.F., Tol, R.S..J. (2001), Estimates of the economic effects of sea level Rise, Environmental and Resource Economics 19 pp113-129

Dossou,K.M.R. and B.Glehouenou-Dossou (2007), The vulnerability to climate change of Cotonou (Benin): the rise in sea level. Environment and Urbanization 19:65,DOI:10.1177/0956247807007149

Greenfieldboyce, N. (2007), Study: 634 million people at risk from rising seas. In http://www.npr.org/templates/story/story.php?storyId=9162438&sc=emaf

http://ecology.org/downloads/publications/blue-environment.pdf

http://www.wepa-db.net/policies/state/philippines/seaareas.htm

IPCC (1990), Strategies for adaptation to sea level rise. Report of the Coastal Zone Management Subgroup

Jackson, S.L. (2012), Research methods and statistics: a critical thinking approach, 4th ed., USA, WADSWORTH Cengage Learning

Lal, M., Harasawa, H., Murdiyarso, D., Adger, W.N., Adhikary, S., Ando, M., Anokhin, Y., Ilyas, M., Kopaliani, Z., Lansigan, F., Li, C., Patwardhan, A., Safriel, U., Zhang, X. (2007), Asia, In www.grida.no/climate/ipcc-tar/wg2/pdf/wg2TARchap11.pdf

Nicholls, R.J. and A. Cazenave (2010), Sea-level rise and its impact on coastal zone. Science 328,1517;DOI:10.1126/science.1185782

Palis, H.G., Taguiam,C.G. (2011), Adaptation responses to coastal perturbations: the case of Prieto Diaz and Baler coastal dwellers in the Philippines. *In* Proceedings of the Asia and the Pacific Symposium – *Vulnerability Assessment to Natural and Anthropogenic Hazards*, IUFRO World Series Vol. 29, pp 35-37

Rietbergern-McCraken, J., Narayan, D. (1998), Participation and social assessment:Tools and techniques. Washington D.C., The International Bank for Reconstruction and Development/ The World Bank

Turner, R.K., Subak, S., Adger, W.N. (1996), "Pressures, trends, and impacts of coastal zones: Interactions between socioeconomic and natural systems", Environmental Management 20(2), pp 159-173

UNFCCC, (2006), Technologies for adaptation to climate change, UN Headquarters, New York, pp 1-65

Vincent, J.R., Forstian, C.F., Das, S. (2009), Mangrove forests save lives in storms study of 1999 super cyclone finds, available from http://www.sciencedaily.com/releases/2009/04/090407292

Walters, B.B. (2004), "Local management of mangrove forests in the Philippines: successful conservation or efficient resource exploitation", Human Ecology 32(2) pp 177-194.

**Coastal Management: Changing coast, changing climate, changing minds**
ISBN 978-0-7277-6114-9

ICE Publishing: All rights reserved
doi: 10.1680/cm.61149.733

# How to Teach Engineers to Interact in a Political Decision Making Environment

**Mireia López Royo**, Unesco-IHE, Delft, The Netherlands (Offshore Independents)
**Madelon Burgmeijer**, Delft University of Technology, Delft, The Netherlands
**Henk Jan Verhagen**, Delft University of Technology, Delft, The Netherlands

## Introduction

Sectoral professionals, like coastal engineers, need to know how to cooperate in a multi-disciplinary environment, as well as how to present substantial information to (political) decision makers. In order to train (coastal engineering) students in this matter, Unesco-IHE and TU Delft have in their curriculum an Integrated Coastal Zone Management Seminar. Because there is also a demand for such a seminar from outside the normal student population, this seminar is also open for external participants, like tourist resort developers, ecologists as well as policy support officials, who are in need to learn how to present to decision makers the various development options in a coastal zone.

This paper will give an overview of the nearly 25 years of experience we have collected in giving this seminar in Delft and Overseas. It is written from the learning perspective of a professional who attends the ICZM Seminar, while remarking the challenges and goals expected to be achieved.

## Broadening professionals 'minds

As pointed out in the previous paper Training of coastal engineers to work in a non-engineering environment (Burgmeijer *et al.*, 2011), ICZM is not really a science, but a matter of good cooperation in planning and decision making. The intention of the seminar is therefore to focus on the process of decision making in the coastal zone and on how important communication and interaction between stakeholders are when making the final decision. A conclusion that the participants should get to should be that due to the weight of non-technical arguments in the political discussion, what seems to be the "best" solution to solve a specific problem from an engineering perspective is not always the chosen alternative by the ones who have to decide.

Strange as it may seem, engineers are not taught how to tackle this kind of situations, they are only trained to find the best solution from a technical point of view. In the seminar it is said that they should be trained also as decision makers or be given the tools to participate in the decision making process, as in many developing countries engineers are in charge of taking the decisions as well as finding the technical optimal solution to a problem.

Hence, along the seminar technical lectures are combined with interactive activities and followed by a role play that immensely helps in accomplishing the main goal.

The seminar succeeds in broadening the minds of engineers and the other involved professionals so that they are taught not to only tackle the situation from an engineering perspective, as suggested previously, but to also be capable of understanding the implication of other sectors and cooperate with them to achieve the "best" solution for the given situation. The multicultural environment in the class is also decisive in achieving the established goal of the seminar, as interacting with people from other countries may help in widening one's mind even more. Each one's expertise and professional experience are for sure helpful in that respect.

Since all attendants have a good knowledge of the physical processes in the coastal zone, there is no need to go into much detail in these matters, because the main line of the seminar is the process of decision making in the coastal zone. However, the technical lectures addressing important topics are also valuable for professionals in this sector. Two of the specific lectures in this seminar are about setback lines and climate change effects on the coast, but other important topics are addressed, such as policy analysis, and ecology in the coastal zone among others. In both lectures a direct link can be given between the technical aspects and the implications for decision makers. Also a lecture on ecosystem services can be considered as a "technical lecture", as it emphasizes the services that coastal ecosystems provide to people, which can be material but also non-material and which can be classified into different groups.

The process of Integrated Coastal Zone Management is explained in general terms and it is pointed out that it should be cyclic, starting by an identification of the main issues, a preparation of a program, a formal adoption and funding, its implementation and a final evaluation, which leads at the same time with the beginning of the cycle (*identify-implement-evaluate-identify*). Development of the ICZM plan must take into account the long term goals but also the short term ones. Figure 1 shows the ideal development of an Integrated Coastal Zone Management Plan.

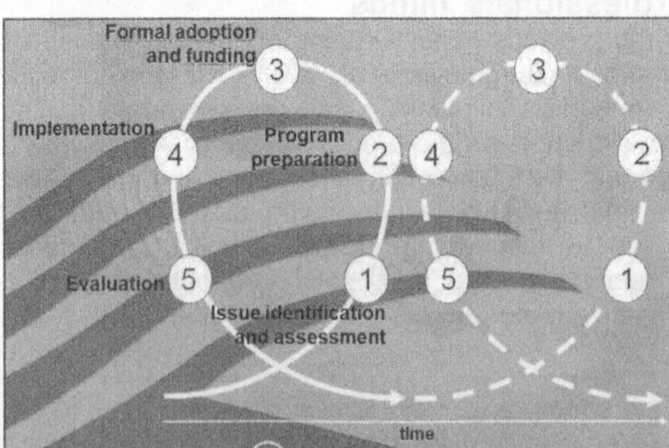


Figure 1: Integrated Coastal Zone Management Cycle. (UNESCO-IHE)

In this cycle, public participation should take part in the early stages of the project. Otherwise it might happen that the achieved engineering solution cannot be put into practice due to the final public opposition.

## First-hand Expert Knowledge and Experiences

As it has been already mentioned, in order to provide the best seminar possible several guest speakers are invited to give some lectures. These guests are experts in their fields and provide, by means of examples and study cases from personal experience, valuable knowledge for professionals of the field.

A very specific lecture is on "Building with nature" that presents a twist towards a more nature-friendly and conscious approach of coastal engineering projects. Through a series of case studies, the guest speaker provides examples of a successful implementation of the exposed approach, which may serve as an inspiration for the attendants of the seminar in their future professional challenges. But apart from the fact that the lecturer in this case is quite knowledgeable on the topic, he has a political background. Therefore his lecture is a good starting point for several discussions with the student.

The decision making process is approached by lectures on "Game theory" and "Policy Analysis". They give useful tools for taking decisions. As illustration a number of case studies are discussed with the students. Other guest lecturers present case studies that illustrate either successful or unfortunate examples of existing situations of other relevant topics in the coastal zone in order to teach a lesson from experience. With the help of these several case studies this working cycle is exposed. Examples of case studies are Coastal Development in Mozambique, Tsunami response in Japan, sustainable expansion of the port of Rotterdam, Dune development near The Hague (Spanjaards dune) and de-poldering of an area along the Westerschelde (Hedwige Polder).

The lecture about socio-economic aspects of ICZM points, among other things, that the gender aspect may be present. The lecture combines scientific knowledge with non-scientific facts that are certainly involved in the process.

## Main Outcome from the Fieldtrip

In the last edition of the seminar the one-day fieldtrip consisted in visiting the border that separates the Dutch coastline from the Belgium coastline and noticing how a different management can make two similar coastlines look completely unlike due to different kinds of management. The consequences of these different approaches are visible and may lead to complete different risk exposures, such as the risk to inundation of the Belgium coast, for example.

From the experience of a student who attends the seminar, the fieldtrip is as important and instructive as the lectures, as what one sees in the fieldtrip complements totally to what is learned in the lectures that precede the fieldtrip.

The fieldtrip contributed together with the role play to make students interact with each other all the time, and it was a first fruitful step to carry on some networking among the participants, which would be very useful for the following activity.

## Role Play – Step by Step

The role play is the key part of the seminar as it certainly triggers the successful achievement of the main goal. Although the role play itself is held in the last two days of the seminar, there are several activities before it starts that are directly related to it and which outcomes are used in the main part of it.

Jesew, a computer program for "Joint Ecological and Socio-economic Evaluation of Water resources development (Zitman, 2003), is rewritten as a spreadsheet application and it is used as a first step to help students elaborate a report they have been hired to do as consultants for a certain client with very specific interests. The main task of each group is to assess the client on how to invest the money available and set a strategy on employment, income, pollution, ecological quality, etc. to find the optimum solution. The assignment that every group gets is basically the same, with the difference that each report has to be done for a different client- who is a stakeholder of the coastal zone, and therefore the objectives of each one of them are different.

In the second phase roles are changed, as well as groups, and students play each one a role as a stakeholder in the decision making process (one can be the Mayor of the town, the Minister of Finance, the representative of the fishermen, etc.). The role each one represents is given a list of the main interests to care for and a report made by a consultant (reports from the previous activity made by another group; one cannot receive his own assignment), and with this indications he/she has to work to achieve their accomplishment. During the discussions the strengths and weaknesses of the participants are uncovered as well as the quality and usefulness of the reports previously made by others. This has a clear impact on the results of the role-play and the decision taken by the stakeholders involved.

One of the main goals of this role play is to acclaim the similarities with the real situation by, for instance, not giving time to think- as politicians receive reports on their way to a meeting where a decision needs to be taken, with just the sufficient time to read it. The reports are supposedly done to fulfill the requirements of each client so each stakeholder should treat each report like self-made. The pressure each stakeholder makes in the decision process is decisive.

Needless to say, decision making is very expensive in real life, as many hours from stakeholders are included. Although often these hours are not paid by the project, but by the stakeholders themselves, it is a considerable financial burden to society. Therefore it is important to make meetings with stakeholders efficient. The direct project expenses for a project may run up to 0.5% of the total project expenditures.

One important remark is that the groups are intentionally designed as multicultural as possible and coordinators try not to put friends in the same group, with the clear intention of promoting interaction and performing the real situation as close as possible.

The final part of the role play consists of discussions and making a decision with all the stakeholders involved in a group. The process itself is closely observed by the coordinators as it is recommendable that certain behaviours are avoided, such as the philosophy "one man, one vote". The process needs to involve all stakeholders, who are invited to give their opinion and propose a strategy. One other significant remark is that is it always easier to convince other stakeholders if one's strategy involves investments in different sectors and therefore benefits others apart from oneself.

In the role play it is important to analyse to what extent the consultants 'reports are being used by the clients who paid for them, and how much the clients believe in them, and make of them their own stance.

The decisions taken are influenced by the personality of the different stakeholder's, as if a certain stakeholder was represented by someone else, the way this person interacts with the other stakeholders might be different and so the decision making process might lead to another result.

The daily press is delivered by the organisers of the seminar with the clear intention to alter the decision making process by inventing some breaking news that could also happen in reality if that was the case. Under these unexpected events the participants of the seminar tend to act the same way- although it is important to remark that sometimes, due to the great persuasive attitude of some participants, this bad news do not affect the sector they are representing, even if it is damaging for many other sectors. This happened in one of the groups in last year's edition of the seminar. On the other hand, even the green associations in favour of a sustainable development might not get the necessary amount of money if that certain person does not make himself/herself heard in the decision making process.

Each coastal zone, however, might have different stakeholders and therefore the management of the coastal zone, despite following the same guidelines and practices, may lead to completely different developments on the coastal zone and a result its appearance look completely different. Moreover, the boundary conditions from one region to another define each specific case.

## Lessons learned

One of the main lessons a participant can learn from this seminar is the importance to broaden his/her perspective and comprehend a problem in the coastal zone as a multidisciplinary matter involving a series of stakeholders, some of which may be "non-professional", which refers to groups whose main profession is not the management of the coastal zone, like for instance fishermen, but also local politicians.

Year by year the same behavioural pattern is observed by the seminar coordinators, so one could definitely say that the way participants deal with the situation follows the same pattern. However, cultural differences are captured in the role play as the way people from different cultures may deal with negotiations and meetings might be quite different.

On the other hand, the seminar is very useful for participants to note the similarities that the real situation has with the activity they are confronted with, as there is not much time to think, some external influences or news might suddenly change the situation completely and the consultancy report is their main source of information for either taking the decision or supporting one. This so called consultancy report has therefore to be produced in an appropriate way, one which can easily be understood and visually apprehended.

## Conclusion and Recommendations

The main goal of the seminar, which is training engineering students to focus on the fact that other disciplines may add useful additional material for decision makers but also broadening their minds, is highly achieved mainly due to student's implication in the role play. The fact that it is so interactive and well-structured makes it even more easy and interesting to follow.

From a student's point of view, the seminar is structured in such a way that it becomes clear what is required in many situations on a professional career such as in decision making processes. There is not the pressure of an exam but it is a continuous personal challenge of having to interact with other students while defending one's interests to others and trying to make them support one's own plan.

This seminar has been inspiring other universities to come and further develop it, which is for example the case of Vietnam.

A recommendation for a further improvement of the seminar is the introduction of the Integrated Water Resources Management (IWRM) concept aligned with ICZM. Participants should get the idea that what happens upstream in a river influences that happening downstream and entering the sea. Rivers flow into the sea and it seems quite obvious that the fact that they carry less or more sediments, and changes in this particular load, will influence the morphology of the coastline, caused by, for instance, the construction of a dam. So the seminar itself should not need to be modified but it would be interesting to reflect this reality and introduce the concept of IWRM.

To sum up, it is all about changing the perspective and learning that to achieve a proper coastal zone management not only scientific and technical knowledge are important, but also being able to interact with other stakeholders and parties, as that is actually a significant part of the process.

## References

Burgmeijer, M.A.; Verhagen, H.J.; Van der Wegen, M. (2011) Training of coastal engineers to work in a non-engineering environment. Coastal Management 2011: Innovative Coastal Zone Management: Sustainable Engineering for a Dynamic Coast, Belfast, UK.

Zitman, T.J. (2003) Jesew user guide, TU Delft report

**Coastal Management: Changing coast, changing climate, changing minds**
ISBN 978-0-7277-6114-9

ICE Publishing: All rights reserved
doi: 10.1680/cm.61149.739

# East Cliff, Lyme Regis: The Challenges of Coastal Protection and Slope Stabilisation of an Historic Town and Community within a World Heritage and Special Area of Conservation Site

**Geoff Davis**, CH2M HILL, Birmingham, UK
**Roger Moore**, CH2M HILL, Birmingham, and University of Sussex, UK
**Jon Cartwright** Balfour Beatty Construction Services UK, Southampton, UK
**Dimitri Daskalopoulos**, AECOM, Basingstoke, UK
**Nick Browning**, West Dorset District Council, Dorchester, UK

## Introduction

The Lyme Regis Phase IV East Cliff scheme completed in 2015 is one of the largest and most complex coast protection schemes attempted in the UK. Thirty years in planning, the ground engineering, environmental, community and financial challenges of the scheme provide an exemplar that will be followed by the industry for decades to come (Watson 2015).

The historic coastal town of Lyme Regis is situated in one of the most unstable geological settings in the UK and has suffered severely from the effects of coastal erosion and landsliding. Over the years, landslide activity and cliff retreat has damaged or destroyed many properties throughout the town together with roads, land and infrastructure. In the absence of engineering intervention the future for the town was bleak, with existing coast protection structures reaching the end of their useful life and landslide activity increasing in response to climate change and a higher frequency of wet winters over the last two decades.

In order to secure the future of Lyme Regis, West Dorset District Council has, since the late 1980s, promoted a long-term programme of phased investigations and engineering schemes to protect and stabilise the coastal frontage of the town. The East Cliff project is the latest phase in this programme which addresses the coastal erosion and landslide threat to the east flank of the town through:

- The construction of a new seawall to replace an existing dilapidated wall
- Slope stabilisation works
- Extensive landscaping and environmental mitigation works

The objectives of the project are to:

- Protect houses, infrastructure and other assets in the long term – up to 480 homes will be saved from damage or loss of access
- Provide improvements to the environment in terms of landscape, geology, ecology and for public amenity

The project required some innovative approaches in order to reconcile the challenging technical objectives of the projects with considerable environmental constraints.

Figure 1: The location of the site with the Black Ven landslide in the foreground

## Background

Lyme Regis is located on an actively eroding soft rock coastline in southern England and much of the town has been constructed on old coastal landslides which are subject to ongoing slow movements, and, from time to time, destructive landslide activity (Figure 1). West Dorset District Council, in its role as coast protection authority, has over many years shown vision, commitment and persistence in promoting coast protection and slope stabilisation schemes to protect the town. In particular the Council has been proactive in identifying and dealing with coastal erosion and landslide problems in a strategic manner, rather than reacting to disasters once they have occurred. Central to the success of this strategy has been extensive long-term consultations with the local community and other stakeholders (Davis and Cole 2002).

Since the late 1980s, West Dorset District Council has promoted a long-term programme of engineering projects covering the whole of the frontage at Lyme Regis, all within an overarching coastal management strategy and with each project typically comprising:

- Desk studies
- Geomorphological and geological mapping of the cliffs, landslides and beaches
- Phased ground investigations
- Installation of ground instrumentation and monitoring
- Preparation of conceptual ground models and process models for the cliffs and beaches
- Local emergency works

- Full engineering schemes to stabilise and protect the coastal areas of the town
- Ongoing extensive consultations with residents, the local community and other stakeholders

This holistic, proactive approach is considered to have many benefits, particularly the economic case that prevention is better than having to deal with the aftermath of destructive erosion or landsliding (Cole *et al.* 2002; Moore and Davis 2014;Moore and McInnes 2011; Moore *et al.* 2010).

The main driver of the present phase of the scheme was that the eastern flank of Lyme Regis was under severe threat from active landsliding which, in absence of stabilisation works, had the potential to expand rapidly inland into the built-up areas of the town, including the A3052, the main strategic access road into the town from the east. Although the benefit cost ratio was less than that for the central areas of the town's frontage, where a high density of property and businesses are present, the economic justification for engineering intervention at East Cliff was met in accordance with the rules for central government grant in aid for coastal protection works under the Coast Protection Act 1949 (Moore and Davis 2014).

## Coastal Management Challenges Facing the Eastern Flank of Lyme Regis

### Challenge 1 - Countering the Threat of Coastal Erosion and Landsliding

The main objective of the project is to secure the long-term future of the eastern part of Lyme Regis by providing protection against coastal erosion and landsliding. This was particularly challenging because:

- Much of the eastern part of the town is underlain by pre-existing landslides in the Lower Lias mudstones and weak, periglacially disturbed material susceptible to reactivation of ground movement
- The coastal landslide systems to the east of the town are some of the largest and most active in Europe and are expanding towards the town (Figure 1).

### Challenge 2 - International Nature Conservation and Landscape Designations

The site is internationally important for geology, geomorphology and wildlife. Severe constraints were placed on the engineering aspects of the project from numerous heritage and nature conservation designations reflecting the considerable geological, nature conservation, heritage and landscape value of the eastern side of Lyme Regis. These designations include:

- UNESCO World Heritage Site status (the "Jurassic Coast")
- European Special Area of Conservation (SAC)
- Site of Special Scientific Interest, for wildlife, geomorphology and geology
- Area of Outstanding Natural Beauty
- Heritage Coast
- Heritage Conservation Area

Coastal erosion and landsliding is of critical importance to the geological and biodiversity value of the World Heritage Site, particularly for the ongoing exposure of abundant fossils, and this need had to be reconciled with the clear but conflicting requirement to protect people, property and infrastructure from the same erosion.

The main implications were:

- Stabilisation of the coastal landslide system was not permitted over large areas due to nature conservation designations
- Seawall works were not allowed to extend a significant distance beyond the footprint of the existing defences, due to the geological importance of the foreshore ledges and cliffs

**Challenge 3 - Physical Constraints**
The site is in an exposed location which is difficult to access over steep unstable slopes, with a foreshore which is completely submerged at high tide and in area where landslides are frequently active during the winter.

## Meeting the Challenges

**Design and Build Procurement Approach**
A design and build contract was let in order assist the rapid development of the project to construction stage and to encourage cost saving and innovation by the principal contractor and their designers, whilst building upon the knowledge gained in the preliminary studies stages and community engagement over several years.

**Developing a Detailed Conceptual Ground Model**
A detailed understanding of current and potential landslide mechanisms and erosion processes was gained through several phases of geological and geomorphological mapping, ground investigation and monitoring conducted at feasibility stage and over the period 1990-2011(Brunsden 2002, Sellwood *et al.* 2000) and specifically for the detailed design of the works. This allowed the position of landslide-controlling strata to be located with a high degree of precision and the design of the slope stabilisation works to be designed to suit local ground conditions with a high degree of efficiency. The geomorphology mapping was key to defining the extent and nature of landsliding, providing the spatial framework and conceptual understanding for ground investigation and design of solutions (Brunsden and Moore 1999).

**Designing the Geotechnical Works to Prevent Expansion of Landslide Systems, Rather Than Complete Stabilisation**
An overall design concept was adopted where slope stabilisation works are intended to prevent expansion of destructive landsliding inland, rather than stabilising all the areas of existing landslides. This required several different interrelated geotechnical engineering solutions to suit the varied ground conditions and different landslide mechanisms, with the main elements involving (Figure 2):

- Soil nails between 6 and 22m long, with associated plates and facing mesh - 34km of soil nail bar were installed in total
- A piled wall comprising 900mm diameter piles up to 27m long, tied back with ground anchors between 35m and 46m in length (Figure 3).
- Dowel piling 300mm in diameter and between 8 and 12m long
- Drainage works including sub-horizontal drilled drains and cut-off trench drains

This had the benefit that the most environmentally sensitive areas of the active landslide were left untouched thereby allowing continuing erosion and avoiding unacceptable environmental impacts. Despite the challenges, cost savings were made through value engineering, bespoke geotechnical design and working with a variety of different procurement routes which allowed the project to be completed within the available budget.

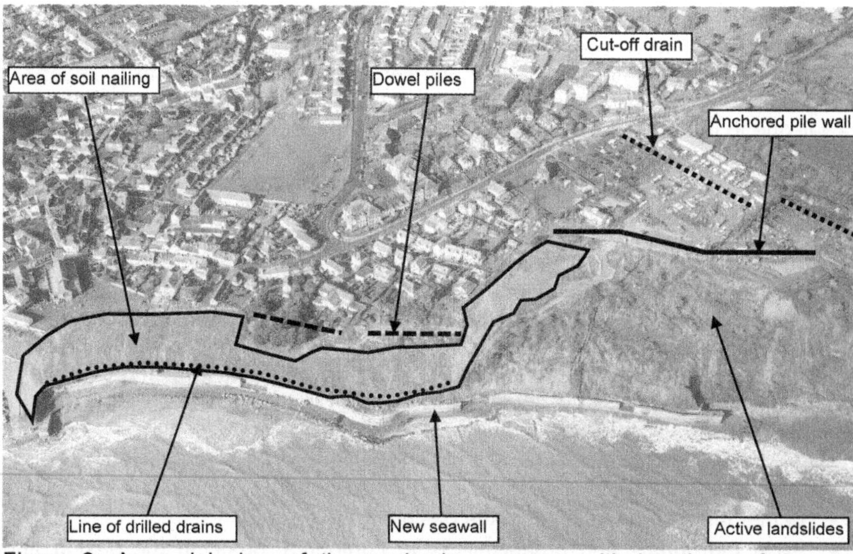


Figure 2: An aerial view of the works in progress, with locations of the main elements.

Figure 3: Installing ground anchors through the capping beam for part of the anchored bored pile wall. The active cliff edge is just beyond the Heras fencing on the right. The owners of the cottage will have their garden reinstated and sympathetically landscaped.

**Minimising the Footprint of the New Seawall**

The footprint of the new works was kept to a minimum by constructing a new 390m long seawall in front of the existing dilapidated wall but with no protective rock armour in front of it, nor extending much beyond the length of the existing wall. This had the benefit of leaving the foreshore ledges in front of the wall and the sea cliff beyond the end of wall exposed to the waves so that they could continue to erode through natural processes, thereby preserving the internationally important geological interest of the site. The possibility of long-term undermining of the new seawall due to foreshore lowering was mitigated by deepening the toe of the wall into strong limestone beds, using detailed geological knowledge of the foreshore strata to determine the correct depth. Potential outflanking of the new seawall by landsliding was mitigated by providing heavy piling at the cliff top above the seawall termination.

**Community Involvement**

The community of Lyme Regis has a long tradition of managing its own affairs and historically has tended to resist actions which were deemed to have been imposed upon it by 'outside' agencies. Extensive liaison and consultations were therefore held with community organisations, landowners and residents over many years in order to develop a successful scheme (Davis and Cole 2002). The principal aims of the consultations were to:

- keep the public, landowners, businesses and other organisations with an interest in the project up to date with developments
- gather information relevant to the design of the works
- allow individuals and organisations to contribute ideas on the design and implementation of schemes
- discuss, answer questions and receive feedback on scheme proposals
- encourage debate on critical issues
- reduce potential sources of conflict
- encourage a spirit of openness and cooperation.

A valuable method of encouraging public participation was a regular coastal forum hosted by Lyme Regis Town Council and held at a venue close to the Phase IV site. A typical forum meeting would include an update on the overall progress of the project, a detailed presentation on a particular topic by invited speakers from consultants or contractors, followed by a question-and-answer session and discussions. The coastal forum has allowed the townspeople to have a direct input into the development of the scheme. The comments of members of the forum on specific ideas for stabilisation or coast protection helped to identify, at an early stage, which proposals would be acceptable, or unacceptable to the town. For example, it was established early in the conceptual design stage that works elements which would give a "heavily engineered" appearance would be unacceptable to the majority of townspeople on visual grounds alone. As another example, there was a strong opinion within the town to include as part of the scheme a new footpath from the town's main car park direct to the seawall, which was implemented despite the additional technical challenges it presented. Scheme concepts were essentially agreed upon with the local community and stakeholders in advance, helping the scheme to progress relatively smoothly through the planning process. A high level of interest and community engagement has been maintained throughout the detailed design and construction stages, culminating in the coastal forum's input to arrangements for the opening ceremony and the naming of the new walkway on top of the seawall.

### Environmental Mitigation Strategy

Prior to and during construction, a comprehensive mitigation strategy was developed and implemented to minimise and ameliorate adverse effects on the environment, critically the natural geological features, species and habitats specific to the locality. This included measures to replace and improve coastal cliff habitat that would support a diverse range of typical species, including rare and protected animals, such as reptiles, dormice and maritime grassland invertebrates. Both soft and hard landscaping was designed to physically and visually integrate the scheme into the local landscape setting, whilst balancing the needs and securities of local residents, visitors and users of the coastal footpath. The landscaping and environmental mitigation works will conceal nearly all of the geotechnical engineering works from view and thereby conserve the quality of the coastal views which form an essential characteristic of the Dorset Area of Outstanding Natural Beauty.

### Environmental Stakeholders Group

A major contributing factor to the success of the mitigation strategy was the establishment of the scheme Environment Group, a multi-disciplinary group of stakeholders including:

- Natural England
- The Dorset County Ecologist
- The Jurassic Coast World Heritage Team
- Dorset Area of Outstanding Natural Beauty Team
- Local authority landscape officers
- Planning officers
- The National Trust

Group meetings provided a regular dialogue and successful working relationship between the stakeholders, the client, contractor and the designers. The group proved to be vital in enabling appropriate and coordinated solutions to be efficiently and promptly agreed, where environmental issues and challenges arose during design and construction.

### Working Restrictions and Ground Monitoring

Weight restrictions were imposed on plant operating on potentially unstable slopes and weather restrictions on some types of activity, in order to reduce the risk of landsides being triggered by the stabilisation works themselves. Much of the soil nailing was installed using rope-access. Monitoring of groundwater and slope movements was carried out, much of it using real-time automated instrumentation systems uploading to a website with associated alarm thresholds, to provide early warning of any potential problems during construction.

## Main Benefits of the Project

The main benefits of the project to Lyme Regis may be summarised as follows:

- The long-term protection of some 480 properties with associated land and infrastructure. The scheme has a design life of 60 years.

- The associated economic benefit of the protection, together with spin-off benefits to the tourist economy of the town by making the area more attractive to visitors.
- Continuation of natural landslide activity and erosion processes, thereby preserving the universal/international geological and wildlife value of the site.
- Significant enhancements to the natural environment, such as providing coppiced hazel to encourage dormice and wild flower seeding to provide food plants for rare insects.
- Enhancing the educational value of the site by:
  - Preserving geological outcrops and geomorphological features (Figure 4)
  - Providing safe access to these features which previously had been largely inaccessible, and to fossil collecting sites on the beach.
  - Providing interpretation boards around the site explaining the engineering scheme, wildlife and geological interest.
- Improvements to amenity by providing:
  - A new scenic footpath from the town's major eastern car park direct to the seawall, so that pedestrians no longer have to access the seafront via the narrow and busy main road. This footpath allows safe viewing of the natural features in the area which were previously hidden or dangerous to access.
  - Providing an elevated, level walkway on top of the seawall suitable for wheelchair and pushchair users, providing new and spectacular views of the coast (Figure 4.)

Figure 4: An aerial view the seawall, showing the preservation of geomorphological features and geological outcrops above. Nearly all of the geotechnical works above the wall will be concealed by landscaping and environmental mitigation works.

- Significant health and safety improvements to the foreshore area – the old dilapidated seawall apron was slippery and the existing groyne field difficult to negotiate on foot. There were frequent instances of walkers slipping over or getting cut off by the tide. The new seawall provides much safer access to walkers and allows escape from the tide at several different locations. The old timber groynes were recycled to form seating on the new walkway.

## Main Achievements of the Project

The main achievements of the project are considered to be:

- Securing the long-term future of Lyme Regis in the face of a considerable and growing threat from aggressive coastal erosion and landsliding in one of the most unstable geological settings in the UK.
- Engaging the local community throughout the scheme development process, with high levels of participation helping to progress the project in a positive and constructive manner, thereby producing a sense of public ownership and a high level of consensus across the town.
- Delivery of a technically demanding engineering scheme both on time and within budget, whilst meeting all of the environmental requirements in a sensitive paleontological and nature conservation area.
- A varied selection of procurement routes within the design and build contract, with subcontractors providing specialist advice and the full or partial design of some of the geotechnical elements, within the overall framework provided by the principal designer.
- Working safely through one of the most extreme winters on record (2013-2014) both in terms of prolonged heavy rainfall and sea state, whilst still remaining on programme. Active landslide movements took place over several months in early 2014 in the immediate location of the works, with landslide displacements of up to 40mm per hour occurring at the worst times and with the resulting debris falling into the area of seawall construction on the foreshore. The retreating cliff top came to within a few metres of construction operations for the anchored piled wall which could be described as being built "just in the nick of time."
- Successfully balancing the technical requirements of the project, the needs of the local community and environmental issues, to the extent that the scheme has been well received from a grateful community and environmental stakeholders alike.
- Raising the reputation of civil engineering amongst the residents of West Dorset. A journalist from a local newspaper wrote (Boothroyd 2014):

    > *"I love engineers. They're logical, methodical, business-like and attentive to detail – for physical and material forces are unforgiving of sloppiness, and one small technical error in an engineering project can be catastrophic. There's no ego and no waffle; they calmly analyse, plan and do it. Yet they're also imaginative and adaptable as circumstances change, and easy to work with. What a shame it is that, as a country, we tend in our public life not to promote the values and qualities of the engineer. How lucky we are in Lyme they've done us proud."*

## Conclusion

It is considered that there would be great benefits in using the coastal management methodologies applied to the Lyme Regis scheme in future projects facing similar challenges. There are many locations in the UK and internationally where erosion and landsliding in soft rock cliffs are threatening communities and where there are also considerable environmental constraints and opportunities. Employing the philosophies outlined in this paper could make the difference between protecting a community, despite the considerable challenges involved, and allowing assets to fall into the sea along with the associated property blight and stress this causes.

## Acknowledgements

West Dorset District Council is the client for the project and the scheme promotor. CH2M HILL is the client's consulting engineer. Balfour Beatty Construction Services UK is the design and build contractor. AECOM is Balfour Beatty's principal designer. The scheme has been funded by EA/Defra grant in aid under the Coast Protection Act 1949, by West Dorset District Council and by Dorset County Council in their role as highways authority. The authors are grateful for the support of colleagues and consultants involved in the development of the scheme since the late 1980s, including High-Point Rendel who led the investigation and design in the early phases.

## References

Boothroyd C., 2014. Continuing a fine tradition. *View from Lyme Regis.* Issue 403. Page 12.

Brunsden D. 2002. The Fifth Glossop Lecture – Geomorphological roulette for engineers and planners: some insights into an old game. *Quarterly Journal of Engineering Geology and Hydrogeology*, 34.

Brunsden D. and Moore R. 1999.Engineering geomorphology on the coast: lessons from West Dorset. *Geomorphology* No. 834; pp1-19.

Cole K., Davis G.M., Clark A.R. and Fort D.S., 2002. Managing coastal instability - a holistic approach. Instability - Planning and Management, 679-686. Thomas Telford. London.

Davis G.M. and Cole K. 2002. Working with the community - public liaison in instability management at Lyme Regis, Dorset, England. Instability - Planning and Management, 695-700. Thomas Telford. London.

Moore R. and Davis G. 2014.Cliff instability and erosion management in England and Wales. *J Coastal Conservation: planning and management.* DOI 10.1007/s11852-014-0359-3.

Moore R. and McInnes R. 2011.*Cliff instability and erosion management in Great Britain: a good practice guide*. Published by Halcrow Group Ltd.

Moore R., Rogers J., Woodget A.& Baptiste A. 2010.Climate change impact on cliff instability and erosion in the UK.*FCRM>10, Proc. of the 45th EA Conference of River and Coastal Engineers*, 079, 2010.

Sellwood M., Davis G.M., Brunsden D. and Moore R. 2000. Ground models for the coastal landslides at Lyme Regis Dorset, UK. Landslides in research, theory and practice, 3, 1361-1366. Thomas Telford. London.

Watson N., 2015. Environment Agency. Personal Communication.

**Coastal Management: Changing coast, changing climate, changing minds**
ISBN 978-0-7277-6114-9

ICE Publishing: All rights reserved
doi: 10.1680/cm.61149.I1

# Author Index